Environmental Management of River Basin Ecosystems

Springer Earth System Sciences

More information about this series at http://www.springer.com/series/10178

Mu. Ramkumar · K. Kumaraswamy
R. Mohanraj
Editors

Environmental Management of River Basin Ecosystems

Springer

Editors
Mu. Ramkumar
South East Asia Carbonate Research Laboratory (SeaCarl)
Universiti Teknologi Petronas
Tronoh
Malaysia

and

Depatment of Geology
Periyar University
Salem
India

K. Kumaraswamy
Department of Geography
Bharathidasan University
Tiruchirappalli
India

R. Mohanraj
Department of Environmental Management
Bharathidasan University
Tiruchirappalli
India

ISSN 2197-9596 ISSN 2197-960X (electronic)
Springer Earth System Sciences
ISBN 978-3-319-13424-6 ISBN 978-3-319-13425-3 (eBook)
DOI 10.1007/978-3-319-13425-3

Library of Congress Control Number: 2014957299

Springer Cham Heidelberg New York Dordrecht London

Printed on acid-free paper

Springer International Publishing AG Switzerland is part of Springer Science+Business Media (www.springer.com)

Preface

Rivers have been the prime sources of sustenance for mankind. Man has continued to reap the benefits provided by the rivers for centuries, largely without understanding much of how the river ecosystem functions and maintains its vitality. Rivers play a significant role in the biogeochemical cycle and in the provision of water for domestic, agricultural, recreational, navigational, and industrial purposes and sediment and nutrients for the sustenance of natural ecosystems. A river basin is a basic geographic and climatological unit within which the vagaries of natural processes act and manifest at different spatiotemporal scales. The channels, the water, and the sediment transported and distributed within the river basin act as a unified system connected to each other through a delicate environmental equilibrium. This equilibrium is influenced by the vagaries of nature including climate, lithology, slope, etc. Left undisturbed, the river system has its own dynamism of maintaining its natural health and equilibrium. Any change in the factors that affect the equilibrium would result in recognizable changes in the system, often to the detriment of the natural dynamism of the rivers and thus the sustenance of mankind.

Given cognizance to the importance of holistic study, integrated management practices, and sustenance of environmental flow of river systems for nourishment of river basin ecosystems, there are a number of international efforts in terms of the Stockholm conference on the Human Environment (1972), United Nations conference on Environment and Development (1992), International Geosphere-Biosphere Program, International Human Dimension Program, and more recently, the EU water framework directive, and flood directive. There are many other programs that provide a comprehensive guideline for environmental management of river basins. As a corollary to these efforts, this volume presents specific thematic papers covering the various facets of river basin ecosystems, methods of study, myriad varieties of influences on natural environmental processes, anthropogenic interventions and resultant impacts, methods of reclamation, remediation, management, and nourishment. Though the interconnectedness of river basins as a unified system and its delicate balance between litho-bio-hydro-atmospheric processes are known to the scientific community, the importance of sustenance of this delicate equilibrium is exemplified through various case studies and methodology papers in this volume.

The publication of this volume is intended to enlighten academicians, researchers, administrators, and planners. This intention would be served if the readers spread awareness among the common people and those concerned for the wellness of the Earth and sustainability of its resources.

Mu. Ramkumar
K. Kumaraswamy
R. Mohanraj

Acknowledgments

The seeds for compiling a special volume emphasizing the multifaceted nature of river basin ecosystems were sown by the discussions held between the editors of this volume during the national seminar on River Basins organized under the aegis of SAP—DRS Phase-I sanctioned to the Department of Geography, School of Geosciences, Bharathidasan University, Tiruchirapalli, India. The grant sanctioned by the University Grants Commission, New Delhi is gratefully acknowledged. A casual pep-talk that originated at tea-break during the seminar among the editors has culminated in the publication of this special volume.

We profusely thank Dr. Johanna Schwarz and her editorial team of Earth Sciences, Springer-Verlag GmbH, Heidelberg, Germany, for agreeing to publish this special volume of papers, and for the professional support extended to us. We thank those academicians and researchers who have actively involved in reviewing the articles and helped the authors to present the data in a better way. Ms. Radhika Sree and her team at production unit have done excellent work and are thanked for their professional and flawless handling.

R. Mohanraj thanks Dr. P.A. Azeez, Director, Salim Ali Centre for Ornithology and Natural History, Coimbatore, India, for his suggestions and valuable inputs.

Mu. Ramkumar thanks Smt. A. Shanthy (wife), Selvi Ra. Krushnakeerthana (daughter) and Selvan Ra. Shreelakshminarasimhan (son) for their support and understanding.

Above all, we submit our thankfulness to The Lord Shree Ranganatha, who by His choice and design chose to reside on a river island (Srirangam), for his boundless mercy showered on us.

Mu. Ramkumar
K. Kumaraswamy
R. Mohanraj

Contents

Land Use Dynamics and Environmental Management of River Basins with Emphasis on Deltaic Ecosystems: Need for Integrated Study Based Development and Nourishment Programs and Institutionalizing the Management Strategies

Mu. Ramkumar, K. Kumaraswamy and R. Mohanraj

Abstract The deltaic systems maintain a delicate equilibrium with the litho, hydro, bio and atmospheric processes and provide sustenance to the biosphere. Located at the receptive end, the lower reaches of the river basins *a la* deltas are fragile ecosystems and are susceptible to the changes in the upstream regions. These ecosystems survive at the mercy of processes not only in their vicinity, but also the processes that act at the catchments of the river systems. With the growing population and economic development, they are under multiple pressures emanating from anthropogenic activities. Unlike the natural processes, the anthropogenic interventions with the deltaic ecosystems are highly complex and varied. These range from modification of physical surface characteristics, introduction of physical, chemical and biological pollutants from point and non-point sources, in the name of developmental activities including but not limited to urbanization, commercial, industrial, recreational and agricultural endeavors, culture, geopolitics, racism, quality of life, economic might, and mismanagement and unsustainable rates of exploitation of natural resources. All these could be comprehensively

Mu. Ramkumar (✉)
South East Asia Carbonate Laboratory (SeaCarl), Universiti Teknologi Petronas, Tronoh, Malaysia
e-mail: muramkumar@yahoo.co.in

Mu. Ramkumar
Department of Geology, Periyar University, Salem 636011, India

K. Kumaraswamy
Department of Geography, Bharathidasan University, Tiruchirapalli 620024, India

R. Mohanraj
Department of Environmental Management, Bharathidasan University, Tiruchirapalli 620024, India

Mu. Ramkumar et al. (eds.), *Environmental Management of River Basin Ecosystems*, Springer Earth System Sciences, DOI 10.1007/978-3-319-13425-3_1

termed into conversion of natural land cover into land use. In this paper, an attempt is made to enlist the human endeavors that convert the natural land cover into anthropogenic land use and resultant adverse impacts and irreversible reactions. These are then summed up into a scheme of comprehensive study for integrated management. A plea is also made to institutionalize the management practices for effective implementation and sustainable development.

Keywords River basin management · Deltaic ecosystem · Integrated model study · Strategies

1 Introduction

The rivers have been the prime sources of sustenance for the mankind since the advent of civilization. Rivers play significant roles in the provision of water for domestic, agricultural and industrial activities (Ayivor and Gordon 2012), and generate sediments and nutrients for the sustenance of the natural ecosystem. Humans continue reaping the benefits provided by rivers for centuries, without understanding much on how the river ecosystem functions and maintains its vitality (Naiman and Bilby 1998). The bludgeoning population and its ever-increasing dependence on the natural resources for agriculture, transport, water and land for domestic, commercial and industrial activities, make nourishing the river systems an all the more important and urgent task.

Within a river basin, the deltaic regime, being a receiving basin acts as an interface between fluvial, oceanographic, atmospheric and anthropogenic dynamics. These traits make this important macrogeomorphic unit ecologically fragile and susceptible to environmental deterioration very easily, even by changes in the catchment (pollution, siltation, flooding, etc.), and ocean (inundation, erosion, accretion, etc.,) and atmospheric equilibrium. For this reason, understanding the deltaic processes has been the endeavor of researchers of various disciplines. Deltas are one of the most frequently and intensively studied natural ecosystems as they host a wide variety of depositional, geomorphic and oceanographic settings. These ecosystems support a wide variety of economic, agricultural and transport avenues.

In this paper, we reviewed the current trends of research on the river basins, particularly the lower reaches (deltas of river basins) with specific emphasis on the land use dynamics and land use-land cover-human interactions, and present a multidisciplinary schematic procedure as an idealistic methodology for understanding the deltaic systems for better management and ecosystem nourishment.

2 A Review on the State of the Art

2.1 Nature of the River Basins

In a classic review, Barrow (1998) stated that except for the most arid and cold regimes, the World's landscape can be divided into distinctly mappable river basins of various scales (large, medium and small) and the river basins themselves can be sub-divided into upper, middle and lower basin, based on the hydrologic and geomorphic characteristics. Since each river basin acts as a holistic system in tune with the climatic, geological and anthropogenic interactions, any study on the river system should understand the dynamics of the river basin as a whole. A river basin is a basic geographic and climatological unit within which the vagaries of natural processes act and manifest at different spatio-temporal scales. However, even if juxtaposed, no two river basins respond to natural processes in a similar way and thus, each river basin is unique. Hence, any developmental activity or conservation effort has to be uniquely designed and implemented for each river basin.

The actions of the rivers include transport of water and fertile soil/sediment from their catchment areas and distributing them in the lower reaches that form the basis for cultivation, settlements, etc. The channels, the water and the sediment transported and distributed within the river basin act as a unified system connected to each other through a delicate environmental equilibrium. This equilibrium is influenced by the vagaries of nature including climate, lithology, slope, etc. Left undisturbed, the river system has its own dynamism of maintaining its equilibrium. Any change in the factors that influence the equilibrium results in recognizable changes in the system, including adverse impacts such as flooding, erosion (Walling 1999), and desertification, which in turn, may cause loss of critical resources that provide sustenance to human race such as land, agricultural produce and other commercial activities. In addition, it also alters the nutrient availability in the deltaic region and contributes to the proliferation of exotic species in coastal regions, that adversely affect the people who depend on the normal natural processes at downstream regions (*for example*, Wu et al. 2008). Hence, a proper understanding on the natural processes, human endeavors and interactions between litho-hydro-eco-atmospheric processes at various scales in a basin-wide ambit is necessary for sustainable future. However, the existing pattern of research, either theoretical, laboratory and or field based studies are segmented. In order to understand the basin-scale processes, transcending subject barriers and conducting interdisciplinary studies are necessary.

Though, for thousands of years, the human beings have been modifying land for food, shelter and other essentials of life, the current rates, extents and intensities of such modifications are far greater than ever in the history (Efiong 2011). In view of the prevailing favorable conditions for habitation, cultivation and industry, explosive growth of human population and resultant pressure on natural environment is high in river basins (Zarea and Ionus 2012). Traditionally, the river basins are treated as treasure troves for natural resources, but the human needs have taken precedence over environmental concerns (Triedman 2012). Nevertheless, the land use dynamics continue to impact on the river catchments (Wu et al. 2008) which have negative repercussions for river health. The impacts of change of natural land cover, particularly the urban expansion is detriment to the local and regional hydrology, and water quality (Li et al. 2008; Koch et al. 2011) and increases the non-point sources of pollution (Tang et al. 2005). The impacts of the human activities translate into catchment erosion (Walling 1999), sedimentation within river channel, surface and ground water pollution, increased levels of turbidity, acidity, mean dissolved oxygen and nutrient loads (Ayivor and Gordon 2012; Memarian et al. 2013). Catchments of many river basins are subjected to irreversible damages particularly by mining and hydro-electric projects. In addition to landscape modifications, such developmental projects cause acid mine drainage, metal pollution and excess sedimentation in the river basin and invigorate a chain of secondary detrimental effects.

2.2 Land Use Dynamics and Their Impacts on the Deltaic Ecosystems

When compared with other ecosystems, the deltaic regions support a high population density and provide sustenance to fauna and flora. Johannes et al. stated that rapid population growth in deltaic regions inevitably brings in recreational, commercial and industrial use of these lands and waters, causing environmental deterioration. To assess the impacts of these activities, baseline information is a prerequisite, with which the ongoing and future trends of sustainability of ecosystems could be adjudicated to strategize the management programs for wellness of the human community and economic prosperity. The availability of a variety of floodplain environments in large lowland river valleys provides opportunities and constraints for a host of human activities (Hudson et al. 2006). One of the visible signs of pressure on the natural environment is the change in land use/land cover which often if not most probably results as a function of population growth (Qi and Luo 2006). Being riparian, the deltaic regimes are subjected to ecological deterioration and experience geohazards not only due to the changes in the immediate vicinity, but also under the vagaries of changes in the catchment regimes, such as rainfall, land use, deforestation, etc. (*for example*, Jessel and Jacobs 2005; Bathurst et al. 2011). Brush demonstrated that the water quality, climatic changes, land use dynamics, imprints of forest fire, cultivation, human settlement and industrialization are all recorded in deltaic sediments. Deterioration of water quality due to conversion of forest land cover into urban and agricultural land use on a basin scale has been documented by Li et al. (2008), Kim et al. (2011) and Koch et al. (2011). Hesslerova et al. (2012) are of the opinion that the higher human activity, particularly the conversion of natural land cover into land use increases the surface temperature that results in water loss on a basin scale.

Murty et al. attempted linking land use patterns along river systems and the quality of the environment. Mutie et al. (2006) are of the opinion based on the study of land use patterns of Mara River basin that, the conversion of natural land cover into mining, agricultural, pastoral and game parks has resulted in severe constraints on the long-term sustainability of environmental health of the river system and dependent wildlife and human society. Similar conclusions were drawn by Siahaya and Hermana (2013) based on their study of land use changes and resultant water resource potential in urban watershed. Wijesekara et al. (2010) observed that conversion of 5 % of watershed area into urban lands resulted in 2.6 % increase of overland flow, 2.3 % reduction of evapotranspiration and 11 % increase of combined overland and base flow into Elbow River, southern Alberta, signifying the exponential nature of flooding of rivers due to the growth of urban sprawl. Coulthard and Macklin (2001) observed that the British river systems have been sensitive to short term (ca. 102 years) climate fluctuations, land-use change and sediment supply. They further stated that this observation has important implications for forecasting river response to future climate and land-use changes.

Li et al. (2008), Tong and Chen (2002), Koch et al. (2011) and Yaakub et al. (2012) demonstrated the existences of significant positive relationship between physico-chemical properties of river water quality and the land use/land cover patterns. According to Zwolsman (1994), there has been much concern about the effects of increasing nutrient inputs into natural environments as the rivers draining densely populated and industrialized areas are being loaded with very high concentrations of phosphate, nitrate and ammonium. Enhanced nutrient inputs to the river basin ecosystems often bring in exotic biota that destruct the native species and eventually create ecological imbalance. Increase of nutrient inputs, especially nitrogen in river water as a result of changing of forest land cover to agricultural land use in riparian areas of various watersheds in Hokkaido, Japan was reported by Okazava et al. (2010). Based on their study, these authors have suggested a management strategy to curtail the detrimental effects of nitrogen pollution. According to them, when cropland covers much of the catchment region and increases the nitrogen concentration in the river water, the elevated levels can be reduced significantly through conversion of much of the riparian region with forest cover. Kroeze and Seitzinger (2002) analyzed the nitrogen loading in rivers and estuaries in relation with changing land use scenarios and concluded that significant reduction of these nutrient-pollutants could be achieved by reducing the use of fertilizers and NO_2 emissions.

Ramkumar (2003a, b) emphasized the link between land use dynamics with that of coastal environments, particularly the mangrove swamps. The coastal deltaic lands serve as repositories of rich biodiversity and as nurseries to endemic avian, aquatic and other species, including mangroves. It was Balls who had established a firm link between nutrient accumulation in natural sedimentary environments, particularly in estuary and adjoining regions and transport of nutrients to coastal marine regions, as a function of land use change based on data from nine rivers located along the east coast of Scotland. Padma and Periyakali analyzed physicochemical and geochemical characteristics of the coastal lakebed sediments and reported the impact of industrial contaminants over sediment textural and geochemical properties. As the coastal environments are homes to a vast array of primary producers, and act as nurseries, which in turn are dependent on the physical and chemical characteristics of the bed-sediments, significant alteration of these traits by industrial contaminants affects the fundamental units of the food-chain and primary producers. A similar study by Solaraj et al. (2010) in the deltaic regions of the Cauvery River basin, Southern India, reported high concentrations of dissolved salts, organic pollutants and phosphate exceeding the permissible levels at certain sampling stations as a result of agricultural runoff from the vicinity, sewage, and industrial effluents.

Employing geochemical analyses and dating of core samples, agricultural and other land use records collected from governmental records, Heathwaite was able to accurately document the link between land use changes and geochemistry of ensued sediments. Based on geochemical data of the overbank deposits of the English and Welsh flood plains, Macklin stated that geochemical profiles show apparent changes across records of the industrial revolution, forest fire, agriculture and human settlements. Supplemented with carbon, pollen and thermoluminescence dating, it was made possible by their study to assess the impacts of anthropogenic activities over the natural ecosystems. Passmore and Macklin recognized coeval shifts in geochemical profiles of ancient sediments coinciding with the periods of conversion of natural vegetation into agriculture, industrialization, and mining land uses.

Qi and Luo (2006) observed the land use changes on a basin scale in the Heihe River basin and found that the observed changes caused severe environmental problems in terms of surface water runoff change, decline of groundwater table and degeneration of surface water and groundwater quality), land desertification and salinization (Zhang and Zhao 2010), and vegetation degeneracy. Glendenning and Vervoort (2010) evaluated the effect of rain water harvesting structures on a catchment-basin scale in terms of water balance models, intensity of surface run off, water variability in storage capacity, etc., to suggest suitable eco-friendly developmental activities. The nexus between agricultural land use, effluent disposal and the quality of river water have been examined by Tafangenyasha and Dzinomwa (2005) and the results have shown that the land use change imposes beneficial as well as detrimental effect on the natural environment and biodiversity.

The abilities of the estuaries and adjacently located microenvironments the deltas to receive, assimilate and disperse nutrient and contaminant inputs from natural and anthropogenic sources is well known. Throughout the World, this capacity of the deltaic systems has been exploited as a means of disposing such waste materials. However, this capacity is finite and if exceeded, environmental quality may be unacceptably compromised (Balls et al. 1996). As the estuaries and coastal environments are valuable ecosystems characterized by high primary productivity and large stocks of juvenile fish (Robertson and Duke 1987), assessing the status of natural environments in these regions is important in terms of academic, economic and ecological points of view (Zwolsman 1994; Ramkumar et al. 1999; Ramkumar 2007; Ramkumar and Neelakantan 2007). Estuaries are characterized by naturally derived organic matter that originates from the autochthonous production, the open sea, surrounding salt marsh and mangroves and river drainage (Cifuentes et al. 1990). In addition to these natural sources, the rivers are also recipients of the organic matter from anthropogenic activities and hence, the documentation of the source, quantum, and the nature of interaction of organic matter in estuarine waters is essential.

The tendency of organic matter to attract heavy metals, available from natural sources in trace amounts but mostly released from the industrial and urban wastes, aggravates the ecological vulnerability as the organic matter in coastal regions form the basic units of energy source and is actively consumed by lowermost link of food web, through which, the toxic metals reach top of the food pyramid often to the detrimental effects on humans (Rajaram and Devendran 2013). The very nature of mushrooming industrialization, urban sprawl, commercial and recreational activities along river channels, deltaic and coastal plains generates newer domestic, exotic and toxic wastes and nutrients and stresses further the already fragile ecosystems leading to environmental deterioration (Rao 1998). This observation is affirmed by the study of Hampson et al. (2010) that has documented the microbial pollution of river systems of UK due to land use change.

2.3 Geohazards Associated with the Deltas

Geohazards occur everywhere and no region of this earth is safe. With the opening of global economy and consequent expansion of domestic, industrial, commercial, recreational and aesthetic land use in every available piece of land by the human race result into intervention of natural geological processes leading to the potential loss of life and property. Owing to their vast expanses of monotonously low altitude topography, the lower reaches of the river basins are home to many geohazards. However, the unscientific anthropogenic developmental activities in the deltaic regions often aggravate the intensities of geohazards in terms of susceptibility to flooding (Hickey and Salas 1995), coastal inundation, river and coastal erosion and sedimentation, tsunami, storms, cyclones and quick sand (Ramkumar 2009). As these areas are thickly populated, the resultant damages often reach catastrophic levels even during slightest change in existing equilibrium of coastal geological processes (Jayanthi 2009).

While the developmental activities could not be contained for the sake of non-intervention with geological processes, a well thought-out planning and developmental activities with minimal intervention with geological processes would thwart many a potential hazardous events. However, the very nature of the geohazards being unpredictable, it would be advisable to classify the land and water regions that are being and/or planned to be utilized, according to vulnerability levels to various potential geohazards, and not to engage in potential catalytic activities. In addition, should there be any eventuality, being ready with contingent plans for relief and rescue operations would minimize the potential loss. Ramkumar and Neelakantan (2007) have evolved an idealistic model for geohazard mitigation,

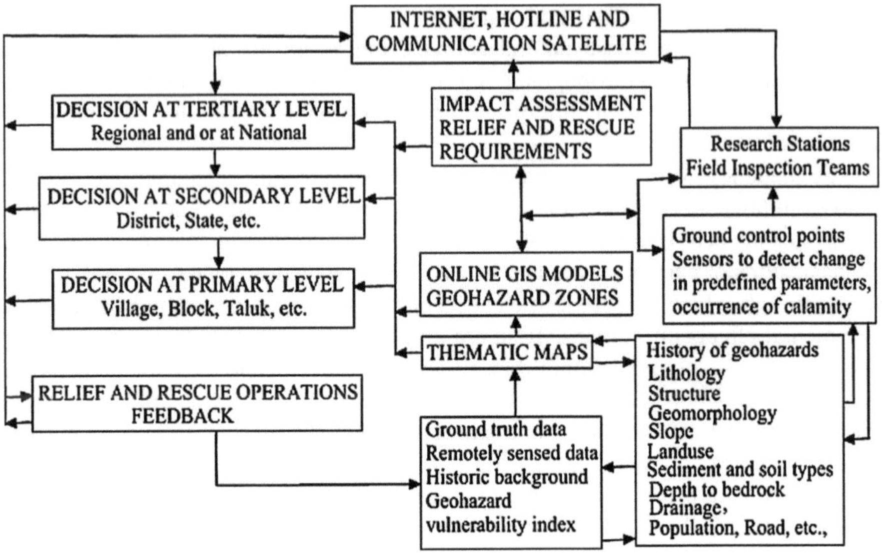

Fig. 1 Scheme for geohazard mitigation, information dissemination and coordination

information dissemination and coordination of rescue and reclamation efforts (Fig. 1).

Among all the geohazards, flooding is more typical and recurrent in the lower reaches of the delta. It occurs from riverine and marine sources and is also influenced by other geohazards such as storm surges, cyclones, etc. Although flooding is a natural phenomena that plays a major role in replenishing wetlands, recharging groundwater and supports agriculture, extreme demands on the natural resources due to the population growth people and their property move closer to water bodies, which in turn increases the risk of life and property loss due to flooding (World Meteorological Organization 2005). Land use change in the catchment regions in terms of deforestation, may cause severe flooding in the riparian regions, in terms of increase of peak discharge and shorter return period of flood events, especially when deforestation exceeded 20–30 % (Bathurst et al. 2011). Studies on fluvial geomorphology of various river basins including Yellow River, Sacramento River, Strickland River, and Brahmaputra River indicate that rivers maintain valley-scale controls on channel alignment, conveyance of flood waters and transport and deposition of sediments (Rogers et al. 1989; Gupta 1988, 1995; Shu and Finlayson 1993; Fischer 1994; Jain and Sinha 2003; Singer et al. 2008; Sacramento River Flood Control Project 2012). Hence, valley-scale documentation of these characteristics and designing flood control measures with site and measure specifics is necessary.

2.4 Sand Mining and Their Impacts on the Deltaic Ecosystems

With the bludgeoning population, economic liberalization, integration of national economy with global systems, and increase in the purchasing capacity of people of the fast growing economic countries such as Brazil, Russia, India and China, construction activity is fast apace not only for the human dwellings but also for commercial activities. The fast pace of economic developments, rise in foreign remittances and liberalized housing schemes for building constructions, mainly from banking sector are some of the causative factors responsible for unabated sand mining from river beds (Padmalal et al. 2008; Sreebha and Padmalal 2011). The environmental impact of river sand extraction becomes increasingly well understood and linked with the globalization in developing countries such as China and India (De Leeuw et al. 2010). There were arguments in favor of linking sand mining with globalization and resultant global environmental change (Sonak et al. 2006). The rivers in which sand is harvested at rates in excess of natural replenishments often undergo channel degradation, causing incision of the entire river system including its tributaries (Ashraf et al. 2011). Striking cases of excessive removal of river sediment removal are summarized by many researchers (Bull and Scott 1974; Sandecki 1989; Kondolf and Swanson 1993; Kondolf 1997; Macfarlane and Mitchell 2003; Hemalatha et al. 2005; Ramkumar et al. this volume). Though mining of building stones and limestone create permanent scars in the vicinity of mining that are located in a variety of geomorphic settings, the impacts are felt only in a geographically restricted region. However, the very nature of unified system on a valley-scale (Singer et al. 2008) and quick reaction time of fluvial system (Kale et al. 2010), mining of river sands from channels advocates a cause of serious concern. Though such generalizations are in the common knowledge, the environmental and geomorphic consequences are not fully understood due to the paucity of requisite data (Macfarlane and Mitchell 2003; Kale 2005; Ashraf et al. 2011; De Leeuw et al. 2010). Ramkumar et al. (this volume) attempted documenting the effects of sand mining on geomorphology, sediment texture and long-term sustainability of the natural fluvial system.

2.5 Socio-economic-Cultural-Racial-Geopolitical Impediments in Managing the Land Use Dynamics

Soini (2005, 2006) examined the interactions between the land use change and the livelihood, environment, socio-economic conditions and biodiversity in the context of population pressure, cultural and regional historical factors, and climate change in selected regions of East Africa for improving the existing understanding on their dynamics and to design better management strategies. The results have emphasized the intricate and complex nature of land use change. Not only the culture, but race and socio-economic conditions were also found to have influenced the land use planning and river basin environmental management and the tendency was termed as "environmental racism" by Miller (1993). Verbatim, the statement of this author reads—"*this essay traces the political and economic framework of New York City's land use and planning processes for North River, discusses how issues of race and socio-economic status are integral elements in this process, and examines one community's actions against environmental racism*". Although not exactly similar, but related issue was documented by Wu et al. (2008) who have recorded the environmental and socio-political pressures from upper reaches of the Yangtse River, China and traced their impacts on the dynamics of land use/land cover changes. In an exhaustive study covering many countries and continents, Englund et al. (2011), assessed the impacts of land use change from agriculture to cultivation of feedstock for biofuels on air and water quality and biodiversity. This is one of the test cases where the overriding influence of land use on environmental, social, economic and livelihoods of people was demonstrated unequivocally by scientific data, leaving no room for subjectivity. In recent years, there are attempts such as that of Costello (2010) that exhibited the link between consumption pattern of agricultural products and the land use dynamics.

The study of Weng (2000) is a classic work that documented the origin and propagation of agriculture in relation with the climate and sea level changes during the Holocene era, and the development and widespread practice of rice cultivation, horticulture and dyke-pond system of human-environment interaction. Based on these data, the study has also exemplified the adverse effect of imprudent use of technology over natural environment that manifested through increased incidents of flooding and increase in flood intensity, which in turn were found to be detrimental to the economic advancements obtained from agricultural innovations.

In one of the pioneering integrated spatial and statistical study of land use, land management practices, soil, geomorphic and hydrological parameters and their relationship with river water quality data of more than hundred basins of Ontario, Ongley and Broekhoven (1978) concluded that the land use exhibited close relationship with water quality and the relationship was found to be more influential than the hydrological parameters! Wolanski et al. observed that the floodplains of Mekong River delta are extensively utilized for rice cultivation and aquaculture while the mangrove swamp regions supply wood and fish stocks providing sustenance to millions of people. These economic benefits are increasingly being curtailed by rising sea level, resultant salt-water intrusion into the river channel and deposition of sediments within estuarine channel itself and hinter inland water transport and felt that it is necessary to understand the depositional dynamics to plan for developmental activities.

2.6 Other Impediments for Effective Management Programs

Despite endowed with vast expanses of coastal lands, many of the developing countries lack accurate information on comprehensive scientific data on the coastal land use, their areal extent, and condition and utilization status of them in the form of maps and statistical data for use in policy decisions on effective coastal zone management plan. These authors have also commented that data may be available in few cases, but there seem to be a long time-gap between data collection and publication that render the data not suitable for effective management plan. While reviewing the river basin-scale environmental management programs world over, Coppola (2011) observed that the failure to achieve many of the plans was mainly due to the fragmentary approach, instead of employing a comprehensive and integrated multi-disciplinary approach. Seto et al. (2002), Tong and Chen (2002), Moss (2004), Qi and Luo (2006), Ranade (2007), Li et al. (2008), Zhang and Zhao (2010), Yueqing et al. (2011), Sharma et al. (2011), and Kotoky et al. (2012) suggested the use of remote sensing and GIS for monitoring and mapping land use/ land cover changes for environmental management planning. Applying Spatial Auto Regression model on salinization, environmental quality, and land use pattern etc., Zhang et al. (2011) demonstrated a methodology for strategizing deltaic environmental management plans.

The fallacy of one dimensional modeling of effluent dispersal in estuaries and adjacent coastal waters has been explained by the studies of Ramkumar (2000a, 2001, 2003a, b, 2004a, b, 2007), and Ramkumar et al. (2001a, b). These studies, together with Rajani Kumari et al. (2000), Ramkumar (2000b), Ramkumar and Pattabhi Ramayya (1999), Ramkumar and Gandhi (2000), Ramkumar and Vivekananda Murty (2000), Ramkumar and Neelakantan (2007) and Ramkumar et al. (1999, 2000a, b, c), have demonstrated the methods to discriminate the natural and anthropogenic processes prevalent in the deltaic ecosystems and integrate spatio-temporal variations of geomorphology, land use, sediment textural, geochemical and remotely sensed data through conventional methods and geostatistical and GIS analyses for modeling deltaic systems for designing better management plans.

3 Strategizing and Institutionalizing the Management Programs

It was during the year 1972 at the Stockholm conference on the Human Environment, a clarion call was made to study the land use changes. Similar calls were made at the United Nations Conference on Environment and Development (UNCED), International Geosphere and Biosphere Program (IGBP) and International Human Dimension Program (IHDP) to foster research on land use—land cover changes and to integrate other environmental parameters for establishing developmental planning (Gajbhiye and Sharma 2012). Koch et al. (2011) advocated that since land-use change strongly affects the water quantity and quality as well as biodiversity and ecosystem functioning, analyzing the land-use change scenarios should form an essential part of these regional development scenarios. According to Crawford (2005) and Ganf and Oliver (2005), owing to the conversion of flood plains and river courses into urban, commercial, agricultural and recreational land uses, the consumption of water from rivers and adjoining regions multiplies and as a result of which the estuarine regions and dependent ecosystems are affected severely. Hence, determination of environmental water flow for the sustenance of riparian regimes and acting accordingly is an essential task in any river management plan.

In recent years, a term "environmental flow" (*sensu* Cottingham et al. 2003; Scoccimarro and Collins 2005; Speed et al. 2012; Morrison 2013) and "environmental water requirement" (Robson et al. 2005; McNeil and Fredberg 2011) are being emphasized for planning river basin management planning. It refers to the minimum water and sediment transport by a river system at various points of its traverse, so that the natural ecosystem on a basin scale (Ganf and Oliver 2005) is sustained without any adverse impact either on its own and/or on the developmental activity that is being made in the basin. Accordingly, planners and administrators are increasingly becoming aware of the economic and ecological benefits of the environmental flow and take it into consideration while strategizing developmental *vis-a-vis* environmental management plans. Assessment of environmental flow is unique to each river basin (Speed et al. 2012) depending on its climate, land use, soil, geomorphic, demographic and many other traits.

Faced with pressures from the population, urban, aquaculture, recreational, commercial and other utilities on coastal environments, the Tasmanian government has evolved a policy of sustainable development following the International practices (Anutha and Johnson 1996 and references cited therein) of conservation and management. The policy states that "*sustainable developmental plan means managing the use, development and protection of natural resources in a way, or at a rate, which enables people and communities to provide for their social, economic and cultural well-being and for their health and safety while sustaining the potential of natural resources to meet the reasonably foreseeable needs of future generations, safeguarding the life-supporting capacity of air, water, soil and ecosystems and avoiding, remedying or mitigating any adverse effects on the environment*". According to this policy, a holistic environmental management system has to be established through an integrated study of status of the ecosystem based on which, remedial and conservation efforts have to be taken up and effectiveness of the system has to be monitored periodically through measurable attributes. Leggett et al. (2003) and Memarian et al. (2013) are of the opinion that different strategies and methods of natural resource management themselves need to be evaluated for their effectiveness and implemented accordingly. A combination of mathematical and GIS tools for effective monitoring mechanisms and assessment of effectiveness of environmental management strategies has been suggested by these authors. Jessel and Jacobs demonstrated a method to evaluate the effectiveness of management programs based on stakeholder responses. Collection of socio-economic and environmental data from stakeholder interviews has led Albinus et al. (2008) to interpret the importance of understanding land use changes over the community lives and ecological balance. The study has also evinced the importance of understanding the cause-effect processes on a basin scale.

To preserve the coastal environment and to regulate the use of land near the Indian coastline, Coastal Regulation Zone (CRZ) was introduced in 1991 in India and amended periodically. The main aim of the CRZ is to ensure the developmental activities along the Indian coast in a sustainable manner based on scientific principles taking into account the dangers of natural hazards in the coastal areas and sea level rise due to global warming (http://moef.nic.in/downloads/public-information/CRZ-Notification-2011.pdf).

In order to judicially utilize the natural resources and to reduce the developmental pressure on land use change, based on the land use reform practices of Hudson River valley, Knudson (2011), suggested involving non-profit organizations for regional and local planning and for implementing the management strategies. Fallon and Neistadt (2006) suggested practical measures for planning, implementing and monitoring the physical, mental and social well-being of the society by involving the local boards. In his review, Barrow (1998) stated that the river basins have been used for developmental planning since the 1930s, but the results have been disappointing. According to Moss (2004), despite the involvement of many factors, including paucity of data, lacuna in scientific understanding of natural and anthropogenic processes that are in operation at various scales, the lack of appreciation for integrated study and institutionalizing the efforts are the reasons for these disappointments. Realizing the importance of protection of fluvial environments, soils, water, and landscape contained in them, EU water framework directive (Moss 2004) and flood directive that include establishment of baseline database through holistic study have been designed and are being implemented in many parts of the World (*for example*, Fleta et al. 2008; Clapcott et al. 2011; Appiah-Opoku 2012 and the papers contained in the edited volume) and these directives caused a paradigm shift towards the integrated management practices (Hampson et al. 2010).

4 Conclusions

From the foregoing review, a broad consensus as concluding remarks can be made.

- The rivers act as a holistic system wherein any change at any part of the basin has repercussions in other parts of the basin and wellness of the system. Hence, any management or nourishment program should take into account the basin scale processes and basin-wide responses of the river system for effective management.

- The rivers have a self-healing ability to the anthropogenic interventions. However, this ability is limited and if exceeded, the resultant damage would be irreversible. Hence, the reclamation, management and nourishment programs of river basins should also include objectives based on these "critical/threshold limits" of the river systems. These limits have to be assessed with more realistic, recent and comprehensive data.
- The growing population and resultant anthropogenic interventions in the name of "developmental activities" translate into conversion of natural land cover into anthropogenic land use. The land use in turn, is influenced by a variety of factors, including geopolitics, racism, culture, etc., which are beyond the realm of any reasonable scientific study. Hence, it is suggested that in addition to setting objectives based on "limits" of the river systems, these factors should also be taken into account for effective management programs.
- As observed by many of the publications cited in this paper and references cited therein, paucity of baseline information about the river basins and their deltaic systems hampers the understanding of scientific community, leave alone the planners and administrators. If at all databases exist, they were found to be fragmentary and also that there is a big gap between acquiring the data and their availability for stakeholders and users including the academics and researchers. This is a serious impediment that hampers proper understanding, designing effective management program and strategizing implementation of the programs.
- Owing to the large quantum and varied types of data involved in the study of basin scale processes, many studies use temporal remotely sensed data and apply a combination of spatial, statistical and mathematic modeling, together with conventional analyses for better understanding. From the review, it is also revealed that there is an improved awareness among the planners, administrators and political class about the complex interactions of land use dynamics and tenuous nature of the deltaic ecosystems.
- An idealistic methodology for integrated study of river basin and deltaic ecosystems (Fig. 2) is presented in this paper. In order to circumvent the racial, geopolitical and cultural and other factors that hamper effective implementation of management programs and monitoring the nourishment efforts, institutionalizing the programs is suggested. It may include policy guidelines and framework for implementation and monitoring, such as those of EU Water Framework directive.

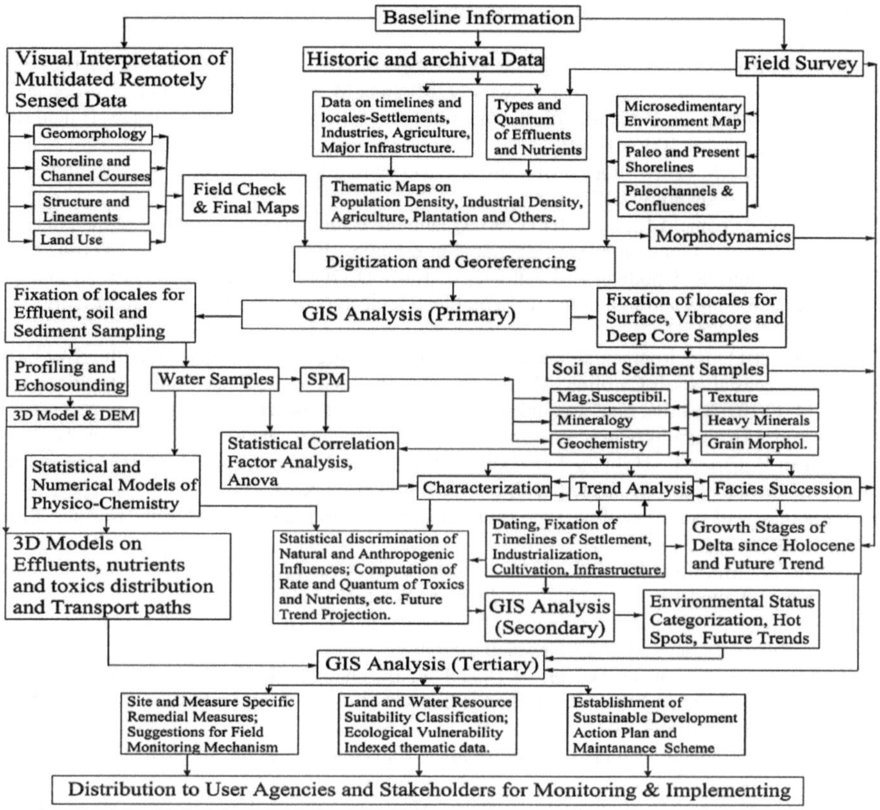

Fig. 2 Model scheme for integrated study of river basin ecosystem in regard to land use

Acknowledgments We thank the authors whose articles are cited in this paper. We thank the scientific databases and websites from where published and unpublished reports are made accessible for the present review.

References

Albinus MP, Makalle JO, Bamutaze Y (2008) Effects of land use practices on livelihoods in the transboundary sub-catchments of the Lake Victoria Basin. Afr J Environ Sci Tech 2:309–317

Appiah-Opolu S (ed) (2012) Environmental land use planning. Intech, Croatia, 232 pp. http://dx.doi.org/10.5772/2728

Ashraf MA, Maah MJ, Yusoff I, Wajid A, Mahmood K (2011) Sand mining effects, causes and concerns: a case from the Bestari Jaya, Selangor, Peninsular Malaysia. Sci Res Essays 6:1216–1231

Ayivor JS, Gordon C (2012) Impact of land use on river systems in Ghana. West Afr J Appl Ecol 20:83–95

Balls PW, Brockie N, Dobson J, Johnston W (1996) Dissolved oxygen and nitrification in the upper forth estuary during summer (1982–1992): patterns and trends. Est Coast Shelf Sci 42:117–134

Barrow CJ (1998) River basin development planning and management: a critical review. World Dev 26:171–186
Bathurst JC, Iroume A, Cisneros F, Fallas J, Iturraspe R, Novillo MG, Urciuolo A, de Bievre B, Borges VG, Coello C, Cisneros P, Gayoso J, Miranda M, Ramirez M (2011) Forest impact on floods due to extreme rainfall and snowmelt in four Latin American environments 1: field data analysis. J Hydrol 400:281–291
Bull WB, Scott KM (1974) Impact of mining gravel from urban stream beds in the southwestern United States. Geology 2:171–174
Cifuentes LA, Schemel LE, Sharp JH (1990) Quantitative and numerical analyses of the effects of river inflow variations on mixing diagrams in estuaries. Est Coast Shelf Sci 30:411–427
Clapcott J, Young R, Goodwin E, Leathwick J, Kelly D (2011) Relationships between multiple land-use pressures and individual and combined indicators of stream ecological integrity. DIC Res Dev Ser 326:57
Coppola H (2011) Environmental land use planning and integrated management at the river basin scale in coastal North Carolina. Master's thesis submitted to the Duke University, 47 pp
Costello C (2010) Relating land use and select environmental impacts to U.S. Consumption with a focus on agricultural products. Dissertation submitted to the Carnegie Mellon University, 46 pp
Cottingham P, Crook D, Hillman T, Roberts J, Rutherfurd I, Stewardson M (2003) Flow related environmental issues associated with the Goulburn river below Lake Eildon. A report to the Department of sustainability and environment, Victoria and the Murray Darling Basin Commission, pp 73–185
Coulthard TJ, Macklin MG (2001) How sensitive are river systems to climate and land use changes ? A model-based evaluation. J Quater Sci 16:347–351
Crawford C (2005) Water use across a catchment and effects on estuarine health and productivity. River Riparian Lands Manage Newslett 29:7–9
De Leeuw J, Shankmann D, Wu G, de Boer FW, Burnham J, He Q, Yesou H, Xiao J (2010) Strategic assessment of the magnitude and impacts of sand mining in Poyang Lake. Chin Reg Environ Change 10:95–102
Efiong J (2011) Changing pattern of land use in the Calabar river catchment, Southeastern Nigeria. J Sustain Dev 4:92–102
Englund O, Berndes G, Lundgren L, Palm M, Engstörm L, Bäckman C, Marquardt K, Stephansson E (2011) Producing feedstock for biofuels: land-use and local environmental impacts. Technical report for the EU biofuel baseline project. Report No.FRT 2011: 06, 195 pp
Fallon LF, Neistadt J (2006) Land use planning for public health: the role of local boards of health in community design and development. A report by Atlanta Regional Health Forum and Atlanta Regional Commission, 33 pp. www.nalboh.org
Fischer KJ (1994) Fluvial geomorphology and flood control strategies: Sacramento River, California. In: Schumm, SA, Winkley BR (eds) The variability of large alluvial rivers. ASCE Press, New York, pp 115–138
Fleta J, Garcia E, Gode L, Gracia A, Moxo D, Verdu J (2008) Fluvial environment planning (FEP) in Catalonia: Harmonizing river and floodplain restoration with human developing. In: Proceedings of 4th international symposium on flood defence: managing flood risk, reliability and vulnerability. Toronto, Canada, pp 1–7, 6–8 May 2008
Gajbhiye S, Sharma SK (2012) Land use and land cover change detection of Indra river watershed through remote sensing using multi-temporal satellite data. Int J Geomatic Geosci 3:89–96
Ganf G, Oliver R (2005) Water allocation to river Murray wetlands: a basin wide modelling approach. River Riparian Lands Manage Newslett 29:17–18
Glendenning CJ, Vervoort RW (2010) Hydrological impacts of rainwater harvesting (RWH) in a case study catchment: the Arvari River, Rajasthan, India. Part 1: field scale impacts. Agr Water Manage 98:331–342
Gupta A (1988) Large floods as geomorphic events in the humid tropics. In: Baker VR, Kochel RC, Patton PC (eds) Flood geomorphology. Wiley, New York, pp 151–177
Gupta A (1995) Magnitude, frequency and special factors affecting channel form and processes in the seasonal tropics. In: Costa JE, Miller AJ, Potter KW, Wilcock PR (eds) Natural and

anthropogenic influences in fluvial geomorphology. vol 89, American Geophysical Union Monograph, Washington, pp 125–136
Hampson D, Crowther J, Bateman I, Kay D, Posen P, Stapleton C, Wyer M, Fezzi C, Jones P, Tzanopoulos J (2010) Predicting microbial pollution concentrations in UK rivers in response to land use change. Water Res 44:4748–4759
Hemalatha AC, Chandrakanth MG, Nagaraj N (2005) Effect of sand mining on groundwater depletion in Karnataka. In: Proceedings of V international R&D conference of the Central Board of irrigation and power, Bangaluru, pp 1–15
Hesslerova P, Chmelova I, Pokorny J, Sulcova J, Kropfelova L, Pechar L (2012) Surface temperature and hydrochemistry as indicators of land cover functions. Ecol Eng 49:146–152
Hickey JT, Salas JD (1995) Environmental effects of extreme floods. U.S.-Italy Research Workshop on the Hydrometeorology, Impacts, and Management of Extreme Floods, Perugia, pp 1–23
Hudson PF, Colditz RR, Aguilar-Robledo M (2006) Spatial relations between floodplain environments and landuse-land cover of a large lowland tropical river valley: Panuco basin. Mexico Environ Manage 38:487–503
Jain V, Sinha R (2003) Geomorphological manifestations of the flood hazard: a remote sensing based approach. Geocarto Int 18:51–60
Jayanthi M (2009) Mitigation of geohazards in coastal areas and environmental policies of India. In: Ramkumar M (ed) Geological hazards: causes, consequences and methods of containment. New India Publishers, New Delhi, pp 141–147
Jessel B, Jocobs J (2005) Land use scenario development and stakeholder involvement as tools for watershed management within the Havel River Basin. Limnologica 35:220–233
Kale VS (2005) Fluvial hydrology and geomorphology of monsoon-dominated Indian rivers. Revista Brasileira Geomorfologia 6:63–73
Kale VS, Achyuthan H, Jaiswal MK, Sengupta S (2010) Palaeoflood records from upper Kaveri river, southern India: evidence for discrete floods during Holocene. Geochronometria 37:49–55
Kim I, Jeong GY, Park SJ, John T (2011) Predicted land use change in the Soyang river basin, South Korea. In: Proceedings of Terreco science conference. Karlsruhe Institute of Technology, Karlsruhe, Germany, pp 17–21, 2–7 Oct 2011
Knudson PT (2011) Building regional capacity for land use reform: environmental conservation and historic preservation in the Hudson river valley. Human Ecol Rev 18:53–66
Koch J, Onigkeit J, Schaldach R, Alcamo J, Köchy M, Wolff H, Kan I (2011) Land use scenarios for the Jordan River region. Int J Sustain Water Environ Syst 3:25–31
Kondolf GM (1997) Hungry water: effects of dams and gravel mining on river channels. Environ Manage 21:533–551
Kondolf GM, Swanson ML (1993) Channel adjustments to reservoir construction and instream gravel mining, Stony Creek. Calif Environ Geol Water Sci 21:256–269
Kotoky P, Dutta MK, Borah GC (2012) Changes in land use and land cover along the Dhansiri River channel, Assam: a remote sensing and GIS approach. J Geol Soc India 79:61–68
Kroeze C, Seitzinger SP (2002) The impact of land use in Europe on N inputs to rivers and estuaries and related NO_2 emissions: a scenario analysis. In: Proceedings of symposium on agricultural effects on ground and surface waters: research at the edge of science and society, IAHS Publication. No.273, pp 355–360, Oct 2000
Leggett K, Fennessy J, Schneider S (2003) Seasonal vegetation changes in the Hoanib River catchment, north-western Namibia: a study of a non-equilibrium system. J Arid Envron 53:99–113
Li S, Gu S, Liu W, Han H, Zhang Q (2008) Water quality in relation to land use and land cover in the upper Han River Basin, China. Catena 75:216–222
Macfarlane M, Mitchell P (2003) Scoping and assessment of the environmental and social of river mining in Jamaica. MERN Working Paper, University of Warwick, 87 pp
McNeil DG, Fredberg J (2011) Environmental water requirements of native fishes in the Middle river catchment, Kangaroo Island, South Australia. Report submitted to the SA department for water, SARDI research report series 528, 50 pp

Memarian H, Tajbakhsh M, Balasundaram SK (2013) Application of SWAT for impact assessment of land use/land cover change and best management practices: a review. Int J Adv Earth Environ Sci 1:36–40

Miller MD (1993) Planning, power and politics: a case study of the land use and sitting history of of the North River water pollution control plant. Fordham Urban Law J 21:707–722

Morrison A (2013) Sustaining environmental flows in southern New England rivers, effects of watershed factors and land use. Dissertation submitted to the University of Rhode Island, U.K, 148 pp

Moss T (2004) The governance of land use in river basins: prospects for overcoming problems of institutional interplay with the EU water framework directive. Land Use Policy 21:85–94

Mutie SM, Mati B, Home P, Gadain H, Gathenya J (2006) Evaluating land use change effects on river flow using USGS geospatial stream flow model in Mara River basin, Kenya. In: Proceedings of 2nd workshop of the EARSeL SIG on land use and land cover, Bonn, pp 141–148, 28–30 Sept 2006

Naiman RJ, Bilby RE (1998) River ecology and management in the Pacific coastal ecoregion. In: Naiman RJ, Bilby RE (eds) River ecology and management: lessons from the Pacific Coastal ecoregion. Springer, New York, pp 1–22

Okazava H, Yamamoto D, Takeuchi Y (2010) Influences of riparian land use on nitrogen concentration of river water in agricultural and forest watersheds of northeastern Hokkaido. Jpn Int J Environ Rural Dev 1:1–2

Ongley ED, Broekhoven LH (1978) Land use, water quality and river mouth loadings: a selective overview for Southern Ontario. Technical report submitted to the reference group on Great Lakes pollution from land use activities, International Joint Commission, Canada, 110 pp

Padmalal D, Maya K, Sreebha S, Sreeja R (2008) Environmental effects of river sand mining: a case from the river catchments of Vembanad lake. Southwest India Environ Geology 54:879–889

Qi S, Luo F (2006) Land use change and its environmental impact in the Heihe River basin, arid northwestern China. Environ Geol 50:535–540

Rajani Kumari V, Pattabhi Ramayya M, Ramkumar M (2000) Physico-chemical quality conditions of ground water in coastal regions of the Krishna delta during summer season. J Geol Assoc Res Centre 8:61–66

Rajaram R, Devendran S (2013) Ichthyofauna as a tool to assess the heavy metal pollution in the Cuddalore coast, Southeast India. In: Ramkumar M (ed) On a sustainable future of the Earth's Natural resources. Springer, Heidelberg, pp 411–424

Ramkumar M (2000a) Estuarine water and suspended sediment sampling: a new instrument and etiquettes in sampling. J Geol Assoc Res Centre 8:1–3

Ramkumar M (2000b) Recent changes in the Kakinada spit, Godavari delta. J Geol Soc Ind 55:183–188

Ramkumar M (2001) Sedimentary microenvironments of modern Godavari delta: characterization and statistical discrimination—towards computer assisted environment recognition scheme. J Geol Soc India 57:49–63

Ramkumar M (2003a) Geochemical and sedimentary processes of mangroves of Godavari delta: Implications on estuarine and coastal biological ecosystem. Indian J Geochem 18:95–112

Ramkumar M (2003b) Progradation of the Godavari delta: a fact or empirical artifice? Insights from coastal landforms. J Geol Soc India 62:290–304

Ramkumar M (2004a) Dynamics of moderately well mixed tropical estuarine system, Krishna estuary, India: part IV zones of active exchange of physico-chemical properties. Indian J Geochem 19:245–269

Ramkumar M (2004b) Dynamics of moderately well mixed tropical estuarine system, Krishna estuary, India: part III nature of property-salinity relationships and quantity of exchange. Indian J Geochem 19:219–244

Ramkumar M (2007) Spatio-temporal variations of sediment texture and their influence on organic carbon distribution in the Krishna estuary. Indian J Geochem 22:143–154

Ramkumar M (2009) Types, causes and strategies for mitigation of geological hazards. In: Ramkumar M (ed) Geological hazards: causes, consequences and methods of containment. New India Publishers, New Delhi, pp 1–22
Ramkumar M, Gandhi MS (2000) Beach rocks in the modern Krishna delta. J Geol Assoc Res Centre 8:22–34
Ramkumar M, Neelakantan R (2007) GIS technology based geohazard zonation and advance warning system for geohazard mitigation and information dissemination towards relief and rescue operations. J Earth Sci 1(3):65–70
Ramkumar M, Pattabhi Ramayya M (1999) Low cost water sampler for shallow water bodies. J Geol Soc India 54:93–96
Ramkumar M, Vivekananda Murty M (2000) Distinction of sedimentary environments of the Godavari delta using geochemical and granulometric data through analysis of variance (ANOVA). Indian J Geochem 15:69–84
Ramkumar M, Pattabhi Ramayya M, Swamy ASR (1999) Changing landuse/land cover pattern of coastal region: Gautami sector, Godavari delta, India. J A.P Acad Sci 3:11–20
Ramkumar M, Rajani Kumari V, Pattabhi Ramyya M, Vivekananda Murty M (2000a) Geochemical characteristics and depositional conditions of coastal sedimentary environments of the modern Godavari delta. Indian J Geochem 15:31–44
Ramkumar M, Sudha Rani P, Gandhi MS, Pattabhi Ramayya M, Rajani Kumari V, Bhagavan KVS, Swamy (2000b) Textural characteristics of coastal sedimentary environments of the modern Godavari delta. J Geol Soc India 56:471–487
Ramkumar M, Pattabhi Ramayya M, Gandhi MS (2000c) Beach rock exposures at wave cut terraces of modern Godavari delta: their genesis, diagenesis and indications on coastal submergence and sealevel rise. Indian J Mar Sci 29:219–223
Ramkumar M, Rajani Kumari V, Pattabhi Ramayya M, Gandhi MS, Bhagavan KVS, Swamy ASR (2001a) Dynamics of moderately well mixed tropical estuarine system, Krishna estuary, India: part I spatio-temporal variations of physico-chemical properties. Indian J Geochem 16:61–74
Ramkumar M, Pattabhi Ramayya M, Rajani Kumari V (2001b) Dynamics of moderately well mixed tropical estuarine system, Krishna estuary, India: part II flushing time scales and estuarine mixing. Indian J Geochem 16:75–92
Ranade P (2007) Environmental impact assessment of land use planning around the leased limestone mine using remote sensing techniques. Iran J Environ Health Sci Eng 4:61–65
Rao GK (1998) Geo-hazard with oil and gas production in Krishna-Godavari Basin. Curr Sci 74:494
Robson B, Austin C, Chester E (2005) Environmental water allocation required to sustain macroinvertebrate species in ephemeral streams. River Riparian Lands Manage Newslett 29:13–15
Robertson AI, Duke NC (1987) Mangroves as nursery sites: comparisons of the abundance of fish and crustaceans in mangroves and other nearshore habitats in tropical Australia. Mar Biol 96:193–205
Rogers P, Lydon P, Seckler D (1989) Eastern waters study: strategies to manage flood and draught in the Ganga-Brahmaputra basin. ISPAN, USAID, Washington
Sacramento River Flood Control Project (2012) Flood control and geomorphic conditions.112 pp Report Accessed on 07 July 2012
Sandecki M (1989) Aggregate mining in River systems. Calif Geol 42:88–94
Scoccimarro M, Collins D (2005) Using buy backs to secure water for environmental flows. River Riparian Lands Manage Newslett 29:5–6
Seto KC, Woodcock CE, Song C, Huang X, Lu J, Kaufmann RK (2002) Monitoring land-use change in the Pearl River delta using Landsat TM. Int J Remote Sens 23:1985–2004
Sharma P, Deka D, Saikia R (2011) An analysis of changing land use pattern and its effect on Umtrew basin, Northeast India. Hung Geogr Bull 60:67–78
Shu L, Finlayson B (1993) Flood management on the lower Yellow River: hydrological and geomorphological perspectives. Sed Geol 85:285–296
Siahaya FZ, Hermana J (2013) Environmental impact assessment of land use due to the change of water resources potential in urban watershed: a review. J Basic Appl Sci. Res 3:420–425

Singer MB, Aalto R, James LA (2008) Status of the lower sacramento valley flood-control system within the context of its natural geomorphic setting. Nat Hazards Rev 9:104–115

Soini E (2005) Land use change patterns and livelihood dynamics on the slopes of Mt. Kilimanjaro. Tanzania Agric Syst 85:306–323

Soini E (2006) Livelihood, land use and environment interactions in the highlands of East Africa. Dissertation report submitted to the University of Helsinki, Finland, 39 pp

Solaraj G, Dhanakumar S, Rutharvel Murthy K, Mohanraj R (2010) Water quality in select regions of Cauvery Delta River basin, southern India, with emphasis on monsoonal variation. Environ Monit Assess 166:435–444

Sonak S, Pangam P, Sonak M, Mayekar D (2006) Impact of sand mining on local ecology. In: Sonak S (ed) Multiple dimensions global environmental change. Teri Press, New Delhi, pp 101–121

Speed R, Gippel C, Bond N, Bunn S, Qu X, Zhang Y, Liu W (2012) Assessing river health and environmental flow requirements in Chinese rivers. International Water Centre, Brisbane 63p

Sreebha S, Padmalal D (2011) Environmental impact assessment of sand mining from the small catchment river in the southwestern coast of India. Environ Manage 47:130–140

Tafangenyasha C, Dzinomwa T (2005) Land-use impacts on river water quality in lowveld sand river systems in south-east Zimbabwe. Land Use Water Resour Res 5:3.1–3.10

Tang Z, Engel BA, Pijanowski BC, Lim KJ (2005) Forecasting land use change and its environmental impact at a watershed scale. J Environ Manage 76:35–45

Tong STY, Chen W (2002) Modeling the relationship between land use and surface water quality. J Environ Manage 66:377–393

Triedman N (2012) Environment and ecology of the Colorado River basin. In: The 2012 state of the Rockies report card, pp 89–107

Walling DE (1999) Linking land use, erosion and sediment yields in river basins. Hydrobiologia 410:223–240

Weng Q (2000) Human-environment interactions in agricultural land use in a South China's wetland region: a study on the Zhujiang delta in the Holocene. Geo J 51:191–202

Wijesekara GN, Gupta A, Valeo C, Hasbani JG, Marceau DJ (2010) Impact of land use changes on hydrological processes in the Elbow River watershed in Southern Alberta. In: Swayne DA, Yang W, Voinov A, Rizzoli A, Filatova T (eds) Proceedings of international Congress on environmental modeling and software modeling for environment's sake, Ottawa, Canada. http://www.iemss.org/iemss2010/index.php?n=Main.Proceedings

World Meteorological Organization (2005) Economic aspects of integrated flood management. WMO, Kobe, 69 pp

Wu X, Shen Z, Liu Z, Ding X (2008) Land use/cover dynamics in response to environmental and socio-political forces in the upper reaches of Yangtse River. China Sens 8:8104–8122

Yaakub A, Norulaini N, Rahman NA (2012) Water quality status of Kinta River tributaries based on land use activities. In: Proceedings of international conference on environment, energy and biotechnology, Singapore, pp 178–182

Yueqing X, Ding L, Jian P (2011) Land use change and soil erosion in the Maotiao River watershed of Guizhou province. J Geogr Sci 21:1138–1152

Zarea R, Ionus O (2012) Land use changes in the Basca Chiojdului River basin and the assessment of their environmental impact. Forum Geogr 11:36–44

Zhang T, Zhao B (2010) Impact of anthropogenic land-uses on salinization in the Yellow River delta, China: using a new RS-GIS statistical model. Int Arch Photogram Remote Sens Spat Inf Sci 38:947–952

Zhang T, Zeng S, Gao Y, Ouyang Z, Li B, Fang C, Zhao B (2011) Assessing impact of land uses on land salinazation in the Yellow River delta, China using an integrated and spatial statistical model. Land Policy 28:857–866

Zimmermann B, Elsenbeer H, de Moraes JM (2006) The influence of land-use changes on soil hydraulic properties: implications for runoff regeneration. For Ecol Manage 222:29–38

Zwolsman JJG (1994) Seasonal variability and biogeochemistry of phosphorus in the Scheldt estuary, southwest Netherlands. Est Coast Shelf Sci 39:227–248

Spatio-temporal Analysis of Rainfall Distribution and Variability in the Twentieth Century, Over the Cauvery Basin, South India

Sawant Sushant, K. Balasubramani and K. Kumaraswamy

Abstract Knowledge on the spatial variability and temporal trends of mean rainfall is essential for efficient management of water resource and agriculture. We have analyzed the rainfall data of the Cauvery river basin, a larger river basin in Southern India which plays a significant role in agricultural development and consequently in the overall growth of Karnataka, Tamil Nadu, Pondicherry and some parts of Kerala. The analysis includes distribution, variability and trends in rainfall over the Cauvery basin during twentieth century (1901–2002). The impact of climate change on temporal and spatial patterns of rainfall over smaller spatial scales is clearly noticed in this analysis. It is also observed that the coefficient of variation shows significant fluctuations during the winter than other seasons. Long term changes in rainfall have been determined by. Significant decreasing trend in the winter rainfall and increasing trend in the post-monsoon season with insignificant levels have been inferred based on the Mann-Kendall rank statistics and linear trend. Overall, insignificant decrease in annual rainfall over the Cauvery river basin has observed during twentieth century.

Keywords Rainfall distribution · Variability · Trend · Mann-Kendall rank statistics · Cauvery basin

1 Introduction

River basins are important from hydrological, economic and ecological points of view. They absorb and channel the run-off from snow-melt and rainfall which, when wisely managed, can provide fresh drinking water as well as access to food, hydropower etc. Experts agree that the best approach to conserving the world's freshwater resources is through managing river basins sustainably. We need to make wise choices about the use of available resources, based on an understanding

S. Sushant (✉) · K. Balasubramani · K. Kumaraswamy
Department of Geography, Bharathidasan University, Tiruchirappalli 620 024, India
e-mail: sushantsawant13@gmail.com

Mu. Ramkumar et al. (eds.), *Environmental Management of River Basin Ecosystems*, Springer Earth System Sciences, DOI 10.1007/978-3-319-13425-3_2

of how to sustain the dynamic, living systems in the long term. Unfortunately, the close inter-linkages between the hydrological, ecological and socio-economic components of river basins have rarely been given adequate consideration by decision-makers such as politicians, land use planners and water engineers. As a result, river basins all over the world are not being managed scientifically. The need to conserve and manage freshwater and its ecosystems at the basin scale is increasingly being recognized by governments and NGOs. The principle of integrated river basin management is included in many international agreements. However, far too little is being done to put words into action.

Given cognizance to the recent reports of climate change and its extreme variability, an understanding of temporal and spatial characteristics of rainfall is crucial to planning and management of water resources especially on a basin-scale. Such information is important in agricultural planning, flood frequency analysis, flood hazard mapping, hydrological modeling, water resource assessments, climate change impacts and other environmental assessments (Michaelides et al. 2009). Precipitation is one of the most variable climatic elements, both spatially and through time scales ranging from daily to decadal and longer-term fluctuations (Juan-Carlos et al. 2009). Several researchers have studied the distribution, variability and trends of rainfall at global, regional and basinal scales (for example, Jagannathan and Parthasarathy 1973; Mooley and Parthasarathy 1983; Thapliyal and Kulshrestha 1991; Parthasarathy et al. 1995; Serrano et al. 1999; Smith 2000; Guhathakurta and Rajeevan 2008; Soltani et al. 2007; Werner 2009; Taschetto and Matthew 2009; Zhang et al. 2009). Krishnakumar et al. (2008) studied temporal (monthly, seasonal and annual) rainfall trends in twentieth century over Kerala, India and reported a significant decrease of rainfalls during the southwest monsoon and an increase during the post-monsoon season. Sivapragasam et al. (2013) studied rainfall trend at basin scale on trends in rainfall patterns over the Tamarabarani basin in Tamil Nadu, India. Anandakumar et al. (2008) studied spatial variation and seasonal behavior of rainfall pattern in lower Bhavani river basin, Tamil Nadu, India and attempt has been made to analyze the occurrence and distribution of rainfall in the basin. Chakraborty et al. (2013) made an attempt to study the spatial and temporal variability of rainfall at Seonath sub basin in the Chhattisgarh State (India) for 49 years (1960–2008). In order to study rainfall characteristics they used different statistic methods. Modified Mann-Kendall (MMK) (non-parametric) and Spearman's rho test (parametric) were applied to detect the trend. Sen's slope was used to detect trend magnitude. The CUSUM and cumulative deviations test were applied to detect change points. The Coefficient of Variation was used for variability analysis. According to both the tests decreasing trend was found in annual and seasonal rainfall series for the whole river basin. Mondal et al. (2002) applied Mann-Kendall test to document the changing trend of rainfall of a river basin of Orissa near the coastal region. Despite providing sustenance to a large geographic area, the rainfall trends of the Cauvery Basin have not yet been studied in the light of changing climatic scenarios and hence in this paper, we have constructed monthly, seasonal and annual rainfall series for the Cauvery River Basin. Statistical features of the temporal rainfall series of each district contained within the Cauvery Basin followed by the study of rainfall distribution, variability and trends of the monthly, seasonal and annual total rainfall for each of the districts.

2 Study Area

The Cauvery Basin lies between 75°27′–79°54′ East longitudes and 10°9′–13°30′ North latitudes (Fig. 1), and extends over the States of Tamil Nadu, Karnataka, Kerala and Union Territory of Pondicherry. It is draining an area of 81,155 km^2 that forms 2.7 % of the total geographical area of the country. It has a maximum length of about 560 km and width of 245 km. It is bounded by the Western Ghats on the west, by the Eastern Ghats/Bay of Bengal on the east and the south and by the ridges separating it from Krishna Basin and Pennar Basin on the north. The major part of basin is covered with agricultural land accounting to 66.21 % of the total area. The Cauvery Basin is heavily dependent on monsoon rains, and thereby is prone to droughts when the monsoons fail. The climate of the basin ranges from dry sub-humid to semi-arid. As per the Indian Metrological Department (IMD) classification, the basin is under the influence of four distinct seasons namely, Winter (January–February), Pre-monsoon with dry season (March–May), South-west monsoon with strong southwest winds (June–September) and Post-monsoon with dominant north-east retreating winds (October–December). Owing to these seasonal

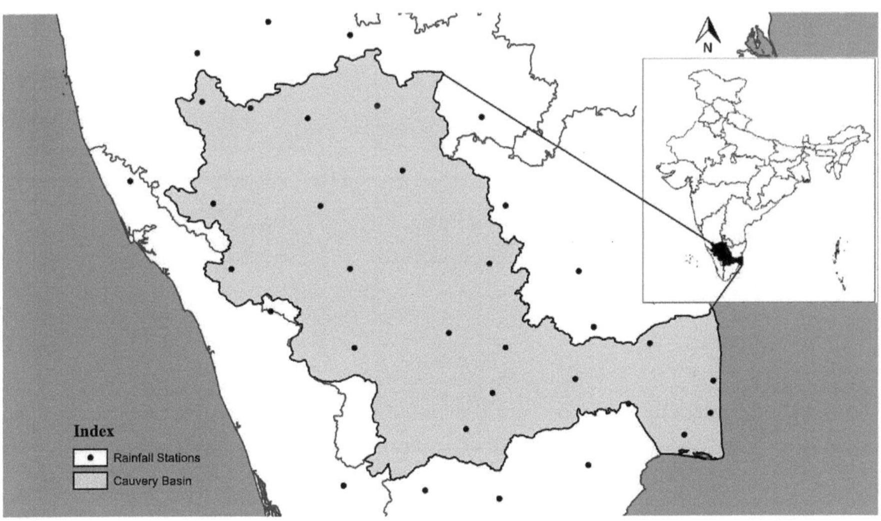

Fig. 1 Location map of the study area and rain gauge stations

influences and the location of the basin in varying topographic and geographic sprawl, the Cauvery Basin region experiences wet and dry periods, resulting in coeval prevalence of heavy rainfall and some areas receive very less rainfall.

3 Materials and Methods

The methodology has been divided into three major parts as (1) data collection and input, (2) data processing function, and (3) presentation of output. The following schematic diagram explains the steps followed in this study (Fig. 2).

The rainfall (mm) data over Cauvery Basin for the period from 1901 to 2002 has been obtained from the Indian Meteorological Department (IMD) and India Water Portal (http://www.indiawaterportal.org/met_data, Accessed date: 30/04/2013). Seasonal, annual and century rainfall data series of Cauvery basin were computed using monthly rainfall data of rain gauge stations located in and around Cauvery river basin area. It was followed by computation of mean monthly, seasonal and annual Standard Deviation (SD) and Coefficient of Variation (CV).

The coefficient of variation indicates the amount of fluctuation in rainfall recorded over a long period of time from the mean values. The coefficient of variation of annual precipitation is an index of climatic risk, indicating a likelihood of fluctuations in reservoir storage or crop yield from year to year. Agriculturally it is, perhaps, a more crucial statistics for marginal areas than in either very dry areas,

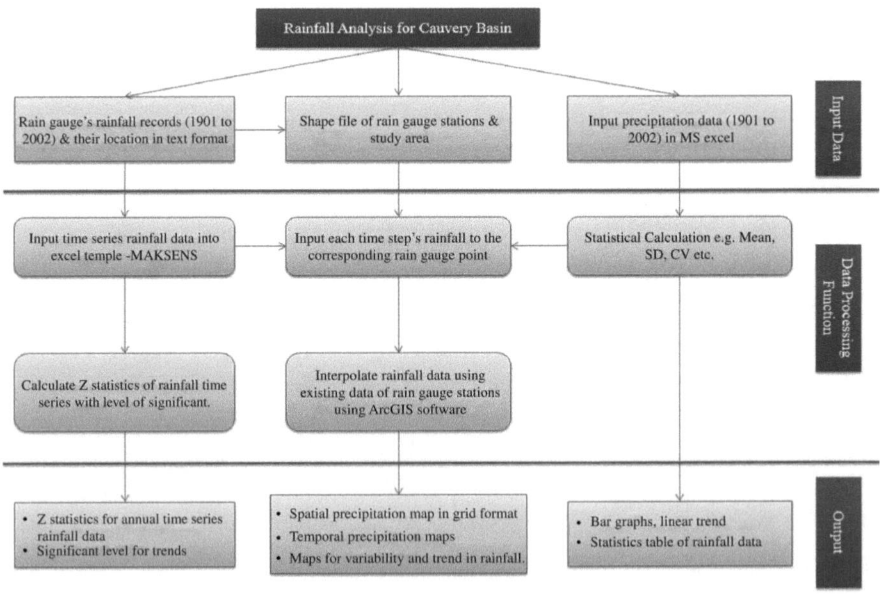

Fig. 2 Schematic diagram representing the methodology of the study

where farming practices have adapted to variability, or in wet areas, where relatively lower inter-annual variability are generally expected. Contrary to these general perceptions, no perceptible spatio-temporal variations were observed, that prompted the authors to find out the long term trends with the help of average rainfall graphs constructed with the trend line. Temporal changes in the monthly, seasonal and annual rainfall were also analyzed by Mann-Kendall test to confirm the significance of the observed trends.

Mann-Kendall test is a statistical test widely used for the analysis of trend in climatologic and hydrologic time series (Milan et al. 2013; Cannarozzo et al. 2006). There are two advantages of using this test. First, it is a non-parametric test and does not require the data to be normally distributed, suiting perfectly for the nature of distribution observed in the Cauvery Basin. Second, the test has low sensitivity to abrupt breaks due to non-homogeneous time series. In this test, the number of annual values in the studied data series is denoted by 'n'. If 'n' is at least 10, the normal approximation test is used. However, if there are several tied values (i.e. equal values) in the time series, it may reduce the validity of the normal approximation when the number of data values is close to 10. First, the variance (S) is computed by the following equation which takes into account that ties may be present.

$$VAR(S) = \frac{1}{18}\left[n(n-1)(2n+5) - \sum_{p=1}^{q}\left(t_p - 1\right)\left(2t_p + 5\right)\right] \quad (1)$$

Here q is the number of tied groups and t_p is the number of data values in the pth group.

The values of S and $VAR(S)$ are used to compute the test statistics Z as follows:

$$Z = \begin{cases} \frac{S-1}{\sqrt{Var(S)}} & if\ S > 0 \\ 0 & if\ S = 0 \\ \frac{S+1}{\sqrt{Var(S)}} & if\ S < 0 \end{cases} \quad (2)$$

Presence of a statistically significant trend is evaluated using the Z value and here the statistics Z has a normal distribution. Significance level α is used for testing either an upward or downward monotone trend (a two-tailed test). If Z appears greater than $Z\alpha/2$ where α depicts the significance level, then the trend is considered as significant. The value for $Z\alpha/2$ is obtained from the standard normal cumulative distribution tables for the significance levels (α) 0.001, 0.01, 0.05 and 0.1 (Timo et al. 2002).

4 Rainfall Characteristics of the Cauvery Basin

Rainfall characteristics of the Cauvery basin are presented in the Table 1 and in the Figs. 2, 3, 4 and 5. The observations and inferences from these are discussed herein.

The average annual normal rainfall over the Cauvery Basin from 1901 to 2002 is 1389.20 mm with a standard deviation of 643.4 mm. The coefficient of variation of annual rainfall is 46.3 % indicating that it is moderately variable. Rainfall during July is the highest (250.07 mm) and contributes 18.00 % of annual rainfall (1389.20 mm), followed by June (14.85 %), August (14.10 %) and October (13.94 %). Contribution of the seasonal rainfall to the annual rainfall is highest during the monsoon period (56.81 %), followed by post-monsoon period (27.80 %), pre-monsoon period (13.71 %) and winter period (1.68 %) in the decreasing order. Least amounts of rainfall are observed during the month of February (10.70 mm) followed by January (12.69 mm), which contribute only 0.77 and 0.91 % to the annual rainfall respectively. The coefficient of variation is highest in July (100 %), followed by June (97.96 %), December (97.96 %) and January (78.48 %) and the least during October (15.19 %) and September (23.77 %). Significant relationship between SD and CV has been observed during the months of highest rainfall (July and June).

Table 1 Mean monthly, seasonal and annual rainfall statistics of Cauvery basin (1901–2002)

Month/season	Rainfall (mm)	Standard deviation	Coefficient of variation (%)	% Contribution to annual rainfall
January	12.69	10.0	78.5	0.91
February	10.70	7.0	65.2	0.77
March	15.95	7.0	43.9	1.15
April	60.30	25.4	42.2	4.34
May	114.20	48.1	42.1	8.22
June	206.27	202.1	98.0	14.85
July	250.07	250.2	100.0	18.00
August	195.91	144.9	74.0	14.10
September	136.96	32.6	23.8	9.86
October	193.68	29.4	15.2	13.94
November	133.48	56.1	42.0	9.61
December	58.99	46.3	78.5	4.25
Winter (January–February)	23.39	15.8	67.5	1.68
Pre-Mon (March–May)	190.45	73.4	38.6	13.71
Monsoon (June–September)	789.22	621.4	78.7	56.81
Post-Mon (October–December)	386.14	110.7	28.7	27.80
Annual	1389.20	643.4	46.3	100

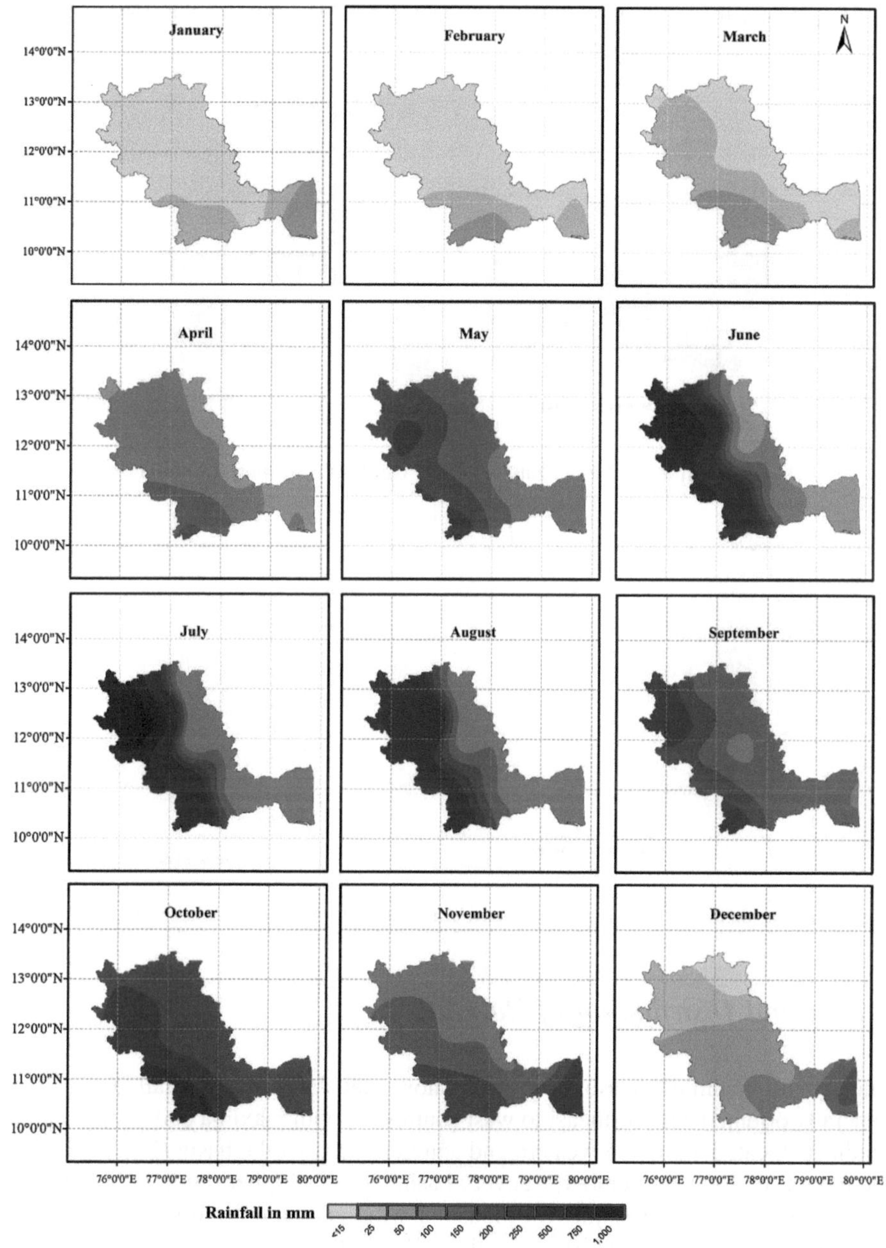

Fig. 3 Spatial distribution of the mean monthly rainfall over Cauvery basin (1901–2002)

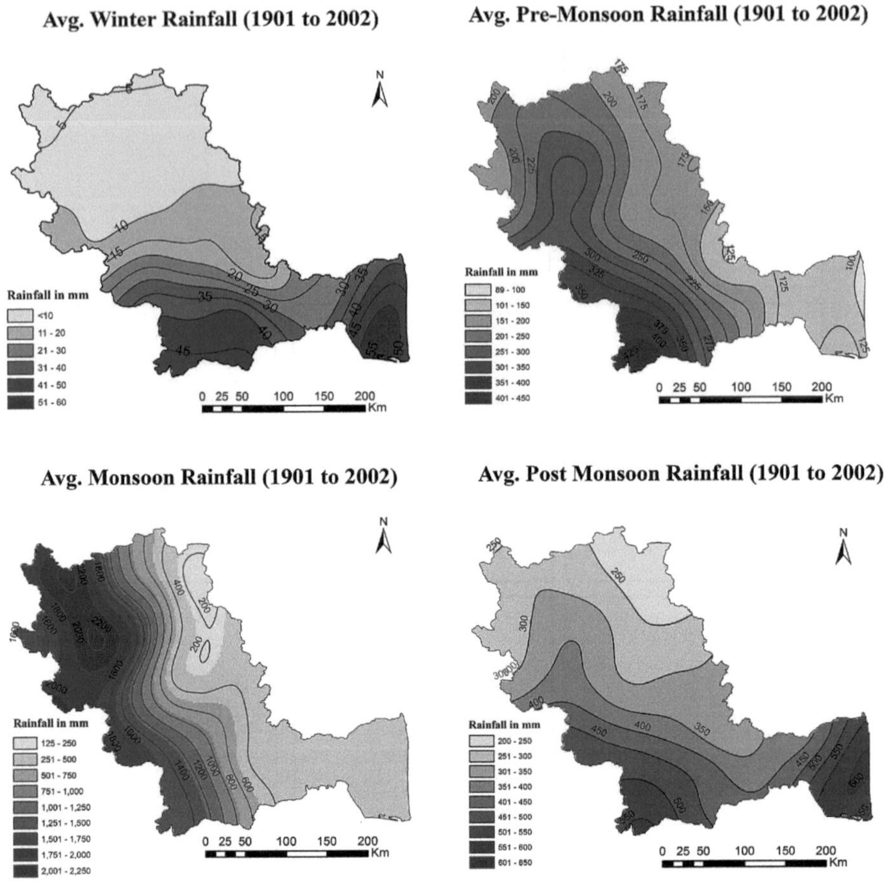

Fig. 4 Spatial distribution of the mean seasonal rainfall over Cauvery basin (1901–2002)

4.1 Spatial Distribution of Rainfall

The monthly rainfall distribution maps show that the major rainfall patterns are generally oriented east northeast to west southwest, with maxima and strong rainfall gradients located along the western and southern parts of the basin (Fig. 3). During the months of June–September, Cauvery basin receives higher rainfall (789.99 mm) as a result of prevalent southwest monsoon wind. Though the northeast monsoon (October–December) provides only the half of the total amount of southwest monsoon (386.14 mm), it remains important since it satisfies the agricultural activities especially in the Cauvery deltaic region. Generally, highest rainfall (up to

Fig. 5 Spatial distribution of the mean annual rainfall over Cauvery basin (1901–2002)

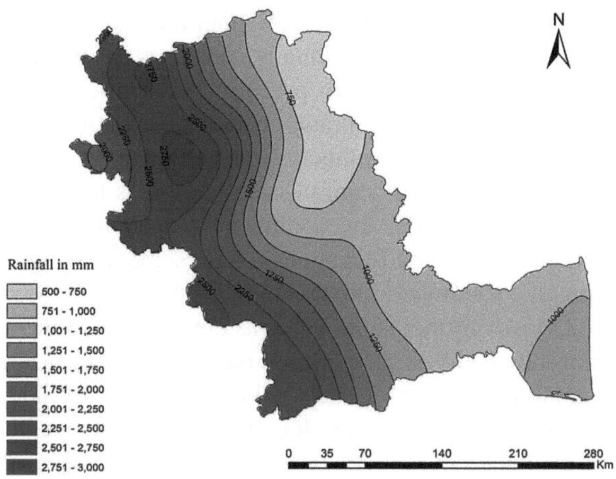

250 mm) pattern is observed in the western boundary of the study area during May–September. However, in the months of October, November and December, the pattern changes from western boundary to extreme eastern side (Cauvery Delta). This seasonal shift of rainfall maxima zones could be clearly identified from the maps (Fig. 3). Overall, the month of January, February, March, April and May experience lower amount of rainfall thus known as dry season.

4.2 Seasonal Rainfall Distribution

On the basis of the percentage of contribution to the annual rainfall, the basin experiences four distinct periods of rainfall (Fig. 4) namely, the pre-monsoon rainfall season from March to May (Hot summer), the southwest monsoon rainfall season with strong southwest wind (June–September), the post-monsoon with dominant northeast wind during October–December (Northeast monsoon), and the winter season (January–February).

During the pre-monsoon, the rainfall ranges from 90 to 450 mm. Isohyets of low rainfall values (100–200) are located in the eastern and central parts of the basin and are aligned north–south. Contrary to this pattern, isohyets of the higher rainfall values are located in the southwest part of the basin. Western and central west areas of the basin experience moderate rainfall. The basin receives copious rainfall (57 % of annual rain fall) during the monsoon season. This season has maximum number of rainy days and is called the 'wet season'. Rainfall during this season is caused by the southwest monsoon winds blowing from the Indian Ocean. Since a part of the basin is shared by western face of the Western Ghats section, the western part of the basin gets higher rainfall throughout the monsoon season, especially northwest part of the basin. Rainfall gradually decreases eastwards due to rainshadow effect. Thus eastern and south-eastern parts of the basin experience lower rainfall (<300 mm).

The monsoon withdraws from the peninsula by October and from the extreme southern tip by December. Due to retreat of the monsoon, the rainfall is called retreating monsoon rainfall, also known as the northeast monsoon rainfall. By the end of September, the southwest monsoon becomes weak as the low pressure trough of the Indo-Gangetic plain starts moving southward in response to the apparent southward movement of the Sun. By October, it reaches the Bay of Bengal and moves further southwards as the season advances. The weather in the retreating monsoon is dry in the northern part of the basin (Karnataka) but it is associated with rainfall (>400 mm) in the southern part (Tamil Nadu) of the basin. The widespread rain in this season in the deltatic region of the basin is associated with the passage of cyclonic depressions which originate over Bay of Bengal. Southeastern part of the Cauvery basin gets more rainfall during this season than in any other seasons. The winter season remains dry and contributes very low rainfall (0–60 mm) to the annual total. South and southwest part of the basin receive about 40–60 mm and remaining area comes under dry climate with less than 30 mm rainfall.

4.3 Annual Rainfall

Spatial distribution of the annual rainfall is shown in the Fig. 5. This map clearly depicts the decreasing trend of rainfall from western part to eastern part of the basin. The western margin of the study area receives higher rainfall (>2,000 mm) during June–September due to orographic effect whereas the eastern margin of the basin remains dry due to its location on the leeward side of the Western Ghats. However, the southeast part of the Cauvery basin gets moderate amount of rainfall from northeast monsoon. Most of the drought phenomena occur in the eastern part of the basin due to the low rainfall.

5 Spatial Variability of Rainfall as Deduced from Coefficient of Variation

Thought the distribution characteristics of the raw data show certain patterns, in order to understand precisely the spatial variability of rainfall in the Cauvery basin, Coefficient of Variation (CV) has been computed using mean seasonal and annual rainfall values as these showed higher variability than on an annual basis (Schulze 1983). It is observed that the pre-monsoon, monsoon and post-monsoon seasons

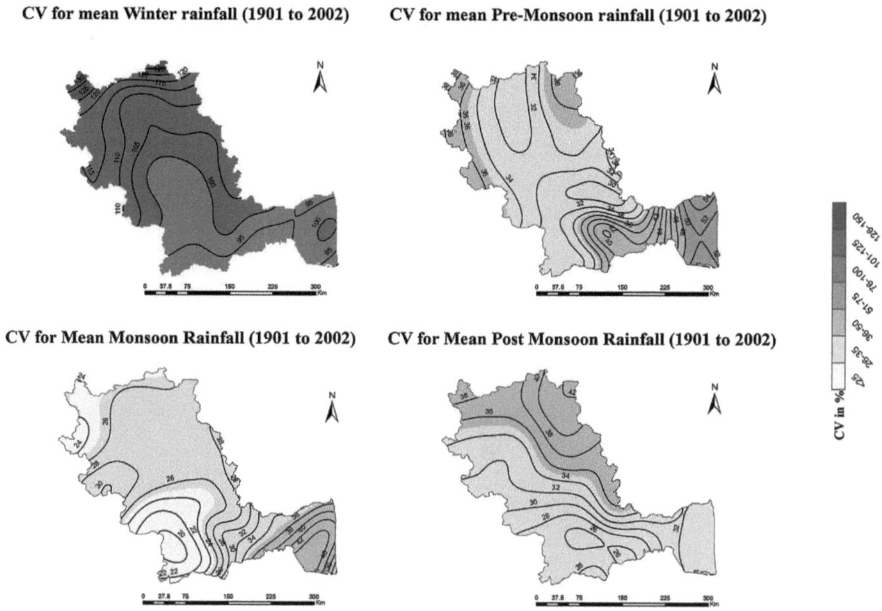

Fig. 6 Coefficient of variation in mean seasonal rainfall in the Cauvery basin (1901–2002)

have CV less than 55 % which indicates less to moderate variability. On the contrary, the winter season has significant variability (139 %). In the winter season, the west, north and eastern part of the Cauvery region has higher rainfall variability and isolines with the value higher than 100 % have been concentrated over this region (Fig. 6). A rule of thumb established already by Conrad (1941), from analyses of precipitation worldwide, is that higher the mean average precipitation the lower its inter-annual variability. In other words, areas with a low annual rainfall are likely to be doubly worse off, because they will additionally suffer from high deviations around their already low average rainfall. In the following Table 2, districts which experience highest and lowest CV in different seasons are listed.

The map of inter-annual CV is a "best case" scenario of rainfall variation. The CV for annual rainfall ranges between 15 and 23 %. Surprisingly, the annual spatial variation in the northwest and southeast parts of the basin is the highest (20–25 %) even though both these regions receive comparatively higher annual mean rainfall (Fig. 7). This is mainly due to the uncertainty in pre-monsoon rainfall and it is clearly noticed in the map of CV of the pre-monsoon period (Fig. 6). The figure also depicts the occurrence of lower CV (<20 %) all over the basin excluding the northwest and southeastern regions.

Table 2 Range of coefficient of variation (in percent) and associated districts

Season	Pre-monsoon		Monsoon		Post-monsoon		Winter		Annual	
	Min.	Max.	Min.	Max.	Min.	Max.	Min.	Max.	Min.	Max.
District	Erode	Cuddalore	Kodagu	Pudukkottai	Dindigul	Bangalore (rural)	Idukki	Tumkur	Coimbatore	Karaikal
CV Values	30.80	55.35	21.09	42.61	27.30	40.86	90.42	139.26	15.11	22.69

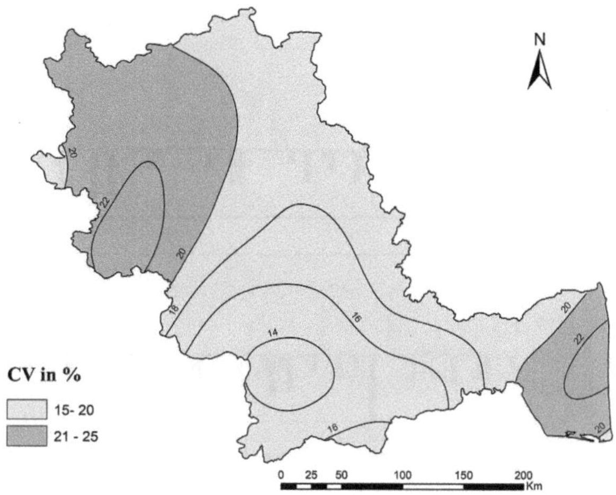

Fig. 7 Coefficient of variation for mean annual rainfall (1901–2002)

6 Deviation in Rainfall from Mean

The deficient or excess rainfall years are defined when rainfall of that year departs from the mean rainfall. In the twentieth century (1901–2002), there were 56 years which recorded annual rainfall below average (Fig. 8). Guhathakurta and Rajeevan (2008) studied the rainfall pattern over India and found 30 years of alternating sequences of dry and wet periods. They delineated the twentieth century into (a) 1901–1930 as dry period (b) 1931–1960 as wet period (c) 1961–1990 as dry period and (d) 1991–2020 as likely wet period. During the multi-decadal dry period 1901–1930 and 1961–1990, there were 12 years (6 years in each spell) of negative deviation of annual rainfall with more than 200 mm over the Cauvery Basin. Similarly, during wet period 1931–1960, there were 6 years of positive deviation of annual rainfall with more than 200 mm.

During the winter season, especially after 1961, more negative deviations of rainfall from the mean have been identified. Winter is the only season in which decreasing trend with significant level has been detected (Fig. 9). This is also validated by the Mann-Kendall trend analysis. For agricultural reason, this declining trend is more important since the agriculturalists grow more crops grow during this season traditionally. Akin to the winter season, rainfall deviation during the pre-monsoon season was higher after 1961 but the number of years was less than the winter season. In contrast, during post-monsoon season the numbers of years with negative (positive) deviation from mean rainfall is higher before (after) 1961. Unlike other seasons, monsoon period had alternate positive and negative deviations from mean rainfall but amount of deviation was higher in positive direction (14 years recorded positive deviation of >200 mm). Notwithstanding, negative deviation years were more (61 years) than the positive deviation years during the monsoon period (Fig. 8).

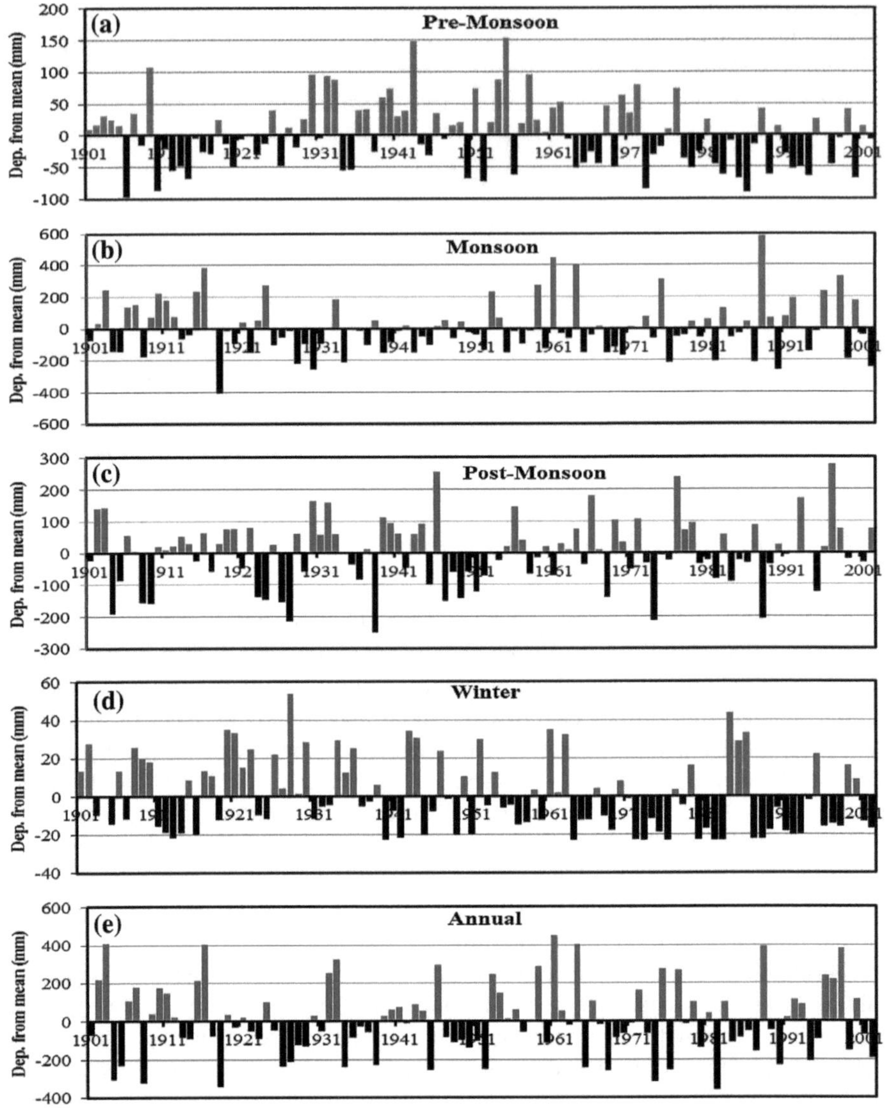

Fig. 8 Rainfall deviation from mean over Cauvery basin (1901–2002)

7 Trends in the Rainfall

Figure 9 shows the 5-year-running mean and trend line drawn based on the linear fit for the seasonal and annual rainfall series of the Cauvery Basin. The results of the Mann-Kendall trend analyses for annual, seasons and all months are summarized in

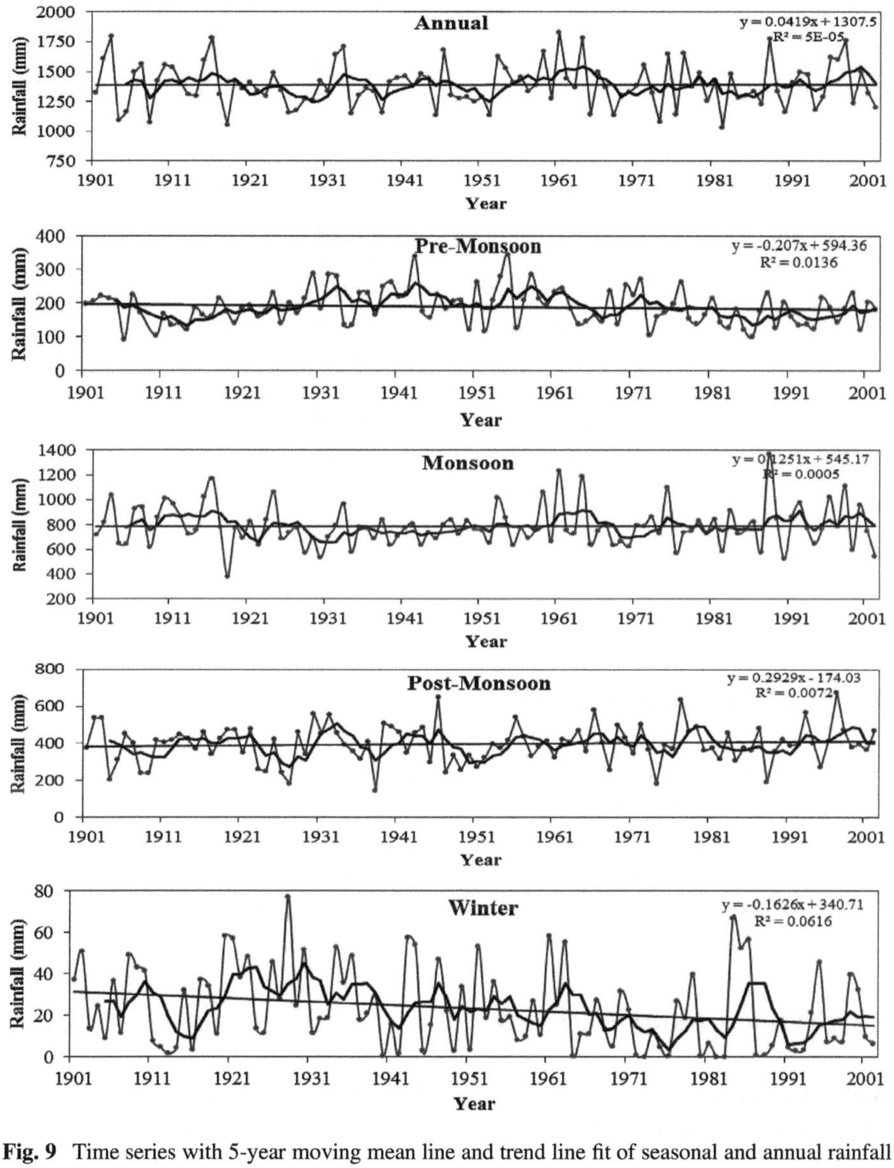

Fig. 9 Time series with 5-year moving mean line and trend line fit of seasonal and annual rainfall over the Cauvery basin (1901–2002)

the Table 3. The results of district-wise Mann-Kendall trend analyses are presented in the Table 4 and the same have been depicted in the Fig. 11. The observations and inferences drawn from these tables and figures are presented herein.

Table 3 Mann-Kendall trend statistics of monthly and seasonal rainfall over the Cauvery Basin

Month/season	Test Z	Significant level (%)
January	−2.32758	95
Febrauary	−1.47188	<90
March	−1.13053	<90
April	−0.27757	<90
May	−0.72284	<90
June	0.38744	<90
July	−0.38166	<90
August	−0.00578	<90
September	0.69393	<90
October	1.68277	90
November	−1.05246	<90
December	0.87319	<90
Winter	−2.76414	99
Pre-monsoon	−1.13341	<90
Monsoon	0.01735	<90
Post-monsoon	0.55514	<90
Annual	−0.1677	<90

First year = 1901; last year = 2002; n = 102

7.1 Monthly Rainfall Trends

The rainfall during January shows negative trend which is significant at 0.05 level. Rainfall in the months of February–May, July, August and November also show negative trends but all are insignificant. Rainfall in October shows positive trend, which is statistically significant at 0.10 level. Rainfall in the month of June, September and December show insignificant positive trends.

7.2 Seasonal Rainfall Trends

With reference to the long-term average (1901–2002), a decrease in the pre-monsoon rainfall from 1910 to 1929 followed by an increase up to 1965 with negligible fluctuation, and a declining trend from 1982 to 1999 are observed. There is an insignificant decrease of 21 mm rainfall over the 102 years period compared to the normal rainfall of 190.45 mm. Decreasing trend with −89 % confidence level has been observed over most parts of the Cauvery Basin (Fig. 10), except the Bangalore Urban, Bangalore Rural and Idukki districts which experienced an increasing trend (89 % level) and Karaikal and Nagapattinam experienced a decreased trend (−90 % level).

Table 4 District-wise mean annual rainfall, CV and Mann-Kendall trend statistics (1901–2002)

District ID	District name	Mean annual rainfall (1901–2002)	Annual CV	Annual Z Stat.	Significant level
1	Ariyalur	876.48	18.78	0.79	<90
2	Bangalore	838.66	18.75	1.14	<90
3	Bangalore rural	834.17	19.25	1.49	<90
4	Chamarajanagar	1532.72	18.87	0.25	<90
5	Chikmagalur	2442.02	22.24	−1.14	<90
6	Coimbatore	2239.09	15.11	−0.02	<90
7	Cuddalore	972.39	19.6	0.67	<90
8	Dharmapuri	815.17	19.6	1.78	90
9	Dindigul	1521.33	15.25	0.56	<90
10	Erode	1306.37	15.56	0.39	<90
11	Hassan	2727.19	20.9	−0.6	<90
12	Idukki	2112.96	15.14	−0.01	<90
13	Karaikal	1020.57	22.69	0.34	<90
14	Karur	1068.17	15.72	0.79	<90
15	Kodagu	1991.91	17.05	−1.64	<90
16	Mandya	1741.63	20.75	0.62	<90
17	Mysore	2586.03	21.15	−0.06	<90
18	Nagapattinam	1000.93	21.4	0.35	<90
19	Namakkal	879.64	17.34	1.54	<90
20	Nilgiris	2583.69	21.81	1.5	<90
21	Perambalur	842.1	19.62	0.52	<90
22	Pudukkottai	862.13	17	1.91	90
23	Salem	822.32	19.23	0.56	<90
24	Thanjavur	913.41	19.93	−0.62	<90
25	Thiruvarur	1073.43	21.74	0.14	<90
26	Tiruchirappalli	886.38	17.19	1.39	<90
27	Tumkur	1017.63	18.2	0.34	<90
28	Wayanad	2514.97	21.63	−3.02	99

Both the Mann-Kendall statistics and trend line manifested an insignificant trend of deviation in the monsoon season from the 5-year moving average value line. These have also shown the cyclic nature of the increases and decreases of the rainfall trend with roughly a period of 15 years. Increase in the rainfall during monsoon period is insignificant (0.017348). Increase in rainfall trend with 90 % significant level has been observed especially in the eastern part of the study area. Besides, the western and south eastern parts show the occurrence of decrease in trend with −89 % significant level (Fig. 10). Interestingly, significant level of −99 % decreasing trend had been observed in Waynad.

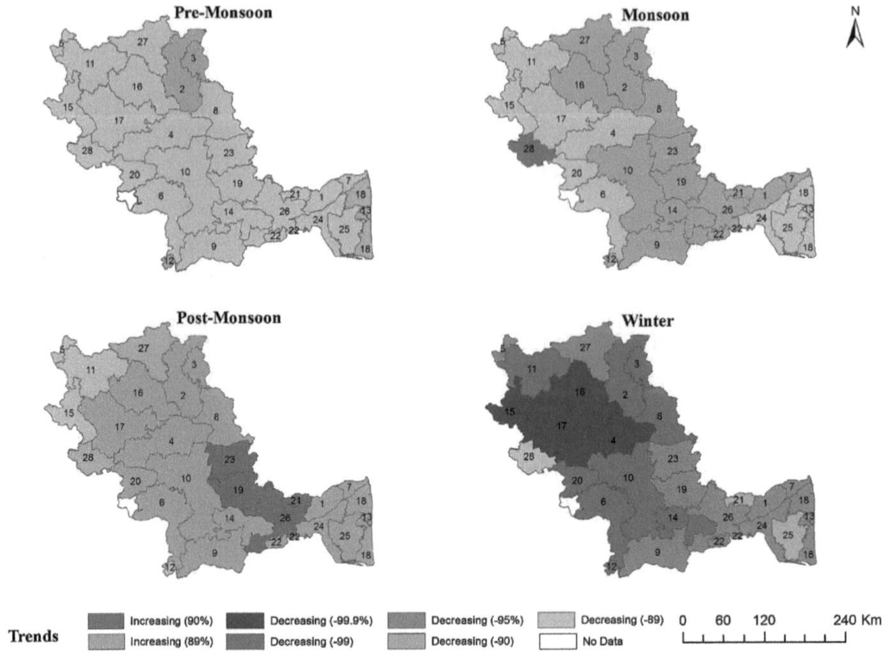

Fig. 10 District-wise trends in the seasonal rainfall over the Cauvery basin (1901–2002)

Rainfall in the post-monsoon season fluctuated moderately from mean. Similar to the monsoon period, post-monsoon is also showing cyclic nature of wet and dry periods as a result of recurrent cyclones over the Bay of Bengal. Overall, during 20th century, the post-monsoon rainfall had increased at insignificant level. This increase was about 0.6 mm/year during the study period (1901–2002). Increasing trend with 89 % significant level is observed in 18 districts and with 90 % significant level is observed in four districts (Fig. 10).

The winter rainfall has decreased significantly from 1901 to 2002. Out of four seasons, only in winter season decreasing trend is with 0.01 significant level is observed. Though, there is a significant level of decrease, the actual rainfall is not significant since the average rainfall in this season is only 23.4 mm. The decrease of 0.16 mm rainfall per year has a negligible contribution in the overall trend. Compared to all other seasons, the winter season manifested significant rainfall trend. Almost all the districts recorded decreasing rainfall trend with different significant levels especially those located in the central, north eastern and western parts of the basin with −99 and −99.99 % significant levels (Fig. 10). There is only one exception of Idukki district which recorded an increase rainfall trend.

7.3 *Annual Rainfall Trends*

The mean annual rainfall over the Cauvery Basin shows a long term insignificant declining trend. However, the declining trend (−1.02051) in annual rainfall was significant if the annual rainfall is considered from 1901 to 1950. The annual rainfall after 1950 has declined with insignificant level (−0.48136). The 5-year moving average shows the cyclic nature of the wet and dry periods (Fig. 9). During the period of 50 years from 1901 to 1950, a decrease of 150 mm was noticed which is highly appreciable quantity. During 1951–2002, a negligible decline of 33.75 mm rainfall has been recorded. As most parts of the basin in the Tamil Nadu side receive rainfall not only from monsoon season (56.81 %) but also from the post-monsoon season (27.80 %), such a trend is exhibited. Out of 28 districts in the study area, 20 have recorded positive rainfall trend and 8 have negative trend.

8 Conclusions

Analysis of the distribution, variability and trends of rainfall data of more than 100 years over the Cauvery Basin revealed the following.

Over the western part of the basin the amount of rainfall is high since this part experiences southwest monsoon rainfall. The coastal southeastern part of the basin gets more rainfall during the retreating monsoon. The CV ranges between 15 and 23 % for the Cauvery Basin as a whole. The CV in the northwestern and southeastern parts of the basin are the highest (20–25 %). Lower CV are observed all over the basin excluding the northwest and southeast regions. The results of the Mann-Kendall trend analyses show a significant decreasing trend in the winter rainfall over the Cauvery Basin. On the other hand, increasing trend in the post-monsoon season with insignificant level has been observed. The Dharmapuri and Pudukottai have highest positive Z values i.e. increasing trend of rainfall while Wayanad has negative Z values i.e. decreasing trend of rainfall. The maps (Figs. 10 and 11) showing

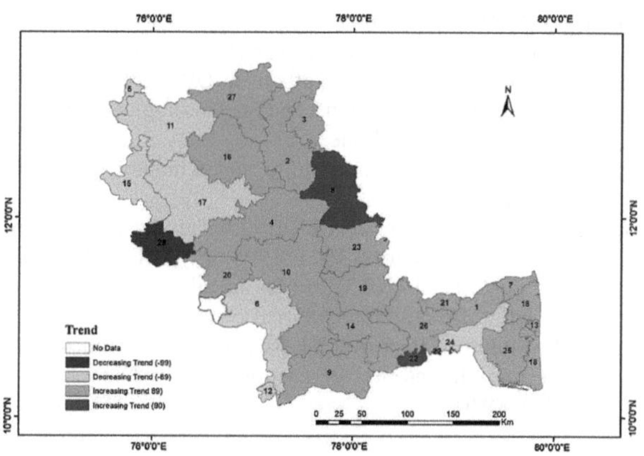

Fig. 11 District-wise trends in the annual rainfall over the Cauvery basin (1901–2002)

district-wise Z values for seasonal and annual rainfall clearly exhibit the trends in spatial dimension over the basin. These results can help the planners, administrators and the stake-holders (principally the agriculturalists) to strategize the development, management and utilization activities.

Acknowledgments This research was supported by UGC-SAP-DRS Phase-I Program of the Department of Geography, Bharathidasan University, Tiruchirappalli and the support is gratefully acknowledged. The authors also thank Indian Metrological Department (IMD) and India Water Portal for providing necessary data.

References

Anandakumar S, Subramanp T, Elango I (2008). Spatial Variation and Seasonal behaviour of rainfall pattern in Lowerbhavani river basin, Tamil Nadu, India. Int Biannu J Environ Sci (The Ecoscan 2):17–24

Cannarozzo M, Noto LV, Viola F (2006) Spatial distribution of rainfall trends in Sicily (1921–2000). Phys Chem Earth 31(18):1201–1211

Chakraborty S et al (2013) Trend and variability analysis of rainfall series at Seonath River Basin, Chhattisgarh (India). Int J Appl Sci Eng Res 2(4):425–434

Conrad V (1941) The variability of precipitation. Mon Weather Rev 69:5–11

Guhathakurta P, Rajeevan M (2008) Trends in the rainfall pattern over India. Int J Climatol 28:1453–1469

Jagannathan P, Parthasarathy B (1973) Trends and periodicities of rainfall over India. Mon Weather Rev 101:371–375

Juan-Carlos A, Brian HL (2009) Spatio-temporal rainfall patterns in Southern South America. Int J Climatol 29:2106–2120

Krishnakumar KN, Prasada Rao GSLHV, Gopakumar CS (2008) Rainfall trends in twentieth century over Kerala, India. Atmos Environ 43:1940–1944. doi:10.1016/j.atmosenv.2008.12.053

Michaelides SC, Tymvios FS, Michaelidou T (2009) Spatial and temporal characteristics of the annual rainfall frequency distribution in Cyprus. Atmosph Res 94(4):606–615

Milan G, Slavisa T (2013) Analysis of changes in meteorological variables using Mann-Kendall and Sen's slope estimator statistical tests in Serbia. Global Planet Change 100:172–182

Mondal A, Kundu S, Mukhopadhyay A (2002) Rainfall trend analysis by Mann-Kendall test: a case study of north-eastern part of Cuttack district, Orissa. Int J Geol Earth Environ Sci 2 (1):70–78

Mooley DA, Parthasarathy B (1983) Variability of the Indian summer monsoon and tropical circulation features. Mon Weather Rev 111:967–968

Parthasarathy B, Munot AA, Kothawale DR (1995) Monthly and seasonal rainfall series for all India homogeneous regions and meteorological sub-divisions: 1871–1994. Research report IITM, vol 65, p 113, Pune

Schulze RE (1983) Agrohydrology and Climatology of Natal. Water Research Commission, Pretoria, p 137

Serrano A, Mateos VL, Garcia JA (1999) Trend analysis of monthly precipitation over the Iberian Peninsula for the Period 1921–1995. Phys Chem Earth 24(1–2):85–90

Sivapragasam C et al (2013) Trends in rainfall patterns over the Tamarabarani basin in Tamil Nadu, India. In: 20th international congress on modelling and simulation, Adelaide, Australia

Smith IN, McIntosh P, Ansell TJ, Reason CJC, McInnes K (2000) Southwest Western Australian winter rainfall and its association with Indian Ocean climate variability. Int J Climatol 20:1913–1930

Soltani S, Modarres R, Eslamian SS (2007) The use of time series modeling for the determination of rainfall climates of Iran. Int J Climatol 27:819–829

Taschetto AS, Matthew HE (2009) An analysis of late twentieth century trends in Australian Rainfall. Int J Climatol 29:791–807

Thapliyal V, Kulshrestha SM (1991) Climate changes and trends over India. Mausam 42:333–338

Timo S, Anu M, Pia A, Tuija Ruoho-Airola, Toni A (2002) Detecting trends of annual values of atmospheric pollutants by the Mann-Kendall test and Sen's slope estimates—the Excel template application MAKESENS. Finnish Meteorological Institute Publications on Air Quality, Helsinki, p 31

Werner N (2009) Rainfall trends in the KwaZulu-Natal Drakensberg region of South Africa during the twentieth century. Int J Climatol 29:1634–1641

Zhang Q, Xu CY, Zhang Z, Chen YD, Liu CL (2009) Spatial and temporal variability of precipitation during 1951–2005 over China. Theor App Climatol 95:53–68

Selection of Suitable General Circulation Model Precipitation and Application of Bias Correction Methods: A Case Study from the Western Thailand

Devesh Sharma

Abstract The General Circulation Models (GCMs) precipitations are generally characterized by the biases and low spatial resolution. These two are the major limiting factors for direct application of GCMs scenarios in the studies of climate change impact assessment. Based on 17 experiments over the two river basins of the western Thailand, Six GCMs were analyzed for their ability to simulate the magnitude and spatial variability of current precipitation. Monthly precipitation scenarios from six GCMs (17 experiments) are downloaded from the IPCC data centre. Three bias-correction techniques namely, scaling, empirical-gamma and gamma-gamma transformations were applied on a daily scale of 9 years (1991–1999) to improve the quality of the selected ECHAM4/OPYC SRES A2 and B2 precipitation for the Mae Ping and Mae Klong River Basins in the Western Thailand. All the three bias correction methods have been compared with observed precipitation based on statistical parameters. Gamma-gamma transformation method is found to be effective in correcting the rainfall frequency and intensity simultaneously as compared to other methods. The bias corrected daily precipitation is useful in studies related to climate change and water resources management at basin level.

Keywords Climate models · Bias-correction · Frequency · Intensity · River basin

1 Introduction

The global average surface temperature has increased by 0.6 ± 0.2 °C since the late 19th century (IPCC 2001). The most important impacts of the climate change and variability are their effects on the hydrologic cycle and water resources. Impact of climate change on hydrology of various river basins in different regions of the

D. Sharma (✉)
Department of Environmental Science, Central University of Rajasthan, NH-8, Bandarsindari, Kishangarh, Ajmer 305817, Rajasthan, India
e-mail: devesh.water@gmail.com

Mu. Ramkumar et al. (eds.), *Environmental Management of River Basin Ecosystems*, Springer Earth System Sciences, DOI 10.1007/978-3-319-13425-3_3

world indicate the changes in total annual flows that directly affected the seasonal aspects of water supply and demand (Prayuth et al. 2010; Sharma and Babel 2013; Shrestha 2014). These changes will also have significant impacts on the regional water resources availability. The effects of climate change on water availability are likely to place further stresses on the water resource management, due to ever increasing demand, causing serious problems in many parts of the world.

Global circulation models (GCMs) predict an increase in temperature throughout the World, model predictions indicate regionally varying (positive/negative) changes in precipitation and runoff. GCMs are important tools to project the expected future scenarios of climatic parameters but the model predictions also contain biases when compared to the observed data due to their parameterization systems and large grid size ($\sim$300 km $\times$ 300 km). These types of errors are considered less important when applying for the estimation of the impacts of climate change impact at large areas. However, at basin-scale, the model predictions tend to be flawed and any management strategy based on the biased climate model data in hydrologic model at basin scale tend to result in poor/mismanagement. Bias-correction methods, when applied with statistical parameters of historical data can largely eliminate these kinds of problems. The simplest kind of bias-correction corrects for a systematic discrepancy in the mean by "rescaling" the mean of the simulations to match the observations. Similarly, a discrepancy between the variance of the simulations and the observations can be corrected by assuming a probability distribution and mapping normalized anomalies between the simulated and observed populations. In many cases, however, the true forms of the probability distributions of the simulated and observed data are not known with any certainty and the two probability distributions are not necessarily of the same form or statistically well behaved. A 'quantile-based' bias-correction approach is useful to statistically transform the rainfall simulated by GCMs to bias corrected data and make it more applicable for use in the impact assessment models (Wood et al. 2002; Hamlet et al. 2003; Ines and Jansen 2006).

Variability of the rainfall generally depends on its frequency and intensity and it is difficult to estimate average rainfall at a particular region. In the present scenarios, many climate models, generally mathematical models, comprising the dynamics of the climate system, physics of the climate system and other factors such as air-sea interaction, topography, and vegetation parameters based on known physical laws and empirical relations are used. However, the actualistic and model predicted data on the rainfall pattern of climate system never match at reasonable levels of margin of error (Ines 2006).

Wood et al. (2002) used a "quantile-based" bias-correction scheme to transform the simulated and observed populations. In this scheme, the simulated and observed data covering the same period of record are used to develop a "quantile map" of each population using an unbiased quantile estimator applied to ranked data. Ines and Jansen (2006) used a procedure to calibrate both the frequency and the intensity distribution of daily GCM rainfall relative to a target station, and demonstrated its application to maize yield simulation at a location in semi-arid Kenya. They used the simple multiplicative shift, empirical-gamma distribution and gamma-gamma transformation techniques to correct the frequency and intensity distribution of daily GCM rainfall. The results have done a better job of correcting monthly and seasonal rainfall totals, but were less effective in improving the frequency and intensity bias. The empirical-gamma distribution and gamma-gamma transformation techniques corrected both the mean and variance of monthly and seasonal GCM rainfall total, frequency and mean intensity. The study concluded that the proposed correction methods provide an option for using the daily output of dynamic climate prediction models for impact studies in a manner that conserves any useful predictive information about rainfall variability within the season.

Hashino et al. (2006) evaluated three bias-correction methods for ensemble streamflow volume forecasts of the monthly flow volumes for the Des Moines River, issued sequentially for each month over a 48 year record. All the three adjusted the ensemble traces using a transformation derived with simulated and observed flows from a historical simulation. The quality was then evaluated using a distributions-oriented verification approach. The results showed that all the three bias-correction methods significantly improved the forecast quality by eliminating unconditional biases.

The aim of this paper is to select suitable GCM and ways to improve the rainfall scenarios to make them more realistic for further application. The methodology adopted in this study can be linked with hydrological models to study the impact of climate change on water resources of river basins.

2 Study Area

The study area comprised of two main river basins of the Thailand, namely, Mae Klong River Basin and Ping River Basin in the western Thailand (Fig. 1). The Mae Klong River Basin is located in the western part of Thailand, stretches from latitude 16°23′ N to 13°10′ N and from longitude 98°15′ E to 100°17′ E, with catchment area of 30,840 km^2. The Mae Klong River Basin has been divided into five sub-basins, namely, the Khwae Yai River, the Khwae Noi River, the Lam Tapoen River, the Lam Pachee River and the Mae Kong River plain. About 38.5 % of the total area is under the elevation range of 500–1,500 m. Based on annual mean, the average rainfall for the lower part of the basin is roughly estimated to be 1,000 mm, and it increases to roughly 2,200 mm in the upper part of the basin. The rainfall increases from east to west in the basin. The temperature in the basin is higher in the upper stream and almost uniform throughout the year with little seasonal variation.

The Ping River Basins is one of eight sub-basins in Chao Phraya basin. It stretches from latitude 19°45′ N to 15°45′ N and from longitude 98°05′ E to 100°10′ E, with catchment area of 34,856 km^2. About 55.5 % of total area is under the elevation range of 500–1,500 m. The weather is influenced mainly by the southwest and northeast monsoon. It is also influenced by the depression from the South China Sea during July and September resulting in abundant rain from May to October. The climate is characterized by average annual precipitation of 1,097 mm and average annual temperature is 26.7 °C.

Rainfall data of 99 stations are used in this paper study. Location and distribution of these stations are shown in the Fig. 1. Grid 1 and grid 2 represent the Ping River Basin and Mae Klong Basin, respectively.

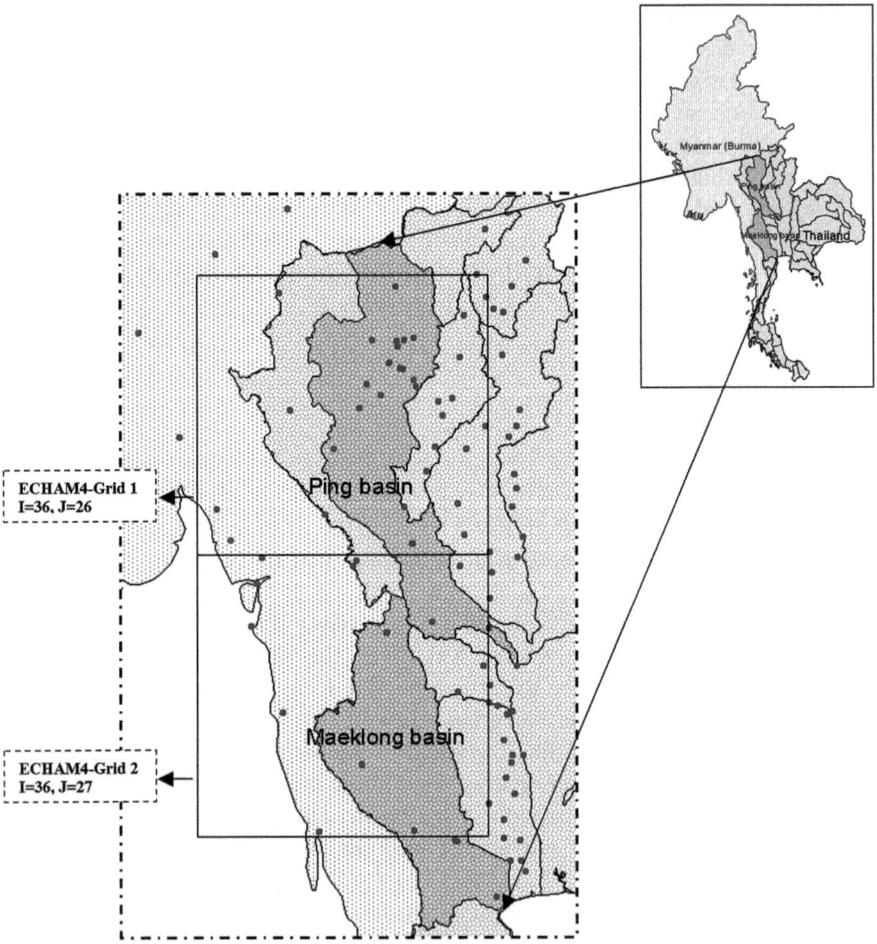

Fig. 1 Location of the study area and the grid layout of general climate model (after Sharma 2007)

3 Methodology

3.1 Selection of General Circulation Model

The main idea behind this analysis is to assess the GCMs for their ability to simulate the magnitude and spatial variability of current precipitation over the study domain at or near the realistic levels. The outcome of this analysis helps in identifying the best GCM for further use in the study framework. Monthly precipitation scenarios from six GCMs (17 experiments) were downloaded from the IPCC data centre (http://ipcc-ddc.cru.uea.ac.uk) and preprocessed to use in the analysis. Download data are then converted in '*ascii*' format using '*grbconv*' program (http://cera-www.dkrz.de/IPCC_DDC/GRIBGZIP.html).

Smith and Hulme (1998) suggested four essential criteria for the selection of suitable GCM experiment, namely, vintage, resolution, validity and representativeness of the results, which in turn were give consideration in the present study. Statistical parameters including the coefficient of determination, the root mean square error and the standard deviation between model and observed values have been calculated. The coefficient of determination values and root mean square error (RMSE) values provide complementary statistical information describing the correspondence between two patterns. A high coefficient of determination is associated with a low RMSE and vice versa. The closeness of these two parameters shows that a model is good at simulating the magnitude of the variability. The Coefficient of Determination (R^2) is the square of correlation coefficient (R) and gives the proportion of the variance of one variable (x) that is predictable from the other variable (y).

$$R = \frac{n\sum xy - (\sum x)(\sum y)}{\sqrt{n(\sum x^2) - (\sum x)^2}\sqrt{n(\sum y^2) - (\sum y)^2}} \tag{1}$$

The Root Mean Square Error (RMSE) is useful to quantify pattern similarity between modeled and observed values.

$$RMSE = \left[\frac{1}{N}\sum_{n=1}^{N}(y-x)^2\right]^{1/2} \tag{2}$$

The Standard Deviation (σ) shows the magnitude of variability of the model with the observed data with respect to mean $(\overline{x})$

$$\sigma = \left[\frac{1}{N}\sum_{n=1}^{N}(x-\overline{x})^2\right]^{1/2} \tag{3}$$

3.2 Bias-Correction Methods

Three methods namely, scaling, empirical-gamma and gamma-gamma transformation are commonly applied for improving the quality of raw GCM precipitation. The scaling procedure considers the bias-correction in the mean monthly GCM rainfall amount, R_{mGCM} (Eq. 4). This method does not assure, however, the corrections of both rainfall intensity and frequency.

$$aR_{mGCM} = R_{mhis} \tag{4}$$

The empirical-gamma distribution and gamma-gamma transformation techniques used for the correction of the frequency and intensity distribution of the daily GCM rainfall are shown in the Fig. 2. These methods are applied to reduce the gap between the daily GCM rainfall and the local condition (Ines 2004). The basic principle of applying bias-correction method on daily GCM data is that the mean rainfall amount (P_m) is a product of the rainfall intensity (I), and frequency (F). It means that by correcting the biases of the two rainfall components namely, frequency and intensity, the monthly rainfall corrects itself.

The rainfall frequency correction establishes the empirical distributions, F(x), by classifying first the long-term daily rainfall data for any particular month of interest based on the positions of the ordered datasets. The empirical distribution function F(x) is selected using Weibull probability plotting method as given in Eq. (5):

$$F(x) = \frac{n}{m+1} \tag{5}$$

where n is the position of x in the ordered array, and m is the total number of data in the array. It is followed by calculation of a threshold value ($\tilde{x}_{GCM}$), derived from the

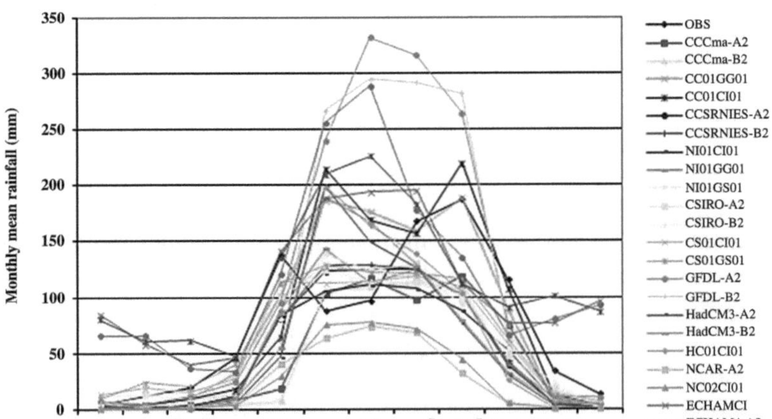

Fig. 2 Trend in mean monthly GCM precipitation for Ping Basin (Grid 1)

empirical distribution of daily historical rainfall, and truncation of the empirical distribution of the raw daily GCM rainfall for that particular month. Eventually, determination of F (x_{his} = 0.0) followed by mapping it to the daily GCM rainfall distribution have to be attempted.

Two approaches, namely gamma-gamma transformation (GG) and empirical-gamma transformation (EG) were used for the correction of rainfall intensity. For the gamma-gamma transformation, fitting the truncated daily GCM rainfall and historical rainfall data to a two-parameter gamma distribution (Eq. 6) and then mapping the cumulative distribution (Eq. 7) of the truncated daily GCM rainfall to the cumulative distribution of the truncated historical data (Eq. 8) have to be attempted.

$$f(x;\alpha,\beta) = \frac{1}{\beta^{\alpha}\Gamma(\alpha)} x^{\alpha-1} \exp\left(-\frac{x}{\beta}\right)_{;\, x \geq x_{Trunc}} \tag{6}$$

$$F(x;\alpha,\beta) = \int_{x_{Trunc}}^{x} f(t)\, dt \tag{7}$$

$$F(x_{GCM};\alpha,\beta|_{GCM}) \Rightarrow F(x_{His};\alpha,\beta|_{His}) \tag{8}$$

The corrected GCM rainfall amount for that day can be calculated by taking the inverse of (Eq. 8) such that

$$x'_{GCM} = F^{-1}\{F(x_{His};\alpha,\beta|_{His})\} \tag{9}$$

The shape and scale parameters α and β for each gamma distribution are determined using Maximum Likelihood Estimation. For the empirical-gamma transformation, the procedure is the same as described so far, but for the use of an empirical distribution to truncate the daily GCM rainfall.

The ability of these methods to correct the bias from the GCM rainfall data was explained by statistical parameters, i.e., coefficient of determination, root mean square error, standard deviation and index of agreement (d-statistics). The first three parameters are explained in a previous section. Index of agreement (d) is formulated as:

$$d = 1 - \frac{\sum_{i=1}^{n} (\hat{y} - y_i)}{\sum_{i=1}^{n} (|\hat{y} - \bar{y}| + |y_i - \bar{y}|)^2} \tag{10}$$

where $\hat{y}$ is bias-corrected rainfall, $\bar{y}$ is observed mean rainfall and y_i is observed rainfall at time step i.

A model result closer to the observed value should indicate that it has a similar standard deviation, a low RMSE, high coefficient of determination and high value of d-statistics. The quality of bias-corrected rainfall on monthly scale is then evaluated using the mean square error (*MSE*) skill score (SS) with the raw GCM as a reference and formulated as:

$$SS = 1 - \frac{MSE_{corrected}}{MSE_{raw}}, \tag{11}$$

where $MSE_{corrected}$ and MSE_{raw} are mean square errors of the bias-corrected GCM and raw GCM data with respect to observed values. The range of MSE skill score varies from negative infinity to 1, and 0 indicates no skill when compared to raw GCM data.

4 Results and Discussion

4.1 Selection of GCM

Monthly precipitation scenarios from six GCMs (17 experiments) are downloaded from the IPCC data centre. Summary of the selected GCMs including their type, spatial resolution and data period are presented in the Table 1. All the GCMs are characterized by their coarse resolution. It may result in the occurrences of biasness in their precipitation scenarios.

Figures 2 and 3 show the trends in the monthly mean precipitation from GCMs and observed precipitation for both the basins. Observed precipitation was calculated by using the Inverse Distance Weighing (IDW) method based on the data from 99 rainfall stations distributed in both the river basins.

Table 1 Characteristics of GCMs

GCM name	Scenario type	Model resolution	Data period
HadCM3A2	SRES-A2	2.5° × 3.75°	1960–2002
HadCM3B2	SRES-B2	2.5° × 3.75°	1960–2002
HC01CI01	Control integration	2.5° × 3.75°	1960–2002
CC01CI01	Control integration	3.75° × 3.75°	1960–2002
CC01GG01	Greenhouse gas integration	3.75° × 3.75°	1960–2002
CCCma-A2	SRES-A2	3.75° × 3.75°	1960–2002
CCCma-B2	SRES-B2	3.75° × 3.75°	1960–2002
CSIRO-MK2 (A2)	SRES-A2	3.2° × 5.6°	1960–2002
CSIRO-MK2 (B2)	SRES-B2	3.2° × 5.6°	1960–2002
CS01CI01	Control integration	3.2° × 5.6°	1960–2002
CS01GS01	Greenhouse gas integration	3.2° × 5.6°	1960–2002
GFDL-R30 (A2)	SRES-A2	3.75° × 2.25°	1960–2002
GFDL-R30 (B2)	SRES-B2	3.75° × 2.25°	1960–2002
NCARPCM (A2)	SRES-A2	7.5° × 4.5°	1990–2002
NC02CI01	Control integration	7.5° × 4.5°	1960–2002
ECHAM4A2	SRES-A2	2.8° × 2.8°	1990–2002
ECHAM4B2	SRES-B2	2.8° × 2.8°	1990–2002

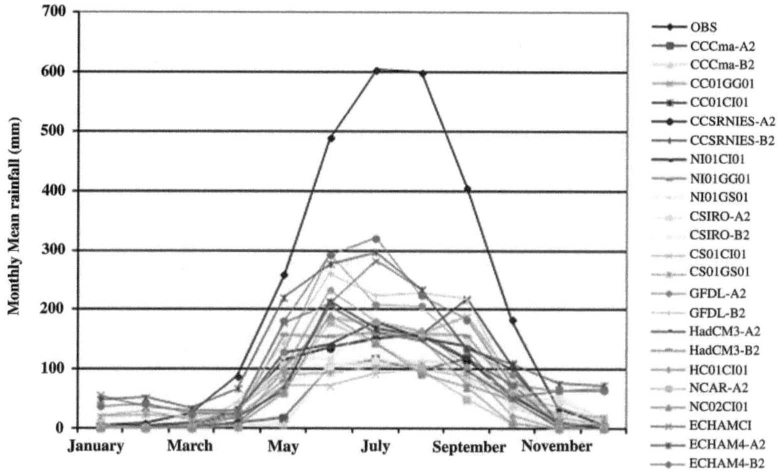

Fig. 3 Trend in mean monthly GCM precipitation for Mae Klong Basin (Grid 2)

Data from 15 stations were used to compare the observed values and GCMs data at respective stations. GCMs ability is checked for precipitation using statistical parameters, namely, coefficient of determination, root mean square error and standard deviation. Precipitation values at respective station locations were removed from the processed GCMs scenarios based on their location (latitude, longitude). Statistical characteristics of the precipitation data are summarized in the Table 2.

The results reveal a considerable variability pertaining to the GCM simulations for the observed climate. It was observed that, while nearly all the models substantially overestimate the magnitude of total temperature on annual basis, the HadCM3, ECHAM4, GFDL-R30 and US NCAR models are good at simulating the magnitude and spatial variability of mean temperature. Every model has a good R^2, indicating that they collectively represent the spatial pattern of variation. But a wider range in the standard deviations of Canadian and Australian model suggests differing abilities in simulating the magnitude of this variability. The precipitation comparisons show much higher variations than those for mean temperature for all the models and for all the stations. The Australian (CSIRO-Mk2) model best represents the observed magnitude and spatial variability of the precipitation. The German (ECHAM4) model is good in simulating the magnitude of spatial variability over the study area. Nearly all the models except ECHAM4 (low value of standard deviation) over-predicted the rainfall. The RMSE, ECHAM4 and CSIRO-Mk2 showed the lowest value, whereas the GFDL provided the highest errors. Coefficient of determination values ranged between 0.4 and 0.65, whereas the CSIRO-Mk2 and ECHAM4 models were closer to the observed standard deviation.

Finally, ECHAM4_OPYC has been selected based on the statistical characteristics, spatial resolution, areal coverage and data accessibility. The detailed description of ECHAM4 model and simulation of present-day climate is given in Roeckner et al. (1996).

Table 2 Statistical summary of monthly GCM precipitation

GCM Name	7052			7092			7142			12042		
	RMSE	R^2	SD	RMSE	R^2	SD	RMSE	R^2	SD	RMSE	R^2	SD
Observed	–	–	3.19	–	–	2.82	–	–	3.84	–	–	3.78
HadCM3A2	3.67	0.33	4.37	6.91	0.17	7.02	3.98	0.29	4.37	4.68	0.27	5.04
HadCM3B2	3.66	0.33	4.38	6.96	0.15	6.99	3.99	0.29	4.38	6.78	0.17	6.99
HC01CI01	4.3	0.35	5.04	7.63	0.18	7.57	4.41	0.32	5.04	7.39	0.19	7.57
CC01CI01	12.58	**0.45**	12.25	8.96	0.28	8.64	12.46	0.34	12.25	6.58	0.18	6.68
CC01GG01	12.51	0.44	12.14	9.35	0.27	8.95	12.45	0.32	12.14	5.94	0.25	6.3
CCCma-A2	7.08	0.34	7.68	5.76	0.2	6.11	7.07	0.28	7.68	4.52	0.18	**4.54**
CCCma-B2	6.98	0.36	7.64	5.68	0.19	6	7.02	0.28	7.64	4.74	0.15	4.64
CSIRO-MK2 (A2)	3.6	0.38	3.49	3.92	0.27	3.49	3.78	0.34	3.49	4.08	0.25	3.49
CSIRO-MK2 (B2)	3.57	0.38	**3.4**	3.93	0.25	**3.4**	3.74	0.35	3.4	4.07	0.24	3.4
CS01CI01	3.48	0.43	3.57	3.82	**0.32**	3.57	3.58	**0.41**	3.57	3.85	**0.32**	***3.57***
CS01GS01	3.54	0.43	3.6	3.95	0.28	3.6	3.74	0.37	***3.6***	4.07	0.27	3.6
GFDL-R30 (A2)	15.31	0.4	14.55	16.08	0.2	14.55	15.15	0.31	14.55	12.7	0.22	11.64
GFDL-R30 (B2)	15.07	0.36	14.41	15.83	0.17	14.41	14.8	0.32	14.41	12.76	0.22	11.78
NCARPCM (A2)[a]	2.66	0.39	***3.16***	3.33	0.14	3.16	3.32	0.29	3.16	4.96	0.12	4.91
NC02CI01	2.85	0.37	3.1	3.26	0.16	3.1	3.33	0.35	3.1	4.71	0.16	4.63
ECHAM4A2[a]	**0.53**	0.28	2.88	**0.53**	0.16	2.59	**0.98**	0.2	**4.15**	**0.85**	0.23	***3.57***
ECHAM4B2[a]	0.73	0.28	2.88	0.67	0.24	***2.63***	**0.98**	0.27	**4.15**	**0.85**	0.21	***3.57***
	0.53	***0.45***	***0.21***	**0.53**	***0.32***	***0.58***	**0.98**	***0.41***	***0.24***	**0.85**	***0.32***	***0.18***
GCM Name	13013			17042			26013			63013		
	RMSE	R^2	SD	RMSE	R^2	SD	RMSE	R^2	SD	RMSE	R^2	SD
Observed	–	–	2.94	–	–	3.06	–	–	2.99	–	–	4.81
HadCM3A2	5.04	0.17	5.04	6.45	0.27	7.02	4.39	0.36	5.04	5.59	0.41	7.02

(continued)

Table 2 (continued)

GCM Name	13013			17042			26013			63013		
	RMSE	R^2	SD	RMSE	R^2	SD	RMSE	R^2	SD	RMSE	R^2	SD
HadCM3B2	5.2	0.15	5.11	6.44	0.28	6.99	6.64	0.23	6.99	5.51	0.43	6.99
HC01CI01	5.86	0.13	5.66	6.99	0.35	7.57	7.22	0.28	7.57	5.86	**0.49**	7.57
CC01CI01	7.25	0.11	6.55	8.5	0.39	8.64	6.35	0.26	6.68	7.85	0.37	8.64
CC01GG01	7.76	0.14	7.03	8.9	0.38	8.95	5.8	0.33	6.3	7.9	0.43	8.95
CCCma-A2	6.2	0.1	5.93	5.39	0.3	6.11	4.13	0.22	4.54	5.18	0.34	6.11
CCCma-B2	6	0.09	5.71	5.28	0.29	6	4.28	0.2	4.64	5.05	0.35	**6**
CSIRO-MK2 (A2)	3.72	0.26	3.54	3.61	0.36	3.49	3.16	0.46	**3.54**	4.07	0.36	3.49
CSIRO-MK2 (B2)	3.74	0.27	3.58	3.6	0.35	3.4	3.21	0.46	3.58	4.07	0.36	3.4
CS01CI01	3.67	**0.3**	3.63	3.48	**0.42**	3.57	3.14	**0.48**	3.63	3.74	0.46	3.57
CS01GS01	3.83	0.27	3.74	3.57	0.4	3.6	3.26	0.47	3.74	3.9	0.42	3.6
GFDL-R30 (A2)	11.5	0.14	10.29	15.49	0.34	14.55	12.85	0.27	11.64	14.04	0.39	14.34
GFDL-R30 (B2)	12.29	0.14	10.99	15.3	0.29	14.41	12.93	0.27	11.78	13.93	0.46	14.37
NCARPCM (A2)[a]	6.9	0.02	6.45	2.95	0.31	3.16	4.52	0.2	4.91	4.77	0.47	6.45
NC02CI01	6.56	0.05	6.28	2.91	0.32	**3.1**	4.25	0.22	4.63	4.93	0.41	6.28
ECHAM4A2[a]	**0.76**	0.16	**3.11**	**0.75**	0.26	***2.9***	**0.74**	0.33	***2.86***	**1.15**	0.27	**4.88**
ECHAM4B2[a]	**0.76**	0.26	**3.11**	**0.75**	0.26	***2.9***	**0.74**	0.31	***2.86***	**1.15**	0.27	**4.88**
	0.76	***0.3***	***0.17***	**0.75**	***0.42***	***0.04***	**0.74**	***0.48***	***0.55***	**1.15**	***0.49***	***0.07***
GCM Name	Chiang Mai			Mae Sot			Tak			Kanchanaburi		
	RMSE	R^2	SD	RMSE	R^2	SD	RMSE	R^2	SD	RMSE	R^2	SD
Observed	–	–	3.29	–	–	5.02	–	–	3.21	–	–	3.26
HadCM3A2	6.4	0.26	7.02	5.53	0.42	7.02	3.9	0.26	4.37	7.16	0.1	7.02
HadCM3B2	6.34	0.28	6.99	5.45	0.44	6.99	3.88	0.27	4.38	7.15	0.11	6.99
HC01CI01	6.96	0.32	7.57	5.75	0.5	7.57	4.57	0.26	5.04	7.67	0.15	7.57

(continued)

Table 2 (continued)

GCM Name	Chiang Mai			Mae Sot			Tak			Kanchanaburi		
	RMSE	R^2	SD	RMSE	R^2	SD	RMSE	R^2	SD	RMSE	R^2	SD
CC01CI01	12.52	0.41	12.25	7.69	0.38	8.64	12.69	0.38	12.25	8.9	0.25	8.64
CC01GG01	12.4	0.42	12.14	7.81	0.43	8.95	12.55	0.4	12.14	9.22	0.26	8.95
CCCma-A2	6.98	0.34	7.68	5.22	0.34	6.11	7.1	0.32	7.68	5.8	0.19	6.11
CCCma-B2	6.94	0.34	7.64	5.12	0.34	6	7.04	0.32	7.64	5.78	0.17	6
CSIRO-MK2 (A2)	3.53	0.38	3.49	4.11	0.38	3.49	3.66	0.34	3.49	4.06	0.24	3.49
CSIRO-MK2 (B2)	3.52	0.37	3.4	4.13	0.37	3.4	3.66	0.33	3.4	4.08	0.22	3.4
CS01CI01	3.38	0.43	3.57	3.76	0.48	3.57	3.55	0.39	3.57	3.94	0.28	3.57
CS01GS01	3.46	0.42	3.6	3.94	0.43	3.6	3.59	0.39	3.6	4.09	0.25	3.6
GFDL-R30 (A2)	15.25	0.36	14.55	13.82	0.41	14.34	15.41	0.33	14.55	15.75	0.17	14.34
GFDL-R30 (B2)	15.04	0.32	14.41	13.7	0.48	14.37	15.19	0.29	14.41	15.95	0.15	14.37
NCARPCM (A2)[a]	2.94	0.32	3.16	4.59	0.5	6.45	3.02	0.28	3.16	3.82	0.1	3.16
NC02CI01	3.01	0.34	3.1	4.87	0.42	6.28	3.04	0.31	3.1	3.69	0.11	3.1
ECHAM4A2[a]	0.77	0.2	2.92	1.42	0.26	4.8	0.77	0.16	3.17	0.74	0.18	2.96
ECHAM4B2[a]	0.77	0.24	2.92	1.14	0.26	4.8	0.77	0.15	3.17	0.74	0.26	2.96
	0.77	0.43	0.11	1.14	0.5	0.98	0.77	0.4	0.19	0.74	0.28	0.14

GCM Name	Thong			Suphan			7013		
	RMSE	R^2	SD	RMSE	R^2	SD	RMSE	R^2	SD
Observed	–	–	2.98	–	–	4.94	–	–	3.28
HadCM3A2	4.99	0.18	5.04	5.56	0.46	7.28	4.83	0.23	5.04
HadCM3B2	5.16	0.15	5.11	5.49	0.47	7.24	4.95	0.21	5.11
HC01CI01	5.82	0.14	5.66	6.47	0.51	8.85	5.54	0.21	5.66
CC01CI01	7.23	0.1	6.55	4.57	0.41	5.6	7.03	0.15	6.55
CC01GG01	7.71	0.15	7.03	5.11	0.38	6.06	7.61	0.17	7.03

(continued)

Table 2 (continued)

GCM Name	Thong			Suphan			7013		
	RMSE	R^2	SD	RMSE	R^2	SD	RMSE	R^2	SD
CCCma-A2	6.17	0.1	5.93	4.48	0.36	4.93	5.97	0.15	5.93
CCCma-B2	5.95	0.1	5.71	4.4	0.37	4.83	5.84	0.13	5.71
CSIRO-MK2 (A2)	3.66	0.28	3.54	3.3	0.55	3.54	3.48	0.36	3.54
CSIRO-MK2 (B2)	3.68	0.29	3.58	3.26	0.57	3.58	3.52	0.36	3.58
CS01CI01	3.58	0.32	3.63	3.4	0.53	3.63	3.41	0.39	3.63
CS01GS01	3.76	0.29	3.74	3.29	0.56	3.74	3.59	0.36	3.74
GFDL-R30 (A2)	11.44	0.15	10.29	10.27	0.38	11.26	11.18	0.19	10.29
GFDL-R30 (B2)	12.22	0.15	10.99	9.9	0.42	11.1	12.03	0.18	10.99
NCARPCM (A2)[a]	6.88	0.03	6.45	4.71	0.47	6.45	5.11	0.07	4.91
NC02CI01	6.52	0.05	6.25	4.81	0.43	6.28	4.79	0.11	4.63
ECHAM4A2[a]	1.26	0.28	4.88	0.74	0.15	2.96	0.77	0.21	2.97
ECHAM4B2[a]	1.26	0.27	4.88	0.74	0.25	2.96	0.77	0.25	2.97
	1.26	0.32	0.56	0.74	0.57	0.01	0.77	0.39	0.25

[a] Analyses data period is only from 1990–2002

Table 3 Description of ECHAM4_OPYC model

Scenarios	SRES-A2, SRES-B2
Spatial resolution	2.8° × 2.8°
Grid geometry	64 × 128
Direction of data	W–E, N–S
Temporal resolution	6-h (SRES A2, SRES B2)
	12-h (EH4OPYC_22723GSDIO)
Data period	1990–2100 (SRES A2, SRES B2)
	1860–1990 (EH4OPYC_22723GSDIO)
Data format	Grib (GRIB converter, file splitter)
Grid identity	Grid 1 (I = 36, J = 26) and Grid-2 (I = 36, J = 27)

Daily precipitation data of the ECHAM4/OPYC model scenarios were obtained from the research center (Max Planck Institute for Meteorology, Hamburg). The model description is presented in the Table 3. Schematic and location detail of the selected grids are shown in the Fig. 1.

Figure 4 illustrates the variation of annual standardized anomalies of the ECHAM4/OPYC3 precipitation scenarios. It shows that there is an inclination towards less abundant rainfall (rainfall close to mean) in future periods. Changes in the annual precipitation across the region may vary between −0.1 to 0.2 mm/day by 2020s and +0.1 to 0.2 mm/day by the 2050s, according to ECHAM4 for the high emissions scenario (A2). In the case of the low emissions scenario (B2), the variations may be between 0.0–0.2 mm/day by 2020s and 0.0–0.5 mm/day by 2050s according to ECHAM4.

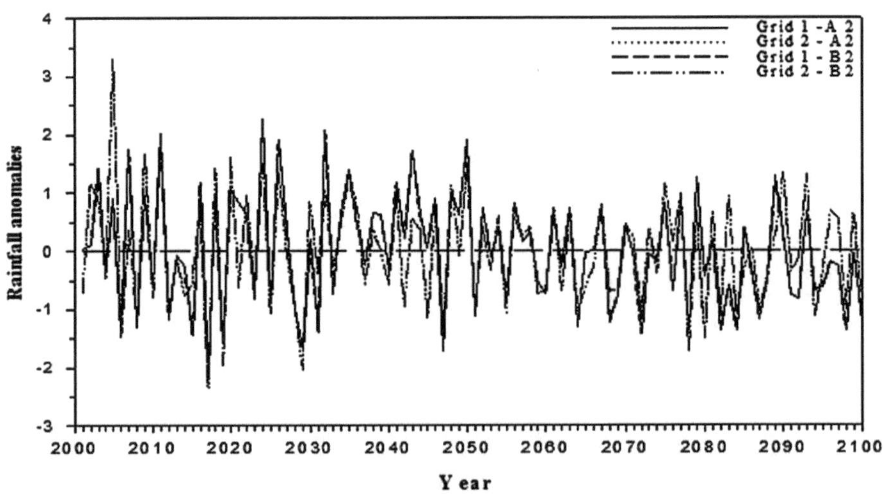

Fig. 4 Annual variability of the standardized rainfall anomalies. Standardized anomaly = (predicted value − mean value)/standard deviation

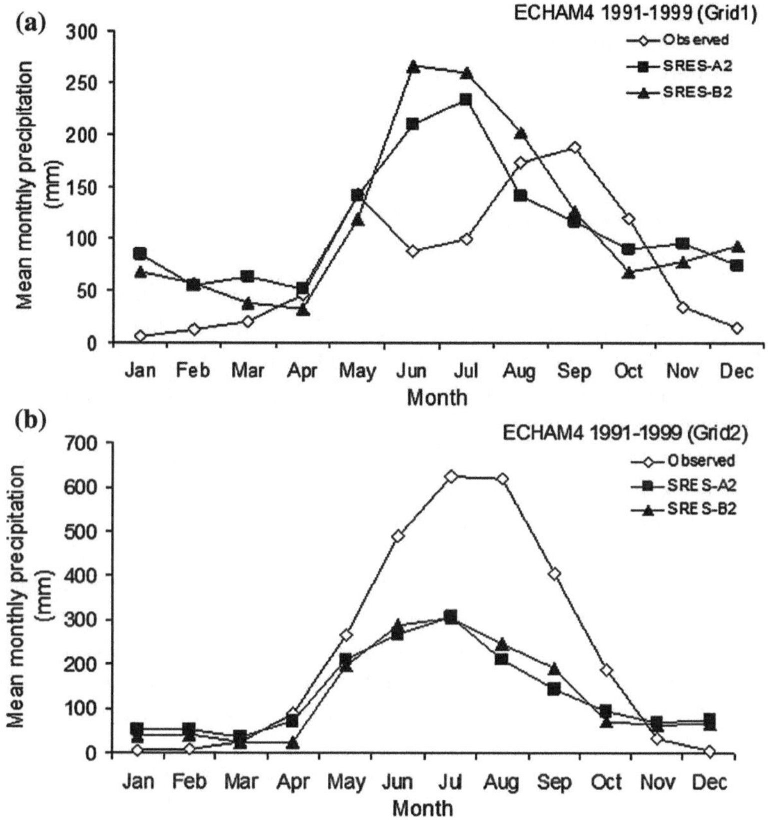

Fig. 5 Mean monthly rainfall of ECHAM4 and observed (1990–1999) for **a** Grid 1, and **b** Grid 2

The Fig. 5a, b shows the mean monthly trend of the ECHAM4 precipitation and observed rainfall. Observed monthly mean rainfall was calculated by the IDW method using 99 rainfall stations. The selected model over-predicts the mean monthly precipitation at grid 1 and under predicts at grid 2 for the wet period (May–October). In the case of the dry period (November–April), the ECHAM4 model over-predicts the rainfall for the whole domain. It is observed that the ECHAM4 simulates continuous wet days (rain ≥ 0.1 mm) with similar trend throughout the year. This trend of simulated rainfall lacks the demarcation in seasons and results in a large difference in the mean frequency of wet days between the model and observed data. Mean rainfall intensity is also found to be low when compared to that observed in the study domain, except for the period May–August at grid 1. This is because of difference in mean rainfall amount as compared to mean frequency. These drawbacks may cause serious problems in direct application of the GCM data in impact assessment studies. Overall, it shows that there is a tendency towards less abundant rainfall (rainfall close to mean) in future periods. A comparison of these patterns of trend suggested that the warmer temperatures and less rainfall will soon affect the present situation of water resources in the study area.

4.2 Bias-Correction of ECHAM4_OPYC Precipitation

It is observed that the GCMs daily precipitation was not accurate enough to predict the existing and future trend at the basin level. This is mainly due to the lag between the intensity and frequency when compared to the observed data. It is observed that ECHAM4/OPYC3 over-predicts the mean monthly rainfall observed at the grid 1 which form the major portion of the Ping basin while the GCM under-predicts for the grid 2 for the wet period (May–October). In the case of the dry period (November–April), the model overestimates the rainfall for the whole domain. It is also observed that the GCM simulates continuous rainfall events (rain ≥ 0.1 mm) throughout the year with more or less same pattern and fails to distinguish different seasons. There is large difference of mean frequency of wet days between the model and observed data. Mean rainfall intensity is also found to be low when compared to that observed in the study domain, except for the period May–August. This is because of the difference in mean rainfall amount as compared to mean frequency. Thus, the direct application of GCM output may introduce uncertainty in climate change studies.

The bias correction methods were applied on the ECHAM4/OPYC3 SRES A2 and B2 scenarios to reduce the lag between the frequency and intensity with respect to grid mean precipitation. Mean observed rainfall at grid nodes was calculated by IDW spatial interpolation method to use in bias-correction along with raw GCM scenario. The Figs. 6 and 7 show the comparative trends in mean monthly rainfall amount and frequency for the observed data, raw GCM scenarios and bias corrected GCM scenarios at each of the grid for control, SRES A2 and SRES B2 scenarios. Out of all the three bias-correction methods applied to the raw GCM, gamma-gamma transformation method was better than the other two methods. The improvements by the bias-correction methods also vary by month-wise values.

The statistical parameters indicating the correspondence of the raw GCM and bias-corrected GCM scenarios with the data from field observation for monthly mean rainfall are provided in the Tables 4 and 5. The standard deviation of

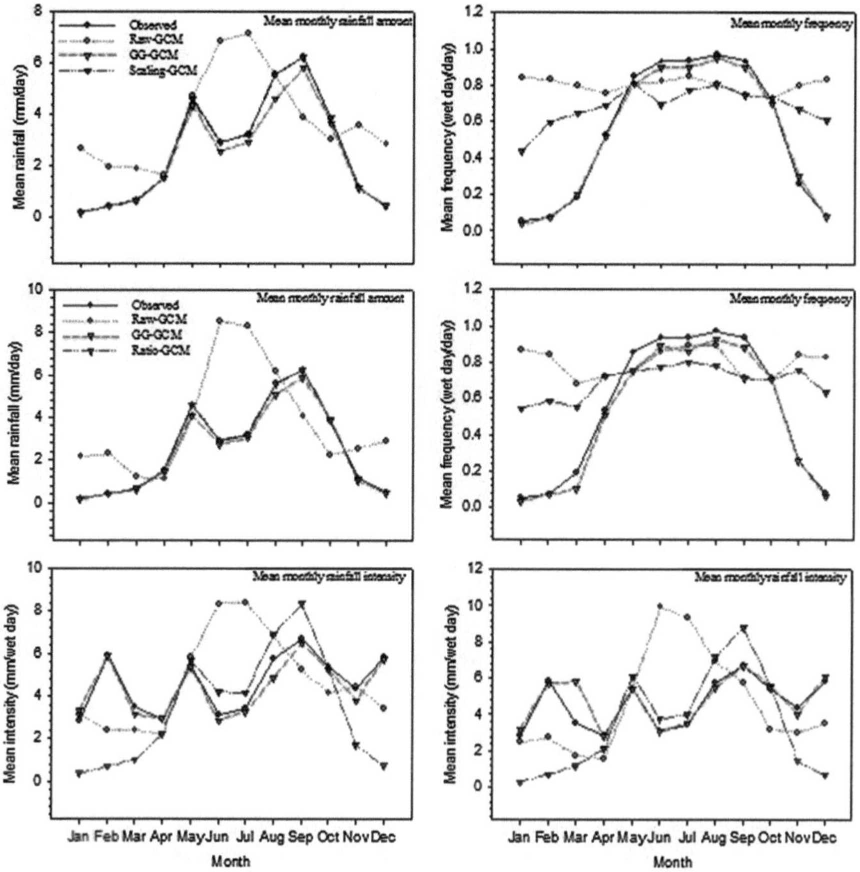

Fig. 6 Trend in mean monthly amount and frequency observed, raw GCM and bias-corrected rainfall at Grid 1

GG-GCM precipitation data, when compared with the raw GCM data, was closer to the standard deviation derived for the observed values. An increase in the correlation coefficient under all the bias-corrected scenarios has also been inferred. For example, there is an increase in correlation coefficient value from 0.32 to 0.64 in SRES A2 scenario of grid 1. A correlation coefficient of the order of 0.56–0.66 has been achieved for the grid 1, while the correlation for grid 2 lies in the range of 0.63–0.77. RMSE for the grid 1 varied from 2.05 to 2.42 and for grid 2 varied from 5.55 to 7.35. Similarly, an increase in d-statistics under all the methods as compared

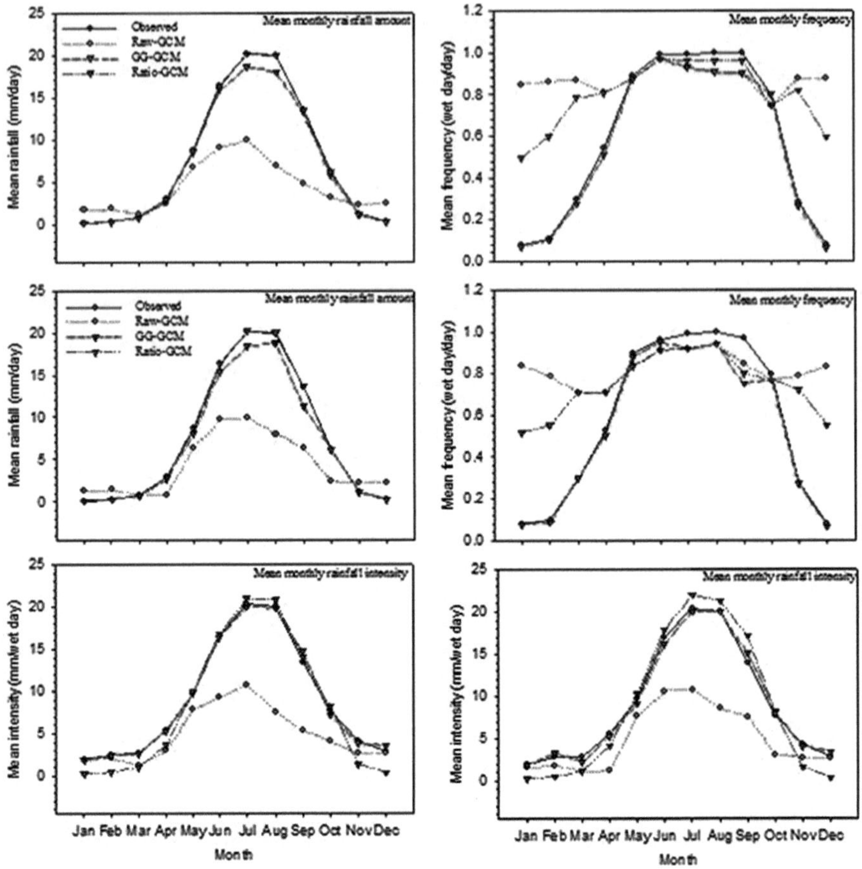

Fig. 7 Trend in mean monthly amount and frequency for observed, raw GCM and bias-corrected rainfall scenarios at Grid 2

to raw GCM data is also observed. The result shows the effectiveness of GG transformation in reducing the biases from the raw GCM datasets.

The effect of the bias-correction method was also checked using the MSE skill score, as shown in the Fig. 8. Positive value of the skill score reveals that the corrected data are better than the raw data, but the improvements vary by month-

Table 4 Performance of bias-correction methods for Grid 1

	Observed	Raw-GCM	GG-GCM	EG-GCM	Scale-GCM
Control					
Stdev	2.48	2.85	2.34	2.36	2.70
R		0.29	0.64	0.61	0.62
RMSE		3.35	2.05	2.15	2.26
D		0.50	0.79	0.77	0.78
SRES-A2					
Stdev	2.48	3.28	2.52	2.55	3.06
R		0.32	0.66	0.64	0.63
RMSE		3.64	2.06	2.14	2.42
D		0.74	0.81	0.79	0.78
SRES-B2					
Stdev	2.48	3.47	2.30	2.30	2.77
R		0.27	0.60	0.56	0.60
RMSE		3.87	2.15	2.24	2.35
D		0.47	0.76	0.74	0.76

Table 5 Performance of bias-correction methods for Grid 2

	Observed	Raw-GCM	GG-GCM	EG-GCM	Scale-GCM
Control					
Stdev	8.48	3.42	8.06	8.11	8.79
R		0.72	0.77	0.77	0.74
RMSE		7.16	5.55	5.45	5.94
D		0.55	0.86	0.87	0.86
SRES-A2					
Stdev	8.48	4.08	8.24	8.31	9.39
R		0.64	0.73	0.73	0.68
RMSE		7.02	6.00	6.01	6.85
D		0.83	0.84	0.84	0.82
SRES-B2					
Stdev	8.48	4.42	8.20	8.28	9.33
R		0.62	0.68	0.68	0.63
RMSE		7.06	6.45	6.51	7.35
D		0.61	0.81	0.81	0.78

wise data. The improvement is relatively high during the dry months compared to rainy months. The high skill score in dry months is the effect of reduction in wet days and in the amount of rainfall that happened to be the characteristics of raw GCM data as explained earlier. The GG-transformation is, therefore, effective in reducing the biases from the raw GCM precipitation when compared with the observed data.

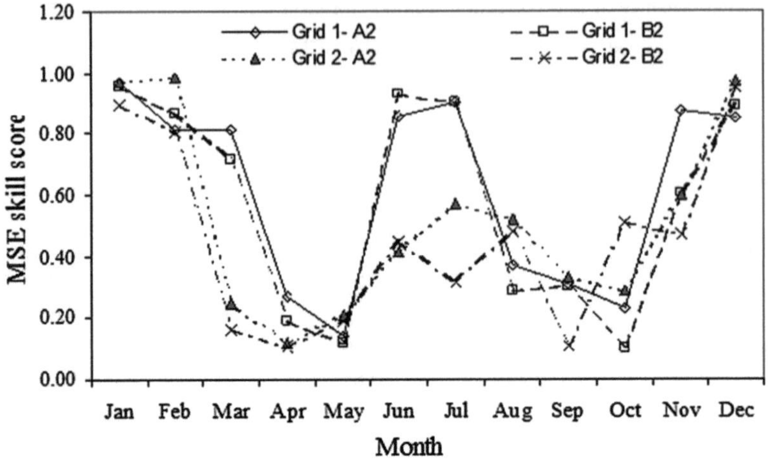

Fig. 8 Monthly variation in MSE skill scores for ECHAM4/OPYC rainfall with bias correction method

5 Conclusions

- A considerable variability is observed in different GCMs simulations for the observed climate. Among the models assessed, the HadCM3, ECHAM4, GFDL-R30 and US NCAR models were good in simulating the magnitude and spatial variability of mean temperature. By considering the statistical characteristics, the Australian model (CSIRO Mk2) and German model (ECHAM4) represented the observed magnitude better than other models.
- The ECHAM4 SRES-A2 and SRES-B2 scenarios were selected for further application due to their statistical characteristics, high spatial resolution and daily data availability. Three bias-correction techniques, namely, scaling, gamma-gamma (GG) transformation and empirical-gamma (EG), are applied to improve the quality of raw ECHAM4/OPYC SRES A2 and B2 precipitation scenarios at the grid nodes. Out of three techniques, the GG transformation method is found to be more effective in correcting the rainfall frequency and intensity simultaneously.
- The results strongly suggest that the direct application of GCM output is a source of uncertainties in climate change studies. Therefore, it is advisable to apply bias-correction method to raw GCM datasets to reduce the error and uncertainty in impact assessment studies.

References

Hamlet AF, Snover A, Lettenmaier DP (2003) Climate change scenarios for Pacific Northwest water planning studies: motivation, methodologies, and a user's guide to applications. Technical document. http://www.hydro.washington.edu/Lettenmaier/permanent_archive/hamleaf/bams_paper/technical_documentation.pdf

Hashino T, Bradley AA, Schwartz SS (2006) Evaluation of bias-correction methods for ensemble streamflow volume forecasts. Hydrol Earth Syst Sci Discuss 3:561–594

Ines AVM (2004) GCM bias correction tool. Version 0.3a. IRI-Columbia University, New York

Ines AVM, Hansen JW (2006) Bias correction of daily GCM rainfall for crop simulation studies. Agric Forest Meteorol (In press)

Intergovernmental Panel on Climate Change (2001) Climate change 2001: impacts, adaptation and vulnerability. Third assessment report (TAR), Cambridge University Press. http://www.grida.no/climate/ipcc_tar/wg2

Prayuth G, Pongput K, Tangtham N, Gassman PW (2010) Hydrologic evaluation and effect of climate change on the At Samat watershed, Northeastern Region, Thailand. Int Agric Eng J 19 (2):12–22

Roeckner E, Arpe K, Bengtsson L, Christoph M, Claussen M, Dümenil L, Esch M, Giorgetta M, Schlese U, Schulzweida U (1996) The atmospheric general circulation model ECHAM4: model description and simulation of present-day climate. Max-Planck-Institut für Meteorologie, Hamburg, Report no 218

Sharma D (2007) Downscaling of general circulation model precipitation for assessment of impact on water resources at basin level trends in extreme rainfall and temperature indices for two river basins of Thailand. Ph.D. dissertation, Asian Institute of Technology, Thailand

Sharma D, Babel MS (2013) Application of downscaled precipitation for hydrological climate-change impact assessment in the upper Ping River Basin of Thailand. Clim Dyn 41(9–10): 2589–2602

Shrestha S (2014) Assessment of water availability under climate change scenarios in Thailand. J Earth Sci Clim Change 5:184. doi:10.4172/2157-7617.1000184

Smith JB, Hulme M (1998) Climate change scenarios (Chap. 3). In: Feenstra J, Burton I, Smith JB, Tol RSJ (EDS) Handbook on methods of climate change impacts and adaptation strategies. UNEP/IES, Version 2.0, Amsterdam

Wood AW, Maurer EP, Kumar A, Lettenmaier DP (2002) Long range experimental hydrologic forecasting for the Eastern U.S. Geophys Res 107(D20):4429

Hydrological Regime Responses to Climate Change for the 2020s and 2050s Periods in the Elbow River Watershed in Southern Alberta, Canada

Babak Farjad, Anil Gupta and Danielle J. Marceau

Abstract The Elbow River watershed, located in southern Alberta, drains approximately 1,235 km^2 area, and supplies the Glenmore Reservoir that provides water to nearly half of Calgary, a fast growing city of 1.1 million inhabitants. The watershed is characterized by a complex hydrological regime and is typical of snow dominated basins, with a spring freshet driven by snow melt and rainfall in late spring and early summer. This context creates favorable conditions for springtime flooding, which resulted in extensive damage in 2005 and 2013. Therefore, understanding how future climate changes might influence the watershed hydrological regime is critical. This research was conducted to investigate the hydrological regime responses of the watershed to climate change for the period of 2020s (2011–2040) and 2050s (2041–2070), relative to 1961–1990. The physically-based, distributed MIKE SHE/MIKE 11 model was used to simulate hydrological processes based on a warmer and drier (CCSRNIES A1FI) climate scenario. Results reveal that the average annual overland flow, baseflow, and river flow will decrease over the next 60 years and that this decrease combined with a drastic increase in evapotranspiration might increase water scarcity. Peak flows will increase in the winter and early-mid spring while decreasing in the summer and early fall. This enhances the risk of flooding in the spring, especially in the month of April which exhibits a significant increase in rainfall coinciding with the highest increase in spring freshet.

Keywords Hydrological modeling · MIKE SHE/MIKE 11 · Elbow River watershed · Alberta · Climate change · Flood

B. Farjad (✉) · D.J. Marceau
Department of Geomatics Engineering, University of Calgary,
2500 University Drive NW, Calgary, AB T2N 1N4, Canada
e-mail: bfarjad@ucalgary.ca

A. Gupta
Alberta Environment and Sustainable Resource Development (AESRD),
Calgary, AB T2E 7L7, Canada

Mu. Ramkumar et al. (eds.), *Environmental Management of River Basin Ecosystems*,
Springer Earth System Sciences, DOI 10.1007/978-3-319-13425-3_4

1 Introduction

The Elbow River watershed, located in southern Alberta, drains approximately 1,235 km^2 (Fig. 1) area. It belongs to the Canada's Western Prairie Provinces (WPP), which lie in the rain shadow regions of the Rocky Mountains and are the driest of southern Canada (Schindler and Donahue 2006). This area has experienced several severe droughts in the 20th century. In one of the worst events in the 1930s, referred to as the "dirty thirties", 7.3 million hectares of agricultural land were damaged and 250,000 people left the Canadian prairies (Gan 2000).

The Elbow River watershed is one of the regions in the WPP that is most affected by climate change. Valeo et al. (2007) performed statistical analysis on historical temperature data in the watershed and found that the annual average temperature has increased by 0.056 °C/yr between 1965 and 2004 in the west part of the watershed and by 0.007 °C/yr between 1885 and 2004 in the east part. Future projected data (provided by Alberta Environment and Sustainable Resource Development) for the CCSRNIESA1FI climate model project that the temperature may increase by approximately 4 °C by 2050, relative to 1990, in the watershed. This considerable change can lead to extreme hydrological events in the future, such as droughts (Schindler and Donahue 2006) and floods (Valeo et al. 2007).

Chen et al. (2006) investigated future climate trends and river water resources availability based on historical climate, streamflow, and population data of the Calgary region. They indicated that Calgary might face significant water supply

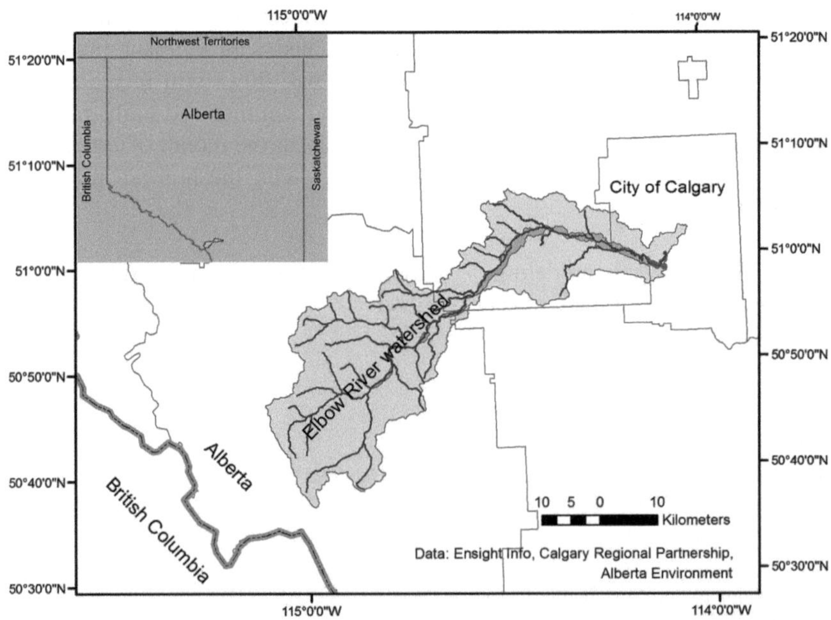

Fig. 1 Location of the Elbow River watershed

challenges in the future. For this city to maintain a sustainable water supply, it will require water conservation efforts to reduce the per-capita water consumption to less than 50 % of the current level by 2064. Even then, in the hot and dry projected periods, water demand could exceed the supply allotments (Chen et al. 2006). As a result, the Province of Alberta has stopped accepting new applications for the allocation of water since August 2006 in the Bow River basin of which the Elbow River is an important multi-use tributary (Pernitsky and Guy 2010).

Flooding is the other great stress endured in the Elbow River. In June 2013, the Elbow River was flowing through Calgary at 12 times the regular rate causing $400 million of damages, and the evacuation of 110,000 people (City of Calgary 2013). Two other floods of less magnitude occurred in 1995 and 2005. These floods happened in the month of June, which coincided with Albertans experiencing the driest years in climate history (Valeo et al. 2007). This is an indication of how climate change can alter the frequency and severity of extreme and contrasting events in a watershed, which depend not only on the magnitude of the change but also on the watershed characteristics, and its vulnerability to climate change. To better understand the vulnerability of the Elbow River watershed to climate change, its main characteristics are presented in the next section.

1.1 Characteristics of the Elbow River Watershed

The Elbow River originates at Elbow Lake at an elevation of 2,095 m above sea level and flows 120 km eastward through the alpine, subalpine, boreal foothill, and aspen parkland before joining the Bow River at 1,033 m above sea level in downtown Calgary (Beers and Sosiak 1993). In terms of land-use (Fig. 2), the watershed is comprised of urban area (5.9 %), agricultural land (16.7 %), rangeland/parkland (6.2 %), evergreen forest (34 %), deciduous forest (10 %), and clear-cut (1.8 %), (Wijesekara et al. 2012).

1.1.1 Meteorological Characteristics of the Watershed

Climate data (obtained from the Alberta Environment and Sustainable Resource Development (AESRD)) indicate that the average annual air temperature is 2.5 °C in the watershed. The warmest month is July with an average temperature of 13.2 °C, while the coldest month is January with an average temperature of −9 °C. The average total annual precipitation is 690 mm, of which almost 67 % falls between the months of April–September. The month of June is the wettest month with an average precipitation of 99.6 mm. The average annual potential evaporation is 552.5 mm, with the highest rate of evaporation (101 mm) recorded in July.

There is a significant difference in climate between the eastern and western portions of the watershed, since it lies between almost 2,100 m difference in elevation (Appendix A.1). To take this difference into consideration, the watershed

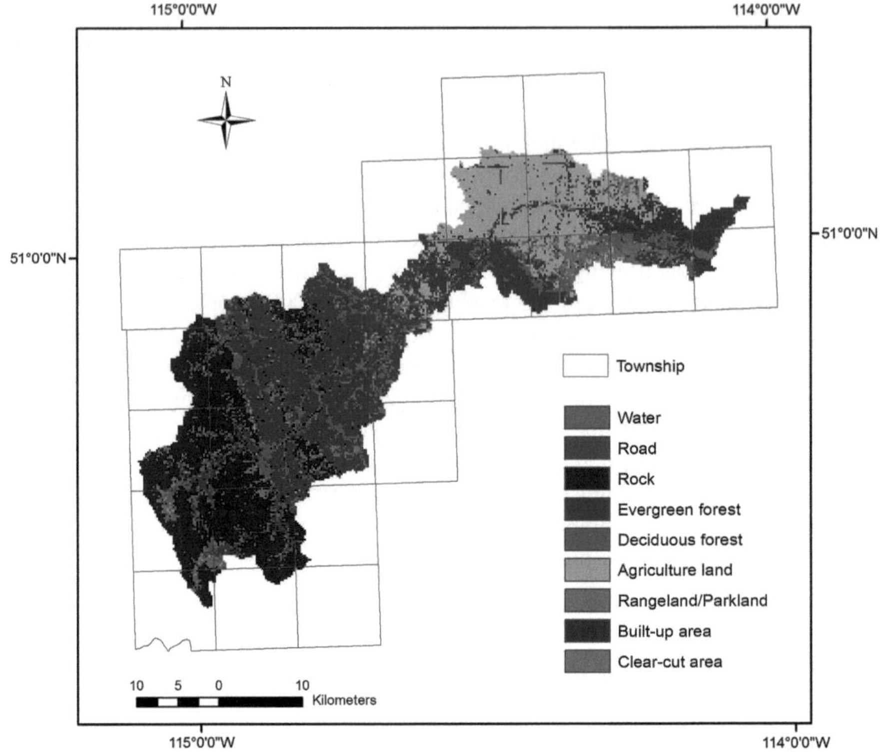

Fig. 2 Land-use map of the Elbow River watershed for the year 2010 *(Data source* Geocomputing Laboratory, University of Calgary)

was delineated into two sub-catchments (based on a digital elevation model): the west sub-catchment which is upstream of the 05BJ004 station (Bragg Creek), and the east sub-catchment which is downstream of that station (Fig. 3).

To estimate the average precipitation for the eastern and western sub-catchments, two precipitation gauges were selected for each sub-catchment in different locations; since precipitation varies spatially, it was necessary to use the data from gauges located at different locations. For the eastern sub-catchment, the temperature index station is Calgary, and the precipitation index stations are 3031875 and 3031090 (Fig. 3). For the western sub-catchment, the temperature index station is Nakiska, and the precipitation index stations are 305LRKB and 353602 (Fig. 3).

In the east sub-catchment, the annual average temperature is 4.1 °C at the Calgary station while in the western sub-catchment the annual average temperature is −1.9 °C at the Nakiska station. From April to September, precipitation reaches 399 mm in the eastern sub-catchment and 332 mm in the western sub-catchment (Figs. 4 and 5). However, the western sub-catchment gets significantly higher precipitation (272 mm) than the eastern sub-catchment (98 mm) between the months of October and March.

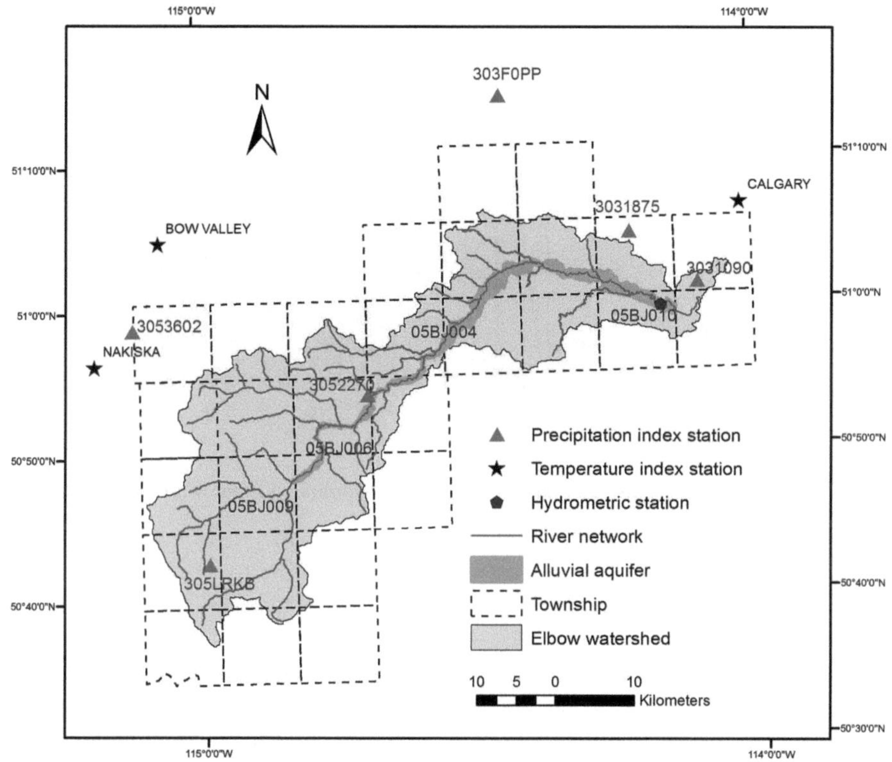

Fig. 3 Location of the climate index and hydrometric stations in the Elbow River watershed (*Data source* Alberta Environment and Sustainable Resource Development)

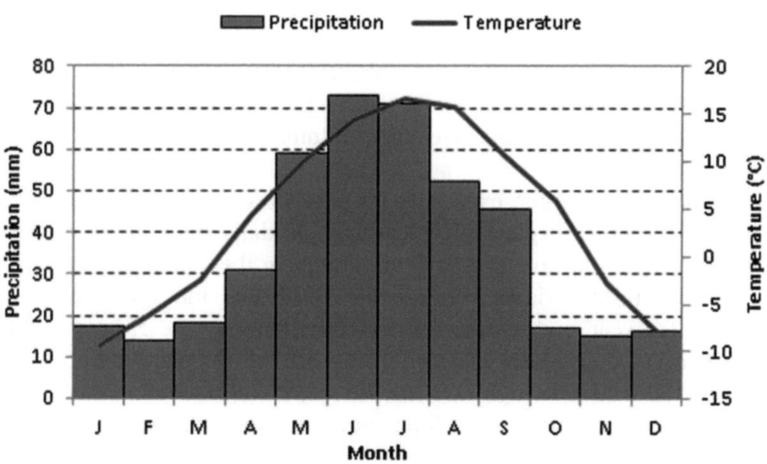

Fig. 4 Average monthly precipitation and temperature distribution in the eastern sub-catchment of the Elbow River watershed

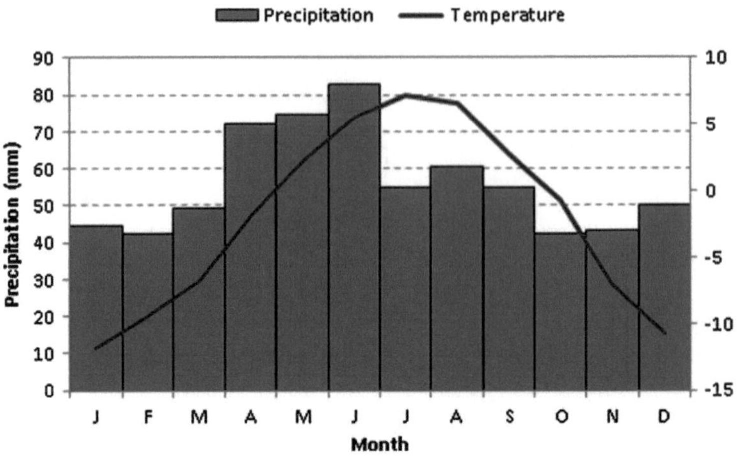

Fig. 5 Average monthly precipitation and temperature distribution in the western sub-catchment of the Elbow River watershed

1.1.2 River Regime

Four hydrometric stations, 05BJ009, 05BJ006, 05BJ004, and 05BJ010 (Fig. 3), measure discharge rates along the river. From west to east, the 05BJ009 and 05BJ006 stations cover 129 and 437 km^2 drainage areas in the front ranges of the Rocky Mountains in the western sub-catchment. The average annual discharge is 3.33 and 6.44 m^3/s at these stations respectively. The volume of water flowing down the river increases at the 05BJ004 station with the average annual discharge of 8.13 m^3/s. This station represents the outlet of the western sub-catchment and measures discharge rates of 791 km^2 drainage area upstream of the hamlet of Bragg Creek. The river flows to the lowlands areas in the eastern sub-catchment and drains a cumulative area of almost 1,200 km^2 at the 05BJ010 station upstream of the outlet of the watershed. The average annual discharge rate at this hydrometric station, which can be considered as the river flow volume discharging into the Glenmore reservoir, is 10 m^3/s.

The discharge rates differ, especially for peak flows, from month to month, and year to year. Figure 6 displays a typical average daily hydrograph of the Elbow River based on 72 year average daily discharge in the watershed. Generally, the flow of the river starts rising between the 100th and the 130th days from the beginning of the year, and reaches its peak flow between the 150th and the 180th days and then gradually starts decreasing between the 181st and the 210th days. In fact, overland flow and through flow are the major contributors to the river flow in the days between rising and falling limbs of the hydrograph and the baseflow is the main contributor in the remaining of the days during a year. In terms of monthly discharge variability (Fig. 7), the high flow period occurs in May, June, and July.

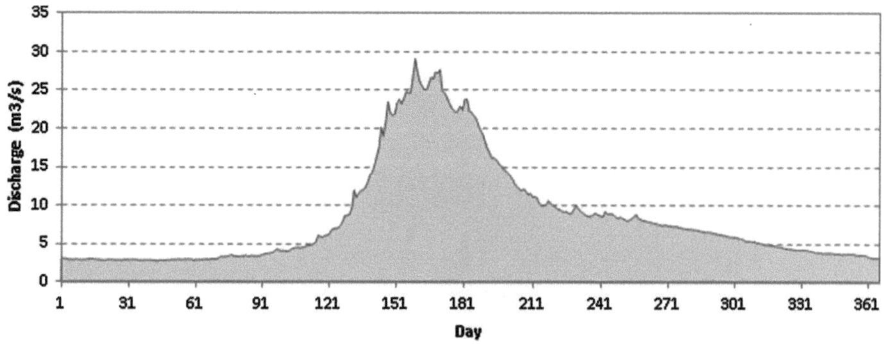

Fig. 6 Average daily hydrograph in the watershed (*Data source* Alberta Environment and Sustainable Resource Development)

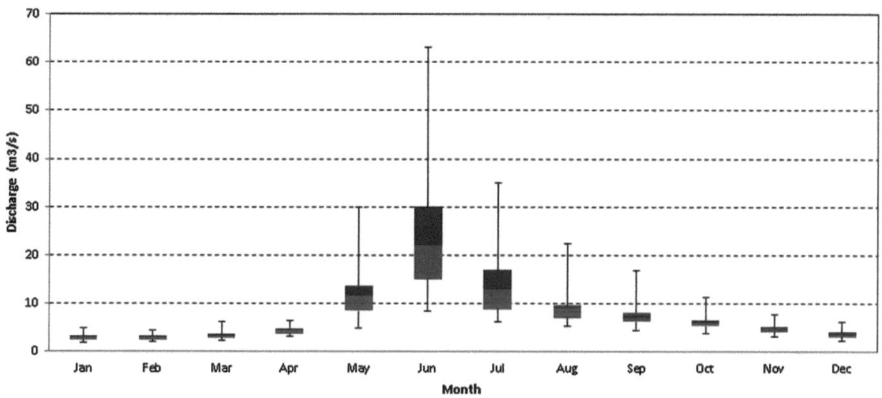

Fig. 7 Box plot of average monthly discharge illustrating the minimum, the 25 percentile, the median, the 75 percentile, and the maximum discharge values

The average peak flow for these months are 30, 63.1 and 35 m^3/s while the low flow are 4.83, 8.51 and 6.27 m^3/s for the period of 1978–2011, respectively.

The historical variability in low flow and high flow were investigated to understand the vulnerability of the watershed to droughts and floods. Daily historical low flow, peak flow, and instantaneous discharge are available for the periods of 1978–2011, 1935–2011, and 1950–2011, respectively. A trend analysis of the streamflow exhibits a moderate (0.30 $m^3s^{-1}y^{-1}$) and slight (0.11 $m^3s^{-1}y^{-1}$) increase in the instantaneous discharge and peak flow for the periods of 1950–2011 and 1935–2011, respectively (Fig. 8). In addition, low flow increases slightly (0.01 $m^3s^{-1}y^{-1}$) for the period of 1978–2011 (Fig. 9). An increasing trend in both low flow and high flow is a clear indication of increasing risk of droughts and floods in the past decades in the watershed, which might be associated with climate and/or land-use changes in the watershed.

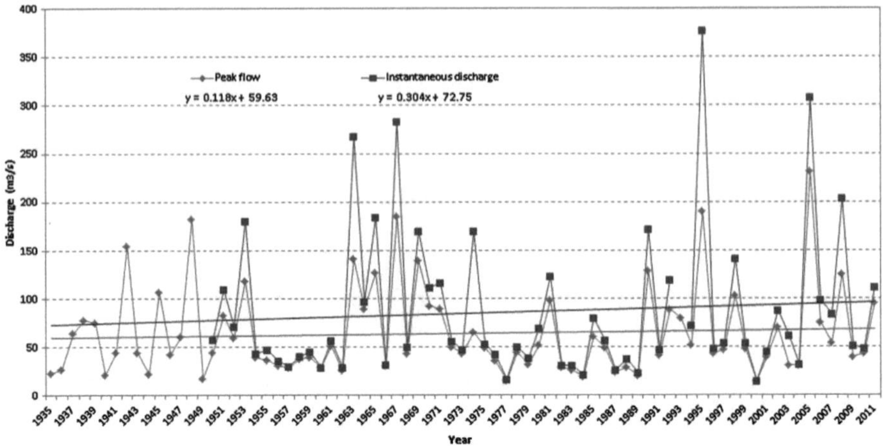

Fig. 8 Peak flow and instantaneous discharge series in the watershed (*Data source* Alberta Environment and Sustainable Resource Development)

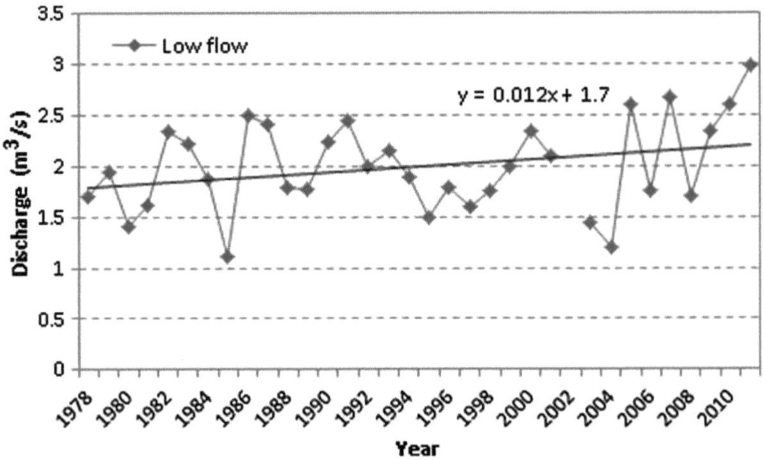

Fig. 9 Low flow discharge series in the watershed (*Data source* Alberta Environment and Sustainable Resource Development)

The highest peak flows of the watershed that have been recorded in history (at the 05BJ004 station) occurred in 1879 (Waterline 2011) and in 2013 with discharge values of 980 and 959 m^3/s, respectively. Other high peak flows of lesser magnitude are 836 m^3/s in 1932 and 489 m^3/s in 1929 while the recurrence of 20 and 100 year of flood events are 340 and 758 m^3/s, respectively (Waterline 2011). It is not possible to directly relate the flood events (such as the recent ones of 1995, 2005, and 2013) to climate change. However, the observed trends in increased peak flows and temperature (which results in increased snow melting) are affirmative signs of increasing vulnerability to floods of higher magnitudes and frequencies.

1.1.3 Geomorphological Characteristics of the Watershed

The geomorphological characteristics of the watershed can influence the flow regime, especially during flooding. For example, the time of concentration, which describes the speed and intensity of the watershed response to storms, changes with the different morphological characteristics. The geomorphological characteristics of the Elbow River watershed, namely the stream patterns, shape, drainage density, stream order, and topography vary considerably from west to east (Fig. 10). The western sub-catchment mostly lies in the highland areas with less recession time of overland flow. The stream patterns of the western sub-catchment are similar to trellis patterns that are characterized by long main streams intercepted by numerous shorter right-angle tributaries. Trellis patterns are commonly found in regions of folded or tilted strata. However, the stream pattern of the eastern sub-catchment is dendritic, which is characterized by gentle regional slope, and relatively uniform lithology (Mejía and Niemann 2008).

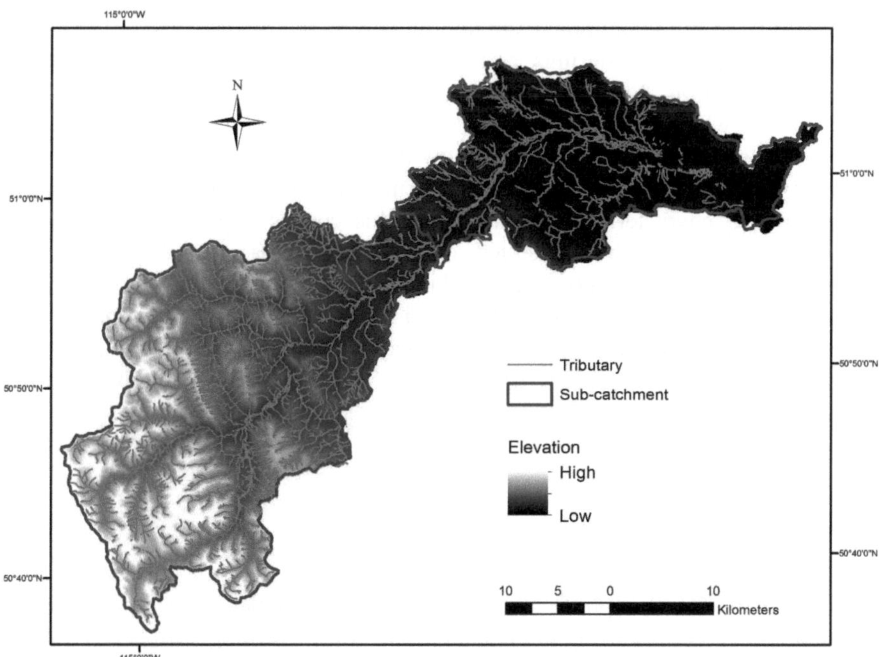

Fig. 10 Geomorphological characteristics of the Elbow River watershed (*Data source* Alberta Environment and Sustainable Resource Development)

These physical characteristics in the western sub-catchment along with the orographic precipitation and snowmelt are the major factors explaining that about 80–90 % of streamflow originates upstream of the station 05BJ004.

1.1.4 Geological and Hydrogeological Characteristics of the Watershed

The surficial geology of the Elbow River watershed is dominated by glacial deposits and recent alluvial deposits (Manwell et al. 2006). The watershed contains the following aquifers (Waterline 2011): Unconsolidated Glacial Overburden aquifers, Elbow River alluvial aquifer, Porcupine Hills Formation aquifers (multiple aquifers with depth), Coalspur Formation aquifer, Brazeau Formation aquifer, and the karstic Paleozoic carbonate aquifer(s).

The alluvial aquifer plays an important role in the hydrological regime of the watershed since it is generally very permeable and hydraulically connected to the Elbow River. The aquifer lies along the Elbow River for approximately 5 % of the watershed (61 km^2) which extends from the near headwaters in the west to the Glenmore reservoir in the east (Waterline 2011).

The hydraulic conductivity of the aquifer is on the order of 1×10^{-3} m s^{-1} (Manwell et al. 2006; Meyboom 1961), and the direction of groundwater flow is generally from the west to the east along the axis of the watershed (Waterline 2011). Discharge from groundwater to river flow, between August and April, is approximately 40 % of the total annual streamflow (Beers and Sosiak 1993).

Different geological settings and the alluvial aquifer along the river create complex interactions between surface water and groundwater. This complexity can be enhanced when there are different climate conditions in the watershed. Furthermore, the watershed is typical of snow dominated basins, with a spring freshet driven by snow melt and rainfall in late spring and early summer. Along with the morphological characteristics of the basin, this creates favorable conditions for springtime flooding, which resulted in extensive damage in 2005 and 2013. Therefore, understanding how future climate change might influence the hydrological regime of the Elbow River watershed is of critical importance. The objective of this study is to investigate the hydrological regime responses to climate change for the period of 2020s (2011–2040) and 2050s (2041–2070), relative to 1961–1990.

2 Methodology

MIKE SHE/MIKE 11, a physically-based, distributed model, capable of simulating the entire processes occurring in the land phase of the hydrologic cycle (DHI 2009), was used to simulate the hydrological processes in the watershed.

Most of the equations in the MIKE SHE/MIKE 11 model are based on the interchangeable types of mechanical energy (kinetic energy, potential energy, and pressure energy) for moving a water particle along a streamline. MIKE SHE quantified overland flows using the two-dimensional Saint-Venant equation, unsaturated zone flows using the two-layer water balance method, saturated zone flows using the Darcy equation, and evapotranspiration using Kristensen and Jensen method. MIKE 11 is a fully dynamic and one-dimensional hydraulic model that simulates flows, rivers, channels, and other surface water bodies based on the complete dynamic wave formulation of the Saint Venant equations (Singh 1995).

The methodological framework used in this study is illustrated on Fig. 11. The coupled MIKE SHE/MIKE 11 model requires a large amount of data. Some of these data, such as the physical characteristics of the surface/subsurface (Appendix B.1), were provided by earlier research (Wijesekara et al. 2014). Land-use maps of the years 1985, 1992, 1996, 2001, and 2006 were generated from Landsat Thematic Mapper imagery at the spatial resolution of 30 m (Hasbani et al. 2011). Climate data were provided by the Alberta Environment and Sustainable Resource Development (AESRD). The observed temperature data are from the period 1961–1990, acquired at an hourly basis for three index stations: Calgary, Nakiska, and Bow valley. The observed precipitation data were acquired for the same period of time (1961–1990), on a daily basis at six index stations: 3031090, 3031875, 303F0PP, 3052270, 353602, and 305LRKB.

Future climate data were selected for the warmer and drier climate scenario (A1FI), and were projected for the period of 2011–2040 (2020s) and 2041–2070 (2050s) relative to the baseline period of 1961–1990 using the CCSRNIES model (Barrow and Yu 2005). These data include precipitation and temperature which have been downscaled using the delta method. The delta method is widely used in hydrological studies of climate change (Hay et al. 2000; Snover et al. 2003; Buytaert et al. 2009; Xu et al. 2009; Buytaert et al. 2010). This method applies change values, obtained from simulation of baseline and future, to perturb the historical daily climate data (1961–1990). The perturbed precipitation and temperature data were generated for the mentioned index climate stations (Golder Associates 2010).

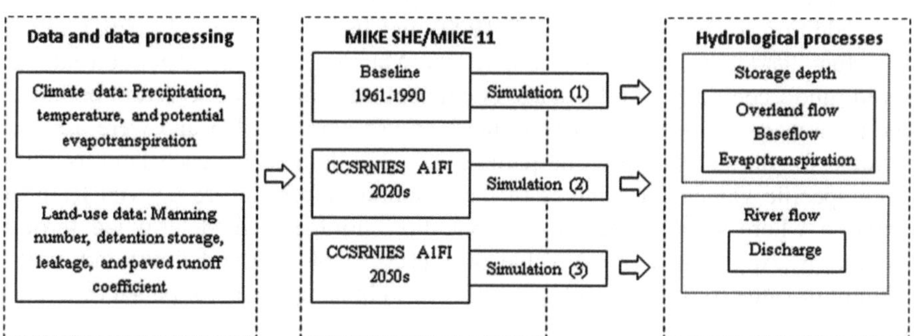

Fig. 11 Methodological framework

The index-based precipitation data were used for configuring the MIKE SHE model. However, the index-based temperature data were converted into township-based temperature (the MIKE SHE model performed better based on township-based temperature rather than index-based temperature during the calibration process as revealed in previous studies (DHI Water and Environment 2010; Wijesekara et al. 2014)).

A programming code was developed in Matlab to interpolate the temperature data using the following procedure. First, Thiessen polygons were created to identify the area of influence for the temperature index stations. Second, a digital elevation model was used along with an elevation lapse rate to interpolate temperature at the resolution of 80 m. The orographic correction factor that was selected is 0.75 °C/100 m, a value that was recommended in previous studies (DHI Water and Environment 2010; Wijesekara et al. 2014). Finally, the interpolated temperature data were split into 29 townships (each 6 by 6 square mile) and employed for configuring the MIKE SHE model.

Potential evapotranspiration (PET) is the other primary climate dataset required in MIKE SHE. It must be estimated using the available projected data (temperature) for the future periods. A literature review was conducted to identify models that could be used for calculating future PET in the watershed. The following criteria were considered for the selection of a model: (a) it must be based on air temperature for calculating PET, (b) it must have been applied in the study area or in a similar climate, and (c) it must be widely used (accepted). Based on these criteria, the following models were chosen: (a) the Hargreaves-Samani (Hargreaves and Allen 2003), (b) the Thornthwaite (Sentelhas et al. 2010), and (c) the Blaney-Criddle models (Espadafor et al. 2011). These models were compared using PET rates provided by AESRD for the period of 1961–2005. These rates have been calculated using the Priestley-Taylor model (Priestley and Taylor 1972) used by Alberta Agriculture and Food for the Elbow River watershed. The performances of the models were evaluated using linear regression analysis, Root Mean Square Error (RMSE), and Mean Bias Error (MBE). The Hargreaves model showed the best performance and was selected for calculating PET. This model was calibrated for the period 1961–1990 and validated for the period 1991–2005.

In addition to the baseline (1961–1990) and future (2020s and 2050s) climate data, four land-use parameters, namely, the Manning number, detention storage, paved runoff coefficient, and leakage coefficient were also used to setup the model. The parameters were extracted from the land-use map of 1985 and were assumed constant for all simulations. The parameters were defined for each land-use class using specific values. The values of the Manning number were derived from the literature whereas the values of the detention storage were determined through calibration (Appendix C.1). A value of 1 (100 % of overland flow) and 1e-013 (minimum value for infiltration) was assigned to built-up areas for paved runoff and leakage coefficient, respectively.

The water balance of the watershed was determined in the MIKE SHE model, since it is an important indicator to assess the responses of hydrological processes to climate change (Hendriks 2010). The water balance corresponds to the amount of water that is taken into or released for storage within the watershed. The hydrological processes such as evapotranspiration (ET), baseflow, and surface flow were used as the major water balance components for estimating the storage depth in the watershed. They were simulated for the entire watershed and for the western and eastern sub-catchments on an average monthly and annual basis for the 2020s, 2050s, and the baseline period. Apart from the water balance assessment, the daily absolute value of discharge at hydrometric stations in m^3/s was obtained through the coupling of MIKE SHE and MIKE 11. For the purpose of this study, only discharge values for station 0BJ010 were evaluated since this hydrometric station is located near the watershed outlet and represents most of the drainage area.

The model was calibrated and validated using a rigorous procedure described in detail in Wijesekara et al. (2014). The calibration was conducted for the period of 1981–1991 with the land-use map of 1985. Four time periods (1991–1995, 1995–2000, 2000–2005, and 2005–2008) were used for validation with their corresponding land-use maps (1992, 1996, 2001, and 2006). The goodness-of-fit was evaluated by comparing observed data and simulated data of total snow storage and stream flow.

3 Results

In this section, results from the simulation of hydrological processes and streamflow in response to climate changes are presented.

3.1 Simulation of Hydrological Processes

Table 1 and Fig. 12 illustrate the average annual storage depth and percentage of change of overland flow, baseflow, and evapotranspiration (ET) for the periods 2020s and 2050s, relative to the baseline. The average annual overland flow decreases by 10.5 % in the 2020s and by 10.6 % in the 2050s. These decreases correspond to the changes in the projected climate variables namely, temperature and precipitation (Table 2). In the 2020s, the considerable decrease in overland flow is primarily associated with the average annual precipitation that reveals a 4.7 %

Table 1 Average annual storage depth and percentage of change (relative to the baseline) for overland flow (OL), baseflow (BF) and evapotranspiration (ET)

Hydrological processes	Average annual storage depth (mm)		
	Baseline	2020 s	2050 s
OL	141.24	126.43	126.27
BF	86.43	80.46	83.34
ET	415.47	411.55	444.28

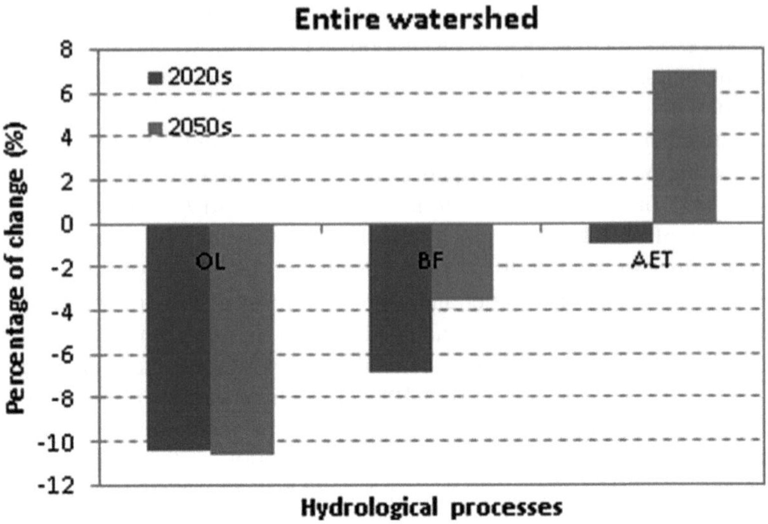

Fig. 12 Percentage of change in overland flow (OL), baseflow (BF) and evapotranspiration (ET) relative to the baseline for the entire watershed

Table 2 Projected average annual precipitation and temperature

Climate variable	Baseline	2020s	2050s
Precipitation (mm)	689.64	657.26	689.60
Temperature (°C)	2.53	2.93	6.63

decline, and to the average annual temperature, which increases by 0.4 °C. The decline in average annual overland flow in the 2050s is almost the same as in the 2020s. However, the average annual precipitation in the 2050s remains stable relative to the baseline. Therefore, the decline in the average annual overland flow in the 2050s is mainly linked with the 4.1 °C projected changes in the average annual temperature.

The average annual baseflow decreases by 6.9 % in the 2020s, which is related to the 4.7 % decrease in precipitation and 0.4 °C increase in the average annual temperature. In the 2050s, the percentage of decrease in baseflow (3.6 %) is mainly due to the 4.1 °C increase in the temperature (that results in increased evaporation loss from surface and subsurface), while the precipitation is invariable relative to the baseline.

In response to the changes in precipitation and temperature, the average annual ET declines by 0.9 % in the 2020s, but increases by 6.9 % in the 2050s. This increase occurs when there is a 4.1 °C increase in temperature and constant precipitation relative to the baseline period. This indicates that ET is more associated with the projected temperature changes than the projected precipitation changes compared to overland flow and baseflow in the 2050s.

The impact of climate change on the hydrological processes is more important in the eastern sub-catchment than the western sub-catchment, except for ET in the 2050s (Figs. 13 and 14). In the 2020s, overland flow and baseflow decrease in the western and eastern sub-catchments. They significantly drop in the eastern sub-catchment by 20.7 and 12.9 %, respectively, while the decline in overland flow

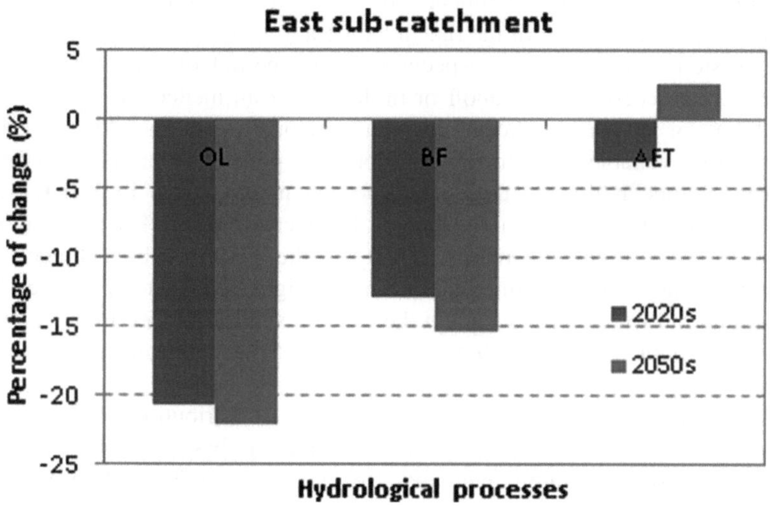

Fig. 13 Percentage of change in overland flow (OL), baseflow (BF) and evapotranspiration (ET) in the eastern sub-catchment

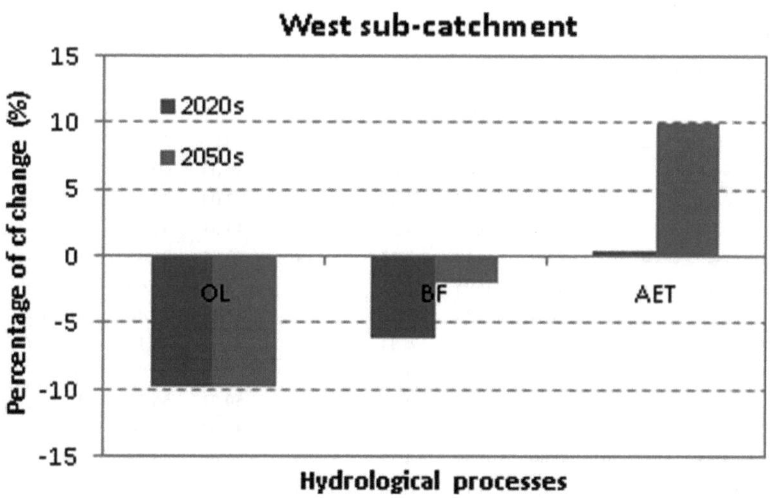

Fig. 14 Percentage of change in overland flow (OL), baseflow (BF) and evapotranspiration (ET) in the western sub-catchment

(9.7 %) and baseflow (6 %) is modest in the western sub-catchment, compared to the eastern sub-catchment. Although, the impact of climate changes is more perceptible in the eastern sub-catchment, these changes are offset by the western sub-catchment. In the 2050s, the significant rise in temperature (Table 2) results in even larger percentage decreases in the surface flow (22.1 %) and baseflow (15.3 %) in the eastern sub-catchment compared to the surface flow (9.8 %) and baseflow (2.0 %) in the western sub-catchment. This is mainly due to more evaporation and loss of soil water as a result of the rise in temperature. However, it is less noticeable in the western sub-catchment, especially in the period of time when only the highland areas receive either runoff or infiltration from melted snow.

In terms of ET, a significant drop in precipitation in the 2020s (Table 2) decreases the available water for evaporation from ponded water and soil (these are the main factors in the Kristensen and Jensen method used in MIKE SHE to calculate ET). In response to these changes, ET decreases by 3 % in the eastern sub-catchment, but it slightly increases (0.4 %) in the western sub-catchment due to sublimation and snow melting (caused by a slight rise in temperature) in the summer and fall which only occurs in the western sub-catchment. In the 2050s, ET increases in both the west (9.9 %) and east (2.5 %) sub-catchments due to a significant increase in temperature by 4.1 °C.

Table 3 and Fig. 15 present how the seasonal distribution of the projected temperature and precipitation change compared to the baseline. In the 2020s, the average monthly temperature increases for every month (0.1–1.5 °C), with the exception of January and February that experience a decrease of −1.3 and −0.6°C, respectively. The significant increase of temperature occurs in the spring and summer (0.4–1.5 °C) with the highest rise occurring in April (1.5 °C). Contrary to

Table 3 Average monthly projected temperature and precipitation

Month	Temperature (°C)			Changes in temperature (°C)		Precipitation (mm)			Changes in precipitation (%)	
	Baseline	2020s	2050s	2020s	2050s	Baseline	2020s	2050s	2020s	2050s
J	−9.05	−10.37	−5.37	−1.32	3.68	36.96	34.03	40.98	−7.93	10.88
F	−6.24	−6.81	−1.78	−0.58	4.46	34.23	32.04	38.37	−6.41	12.09
M	−3.07	−2.69	1.44	0.38	4.51	44.05	43.31	52.72	−1.68	19.68
A	2.66	4.15	7.79	1.48	5.13	68.49	73.31	81.49	7.03	18.98
M	7.60	8.55	10.45	0.95	2.84	83.16	73.98	77.66	−11.03	−6.61
J	11.32	11.71	14.42	0.39	3.10	99.63	81.19	90.44	−18.51	−9.23
J	13.20	14.07	17.54	0.87	4.34	75.60	74.96	78.11	−0.85	3.33
A	12.53	13.57	16.78	1.04	4.25	73.61	74.65	62.98	1.40	−14.45
S	8.05	8.12	10.81	0.07	2.76	63.04	59.71	56.50	−5.28	−10.39
O	3.97	4.09	7.25	0.12	3.29	37.27	36.97	33.63	−0.79	−9.75
N	−3.38	−2.79	0.99	0.59	4.37	35.05	36.00	34.35	2.69	−2.02
D	−7.67	−6.90	−1.19	0.77	6.48	38.55	37.12	42.38	−3.73	9.91

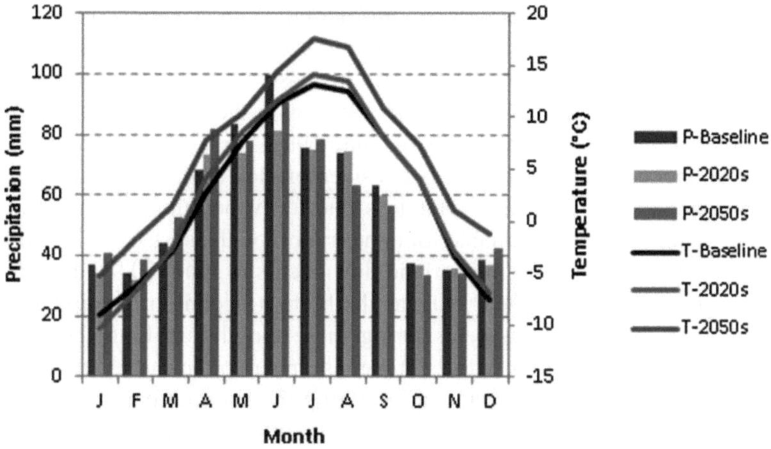

Fig. 15 Monthly average projected temperature and precipitation

the temperature, the average monthly precipitation decreases (0.8–18.5 %) for all the months, except in April, August, and November. In the 2050s, the average monthly temperature increases in every month (2.8–6.5 °C), while precipitation increases considerably in the winter and early and mid-spring.

In response to the seasonal distribution changes, overland flow significantly increases in the mid- and late spring in the 2020s and in the winter, and early and mid-spring in the 2050s (Fig. 16). This is mainly due to snow melt and a higher rain/snow ratio caused by higher temperatures. In addition, during this period of time, a unit of rainfall generally produces more overland flow when it falls on wet soils in early spring compared to the summer where rains often fall on very dry soils and generates less overland flow relative to the intensity of the events. The highest increase in overland flow occurs for both the 2020s and 2050s in the month of April

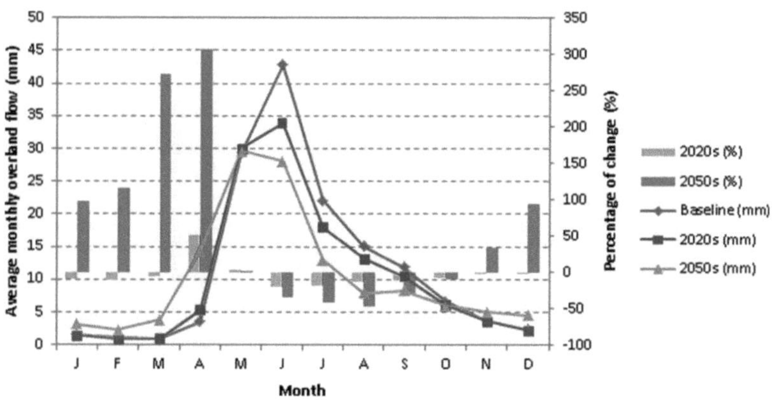

Fig. 16 Monthly average overland flow and percentage of change relative to the baseline

since snow starts melting usually in this month, and a significant increase in temperature amplifies sublimation and melting snow packs. In the 2050s, the precipitation in April is another factor that can increase the intensity of overland flow. April is the second wettest month in the 2050s with the average monthly precipitation of 81.4 mm after the month of June (99.4 mm). However, the average monthly temperature in this month is considerably lower than the month of June, and consequently, there is less water loss due to evaporation. Therefore, a large portion of water storage is available for either runoff on the surface or for infiltration into the soil. The lowest drop in overland flow occurs in June (20.9 %) in the 2020s and in August (46.9 %) in the 2050s, as the highest decline in precipitation (18.5 and 14.4 %, respectively) happens during these months. In both the 2020s and 2050s, overland flow declines in the summer as a result of an earlier and less intense snowmelt and an increase in evaporation during the summer period.

In the 2020s, baseflow decreases for every month, except in April and May in which case there is a slight increase (0.24 mm) relative to the baseline (Fig. 17). Baseflow during these months is mainly driven by the spring time snowmelt due to the considerable increase in temperature. In addition, the rising of temperature in early spring increases the period of time available for infiltration, and eventually increases the baseflow. The highest rise in baseflow occurs in April due to the highest increase in precipitation and temperature which ultimately leads to an increase in snow melting. In the 2050s, baseflow increases in winter and spring which is associated with more snow melting as a result of the increase in temperature; it decreases in the summer and fall due to an increase in temperature (and an increase in evaporation from unsaturated zone), and a decrease in precipitation.

In the 2020s, ET decreases in the summer and early fall (1.5–4.8 %) (Fig. 18), due to an increase in temperature and a reduction in soil water which result in less available water to the roots and a drop in transpiration. The lowest decrease in ET (14.9 %) happens in January in relation with the lowest decline in precipitation. However, the highest increase in ET (11.23 %) occurs in April, which is associated with the highest rise in precipitation (7.9 %), and additional available water to the

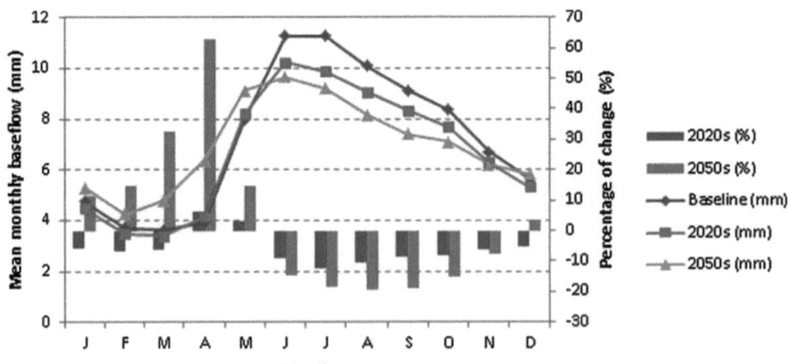

Fig. 17 Monthly average baseflow and percentage of change relative to the baseline

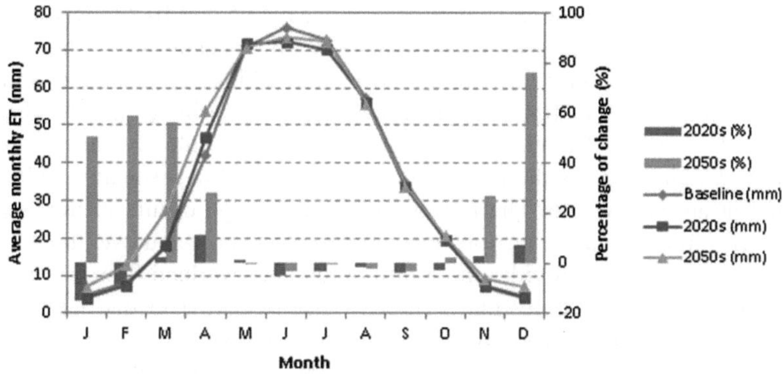

Fig. 18 Monthly average evapotranspiration and percentage of change relative to the baseline

roots due to a significant sublimation and snow melting during this month. The lowest decrease in ET (14.9 %) happens in January due to the lowest decline in precipitation. In the 2050s, evapotranspiration increases significantly in the winter and early spring due to snow melting as a result of an increase in temperature during this period.

3.2 Simulation of Streamflow

Simulations of 30 year average daily hydrograph for the 2020s, 2050s, and the baseline are shown in Fig. 19. Streamflow simulation reveals that the average annual stream flow decreases by 8.47 and 7.29 % in the 2020s and 2050s, respectively.

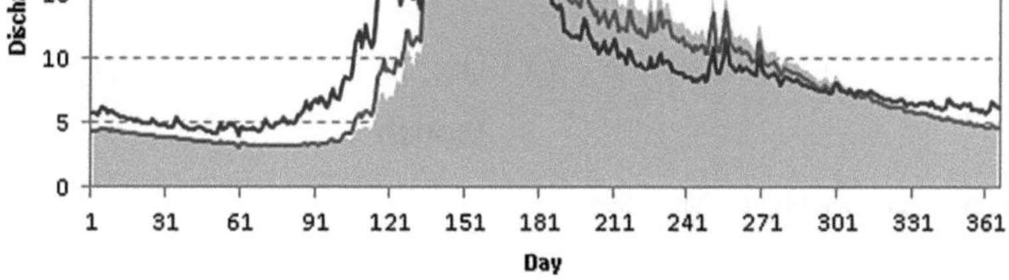

Fig. 19 Simulated 30 year average daily hydrograph for the 2020, 2050s, and the baseline

In the 2020s, both high flow and low flow decrease by 3.5 and 10.7 % whereas in the 2050s, the high flow increases by 1.4 % and low flow decreases by 7 %.

The streamflow declines for both the 2020s and 2050s in the summer and early fall, and increases in the 2050s in winter and early-mid spring (Fig. 20). There is a shift in peak river flow from late spring-early fall to the middle of spring-summer in the 2050s which indicates the possibility of spring flood in association with snow melt and at an earlier date. The highest rise in discharge occurs in the month of April. A rise in temperature in April increases rain-on-snow and eventually rainfall on a melting snowpack provides more infiltration and overland flow. Furthermore, the high flow season becomes much shorter. Historically, river discharge exceeds 10 m^3/s during 5 months (from May to September) of a year; however, it only lasts three months (May, June, and July) in the 2050s. The predicted average monthly winter temperature stays below 0 °C for only 3 months (December, January and February), while historically, five months of a year were below 0 °C. This results in a large shift away from winter snowfall to rainfall that will cause more river flow during the winter months in return of lower flows during the spring and summer months. Rising temperature above 0 °C in winter, increases the sublimation of snow packs before the melting occurs, which results in a reduction of water storage in winter and of contribution to river flow in summer.

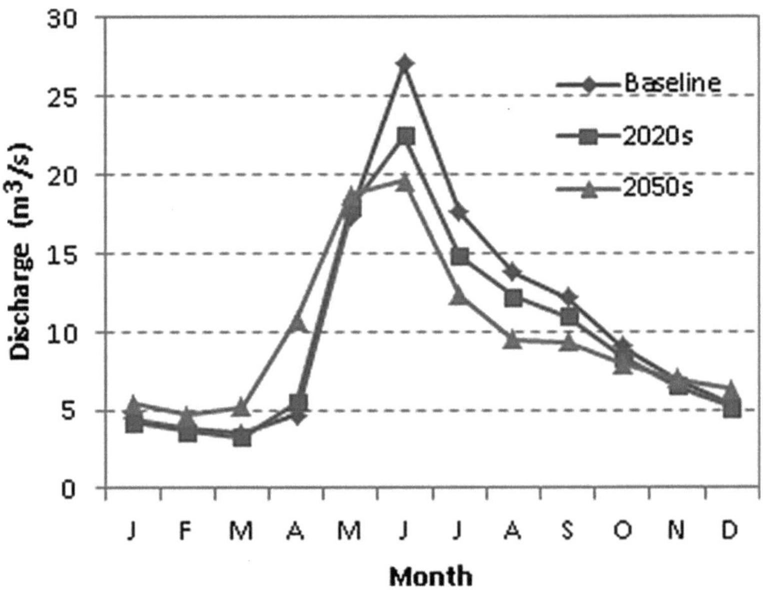

Fig. 20 Average monthly discharge flow

4 Conclusions

- This study reveals that the climate changes projected for the 2020s and 2050s might induce significant modifications on the Elbow River watershed hydrological regime. These changes are more noticeable in the eastern sub-catchment compared to the western sub-catchment. However, the induced changes in the eastern sub-catchment are offset by the western sub-catchment which almost governs the hydrology of the entire watershed due to its characteristics.
- The average annual overland flow, baseflow, and river flow will decrease over the next 60 years, while evapotranspiration will considerably increase, resulting in water scarcity. The situation might become critical due to projected increase in water supply demands.
- An increase in temperature during winter and spring (especially in April) will drive most of the changes affecting the hydrological processes and river flow. It will induce an important increase in the proportion of rainfall. Furthermore, the increase in number of months having a temperature above 0 °C will enhance the sublimation of snow packs before the melting occurs, which will result in an increase in winter and spring runoff and a reduction in the volume of water stored in the snowpack relative to the baseline period.
- Overall, climate change in the watershed will lead to an increase in overland flow, baseflow, evapotranspiration, and river flow in the winter-spring season as a result of an earlier and intense snowmelt, and a decrease in the summer-fall due to less intense snowmelt and an increase in evaporation. This implies a smaller difference in average monthly discharge between these two periods (summer-fall and winter-spring).
- The shift in peak river flow from late spring-early fall to the middle of spring-summer (which is mainly associated with earlier spring melt) will enhance the risk of flooding, especially in the lowlands in the eastern sub-catchment. The risk of flooding will increase in the month of April, which exhibits a significant increase in rainfall coinciding with the highest increase in spring freshet. This might even increase the risk of flooding in the following months (May and June), since the soil moisture reaches field capacity in early spring (March–April), and water released by snowmelt in May and June primarily contributes to runoff. This situation, in May and June, would be intensified once all the rainfall becomes runoff due to soil moisture field capacity.
- All the significant changes, such as the graduated shift from snow to rain, the snowpack reduction, along with the changes in hydrological processes will influence the timing and frequency of low and high discharge in the Elbow River, and eventually will affect the operation strategy of the Glenmore reservoir. The future modifications of the Elbow River watershed will not only be related to the hydrological regime, but will also cause important modifications in the watershed ecology.

Work is currently in progress to incorporate four additional climate change scenarios to cover the range of possible future climate conditions, along with projected land-use changes to fully evaluate the impact of future climate change and land-use changes on the hydrology of the watershed.

Acknowledgments This project was funded by a research grant awarded to Danielle Marceau by Tecterra and Alberta Environment and Sustainable Resource Development (AESRD), and by an Alberta Innovates Technology Futures graduate student scholarship awarded to Babak Farjad. We thank Dr. Shawn Marshall, Department of Geography, University of Calgary for his useful comments.

Appendix A

See Table 4.

Table 4 Physiography of east and west sub-catchments of the Elbow River watershed

Sub-catchment	Min-elevation (m)	Max-elevation (m)	Mean elevation (m)	Area (km^2)
West	1301.07	3208.38	1938.60	791.3
East	1039.78	1495.70	1209.06	443.7

Appendix B

See Table 5.

Appendix C

See Table 6.

Table 5 Datasets and parameters used for the setup of MIKE SHE in the Elbow River watershed

Data	Description
Topography	A digital elevation model was obtained from GeoBase at the spatial resolution of 80 m and was re-sampled at 200 m
Root depth (RD) and leaf area index (LAI)	These data were obtained from the literature. Root depth values are based on average depths of actual root zone of the vegetation, and leaf area index values vary between 0 and 7 in different seasons
Manning number	These data were obtained from the literature for each land-use class.
Detention storage	Detention storage values were obtained based on the sensitivity analysis and calibration for each land-use class
Channel flow	Water level data were obtained from field survey and LiDAR generated cross sections located across the Elbow River and its branches
Soil hydraulic properties	Soil hydraulic values were acquired based on physical properties of different soil classes. A soil classification map was obtained from the Agricultural Region of Alberta Soil Inventory Database and the Canadian Soil Information Service Data sources
Land-use maps	The land-use maps of 1985 indicate other years were generated from Landsat Thematic Mapper imagery at 30 m resolution

Wijesekara et al. 2014

Table 6 The assigned values for the manning number and detention storage parameters

Parameter	Assigned value
Manning number	Water: 25.04, road: 76.9, rock: 40.0, evergreen: 10.0, deciduous: 10.0, agriculture land: 28.57, rangeland/parkland: 33.33, built-up area: 90.9, clear-cut area: 90.9
Detention storage	Water: 0 mm; road: 5 mm; rock: 10 mm; evergreen: 20 mm; deciduous: 20 mm; agriculture land: 20 mm; rangeland/parkland: 20 mm; built-up area: 0 mm; clear-cut area: 0 mm

Wijesekara et al. 2014

References

Barrow E, Yu G (2005) Climate scenarios for Alberta. Prairie adaptation research collaborative, Regina

Beers C, Sosiak A (1993) Water quality of the Elbow River. Environmental Quality Monitoring Branch, Alberta Environmental Protection

Buytaert W, Célleri R, Timbe L (2009) Predicting climate change impacts on water resources in the tropical andes: effects of GCM uncertainty. Geophysl Res Lett 36(7):L07401

Buytaert W, Vuille M, Dewulf A, Urrutia R, Karmalkar A, Celleri R (2010) Uncertainties in climate change projections and regional downscaling in the tropical Andes: implications for water resources management. Hydrol Earth Syst Sci 14(7):1247–1258

Chen Z, Grasby SE, Osadetz KG, Fesko P (2006) Historical climate and stream flow trends and future water demand analysis in the Calgary region, Canada. Water Sci Technol 53(10):1–12
City of Calgary (2013). Calgary flood 2013. http://www.calgary.ca/General/flood-recovery/Pages/Calgary-flood-2013-infographic-recap.aspx
DHI (2009) MIKE SHE. User manual vol 2. Reference guide
DHI Water and Environment (2010) Elbow River watershed hydrology modeling. Report submitted to Alberta Environment
Espadafor M, Lorite IJ, Gavilán P, Berengena J (2011) An analysis of the tendency of reference evapotranspiration estimates and other climate variables during the last 45 years in Southern Spain. Agric Water Manag 98(6):1045–1061
Gan TY (2000) Reducing vulnerability of water resources of Canadian prairies to potential droughts and possible climatic warming. Water Resour Manage 14(2):111–135
Golder Associates (2010) Hydro-climate modeling of Alberta south saskatchewan regional planning area, 09-1326-1006, pp 82. Report submitted to Alberta environment
Hargreaves GH, Allen RG (2003) History and evaluation of Hargreaves evapotranspiration equation. J Irrig Drainage Eng 129(1):53–63
Hasbani J-G, Wijesekara N, Marceau DJ (2011) An interactive method to dynamically create transition rules in a land-use cellular automata model. In: Salcido A (ed) Cellular automata simplicity behind complexity, InTech. http://www.intechopen.com/books/cellular-automata-simplicity-behind-complexity/aninteractive-method-todynamically-create-transition-rules-in-aland-use-cellular-automata-model
Hay LE, Wilby RL, Leavesley GH (2000) A comparison of delta and downscaled GCM scenarios for three mountainous basins in the United States. JAWRA J Am Water Resour Assoc 36(2):387–397
Hendriks MR (2010) Introduction to physical hydrology. Oxford University Press, New York
Manwell BR, Ryan MC (2006) Chloride as an indicator of non-point source contaminant migration in a shallow alluvial aquifer. Water Qual Res J Can 41(4):383–397
Mejía AI, Niemann JD (2008) Identification and characterization of dendritic, parallel, pinnate, rectangular, and trellis networks based on deviations from planform self-similarity. J Geophys Res 113(F2):F02015
Meyboom P (1961) Groundwater resources in the City of Calgary vicinity. Research council of Alberta. Bulletin 8, Edmonton
Pernitsky DJ, Guy ND (2010) Closing the south saskatchewan river basin to new water licences: effects on municipal water supplies. Can Water Resour J 35(1):79–92
Priestley CHB, Taylor RJ (1972) On the assessment of surface heat flux and evaporation using large-scale parameters. Mon Weather Rev 100(2):81–92
Schindler DW, Donahue WF (2006) An impending water crisis in Canada's western prairie provinces. Proc Natl Acad Sci 103(19):7210–7216
Sentelhas PC, Gillespie TJ, Santos EA (2010) Evaluation of FAO Penman-Monteith and alternative methods for estimating reference evapotranspiration with missing data in Southern Ontario Canada. Agric Water Manag 97(5):635–644
Singh VP (1995) Computer models of watershed hydrology. Water Resources Publications, Highlands Ranch
Snover AK, Hamlet AF, Lettenmaier DP (2003) Climate-change scenarios for water planning studies: pilot applications in the Pacific Northwest. Bull Am Meteorol Soc 84(11):1513–1518
Valeo C, Xiang Z, Bouchart FC, Yeung P, Ryan MC (2007) Climate change impacts in the Elbow River watershed. Can Water Resour J 32(4):285–302
Waterline (2011) Groundwater elevation and monitoring plan Elbow River watershed sub-region TWPS 018–024, RGES 29W4–09W5 Alberta.WL10–1716. Alberta Environment. p 100
Wijesekara GN, Gupta A, Valeo C, Hasbani JG, Qiao Y, Delaney P, Marceau DJ (2012) Assessing the impact of future land-use changes on hydrological processes in the Elbow River watershed in southern Alberta, Canada. J Hydrol 412:220–232

Wijesekara GN, Farjad B, Gupta A, Qiao Y, Delaney P, Marceau DJ (2014) A comprehensive land-use/hydrological modeling system for scenario simulations in the Elbow river watershed, Alberta Canada. Environ Manage 53(2):357–381

Xu ZX, Zhao FF, Li JY (2009) Response of streamflow to climate change in the headwater catchment of the Yellow River basin. Quatern Int 208(1):62–75

[illegible]

Climate Effects on Recharge and Evolution of Natural Water Resources in middle-latitude Watersheds Under Arid Climate

Bingqi Zhu, Jingjie Yu, Patrick Rioual, Yan Gao, Yichi Zhang and Heigang Xiong

Abstract This paper analyzes the physico-chemical characteristics of natural waters in middle-latitude drainage systems of central Asia, including the climatic, lithological and geomorphological conditions in which water flows and resides. This analysis allowed the identification of the geological evolution and recharge mechanism of the water resources in an arid environment. The studied waters at various sites are different in mineralization but similar to the majority of large rivers on earth, which are typically alkaline. However, no Cl-dominated water type occurs in the study area, indicating that these natural waters are still at an early stage of evolution. The regolith and geomorphological parameters controlling ground-surface temperature may play a large role in rock weathering regime and so in the geological evolution of water. Three main morphological and hydrological units are reflected in water physico-chemistry: the montane areas (recharge area) with silicate and carbonate weathering, the piedmonts and sedimentary platform (runoff area) with carbonate weathering, and the desert plains (discharge area) with evaporite dissolution. Climate influences the salinization of natural waters substantially. Direct recharge from seasonal snow and meltwater and infiltration of rainfall into the ground are thought to be significant recharge processes for natural waters in the

---

B. Zhu (✉) · J. Yu · Y. Zhang
Key Laboratory of Water Cycle and Related Land Surface Processes, Institute of Geographic Sciences and Natural Resources Research, Chinese Academy of Sciences, Beijing, China
e-mail: zhubingqi@igsnrr.ac.cn

P. Rioual
Key Laboratory of Cenozoic Geology and Environment, Institute of Geology and Geophysics, Chinese Academy of Sciences, Beijing, China

Y. Gao
Centro de Investigaciones en Geografia Ambienta, Universidad Nacional Autonoma de Mexico, Campus Morelia, Morelia, Mexico

H. Xiong
Key Laboratory of Oasis Ecology, Ministry of Education, Urumqi, China

Mu. Ramkumar et al. (eds.), *Environmental Management of River Basin Ecosystems*, Springer Earth System Sciences, DOI 10.1007/978-3-319-13425-3_5

study area, while recharge from potential deep groundwater may be much less important. The chemistry of lakes is generally consistent with those of large lakes in the world, but the enrichment of the ions in the lakes has been caused mainly by evaporation, rather than through the quality of the recharged water.

Keywords Natural water · Geological evolution · Recharge mechanism · Arid environment · Climatic effect · Central Asia

1 Introduction

Natural water in the Earth system is not only one of the key components but also a major agent of the biogeochemical cycles that link the lithosphere, biosphere and atmosphere. Unlike seawater, which can be spatially approximated with an artificial standard, continental natural water varies considerably in its geochemical composition and other properties (Livingstone 1963). Many factors influence the recharge and evolution of continental natural water, causing spatial geochemical variation. The dissolution and weathering of rocks are commonly considered to be the major determinants of natural water chemistry, but are under the influences of local geology and climate, magnitudes of inputs via other pathways including precipitation, volcanic activity, and pollution. Under arid climate the input materials are concentrated by evaporation and altered by chemical and biological interactions, and add another complexity to the evolution of natural waters.

Over the past century, the world has experienced an undeniable temperature increase and thereby inevitably is faced with environmental problems, such as glacier retreat, permafrost melting, sea level rising, droughts, storms, floods, heat waves and cold spells. In addition to the direct impacts, these environmental problems also accelerate in global water cycle processes (Menzel and Burger 2002). Recently, there has been a spurt in the studies on recharge and recycling of the water resources and their response to climate change (White and Blum 1995; Arnell 1999; Barnett et al. 2005; Burns et al. 2007; Hagg et al. 2007). In the Himalaya-Hindu Kush region, home to the third largest ice reservoir on the earth, the mass of glacier has been observed to decrease quickly during the last two decades (Meier and Dyurgerov 2002), which led to an increase in glacier melting and runoff by 33–38 % (Singh and Kumar 1997) thereby an increase not only in the available water supply, but also in the frequency and intensity of extreme floods. In the Rhine River basin in Europe, the projected changes in temperature will shift the basin from a combined rainfall and snow melt regime to a more rainfall-dominated regime, resulting in an increase in winter discharge and a decrease in summer discharge (Middelkoop et al. 2001). In the Columbia River basin, USA, a seasonal shift in snow melt runoff is associated with reduced winter snow accumulation, earlier peak snow melt and higher winter runoff (Lettenmaier and Sheer 1991). In the Tarim River Basin, China, long-term change of seasonal snow cover has greatly affected the river runoff

(Xu et al. 2009), which was replenished by snow melt water and rainfall but now is related primarily to summer precipitation. These case studies suggest that the climate-induced changes in water resources have been occurring in regions all over the world. The consequences of these hydrological changes for future water availability —predicted with high confidence and already diagnosed in some regions—are likely to be severe (Barnett et al. 2005), especially in the arid zones.

Beside the topics of response of water resources to climate change, the keen interest of scientists in past and present global climate change has renewed the efforts to quantitatively understand the feedback mechanisms between climate and river chemical weathering at the watershed scale (White and Blum 1995). Discovering how climate, weathering, and erosion are related in continental watersheds is essential for understanding how geomorphology and tectonics affect Earth's long-term climatic evolution (Raymo et al. 1988; Molnar and England 1990, Riebe et al. 2001). Various rivers of the world, particularly, those draining the highlands of Asia (such as Himalaya, Tibetan Plateau) have received special attention in recent years because of possible connection between the Himalayan uplift and Cenozoic climate change (Raymo and Ruddiman 1992; Edmond 1992; Blum and Erel 1995). Raymo and Ruddiman (1992) ascribed the Cenozoic deterioration in global climate to the rise of the Himalayas and the resulting increase in mechanical and chemical weathering rates and consumption of atmospheric CO_2. In this "tectonics-weathering-climate" hypothesis, the uplift of the highlands of Asia was proposed as a major driver of Cenozoic cooling as it could have promoted rapid silicate weathering and thereby enhanced CO_2 drawdown from the atmosphere. However, quantifying the feedback and its response to the multiple controlling processes has proved to be difficult and controversial (e.g. Berner et al. 1983; Sundquist 1991; Raymo 1994). For instance, Huh and Edmond (1999) concluded that there were no primary climatic effects on weathering rates on the present day Earth. Similarly, Noh et al. (2009) proposed that at a regional scale, the climatic factors exert only a weak effect on the weathering rates, in particular, the net CO_2 consumption rate by silicate weathering was a function primarily of runoff and secondarily of relief but not of climate (temperature and precipitation). On the contrary, some other studies (White and Blum 1995; White et al. 1999) provide strong support for significant temperature effects on the rates of chemical weathering for granitoid rocks, and few others consider that chemical weathering is greatly climate-dependent and is a function of temperature and precipitation/runoff (e.g. Berner et al. 1983; Velbel 1993; White and Blum 1995; White et al. 1998; Dessert et al. 2001; Zhu et al. 2013b). In general, the climatic forcing has remained elusive in active orogens because of the strong lithological dependence of water chemistry (Bluth and Kump 1994; Gaillardet et al. 1999). At present, climatic effects on natural water evolution are evaluated by correlating variations in solute concentrations and fluxes (particularly the CO_2 flux) with temperature, precipitation, runoff, and evapotranspiration (ET) for a worldwide distribution of large watersheds underlain by different rock types (Bluth and Kump 1994; Whiter and Blum 1995; Gaillardet et al. 1999; White et al. 1999; Dessert et al. 2001; Noh et al. 2009; Xu et al. 2013; Zhu et al. 2011, 2012a, 2013a, b). Among these watersheds, stream solute concentrations are strongly correlated with proportional ET loss, and evaporative concentration makes stream solute concentration an inappropriate surrogate for chemical weathering (Whiter and Blum 1995). However, arid watersheds produce anomalous weathering rates relative to warm and wet watersheds (Zhu et al. 2013b).

In middle-latitude zone of centrall Asia, high mountains, particularly the Tibetan Plateau and Tianshan Ranges, function as barriers for atmospheric circulation and keep moisture from reaching an extensive region in western China, causing arid and hyperarid climatic conditions (Domros and Peng 1988; Sun et al. 2010). Northwestern China acts as a significant repository of information relating to the hydrological evolution and climatic changes in central and eastern Asia. Northern Xinjiang in northwestern China is the geographical centre of the Asian Continent and an extremely arid and ecologically fragile region. Most of the rivers in the northern Xinjiang drainage system originate from the peripheral mountain glaciers and snow and drain the areas of diverse geological and climatic conditions, with their lower reaches located in a desert environment. Over the last 50 years, climate of the entire Xinjiang has experienced an increase both in temperature and precipitation, most notably since the mid-1980s, and northern Xinjiang was the area that changed most significantly (Xu et al. 2010). These variations were closely related with the replenishment types of rivers. Climate warming has had an effect on the regional hydrological cycle. Studies on the long-term change of seasonal snow cover in Xinjiang (e.g. Xu et al. 2009) proved that the entire catchment in Xinjiang was mainly affected by one-large-scale weather system. The runoff of the rivers replenished during the past by snow melt water and rainfall in Xinjiang is now related primarily to summer precipitation, followed by summer temperatures or the maximum snow depth in the cold season (Xu et al. 2009). These changes suggest that due to the climate change, snow is no longer the principal factor that contributes to the runoff increase in the headstreams in Xinjiang, although there was a slow increase in snow depth. Until now, however, it is not clear how the climatic factors are responsible for the steady and continuous water increase in these areas.

Geochemical studies of waters are of considerable importance for a better understanding of the sources and evolution of the natural water resources and the influencing factors. Rivers carry the imprints of chemical erosion on the continents in the forms of dissolved materials. Fresh water in arid regions is definitely a non-substitutable resource upon which humans and ecosystems depend. An understanding of the source of natural waters and their evolution in the middle-latitude northern Xinjiang watersheds is significant not only to the policy makers for regional planning but also to the scientists interested in hydrological cycles in arid environments and climate change in global arid regions (Meyer et al. 1988; Kimbadi et al. 1999; Zhu et al. 2011). Until now, little work focusing on these questions has been done in arid middle-latitude drainage systems. Using northern Xinjiang as an example, this paper provides insight into the physicochemistry of natural waters in arid environment.

2 Regional Setting

The northern Xinjiang (China) is bounded between 78°–90°E and 42°–50°N covering approximately an area of 603,000 km^2 and is bordered by Tianshan Mountains in the south and the Altai Mountains in the north. Topographic variations are from less than 500 m above sea level (asl) in the centre of the Zhungarer Basin to more than 3,000 m asl in the south and north (Fig. 1c). The topography is generally flat in the central plain and is rather cragged in the peripheral mountainous areas. The wide piedmonts and pediment plains are marked by the Gobi desert, grassland and oasis areas. An arid temperate continental climate predominates the region. The mean annual air temperature is about 5 °C, with a minimum of −10 to −20 °C in January and a maximum of 28–33 °C in August. The precipitation is mainly from westerlies, with a mean annual precipitation of 60–150 mm in the central plain and 200–500 mm in the surrounding mountainous areas (Fig. 1b). While the potential evapotranspiration is approximately 1,000–3,500 mm/year, it shows seasonal variation as well as geographic variation along the elevation gradient across the northern Xinjiang, due to the distribution patterns of seasonal precipitation, temperature and relative humidity and the orientation of delivery of moisture by the westerlies.

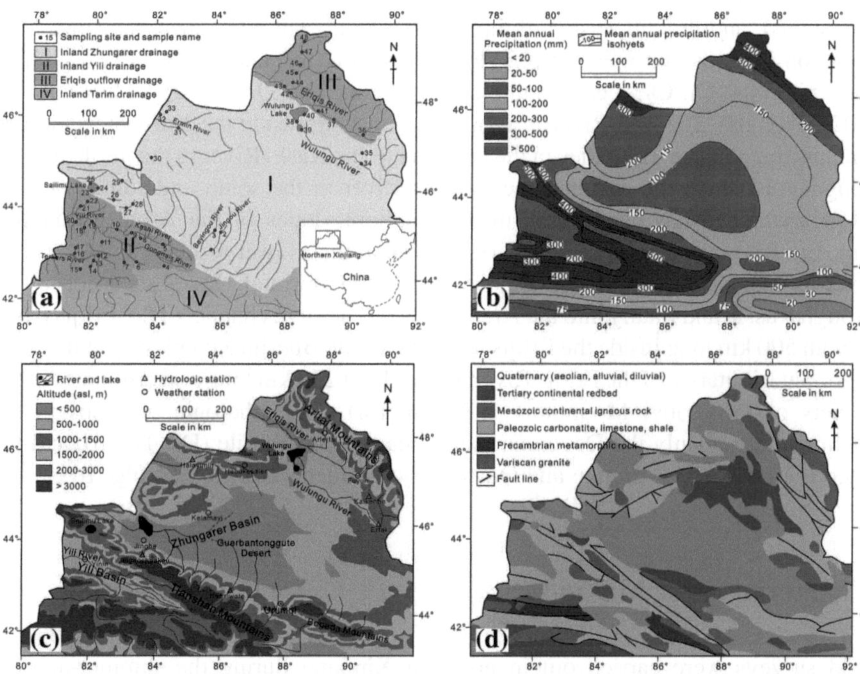

Fig. 1 **a** Hydrological setting and sampling locations map, **b** mean annual precipitation isohyets distribution map, **c** topographical map, and **d** lithological-distribution map of the northern Xinjiang (Central Asian) in China

The northern Xinjiang drainage system includes three watersheds: the Zhungarer, the Yili and the Erlqis (Ma 2002) (Fig. 1a). The Zhungarer and Yili watersheds are inland continental watersheds and the Erlqis watershed drains into the Arctic Ocean. Water resources in the northern Xinjiang are mainly distributed in Yili (170.9×10^8 m^3), Arltai (129×10^8 m^3) and Tacheng districts (61.6×10^8 m^3) and the total volume of the surface water in the northern Xinjiang is about 435.7×10^8 m^3. The temporal distribution of runoff in a year is extremely heterogeneous with 50–70 % in summer, 10–20 % in both spring and autumn, and less than 10 % in winter. The uneven spatial-temporal distribution of water resources and dry climate determine the dependence of oasis agricultural activities on irrigating system. The Zhungarer Basin (379,000 km^2) is located in the central part of the northern Xinjiang. The basin is a structural depression filled with 500–1,000 m thick unconsolidated Quaternary and Tertiary sediments (XETCAS et al. 1978), and is an extension of the Palaeozoic Kazakhstan block surrounded by Palaeozoic folded mountains. Aeolian deposits are widespread in this basin. The Gulbantonggute Sand Sea (48,800 km^2) is located in the central part of the basin. A large geographic distance from the surrounding oceans and the presence of the rain-shadow effect due to the surrounding orographic conditions is responsible for the arid climate of the Zhungarer Basin. Most of the rivers in the Yili watershed converged into the Yili River at their lower reaches. The Yili River is a large international river running from northeast Borohoro Mountains and southeast Halik Mountains (both form part of the eastern Tianshan Mountains), and flowing eastward through the Yili basin into Kazakstan (Fig. 1). It drains on igneous and metamorphic Precambrian (An€) and Variscan (γ_4) granite, Carboniferous (C_1, C_{2+3}) carbonatite and limestone and Quaternary sediments (Ma 2002) (Fig. 1d). The Yili River is about 430 km long inside the Yili watershed and contains three major tributaries in the catchment: the Kashi River in the northeast, the Gongnais River in the east and the Terkers River in the south and southeast. The Erlqis watershed has headwaters in the southern slopes of the Arltai Mountains (Fig. 1a). Rivers converge into the Erlqis River at their lower reaches, and the Erlqis River is also a large international river, flowing through Kazakstan northward through Russia and finally into the Arctic Ocean. The stem channel of the Erlqis River is about 500 km long inside the Erlqis watershed. The Buerjin River is one of the most important tributaries of the Erlqis River and is about 250 km long and originates in the glaciers of the Youyi Mountains (4,373 m asl). The lithologic outcrops in the catchment are mainly the Devonian marine carbonate, clastite (D2, D2+3), and the igneous Variscan (γ_4) granite and Quaternary sediments (Ma 2002) (Fig. 1d).

3 Materials and Methods

Field surveys were carried out in northern Xinjiang during the summer-autumn season of 2008. Chinese 1:50,000 scale topographic maps, a prismatic compass, a Thrommen altimeter and a Garmin GPS were used for orientation and fixation of location of water sampling points. All locations and elevations were recorded using

GPS and topographic maps. Water samples were collected under natural flow conditions in one-liter polyethylene bottles from various parts of the Yili, Zhungarer, and Erlqis watersheds (Fig. 1a), including river stems, stream channels, hill slopes, wells, lakes, ponds, man-made trenches and reservoirs. Taking into account that the tributaries reflect a much broader variety of geological, biological, and population patterns than main stem rivers (Pawellek et al. 2002), it was of interest to sample tributary water to look for common features reflected in their hydrochemistry. Some pristine streams draining forested catchments were chosen in the Tianshan and the Arltai Mountains. All the water samples were surface-sampled except for the well water which was collected from the pump shafts. Most water samples were colorless and turbid-free, but a few were yellow, gray and turbid due to dissolved iron contents or suspended solid particles. The colored river water samples primarily drained through the south and north piedmont areas of the Tianshan Mountains, with deep coves of loess and loess-like soil. Because no river and spring exist in the hinterland of the Gulbantonggute Desert, no surface water samples was collected for comparison with samples from other areas.

Water samples were filtered through a 0.45 μm Millipore membrane filters in the field and analyzed in the institute of Geology and Geophysics, Chinese Academy of Sciences (IGGCAS). Analysis of water samples included total dissolved solids (TDS), pH, and ion concentrations. Temperature (T), pH, and TDS were measured in situ using a portable Multi-Parameter Analyzer (Eijkelkamp 18.28). The water samples for cation analysis were acidified with hyper-pure HNO_3 below pH 4.5. Cations and anions were determined by ion chromatography (IC, Dionex 600) with deionized water (EC < 2.1 μS/cm) as the dilute base. The eluents used were 20 mmol/L MSA (methanesulfonic acid) at pH 4.5 and conductivity of approximately 400 μS/cm for cation analysis and 3.5 mmol/L Na_2CO_3, 1.0 mmol/L $NaHCO_3$ at pH 8.5 for anion analysis, respectively. One replicate for each sample was taken for ion analysis, and the instrument was recalibrated after analyzing every five samples. The alkalinity was measured with a Hach digital titrator using the Gran method (Wetzel and Likens 2000) within 3 days after sampling.

4 Results

Statistical characteristics of the physio-chemical and chemical properties of the water samples are listed in the Table 1. The data show that the natural waters at various sites are different in terms of mineralization, with TDS ranging from 24.6 to 6,200 mg/L, the mean being 580 mg/L. More than half of the analyzed samples belong to fresh water (TDS < 1,000 mg/L), and the remainder to brackish water (TDS 1–10 g/L). All natural waters are alkaline, and pH values range from 7.0 to 9.81, with an average value of 7.87. In the Piper's diagram, the natural waters are distributed in the fields where alkaline earths exceed alkalis and strong acids and weak acids dominate partially in composition (Fig. 2). There are four main water types: Ca-HCO_3, Ca-NDA (non-dominant anion), Ca-SO_4 and NDC-NDA (non-dominant cation) or

Table 1 Statistical summary of the parameters determined in the studied water samples

Parameter	Maximum	Minimum	Average	Median	10th	90th	SD
pH	9.81	7.00	–	7.85	7.33	8.24	0.47
Eh	−21.0	−187	−72.6	−72.5	−94.6	−41.8	27.6
EC(μs/cm)	5,380	46.3	864	352	162	3,426	1,258
TDS(mg/L)	6,200	24.6	580	186	86.1	1,843	1,071
Li^+(mg/L)	0.13	nd	0.02	0.01	nd	0.04	0.02
Na^+(mg/L)	1,673	1.66	105	6.86	2.60	272	281
NH_4^+(mg/L)	0.17	nd	0.02	nd	nd	0.03	0.04
K^+(mg/L)	77.7	0.49	8.48	1.63	0.82	25.9	17.4
Mg^{2+}(mg/L)	190	0.63	27.2	6.49	2.44	81.8	49.5
Ca^{2+}(mg/L)	91.9	3.05	29.1	26.4	14.0	48.0	16.7
F^-(mg/L)	3.04	0.02	0.31	0.18	0.06	0.54	0.51
Cl^-(mg/L)	648	1.83	74.5	6.91	2.97	223	160
NO_3^-(mg/L)	36.5	–	4.96	2.88	0.72	8.89	6.62
SO_4^{2-}(mg/L)	2,655	2.52	205	45.3	9.01	682	437
HCO_3^-(mg/L)	799	9.76	117	69.4	29.7	283	142
TH(mg/L)	824	10.2	184	102	47.9	436	209
SAR	28.2	0.058	1.97	0.278	0.094	4.56	4.85

(TH, total hardness, SAR, sodium adsorption ratio)

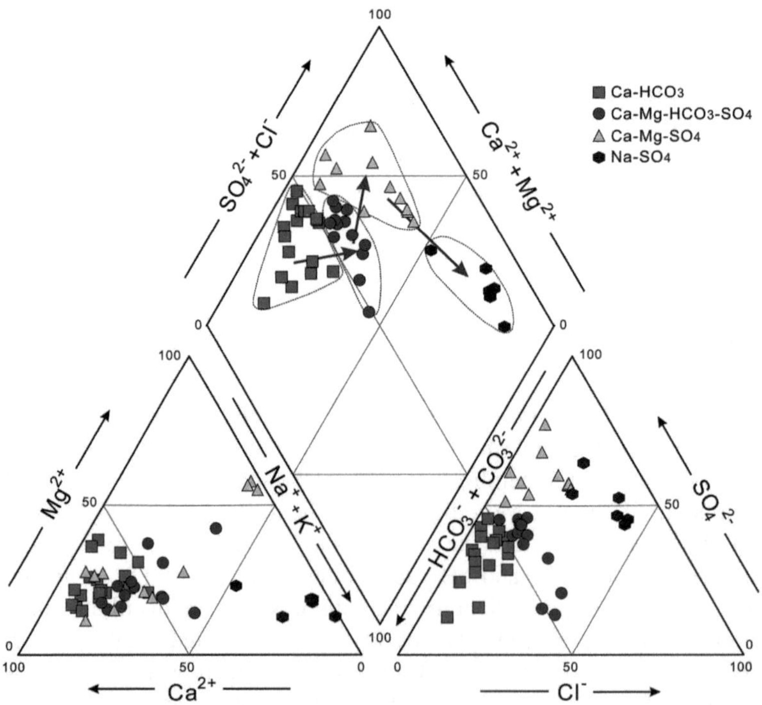

Fig. 2 Piper plot of the studied water samples

Na-NDA. The Ca-type water is the most widely distributed in the study area. The Ca-HCO_3 and Ca-NDA types mainly occur in the montane areas and the low-pitched piedmont zones. The Ca-SO_4 and Na-SO_4 types are mostly distributed in the transition band between oasis and desert plain. Mg-type water exists only in the Sailimu Lake. No Cl-dominated water type occurs in the study area (Fig. 2).

5 Discussion

5.1 Geochemical Evolution and Distribution

The natural waters in northern Xinjiang are similar to the majority of the large rivers on the earth, which are typically alkaline in nature (Meybeck 1987). Compared with the upper ranges of typical values for major world rivers and the lower ranges for major pristine rivers (most common natural concentrations—MCNC) (Fig. 3), such as the Amazon and Mackenzie Rivers (Meybeck and Helmer 1989; Meybeck 1996), the mean major-ion concentrations of the waters studied are almost in excess of typical natural values of the major world rivers (Fig. 3), except for Ca^{2+} and HCO_3^-. This suggests that the natural waters in northern Xinjiang fall close to the upper limit of the concentrations in major world rivers, classified as "salted" by Meybeck (1996). Furthermore, the studied waters are notably characterized by an enormous range for concentrations and yields, over three orders of magnitude compared to that of global rivers (Fig. 3). This indicates a definable spatial variation of water chemistry and quality in the studied watersheds.

Cl-type water is usually one of the major water types in the arid environment in northern China, such as in the Taklamakan Desert (Zhang et al. 1995; Zhu and

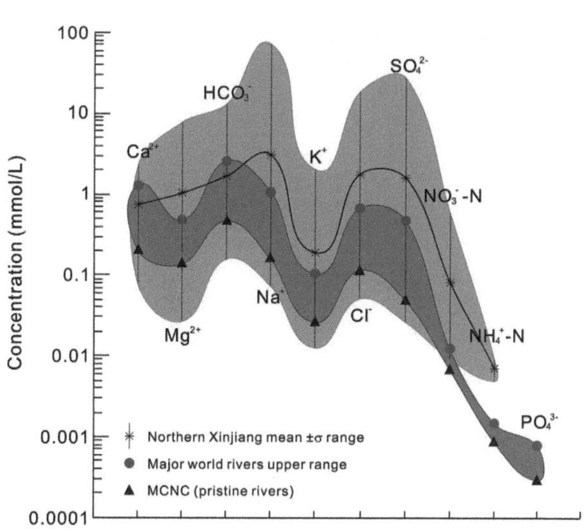

Fig. 3 Typical ranges for major ions in the natural waters in northern Xinjiang. Also marked are the upper limits for these components in major world rivers and the median concentrations in pristine rivers (MCNC) (Meybeck and Helmer 1989; Meybeck 1996)

Yang 2007) and the Badanjilin Desert (Yang and Williams 2003; Zhu and Yang 2010; Zhu et al. 2012b). Cl-dominated water type does not exist in the study area (Fig. 2), indicating that the natural waters in northern Xinjiang are still at the early stage of water evolution.

The large variation in Ca^{2+} and Na^{+} concentrations in the natural water samples from the study area (Table 1 and Fig. 2) may theoretically reflect local mineralogical changes in the sediments and carbon dioxide produced by biological processes in their surface layers. However, biogenic weathering might not be a key factor, because the natural water in the study area is alkaline in general. The CO_2 reactions in the natural water are possibly not very active at many of the studied sites because of alkaline conditions and limited vegetation on the surface. The heterogeneity of the ionic compositions (Fig. 2) may be related to the local variations in the mineralogy of the natural water reservoir and geochemical processes occurring in the aquifer. The amount of sodium is sympathetic with the chloride content in the studied samples (Fig. 4), supporting the view that geochemical processes play a role in the chemistry of natural water. Regarding to the Na/Cl ratios, part of the samples (mostly from mountainous areas) with lower TDS lie very close to the 1:1 ratio line (Fig. 4), indicating that sulphate and other salts are relatively abundant in the natural water with higher TDS (mainly under fluvial-plains and desert environments). The regolith and geomorphological parameters controlling the ground-surface temperature may play an important role not only in rock weathering regime but also in the water chemistry, because solution of lithological components is controlled to a large extent by the ground temperature in a periglacial landscape (Beylich et al. 2004). Such as in areas of shade, cold ground and thin regolith, the TDS of natural water (surface water and shallow groundwater) is low, whereas the TDS is high in areas exposed to intensive solar radiation. Data from the arctic-oceanic drainage basin of northern Sweden (Latnjavagge) show that

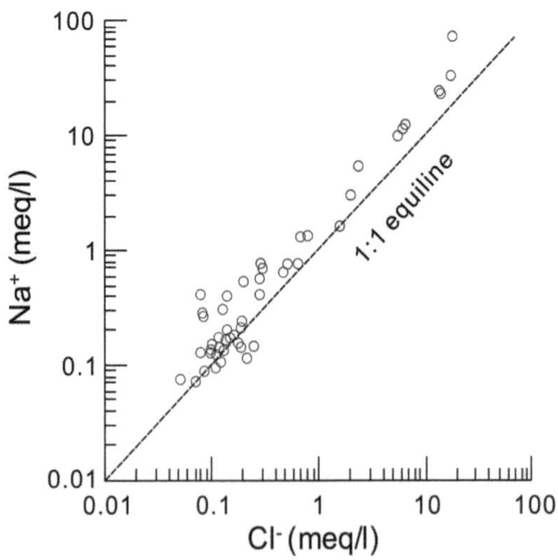

Fig. 4 Scatter diagram of the concentrations of Na^{+} versus Cl^{-} of the study waters

water chemistry may vary considerably in a small area with very homogeneous lithology but great slope segments (Beylich et al. 2004). The regolith in northern Xinjiang area varies in thickness from less than a meter to more than 5 m between upper and lower slope segments.

The water samples from the oasis areas in the Zhungarer basin where the local communities are located show much higher TDS than samples from more pristine areas. This is probably caused by two processes: (a) direct pollution from the domestic and agricultural activities and (b) re-seepage of wastewater emanating from the former sources. It appears that the wastewater can infiltrate into the natural water systems quickly, because of favorable proximity, and hydrological and topographic conditions. For this reason in the Zhungarer watershed, the nitrate content is clearly much higher in groundwater samples (22–37 mg/L) than in the surface waters (0.5–7.7 mg/L). The high concentration of nitrate in groundwater samples is probably related to the decomposition process of untreated waste. In fact, the nitrate concentration is rather high in the majority of the groundwater samples from the entire study area. In the arid environment of the desert, the life of both humans and their live stocks depends on the groundwater (wells). Therefore human influence is quite strong around the wells and the dung depot is often located not far from the wells. This explains the high concentration of nitrate in the water samples. Despite such precarious conditions, neither a waste water treatment plant been constructed in this region, nor any plan to contain the pollution effects of domestic and irrigation practices.

5.2 Recharge Mechanisms of the Natural Waters

The possible mechanisms for natural water recharge (especially river and shallow groundwater) in northern Xinjiang may be: (a) direct infiltration of rainfall (summer and winter precipitation); (b) direct recharge from seasonal snow and meltwater; (c) recharge from potential deep groundwater. The first two processes are thought to be more important than the third one, because the northern Xinjiang watersheds contain varying topographic ranges in a desert landscape. Based on the tectonic structure of mountain-basin couples, it is inferred that the recharge of the shallow groundwater is mainly through infiltration of seasonal snow and ice-melt water from mountain areas and local precipitation that infiltrates downwards quickly due to the loose nature of the regolith. Because of the large temporal and spatial variability of rainfall, infiltration is related to the amount of precipitation of each event. From the observation of dried channels in the field, it is assumed that there may be ephemeral streams and ponds in the study area when it rains. As a result, the overall recharge process is rather heterogeneous. The difference in ion concentrations suggests that the hydraulic relationship among the sampling sites is weak.

In fluvial plains of Zhungarer watersheds, we consider that the precipitation directly into the rivers, lakes and reservoirs plays a minor role in the recharge today, because the amount of precipitation is much less than the prevailing evaporation.

Field observations indicate that currently all the rivers in these areas are recharged mainly by mountain stream water and partially by groundwater. Owing to the isolated geomorphological basin structures, it is unlikely that the recharge mechanisms might have fundamentally changed in the past. The large range in chemical composition of water samples suggests not only the heterogeneity of the recharge conditions in the study areas, but also a possible mixing with enriched waters that could have been drawn from different aquifers. However, the input from other regions (via surface runoff or groundwater inflow) seems to be rather insignificant in the local water cycle. Besides, a large proportion of rain is probably added to the shallow groundwater system because of rapid infiltration in the loose and sandy sediments. Field observations and interviews with local inhabitants suggest that the water from the shallow aquifer can meet the demands of the inhabitants with traditional life styles and reasonable population of livestock in the desert environment of northern Xinjiang.

The highest TDS concentrations were observed at some lake depressions (such as the Salimu Lake) in the central Zhungarer watershed. From field observation, it is assumed that the Salimu Lake is mainly recharged by glacier/snow waters, because it is located in a highland depression surrounded by high mountains and there is surficial inflow but no sign of outflow. The major ions in the lake are Mg^{2+} for cations and SO_4^{2-} for anions. At present, Salimu is one of the largest lakes in northern Xinjiang and has relatively high TDS (1,850 mg/L). Compared with the chemistry of other lake samples, the proportions of sodium and chloride are much lower in the Salimu lake water. From a global point of view, the dominant ions in most saline lakes are Na^+ and Cl^-, and occasionally in some less saline lakes, Na^+ and HCO_3^-/CO_3^{2-} (e.g. Day 1993). It is reasonable to believe that the brines of these lakes were formed by long and intensive evaporation, because Na^+ and Cl^- are often associated with evaporites. A halite origin can be excluded for the Salimu Lake. However, Mg^{2+} and SO_4^{2-} are also often associated with evaporites. Dissolution of sulphates (such as gypsum and anhydrite) may be an important source. In general, major ions in saline waters were derived from the catchment or the substratum (e.g. Kilham 1990). However, potentially, precipitation should be a significant source of ions (Eriksson 1985) in the waters of the study area because of frequent dews and rimes. In many large saline lakes of northern China and Mongolia, Cl^-, SO_4^{2-} and HCO_3^- are almost equally predominant anions, while Na^+ and K^+ are dominant cations. The salts of these lakes were thought to have originated from high evaporation and soil salinization (Egorov 1993). The salinity of the Chinese lakes increases westwards, consistent with the trend of increasing aridity. Both natural climatic changes and irrigation activities have caused decadal variations in the salinity in these lakes. From limnological analysis of different regions in the world, it was concluded that large, deep, saline lakes often contain alkaline, sodium rich waters with considerable chloride, sulphate, and carbonate plus bicarbonate (Melack 1983). The chemistry of lakes in the northern Xinjiang drainage basins is, to a considerable extent, consistent with those of large lakes. Also, from this point of view, it appears that the ionic enrichment in the lakes has been caused mainly by evaporation, rather than through the quality of the recharged water.

5.3 Comparison Between Watersheds in Arid Environment

Investigation of the natural water recharge and evolution has been the focus of many researchers in arid lands of the world (Zhu et al. 2007; Vanderzalm et al. 2011; Abdesselam et al. 2013; Martos-Rosillo et al. 2013; Sener et al. 2013). Abdesselam et al. (2013) investigated the evolution of groundwater in arid regions and they emphasized the anthropogenic impact on ground waters and its increase since the last decades. In the Egirdir watershed in Turkey, contaminants in the catchment affect the natural water quality negatively (Sener et al. 2013). The major factors that control the quality of the Egirdir waters are firstly the agricultural activities in the basin, secondly water-rock interaction and domestic and industrial wastes (Sener et al. 2013). This indicates that anthropogenic processes have greatly influenced the recharge and evolution of natural water resources in the basin. This is different from the northern Xinjiang watersheds, which, for large parts are basically controlled by natural processes compared to the Egirdir watershed. Pertaining to the Zhungarer basin in the northern Xinjiang, although the combined effects of the anthropogenic and geologic sources are observed, restriction of these sources and influences within geographically limited regions has also been inferred from the highly heterogeneous nature of ionic concentrations. However, given cognizance to the shallow aquifers, high reactivity of loose-unconsolidated sediments and regoliths, aided by the general aridity and rising temperature, it is suggested that suitable management plans and programs need to be implemented.

In the arid lands of central Australia, natural water in the surficial alluvial aquifer system is recharged from multiple sources, including infiltration from ephemeral stream flow in the Todd River, groundwater throughflow between connected alluvial basins, regional groundwater flow from the underlying Tertiary aquifer, and diffuse recharge (Vanderzalm et al. 2011). The complicated recharge system is similar to that of northern Xinjiang, but the latter has different recharge sources from snow and glacier melt waters and no deep groundwater sources. This may indicate a difference in natural water recharge sources between the middle-/high-latitude and the low-latitude zones. In central Australia, the contribution of the multiple recharge sources varied spatially with proximity to the recharge source. Effluent recharge is evident in most of these basins. The long-term impact of effluent recharge is experiencing a shift from sodium and calcium as co-dominant cations in the groundwater to dominance by sodium alone. It is very different from the groundwaters in the northern Xinjiang, which are calcium dominated and recharged by influent water, indicating that natural water recharge in most areas in northern Xinjiang are controlled by geological process, without evident recycle recharge of natural water.

In the hyperarid Minqin Basin of northwestern China, the shallow groundwater is largely evaporitic in origin and is older than 1 ka, representing paleowaters mixed with a limited quality of modern recharge (Zhu et al. 2007). The rain-fed direct recharge of groundwater ranges from 1.55 to 1.64 mm/year, which is about 1.5 % of local rainfall (Zhu et al. 2007). Compared with the water in the Minqin Basin, no radiocarbon residence time can be estimated from water samples in the present study in the northern Xinjiang. However, estimates on the contribution of water recharge from atmospheric precipitation can be made. Na^+ and Cl^- in water are conservative ions. Cl^- in stream water in wet conditions is primarily influenced by climate (atmospheric precipitation) and not by geological weathering (Feller 2005). Thus, the excess of Na^+ concentration in comparison to Cl^- in stream water relative to the Na^+ concentration in local rainfall can be regarded as the contribution from regional weathering and anthropogenic sources (Wu and Gibson 1996). According to this consideration, contributions of various ions from regional rock weathering and precipitation in the stream water can be evaluated from the equations in Dunne (1978). Although some degree of uncertainty may exist in this approach, it effectively assesses the relative roles of precipitation and weathering effects (Mast and Drever 1990; Ahmad et al. 1998; Chae et al. 2004). Based on the formulas from Dunne (1978), the ionic contributions resulting from atmospheric precipitation to the natural waters in the basins of northern Xinjiang are estimated to be between 2 and 39 % (13 % in average) of the total dissolved solids. This indicates that the contributions of rainfall to water chemistry in northern Xinjiang are more effective than those in the Minqin Basin. It implies that as could be expected elsewhere, the direct recharges by rainfall in arid lands are regionally variable. This pattern in the study area is not only influenced by the regional variability of rainfall, but also by the lithological, geomorphological and orographic characteristics.

5.4 *Response to Climate*

The regional distribution pattern of chemical components in natural waters (Fig. 2) reveals that a transitional trend exists among natural waters in the northern Xinjiang drainage basins. A general transformation of the water types in the order of Ca-HCO_3, Ca-Mg-HCO_3, Ca-Mg-SO_4 and Na-SO_4, corresponding to the altitude change of landscapes in the order of mountains, pediments, piedmont plains and desert plains could be discerned from the study area. It indicates a potential control of a vertical zonality of temperature and relevant geological effects on the natural water evolution. Besides this trend, another kind of zonal distribution of hydrogeochemistry also exists in the northern Xinjiang drainage basins. For example, the relatively wet zones, with AMP > 200 mm in three watersheds, are characterized by Ca-HCO_3, Ca-NDA and NDC-NDA water types. In the relatively dry zones, with AMP < 200 mm, however, the Ca-SO_4 and Na-SO_4 are dominant components. The Ca-HCO_3-dominated chemistry of river water in the relatively wet zones couples well with the regional carbonate lithological distributions (Fig. 1d). It confirms that rock weathering and the regional lithologic distributions have a major effect on the evolution of river chemistry in the wet zones. The precipitation processes can accelerate the rock weathering processes (White and Blum 1995). The water types in dry zones tend to be relatively uniform. Because the lithological distributions are not homogeneous, evaporation processes, which can result in the selective precipitation of solute fraction ($CaCO_3$) and subsequent changes in water chemistry as water moves downstream in the relatively arid zones (Gibbs 1970; Kilham 1990; Feller 2005), could be responsible for the accumulation of sodium and sulphate salts. It indicates a major effect of climate rather than geology on the evolution of water chemistry in dry zones.

The strong relationship between the annual mean summer precipitations (AMP) in northern Xinjiang and the altitudes of the sampling sites in this study can be expressed as AMP = 0.14 × altitude + 101.3 ($r = 0.88$, $p < 0.005$). The total dissolved solid contents in most of the samples decreased with the increasing altitude (Fig. 5). It indicates, together with the sympathetic nature between the rainfall and altitude, a potential influence of precipitation on dissolved materials in the water. The fact that rock types are not distributed according to the altitude (Fig. 1c, d) implies that topography and temperature, in addition to rock type and weathering, affects water chemistry indirectly by influencing AMP/evaporation processes.

In the central parts of the studied watersheds, evaporation probably influences the salinization of surface and ground water substantially, because the TDS values of river waters clearly vary along their channel courses, such as the Erlqis, Wulungu, Jingou, Gongnais and Yili rivers (Fig. 5). The river waters in the middle reaches of the Jingou, Gongnais and Wulungu have TDS values of 90, 95 and 321 mg/L, respectively, and gradually change to higher values of 112, 147 and 2,000 mg/L after flowing about 70, 100 and 300 km downward, respectively. The Jingou, Gongnais and Wulungu rivers have few tributaries over their water courses (Fig. 1a), so the waters flowing along the lengths of these courses do not mix extensively with water from different tributary sources. In addition, the seasonal

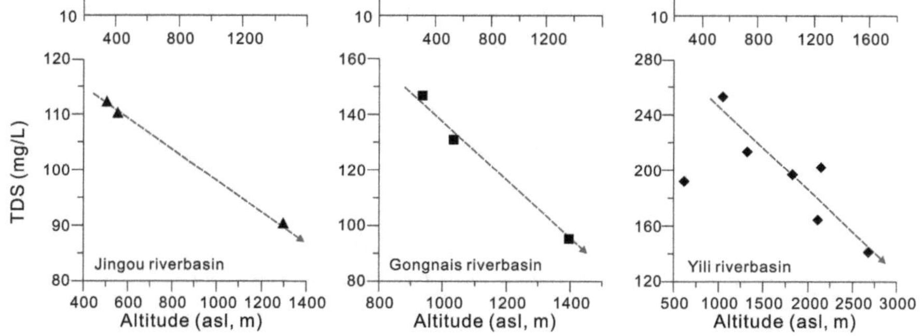

Fig. 5 TDS versus altitude (m, asl) of the studied water samples

changes of discharge and sources cannot be the reasons for the variation in the dissolved solid concentrations for samples collected within a day from a single river. Accordingly, the increase of TDS from upper reaches toward lower reaches in these rivers can be interpreted in terms of evaporation and dissolution of soil salts along their river courses. Due to the relatively fixed channel course at present, the dissolution of soil salts along the river bed does not seem to vary greatly. So if it is the evaporation processes that causes the change of concentration of dissolved material during water flow, and given that there is no change in the flow speed of ~5 km/h, it indicates that it is intense enough in these drainage basins that runoff can decrease by one-third to one-fifth during one day.

6 Conclusions

- The physico-chemical characteristics of the natural waters in the middle-latitude drainage systems, as well as the climatic, geological and topographical context where they flow and reside, allowed the identification of hydrogeochemical evolution and recharge mechanism of the natural water resources in an arid environment.
- The natural waters at various sites in the study area are different in terms of mineralization but are similar to the majority of the large rivers on earth in general, which are typically carbonate rivers and alkaline in nature. However, there is no Cl-dominated water type that occurs in the study area, indicating that the natural waters in northern Xinjiang are still at the early stage of water evolution. The regolith and geomorphological parameters play a large role in water chemistry and its evolution. The high concentration of nitrate in oasis water is probably related to the decomposition process of untreated waste.
- Direct recharge from seasonal snow and ice-melt water and infiltration of rain into the ground are thought to be significant recharge processes for natural water in the study area, while recharge from potential deep groundwater may be much less important.
- The chemistry of lakes present in the study area is largely consistent with those of large lakes in the world but the ionic enrichment in the lakes has been mainly caused by evaporation, rather than through the quality of the recharged water.
- The effect of temperature-and-precipitation-dependent geological weathering and lithological distributions on water evolution in the wet zones and the preponderant role of climate (evaporation) in dry zones are recognized.

Acknowledgements This work was financially supported by the National Natural Science Foundation of China (Grant No.: 41371060, 41271049) and the Kezhen Young Talent Project of the Institute of Geographic Sciences and Natural Resources Research, CAS (Grant No.: 2013RC101).

References

Abdesselam S, Halitim A, Jan A, Trolard F, Bourrie G (2013) Anthropogenic contamination of groundwater with nitrate in arid region: case study of southern Hodna (Algeria). Environ Earth Sci 70:2129–2141

Ahmad T, Khanna PP, Chakrapani GJ, Balakrishnan S (1998) Geochemical characteristics of water and sediment of the Indus River, Trans-Himalaya, India: constraints on weathering and erosion. J Asian Earth Sci 16:333–346

Arnell NW (1999) The effect of climate change on hydrological regimes in Europe: a continental perspective. Glob Environ Change 9:5–23

Barnett TP, Adam JC, Lettenmaier DP (2005) Potential impacts of a warming climate on water availability in snow-dominated regions. Nature 438:303–309

Berner RA, Lasaga AC, Garrells RM (1983) The carbonate silicate geochemical cycle and its effect on atmospheric carbon dioxide over the past 100 million years. Am J Sci 283:641–683

Beylich AA, Kolstrup E, Thyrsted T, Gintz D (2004) Water chemistry and its diversity in relation to local factors in the Latnjavagge drainage basin, arctic-oceanic Swedish Lapland. Geomorphology 58:125–143

Blum JD, Erel Y (1995) A silicate weathering mechanism linking increases in marine $^{87}Sr/^{86}Sr$ with global glaciation. Nature 373:415–418

Bluth GJS, Kump LR (1994) Lithologic and climatologic controls of river chemistry. Geochim Cosmochim Acta 58:2341–2359

Burns DA, Klaus J, McHale MR (2007) Recent climate trends and implications for water resources in the Catskill Mountain region, New York, USA. J Hydrol 336:155–170

Chae GT, Yun ST, Kim KH, Lee PK, Choi BY (2004) Atmospheric versus lithogenic contribution to the composition of first- and second-order stream waters in Seoul and its vicinity. Environ Int 30:73–85

Day JA (1993) The major ion chemistry of some southern African saline systems. Hydrobiologia 267:37–59

Dessert C, Dupre B, Francois LM, Schott J, Gaillardet J, Chakrapani G, Bajpai S (2001) Erosion of Deccan Traps determined by river geochemistry: impact on the global climate and the $^{87}Sr/^{86}Sr$ ratio of seawater. Earth Planet Sci Lett 188:459–474

Domros M, Peng G (1988) The climate of China. Springer, Berlin

Dunne T (1978) Rates of chemical denudation of silicate rocks in tropical catchments. Nature 274:244–246

Edmond JM (1992) Himalayan tectonics, weathering processes and the strontium isotope record in marine limestones. Science 258:1594–1597

Egorov AN (1993) Mongolian salt lakes: some features of their geography, thermal patterns, chemistry and biology. Hydrobiologia 267:13–21

Eriksson E (1985) Principles and applications of hydrochemistry. Chapman and Hall, London

Feller MC (2005) Forest harvesting and streamwater inorganic chemistry in western North America: a review. J Am Water Resour Assoc 41:785–811

Gaillardet J, Dupre B, Louvat P, Allegre CJ (1999) Global silicate weathering and CO_2 consumption rates deduced from the chemistry of large rivers. Chem Geol 159:3–30

Gibbs RJ (1970) Mechanisms controlling world water chemistry. Science 170:1088–1090

Hagg W, Braun LN, Kuhn M, Nesgaard TL (2007) Modeling of hydrological response to climate change in glacierized Central Asian catchments. J Hydrol 332:40–53

Huh Y, Edmond JM (1999) The fluvial geochemistry of the rivers of Eastern Siberia: III. Tributaries of the Lena and Anabar draining the basement terrain of the Siberan Craton and the Trans-Baikal Highlands. Geochim Cosmochim Acta 63:967–987

Kilham P (1990) Mechanisms controlling the chemical composition of lakes and rivers: data from Africa. Limnol Oceanogr 35:80–83

Kimbadi S, Vandelannoote A, Deelstra H, Mbemba M, Ollevier F (1999) Chemical composition of the small rivers of the north-western part of lake Tanganyika. Hydrobiologia 407:75–80

Lettenmaier DP, Sheer DP (1991) Climatic sensitivity of California water resources. J Water Resour Plan Manage 117(1):108–125
Livingstone DA (1963) Chemical composition of rivers and lakes. Geological Survey Paper, 440-G, pp 1–64
Ma L (2002) Geological Atlas of China. Geological Press, Beijing (in Chinese)
Martos-Rosillo S, Rodriguez-Rodriguez M, Pedrera A, Cruz-SanJulian JJ, Rubio JC (2013) Groundwater recharge in semi-arid carbonate aquifers under intensive use: the Estepa Range aquifers (Seville, southern Spain). Environ Earth Sci 70:2453–2468
Mast MA, Drever JI (1990) Chemical weathering in the Loch Vale watershed, Rocky Mountain National Park, Colorado. Water Resour Res 26:2971–2978
Meier M, Dyurgerov M (2002) Deciphering complex changes in snow and ice. Science 297:350–351
Melack JM (1983) Large, deep salt lakes: a comparative limnological analysis. Hydrobiologia 105:223–230
Menzel L, Burger G (2002) Climate change scenarios and runoff response in the Mulde catchment (southern Elbe, Germany). J Hydrol 267:53–64
Meybeck M (1987) Global chemical weathering of surficial rocks estimated from river dissolved loads. Am J Sci 287:401–428
Meybeck M (1996) River water quality: global ranges, time and space variabilities, proposals for some redefinitions. Verhandlungen des Internationalen Verein Limnologie 26:81–96
Meybeck M, Helmer R (1989) The quality of rivers: from pristine stage to global pollution. Palaeogeogr Palaeoclimatol Palaeoecol 75:283–309
Meyer JL, McDowell WH, Bott TL, Elwood J, Ishizaki C, Melack JM, Peckarsky B, Peterson B, Rublee P (1988) Elemental dynamics in streams. J N Am Benthol Soc 7:410–432
Middelkoop H, Daamen K, Gellens D, Grabs W, Kwadijk JCJ, Lang H, Parmet BWAH, Schadler B, Schulla J, Wilke K (2001) Impact of climate change on hydrological regimes and water resources management in the Rhine basin. Clim Change 49:105–128
Molnar P, England P (1990) Late Cenozoic uplift of mountain ranges and global climate change: chicken or egg? Nature 346:29–34
Noh H, Huh Y, Qin J, Ellis A (2009) Chemical weathering in the three rivers region of Eastern Tibet. Geochim Cosmochim Acta 73:1857–1877
Pawellek F, Frauenstein F, Veizer J (2002) Hydrochemistry and isotope geochemistry of the upper Danube River. Geochim Cosmochim Acta 66:3839–3854
Raymo ME (1994) The Himalayas, organic carbon burial, and climate in the Miocene. Paleoceanography 9:399–404
Raymo ME, Ruddiman WF (1992) Tectonic forcing of late Cenozoic climate. Nature 359:117–122
Raymo ME, Ruddiman WF, Froelich PN (1988) Influence of late Cenozoic mountain building on ocean geochemical cycles. Geology 16:649–653
Riebe CS, Kirchner JW, Granger DE, Finkel RC (2001) Strong tectonic and weak climatic control of long-term chemical weathering rates. Geology 29(6):511–514
Sener S, Davraz A, Karaguzel R (2013) Evaluating the anthropogenic and geologic impacts on water quality of the Eğirdir Lake, Turkey. Environ Earth Sci 70:2527–2544
Singh P, Kumar N (1997) Impact assessment of climate change on the hydrological response of a snow and glacier melt runoff dominated Himalayan river. J Hydrol 193:316–350
Sun J, Ye J, Wu W, Ni X, Bi S, Zhang Z, Liu W, Meng J (2010) Late Oligocene-Miocene mid-latitude aridification and wind patterns in the Asian interior. Geology 38:515–518
Sundquist ET (1991) Steady-and-non-steady-state carbonate–silicate controls on atmospheric CO_2. Quatern Sci Rev 10:283–296
Vanderzalm JL, Jeuken BM, Wischusen JDH, Pavelic P, Salle CLGL, Knapton A, Dillon PJ (2011) Recharge sources and hydrogeochemical evolution of groundwater in alluvial basins in arid central Australia. J Hydrol 397:71–82
Velbel M (1993) Temperature dependence of silicate weathering in nature: how strong a negative feedback on long term accumulation of atmospheric CO_2 and global greenhouse warming? Geology 21:1059–1062
Wetzel RG, Likens GE (2000) Limnological analyses, 3rd edn. Springer, New York

White AF, Blum AE (1995) Effect of climate on chemical weathering in watersheds. Geochim Cosmochim Acta 59:1729–1747

White AF, Blum AE, Schultz MS, Vivit DV, Stonestrom DA, Larson M, Murphy SF, Eberl D (1998) Chemical weathering in a tropical watershed, Luquillo Mountains, Puerto Rico: I. Long-term versus short term weathering fluxes. Geochim Cosmochim Acta 62:209–226

White AF, Blum AE, Bullen TD, Davison VV, Schulz M, Fitzpatrick J (1999) The effect of temperature on experimental and natural chemical weathering rates of Granitoid rocks. Geochim Cosmochim Acta 63:3277–3291

Wu Y, Gibson CE (1996) Mechanisms controlling the water chemistry of small lakes in northern Ireland. Water Res 30:178–182

XETCAS (Xinjiang Expedition Team of the Chinese Academy of Sciences), IGCAS (Institute of Geography of the Chinese Academy of Sciences), DGBNU (Department of Geography of Beijing Normal University) (1978) Geomorphology in Xinjiang. Science Press, Beijing (in Chinese)

Xu CC, Chen Y, Hamid Y, Tashpolat T, Chen Y, Ge H, Li W (2009) Long-term change of seasonal snow cover and its effects on river runoff in the Tarim River basin, northwestern China. Hydrol Process 23:2045–2055

Xu CC, Chen Y, Yang Y, Hao X, Shen Y (2010) Hydrology and water resources variation and its response to regional climate change in Xinjiang. J Geog Sci 20:599–612

Xu CC, Chen YN, Chen YP, Zhao R, Ding H (2013) Responses of surface runoff to climate change and human activities in the arid region of Central Asia: a case study in the Tarim River Basin, China. Environ Manage 51:926–938

Yang X, Williams MAJ (2003) The ion chemistry of lakes and late Holocene desiccation in the Badain Jaran Desert, Inner Mongolia, China. Catena 51:45–60

Zhang J, Takahashi K, Wushiki H, Yabuki S, Xiong J, Masuda A (1995) Water geochemistry of the rivers around the Taklimakan Desert (NW China): crustal weathering and evaporation processes in arid land. Chemical Geology 119:225–237

Zhu B, Yang X (2007) The ion chemistry of surface and ground waters in the Taklimakan Desert of Tarim Basin, western China. Chin Sci Bull 52:2123–2129

Zhu B, Yang X (2010) The origin and distribution of soluble salts in the sand seas of northern China. Geomorphology 123:232–242

Zhu B, Yang X, Rioual P, Qin X, Liu Z, Xiong H, Yu JJ (2011) Hydrogeochemistry of three watersheds (the Erlqis, Zhungarer and Yili) in northern Xinjiang, NW China. Appl Geochem 26:1535–1548

Zhu B, Yu J, Qin X, Rioual P, Xiong H (2012a) Climatic and geological factors contributing to the natural water chemistry in an arid environment from watersheds in northern Xinjiang, China. Geomorphology 153–154:102–114

Zhu B, Yang X, Liu Z, Rioual P, Li C, Xiong H (2012b) Geochemical compositions of soluble salts in aeolian sands from the Taklamakan and Badanjilin deserts in northern China, and their influencing factors and environmental implications. Environ Earth Sci 66:337–353

Zhu B, Yu J, Qin X, Rioual P, Zhang Y, Liu Z, Mu Y, Li H, Ren X, Xiong H (2013a) Identification of rock weathering and environmental control in arid catchments (northern Xinjiang) of Central Asia. J Asian Earth Sci 66:277–294

Zhu B, Yu J, Qin X, Rioual P, Liu Z, Zhang Y, Jiang F, Mu Y, Li H, Ren X, Xiong H (2013b) The Significance of mid-latitude rivers for weathering rates and chemical fluxes: evidence from northern Xinjiang rivers. J Hydrol 486:151–174

Zhu GF, Li ZZ, Su YH, Ma JZ, Zhang YY (2007) Hydrogeochemical and isotope evidence of groundwater evolution and recharge in Minqin Basin, Northwest China. J Hydrol 333:239–251

Factors Influencing the Runoff Trend in a Medium Sized River Basin in the Western Ghats, India

P.P. Nikhil Raj and P.A. Azeez

Abstract The present study examines the trends in the annual runoff of a tropical river basin Bharathapuzha, a medium sized river in southern India under the influences of anthropogenic pressures and climate change. The examination of the temporal trends in the rainfall, temperature and river runoff was done using historical datasets. It was supplemented with the data on the land use/land cover (LU/LC) change in the basin based on the LANDSAT TM data. By using a multiple regression model, the influential factors determining the river discharge were identified. The results show that while the rainfall influences the runoff positively, new water bodies, dams and other diversions in the fluvial setup in the basin influence the river runoff negatively.

Keywords Tropical river system · Bharathapuzha · River runoff · Climate change · Land use · Regression model

1 Introduction

Rivers and other water bodies hold enormous ecological values (Schuyt 2005), provide very vital ecological services, and have been the cradles for most of the world's civilizations. Nevertheless, the pressures from human activities and global change in the climatic system are jeopardizing these ecosystems (Turner 1991; Naiman et al. 2002). Human pressures and their manifestations on the river systems are wide and varied and are well known not only to the scientific community, but

P.P.N. Raj · P.A. Azeez
Environmental Impact Assessment Division, Sálim Ali Centre for Ornithology and Natural History (SACON), Coimbatore 641108, India

P.P.N. Raj (✉)
Centre for Sustainable Future, Amrita Vishwa Vidyapeetham, Amritanagar (PO), Coimbatore 641112, India
e-mail: ppnraj@gmail.com

Mu. Ramkumar et al. (eds.), *Environmental Management of River Basin Ecosystems*, Springer Earth System Sciences, DOI 10.1007/978-3-319-13425-3_6

also to the general public. Local alterations (construction of dams and dykes, diversion of the bulk of water flow for agriculture, industrial or human consumption, and waste disposal, and embankment and encroachments across the flow regime) are carried out largely with narrow myopic objectives without giving considerations towards the sustainability of the natural fluvial ecosystems. According to recent statistics 41 % of the global human population resides along river courses of extreme water stress (Bates et al. 2008; CBD 2005). Degradation and alternation of riverine ecosystems have resulted in the extinction or near extinction of about 20 % of the world's fresh water species (CBD 2005). According to the Millennium Assessment report (2005) the world's fresh water ecosystems are reported to be the areas of highest species extinction and several species in this zone fall among the list of threatened species.

Most of the world river basins have been undergoing major change since 19th century. Overexploitation as a result of agriculture and industrial revolutions had modified many of the global rivers. To cater the growing needs of the population, the agriculture required more irrigation facilities and diversion of water. Dams for irrigation and energy production and the extensive inorganic agriculture practices not only altered the natural flow regimes but also polluted the water resources (Varghese 2009). It is noted that 40 % of the world agriculture output purely depends on irrigation (Fischer et al. 2006), and as a result, the water impounded in the reservoirs has quadrupled since 1960 (Millennium Assessment 2005). Industrial revolution had further aggravated the situation by polluting the rivers with industrial, municipal and other forms of wastes. There are many case studies that have unequivocally demonstrated how over-exploitation coupled with other human pressures and climatic changes affect the sustainability of rivers (Dai et al. 2010; Changming and Xiaoyan 2009; Tijiu and Xiaojing 2007; Wilk and Hughes 2002).

Measures to assess the health of the rivers are debatable. According to Changming and Xiaoyan (2009), the continuous and consistent runoff, favourable riverbed and complete drainage system verifies the health of any river. Apparently, the river runoff being directly related to the hydrologic cycle of a basin is an integrated measure of the influence of climate, land cover, and human activities on the hydrologic cycle over a drainage basin (Sharma et al. 2000). Taking the stream flow as a parameter, several river basins located across the globe are examined with reference to the impact of the contemporary environmental changes on the river runoff. Fang et al. (2009) studied the Yellow River, China and appraised the direct influence of rainfall (50 %) and anthropogenic activities (<50 %) on the stream flow in the basin. Sharma et al. (2000) revealed that the anthropogenic pressure coupled with the global warming adversely affected the discharge in the Kosi River basin in the Himalayas. Several authors studied the hydrology of various river basins by focusing on the water quality and runoff (for example, Riedel et al. 2000; Gupta and Chakrapani 2005; Sileika et al. 2006; Quadir et al. 2007; Raj and Azeez 2009a).

During the last few of decades the world has been witnessing drastic changes in land use and land cover (Zhu et al. 2008). Land use change along the catchment of rivers affects the entire river system by altering the river runoff and ground water flow (Zade 2005; Tijiu and Xiaojing 2007; Xu et al. 2007) in general and more precisely by changing the inputs of water, light, and allochthonous materials into the system (Nilsson et al. 2003; Strayer et al. 2003). Studies have been conducted investigating the effect of changes in land use on river basins around the globe (for example, Xiaoming et al. 2007; Galster et al. 2007; Yang et al. 2004; Bhaduri et al. 2000). Expansion of agriculture in river basins is closely associated with diversion of water from the rivers. Nilsson et al. (2003) reported the effect of encroachment in the basin as well as in the riverbeds on the hydrology. Similarly studies on several river basins have documented how the dams altered the natural water flow (Burke et al. 2009; Tukur and Mubi 2002; Cowell and Stoudt 2002) and the natural profile of the river beds (Altaiee 1990).

Untoward variation in climate is one of the determining factors of the health and status of the world's fresh water resources. There are notable studies on meteorology and its relation with the hydrology of the river basins (Yin et al. 2000; DeWit et al. 2007). Drastic and adverse fluctuations in the climatic conditions are likely to have a direct influence on the fresh water resources. The global surface temperature is reported to have risen by 0.74 °C since 1906 and the warming was more rapid during the last 50 years (Bates et al. 2008). The India Meteorological Department (IMD) also reports a 0.913 °C hike in temperature in 2009 than the 1961–1990 average (IMD 2010). However, due to the lack of climate stations and availability of historical climate datasets there is a huge lacuna in documenting the global river flow in relation to the climate change. Seawater intrusion is yet another major crisis observed in most of the river basins particularly those with low physiography. It affects the water quality and the ecosystems in a larger perspective. Sea level rise was observed at a rate of 1.7 ± 0.5 mm/yr for the 20th century (Bates et al. 2008).

The rivers of India are broadly grouped into four classes according to their location and topography: Himalayans, Deccan, Coastal, and rivers of inland drainage basins. The three big rivers, Indus, Ganges, Brahmaputra and their network of tributaries in the Indo-Gangetic plains constitute the Himalayan river systems. The Deccan rivers include the west flowing Narmada and Tapti and east flowing Mahanadi, Krishna, Godavari and Cauvery. The coastal rivers include the small river basins along the eastern and western sides of the peninsula. The small rivers of Rajasthan are known as the inland drainage rivers. The rivers of India could be also classified as major, medium and minor rivers on the basis of catchment size (Jain et al. 2007); the class 'major rivers' includes those with catchment area >20,000 km^2, 'medium rivers' includes those with catchment area between 2,000 and 20,000 km^2, while 'minor rivers' are those with Catchment area <2,000 km^2.

The state of Kerala, located at the southwest corner (8.5°–11°N and 76°–77°E) of the Indian peninsula is unique in its physiography with an undulating terrain bounded by the Western Ghats on its east and the Arabian Sea on the west. The State has 44 rivers with an average length of 64 km. The network of their tributaries and distributaries cover almost 74 % of the total area of the state. The average rainfall of the state is 3,000 mm and the total annual yield from the 44 rivers to the state is 70,323 MCM (http://www.kerenvis.nic.in). However, Nair (2008) states that Kerala's per capita water availability lies far below the more arid states of India such as Rajasthan and Maharashtra. Studies reveal that there is a general declining trend in the annual rainfall in the state (Soman et al. 1998; Kumar et al. 2004; Krishnakumar et al. 2009), and some of the studies reported local aberrations and decline in the annual rainfall in different parts of the state (Soman et al. 1998; Raj and Azeez 2010a).

Although Kerala is covered by a rich river network, studies documenting the environmental status of these rivers are very rare. The present study was carried out on the Bharathapuzha river basin. The Bharathapuzha river system, a major one in the western side of the Western Ghats, is critical as a support system for over 4.5 million people and more than four hundred thousand hectares of agriculture (Raj 2011). In the recent decades, the river in the downstream areas is turning barren with no notable and consistent flow and is experiencing serious water scarcity sometimes even immediately after the monsoons. Apparently unsustainable consumptive water demands, unsustainable exploitation of natural resources such as river-sand and untenable encroachment of the river course are some of the major threats to the river system. Extensive mining of sand and clay have interfered with the flow regime of the main course of the river and most of its tributaries. In addition, imprudent and unhygienic disposal of urban wastes add to the grievous degeneration of the river course (Raj 2011). In this context, the present study attempts to examine the annual runoff in the river in relation to various anthropogenic and climatic parameters.

2 Study Area

The Bharathapuzha River (Fig. 1) is one among the forty-one west flowing rivers of the Kerala State and is bounded by 10°25′–11°15′ north and 75°50′–76°55′ east. Four major tributaries namely, Kalpathy, Chittur, Gayathri and Thootha that originate from the Western Ghats (one of the biodiversity hotspots in India) and join the stem channel of the river which is classified as a seventh order river and is grouped under the median rivers of the country (Jain et al. 2007). It debouches into the Arabian Sea at Ponnani located at the western coast of India. The river flows through high land (>76 m), mid land (76–8 m) and coastal plains (<8 m). The river basin covers almost one-ninth of the total geographical area of the state.

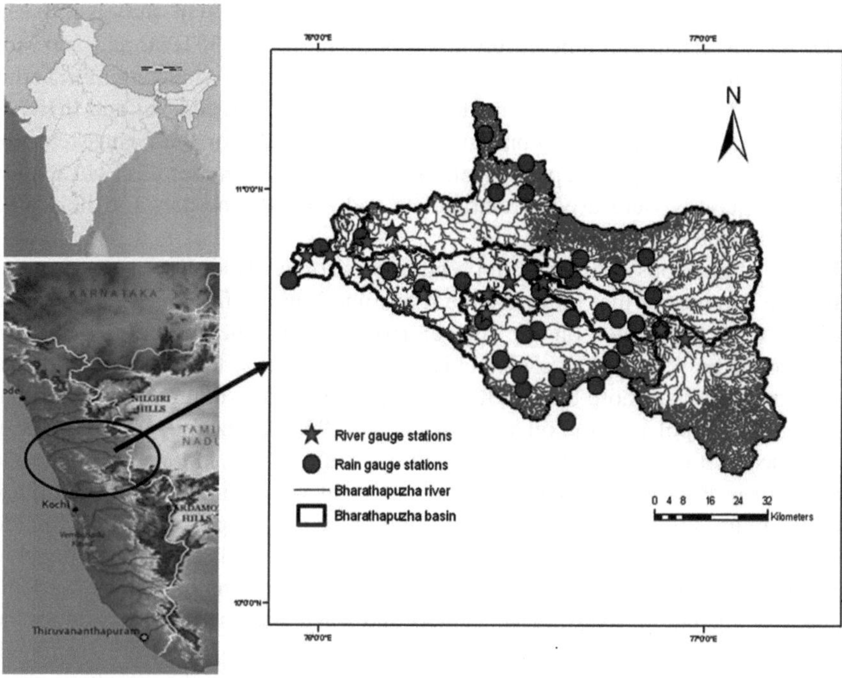

Fig. 1 Map showing the location of the Bharathapuzha river basin and the river and rain gauge stations within the basin

3 Materials and Methods

To study the hydrologic trend of the basin, runoff data for 34 years (1968–2005) from 12 river gauge stations maintained by the Central Water Commission (CWC, Government of India) and Irrigation department (Government of Kerala) were collected. This was followed by examination of the climate trend of the river basin using available meteorological data (annual, monthly temperature and rainfall data) collected from the Indian Meteorological Department (IMD, Pune) for the same period (1968–2005). The basin has a network of 37 rain gauge stations operated by the Irrigation department, Government of Kerala. However, data from only 29 selected stations, that have continuous monthly and yearly data, were used for the study.

The annual runoff, temperature and rainfall trend of the basin were analyzed using the Mann-Kendall rank correlation analysis, since it is a suitable statistical test for a long period of data (Basistha et al. 2007; Krishnakumar et al. 2009; Raj and Azeez 2010a, 2012). The values of t were used as the basis of significance test by comparing it with $Tt = 0 \pm tg\sqrt{[4N + 10/9N(N - 1)]}$, where, tg is the desired probability point of the Gaussian normal distribution. In the present study, tg at 0.01 and 0.05 were considered as the levels of significance.

To understand the land use/land cover changes (LU/LC) in the river basin LANDSAT TM data, with a pixel resolution of 30 m, for 1973, 1990 and 2005 for the whole basin obtained from the Global Land Cover Facility (www.glcf.umd.edu) were utilized. LU/LC analysis was made with the help of Arc GIS 9.3 and ERDAS IMAGINE 8.5 software. Since continuous data on LU/LC for the basin were not available, a method of gross approximation was adopted such that the LU/LC data of the three periods were used to extrapolate and estimate the data for the whole period of the study.

Multiple regressions coupled with partial regression analyses were carried out to identify the parameters that influence the river runoff and to find out the respective weight of each parameter on the annual river runoff. The multiple regression analysis could be regarded as a convenient and robust tool to find the respective weights among two or more factors. It is also applicable in testing hypothesis and in predicting or interpolating the relationships (Zar 1999; Nawaz and Adeloye 1999). It is assumed that both the climatic and anthropogenic factors influenced the trends in the runoff in Bharathapuzha Basin. The factors such as rainfall, number of rainy days, maximum, minimum and mean temperatures were considered as the climatic factors while the land use/land cover changes were considered as the anthropogenic factors. The land use/land cover was categorized into five major classes (agriculture, natural vegetation, roads, urban centers and water bodies) and accordingly the land use changes in the river basin for the time span of 1973–2005 were examined. For generating unbiased results, the area under plantation and agriculture were merged together since the means of utilization of river water is same by both these types of land uses. Altogether, the factors were classified into two; those likely to influence the river runoff positively and those having a negative influence on it. It was assumed that the number of rainy days, rainfall, natural vegetation, roads and water bodies to be positively related with the river runoff and the temperature, urban area and agriculture area to be inversely related with it (Fig. 2).

4 Results

4.1 Trends in the River Runoff, Rainfall and Temperature

It is interesting and useful to examine the annual trend of runoff of a river and the factors influencing it. The data extending to almost four decades show the average runoff in the Bharathapuzha Basin to be 5.39 km^3 (STDEV ± 1.58). Mann-Kendall rank correlation analysis shows a significant ($p < 0.01$) decreasing trend in the runoff as the years proceeds (Table 1, Fig. 3). On examining the temporal trends of the rainfall (Raj and Azeez 2012), it is found that the rainfall shows a statistically significant ($p < 0.01$) decreasing trend (Table 1, Fig. 4). Analysis of the temperature in the basin for the same period shows a significant ($p < 0.01$) increasing trend.

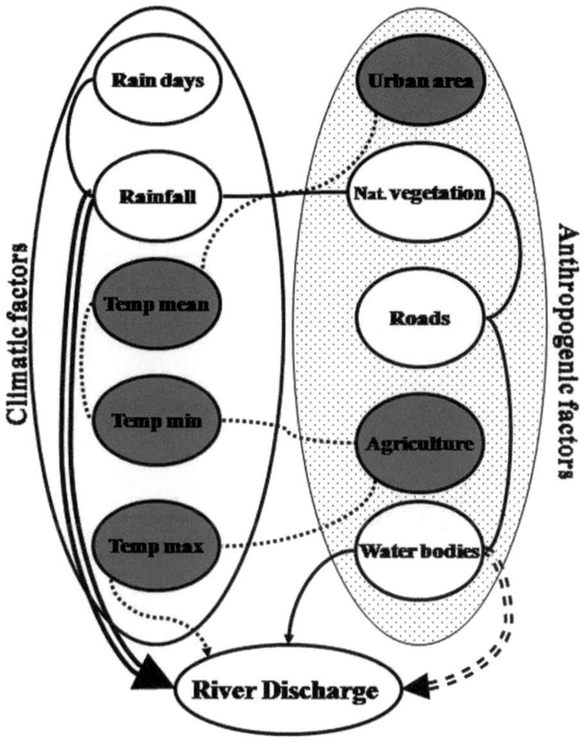

Fig. 2 Factors influencing the river runoff in the Bharathapuzha River Basin (*black single lines* represent the hypothetical scenario, while *black double lines* mark the results after the analysis. *Thick lines* indicate the positive relationship while the *dotted ones* represent the inverse relationships among the factors)

Table 1 Mann-Kendall's rank correlation statistics showing the temporal trend of runoff and other climatic factors in Bharathapuzha river basin

Factors	Coefficient
Annual river runoff	−0. 504[a]
Annual rainfall	−0.496[a]
Annual T max	0.317[a]
Annual T mean	0.404[a]
Annual T min	0.428[a]

[a] Significant at 0.01

Similar trend (Raj and Azeez 2011) is observed for annual maximum (T max), minimum (T min) and mean temperature of the basin (Table 1, Fig. 5).

4.2 Temporal LU/LC in the Basin

During the early period (1973–1990) of the study, land under natural vegetation cover in the basin was 44 % of the total basin area followed by the area under agriculture. During the second half of the study period, the area under agriculture was the highest in extent (41.7 %) followed by the area under urban centre. In 2005, the

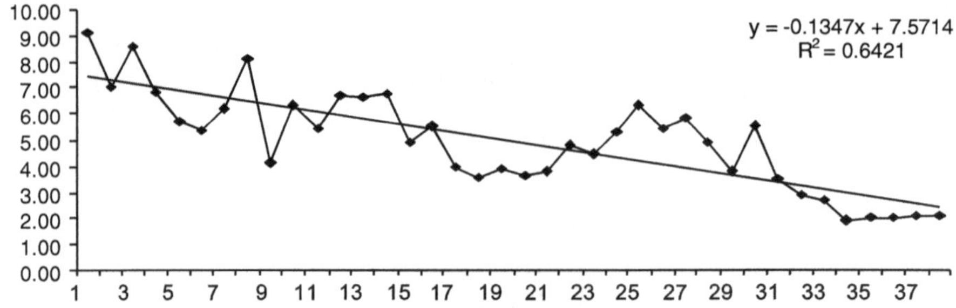

Fig. 3 The temporal trend in annual runoff in the Bharathapuzha river basin

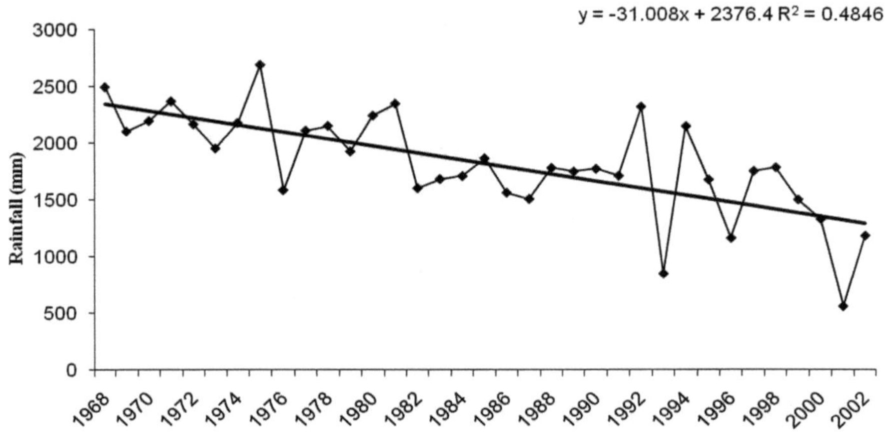

Fig. 4 Temporal trend of rainfall in the Bharathapuzha river basin

area under urban centers was the major land use type in the basin, followed by agriculture at the second position. Thus, the area under the natural vegetation cover consistently showed a declining trend of decline. A positive growth trend of urban centers in the basin was observed during the whole period (Table 2; after Raj and Azeez 2010b).

4.3 Factors Influencing the River Runoff

The trend in runoff of the river was examined to identify other determining factors. The correlation analysis of various parameters on runoff shows significant positive correlations ($p < 0.01$) only with rainfall and natural vegetation ($p < 0.05$). Similarly, the area under water bodies ($p < 0.01$), and the agriculture area ($p < 0.01$ levels) shows negative correlations with the runoff (Table 3).

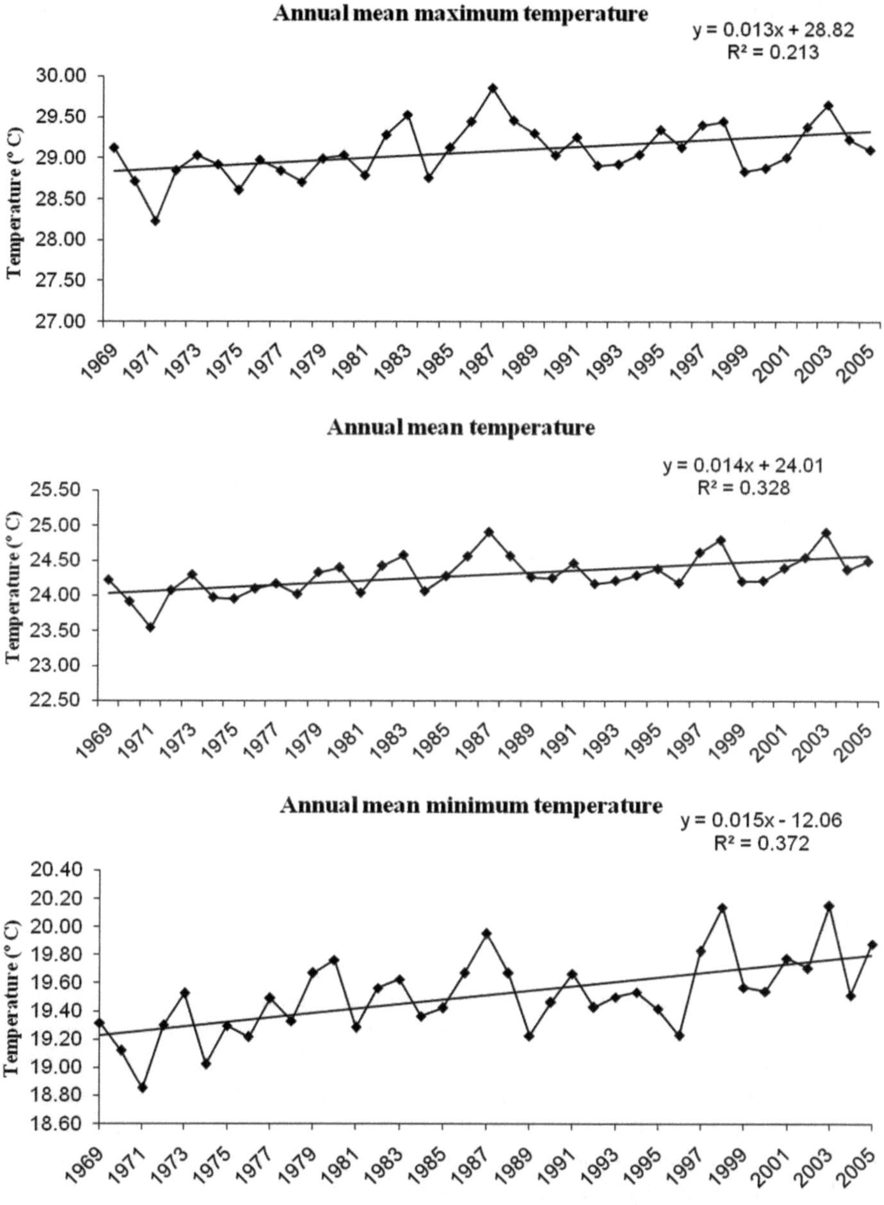

Fig. 5 Temporal trend of T max, T mean and T min in Bharathapuzha river basin

Table 2 Total land cover (in %) as a proportion to the total area, and the net change during the study period

Land use	1973	1990	2005	Change (%) during 1973–1990	Change (%) during 1990–2005	Change (%) during 1973–1005
Agriculture	35.31	41.74	27.80	6.43	−13.94	−7.51
Natural vegetation	43.43	12.07	12.28	−31.36	0.21	−31.15
Road	7.61	8.40	16.24	0.79	7.83	8.62
Urban centers	9.83	32.63	41.76	22.80	9.13	31.93
Water bodies	3.82	5.16	1.93	1.34	−3.23	−1.89

Table 3 Mann-Kendall's rank correlation analysis showing the relationship between runoff and other parameters

Variables	Coefficient
Rainfall	0.442[a]
Rain days	0.17
T max	−0.15
T min	−0.063
T mean	−0.103
Natural vegetation	0.277[b]
Water bodies	−0.459[a]
Roads	0.166
Agricultural area	−0.451[a]
Urban centers	−0.166

[a] Significant at 0.01 [b] 0.05 levels

5 Discussion

Except for the physical structures in the catchment, the rainfall exerts direct control over the river runoff (Pfister et al. 2000) which is true for all river systems dependent on monsoon precipitation (Raj and Azeez 2009). The general rainfall in the basin shows a statistically significant decreasing trend as the year proceeds (Fig. 3, Table 1). At the same time, the temporal trend shows a decrease in the number of rainy days in the basin, which is in conformity with the nation-wide trend reported earlier by Ramesh and Goswami (2007). The past few decades have seen extensive expansion of road transport infrastructure and allied structures. Notwithstanding the flow regimes, the positive relationship (albeit at statistically low levels—Table 3) shown by the areal extent of roads with the river runoff, which is a cause of concern and need to be studied specifically.

The area under water bodies in the basin includes natural water bodies and man-made structures such as dams and ponds. Dams and other embankments reduce or alter the natural flow in the river especially in the lower reaches. There are

Table 4 Summary of the multiple regression model

Model	R	R2	Adjusted R^2	Std. error of the estimate	Change statistics				
					R2 change	F change	df1	df2	Sig. F change
1	0.740	0.547	0.521	3091.44	0.130	10.085	1	35	0.00

Predictors (Constant) Area under water bodies, Rainfall
Dependent Variable Runoff

11 such dams and irrigation projects in the Bharathapuzha Basin. Of these, the Aliyar dam located in the Chittur Basin is the largest and is a part of the Parambikulam Aliyar Project (PAP) and stores water from three river basins, namely, the Bharathapuzha, Periyar and Chalakkudipuzha (Ravi et al. 2004) and may have contributed towards significant negative impact on the total water discharge in the Bharathapuzha River Basin (Sadasivan 2003). The runoff and the agriculture area in the Bharathapuzha river basin show a statistically significant ($p < 0.01$) inverse relationship (Table 3). Change in the vegetation type, conversion of natural forested area into agriculture alters the river runoff due to the change in the vegetation type (Hudson et al. 1997). The over-exploitation of water and diversion for agriculture is a common practice in the Bharathapuzha Basin (Kumar 2001).

Taking these observations as cues, a multiple regression model was attempted, anticipating that the runoff of the basin could be predicted using the above variables. All the factors were utilized through a step-wise and multiple regression that resulted in the model equation $Qw = 13879.29 + (-0.36*Aw) + (4.752*Rf)$, where Qw denotes the runoff, Aw the area under water bodies and Rf the annual rainfall in the basin. The variables namely, maximum, minimum, and mean temperature, natural vegetation, areas under roads, urban centers and agriculture, and total rain days were excluded from this analysis due to their respective lower weights on the river runoff (Table 4). The regression model thus generated is highly significant ($R^2 = 0.521$, Table 4). The multiple regression analysis refined the result such that only two out of the ten factors that were initially considered were found influential. The results have shown that the area under water bodies ($p = 0.000$, $R2 = 0.400$) and rainfall ($p = 0.003$, $R2 = 0.293$) were found to be the influencing factors (Table 5).

Concurrent decreasing trend of rainfall in the basin (Fig. 3, and Table 3) in this context will have a crucial role in determining the health of the river in the

Table 5 Regression Model for predicting runoff in Bharathapuzha river basin

Model	Un-standardized coefficient		t	Significance	Partial correlation coefficient
	B	Std error			
Constant	13,879.29	4,218.85	11.62	0.002	
Area under water bodies (W)	−0.36	0.08	−4.35	0.000	−0.592
Rainfall (RF)	4.752	1.49	3.18	0.003	0.473

coming years. The presence of dams and the other man-made water storage structures have a negative effect on the river runoff. In fact, it is observed that among the four sub-basins of the river, the river Thootha, the only lesser-disturbed sub-basin in the river contributes highly to the annual runoff of the river (Raj and Azeez 2009). Many of the streams in the river basin, except the river Kunthi flowing through the Silent Valley National Park and the Thuppanadupuzha river flowing through the Chenath Nair Reserve Forest have been facing pressures from various corners for harnessing for irrigation as well as power generation, during the last couple of decades. That would have seriously affected the runoff in the main Bharathapuzha River, constraining its other ecological and socio-economic services downstream.

6 Conclusions

- The historical data shows a decreasing trend in the annual runoff in the basin.
- As could be expected in a monsoon dependent fluvial system, the river runoff is predominated by the intensity of rainfall and the number of rainy days, notwithstanding the physical structures in the catchment and resultant changes in the flow and percolation regimes
- Conversion of forestland into agriculture land use has significantly contributed towards enhancing the river runoff while the maximum, minimum and mean temperature, area under plantation contribute negatively to the river runoff.
- Out of ten factors that were initially considered for the study, only two namely, rainfall and areal extent of waterbodies have been found to be significantly influencing the runoff.
- A temporal trend of decrease in rainfall, number of rainy days and increase in maximum, minimum and mean annual temperatures is a cause of concern and this basin warrants urgent adoptive strategies to the changing climatic scenario.

Acknowledgments We are thankful to the IMD, Government of India, Pune, Irrigation Department, Government of Kerala, and the Central Water Commission (Coimbatore and Cochin offices) for providing various data sets on meteorology and river runoff. We thank the Global Land Cover Facility for the RS data. Thanks are also due to the anonymous reviewers for their constructive comments.

References

Altaiee T, Alhamdani A (1990) Morphological variations of a certain Tigris river reach for different periods in Iraq (www.balwois.com)

Basistha A, Goel NK, Arya DS, Gangawar SK (2007) Spatial pattern of trend in Indian subdivisional rainfall. Jalavigyan Sameekha 22:47–57

Bates BC, Kundzewicz ZW, Wu S, Palutikof JP (2008) Climate change and water. In: Technical paper of the intergovernmental panel on climate change, Secretariat, Geneva, p 210

Bhaduri B, Harbor J, Engel B, Grove M (2000) Assessing watershed-scale, long-term hydrologic impacts of land-use change using a GIS-NPS model. Environ Manage 26(6):643–658

Burke M, Jorde K, Buffington JM (2009) Application of a hierarchical framework for assessing environmental impacts of dam operation: changes in stream flow, bed mobility and recruitment of riparian trees in a western North American river. J Environ Manage 90:S224–S236

Changming L, Xiaoyan L (2009) Healthy river and its indication, criteria and standards. J Geog Sci 19:3–11. doi:10.1007/s11442-009-0003-6

Convention on Biological Diversity (CBD) (2005) Inland waters biodiversity introduction, Secretariat of the Convention on Biological Diversity, Montreal. http://www.biodiv.org/programmes/areas/water/default.asp

Cowell CM, Stoudt RT (2002) Dam-induced modifications to Upper Allegheny river stream flow patterns and their biodiversity implications. J Am Water Resour Assoc 38(1):187–196

Dai ZJ, Du J, Chu A, Li J, Chen J, Zhang X (2010) Groundwater discharge to the Changjiang River, China, during the drought season of 2006: effects of the extreme drought and the impoundment of the three Gorges Dam. Hydrol J 18:359–369. doi:10.1007/s10040-009-0538-8

DeWit MJM, Hurk BVD, Warmerdam PMM, Torfs PJJF, Roulin E, Deursen WPA (2007) Impact of climate change on low-flows in the river Meuse. Clim Change 82:351–372. doi:10.1007/s10584-006-9195-2

Fang ZF, Xue XZ, Lu Z, DePeng Z (2009) Stream flow response to climate variability and human activities in the upper catchment of the Yellow River basin. Sci China Ser E–Tech Sci 52 (11):3249–3256. doi:10.1007/s11431-009-0354-3

Fischer G, Tubiello FN, Velthuizen VH, Wiberg D (2006) Climate change impacts on irrigation water requirements: global and regional effects of mitigation. Technol Forecast Soc Chang 74:1990–2080. doi:10.1016/j.techfore.2006.05.021

Galster JC, Pazzagila FJ, Hargreaves BR, Morris DP, Peters SC, Weisman RN (2007) Natural and anthropogenic influences on the scaling of discharge with drainage area for multiple watersheds. Geosphere 3:260–271

Gupta H, Chakrapani GJ (2005) Temporal and spatial variations in water flow and sediment load in Narmada river basin, India: natural and man-made factors. Environ Geol 48:579–589

Hudson JA, Crane SB, Robinson M (1997) The impact of the growth of new plantation forestry on evaporation and stream flow in the Llanbrynmair catchments. Hydrol Earth Syst Sci 1(3):463–475

IMD (2010). www.imd.gov.in/doc/warm2009.pdf

Jain SK, Agarwal PK, Singh VP (2007) River basins of India. Water Science and Technology Library. Hydrology and water resources of India. doi:10.1007/1-4020-5180-8-7

Krishnakumar KN, Rao GSLHVP, Gopakumar CS (2009) Rainfall trends in twentieth century over Kerala India. Atmos Environ 43:1940–1944

Kumar BA (2001) Biodiversity of Bharathapuzha (Nila River), Kerala. Report submitted to UGC, Southern regional office, Bangalore

Kumar RMR, Shenoi SSS, Shankar D (2004) Monsoon onset over Kerala and Pre-monsoon rainfall peak

Millennium Ecosystem Assessment (2005) Ecosystems and human well-being: wetlands and water synthesis

Naiman RJ, Bunn SE, Nilsson C, Petts GE, Pinay G, Thompson LC (2002) Legitimizing fluvial ecosystems as users of water: an overview. Environ Manage 30(4):455–467. doi:10.1007/s00267-002-2734-3

Nair PR (2008) Literacy on water. Kerala Calling, 26–27 Feb 2008

Nawaz NR, Adeloye AJ (1999) Evaluation of monthly runoff estimated by a rainfall-runoff regression model for reservoir yield assessment. Hydrol Sci 44(1):113–134

Nilsson C, Pizzuto JE, Moglen GE, Palmer MA, Stanley EH, Bockstael NE, Thompson LC (2003) Ecological forecasting and the urbanization of stream ecosystems: challenges for economists, hydrologists, geomorphologists and ecologists. Ecosystems 6:659–674

Pfister L, Humbert J, Hoffmann L (2000) Recent trends in rainfall-runoff characteristics in the Alzette river basin, Luxembourg. Clim Change 45:323–337

Quadir A, Malik RN, Husain SZ (2007) Spatio-temporal variations in water quality of Nullah Aik-tributary of the river Chenab, Pakistan. Environ Monit Assess 140:1–3

Raj N, Azeez PA (2009) Spatial and temporal variation in surface water chemistry of a tropical river, the river Bharathapuzha, India. Curr Sci 96(2):245–251

Raj PPN, Azeez PA (2010a) Changing rainfall in the Palakkad plains of south India. Atmósfera 23 (1):81–88

Raj PPN, Azeez PA (2010b) Land use/Land cover changes in a tropical river basin: a case from Bharathapuzha river basin, Southern India. J Geogr Inf Syst 2:185–193. doi:10.4236/jgis.2010.24026

Raj PPN, Azeez PA (2012) Trend analysis of rainfall in Bharathapuzha river basin, Kerala, India. Int J Climatol (Wiley). doi:10.1002/joc.2283

Raj PPN, Azeez PA (2011) Temperature rise in the Bharathapuzha river basin, southern India. Curr Sci 101(4):492

Raj PPN (2011) An analysis of the environmental changes in the Bharathapuzha river basin, southern India. PhD thesis submitted to the Bharathiar University, Coimbatore, India

Ramesh KV, Goswami P (2007) The shrinking Indian summer monsoon. CSIR Centre for Mathematical Modelling and Computer Simulation. Research report RR CM 0709

Ravi SP, Madhusoodhanan CG, Latha A, Unnikrishnan S, Bachan KHA (2004) Tragedy of commons: the Kerala experience in river linking. River Research Centre, Thrissur

Riedel GF, Twilliams SA, Riedel GS, Oilmour CC, Sanders JG (2000) Temporal and spatial patterns of trace elements in the Patuxent river: a whole watershed approach. Estuaries 23:521–535

Sadasivan SN (2003) River disputes in India: Kerala rivers under siege. Mittal Publication, New Delhi

Sharma KP, Moore B, Vorosmarty CJ (2000) Anthropogenic, climatic, and hydrologic trends in the Kosi basin, Himalaya. Clim Change 47:141–165

Schuyt K (2005) Freshwater and poverty reduction: serving people, saving nature: an economic analysis of the livelihood impacts of freshwater conservation initiatives. WWF International, Gland, Switzerland

Sileika A, Lnacke P, Kutra S, Gaigals K, Berankiene L (2006) Temporal and spatial variation of nutrient levels in the Nemunas river (Lithuania and Belarus). Environ Monit Assess 122:335–354

Soman MK, Kumar KK, Singh N (1998) Decreasing trend in the rainfall of Kerala. Curr Sci 57:7–12

Strayer DL, Beighley RE, Thompson LC, Brooks S, Nilsson C, Pinay G, Naiman RJ (2003) Effects of land cover on stream ecosystems: roles of empirical models and scaling issues. Ecosystems 6:407–423. doi:10.1007/s10021-002-0170-0

Tijiu C, Xiaojing T (2007) Impact of forest harvesting on river runoff in the Xiaoxing'an mountains of China. Front Forestry China 2:143–147

Tukur AL, Mubi AM (2002) Impact of Kiri dam on the lower reaches of river Gongola, Nigeria. Geo J 56:93–96

Turner RK (1991) Economics and wetland management. Ambio 20:59–63

Varghese S (2009) Integrated solutions to the water, agriculture and climate crises. IATP, North America

Wilk J, Hughes DA (2002) Simulating the impacts of land-use and climate change on water resource availability for a large south Indian catchment. Hydrol Sci 47(1):19–30

Xiaoming Z, Xinxiao Y, Sihong W, Manliang Z, Jianlao L (2007) Response of land use/coverage change to hydrological dynamics at watershed scale in the Loess plateau of China. Acta Ecol Sinica 27(2):414–423

Xu H, Ye M, Song Y, Chen Y (2007) The natural vegetation responses to the groundwater change resulting from ecological water conveyances to the lower Tarim river. Environ Monit Assess 131:37–48. doi:10.1007/s10661-006-9455-7

Yang Z, Luohui L, Yansui L, Yimei H (2004) Land use change during 1960–2000 period and its eco-environmental effects in the middle and upper reaches of the Yangtze river: a case study in Yiliang county, Yunnan, China. J Mt Sci 1(3):250–263 (ID:1672-6316, 03-0250-14)

Yin Y, Cohen S, Huang GH (2000) Global climate change and regional sustainable development: the case of Mackenzie basin in Canada. Int Assess 1:21–36
Zade M, Ray SS, Dutta S, Panigrahy S (2005) Analysis of runoff pattern for all major basins of India derived using remote sensing data. Curr Sci 88(8):1301–1305
Zar JH (1999) Biostatistical analysis, 4th edn. Prentice-Hall, New Jersey
Zhu W, Graney J, Salvage K (2008) Land use impact on water pollution: elevated pollutant input and reduced pollutant retention. J Contemp Water Res Educ 139:15–21

Morphometric Analysis for Prioritization of Watersheds in the Mullayar River Basin, South India

R. Jaganathan, K. Annaidasan, D. Surendran and P. Balakrishnan

Abstract The Mullayar River is one of the tributaries of the Periyar River, south India. Morphological features and linear and aerial morphometric parameters of the Mullayar River Basin (MRB), located in the Kerala State, India were studied with the help of Geographic Information System and utilizing the Survey of India (SOI) topographic sheets of 1:50,000 scale. The results show that this basin contains 5th order drainage network and drainage pattern mainly of sub-dendritic to dendritic type. It is observed that the drainage density is low indicative of highly permeable soil structure and thick vegetative cover. The circularity ratio value reveals that the basin is highly elongated and is covered by highly permeable homogenous geologic materials.

Keywords Morphometric analysis · Drainage · Geographical information system · River basin · Water resource management

1 Introduction

It is of common knowledge that due to the ever increasing population and economic development and resultant expansion of urban, industrial, commercial and other demands, the available surface and ground water resources are becoming inadequate at the current rates of consumption. The demand for water has increased over the years, due to which the assessment of quantity and quality of water for their optimal utilization has become essential. Identification and outlining of various ground features such as geological structures, geomorphic features and their hydrologic characteristics can serve as both direct and indirect indicators of the

R. Jaganathan (✉) · K. Annaidasan · D. Surendran
Department of Geography, University of Madras, Chennai 600 005, India
e-mail: rjnathan@gmail.com

P. Balakrishnan
Department of Biology and Environmental Sciences, Qatar University, Doha, Qatar

Mu. Ramkumar et al. (eds.), *Environmental Management of River Basin Ecosystems*, Springer Earth System Sciences, DOI 10.1007/978-3-319-13425-3_7

presence of ground and surface water and may help in designing better management practices for sustainable and judicial utilization. The geomorphic conditions are essential prerequisites in understanding the water bearing structures of hard rocks. The role of rock types and geologic structure in the development of stream networks can be better understood by studying the nature and type of drainage patterns by quantitative morphometric analysis.

Morphometric analysis is referred as the quantitative evaluation of linear, areal and Relief forms of characteristics on the earth's surface. This is the most common technique in basin analysis, as river basin is an ideal areal unit for analysis and interpretation of fluvial landforms. The composition of the stream system of a drainage basin is expressed quantitatively with stream numbers, stream order, drainage density, bifurcation ratio and stream length ratio (Horton 1945). It incorporates quantitative study of the various components such as, stream segments, basin length, basin perimeters, basin area, altitude, volume of flow, slope, and profiles of the land.

The morphometric parameters of a watershed are reflective of its hydrological response to a considerable extent and can be helpful in synthesizing its hydrological behavior. A quantitative morphometric characterization of a drainage basin is considered to be the most satisfactory method for the proper planning of watershed management because it enables to understand the relationship among different aspects of the drainage pattern of the basin, and also to make a comparative evaluation of different drainage basins developed in various geologic and climatic regimes. Morphometric method is the measurement and mathematical analysis of the configuration of the earth's surface, shape and dimension of its landform (Clark 1966).

In the present study, Geographic Information System has been used for assessing various terrain and morphometric parameters of drainage basin and its watersheds. Linear, relief and aerial parameters were evaluated for development planning of the sub-watershed of the Mullayar Basin, Kerala State, India.

2 Study Area

The Mullayar River is a tributary of the Periyar River, which is the longest river in Kerala State. It originates at the Kottamalai peak in the Periyar Tiger Reserve. The Mullayar flows westwards through the reserve forest and joins the Periyar at Mullakudy at the beginning of the Periyar lake formed by the Mullaperiyar dam, located at 9° 31′ 43″ N; 77° 8′ 39″ E. It covers a total area of 115.75 km^2. The average annual rainfall varies from 3,265 mm in the western side to 836 mm in the eastern side (Fig. 1).

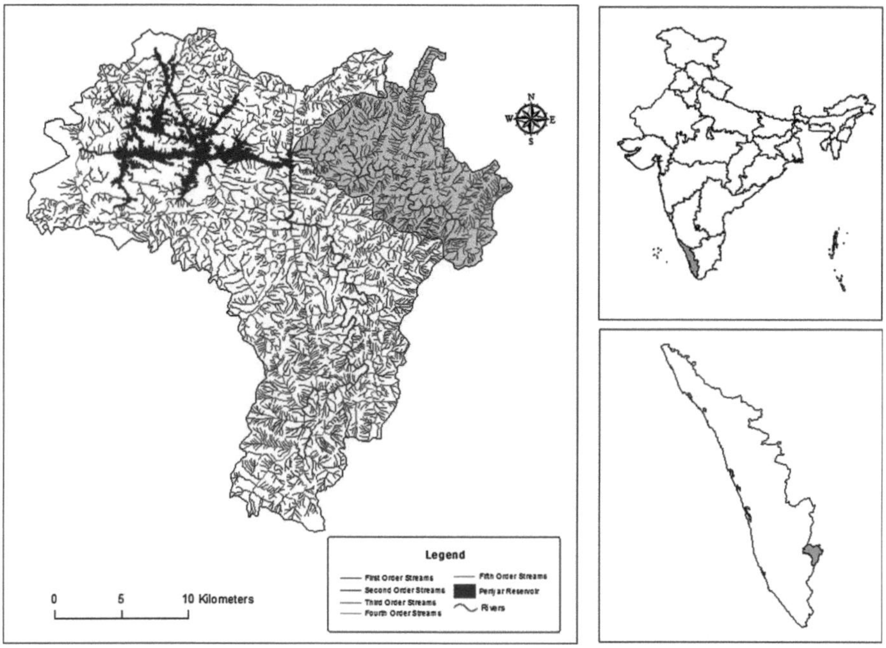

Fig. 1 Location of the Mullayar in the Periyar watershed

3 Materials and Methods

The study includes prioritization of the sub-watersheds based on data on the drainage density, groundwater prospects, areas covered by irrigation, natural forest, wastelands, etc., collected from the Survey of India topographic sheets of 1:50,000 scale and limited field verification. The digitization of the drainage pattern has been carried out in GIS environment (Fig. 2). Assignation of orders to the streams was made using Horton's law. The fundamental parameters namely, stream length, area, perimeter, number of streams and basin length were derived from the drainage layer. The morphometric parameters for the delineated watershed area were calculated based on the formula suggested by Horton (1945), Strahler (1964), Schumm (1956) and Miller (1953) and are listed in the Table 1. The morphometric parameters namely, stream order, stream length, bifurcation ratio, drainage density, drainage frequency, relief ratio, elongation ratio, circularity ratio and compactness constant have been calculated. Prioritization rating of all the nine sub-watersheds of Mullayar watershed was carried out by deriving the compound parameter values.

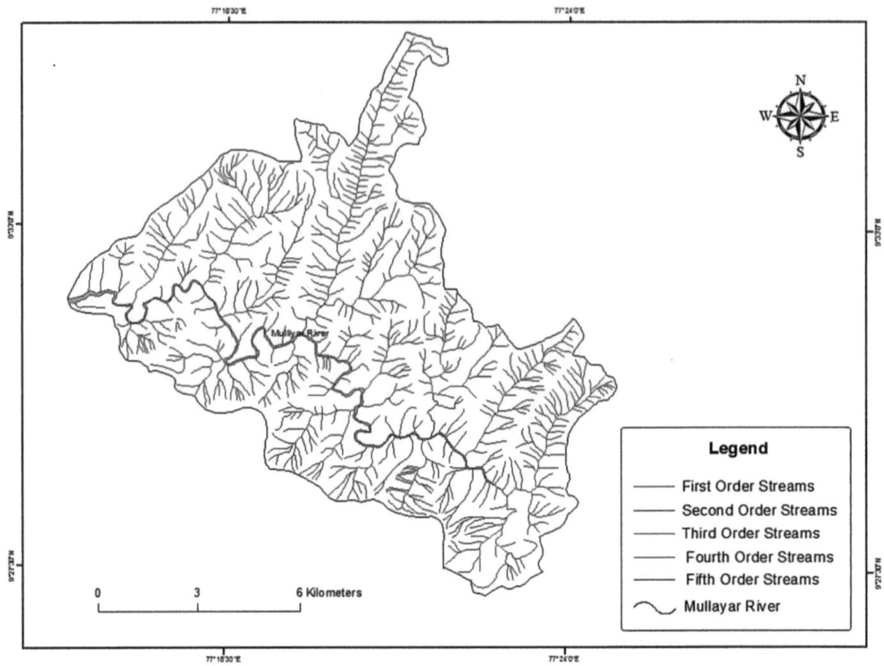

Fig. 2 Drainage network of the Mullayar River basin

4 Results and Discussion

Drainage pattern in the study area is characterized by regular branching of tributaries in many directions with an average angle 90°. The study area has been divided into six sub-watersheds which are designated as SWS1–6 serially. The nature of the streams and their length shows the predomination of structural control over their formation and existence.

4.1 Linear Aspect

The linear aspects of the channel system are stream order (U), stream length (Lu) and stream frequency (Fs). The Mullayar River is a 5th order stream covering an area of 115.75 km^2. The sub-watersheds SWS6 and 5 have 3rd order streams covering an area of 13.27 and 14.87 km^2 respectively. The sub-watersheds SWS2, and 4, have 5th order streams covering an area of 28.01 and 16.21 km^2 respectively. The sub-watershed SWS1 and 3 contain 4th order streams covering an area of 23.08 and 20.30 km^2 correspondingly. The variation in order and the size of the sub-watersheds are largely due to the physiographic and structural conditions of the region.

Table 1 Measurements adopted for computation of morphometric parameters

Morphometric parameters	Measurements	Reference authors
Stream order (u)	Hierarchical rank	Strahler (1964)
Stream length (Lu)	Length of the stream	Horton (1945)
Mean stream length (Lsm)	Lsm = Lu/Nu where, Lsm = Mean stream length Lu = Total stream length of order 'u' Nu = Total no. of stream segments of order 'u'	Strahler (1964)
Stream length ratio (RL)	RL = Lu/Lu − 1 where, RL = Stream length ratio Lu = Total stream length of the order 'u' Lu − 1 = Total stream length of next lower order	Horton (1945)
Bifurcation ratio (Rb)	Rb = Nu/Nu + 1 where, Rb = Bifurcation ratio Nu = Total no. of stream segments of order 'u' Nu + 1 = No. of segments of the next higher order	Schumn (1956)
Mean bifurcation ratio	Rbm = Average of bifurcation ratios of all orders	Strahler (1957)
Relief ratio (Rh)	Rh = H/Lb where, Rh = Relief ratio H = Total relief (Relative relief) of the basin (km) Lb = Basin length	Schumm (1956)
Drainage density (Dd)	Dd = Lu/A where, Dd = Drainage density Lu = Total stream length of all orders A = Area of the basin (km^2)	Horton (1932)
Stream frequency (Fs)	Fs = Nu/A where, Fs = Stream frequency Nu = Total no. of streams of all orders A = Area of the basin (km^2)	Horton (1932)
Drainage texture (Rt)	Rt = Nu/P where, Rt = Drainage texture Nu = Total no. of streams of all orders P = Perimeter (km)	Horton (1945)
Form factor (Rf)	Rf = A/Lb^2 where, Rf = Form factor A = Area of the basin (km^2) Lb^2 = Square of basin length	Horton (1932)
Circularity ratio (Rc)	Rc = 4 * Pi * A/P^2 where, Rc = Circularity ratio Pi = 'Pi' value i.e., 3.14 A = Area of the basin (km^2) P^2 = Square of the perimeter (km)	Miller (1953)
Elongation ratio (Re)	Re = 2/Lb where, Re = Elongation ratio A = Area of the basin (km^2) Pi = 'Pi' value i.e., 3.14 Lb = Basin length	Schumn (1956)
Length of overland flow (Lg)	Lg = 1 / D * 2 Where, Lg = Length of overland flow D = Drainage density	Horton (1945)

Stream length (Lu):

The stream length has been computed on the basis of the rule proposed by Horton (1945), for all the six sub-watersheds and the resultant data are presented in the Table 2. From the table it follows that the total length of the stream segments decreases as the stream order increases.

Stream Length ratio (Rl):

The Horton's law (1945) of stream length states that the mean stream length of each of the consecutive orders of a basin tend to approximate a direct geometric series with stream length increasing towards higher order of streams. The stream length ratio between different sub-watersheds showed a fluctuation from lower order to higher order and in the sub-watersheds of SWS1–6. Change from one order to another order as observed in the study area indicate the late youth stage of the geomorphic development of the streams in the inter basin region.

Stream frequency (Fs):

The stream frequencies of all the sub-watersheds are listed in the Table 3. The data presented in the table along with the observations made in the field show the existences of relationships among low drainage frequency and densely forested region and higher frequency and agricultural lands. While the former could have also been controlled by structure, lithology and percolation characteristics, the latter is clearly influenced by anthropogenic activities, notwithstanding the natural terrain

Table 2 Basic linear aspects of the sub-watersheds of the Mullayar River

Number of streams						Stream length (Km)					Length ratio			
Sub-watershed	I	II	III	IV	V	I	II	III	IV	V	II/I	III/II	IV/III	V/VI
SWS1	72	19	4	1	0	35	9.6	6.8	1.3	–	0.27	0.7	0.19	–
SWS2	125	25	7	1	1	66.4	13.3	8.1	9.5	1.3	0.2	0.61	1.17	0.13
SWS3	64	16	5	1	0	34.8	6.1	5	3.2	–	0.18	0.82	0.64	–
SWS4	40	6	1	2	1	20.8	3.7	1	2.2	0.49	0.18	0.27	2.24	0.21
SWS5	81	27	5	1	0	39.1	12.6	6.2	1.2	–	0.32	0.49	0.19	–
SWS6	70	17	4	0	0	32.3	7.9	2.7	–	–	0.24	0.34	–	–

Table 3 Mulliyar River: basic ratio parameters

Sub watershed	Area (km^2)	Stream frequency	Length of Basin (km)	Form factor	Elongation ratio	Circulatory ratio
SWS1	23.09	5.8	4.44	1.2	2.57	0.56
SWS2	28.02	6.6	9.5	0.3	1.94	0.25
SWS3	20.31	6.35	4.49	1	2.4	0.5
SWS4	16.21	6.8	7.36	0.3	1.68	0.47
SWS5	14.85	7	3.64	1.1	2.28	0.38
SWS6	13.28	6.85	2.23	2.7	2.75	0.42

conditions. It is also observed that the higher drainage frequencies observed in SWS5 and 6 are associated with higher runoff than other sub-watersheds.

Form factor (*Rf*):

The form factor for all the sub-watersheds varies from 0.3 to 2.7 (Table 3), indicating that all the studied sub-watersheds are more or less elongated. The elongated watershed with low value of Rf indicates that the basin will have a high flow for long duration. Flood flows of such elongated basins are easier to manage than the circular basins. In the Mullayar, the sub-watersheds are highly elongated on one side and more of circular pattern on the other side of the view. This heterogeneity may pose certain challenges in basin management as the different regions of the basin, owing to their inherent differences in drainage characteristics require unique management plans and practices.

Elongation Ratio (*Re*):

The elongation ratio for all the sub-watersheds varies from 1.68 to 2.75 which indicates normal relief on top of the basin and gentle slope (Table 3) in the lower reaches. The sub-watersheds SWS3 and 6 are more elongated according to the classification scheme of Schumn (1956).

Circularity Ratio (*Rc*):

The circulatory ratio is controlled by the length frequency of streams, geological structures, land use/land cover, climate, relief and slope of the watershed. In the present study (Table 3), the Rc value varies from 0.25 to 0.56 which shows that the sub-watersheds are almost elongated. The high value of the circularity ratio for SWS1 indicates the late maturity stage of its topography. This difference is due to the diversity of slope, relief and structural conditions prevailing in this sub-watershed.

4.2 Measurement of Intensity of Dissection

Drainage density (*Dd*):

Horton (1932) has introduced drainage density (Dd) as an expression to indicate the closeness of spacing of channels. Drainage density in all the sub-watersheds varies from 2.4 to 2.7 (Table 4). In general, it has been observed over a wide range of geologic and climatic types, the low drainage density is more likely to occur in regions of highly permeable subsoil material under dense vegetative cover. Wherever relief is low high Dd is favored in regions containing weak or impermeable subsurface materials, sparse vegetation and mountainous relief (Nag 1998). The low Dd values of the sub-watersheds SWS1, 3, 6, indicate that the presence of highly resistant, impermeable subsoil material with dense vegetative cover and low relief.

Table 4 Drainage density, texture and bifurcation ratios for Mullayar River

Sub watershed	Perimeter (km)	Drainage density	Drainage Texture	Bifurcation ratio Rb				Mean bifurcation ratio
				I/II	II/III	III/IV	IV/V	
SWS1	22.7	2.4	4.8	3.19	4.8	4	0	2.51
SWS2	37.2	2.6	4.2	5.00	3.6	7	1	3.31
SWS3	22.7	2.4	4.7	4.00	3.2	5	0	2.44
SWS4	20.8	2.8	4.1	6.67	6.0	0.5	2	3.03
SWS5	22	2.7	3.7	3	5.4	5	0	2.68
SWS6	20	2.4	3.5	4.12	4.3	0	0	1.67

4.3 Prioritization of the Sub-watersheds Based on the Morphometric Parameters

The compound parameter values of all the six sub-watersheds of the Mullayar watershed were calculated and the resultant prioritization ratings are presented in the Table 5. The sub-watershed SWS1 with a compound parameter value containing of 2.46 receives the highest priority, followed by SWS2 and 2. Highest priority indicates the greater degree of erosion in the particular micro-watershed and it warrants soil conservation measures. Thus, soil conservation measures can first be applied to sub-watershed SWS1 and then to the other sub-watersheds depending upon their priority. The final prioritized map of the study area and prioritization ranks of the sub-watersheds are shown in the Fig. 3.

Table 5 Prioritization results of morphometric analysis

Sub watershed	Dd	Fs	T	Rf	Rc	C	Rb	Compound parameter	Prioritization
SWS1	2.40	5.80	4.79	1.2	0.24	0.29	2.51	2.46	1
SWS2	2.56	6.60	4.17	0.3	0.07	0.27	3.31	2.47	2
SWS3	2.45	6.35	4.72	1.0	0.21	0.28	2.44	2.49	4
SWS4	2.81	6.80	4.14	0.3	0.07	0.26	3.03	2.48	3
SWS5	2.69	7.00	3.68	1.1	0.31	0.25	2.68	2.53	5
SWS6	2.44	6.85	3.51	2.7	0.76	0.31	1.67	2.61	6

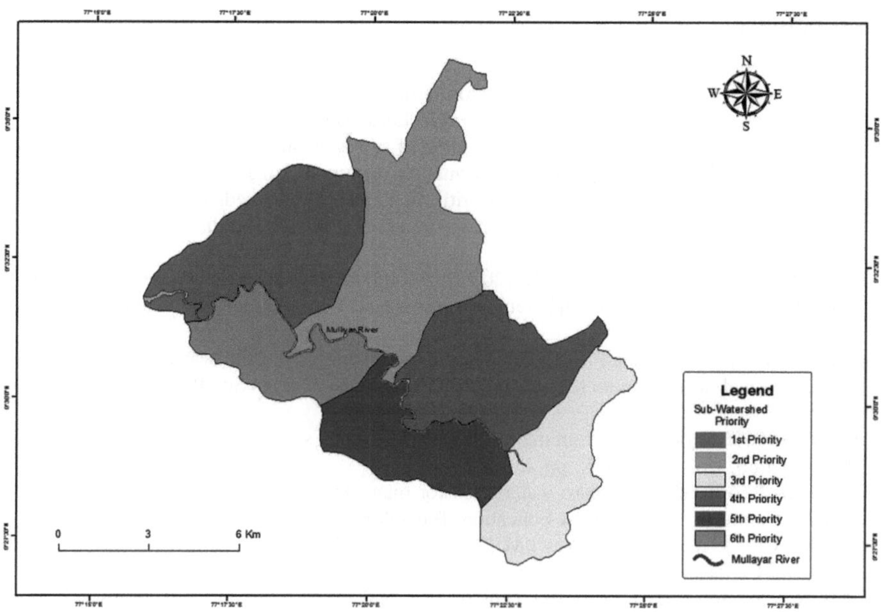

Fig. 3 Mullayar River: prioritized sub-watersheds

5 Conclusions

- The watershed prioritization is important for planning and implementation of river basin and water resource development and management programs.
- The morphometric characteristics of different sub-watersheds show their relative characteristics with respect to the hydrologic response of the watershed.
- Morphometric parameters coupled with thematic maps of drainage density and land use helped in identification of the soil-erosion prone sub-watersheds of the Mullayar Basin.
- Based on the results, it is concluded that the SWS1 sub-watershed of this basin warrants urgent implementation of erosion control measures.

Acknowledgments Authors are grateful to Prof. N. Sivagnanam, Professor, of Col. Colin Mackenzie Chair, Institute of Remote Sensing, Anna University, Chennai for his valuable inputs, encouragement and guidance.

References

Abha M, Dubey DP, Tiwari RN (2011) Morphometric analysis of tons basin, Rewa district, MadhyaPradesh, based on watershed approach. Earth Sci India 4(III):171–180

Abhijit MZ, Nagarajan R et al (2013) Prioritization of sub-watersheds in semi arid region, western Maharastra, India using geographical information system. J Am J Eng Res 02(10):128–135

Biswas S, Sudhakar S, Desai VR (1999) Prioritization of sub-watersheds based on morphometric analysis of drainage basin. A Remote Sensing and GIS approach. J Indian Soc Remote Sens 27 (3):155–166

Chopra R, Dhiman RD, Sharma PK (2005a) Morphometric analysis of sub watersheds in Gurudaspur district, Punjab using remote sensing and GIS techniques. J Indian Soc Remote Sens 33(4):531–539

Chopra R, Dhiman R, Sharma PK (2005b) Morphometric analysis of sub-watersheds in Gurdaspur district, Punjab using remote sensing and GIS techniques. J Indian Soc Remote Sens 33(4):531 –539

Clarke JI (1966) Morphometry from maps. In: Dury GH (ed) Essays in geomorphology. American Elsevier Pub. Co., New York, pp 235–274

CEE (2006) Prioritization of micro watersheds for management in Bijapur district of Karnataka, vol 20. Centre for Environment Education, Bangalore, p 89

Grohmann CH, Riccomini C, Alves FM (2007) SRTM—based morphotectonic analysis of the Pocos de Caldas alkaline massif Southeastern Brazil. Comput Geosci 33:10–19

Hadely RF, Schumm SA (1961) Sediment sources and drainage basin characteristics in upper Cheyenne river basin. United State Geological Survey water-supply paper, 1531-B, pp 137–196

Horton RE (1932) Drainage-basin characteristics transactions, Am Geophys Union 13(1):350–361

Horton RE (1945) Erosional development of streams and their drainage basins; hydrological approach to quantitative morphology. Geol Soc Am Bull 56:275370

Mesa LM (2006) Morphometric analysis of a subtropical Andean Basin (Tucuman, Argentina). Environ Geol 50:1235–1242

Miller VC (1953) A quantitative geomorphic study of drainage basin characteristics in the Clinch Mountain area, Virginia and Tennessee. Project NR 389042, Technical report 3, Columbia University, Department of Geology, ONR, Geography Branch, New York

Ripsarda F (2009) Morphometric and Landsliding "analyses in chain domain: the Roccella Basin, NE Sicily, Italy". Environ Geol 50:1235–1242

Schumn SA (1956) Evolution of drainage systems and slopes in Badlands at Perth Amboy, New Jersey. Geol Soc Am Bull 67:597–646

Singh S, Singh MC (1997) Morphometric analysis of Kanhar river basin. Natl Geogr J India 43 (1):31–43

Smith KG (1950) Standards for grading textures of erosional topography. Am J Sci 248(9):655–668

Strahler AN (1957) Watershed geomorphology. Trans Am Geophys Union 38(6):913–920

Strahler AN (1964) Quantitative geomorphology of drainage and channel networks. In: Fairbridge RW (ed) The encyclopedia of geomorphology, Encyclopedia of earth science studies. Mc. Graw-Hill Book Co., New York, pp 39–73

Sangita M, Nagarajan R (2010) Morphometric analysis and prioritization of using GIS and remote sensing: a case study of Odisha, India. Int J Geomat Geosci 1(2):501–510

Srinivasa VS, Govindaonah S, Home Gowda H (2004) Morphometric analysis of sub-watersheds in the Pawagada area of Tumkur district, South India using remote sensing and GIS techniques. J Indian Soc Remote Sens 32(4):351362

Toy TJ (1977) Hillslope form and climate. Geol Soc Am Bull 88:16–22

Tucker GE, Bras RL (1998) Hillslope processes, drainage density and landscape morphology. Water Resour Res 34:2751–2764

Assessment of the Water Resource of the Yodo River Basin in Japan Using a Distributed Hydrological Model Coupled with WRF Model

K.L. Shrestha and A. Kondo

Abstract The Yodo River basin provides water resource to the highly populated areas of Kinki, Japan. Similar to other river basins located elsewhere, the Yodo River basin is also vulnerable to negative impacts of climate change. Since accurate prediction of extreme events is essential for assessing the impact of climate change, any integrated monitoring and prediction system should be based on the hydro-meteorological system. For this goal, dynamic downscaling of the meteorological data by using coupled mesoscale hydrometeorological modeling approach to simulate the local and regional effects on water resources of the basin, has been attempted. Coupled model consisting of WRF mesoscale meteorological model and distributed hydrological model, along with a simplified dam model, at high-resolution was used to simulate the response of the Yodo River basin to atmospheric forcings in one-way coupling mode. The distributed hydrological model is shown to be capable of simulating the basin hydrology of the Yodo River basin by replacing the atmospheric forcings from observation station data with the high-resolution gridded hydrometeorological variables from WRF mesoscale meteorological model.

Keywords Distributed hydrological model · Yodo River basin · Coupled model · WRF model

1 Introduction

The IPCC (2007) has projected a substantial impact of human-induced climate change on the water resources at global and regional scales. In river basins affecting the urban population, Owing to the urban pressures, the effects of climate change

K.L. Shrestha (✉)
Kathmandu University, Dhulikhel, Kavre, Nepal
e-mail: kundan@ku.edu.np

A. Kondo
Osaka University, 2-1 Yamadaoka, Suita, Osaka, Japan

Mu. Ramkumar et al. (eds.), *Environmental Management of River Basin Ecosystems*, Springer Earth System Sciences, DOI 10.1007/978-3-319-13425-3_8

are more pronounced in river basins as the urban centres are usually concentrated along the river courses and the urban regions have high density of population and a limited supply of water. However, the local impacts of urban and meteorological forcings on the water cycles at the river basin scales are difficult to be estimated using the climate models of the global scale. The spatial resolutions of such models are very coarse and the fine spatial variability of the river basins cannot be represented in such models. The regional models can better represent the complex orographic features and land-use distribution of a river basin. Moreover, regional and local atmospheric phenomena like sea breeze, storm and mesoscale flows can also be represented in the regional scale models.

Basin-scale physically-based distributed hydrological models have been proposed (Freeze and Harlan 1969) for documenting the increasing complexities of land-use changes, anthropogenic activities, and vegetation changes etc. (Abbott et al. 1986). The distributed models require a large set of data related to soil, vegetation, geography, etc. and they are computationally more expensive than the lumped models. Computationally cheaper models have been developed to integrate the hydrological processes with water management strategies that directly affect water resource issues such as water supply, agriculture activities and water quality problems (Arnold et al. 1998). The physically-based distributed hydrological models are important to investigate land-use change scenarios (Cai 1999; Mango et al. 2011; Miller et al. 2002; Cornelissen et al. 2013) and climate change scenarios (Kiem and Verdon-Kidd 2011; Kim et al. 2010; Mango et al. 2011; Sato et al. 2012, 2013; Thompson et al. 2013; Todd et al. 2011). Information on future hydroclimatic changes have been provided by GCM to run different hydrological models (Todd et al. 2011; Thompson et al. 2013) to assess different hydrological changes in the future. Very high resolution GCMs have also been used to provide atmospheric forcing at the resolution of 20 km (Sato et al. 2012, 2013; Kim et al. 2010) to simulate the effect of climate change on river discharge and snow melt. For limited-area applications at regional scale, mesoscale atmospheric models have been coupled with different hydrological models to evaluate the extreme hydrological events as well as the impact of hydroclimatic changes on the hydrological cycle (Cornelissen et al. 2013; Kiem and Verdon-Kidd 2011).

The advances in distributed models have helped in the implementation of better water management plans (Hooper 2011; Liu et al. 2008). The basin-scale models have been simplified and made computationally cheaper by using conceptual hydrological models (Arnold et al. 1998; Malone 2014; Liu et al. 2008; Singh et al. 2005), and their applications in the water quality research at basin scale have contributed to water pollution abatement strategies (Cai 1999; Malone 2014; Abbott et al. 1986). Several regional modeling techniques have been employed to simulate the hydrological cycle and water balance for many purposes like flood forecasting, river discharge forecasting, reservoir operational management, rainfall-runoff relationship, effect of climate change on water resources and ecosystem, groundwater flows, water quality, orographic effect on basin discharge, urban drainage flow, etc. Prediction and forecasting systems for Japanese river basins (Sayama et al. 2005; Tachikawa et al. 2007) and physically-based distributed hydrological models have been used in the Yodo River basin (Shrestha et al. 2005; Tachikawa et al. 2007) and in other urbanized basins (Jia et al. 2001, 2002). Dam operation models have also been incorporated in the hydrological models (Sayama et al. 2005; Tachikawa et al. 2007). The hydrological processes in the paddy fields have also been incorporated in the hydrological models (Sakaguchi et al. 2014). Water quantity and water quality assessments (Ikebuchi et al. 2006; Kojiri et al. 2002, 2008; Nawahda et al. 2005) have been carried out in Japan using various hydrological models.

Many researchers have focused on how to effectively use atmospheric forcings like precipitation in high spatial resolution settings of the hydrological models (Mölders and Raabe 1997; Lin et al. 2006; Verbunt et al. 2006; Seuffert et al. 2002; Hong and Lee 2009). One-way coupling of meteorological and hydrological models is mostly used to drive hydrological models by the hydrometeorological variables like air temperature and precipitation generated by meteorological models (Mölders 2005). Due to the requirement of high resolution spatial and temporal scales for coupling atmospheric and hydrological model, many computational problems arise during such integrated hydrometeorological studies. These computational limitations are also being overcome with the use of more powerful computing resources and better atmosphere-hydrology coupling techniques. Examples of the coupled hydrometeorological models are: Integrated Regional Scale Hydrologic/Atmospheric Model (IRSHAM) (Yoshitani et al. 2001); "System for Prediction of Environmental Emergency Dose Information Multi-model Package" (SPEEDI-MP) with MM5 mesoscale meteorological model as the atmospheric model, Princeton Ocean Model (POM) as the oceanic model, and SOLVEG atmosphere—soil—vegetation model as the land surface model (Chino and Nagai 2003); WRF mesoscale model coupled with a hydrological model (Yoshikane et al. 2005); hydrostatic Swiss Model (SM) and non-hydrostatic Alpine Model (aLMo) coupled with Precipitation Runoff EVApotranspiration Hydrotope (PREVAH) distributed hydrological model (Verbunt et al. 2006); Canadian atmospheric Mesoscale Compressible Community Model (MC2) coupled with the Chinese Xinanjiang hydrological model (Lin et al. 2006); integrated Hydrologic Model System (HMS) (Yu 2000); Hydrologic Model System (HMS) coupled with Regional Climate Model (RCM) (Yu et al. 2002); MM5 mesoscale meteorological model coupled with Distributed Hydrology Soil Vegetation Model (DHSVM) (Westrick and Mass 2001); MM5 mesoscale meteorological model coupled with Soil and Water Assessment Tool (SWAT) (He et al. 2009); MM5 mesoscale meteorological model coupled with distributed Water Simulation Model (WaSiM) (Kunstmann and Stadler 2005); and integrated land surface and atmosphere using MM5 mesoscale model (Kunstmann and Jung 2003).

Japan is facing several water related problems due to the impacts of climate change. The Ministry of Land, Infrastructure, Transport and Tourism (2008), Japan has reported that the snow cover in upstreams of many of the major rivers of Japan will decrease and consequently the water reservoir levels will decrease. The frequencies and intensities of floods are also projected to increase due to the anticipated higher intensities of precipitation. For example, in a future environmental scenario, yearly average maximum daily rainfall is predicted to increase by 11 %

under the A1B scenario (2080–2099 compared with 1979–1998) in Tokyo and surrounding regions.

The Water Resources Development Promotion Law, Ministry of Land, Infrastructure and Transport (Ministry of Land, Infrastructure, Transport and Tourism, Japan 2008) has termed the river systems needing water supply measures due to urban and industrial development as "river systems for water resources development and encouraged application of integrated resource management in these river systems. Seven such river systems have been identified and one of them is the Yodogawa or the Yodo River system (Fig. 1) in the Kinki region of Japan. It is situated in the southcentral region of Honshu, and has a densely populated urban region. The Kinki region includes many big cities like Osaka, Kyoto and Nara. Yodo River originates from the Lake Biwa (670.4 km^2), which is the biggest freshwater lake in Japan. The Yodo River is 75 km long and the Yodo River basin covers an area of 8,240 km^2.

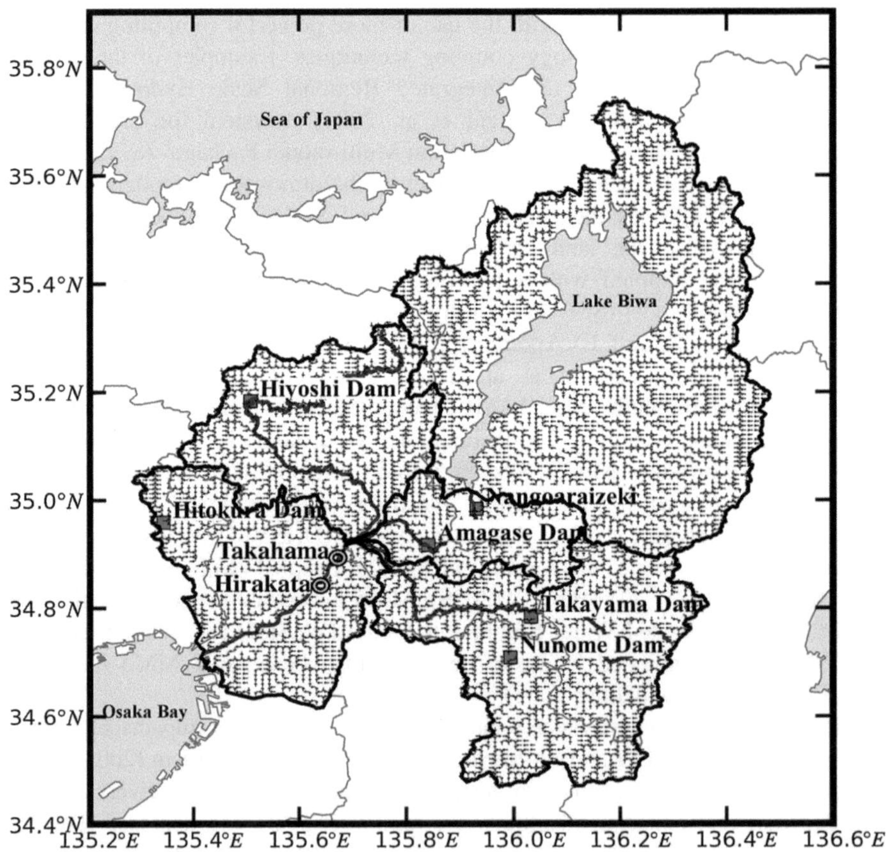

Fig. 1 Yodo River basin grid structure with flow directions. Square symbols are dams and circles represent two observation stations

There are several important factors affecting the water resources of the Yodo River basin. The Yodo River basin experiences many seasonal typhoons and heavy rainfall. Discharge in the Yodo River basin is also affected by snow melting. The Yodo River basin has also witnessed several cases of water supply shortage. Water demand has also increased due to urbanization and regional development. The Kinki region is located mostly below the river water levels and hence the risk of embankment collapse and flood is substantial. Nearly 95 % of the highly urbanized region of Osaka City has risk of flooding. These characteristics necessitate integrated water resource management study of the Yodo River basin. In this paper a hydrometeorological prediction system to simulate water cycle of the Yodo River basin has been attempted through the coupled modeling approach. The coupled model considers the land-use distribution in Yodo River basin and includes a conceptual linear storage model for subsurface transport, a tank model for paddy fields, and a simple dam model to simulate the operations of the complex dam network. This model is also coupled with the WRF mesoscale atmospheric model at high-resolution to enable the application of the model to future hydroclimatic scenarios, water resource assessment and water quality research.

2 Models and Modeling Methods Used in the Study

2.1 WRF Mesoscale Model

The Weather Research and Forecasting (WRF) is a fully compressible and non-hydrostatic mesoscale numerical weather prediction and atmospheric simulation system. It has a terrain-following hydrostatic pressure coordinate system and a flexible, modular and portable code design. It is a successor to the widely used MM5 model (Skamarock et al. 2005). The WRF can be applied at local scales to global scales, and many options of cumulus parameterization, microphysics, radiation, land surface, boundary layer are available in this model. This model was adapted for high-resolution meteorological modeling of the Yodo River basin.

2.2 Hydrological Modeling

2.2.1 Distributed Hydrological Model

The Yodo River basin is a mesoscale river basin. Hence, the impacts of meteorological and climate changes on the water resources can only be studied by high-resolution hydrological modeling. The mesoscale hydrological modeling requires finer spatial and temporal resolutions than the large-scale hydrological modeling. The land-use and precipitation data are also required to have high resolution.

For the present study, high-resolution distributed hydrological modeling approach was used to simulate the water flow and river discharge in response to the changing hydrometeorological variables in the basin. River flow, surface runoff, sub-surface groundwater flow, water intake, and dam reservoir operations in the Yodo River basin were modeled using a high-resolution distributed hydrological model based on a rainfall-runoff model known as Hydrological River Basin Environment Assessment Model (HydroBEAM) (Kojiri et al. 2002).

In the hydrological model (Fig. 2), the basin has been divided into terrain grids (that acted as a single unit basin) with horizontal resolution of 1 km × 1 km. The grids also acted as river flow network (Fig. 1) in which corresponding upstream grids were allocated for each of the downstream grids. The surface runoff in each grid flow into the corresponding downstream river grid. The terrain grid was divided into five land-use categories (fields, forest, urban, paddy and water) (Fig. 3). The basin units or the grids were vertically divided into four soil layers (A, B, C, and D from top to bottom respectively). Surface energy balance model (Sect. 2.2.2) was applied at each grid to estimate the rainfall, snowmelt and evapotranspiration. The surface runoff, runoff from paddy fields, ground water flow, water intake and release, and lateral water flows were simulated by HydroBEAM. The lateral movement of water in soil layers, except the layer D, flow into the river channel. The sewerage network, wastewater and irrigation canals in paddy fields were also included in the model.

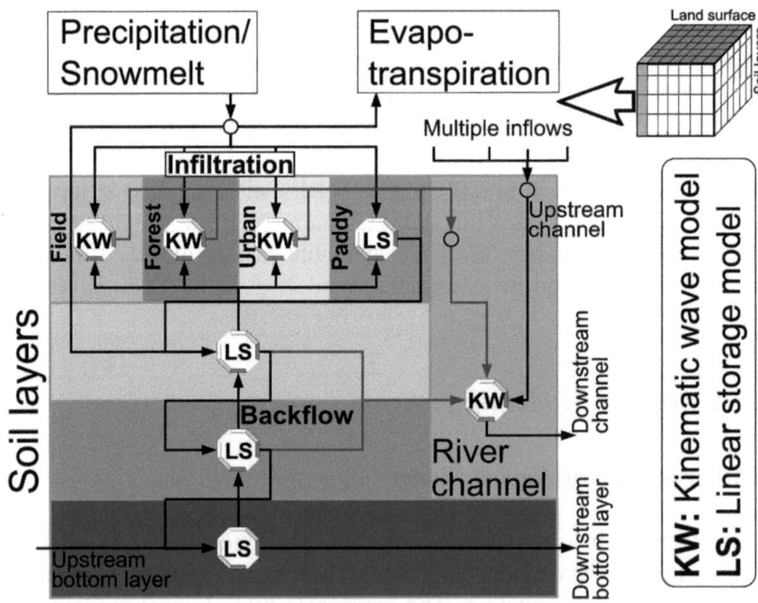

Fig. 2 Rainfall-runoff hydrological model framework

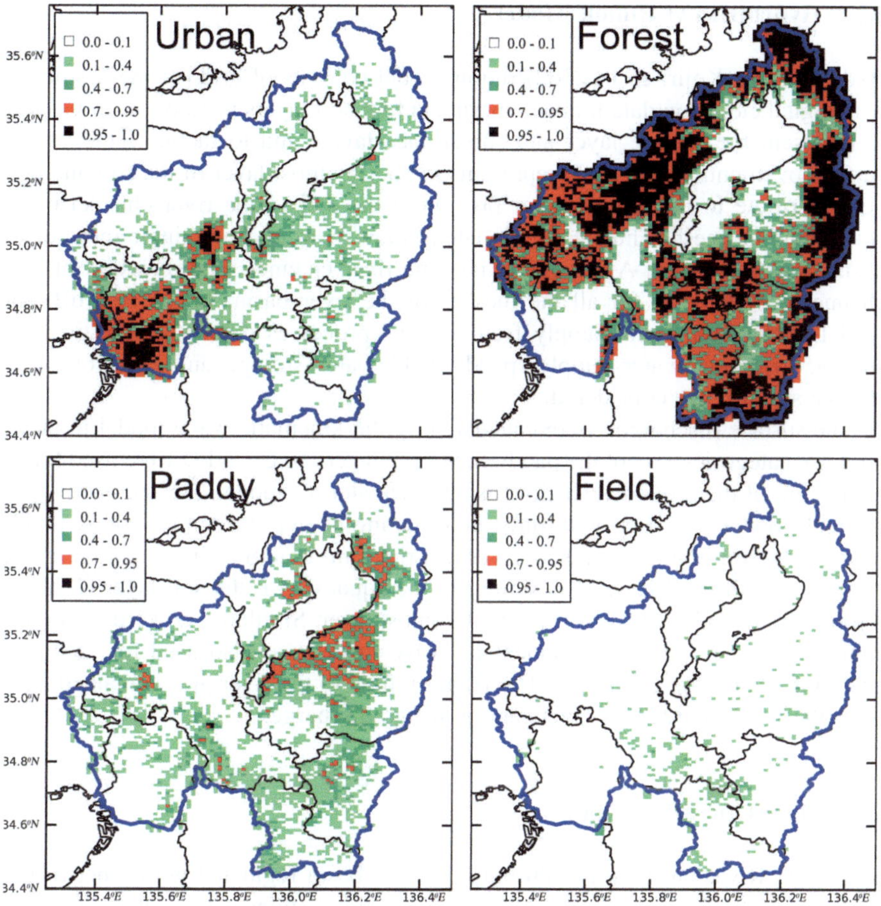

Fig. 3 Land use fraction of different land use categories in Yodo River basin

2.2.2 Surface Energy Balance Model

Using positive sign convention for incoming radiative fluxes and outgoing non-radiative fluxes from the surface, the surface energy balance equation for the basin was modeled with the bulk transfer approach to calculate the surface moisture flux. The ground heat flux was parameterized as having the diurnally averaged value of zero.

Snowfall and snowmelt were modeled by an energy balance at a single layer of snow. When near-surface air temperature was less or equal to the critical temperature, snowfall was assumed.

2.2.3 HydroBEAM Runoff Model

HydroBEAM (Kojiri et al. 2002) runoff model consists of gridded mesh structure with digital elevation data for each mesh and four soil layers namely, A, B, C and D to represent the surface layer and sub-surface layers and to facilitate the vertical water movement. The layer D represents the bottommost layer of the soil and does not contribute to the lateral water flow from the soil to the river channel of the corresponding mesh. Thus water has lateral movement in every single mesh of the basin from soil layers A, B, and C into the river channel. The direction of river channel flow was set for all the meshes of the river basin. In the HydroBEAM model, water intake and supply from various processes namely, domestic water supply, irrigation water supply, paddy field water storage and reservoir water storage systems are considered.

The surface discharges were calculated by the kinematic wave model in all the land-use categories except the paddy field. Soil layers B, C, and D were used for the simulation of the groundwater flow using linear storage model. For the paddy field, the tank model was used with three holes acting as upper ridge overflow, middle lateral ridge infiltration, and lower vertical ridge infiltration. Linear storage model equation for the soil layer D is similar to the equation for the soil layer C, but no return flow was allowed as it is the bottommost layer. Similarly, no vertical outflow from D was considered. The kinematic wave model was also used for each river grid in the hydrological model. The lateral inflow to the river grid contained flows from the A, B, and C soil layers and drainage discharge.

2.2.4 Dam Operation Model

The Yodo River basin is a multi-purpose river basin regulated with a network of dam reservoirs of various sizes that is used for flood attenuation, water supply and power generation. Six large dams have been selected for this study (shown in Fig. 1). The size and capacity of the six dams are shown in the Table 1.

Table 1 Database of six major dams of the Yodo River basin for 2006

Dams	Intake area (km^2)	Total volume ($\times 10^6$ m^3)	Flood control storage[a] ($\times 10^6$ m^3)
Nango-araizeki	3,848	27500.0	2221.0[b]
Amagase	4,200	26.3	20.0
Takayama	615	56.8	35.4
Nunome	75	17.3	6.4
Hiyoshi	290	66.0	42.0
Hitokura	115	30.8	17.5

[a] From normal water level at flooding season to surcharge water level

[b] Capacity of Biwa lake from standard water level of −0.3 m to design high-water level of 1.4 m

The effect of dam reservoirs was included as actual outflow boundary condition for the past scenarios. But this method cannot be used for the future scenarios. To develop a generalized dam model, a simplified operation rule was modeled to simulate the actual dam operation. Since the regulation of water flow in the rivers needs to be considered for correct prediction of the outflow of the rivers, a simple dam model with fluctuating water level was constructed, in which the desired water levels were set according to the normal water levels and flood water levels stipulated by the present dam operation rules (Table 2). The maximum allowed water level was the surcharge level and the water level was not allowed to decrease beyond the minimum level. According to the water levels set for the flooding and non-flooding seasons, the outflows of dams were adjusted to maintain the required water levels. The weather reports and forecasts are referred by dam operators to predict the floods according to the precipitation rate in the catchments of dams. Given cognizance to this, in the simplified dam model, the past and future precipitation forecasts were directly obtained from the meteorological input from the mesoscale model. The water levels were accordingly lowered to adjust for the flood. Though there are several modes of operation depending upon the intensity of precipitation and estimated flood (Sayama et al. 2005), in the present study, only a single operation rule was applied in the simplified model, so that the water level is adjusted if the precipitation intensity predicted by the meteorological model for the next day is more than 50 mm/day in the terrain grid of the dam.

Table 2 Different water levels of six major dams of Yodo River basin for 2006

Dams	Water levels (EL m)[a]	Normal (non-flooding)	Normal (flooding)[b]
	Surcharge		
Nangoaraizeki	1.4	0.3	−0.2 BSL m
(6/16–8/31)			
−0.3 BSL m			
(9/1–10/15)			
Amagase	78.5	78.5	72.0
Takayama	135.0	135.0	117.0
Nunome	287.3	284.0	280.6
			(6/16–8/15)
			279.2
			(8/16–10/15)
Hiyoshi	201.0	191.4	178.5
Hitokura	152.0	149.0	135.3

[a] From Dam database of ministry of land, infrastructure and transport, Japan
[b] Flooding season is 6/16–10/15 if not stated

2.3 Input Data for the Hydrological Model

2.3.1 Observed Precipitation Data

Thiessen polygon method (Thiessen 1911) was used for the regions having precipitation observation network. This method may underestimate the overall rainfall in the regions having a complex terrain and orographic precipitation. There are many relations used to find correlation between the precipitation, elevation (Daly et al. 1994; Drogue et al. 2002; Mölders et al. 1996; Running et al. 1987), and wind (Suzuki et al. 2003). To generate gridded precipitation for distributed hydrological modeling, the SDP (Surface Daily Product) observation station data archived by the Japan Meteorological Agency were processed. The SDP observation data also contained other hydrometeorological variables namely, air temperature, wind speed, water vapor pressure, sunshine duration, etc. Ten observation stations were used to calculate the Thiessen polygons. Then the daily average precipitation data were calculated using the Thiessen polygon method. Terrain height correction method was then used to obtain the areal average precipitation data having 1-km resolution. The corrected SDP precipitation data is termed as 'SDP data' in this paper.

2.3.2 Other Input Data for Evapotranspiration Model

Besides precipitation, near-surface air temperature, near-surface wind speed, surface pressure, water vapor pressure, and incoming shortwave radiation flux at surface are required to run the evapotranspiration model. The input data needed for all these input variables were also gridded into the Yodo River basin domain by using the Thiessen polygon method.

Near-surface air temperature data recorded at the SDP observation stations were provided by the Japan Meteorological Agency. These were used for the calculation of sensible heat flux and latent heat of vaporization. As in the case of precipitation, terrain height correction was applied to air temperature data. The correction coefficient used to consider the effect of terrain on air temperature was 0.65 K per 100 m. Water vapor pressure was also obtained from the SDP observation data, and it was used in the calculation of air density and near-surface specific humidity in the bulk transfer equation for moisture.

Actual sunshine duration obtained from SDP observation data was used to calculate daily incoming shortwave radiation flux at surface. Empirical relationship between incoming solar radiation at surface and extraterrestrial radiation is:

$$R_S\downarrow \; = \; R_{S0}\downarrow(a + b\,N/N_0) \tag{1}$$

where, $R_S\downarrow$: Incoming shortwave radiation $R_{S0}\downarrow$: Extraterrestrial radiation, N: Actual sunshine hours, N_0: Day length, a and b: Empirical numerical parameters.

Table 3 Parameters for evapotranspiration model in Yodo River basin

Land-use	Bulk transfer coefficient	Evaporation efficiency	Albedo
Forest	0.002	0.7	0.1
Paddy—irrigation period	0.005	0.7	0.15
Paddy—non-irrigation period	0.0008	0.3	0.2
Crop field	0.0008	0.2	0.2
Urban	0.0001	0.2	0.3
Water bodies	0.001	1.0	0.06

Irrigation period 120–270th days of a year

The regressional parameters a and b were set at 0.244 and 0.511 respectively. The data on actual sunshine hours were provided in the SDP observation station data. $R_{SO}\downarrow$ is the assumed amount of solar radiation coming to the surface without the influence of the atmosphere. N_0 is day length, or the theoretical maximum duration of the sunshine between dawn and sunset. Near-surface speed was also obtained from the SDP observation data, and it was used in the bulk transfer equation for moisture.

Few important parameters for the five types of land-use surfaces used in the surface energy balance equation of the evapotranspiration model are shown in the Table 3. Since the surface properties of the paddy fields change during irrigation and non-irrigation periods, different sets of parameters were used for these two periods. For the Yodo River basin, the irrigation period was estimated as 120–270 days of a year. Anthropogenic heat flux was set at an average value of 80 Wm^{-2} for the Yodo River basin area.

2.3.3 Data and Parameters for Runoff Model

Precipitation and evapotranspiration quantities simulated by the evapotranspiration model were used as input data in the HydroBEAM runoff model. Water intake and release for water supply, water use, hydroelectric plants, and irrigation purposes were also used as inputs into the runoff model. The sewerage and wastewater flow data were used for the corresponding grids. Dam operation rules (Sect. 2.2.4) were used to control the water level and storage in the six majors dams (Fig. 1). Important soil layer and model parameters used in the runoff model are shown in the Table 4.

2.4 Domain and Grid Structures

The Yodo River basin was gridded into 8,242 square grids each having an unit area of 1 km × 1 km resolution (Fig. 1). Each grid in the basin contains terrain data and

Table 4 Soil layer and model parameters for runoff model in Yodo River basin

Parameter		Value
Equivalent roughness ($m^{-1/3}s$)	Paddy	–
	Crop field	0.300
	Forest	0.700
	Urban	0.030
	River channel	0.035
Direct runoff percentage from surface	Paddy	1.0
	Crop field	0.21
	Forest	0.3
	Urban	0.737
Soil layer depth (m)	A	0.3
	B	1.0
	C	2.5
	D	10.0
Porosity (%)	All soil layers	10.0
Linear storage model constant (1/d)	B soil layer horizontal runoff coefficient	0.03
	C soil layer horizontal runoff coefficient	0.007
	D soil layer horizontal runoff coefficient	0.0039
	B soil layer vertical runoff coefficient	0.11
	C soil layer vertical runoff coefficient	0.013
	D soil layer vertical runoff coefficient	0.0
Tank model constant	A1: runoff coefficient (1/d)	1.0
	Z1: hor. runoff boundary height (mm)	300.0
	A2: hor. runoff coefficient	0.17
	Z2: hor. runoff boundary height	DPD[a]
	A3: hor. runoff coefficient	0.0
	Z3: hor. runoff boundary height	0.0

[a] Desired ponding depth

river network flow direction that channels water flowing in the rivers. The evapotranspiration model was used to solve surface energy balance in each of the basin grids. As the lake model has not been used to simulate the hydrodynamics of the Lake Biwa. So, in the rainfall-runoff model, only 7,557 grids, excluding those of Lake Biwa, were used in the simulation. The tributaries of the Lake Biwa were directly let out from the outlet of the lake.

The distributed hydrological model was used to simulate the response of the Yodo River basin to meteorological forcings for one-year period (the year 2006) using the evapotranspiration model, rainfall-runoff model and dam reservoir model. The simulation period was selected due to the availability of reliable observation data during that period.

2.5 Coupled Hydrometeorological Modeling

2.5.1 Coupling Approach

The modeled hydrometeorological data are the inputs to the hydrological model. One simple way of obtaining the hydrometeorological data is to directly downscale the global climate data obtained from the General Circulation Models (GCMs). However, the very fine spatial resolution required by the hydrological models cannot be realistically obtained from the very coarse GCM outputs. To circumvent this problem, atmospheric models with better spatial and temporal resolutions could be utilized. Regional and mesoscale meteorological models are better suited for providing more accurate and high-resolution input data to the distributed hydrological models. Thus the regional and mesoscale models dynamically downscale the global climate variables to regional and local scales.

Since the Yodo River basin has a mesoscale basin structure, the distributed hydrological modeling at a spatial scale of 1 km × 1 km was found to be satisfactory. The WRF model was used to produce all the meteorological variables required by the hydrological model. The hydrological simulation of the Yodo River basin was then carried out for 2006 by coupling the WRF model with the distributed hydrological model.

WRF mesoscale model was coupled one-way in offline mode with the hydrological model of the Yodo River basin. WRF mesoscale model was first run to simulate the hydrometeorological variables namely, air temperature, precipitation, surface pressure and incoming solar radiation. Then, the required hydrometeorological variables were downscaled and converted into the data format required by the hydrological model.

2.5.2 Parameters and Input Data

The options used in the WRF model are shown in the Table 5. The US NCEP (National Centers for Environmental Prediction) Global Analyses data, available on 1.0 × 1.0° grids continuously at every 6 h since the year 1999 (http://dss.ucar.edu/datasets/ds083.2/), were chosen as boundary and initial conditions for the WRF simulations along with the SST data. NCEP Global Analyses data for the Yodo River basin were dynamically downscaled to 3-km grid domain (Fig. 4).

All the input variables were made available from the WRF output except the water vapor pressure. Air temperature at 2 m was used as near-surface air temperature. Similarly, horizontal wind vector components, u and v, at 2 m were used to calculate the near-surface wind speed. From the WRF output, surface pressure (P) and water vapor mixing ratio at 2 m (q) were extracted, and then water vapor pressure was calculated. In the hydrological model, sunshine duration was used to calculate daily incoming shortwave radiation flux at the surface. Horizontal wind vector components, u and v, at 2-m height were used for the calculation of

Table 5 Specifications of the WRF model

Parameter/option	Value
Nesting	One-way with 3 domains
Domain-1 size	50 × 50 mesh with 27-km mesh size
Domain-2 size	46 × 46 mesh with 9-km mesh size
Domain-3 size	52 × 52 mesh with 3-km mesh size
Vertical grid	35 full levels with top level at 5 kPa
Microphysics	WSM 3-class simple ice
Cumulus	Kain-Fritsch scheme
Planetary boundary layer	YSU scheme
Land surface	Noah land surface model
Longwave radiation	RRTM scheme
Shortwave radiation	Dudhia scheme

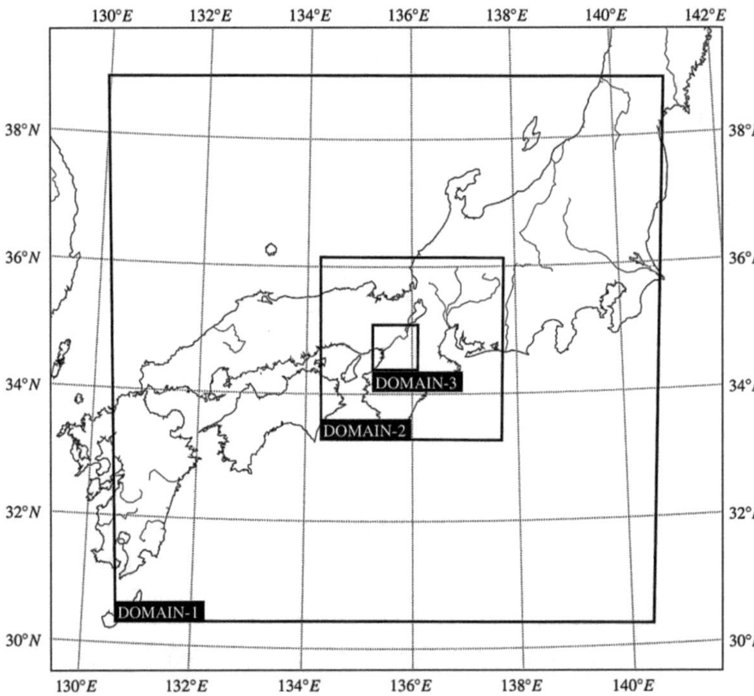

Fig. 4 Nesting of three domains in WRF model (The total area is Domain-1,'d02' is Domain-2 and 'd03' is Domain-3)

near-surface wind speed. First, the 10-m wind vector components were used to find wind speed at 10-m height. Then, using the wind profile based on the Monin and Obukov similarity theory, 10-m wind speed was converted to 2-m wind speed (as followed in the work of Högström 1988).

2.5.3 Data Processing

The grid structure of the Yodo River basin in the hydrological model is represented by nearly 1 km × 1 km GIS grid system, which is also known as Japanese 3D Mesh. In the hydrological model, each grid was given a unique Mesh ID to represent the sequence of channel flow in the basin. Besides a Mesh ID, these grids have 3D Mesh ID as well. A list of Mesh ID and 3D Mesh ID for 8,242 grid mesh of the Yodo River basin was created. Using GRASS GIS (http://grass.itc.it/), geographical coordinates (latitude/longitude) of the basin grid meshes were tabulated against corresponding Mesh ID using the 3D Mesh ID data.

A Fortran program called read_wrf_nc (described in http://www.mmm.ucar.edu/wrf/OnLineTutorial/Tools/read_wrf_nc.htm) is available in the WRF software to look into the NetCDF binary format (http://www.unidata.ucar.edu/software/netcdf/) output data produced by the WRF. This program is slightly modified to obtain X/Y coordinates of WRF for each of the grid mesh of the Yodo River basin. The latitude/longitude information from the previous step for each Mesh ID of the Yodo River basin were used as input to read_wrf_nc program to produce a list containing Mesh ID and corresponding WRF X/Y coordinates.

The required output data from the WRF were extracted from the WRF output files using a Python program that utilizes Climate Data Management System (CDMS, http://www2-pcmdi.llnl.gov/cdat) library to access the NetCDF variables. Since the WRF precipitation data Domain-3 (Fig. 4) were at 3-km resolution, they were downscaled to 1-km resolution required by the hydrological model. Other variables were obtained at basin grids corresponding to the nearest grid in the WRF output data. Then the hourly meteorological data for all the 1-km × 1-km grid mesh of the Yodo River basin were averaged into daily average values and then written into the input data files required by evapotranspiration model of the hydrological model.

3 Results and Discussion

3.1 Validation of the Hydrological Model

3.1.1 Validation of the River Discharge

Table 6 shows the comparison between the modeled and observed annual outflow from the six major dams in the Yodo River basin. The observed outflows from the dams have been provided by Ministry of Land, Infrastructure and Transport, Japan (http://www2.river.go.jp/dam/index.html, in Japanese). The outflow for the Nan-goaraizeki dam has not been included because of data unavailability. The hydrological model reasonably reproduced the actual observed outflows in all the dams. The dam outflows were slightly overpredicted at the Hiyoshi and Nunome dams but some underpredictions of the outflows we observed at the Amagase and Hitokura dams.

Table 6 Comparison of modeled and observed annual outflow from dams in Yodo River basin in 2006

Dam	Outflow ($m^3 \times 10^6$)	
	Observed	From model
Amagase	3523.7	3494.2
Hiyoshi	360.9	396.2
Takayama	474.6	474.3
Nunome	56.5	61.6
Hitokura	95.2	50.3

The slight overpredictions in the annual outflow amount at Hiyoshi and Nunome dams can also be visualized from the hydrographs in Fig. 5 (shown by SDP lines). The overall slight overprediction of outflow at the Hiyoshi and Nunome dams can be attributed to some peaks larger than the observed discharge peak values in the rainy season (June and July). The largest peak discharge at the Nunome has been overpredicted by nearly 20 m^3/s.

At the Amagase dam, though the annual discharge has been slightly underpredicted by the hydrological model, the maximum discharge peak in July has been overpredicted by more than 1,000 m^3/s and the ensuing low discharge periods were also generally overpredicted by the model. Some of the overpredicted peak discharges shown by the hydrological model may be attributed to the under-representation of peak attenuations during actual dam operations at Nangoaraizeki and Amagase dam reservoirs that regulate the outflow from the Lake Biwa into the Uji River. Another reason for the overprediction in the Amagase dam peak discharges may be that the lake hydrodynamics of the Lake Biwa has not been considered in the present model, and hence, the outflow from the Lake Biwa is possibly overpredicted due to the lack of consideration of residence time in the lake. Though the peak discharge in July has also been overpredicted at all the dams except at the Hitokura dam, most of the peak discharges were modeled by the hydrological model within the acceptable limits of error.

The Water Information System database provided online by the Ministry of Land, Infrastructure and Transport, Japan (http://www1.river.go.jp/, in Japanese) contains many observation stations in the Yodo River basin. But, most of the observation stations had missing data in the year 2006. Hence, only two observation stations, viz., Hirakata and Takahama (Fig. 1) were validated with the observed data. In the Hirakata station, only the high discharge peak in July 2006 was available in the observed river discharge data.

Takahama station is situated in the Yodo sub-basin and receives river discharge from Biwa, Uji, Katsura and Kizu sub-basins (Fig. 6). Hirakata station lies further downstream than the Takahama station in the Yodo River. Though the maximum river discharge in July could not be compared with the observed data at the Takahama station, rest of the river discharge in the year 2006 is well-simulated by the hydrological model in the Takahama station with some slight underpredictions in the spring season. The highest peak discharge during July 2006 has been reasonably predicted at the Hirakata station with the observed discharge being overpredicted by nearly 1,000 m^3/s. This overprediction in maximum peak discharge in

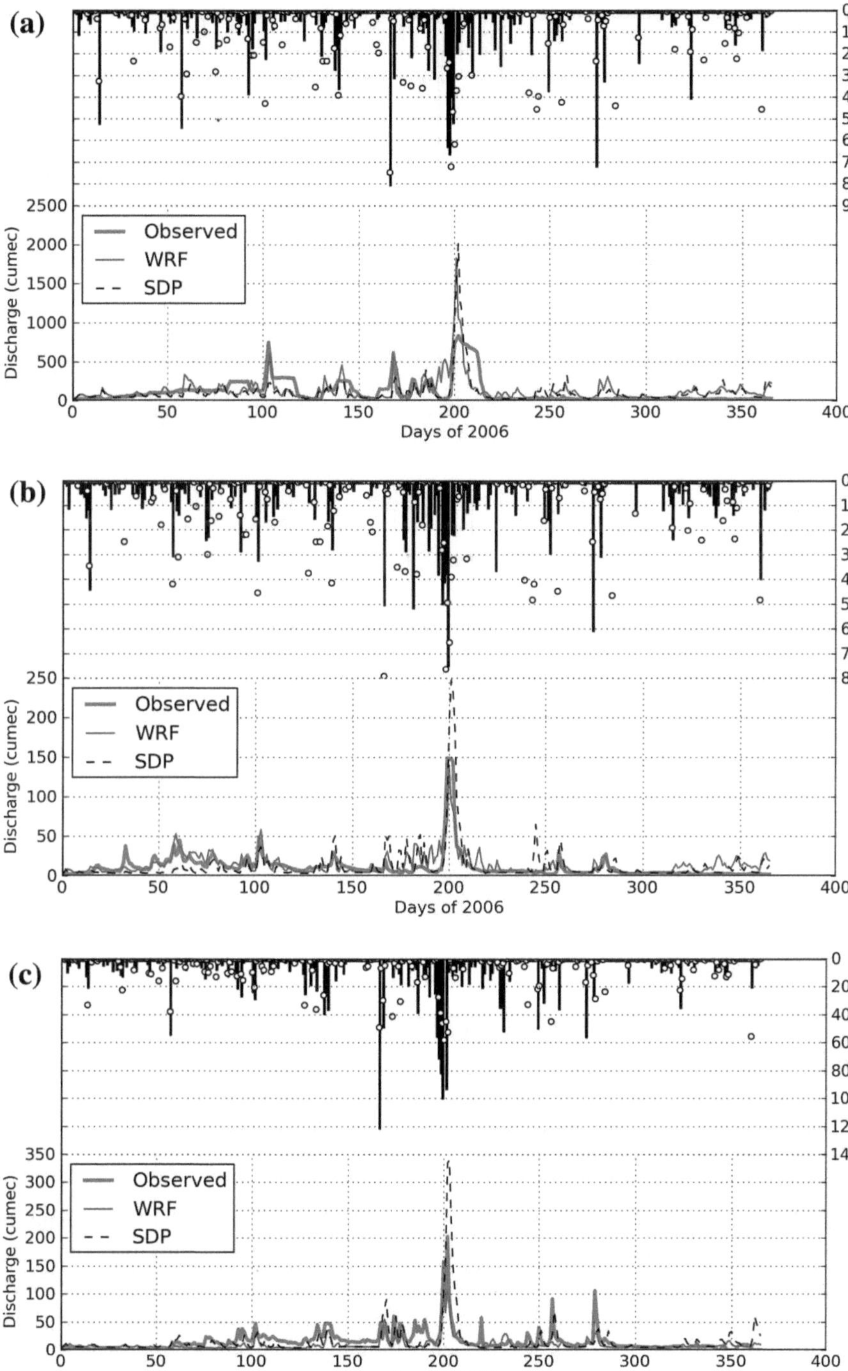

Fig. 5 Hydrographs for dams in Yodo River basin using coupled hydrometeorological model (Circles represent observed precipitation and bars represent precipitation from WRF model.) **a** Amagase dam, **b** Hiyoshi dam, **c** Takayama dam, **d** Nunome dam, **e** Hitokura dam

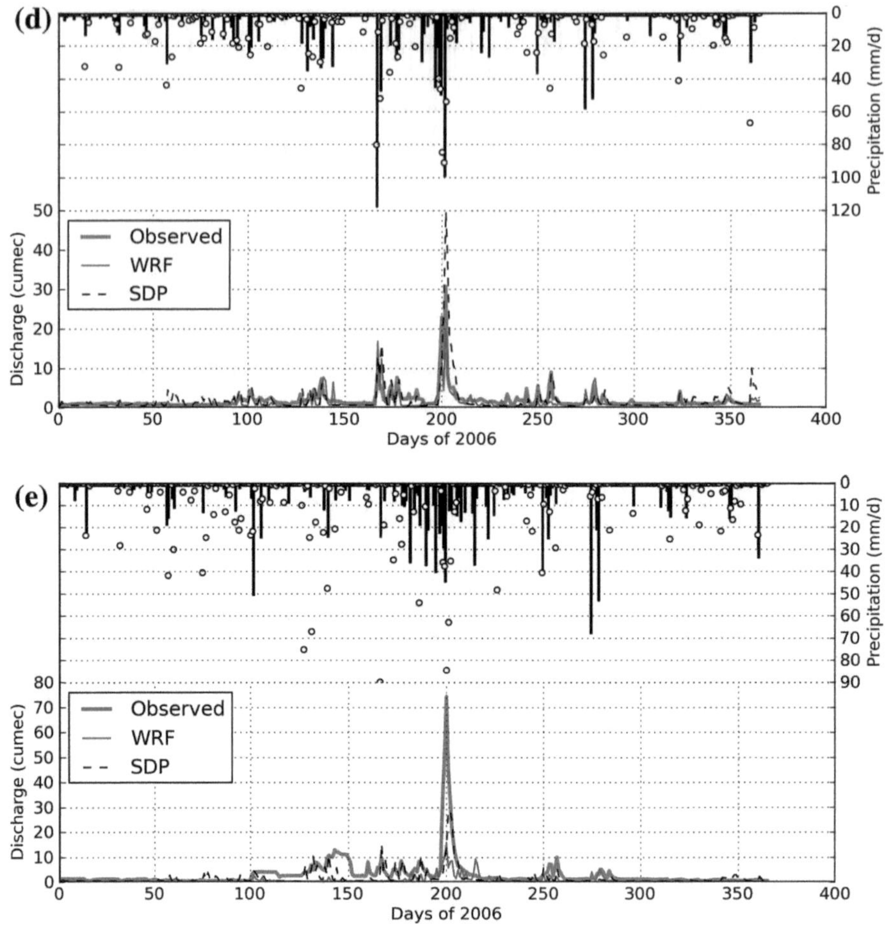

Fig. 5 (continued)

July is most probably due to the overprediction of dam outflow from the Amagase dam located at upstream.

3.1.2 Validation of the Dam Model

To validate the simplified dam model used in the hydrological model to simulate the reservoir operation in six major dams of the Yodo River basin, the observed water levels at those dams (http://www2.river.go.jp/dam/index.html, in Japanese) were compared with the simulated water levels (Fig. 7). The patterns of water level fluctuations in all these dams were well-simulated. The change in water levels during the transition period between the flooding and non-flooding seasons was also predicted with sufficient accuracy. Since the actual dam operations are complex and

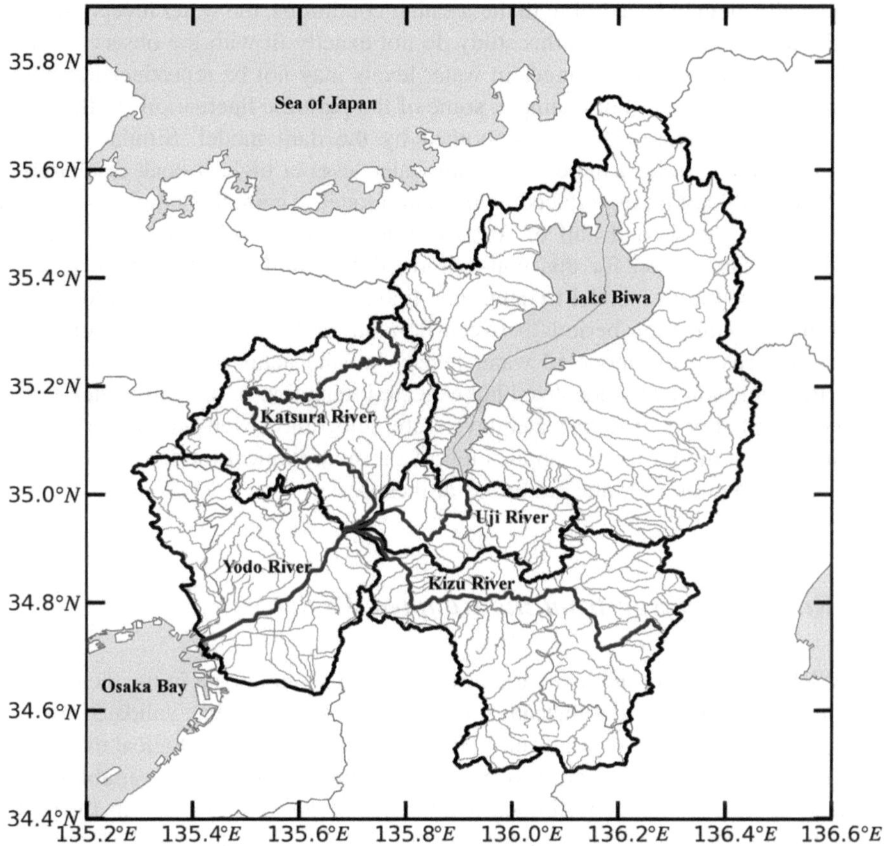

Fig. 6 Sub-basins of Yodo River basin with major rivers (*Thick black lines* are the boundaries of the sub-basins)

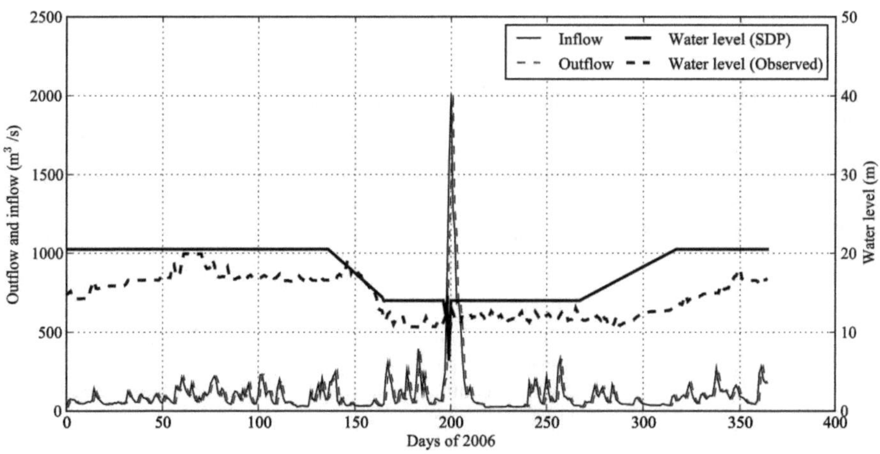

Fig. 7 Inflow, outflow and water level for Amagase dam in 2006

are manually changed according to the weather conditions, the water levels from the simplified dam model used in this study do not exactly fit with the observed water levels. Some short-period changes in water levels may not be reproduced at all by the present dam model. For example, some of the periodic fluctuations in the water levels at the Amagase dam are smoothed by the dam model. Similarly, at the Hiyoshi dam, the sudden increase in dam water level at highest peak discharge in July 2006 could not be reproduced by the dam model. Again, the temporary drop in water level in the second half of August 2006 in the Hiyoshi dam was also not reproduced. The reason for this non-reproducibility might be the simple precipitation limit used in the model to determine flooding (Sect. 2.2.4). It should also be noted that the transition periods between flooding and non-flooding seasons have regular patterns in the modeled water levels, but the actual dam operations may differ from this somewhat idealized but expected transition in the dam water levels.

The overall trend of dam water levels are sufficiently well-reproduced by the simplified dam model for the application of the hydrological model in the assessment and prediction of water resources at mesoscale.

3.2 River Discharge from the Coupled Model

The WRF hydrometeorological variables were used as input data in the the coupled hydrometeorological model instead of observation data. This has validated the use of high-resolution gridded data from WRF to be used in the hydrological model and the future basin scenarios can also be studied using the projected variables from WRF mesoscale model. The limited availability of observation data in the Yodo River basin were overcome by coupling the WRF mesoscale model in one-way mode to drive the basin-scale hydrological simulations.

Since the observation data were limited to dam data and two observation stations, the validation was done only with these available data. Table 7 shows the comparison of total annual outflow discharges from five dams in 2006 between

Table 7 Comparison of annual outflow from dams in 2006 between observed data, hydrological model using SDP data, and coupled hydrometeorological model

Dam	Outflow ($m^3 \times 10^6$)		
	Observed	SDP[a]	WRF[b]
Amagase	3523.7	3494.2	4348.2
Hiyoshi	360.9	396.2	442.7
Takayama	474.6	474.3	302.8
Nunome	56.5	61.6	41.7
Hitokura	95.2	50.3	29.2

[a] "SDP" column is the result of hydrological model using SDP observation data

[b] "WRF" column is the result of coupled hydrometeorological model

observed data, hydrological model using SDP data, and the coupled hydrometeorological model. All the dam outflows are comparatively similar between the hydrological model using SDP data and the coupled hydrometeorological model. The Amagase dam shows slightly larger overprediction in the coupled model than the hydrological model using SDP data. These overpredictions can also be observed from some larger peaks in coupled model output in Fig. 5. The coupled hydrometeorological model is found to be better at predicting the maximum peak discharge of July 2006 at most of the places than the original hydrological model using SDP data. At the Hiyoshi, Takayama and Nunome dams, the maximum peak discharges in July were better represented by the coupled hydrometeorological model. But at the Hitokura dam, the maximum peak discharge in July is found to be highly underpredicted by the coupled model than the hydrological model using SDP data. Moreover, May-June peak discharges at Hitokura dam were also underpredicted by the coupled model than the hydrological model using SDP data.

In the case of two observation stations used in the validation of coupled hydrometeorological model, the maximum peak flow at the Hirakata station and the overall annual discharge at the Takahama station were predicted by the coupled model without any appreciable deviation from the performance of the original hydrological model using SDP data. In fact, the maximum discharge peak in July 2006 at Hirakata has been simulated more precisely by the coupled hydrometeorological model.

4 Conclusions

As a solution to the need of a comprehensive integrated water resource management in the Yodo River basin, a coupled mesoscale hydrometeorological modeling approach was attempted.

The response of the Yodo River basin to atmospheric forcings was simulated for the year 2006 by using a distributed hydrological model containing an evapotranspiration model and a rainfall-runoff model. The hydrometeorological input such as precipitation and air temperature were obtained from the SDP observation data of the Yodo River basin and were gridded into the mesh structure of the hydrological model using Thiessen polygon method. A simplified reservoir model was also used to simulate inflow, outflow and water level at major dam reservoirs in the Yodo River basin. Outflow from the dams, basin discharge and river peak discharge in the observation stations were found to be fairly well-predicted by the hydrological model.

Instead of the traditional observation data used as atmospheric input data in the hydrological model, the high-resolution meteorological data from WRF model were used as the atmospheric forcings in the hydrological model. The coupled mesoscale meteorological—hydrological modeling system overcomes several problems associated with observation data from basin stations. Foremost, the sparse distribution of the observation stations in the basin limits the level of regional and mesoscale features that can be reproduced by the hydrological model. The coupled hydrometeorological modeling system mostly overcomes this spatial problem by producing high-resolution regular gridded meteorological data that is equivalent to a highly dense network of observation stations placed at very close proximity. The upshot of using the coupled hydrometeorological modeling system is that the local and regional variabilities inherently present in the meteorological data are better reproduced by virtue of high-resolution and physically-based modeling of the atmospheric processes. The comparison and validation of the coupled hydrometeorological model results against the observed river discharge and dam outflow data indeed reinforces the notion that regional modeling approach for water resource assessment is imperative for integrated management of river basins.

References

Abbott M, Bathurst J, Cunge J, O'Connell P, Rasmussen J (1986) An introduction to the European hydrological system—systeme hydrologique Europeen, "SHE", 1: history and philosophy of a physically-based, distributed modelling system. J Hydrol 87:45–59

Arnold JG, Srinivasan R, Muttiah RS, Williams JR (1998) Large area hydrologic modeling and assessment part I: model development. JAWRA J Am Water Resour Assoc 34(1):73–89

Cai X (1999) A modeling framework for sustainable water resources management. PhD thesis, The University of Texas at Austin

Chino M, Nagai H (2003) Numerical research of regional environment on IT based laboratory. In: International symposium of the Kanazawa University, Kanazawa University, vol 1, pp 186–189

Cornelissen T, Diekkrüger B, Giertz S (2013) A comparison of hydrological models for assessing the impact of land use and climate change on discharge in a tropical catchment. J Hydrol 498:221–236

Daly C, Neilson RP, Phillips DL (1994) A statistical-topographic model for mapping climatological precipitation over mountainous terrain. J Appl Meteorol 33(2):140–158

Drogue G, Humbert J, Deraisme J, Mahr N, Freslon N (2002) A statistical-topographic model using an omnidirectional parameterization of the relief for mapping orographic rainfall. Int J Climatol 22(5):599–613

Freeze R, Harlan R (1969) Blueprint for a physically-based, digitally-simulated hydrologic response model. J Hydrol 9(3):237–258

He H, Zhang Q, Zhou J, Fei J, Xie X (2009) Coupling climate change with hydrological dynamic in Qinling mountains, China. Climatic Change 94(3):409–427

Högström U (1988) Non-dimensional wind and temperature profiles in the atmospheric surface layer: a re-evaluation. Bound-Layer Meteorol 42(1):55–78

Hong SY, Lee JW (2009) Assessment of the WRF model in reproducing a flash-flood heavy rainfall event over Korea. Atmos Res 93(4):818–831

Hooper BP (2011) Integrated water resources management and river basin governance. J Contemp Water Res Educ 126(1):3

Ikebuchi S, Kojiri T, Hagiwara Y, Tomosugi K, Takemon Y, Tanaka K, Hamaguchi T (2006) Water resources and environment assessment in river basin based on hydro-BEAM. Annuals of Disaster Prevention and Research Institute, Kyoto University, Uji-Shi, vol 49 C, pp 113–118

IPCC (2007) Impacts, adaptation and vulnerability: contribution of working Group II to the fourth assessment report of the intergovernmental panel on climate change. Cambridge University Press, Cambridge

Jia Y, Ni G, Kawahara Y, Suetsugi T (2001) Simulation of hydrological cycle in an urbanized watershed and effect—evaluation of infiltration facilities with WEP model. J Hydrosci Hydraul Eng 19(1):43–52
Jia Y, Ni G, Yoshitani J, Kawahara Y, Kinouchi T (2002) Coupling simulation of water and energy budgets and analysis of urban development impact. J Hydrol Eng 7(4):302–311
Kiem AS, Verdon-Kidd DC (2011) Steps toward "useful" hydroclimatic scenarios for water resource management in the Murray-Darling basin. Water Resour Res 47(12):W00G06
Kim S, Tachikawa Y, Nakakita E, Takara K (2010) Hydrologic evaluation on the AGCM20 output using observed river discharge data. Hydrol Res Lett 4:35–39
Kojiri T, Kinai Y, Park J (2002) Integrated river basin environment assessment on water quantity and quality by considering utilization processes. In: Proceedings of the international conference on water resources and environment research, pp 397–401
Kojiri T, Hamaguchi T, Ode M (2008) Assessment of global warming impacts on water resources and ecology of a river basin in Japan. J Hydro-Environ Res 1:164–175
Kunstmann H, Jung G (2003) Investigation of feedback mechanisms between soil moisture, land use and precipitation in West Africa. IAHS-AISH Publication 280:149–159
Kunstmann H, Stadler C (2005) High resolution distributed atmospheric-hydrological modelling for alpine catchments. J Hydrol 314(1–4):105–124
Lin CA, Wen L, Lu G, Wu Z, Zhang J, Yang Y, Zhu Y, Tong L (2006) Atmospheric-hydrological modeling of severe precipitation and floods in the Huaihe River Basin, China. J Hydrol 330(1–2):249–259
Liu Y, Gupta H, Springer E, Wagener T (2008) Linking science with environmental decision making: experiences from an integrated modeling approach to supporting sustainable water resources management. Environ Model Softw 23(7):846–858
Malone TM (2014) Streamflow modeling of Johnson creek subwatersheds using the precipitation runoff modeling system. Civil and environmental engineering master's project reports. Portland State University, Portland
Mango L, Melesse A, McClain M, Gann D, Setegn S (2011) Land use and climate change impacts on the hydrology of the upper Mara river basin, Kenya: results of a modeling study to support better resource management. Hydrol Earth Syst Sci 15(7):2245–2258
Miller SN, Kepner WG, Mehaffey MH, Hernandez M, Miller RC, Goodrich DC, Kim Devonald K, Heggem DT, Miller WP (2002) Integrating landscape assessment and hydrologic modeling for land cover change analysis. JAWRA J Am Water Resour Assoc 38(4):915–929
Ministry of Land, Infrastructure, Transport and Tourism (2008) Climate change adaptation strategies to cope with water-related disasters due to global warming (policy report). url: http://www.mlit.go.jp/river/basicinfo/english/pdf/policyreport.pdf (Provisional Translation)
Ministry of Land, Infrastructure, Transport and Tourism, Japan (2008) Integrated water resource management addressing climate change and other risks. Technical Report (Interim Report)
Mölders N (2005) Coupled models for the hydrological cycle: integrating atmosphere, biosphere and pedosphere, Springer, Berlin, pp 192–208 (Chap Feedbacks at the hydrometeorological interface)
Mölders N, Raabe A (1997) Testing the effect of a two-way-coupling of a meteorological and a hydrologic model on the predicted local weather. Atmos Res 45(2):81–107
Mölders N, Raabe A, Tetzlaff G (1996) A comparison of two strategies on land surface heterogeneity used in a mesoscale meteorological model. Tellus 48:733–749
Nawahda A, Kojiri T, Kaihotu I (2005) Assessment of global warming impacts on water resources and ecology of a river basin in Japan. J Jpn Soc Hydrolo Water Resour 18(3):293–305
Running S, Nemani R, Hungerford R (1987) Extrapolation of synoptic meteorological data in mountainous terrain and its use for simulating forest evapotranspiration and photosynthesis. Can J For Res 17(6):472–483
Sakaguchi A, Eguchi S, Kato T, Kasuya M, Ono K, Miyata A, Tase N (2014) Development and evaluation of a paddy module for improving hydrological simulation in SWAT. Agric Water Manag 137:116–122

Sato Y, Kojiri T, Michihiro Y, Suzuki Y, Nakakita E (2012) Estimates of climate change impact on river discharge in Japan based on a super-high-resolution climate model. Terr Atmos Oceanic Sci 23(5):527–540

Sato Y, Kojiri T, Michihiro Y, Suzuki Y, Nakakita E (2013) Assessment of climate change impacts on river discharge in japan using the super-high-resolution MRI-AGCM. Hydrol Process 27(23):3264–3279

Sayama T, Tachikawa Y, Takara K (2005) Assessment of dam flood control using a distributed rainfall-runoff prediction system. In: Proceedings of monitoring, prediction and mitigation of water-related disasters, pp 59–64

Seuffert G, Gross P, Simmer C, Wood EF (2002) The influence of hydrologic modeling on the predicted local weather: two-way coupling of a mesoscale weather prediction model and a land surface hydrologic model. J Hydrometeorol 3(5):505–523

Shrestha RK, Sayama T, Tachikawa Y, Takara K (2005) Use of disaggregated rainfall data for distributed hydrological modeling in Yodo River basin. Annuals of Disaster Prevention and Research Institute, Kyoto University (48 B)

Singh J, Knapp HV, Arnold J, Demissie M (2005) Hydrological modeling of the Iroquois river watershed using HSPF and SWAT. JAWRA J Am Water Res Assoc 41(2):343–360

Skamarock W, Klemp J, Dudhia J, Gill D, Barker D, Wang W, Powers J (2005) A description of the advanced research WRF version 2, NCAR technical note. Technical Report, NCAR/TN-468 + STR

Suzuki Y, Nakakita E, Ikebuchi S (2003) Numerical study of rainfall-topography relationships in mountainous regions of Japan using a mesoscale meteorological model. In: Weather radar information and distributed hydrological modelling, international association of hydrological sciences, Sapporo, Japan, vol 282, pp 43–50

Tachikawa Y, Sayama T, Matsuura H, Yamazaki T, Yamaji A, Michihiro Y (2007) Development of a real-time runoff forecasting system using a physically-based distributed hydrologic model and its application to the Yodo River basin. J Nat Disaster Sci 26(2):189–201 (In Japanese)

Thiessen AH (1911) Precipitation averages for large areas. Mon Weather Rev 39(7):1082–1089

Thompson J, Green A, Kingston D, Gosling S (2013) Assessment of uncertainty in river flow projections for the Mekong river using multiple GCMs and hydrological models. J Hydrol 486:1–30

Todd M, Taylor R, Osborn T, Kingston D, Arnell N, Gosling S (2011) Uncertainty in climate change impacts on basin-scale freshwater resources-preface to the special issue: the QUEST-GSI methodology and synthesis of results. Hydrol Earth Syst Sci 15:1035

Verbunt M, Zappa M, Gurtz J, Kaufmann P (2006) Verification of a coupled hydrometeorological modelling approach for alpine tributaries in the Rhine basin. J Hydrol 324(1–4):224–238

Westrick KJ, Mass CF (2001) An evaluation of a high-resolution hydrometeorological modeling system for prediction of a cool-season flood event in a coastal mountainous watershed. J Hydrometeorol 2(2):161–180

Yoshikane T, Ma X, Kimura F, Hara M (2005) Regional climatic simulation for hydrological model using WRF model around Yellow River basin. In: 6th WRF/15th MM5 users' workshop, National center for atmospheric research, 3.11

Yoshitani J, Kavvas M, Chen ZQ (2001) Coupled regional-scale hydrological-atmospheric model for the study of climate impact on Japan. Soil-Veg-Atmos Transfer Schemes Large-Scale Hydrolo Models 270:191–198

Yu Z (2000) Assessing the response of subgrid hydrologic processes to atmospheric forcing with a hydrologic model system. Global Planet Change 25(1–2):1–17

Yu Z, Barron EJ, Yarnal B, Lakhtakia MN, White RA, Pollard D, Miller DA (2002) Evaluation of basin-scale hydrologic response to a multi-storm simulation. J Hydrol 257(1–4):212–225

Remote Sensing—A Fast And Reliable Tool to Map the Morphodynamics of the River Systems for Environmental Management

V. Thirukumaran and Mu. Ramkumar

Abstract River is a dynamic morphometric agent which shapes literally earth's entire surface except the dry deserts and the ice-capped Polar Regions. The morphological variations in the river basins are under the influences of an ensemble of variables including, but not limited to, the precipitation, runoff, slope, soil, and many other processes including tectonics, sedimentation, oceanography and anthropogenic intervention. As the rivers provide sustenance to the fauna and flora including the human race, understanding the river systems is essential from the ecological and economic points of view. The river systems interact with and maintain a delicate equilibrium with the lithosphere–hydrosphere–atmosphere, which make understanding the fluvial dynamics a necessity. The ever-increasing pressures from the anthropogenic interventions, under the climate change scenarios make this task critical for sustainable management which in turn requires extensive data inputs on a spatio-temporal scale. However, the very dynamic nature of the river systems poses constraints on the data acquisition and subsequent analysis. In this paper, we attempt demonstrate the utility of remote sensing and GIS as tools for fast and reliable analysis of morphodynamics of rivers for environmental management.

Keywords Fluvial geomorphology · River dynamics · Remote sensing · Geohazards · Sediment load · LIDAR · Drainage · Neotectonism · Environmental management

V. Thirukumaran (✉)
Department of Geology, Government Arts College (Autonomous), Salem 636007, India
e-mail: mailkumaran75@gmail.com

Mu. Ramkumar
South East Asia carbonate Laboratory (SeaCarl), Universiti Teknologi Petronas, Tronoh, Malaysia

Mu. Ramkumar
Department of Geology, Periyar University, Salem 636011, India

Mu. Ramkumar et al. (eds.), *Environmental Management of River Basin Ecosystems*, Springer Earth System Sciences, DOI 10.1007/978-3-319-13425-3_9

1 Introduction

When compared with other geomorphic systems, the studies on the morphological variation and evolution of the fluvial system are limited by the inaccessible nature of the terrain (either at the catchment or at the coastal swamps and tidal channel networks) and by the dynamic changes in morphological features at spatial and temporal scales. These two major inhibiting factors, along with other logistic restrictions, pose severe constraints in mapping and understanding the morphological variations of river systems that are manifested as a result of natural and anthropogenic processes. However, as the river courses and adjoining regions are the preferred locations of inland transport, agriculture, habitation, commercial, industrial and aesthetic activities, improved understanding on the fluvial geomorphic system is essential, especially in the context of accelerated rate of climate change during the recent years and also in the future scenarios. The river basins are home to a wide variety of natural fauna and flora, and also for the buried georesources such as hydrocarbon and coal. Furthermore, the river basins contain vast expanses of low lying lands and are underlain by thick pile of sediments that are under the influences of tectonic movements, recurrent flooding, wave erosion, and cyclones. Thus, understanding the river systems, particularly the morphodynamics is essential from the ecological, economical and strategic points of view.

2 River System and Its Dynamism

The land and water are linked in a natural system called the catchment that represents an area of land in the upstream region of a drainage basin that collects water either from the rainfall or from the snow melt or both and directs it to a stream. Recently, Thirukumaran (2012) narrated the emergence and actions of rivers from a geomorphic point of view that could be briefly stated thus: From the smallest droplet to the mightiest river, water works to shape the land, taking with it sediment and dissolved materials and delivers them to the coastal lowlands (delta) and into the sea. A river network is an integral component of any landscape that is created as water flows from higher to lower elevations. There is an inherent supply of potential energy in the river systems created by the change in elevation between the beginning and ending points of the river or within any discrete stream reach. This potential energy is expressed in a variety of ways as the river moves through and shapes the landscape, developing a complex fluvial network, with a variety of channel and valley forms and associated aquatic and riparian habitats. Excess energy is dissipated in many ways: contact with vegetation along the banks, in turbulence at steps and riffles in the river profiles, in erosion at meander bends, in irregularities, or roughness of the channel bed and banks, and in sediment, ice and debris transport. The potential energy gets rejuvenated whenever the earth's dynamics revive the elevation and landscapes. Streams are dynamic systems that balance the water flow and the sediment transport. A river's energy must be in balance with the size and volume of the sediment carried by the river. Any aberrations in this balance, either natural, or anthropogenic in origin, results in immediate and extensive reaction from the river flow depending on the nature and intensity of the aberration. The reactions of the river system include, flooding, change of channel course, bank erosion, and channel abandonment, which serve to the detriment of downstream ecosystem as well as services provided by the river to riparian community including humans.

The dynamism of geomorphic agents is relentless in shaping the earth and the agents are the immediate respondents of the geotectonic and geodynamic processes operating on the Earth system. Being fluid, the water responds to the tectonic and other disturbances at a rate faster than other geological agents. Being entirely dependent on this fluid medium, though there are other catalytic factors such as slope and gravity, the fluvial system is inherently dynamic which requires a compatible tool, should the features and responses of the fluvial system are to be studied at their rates and spatio-temporal scales of evolution and modification.

Owing to the provision of synoptic view at spatial, temporal, spectral, and radiometric resolutions of remote sensing systems, the satellite remote sensing (RS) offers an unique tool to map the fluvial dynamics without being in direct physical contact with the ground, which obviates the logistic and inaccessibility constraints posed by the fluvial system. The geographic information systems (GIS) offer the opportunity to gain fresh insights into these earth's surficial features through the provision of analytical and data integration capabilities. These two tool-technologies can be linked together into a synergistic system that is particularly well suited to the examination of landscape conditions through the interrelationships of scale, pattern, and process. There are many automatic pattern recognition functions within the GIS environment that reduce man-hours of analysis, remove operator inconsistency and provide dynamic mapping capability. Thus, this paradigm has gained wide application and prominence in the field of mapping and understanding the fluvial morphological features.

3 Indicators of River Dynamics and Their Mapping and Analysis Through RS and GIS

The river dynamics can be indicated by the frequency, duration, and magnitude of stream discharge, changes in channel morphology, and seasonal variations in sediment load, and formation and modifications in fluvial landforms. The suspension, transport and deposition of sediments are important geomorphic functions in shaping the physical landscape. The river dynamics can be categorically classified into two major types viz., related to sediment load and erosion and response of river system to tectonics or earth dynamics.

3.1 Sediment Load and Erosion

The sediment load of the stream is directly related to the runoff and gradient of the stream course and free supply of weathered materials. Runoff is the function of rainfall and permeability of the subsurface lithology, and gradient is the function of youngness of the terrain. Three types of analyses namely, (a) mapping of areas undergoing erosion due to channel migration over time using multidate imagery, (b) analysis of LiDAR (Light detection and ranging)-derived DEMs (Digital Elevation Models) to quantify stream bank heights to estimate soil volume loss, and (c) combined remote sensing and field observations to estimate sediment loading per eroded feature, reach, and stream, allow efficient quantification of the sediment loading mobilized by stream bank erosion at various spatial and temporal scales. It also provides information on the critical source areas that contribute a disproportionate amount of the total sediment load.

3.1.1 Satellite Reflectance and Sediment Load

Remote sensing holds the potential for monitoring and estimating suspended sediments in surface water. Suspended sediments increase the radiance emergent from the surface waters in the visible and near infrared (NIR) portion of the electromagnetic spectrum. The Landsat TM and ETM+visible and NIR bands, 1–4, are optical bands recording the electromagnetic radiation of 0.45–0.52, 0.52–0.60, 0.63–0.69 and 0.76–0.90 μm, respectively. Cloud-free pixels in the middle of the river width, counting the information only from water surface but not from land, are generally used to estimate the suspended sediment load (Fig. 1). Wang et al. (2012) found overall increase in reflectance with the increases in the sediment concentrations (Fig. 2). To estimate the suspended sediment concentration (SSC) from the satellite images, Digital Counts (DC) are converted to at-satellite radiance and then radiance is converted to reflectance. The equation to convert satellite DCs to satellite radiance is:

$$L = (DC + A)S$$

where L is the satellite spectral radiance for the given spectral band (Wm-2 sr-1 μm-1), DC is the digital count at a given pixel for a given spectral band, A is the offset for a given spectral band in the unit of DC, and S is the slope for a given spectral band in Wm-2 sr-1 μm-1/DC (Chavez 1996). The offset and slope values are taken from the header file.

Atmospheric correction is applied to the satellite spectral radiance and then converted to surface reflectance. Bhatti et al. (2009) have observed sympathetic relationship between reflectance and sediment load when the suspended sediment concentration is in the range of 500–2,000 mg/l. When the suspended sediments in the river is high, and 0 % reflectance or completely dark objects are not likely to exist. The improved dark-object subtraction technique (Chavez 1988, 1989) can be

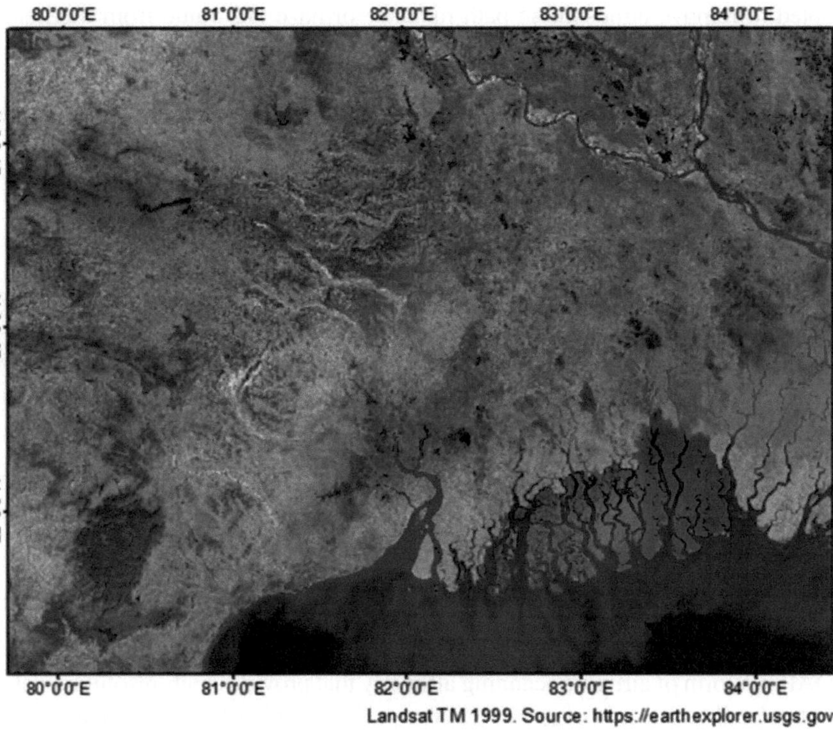

Fig. 1 Suspended sediment in Bay of Bengal showing high reflectance in RGB in landsat TM data 1,999 (*Source* Global landcover facility, www.landcover.org)

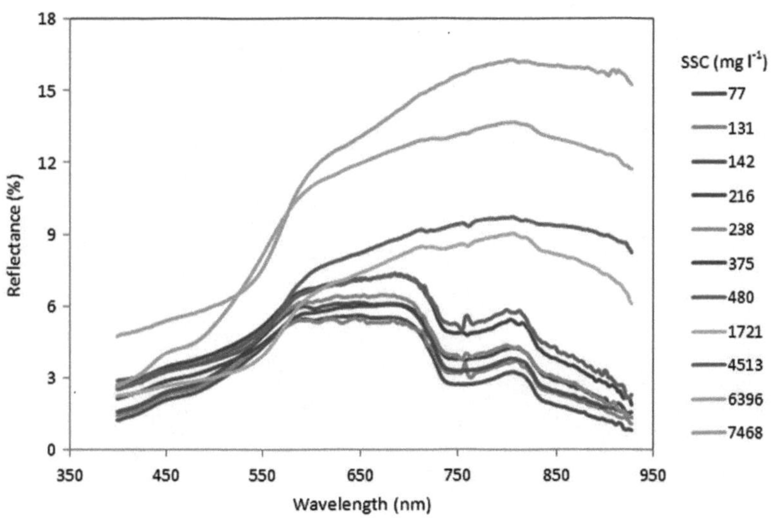

Fig. 2 Spectral reflectance curves versus suspended sediment concentrations (Wang et al. 2012)

adopted to remove atmospheric path radiance or haze DC value from the images. Also, correction for atmospheric transmission coefficient due to the atmospheric absorption can be carried out based on the Cosine of solar zenith angle (COST) model of Chavez (1996). The following equation can be used to calculate reflectance from radiance data.

$$R = \frac{\pi(\mathrm{Ls} - \mathrm{La})}{TAUv(EoCos(TZ)TAUz + Ed)}$$

where R is the spectral reflectance of the water surface, La is the atmospheric path radiance (Wm-2 sr-1 μm-1). TAUv is the atmospheric transmittance along the path from the ground surface to the sensor, TAUz is the atmospheric transmittance along the path from the sun to the ground surface, TZ is solar zenith angle in degree. Eo is solar spectral irradiance on a surface perpendicular to the Sun ray's outside the atmosphere (Wm-2 sr-1 μm-1). Ed is the downwelling spectral irradiance at the water surface due to scattered solar flux in the atmosphere (Wm-2 sr-1 μm-1).

3.1.2 Assessment of Topographic Variations Resulting from Stream Dynamics

LIDAR is a form of airborne scanning altimetry that provides high-resolution data on topography (Fig. 3). It is of great importance for mapping landforms and landform change, especially at relatively local scales. It can be used to produce Digital

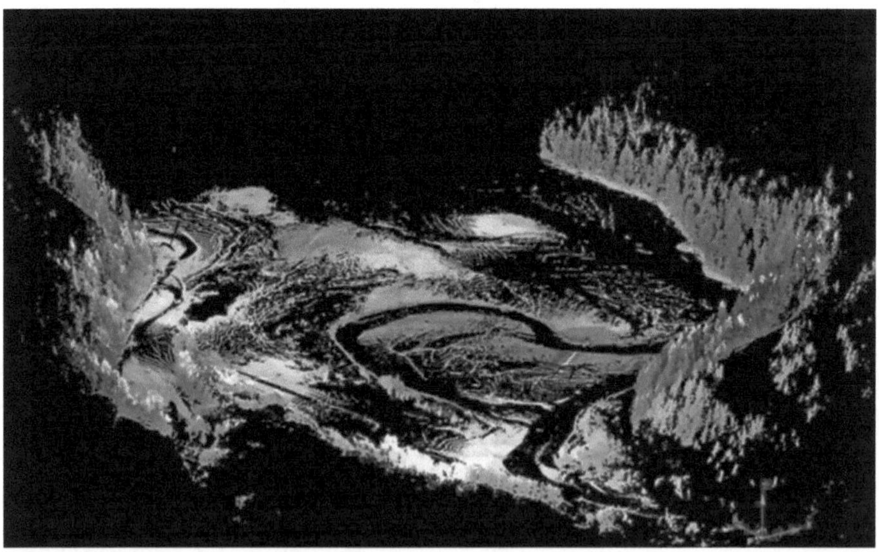

Fig. 3 Lidar image depicting topography of a delta in lake Aldwell (*Source* http://www.venturariver.org)

Elevation Models (DEMs) and is being used in a number of geomorphological applications including cliff and landslide monitoring, the study of tidal channels, assessing subsidence risk, predicting areas subject to storm surges and tsunamis and changes in beach height (Adams and Chandler 2002). Multi-temporal studies may shed light on the changes in channel pattern, sediment accumulation, etc.

3.2 Drainage Anomalies as Observed from the Remotely Sensed Data

The nineteenth—early twentieth-century geomorphologist Davis (1889, 1899) developed an elaborate scheme to describe the components of a river drainage network and related them to various stages of their physiographic development. The river system is the immediate respondent to the earth's dynamics and faithfully obeys the resultant newer landforms and structures. Miller (1937) and Chitale (1970) and a large number of workers have brought out exhaustive information on how these drainages can be used in mapping the lithology and geological structures especially from the aerial photographs.

Pope and Wilkinson (2005) studied the roles of climate and tectonics in the late Quaternary fan development on the Spartan piedmont Greece. Rejuvenated alluvial fan formation along the mountain front is an indication of slope and relief modification brought in by neotectonism. The radial drainages are the drainages that are radiating away in all directions from a central point or drainages converging from the periphery towards a central point. The former is called as centrifugal radial drainage and the latter as centripetal radial drainage. These have been interpreted to be the indicators of recent tectonic movements. However, care has to be taken to stifle through the available evidences on their origin ascribable to topographic or impact origin.

Such radial centrifugal drainages observed in many parts of the world were inferred to be due to active tectonic doming (Twidale 2004). In India too, a lot of inferences are made by many on the possible active tectonics from such radial drainages. In Saurashtra Peninsula, Sood et al. (1982) interpreted a number of such centrifugal radial drainages along with gullying and inferred the role of Post Deccan trap diapirism as the causative factor. Babu (1975) inferred the role of morphostructures related to hydrocarbon in Krishna–Godavari delta as the causative factor for the formation of the anomalous radial cum annular drainages. Ramasamy et al. (2006a) have observed network of radial cum annular drainages in Cauvery delta and inferred them to be the reflection of recent subsurface doming.

A system of co-linear drainages is called as "parallel drainages", whereas, the long and straight flow paths of the drainages are called as rectilinear drainages. The parallel or co-linear drainages are normally observable in the dune fields in between co-linear ridges (Miller 1937). In general, the rectilinear flow of drainages is normally attributed to the faults, whether active or dormant. In fact, most of the lineaments and faults are interpreted by the geoscientists only from such rectilinear flow of drainages (Twidale 2004). In India, the rectilinear flow of Narmada River for over 1,000 km in Central India was inferred to be due to a major crustal dislocation (Oldham et al. 1901; Murty and Mishra 1981) along the Narmada-Son lineament. In parts of Tamil Nadu also, most of the easterly flowing near-rectilinear rivers namely the Palar, the Ponnaiyar, and the Cauvery and parts of the rivers the Vellar, and the Vagai etc. are inferred to flow along WNW–ESE transverse faults, out of which few of them are inferred to be active (Vemban et al. 1977). Chamyal et al. (2003) inferred the closely spaced parallel drainages in Saurashtra Peninsula to be indicative of tectonic uplift.

The slope and lineament controlled drainages which are sharply deflected by the lineaments are interpreted as deflected drainages. The deflected drainages are interpreted from many river systems the world (Bowler and Harford 1966; Panizza 1978; Reid 1992; Saintot et al. 1999; Kusky and El–Baz 2000; Twidale 2004 and many others). For example, Twidale (2004) had observed many such deflections in the Murray River of Australia in association with many lineaments/faults. In India too, a number of workers have utilized the drainage deflection phenomenon as a tool for detecting the land stability and neo-active tectonics. Babu (1975) has observed anomalous drainage deflections in the River Godavari, Andhra Pradesh and attributed these to various lineaments related to tectonic uplift. The rectangular flow of the Bhramaputra River, Assam valley has been explained by NE–SW, NW–SE, E–W and N–S faults (Murty and Mishra 1981). Amalkar (1988) has explained the complex drainage pattern in the Luni Basin, Rajasthan through lineaments of various orientations. Radhakrishna (1992) has observed multiple deflections in the Biligirirangan–Hogenekkal area of the Cauvery River basin and on the basis of such acute and rectangle deflections, identified a spectrum of N–S dextral and sinistral faults. Ramasamy et al. (1992) recorded that the River Cauvery has migrated during 2700–2300 B.P. towards Tiruchirappalli–Thanjavur plains.

An anabranching alluvial river is a system of multiple channels characterized by vegetated or otherwise stable alluvial islands that divide flows at bankful discharges. The islands may be developed from within-channel deposition, excised by channel avulsion from extant floodplain, or formed by prograding distributary channel accretion on splays or deltas. A specific subset of distinctive low-energy anabranching systems associated with mostly fine-grained or organic sedimentation are defined as anastomosing rivers (Smith and Smith 1980; Knighton and Nanson 1993; Makaske 2001). Van Niekerk et al. (1999) found that bedrock anabranching channels on the Sabie River in South Africa have a significantly greater potential to transport sediment than do the other channel types along that river.

Neither of these terms now applies to the braided rivers where divided flow is strongly stage dependent around the bars that are unconsolidated, ephemeral, poorly vegetated and overtopped at less than bankful flow. The hallmark of the braided rivers is the presence of multiple active channels that divide and rejoin to form a pattern of gently curved channel segments separated by exposed bars. Braided rivers are marked equally by temporal dynamism; gradients in sediment flux associated with the complex spatial topography change in local slopes, leading the flow to continually adjust its path as it picks its way through the network. It has been suggested that braiding was the dominant river pattern on the Earth before the first appearance of land plants in late Silurian time (Schumm 1968). The braiding of streams usually happens when the gradients drop near the confluence points. Excluding alluvial fans, if the braiding starts at much earlier stages of the stream then, it could be due to the drop in gradient which is anomalous manifestation of the land subsidence along a fault.

The drainages that flow as a single stream, branch off into two and rarely into four or five, run co-linear or curvilinear and meet after a few hundred meters or kilometers, thus ultimately giving a shape of an eye or biconvex lens are termed as "eyed drainages" because of such morphology. These eyed drainages are invariably found to be either bisected by the orthogonally/obliquely aligned lineaments or confined within two sub parallel/oblique lineaments. Such eyed drainages can be propounded that vertical cutting/incision of drainages and when it happens, it indicates tectonic emergence and the splitting up of the drainages suggests land subsidence. Significant amongst them are the observations made by Smith et al. (1997) in Okavango River, Botswana that the river on reaching a graben split up into four channels and these channels ultimately rejoin after crossing the graben. They have called this tectonically induced phenomenon as "anastomosis" (Fig. 5a).

The types of anomalous occurrences of compressed meanders in otherwise normally flowing drainage systems have been demonstrated to be indicative of active tectonics in such zones of compressions. Many workers explained the anomalous sinuosity of the river Yamuna right from its outlet from Himalayas and up to it's confluence with Ganges in the Indo-Gangetic plain to be due to the active tectonics related to post collision phenomenon. On the contrary, Bakliwal and Sharma (1980) have explained the intense, acute and compressed meandering in the River Yamuna in Agra region of the Indo–Gangetic plains to active scissor fault tectonics along two sub parallel lineaments of the Great Boundary Fault System. Murthy and Sastri (1981), Barooah and Bhattacharya (1989) and many others have interpreted a large number of drainage anomalies in the form of compressed meandering in the Brahmaputra River and explained them to be due to ongoing collision of the Indian plate. The anomalous compressed meandering in otherwise 1,000 km long rectilinear Narmada River in its western end near Broach area, Western India, is demonstrated to be due to the ongoing tectonic activities by three major faults which occur in triangular pattern and caused compressed meandering within them (Ramasamy et al. 1991). Many workers observed the occurrences of compressed meanders [for example, Ramasamy et al. (1995)—in the Vellar River near Cuddalore, Tamil Nadu State (Fig. 4), Valdiya (2001)—Cauvery River in the area west of Bangalore and Jain and Sinha (2005) in the Baghmati River, Himalayan foreland basin] and interpreted them to be the result of active tectonics.

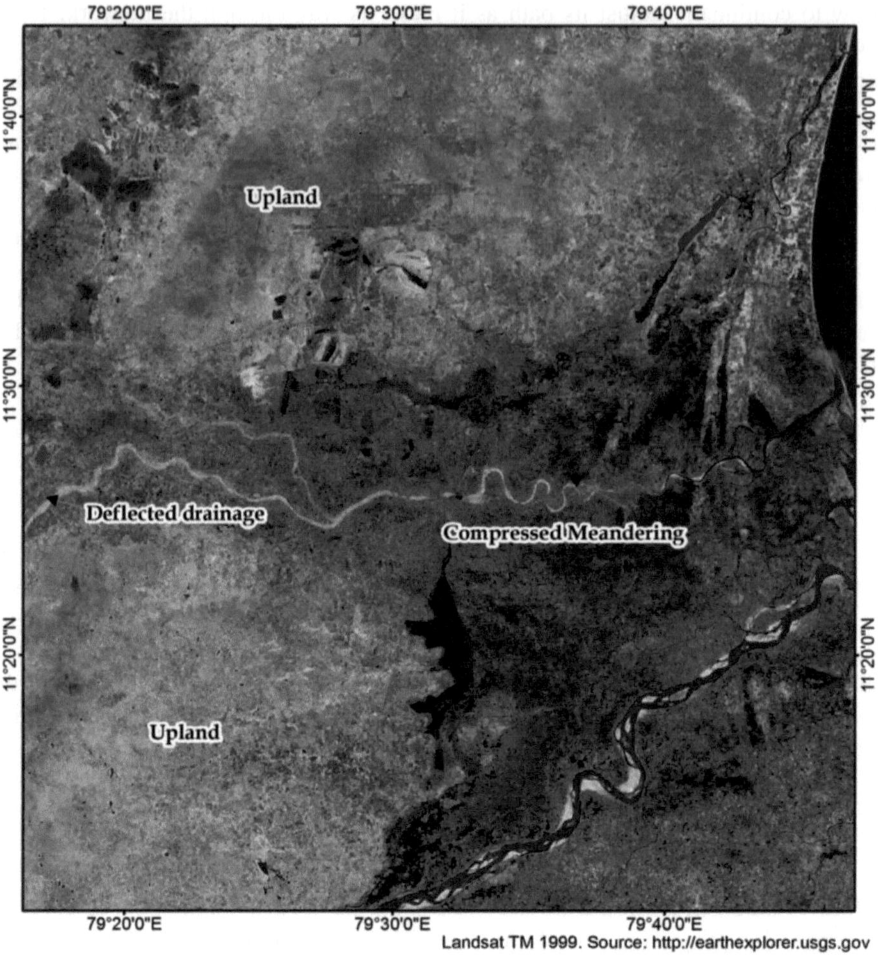

Fig. 4 Compressed meandering and deflected drainage owing to rising upland (Vellar, Tamil Nadu, India)

3.3 Paleochannels

The occurrence of paleochannels indicates that the river has left these traces and migrated away. These Paleodrainages/Paleochannels/buried rivers show typical shapes namely, ribbon, loop, linear, curvilinear and contorted features with black tone in black and white panchromatic aerial photographs and reddish tone in the satellite colour coded FCC images. Mapping the paleochannels and interpretation of their evolutionary pattern provide reliable clues to the past geological and climatological events and processes namely, active tectonic movements, sea level changes, climatic fluctuations, flooding frequency and intensity, etc.

Among these, the active tectonic movements seem to play a greater role in the river migration when these rivers show preferential migrations only in one direction (Ramasamy et al. 1992). Many workers have used this as a tool to understand and map the recent tectonic movements. The phenomenal anticlockwise rotational migration of river Sarasvati and its burial in the northern part of the Great Indian Desert was attributed to the rise of Aravalli Mountains (Yashpal et al. 1980; Valdiya 1997; Gupta et al. 2004).

Ramasamy and Karthikeyan (1998) have observed the southerly migration of the Ponnaiyar River in Pondicherry due to the ongoing activism in the Holocene graben to its south. The anticlockwise rotational migration of the River Cauvery in its deltaic regime in Tamil Nadu was explained by the phenomenon of block faulting of the Mio-Pliocene Sandstone and its uplift during the last 6,000 years (Ramasamy et al. 2006b). Preferential migrations of rivers have been used as an indicators for active tectonic movements around the world viz: the Godavari delta, India (Ramkumar 2003), southerly migration of Echuca River, New South Wales (Bowler and Harford 1966; Pels 1966), the Charwell River in New Zealand (Bull and Knuepfer 1987), the Po River in Italy (Castaldini 1990), and the Murry River in Australia (Twidale 2004), etc.

3.4 Delta

Deltas can be either subaerial or subaqueous accumulations and may have a variety of geometries. The prevailing shape of any given delta depends on the rates of sediment supply by the rivers and the patterns and rates of sedimentation. Deltas vary immensely in both area and volume. The size of a delta depends at the lowest order on the annual sediment discharge of the river but the most extensive deltas also tend to be developed where wide, low gradient continental shelves provide a platform for prolonged sediment accumulation and morphological progradation. Hence, the largest deltas are found on passive (as opposed to active) continental margins (Wright 1985). Despite this fact, active margins are probably equally or more important than the passive margins in supplying river sediment to the sea. Milliman and Syvitski (1992) showed that the numerous small mountainous streams, particularly those of the humid tropics, are collectively the most important sources of terrestrial sediment to the sea (Fig. 5c).

Deltas contain a variety of micro-landforms or sub-environments namely, river channel—tributary and distributary (active, dormant, abandoned and paleochannel), terrace, natural levee, river bank (that can be either straight, curved or any other intermediate form; These can, be either erosional or depositional or transitional, and are termed as cut bank or inner bank of meander), crevasse splay, point bar, braided bar, channel bar, ox-bow lake (Fig. 5b, d), overbank, estuary, tidal channel, mud-flat, marsh, swamp, back water, lagoon, bay, distributary mouth bar, spit, beach-ridge-swale complex, flood plain, chenier plain, tidal flat, beach, prodelta, etc., that

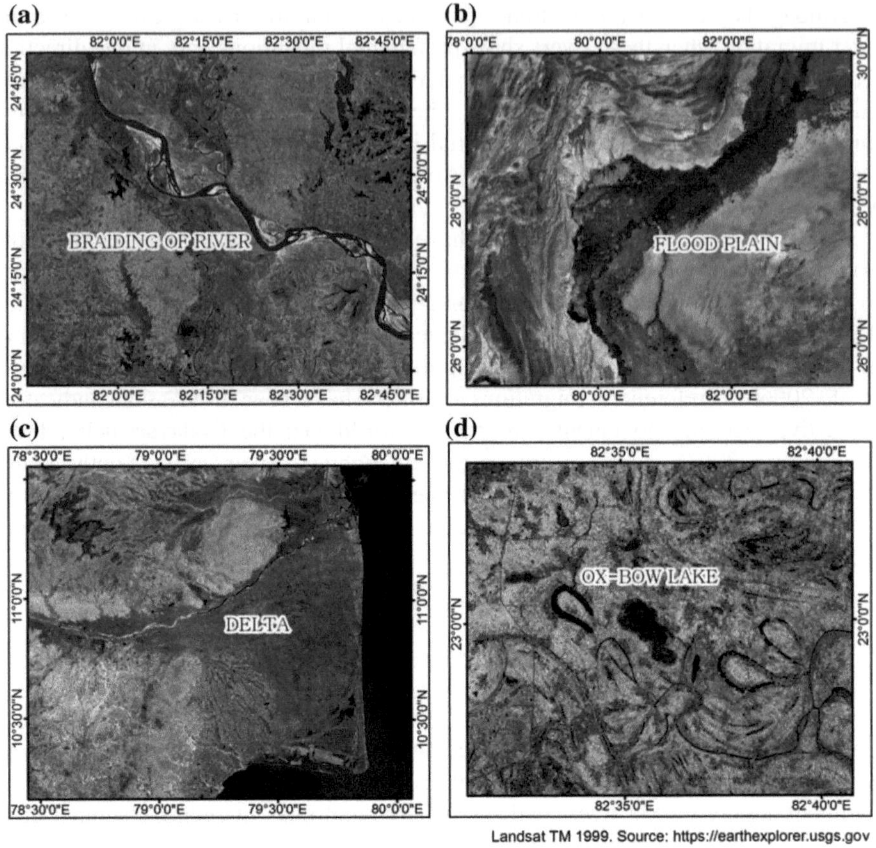

Fig. 5 **a** Anastomosis in Ganges: **b** Flood plain in Indus: **c** Delta of Cauvery: **d** Ox-bow lake along Ganges

are under the influences of a myriad varieties of processes on a spatio-temporal scale. Remote sensing plays a vital role in mapping these micro-landforms for understanding the evolutionary history and future trends.

4 The Road Ahead in Monitoring Approach

Ramakrishna Nemani et al. (2009) have used a modeling framework termed as Terrestrial Observation and Prediction System (TOPS) integrating operational satellite data, microclimate mapping, and ecosystem simulation models to characterize ecosystem status and trends and patterns in landscape. The analysis of coarse or fine resolution satellite-derived normalized difference vegetation index (NDVI) measurements may indicate vegetation condition. The variation in vegetative cover along the channel beds, river islands, sand bars and meandering loops may indicate prevailing/prevalent dynamics of a river. Analyzing MODIS (Moderate Resolution Imaging Spectroradiometer) products (vegetation indices, absorbed radiation, land surface temperature and gross primary production) in conjunction with ground-based measurements, such as runoff, lends additional utility to satellite-based monitoring of dynamic indicators, as together they provide a comprehensive view of the river dynamics. Analyses of satellite products from different seasons are corroborated by observed changes in river runoff patterns. The general overview of changes in future flooding can be created by taking the climate into consideration. This is achieved by combining climate scenarios with hydrologic models.

Laser scanning has enabled highly accurate data gathering with increased horizontal and vertical precision and better availability of detailed spatial data. For example, airborne laser scanning (ALS), ALS systems for bathymetric measurements, fixed-position terrestrial laser scanning (TLS) and mobile laser scanning (MLS), such as boat- and cart-based mapping systems (BoMMS/CartMMS), have revealed new potential in fluvial research. Aerial photography based bathymetry modeling allows us to create depth models at high spatial resolution, based on the connection between water depth and measured reflectance. Establishing protocols for monitoring the river and data from the gauging stations acquired from web-based program archives or real time data feeds as available can be of reliable method of monitoring the fluvial morphodynamics.

5 Conclusions

Knowing that rivers are a dynamic system and are sensitive to change, the geomorphic condition of the stream can be used as both an indicator and predictor of watershed function and health. The tradition of capturing the dynamics with a modern tool—remote sensing has been realized by the scientific community since long back. Widespread application of this tool is being encouraged by the administrators and planners owing to its fast and reliable nature. As integrating the remote sensing and GIS provide an indispensable tool for real-time impact assessment during the times of natural calamities such as flooding this is not only crucial for saving a large population from the misery of river dynamics and floods but is also important for improving the river health (Sinha and Ghosh 2012). The technology of quantifying sediments using spectral band of remote sensors has carried the science to new front in monitoring environmental issues. Hyperspectral remote sensing with high spectral resolution makes easier estimation of sediment load. To sum up, it can be stated that the river dynamics can be monitored, measured and understood better with the help of RS and GIS tool-technologies.

Acknowledgments This review is based on the material and information available freely as open source materials. Authors thank the sources of these data. Anonymous reviewers are thanked for helping the authors to redesign the presentation.

References

Adams JC, Chandler JH (2002) Evaluation of Lidar and medium scale photogrammetry for detecting soft-cliff coastal change. Photogram Rec 17:405–418

Bhatti AM, Rundquist D, Schalles J, Ramirez L, Nasu S (2009) A comparison between above-water surface and subsurface spectral reflectances collected over inland waters. Geocarto Int 24 (2):133–141. doi:10.1080/10106040802460707

Amalkar A (1988) Possible neotectonic activities in the Luni-Jawai plains, Rajasthan. J Geol Soc India 32(12):522–526

Babu PVLP (1975) Morphological evolution of the Krishna delta. J India Soc Remote Sens (Photonirvachak) 3(1):21–27

Bakliwal PC, Sharma SB (1980) On the migration of the river Yamuna. J Geol Soc India 21:461–463

Barooah BC, Bhattacharya SK (1989) A review of basement tectonics of Brahmaputra valley, Assam. Geol Surv India Misc Pub 46:123–128

Bowler JM, Harford LB (1966) Quaternary tectonics of the riverine plain near Echuca, Victoria. J Geol Soc Aust 13:339–354

Bull WL, Knuepfer PLK (1987) Adjustments by the Charwell River, New Zealand, to uplift and climatic changes. Geomorphology 1(1):15–32

Castaldini D (1990) The southern central sector of the po plain (North Itlay): a geomorphological study with examples of evidence of paleorivers. Bull Geomorphol Ankora Turky 18:1–10

Chamyal LS, Maurya DM, Raj Rachna (2003) Fluvial systems of the drylands of Western India: a synthesis of late quaternary environmental and tectonic changes. Quat Int 104(1):69–86

Chavez PS Jr (1988) An improved dark-object subtraction technique for atmospheric scattering correction of multispectral data. Remote Sens Environ 24:459–479

Chavez PS Jr (1996) Image-based atmospheric corrections-Revisited and Improved. Photogram Eng Remote Sens 62:1025–1036

Chavez PS Jr (1989) Radiometric calibration of landsat thematic mapper multispectral images. Remote Sens Environ 55:1285–1294

Chitale SV (1970) River channel patterns. American Society of Civil Engineers. J Hydraul Div 96 (HY1):201–222

Davis WM (1889) The rivers and valleys of Pennsylvania. Natl Geogr Mag 1:183–253

Davis WM (1899) The geographical cycle. Geogr J 14:481–504

Gupta AK, Sharma JR, Sreenivasan G, Srivastava KS (2004) New findings on the course of river Sarasvati. J Indian Soc Rem Sen (Photonirvachak). 32(1):1–24

Jain V, Sinha R (2005) Response of active tectonics on the alluvial Baghmati River, Himalayan foreland basin, eastern India. Geomorphology 70(3–4):339–356

Wang J-J, Lu X-X, Zhou Y, Liew S-C (2012) Remote sensing of suspended sediment concentrations in turbid rivers: a field survey. In: Proceedings of global geospatial conference 2012, Québec City, Canada, pp 14–17 May 2012

Knighton AD, Nanson GC (1993) Anastomosis and the continuum of channel pattern. Earth Surf Proc Land 18:613–625

Kusky TM and El-Baz F (2000) Neotectonics and Fluvial Geomorphology of the northern Sinai Peninsula. J Afr Ear Sci 31(2):213–235

Makaske B (2001) Anastomosing rivers: a review of their classification, origin and sedimentary products. Earth Sci Rev 53:149–196

Miller RP (1937) Drainage lines in bas-relief. J Geol 45:432–438

Milliman JD, Syvitski JPM (1992) Geomorphic/tectonic control of sediment dischargeto the ocean: the importance of small mountainous rivers. J Geol 100:525–544

Murthy KVS, Sastri VV (1981) Tectonic influence of the course of Brahmaputra River. Geol Surv India Misc Pub 46:129–132

Murty TVV, Mishra SK (1981) The Narmada-Son lineament and the structure of the Narmada rift system. J Geol Soc India 22(3):112–120

Oldham RD, Datta PN, Vredenberg EW (1901) Geology of the Son valley in the Rewa state and parts of the adjoining districts of Jabalpur and Mirzapur. Mem Geol Surv India 31(Pt 1):1–178

Panizza M (1978) Analysis and mapping of geomorphological processes in environmental management. Geoforum 9:1–15

Pels S (1966) Late quaternary chronology of the riverine plain of southeastern Australia. J Geol Soc Aust 13:27–40

Pope RJJ, Wilkinson KN (2005) Reconciling the roles of climate and tectonics in the quaternary fan development on the Spartan mount, Greece. In: Harvey AM, Mather AE, Stokes M (eds) Alluvial fans—geomorphology, sedimentology, dynamics. Geological society special publication vol 251. pp 245–264

Radhakrishna BP (1992) Cauvery—its geological past. J Geol Soc India 40(1):1–12

Rajiv Sinha and Santosh Ghosh (2012) Understanding dynamics of large rivers aided by satellite remote sensing: a case study from lower Ganga plains, India. Geocarto Int 27(3):207–219

Ramasamy SM, Bakliwal PC, Verma RP (1991) Remote sensing and river migration in western India. Int Remote Sens 12(12):2597–2609

Ramasamy SM, Karthikeyan N (1998) Pleistocene/Holocene Graben along Pondicherry—Cumbum Valley, Tamil Nadu, India. Geocarto Int 13(3):83–90

Ramasamy SM, Saravanavel J, Yadava MG, Ramesh R (2006a) Radiocarbon dating of some palaeochannels in Tamil Nadu and their significance. Curr Sci 91(12):1609–1613

Ramasamy SM, Saravanavel J, Selvakumar R (2006b) Late Holocene geomorphic Evolution of Cauvery delta, Tamil Nadu, India. J Geol Soc India 67(5):649–657

Ramasamy SM, Venkatasubramanian V, Riaz Abdullah S, Balaji S (1992) The phenomenon of river migration in northern Tamil Nadu—evidence from satellite data, archaeology and Tamil Literature. Man Environ Pune V (XVII)(2):13–25

Ramasamy SM, Balaji S (1995) Remote sensing and Pleistocene tectonics of southern Indian Peninsular. Int J Remote Sens 16(13):2375–2391

Ramkumar Mu (2003) Progradation of the Godavari delta: A fact or empirical artifice? Insights from coastal landforms. J Geol Soc India 62:290–304

Nemani Ramakrishna, Hashimoto Hirofumi, PetrVotava Forrest Melton, Wang Weile, Michaelis Andrew, Mutch Linda, Milesi Cristina, Hiatt Sam, White Michael (2009) Monitoring and forecasting ecosystem dynamics using the terrestrial observation and prediction system (TOPS). Remote Sens Environ 113(2009):1497–1509

Reid JB Jr (1992) The Owens River as a tiltmeter for long valley Caldera, California. J Geol 100:353–363

Saintot A, Angelier J, Chorowicz J (1999) Mechanical significance of structural patterns identified by remote sensing studies: a multiscale analysis of tectonic structures in Crimea. Tectonophysics 313:187–218

Schumm SA (1968) Speculations concerning paleohydrologic control of terrestrial sedimentation. Geol Soc Am Bull 79:1573–1588

Smith ND, McCarthy TS, Ellery WN, Merry CL, Rüther H (1997) Avulsion and anastomosis in the panhandle region of the kavango Fan, Botswana. Geomorphology 20(1–2):49–65

Smith DG, Smith ND (1980) Sedimentation in anastomosed river systems: examples from alluvial valleys near Banff, Alberta. J Sedimen Petrol 50:157–164

Sood RK, Subramanyam V, Baldev S (1982) Quaternary geomorphology of Kathiawar coast. In: Merh SS (ed) Proceedings first national seminar on quaternary environments. pp 3–12

Thirukumaran V (2012) Geoinformatic modelling for certain georesources and geohazards of Attur Valley, Tamil Nadu, India. Ph.D. Dissertation submitted to the Bharathidasan University, India. p 573 (unpublished)

Twidale CR (2004) River patterns and their meaning. Earth Sci Rev 67:159–218

Valdiya KS (1997) River piracy: sarasvati that disappeared. Resonance 1(5):19–28

Valdiya KS (2001) Tectonic resurgence of the Mysore Plateau and surrounding region in cratonic South India. Curr Sci 81(8):1068–1089

Van Niekerk AW, Heritage GL, Broadhurst LJ, Moon BP (1999) Bedrock anastomosing channel systems: morphology and dynamics in the Sabie River, Mpumalanga Province, South Africa. In: Miller AJ, Gupta A (eds) Varieties of fluvial form. Wiley, Chichester, pp 33–51

Vemban NA, Subramanyan KS, Gopalakrishnan K, Venkata Rao V (1977) Major faults/dislocations/lineaments of Tamil Nadu. Geol Sur India Misc Pub 31:53–56

Wright LD (1985) River deltas. In: Davis RA (ed) Coastal sedimentary environments. Springer, New York, pp 1–76

Yashpal Sahai B, Sood RK, Aggarwal DP (1980) Remote sensing of the lost Sarasvati River. Proc Indian Acad Sci (Earth Planet Sci) 69:317–331

Spatio-temporal Analysis of Magnetic Mineral Content as a Tool to Understand the Morphodynamics and Evolutionary History of the Godavari Delta, India: Implications on the Environmental Management of Deltaic Coastal Zones

Mu. Ramkumar

Abstract The fluvial delta acts as a unified system wherein the changes in one part of the system affect the other regions and create unique signatures of such events in terms of geomorphic and sediment textural, mineralogical, geochemical and other characteristics in the ensuing deposits. The spatio-temporal variations of the governing factors of the system introduce proportionate changes, which are normally inferred as sub-environments of a delta. An ability to understand the ongoing dynamics in the sub-environments through sensitive parameters of the sediments and developing a capability to interpret the ancient dynamics would be valuable for application of similar technique in other deltaic systems for environmental management. Magnetic mineral contents of the surface samples and vibracore stations collected from the Godavari delta, India were estimated and analyzed for their spatio-temporal variations together with geomorphologic and other characteristics. The magnetic mineral content showed a general increase from inland to coastline and is at its maximum along the river courses. It is concluded that the concentration of magnetic minerals is controlled by erosion and size sorting within a deltaic system and vary according to the spatio-temporal differences of energy conditions in time and space. As each sub-environment within a deltaic system has unique energy conditions, these primary differences reflect in the sediment characteristics and in the magnetic mineral content. Visualizing these differences on a spatio-temporal scale has helped in inferring paleoshorelines and paleochannels in the study area and reconstruct the evolutionary history of the Modern Godavari Delta. Collation of these inferences together with information on neotectonics,

Mu. Ramkumar (✉)
South East Asia Carbonate Laboratory (SeaCarl), Universiti Teknologi Petronas, Tronoh, Malaysia
e-mail: muramkumar@yahoo.co.in

Mu. Ramkumar
Department of Geology, Periyar University, Salem 636011, India

Mu. Ramkumar et al. (eds.), *Environmental Management of River Basin Ecosystems*, Springer Earth System Sciences, DOI 10.1007/978-3-319-13425-3_10

geomorphology and land use characteristics suggested that the Godavari delta faces multiple and multi-dimensional natural and anthropogenic environmental threats and urgent mitigation and environmental management programs need to be undertaken based on multi-disciplinary and comprehensive studies.

Keywords Fluvial system · Evolutionary history · Sediment size sorting · Energy conditions · Magnetic mineral content · Morphodynamics

1 Introduction

The coastal regions are highly transient environments resulting from a delicate balance between the sediment influx, storage and loss. Subtle modifications affecting any of these individual components can affect their equilibrium, form and existence which, in turn can have serious economic and environmental ramifications including the loss of habitat and reduction in ecological status (Hatfield et al. 2010). The east coast of India is drained by many perennial rivers that have formed extensive deltaic systems. These deltas act as interfaces between fluvial, oceanographic, atmospheric and anthropogenic dynamics and provide vital sustenance to the primary productive ecosystems and to the human populace by supporting the agricultural, navigational, industrial, commercial, domestic and recreational endeavors. As could be observed elsewhere, these traits make the deltas ecologically fragile and susceptible to environmental deterioration by changes in the catchment (pollution, siltation, flooding, etc.), ocean (inundation, erosion, accretion, etc.) and atmosphere. For all these reasons, understanding the morphodynamics and evolutionary history of the deltaic systems is necessary. However, the deltaic systems are dynamic and controlled by a myriad variety of factors and hence are considered to be one of the complex geological features that require extensive and diverse studies to understand the processes that operated and/or operating on them. In addition, environmentally effective management programs of these systems depend upon accurate and comprehensive scientific data on which policy decisions can be based (Nayak et al. 1996). Sediment granulometry, mineralogy and geochemical studies of deltaic sediments can provide excellent information on the processes that are in operation, based on which, interpretations on the morphodynamics and evolutionary history could be made.

In this regard, the statement of Hatfield et al. (2010) that "*grain size and mineralogy analyses are time consuming and mineralogical identification methods and techniques often require specialized training and expensive equipment. Therefore, simplistic assumptions are often made of the effect of sorting processes on the composition of the eroded, transported and deposited load. Compared with complex mineralogical studies, measurement of magnetic properties is relatively rapid, cheap and non-destructive*" is worth mentioning. Owing to these traits, this method has found a wide variety of applications in understanding the geological phenomena primarily dictated by economic and strategic considerations (Ouyang et al. 2013).

Shankar and Karbassi (1991) reported the utility of magnetic mineral content to ascertain the nature and source of shelf sediments. The dependence of magnetic susceptibility of sediments on the dominant magnetite mineral and the other magnetic minerals such as ilmenite and monazite is well known. Hence, the measurements of magnetic susceptibility are always made to detect places of heavy mineral concentrations associated with magnetite (Karbassi and Shankar 1994; Lakshmipatiraju and Rao 1996). Ratcliffe et al. (2004) utilized the stratigraphic variation of heavy mineral contents to deduce the dynamics of sedimentary provenance. With these precedents, it is perceived that the documentation of the change of magnetic mineral content and distribution pattern with reference to the sub-environments of the deltaic system would help infer prevalent morphodynamics and thereby evolutionary history, which in turn could be utilized for environmental management plans. It may also turn out to be an inexpensive, simple, yet effective tool for understanding the deltaic dynamics.

2 Justification for the Study and the Study Area

Though many have established the utility of magnetic mineral content in understanding the depositional conditions and sedimentation in river, estuary, marginal marine and deep sea regimes, its utility in the deltaic coastal regions (Ouyang et al. 2013) has not yet been adequately tested owing to the influence of multiple factors in this region. In this paper, the magnetic mineral content at several depths of sedimentary environments of the Godavari delta is determined and used to distinguish the different environments of the delta, to understand their geographic distribution and to evolve its progradational and evolutionary history. Such a study in the Godavari delta appears to be first of its nature. Rao (1994) reported the occurrence of Ilmenite-magnetite-garnet-zircon assemblage in the Vasishta Godavari region that made the present study area an ideal location.

The Godavari River has built an extensive and uniquely shaped delta (Fig. 1). This delta is known for its diverse sub-environments (Ramkumar 2003) and their dynamic changes. Among the east coast deltas of India, the Godavari delta merits special attention from environmental point of view as the region is under 0–11 m topographic contour and appears to be fragile with a thick pile of sediments dipping towards sea and underlain by faults. It was opined by Rao (1998) that any change in the existing equilibrium would devastate the entire region. This information, along with the recurrent life and property loss, almost on a annual basis due to periodic floods and cyclones owing to its unique geographic location, dense network of river and tidal channels, relief characteristics, intense agricultural activities and dense population, typifies this delta into one of the “endangered” coastal region of the East coast of India. Previous studies on this delta includes, geomorphic features (Rao and Vaidyanadhan 1979a), sediment characteristics (Rao and Swamy 1987; Swamy et al. 1990; Ramkumar et al. 2000a; Ramkumar 2001), geochemistry (Rao and Swamy 1995; Ramkumar and Murty 2000; Ramkumar et al. 2000b), land use

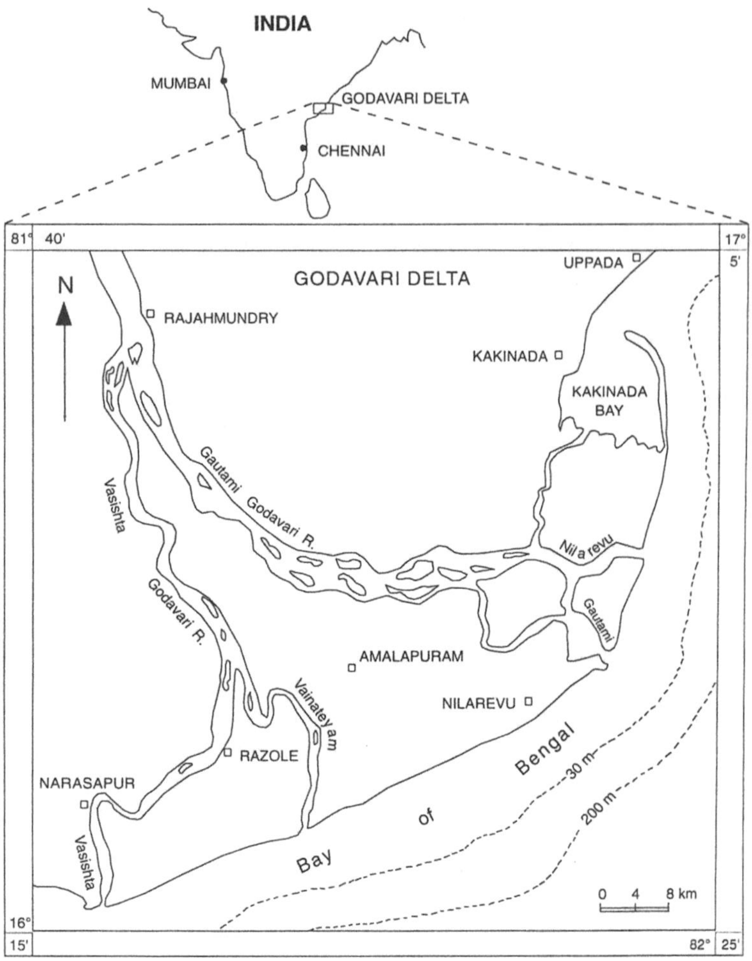

Fig. 1 Godavari Delta and its location along the east coast of India

(Ramkumar et al. 1999a), sea level changes (Ramkumar et al. 1999b) and distribution and dynamics of geomorphic features (Ramkumar 2003). Yet, their magnitude and evolutionary history remain to be explored. This study is an attempt to understand them through documentation of spatio-temporal variation of magnetic mineral contents of various sub-environments of this delta.

3 Location and Setting

The Godavari River is the longest river draining the Peninsular India. It originates in the Western Ghats at an altitude of 1,620 m and flows towards east for a length of about 1,465 km. It has made an extensive arcuate-shaped delta on the east coast of

India, protruding 35 km from the adjoining coast into the Bay of Bengal. The Godavari River deltaic plain starts near Rajahmundry (Fig. 2) and the river branches off into two major distributaries namely, Vasishta and Gautami. The Gautami branches off further into two distributaries namely, Gautami Godavari and Nilarevu. The Vasishta channel also branches into two namely, Vasishta and Vainateyam. There are 21 sedimentary sub-environments in the Godavari Delta (Fig. 2) whose juxtaposition and overlap with the others are controlled by neotectonic and sedimentation processes, which varied with space and time (Ramkumar 1999). The delta building activity of this river commenced during the Lower Cretaceous with the opening of the Bay of Bengal. However, significant sedimentation took place

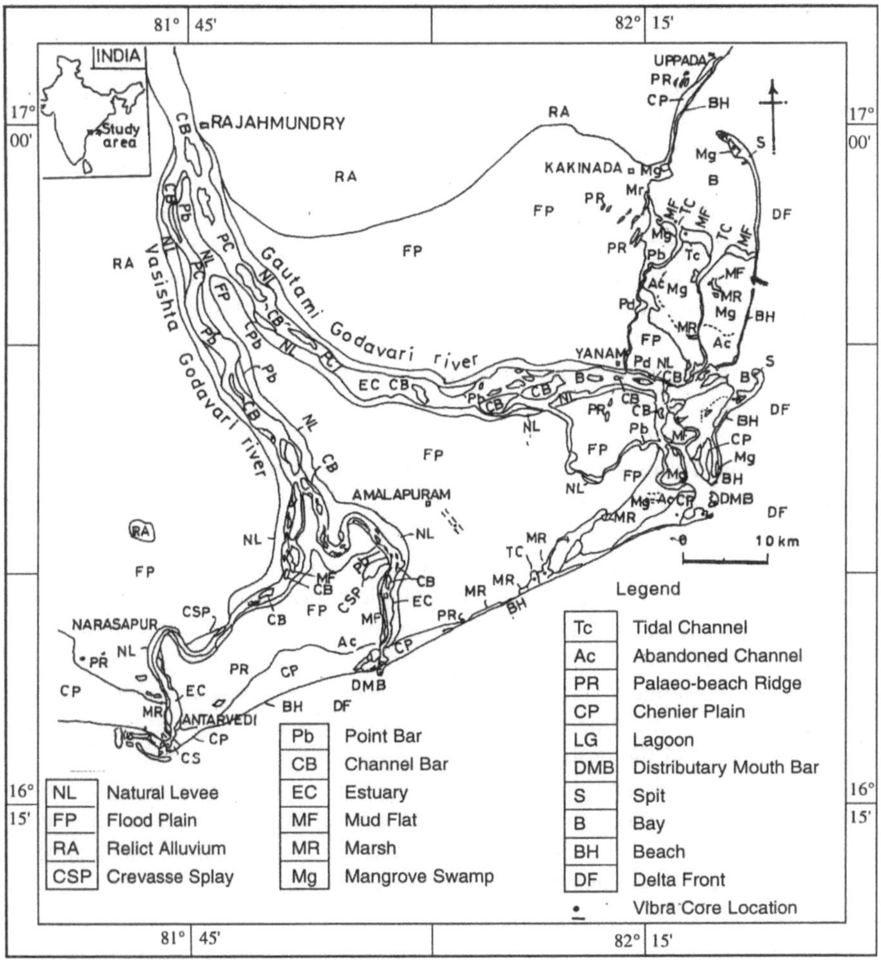

Fig. 2 Distribution of sub-environments within the Modern Godavari delta (After Ramkumar 2003)

during the Neogene (Singh and Swamy 1996) and extensive sedimentation occurred during the 19th and 20th century (Hema Malini and Rao 2004) leading to the development of third largest delta of India. Off Godavari, the continental shelf is 15 km wide and the continental slope starts at 90 m water depth whereas to the north of it, the shelf is much broader. The shelf edge is at about 200 m water depth. The ocean floor gradient is less off Godavari delta due to the significant sediment influx from the four major distributaries of the Godavari River (Ramkumar 2000) that could not be effectively removed either to offshore or to other coastal regions, perhaps due to the fact that the sediment influx outpaces the rate of removal by coastal and marine processes (wave, tide, current).

Significant discharges from the River Godavari commence from June and reach a maximum during August (Suryanarayana 1988). August and September are the months of peak discharge (Sastry et al. 1991). The mean discharge is 3,600 m^3/s and the runoff is 92,245 × 10^6 m^3/year (Subramanian et al. 1987). The freshwater discharge is a potential driving mechanism for coastal circulation though the discharge irregular and non-uniform along the coast. The delta is under the influence of all the three seasons namely, southwest monsoon (June–September), northeast monsoon (October–January) and non-monsoon (February–May) periods. Mean annual rainfall in this delta is 1,100 mm; mean annual temperature is 27.5 °C. The delta is mesotidal (>2 m tidal variations) during normal seasons and may reach up to 3.2 m during spring seasons. Wave width is 30–40 m (near shore). The waves approaching this delta are predominantly swells (responsible for all wider deltas) with significant heights centered around 1 m. These reach 2 m during SW monsoon (Vethamony et al. 1984). From March to September, the waves approach from SE to S direction and during the remaining period, they follow NE-ESE direction. When the wave direction is SE, the associated littoral current is directed up coast while it is directed down coast for the waves approaching NE direction. As wave heights during SW monsoon are higher than those of NE monsoon season, littoral current is stronger during SW monsoon than NE monsoon resulting in net northerly drift all along the coast (Sastry 1958). These coastal currents range from 10 to 30 cm/s with a maximum values recorded up to 200 cm/s.

The river traverses a wide variety of geological formations with the Deccan traps principally in the catchment region occupying 55 % of the total drainage area followed by Archean craton, Khondalite group of rocks of the Eastern Ghats and the Gondwana sediments in the lower reaches in the decreasing order of coverage (Rao 1994). The heavy mineral suite constituting magnetite, ilmenite, zircon, rutile, epidote, etc., found the in the receiving basin (delta) is predominantly drawn from the Deccan traps (Swamy et al. 1990; Sivaprasad 1993; Pattan et al. 2008). It is perceived that as some of these minerals are strongly or weakly magnetic in character, the utility of magnetic mineral content in evaluation of the sedimentary environments could be tested in the Godavari delta under ideal conditions.

4 Methods and Materials

The Modern Godavari delta bounded between N 16° 15′–17° 05′ and E 81° 35′–82° 25′ (Fig. 2) was mapped for landforms *vis-à-vis* sub-environments on a scale of 1:25,000. Field survey was conducted through beach landing craft and trawler boat along the water bodies and by traverses on land to demarcate the neomorphologic features and the changes in occurrence and distribution of older features, and collection of sediment samples from typified sub-environments. A total of 146 uncontaminated surface sediment samples representing typical sub-environments and 119 sub-samples from twenty six vibracore stations were collected initially. The locations of the surface and vibracore sampling are shown in the Figs. 2 and 3. Based on the sand and silt content, 155 samples from these were shortlisted for estimating their

Fig. 3 Locations of the surface and vibracore sampling. Additional sample locations with reference to sub-environments are depicted in the Fig. 2 (After Ramkumar 2003)

magnetic mineral content. The sediments were first desalinized and dispersed with sodium hexametaphosphate. Then, they were air-dried and thoroughly mixed. By coning and quartering method, a portion of the sediment was selected and weighed. The magnetic minerals from this portion was separated with the help of a set of laboratory magnets and weighed and expressed in weight percentage.

5 Results

The magnetic mineral contents expressed in weight percentage are shown in the Table 1 with reference to individual sub-environments. It includes the data related to both surface and sub-surface samples. The geographic distribution of magnetic mineral content is explained through isolines in Figs. 4, 5 and 6. The Fig. 4 explains the distribution of magnetic mineral content at the surface, while the Figs. 5 and 6 show the magnetic mineral distribution at 50 and 100 cm depths respectively.

Table 1 Magnetic mineral content of sub-environments of modern Godavari delta

Environment	N	Minimum	Maximum	Mean	Standard deviation
CB	26	0.119	3.226	1.111	0.905
MF	2	0.025	1.261	0.643	0.874
MR	2	1.699	2.043	1.871	0.243
BY	3	1.148	1.697	1.491	0.299
BH	27	0.289	10.062	2.374	2.852
PR	22	0.599	6.210	1.981	1.212
SP	14	0.444	6.860	1.838	1.929
BW	3	0.540	1.061	0.898	0.311
CP	20	0.161	12.345	2.025	2.591
PB	4	0.899	3.221	2.336	1.082
EC	6	0.060	1.460	0.543	0.493
CSP	1	2.318	2.318	2.318	–
DMB	20	0.439	7.679	1.353	1.647
LG	5	0.387	2.943	1.299	1.094

Note that while all the 21 sub-environments of the Godavari delta (refer Fig. 2 for a complete list of available sub-environments of this delta) have been sampled, due to the paucity of sand and silt content, only 14 sub-environments qualified for magnetic mineral estimation

N Number of samples. Magnetic mineral contents are expressed in weight percentages. *CB* Channel bar; *MF* Mudflat; *MR* Marsh; *BY* Bay; *BH* Beach; *PR* Palaeo sand ridge; *SP* Spit; *BW* Backwater; *CP* Chenier plain; *PB* Point bar; *EC* Estuarine channel; *CSP* Crevasse splay; *DMB* Distributary mouth bar; *LG* Lagoon

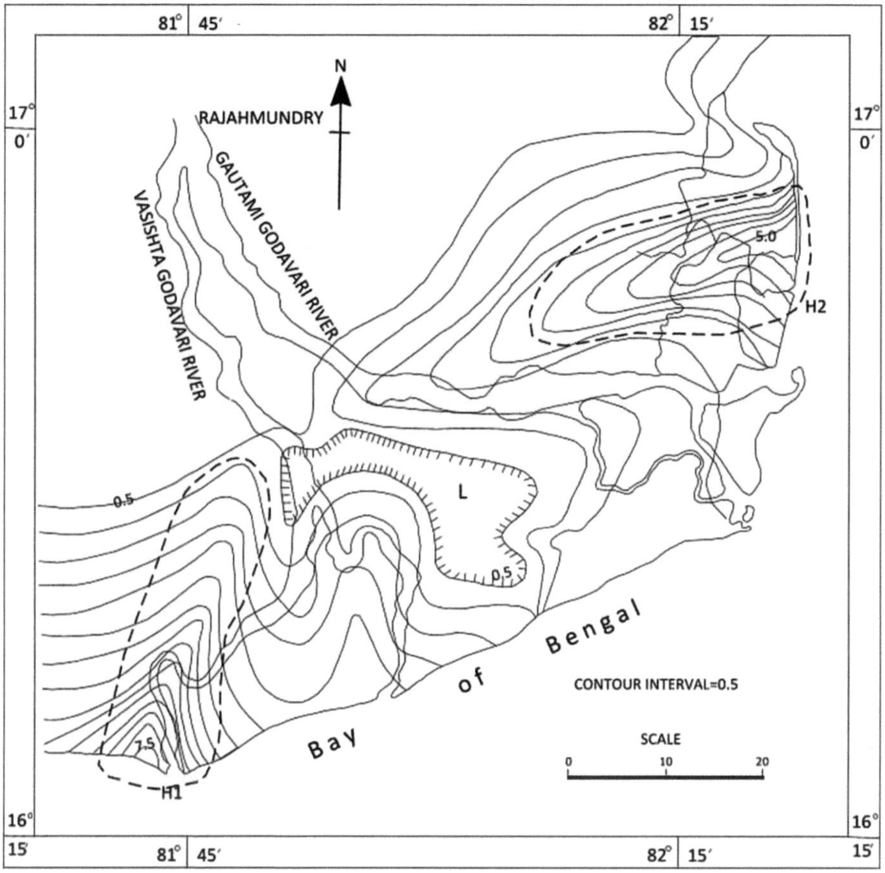

Fig. 4 Isoline map showing magnetic mineral distribution in the surface sediments and the zones of enrichment (H) and depletion (L) within the Godavari delta. While the active delta lobes show the enrichment, the regions that experience low energy and inactive portion of the deltaic regime show the lowest levels of accumulation of magnetic minerals

6 Discussion

6.1 Controls of Magnetic Mineral Content Within Sub-environments of a Deltaic System

As each sedimentary environment is the product of unique set of physical, chemical and biological conditions in a geomorphic setup (Reineck and Singh 1980), each sub-environment within a delta accumulates sediments with unique mineralogical distribution pattern (Ferrell et al. 1998). This uniqueness is produced by the grain size sorting (Sahu 1983; Johan et al. 1990) as a function of prevalent energy, and hydrodynamic conditions (Ramkumar et al. 2000a), and the nature of transportation

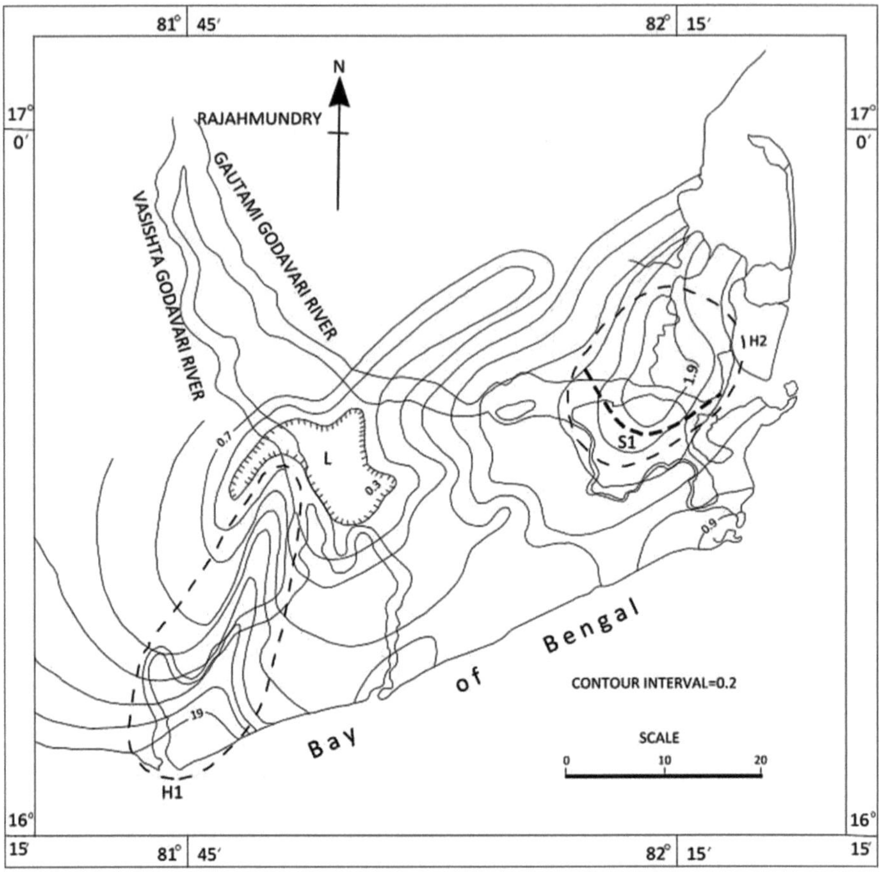

Fig. 5 Isoline map showing the distribution of magnetic mineral content at 50 cm depth. The zones of enrichment and depletion are marked with *dashed line*. *Solid dashed lines* show the configuration and location of inferred palaeo *shorelines* based on maxima (H) and minima (L) zones

and depositional agents, etc. Ferrell et al. (1998) are of the view that the sub-environments in a deltaic system are either dominated by fluvial or marine process. In addition to recognizing these two processes, Ramkumar and Murty (2000) demonstrated the presence of an intermediate process in brackish water regions of the delta wherein both the fluvial and marine processes play roles and create unique groups of sub-environments and thereby unique sets of sediment characteristics owing allegiance to each of those sub-environments and groups of sub-environments. While the groups of deltaic sub-environments forming allegiance to any of these three major processes show significant differences between groups as a result of differences in prevalent energy conditions in terms of their textural, mineralogical and geochemical traits, within each group, they have only minor differences. Ramkumar (2001) demonstrated the presence of unique grain size, mineralogical and

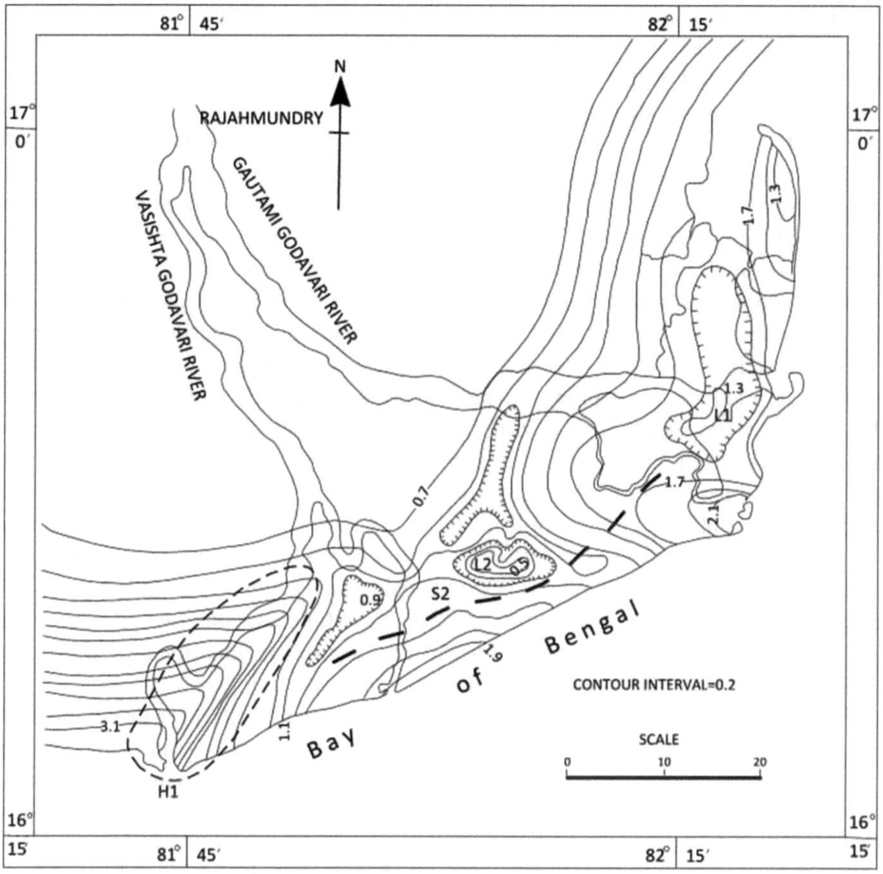

Fig. 6 Isoline map showing the distribution of magnetic mineral at 100 cm depth. *Note* the change in configuration and areal extent of maxima and minima zones. The inferred palaeo shoreline also shows different configuration and location

geochemical traits for the sediments deposited within each of the sub-environments and also their general grouping influenced by three major processes that are operative within the deltaic system. It is perceived that an ability to exploit the uniqueness among the sub-environments of each group and significant difference between groups would help understand the spatial and temporal variations of distribution of sub-environments.

Frihy et al. (1995) utilized the spatial distribution of magnetic mineral in the Nile delta to assess the variations of sediment grain sorting in response to the changes in prevalent hydrodynamic conditions. Frihy and Dewidar (2003) demonstrated the efficacy of using magnetic mineral distribution along with grain size and heavy mineral data to identify littoral sub-cells and with which categorized the study area in terms of erosional/depositional regions. These results have formed the basis for the study of magnetic properties of the coastal sediments of the northwestern Lake Erie, Canada to identify the regions of erosion/deposition (Hatfield et al. 2010). Vijayalakshmi et al. (2011) are of the opinion that the occurrences of beds enriched with heavy minerals are indicative of prevalent erosional episodes. However, Rao et al. (2012) are of the opinion that heavy mineral distribution is principally controlled by selective sorting of the sediments. Combining both of these statements implies that, during erosional episodes, the finer and less denser sediments are winnowed away to enrich the heavier sediment grains to get entrained at the sites of active erosion. Based on their observations on the eastern continental margin of India, Mislanker and Gujar (1996) suggested that reworking of ancient sediments during sea level rise concentrated the heavy minerals. In an experimental field study involving radioactive tracer technique, Gallaway (2012) documented that during low to intermediate wave energy conditions, the magnetic minerals remain below the surface in the swash zone (a zone of active sediment sorting and transport), do not contribute to the net sediment transport and move only during the high energy wave conditions. Horng and Chen (2006) are of the view that the occurrence of magnetic mineral assemblages in the marine regime is influenced by a complicated set of factors. Hatfield et al. (2010) stated that the selective entrainment of dense grains including magnetic minerals and erosion of lighter quartz grains leads to enrichment magnetic grains within the beach sediments, and by studying their magnetic properties, it is possible to ascertain quantitative identification of coastal erosion-prone areas and implement mitigation measures.

From Table 1, it follows that out of 21 sub-environments present in this delta (refer Fig. 2 for a complete list and their distribution), only 14 sub-environments contain noticeable amounts of magnetic minerals. Elucidation of the sub-environments that do not have magnetic minerals reveals that they experience very low energy conditions. Table 1 also shows that among all the sub-environments, only mudflat (Plate 1.1) has least amount of magnetic mineral content while the chenier plain and beach sub-environments showed the maximum. These observations along with the information reviewed in previous paragraphs pertaining to the supportive conditions of heavy mineral occurrence permit to interpret that occurrence and quantum of magnetic mineral in this deltaic system is primarily controlled by the relative energy conditions prevalent. This inference is further affirmed by the higher magnetic mineral content of sub-environments that experience higher energy conditions such as distributary mouth bar, spit, paleo sand ridge, chenier plain and beach (Plates 1.2 and 1.3). The inland sedimentary environments such as point bar and channel bar (Plate 1.4) have magnetic mineral contents next to these higher energy sub-environments.

Sundararajan et al. (2010) are of the opinion that the enrichment of heavy minerals including Ilmenite in the vicinity of river mouths is due to the selective size and density sorting under the control of combined actions of high energy waves, long shore currents and fluvial influx. While analyzing the heavy mineral distribution along the coastal region of the Krishna delta, Reddy et al. (2012) observed that within the heavy mineral suit, towards river mouths, high-specific gravity heavy minerals are found enriched while a trend of heavy minerals of decreasing specific gravity are found at regions away from the river mouths.

1.1
1.2
1.3
1.4
1.5
1.6
1.7
1.8
1.9
1.10

◀**Plate 1** **1.1** Field photograph showing the mudflat sub-environment. *Note* the deposition of very fine silt and clayey sediments as a result of low energy conditions experienced by this sub-environment. **1.2** Beachface located near the mouth of Nilarevu showing the predominant deposition of heavy minerals as a result of very high energy conditions prevailing there. These *dark colored* heavy minerals are dominated by magnetite, ilmenite and rutile. **1.3** Beachface located near the confluence point of Vasishta Channel showing the alternate deposition of *light colored* and *dark colored* sand beds indicative of alternations of relatively high and intermediate energy conditions. **1.4** Field photograph showing the cross sectional exposure of channel bar deposits containing cross bedded coarse sandy sediments. These have magnetic mineral content higher than the mudflat and point bar but less than the beach and chenier plain sub-environments. Location of the photograph: Northeast of Narasapur in the Vasishta estuarine channel. **1.5** Erosional beachface showing the concentration of heavy minerals suggesting the correlation between higher energy conditions and enrichment of magnetic minerals in the ensuing sedimentary record. **1.6** Field photograph showing the erosional cut bank along the upstream part of the Godavari River. Note that the coconut plantations (indicative of stabilized land) are visible within the channel, indicative of the severity of the erosion along stream banks. **1.7** Field photograph showing the development of cut bank along the southern bank of Nilarevu channel. **1.8** Field photograph showing the palaeo sand ridge-swale complex located south of Amalapuram. Excavation of dug pits and trenches at this location have exposed the occurrences of typical river channel deposits. Riverine character has also been affirmed by facies associations, sedimentary structures and sediment textural properties. Recognition of river channel-palaeo sand ridge association suggested debauching of distributary of Godavari and prevalence of shoreline at this location. This inference also paved way for the interpretation of dynamic nature of Godavari, shifting nature of its active lobe, etc. **1.9** Field photograph showing the cut-bank along the northern side of the Vasishta exposing Crevasse splay deposits. Owing to their very nature, the Crevasse splay deposits originate at locales topographically below the bed and flood levels of major channels. However, occurrences of Crevasse splay deposits exposed along the channel bank indicate neotectonic movements (uplift of land) in the vicinity. **1.10** Field photograph showing the wave-cut terrace near Kakinada

Together, all these information suggest that the occurrence and proportion of magnetic mineral content in the deltaic sub-environments are influenced by the relative energy distribution in the depositional environment and also by relative intensity of erosion and/or selective removal of less dense minerals. The geomorphic studies (Ramkumar 1999, 2000, 2003) have shown that the deltaic coastline (Plate 1.5) and banks of major distributary channels (Plates 1.6 and 1.7) are undergoing severe erosion. The superposition of erosional regions and the concentration of magnetic mineral content are observed in the field and clearly indicate the prevailing correlation between them. Thus, it can be interpreted that the quantum of magnetic mineral present in an environment is controlled by energy and size/density sorting of sediments.

6.2 Spatial-Temporal Distribution of Magnetic Mineral Content

Increase of magnetic mineral content on a spatial scale from inland towards present day coastline is explicitly shown by the Fig. 4 that depicts the distribution of magnetic mineral content based on surface samples. It shows a five-fold increase of

Plate 2 **1.1** Beach rock exposures are found scattered all along the deltaic shoreline. Beach rocks are definitively associated with deltaic progradation. Exposures of beach rocks at wave-cut terraces suggest the ongoing shoreline movement towards inland, sea level rise and neotectonic activity. **2.2** Field photograph showing the wave cut terrace on previously prograded and stabilized lands (mangrove swamp) by shifting of active delta lobe from Vasishta to Gautami Godavari. The photograph shows not only the wave cut terrace over mangrove swamps, but also deposition of heavy mineral rich beach sands over them indicative of severity of the active erosion and the rate of shoreline movement towards inland regions. **2.3** Field photograph showing a live freshwater tree amidst sea. Owing to the very gentle slope of the deltaic plain, even slightest change in sea level/neotectonics/fluvial sediment and water influx changes the field conditions drastically at a faster rate. **2.4** Field photograph showing the submergence of chenier plains under tidal waters. Traditionally, occurrences of chenier plains are considered as the indicators of advancement of deltaic land over former marine regions. As the photograph evidences, previously prograded areas of the Vasishta lobe of the Godavari delta are experiencing submergence either due to subsidence, sea level rise, and reduction of fluvial sediment and water influx. **2.5** Field photograph showing the fresh conversion of mangrove swamps into aquaculture ponds. The entire lower deltaic region is under threat due to these ponds. **2.6** Field photograph showing the submergence of mangrove swamp under tidal waters and formation of newer tidal channel. All along the coastline in the vicinity of Vasishta lobe, such newer dense network of tidal creeks and channels are getting formed, which may indicate subsidence of the region

magnetic mineral content from inland (0.05 %) to more than 2.5 % along the present day coastline. While a gradual increase of magnetic mineral content from inland towards coastline could be a normal phenomenon, such five-fold increase, that too all along the coastline regardless of active delta lobe (Gautami Godavari and Nilarevu) and less active lobe (Vasishta and Vainateyam) could only be explained by the relatively subdued nature of the fluvial forces against the marine forces. Hema Malini and Rao (2004) reported dwindling freshwater and sediment influx into the lower deltaic plains and resultant erosional features and associated morphological and sedimentological changes in the lower deltaic plains, particularly along the shoreface. Thus, the predomination of marine forces over the fluvial depositional/erosional characteristics except during the major flood seasons could explain the five-fold increase of magnetic mineral content from inland towards coastline. It may also explain the exponential dissipation of the marine forces from the coastline towards farther inland regions.

It may also due to the fact that, from the delta head (Rajahmundry), the deltaic plain advanced over the former marine regions through four stages (Rengamannar and Pradhan 1991) and at each stage, on the spatial scale, paleo channel-paleo sand ridge-chenier plain and relict alluvium complexes occur (Ramkumar 2003). However, the temporal scale between these four stages is not uniform. According Hema Malini and Rao (2004), very significant deltaic advancement took place only during the 19th and 20th centuries. These information, along with the punctuated sea level rise, alternative activation of the Vasishta and the Gautami channels of the Godavari River (Ramkumar 2000) could have resulted in progressive but exponential increase of selective sorting and erosion at the fluvial-marine interface and resulted in the five-fold increase of the magnetic mineral distribution from inland towards present day coastline. These assumptions are supported by the following magnetic mineral distribution pattern at surface.

The isoline map of magnetic mineral content at surface (Fig. 4) shows two local maxima zones marked as H1 and H2 trending NE-SW and E-W respectively and a minimum zone marked as L located in between the maxima zones. These zones of maxima are parallel to the major distributary channels and hence can be explained by the severe erosion that is taking place along the channels as reported by Ramkumar et al. (1999a, 2000a). The minimum zone located between the maxima zones is naturally due to little or no erosion at that place. Such a zone of least magnetic mineral content is expected because, due to the arcuate shape of this delta, waves always obliquely approach it with the result that erosion takes place along the peripheries and a low energy region prevails at the centre of active delta lobe. In view of this low energy condition, winnowing away of fine sediments does not also take place here restricting the heavy mineral accumulation/enrichment at a reduced level. Similar inferences have been reported earlier by Ramkumar (2000, 2003) based on the erosional and depositional statistics. Examination of magnetic mineral distribution at 50 and 100 cm depths have been examined in the light of these observations and geomorphic information and are presented herein.

Figure 5 retains the NE-SW maxima zone (H1) which has actually undergone a minor change in its trend of N 50° E at the surface to N 45° E at 50° cm depth. This may mean that the Vasishta had an altogether different course during that period. However, the presence of the maxima zone at depth clearly establishes the existence of erosion and higher energy conditions (meaning either beach or distributary mouth bar or chenier plain or sand ridges) during geologic past, when deposition of this strata took place. Even at the 100 cm depth, this maxima zone persists (Fig. 6), but its trend is the same of the present day configuration (N 50° E). This is to imply that the Vasishta channel has undergone changes in its course repeatedly in the past.

The H2 maxima observed at surface in the Gautami sector has undergone a drastic change of trend to NNW-SSE at 50 cm depth (Fig. 5), whereas at the 100 cm depth, this maximum actually turned out to be minimum zone (marked as L1) with more or less the same trend as at the depth 50 cm. The formation of this minimum can only be explained by invoking the presence of a shoreline-paleo channel complex at this place. If so, the Gautami distributary should have had an altogether different course which should be quite north of its present day course. Such a divergent course appears to be justified because a shoreline-paleo channel complex is reported at this location in the archival naval hydrographic charts of 17th century. The high occurring at the same location in Fig. 4 shows that the Gautami distributary has already started moving towards south and moved amply by that time. The shapes of these maxima (at 50 cm depth) and minima (at 100 cm depth) zones also follow the coastal configuration depicted in the hydrographic chart of 17th century, adding credence to the interpretation of the paleo shoreline (marked in the Figs. 5 and 6 as S1) on the basis of magnetic mineral content. The Fig. 6 shows another minima zone (L2) adjacent to the H1 maxima zone. At this location, a paleo channel that met the sea south of Amalapuram was reported earlier by Ramkumar (2003) based on the facies association, sedimentary structures and sediment characteristics.

6.3 Morphodynamics and Evolutionary History

When rivers build land and prograde over the sea by accumulating sediments at the lower deltaic plains, they branch off into many distributaries (Coleman et al. 1970). The present day configuration of the Godavari River shows hierarchical and sequential branching nature and could be the result of prevalent progradational phase of the delta. Occurrence of paleo channels at south of Amalapuram adjoining paleo sand ridges (Plate 1.8), as indicated by the configurations of magnetic mineral maxima and minima zones at varying depths and the occurrences of maxima zones at varying configurations at different depths suggestive of alternating active nature of the Gautami and the Vasishta channels of the Godavari River suggest the prevalent similar sequential branching and abandonment processes of river courses in the geological past. It is interesting to note that while the minimum zone remained almost stable even at different depths, the maxima zones varied considerably. This information, along with the observation that the surface samples show higher magnetic mineral contents along river channels that show intense erosional characteristics could pave way for the affirmation of alternative activation of the Gautami and the Vasishta channels of the Godavari delta in the geologic past too.

While the prevalence of selective sorting in beach, sand ridge and distributary mouth bars could be expected due to the wave reworking and resultant selective removal of finer and less denser sediment grains (Hatfield et al. 2010), leading to the enrichment of heavier and larger grains in the ensuing sediment record, the occurrences of higher concentrations of magnetic mineral in the crevasse splay (Table 1) suggests an interesting feature. The crevasse splay deposits (Plate 1.9) occur at ground level below the maximum flood level and normally contain sediments emanated during flooding events. Sedimentation in this sub-environment occurs during flooding events without sorting. This trait of the crevasse splay sub-environment together with higher magnetic mineral content in the Crevasse splay deposits suggest less than normal supply of sediments during flooding events, concomitant sorting of flood deposits due to the location of the Crevasse splay within estuarine region and continually active neotectonic activity at this location. This inference is supported further by three more observations:

- Occurrences of crevasse splay deposits significantly only adjoining the Vasishta and the Vainateyam estuaries, known for significantly active Cretaceous faults at sub-surface, and general E-W trending nature of crevasse splay deposits,
- occurrences of incision of natural levee to a tune of more than 10 m above highest high water line along the Godavari main channel extending for many kilometers (Ramkumar 2003) and
- a typical erosional cliff section of a channel bar located in the Vasishta estuary has exposed the presence of clay deposits at top followed towards bottom by cross bedded sand and again a clay bed that in turn is followed by a cross bedded sand bed. Below this, river channel sand bed and pebble bed are present. Whole of this sequence indicates cyclic but intermittent deposition. Occurrence of the river channel sand bed and pebble bed 3 m above the highest high waterline definitively affirms the lowering of bed level (incision) under the influence of uplift of land by neotectonic activity.

The ground truth data and measurements based on the hydrographic and topographic sheets published since 17th century (Prakash and Agarwal 1979) have revealed that the shoreline all along the Godavari delta is erosional (Plates 1.10, 2.1, 2.2, 2.3, and 2.4) except the Nilarevu and Gautami mouths (Ramkumar 2000, 2003). The wave cut cliffs exposed all along the coast expose the presence of lithofacies sequences typical of progradational phases of the delta. These suggest that the delta engages progradational phase at one region of the delta lobe while the previously prograded areas undergo adjustments to the newer wave, current and fluvial supply conditions. Given conducive milieu, the adjusted regions become the loci of active erosion and vice versa, resulting in net sediment deposition for the entire deltaic system and progradation of the delta over former marine regions. These inferences have amply been reflected by the configurations and distributions of maxima and minima zones at various depths. Based on these observations and inferences, it may be affirmed that the coastline had changed its configuration in the past. The coastline configuration that could be expected during or prior to the year 1789 is shown in the Fig. 6, marked as S1 and S2. When these configurations are accepted, the zones of maxima and minima of magnetic mineral content stand explained. The results of the geomorphic studies conducted earlier (Mahadevan and Rao 1958; Rao and Vaidyanadhan 1979a, Rengamannar and Pradhan 1991; Ramkumar 1999, 2000, 2001, 2003) also confirm the interpretations made with the help of magnetic mineral content and their geographic distribution during various times of deposition (temporal).

Based on these interpretations on shoreline locations and the best available datum of the year 1789, the rate of deposition can be estimated as 100 cm per 210 years or 0.476 cm/year. Ramkumar et al. (1999b) showed that the Godavari delta, in its marginal section, experienced a combined effect of 134 cm of subsidence and sea level rise, while the global rate of sea level rise is 1.2 mm/year (Barnett 1984; Bird 1985). Since the year 1789, the sea level should then have risen by 25.2 cm and the subsidence should have been 1.09 m.

6.4 Implications on Deltaic Coastal Zone Management

The Godavari deltaic plain is one of the most intensively cultivated regions of India. This region supports dense population by providing sustenance through agricultural, industrial, commercial and recreational avenues. In addition, the delta is home to vast expanses of mangrove swamps that are next only to the great Sundarbans of the Ganges delta. Occurrence of huge reserves of hydrocarbon in this delta (onland and offshore), and the use of the coastal lands for aquaculture have brought far more human populace and all other related infrastructure in this region. On the other hand, owing to the vast expanse of low lying areas, traditionally the Godavari deltaic plain is under the mercy of recurrent floods, cyclones and storms. Ramkumar et al. (1999a) observed rapid shrinking of natural channels and vegetation for the sake of aquaculture ponds (Plate 2.5), salt pans and domestic, industrial and recreational land uses. Furthermore, the Godavari River has one of the longest estuarine channel (extending for about 45 km inland from shoreline) and drains in a very low gradient (the delta head occurs only at 11 m above sea level) and a mere 2 m tidal waters during normal season and 3.2 m tidal waters during spring season transport sea water to the tune of 45 km inland), making the entire region precariously located within the reach of any potential natural calamities such as catastrophic tidal waves, storms, cyclones and flooding. Formation of dense network of tidal channels is observed in the vicinity of Coringa Channel north of Nilarevu and has been interpreted to be the expression of significant subsidence in the region (Ramkumar 2003) based on the criterion suggested by Gould (1970). The Vasishta lobe also shows similar features (Plate 2.6) all along the coastline which may also suggest the commencement of significant subsidence in the vicinity. Added to these is the problem of construction of many dams and other structures at upper deltaic regime that restricts influx of sediment and water to the downstream regions creating ecological imbalance (Hema Malini and Rao 2004).

Rao (1998) cautioned about possible geohazards due to intensive hydrocarbon exploitation from underground reservoirs as the region experiences neotectonic movements and contains thick pile of unconsolidated sediments dipping towards the sea. According to Rao et al. (2004), there are many shallow gas pockets, perhaps associated with deep seated faults and connected through neotectonic lineaments in the central part of lower deltaic plain. These may enhance the vulnerability of the region. According to Rengamannar and Pradhan (1991), the Godavari River has built its vast deltaic plain through four stages of development. Ramkumar (2003) commented that from the third stage of Rengamannar and Pradhan (1991), the deltaic development had been under the influence of alternate northerly and southerly tilting of the underlying faults, due to which the delta prograded through alternative activation of Gautami lobe and Vasishta lobe. These information, together with the fault controlled stream course of Vasishta channel in the lower deltaic plain, the observed uplift and resultant incision of channel bed, occurrences of wave cut terraces all along the deltaic coastline and submergence of previously prograded areas such as mangrove swamps and chenier plains under tidal waters, suggest the essentiality of proper coastal zone management strategy for this region.

7 Conclusions

Deltaic systems have been one of the most complex geological systems that require proper understanding for the sustainable development of ecosystems and for supporting the human endeavors. In this paper, a simple tool of analysis of spatio-temporal variations of magnetic mineral contents of surface and sub-surface sediments was tested for its utility and effectiveness for recognition of morphodynamics and to understand the evolutionary history of the delta. Following are the major conclusions of this attempt.

- The present study had demonstrated that the measurement of magnetic mineral content in deltaic sediments could help delineate sub-environments deposited under relatively low, intermediate and high energy conditions and to identify individual sub-environments from surface and sub-surface sediments.
- The amount of magnetic mineral content in different sub-environments of the deltaic system is influenced by reworking of the sediments brought into the depositional loci. Relative high energy and low rate of sedimentation are the conducive milieu for the prolific occurrence of magnetic mineral content.
- The present study has established the utility of analyzing magnetic mineral content in deltaic sediments towards recognition of paleo shorelines and paleo channels. When landform and tectonic data are added, this helps in enumerating the evolutionary history of the delta.

- Through a conservative estimate, the study has brought out the rate of sedimentation in this delta as 0.476 cm/year since the year 1789. Subsidence of the coastal land is estimated to be 1.09 m, while the sea level rise is 25.2 cm since the year 1789.
- The inherent topographic, fluvial, oceanographic and land use characteristics of the Godavari delta along with the observed geomorphic dynamics, neotectonic activity and the unplanned developmental activities in terms of domestic, industrial, commercial and recreational projects in the vicinity pose severe threat to its sustenance and warrants implementation of comprehensive study based deltaic coastal zone management programs.

Acknowledgments The work was conducted during the author's tenure at the Delta Studies Institute, Andhra University, Visakhapatnam. Partial financial assistance in the form of research grant for this study was provided by ONGC. Prof. I.V. Radhakrishnamurty, (Retd.), Department of Geophysics, Andhra University is thanked for reading earlier version of this manuscript. However, author alone is responsible for the correctness of the data and the interpretations.

References

Barnett TP (1984) The examination of global sea level changes: a problem of uniqueness. J Geophys Res 87:7980–7988

Bird ECF (1985) Coastline changes. Wiley, Chichester 218 pp

Coleman JM, Gagliano SM, Smith WG (1970) Sedimentation in a Malaysian high tide tropical delta. In: Morgan JP (ed) Deltaic sedimentation: modern and ancient, vol 15. SEPM Special publication, pp 185–197

Ferrell RE, Hart GF, Swamy ASR, Bhanumurty P (1998) X-ray mineralogical discrimination of depositional environments of the Krishna delta, Peninsular India. J Sediment Petrol 68:148–154

Frihy OE, Lofty MF, Komar PD (1995) Spatial variations in heavy minerals and patterns of sediment sorting along the Nile Delta. Egypt Sediment Geol 97:33–41

Frihy OE, Dewidar KM (2003) Patterns of erosion/sedimentation, heavy mineral concentration and grain size to interpret boundaries of littoral sub-cells of the Nile Delta. Egypt Mar Geol 199:27–43

Gallaway E 2012 Magnetic mineral transport and sorting in the swash-zone: northern Lake Erie. Masters thesis submitted to the University of Windsor. Electronic theses and dissertations, Paper 152, 59 pp

Gould HR 1970 The mississippi delta complex. In: Morgan JP (ed) Deltaic sedimentation: modern and ancient, vol 15. SEPM special publication, pp 3–30

Hatfield RG, Cioppa MT, Trenhaile AS (2010) Sediment sorting and beach erosion along a coastal foreland: magnetic measurements in point Pelee National Park, Ontario. Canada Sediment Geol 231:63–73

Hema Malini B, Rao KN (2004) Coastal erosion and habitat loss along the Godavari delta front—a fallout of dam construction (?). Curr Sci 87:1232–1236

Horng CS, Chen KH (2006) Complicated magnetic mineral assemblages in marine sediments offshore of Southwestern Taiwan: possible influence of methane flux on the early diagenetic process. Terr Atmos Ocean Sci 17:1009–1026

Johan CS, Majumder TK, Roy MK (1990) Sedimentary environmental discrimination using grain size analysis. J Geol Soc Ind 35:529–534

Karbassi AR, Shankar R (1994) Magnetic susceptibility of bottom sediments and suspended particulates from Mulki-Pavanje River, estuary, and adjoining shelf, west coast of India. J Geophys Res 99:10207–10220

Lakshmipatiraju A, Rao LS (1996) Heavy mineral concentration on the Visakhapatnam-Konada Coast as evidenced by magnetic susceptibility and radioactivity. J Geol Soc India 47:717–724

Mahadevan C, Rao R (1958) Causes of the growth of sand spit north of Godavari confluence. Memoirs in oceanography. VII Andhra University Series, vol 62, pp 69–74

Mislankar PG, Gujar AR (1996) Heavy mineral distribution in the surficial sediments from the eastern continental margin of India and their implications on paleoenvironment. Ind J Earth Sci 23:91–97

Nayak S, Chauhan P, Chauhan HB, Bahuguna A, Nath AN (1996) IRS-1C applications for coastal zone management. Curr Sci 70:614–618

Ouyang T, Appel E, Jia G, Huang N, Zhu Z (2013) Magnetic mineralogy and its implication of contemporary coastal sediments from South China. Environ Earth Sci 68:1609–1617

Pattan JN, Parthiban G, PrakashBabu C, Khadge NH, Paropkari AL, Kodagali VN (2008) A note on the geochemistry of surface sediments from Krishna-Godavari Basin, East coast of India. J Geol Soc Ind 71:107–114

Prakash S, Agarwal J 1979 A coastal geomorphic study of Godavari point sand spit (as evaluated from hydrographic surveys and LANDSAT imagery). Natl Geogr J India XXV:85–92

Rao BN (1994) Significance of heavy mineral ratios in some beach placers of Andhra Pradesh, East coast of India. Curr Sci 67:535–537

Rao GK (1998) Geo-hazard with oil and gas production in Krishna-Godavari basin. Curr Sci 74:494

Rao MP, Swamy ASR (1987) Clay mineral distribution in the Mangroves of the Godavari delta. Clay Research 6:81–86

Rao PS, Swamy ASR (1995) Distribution of trace elements in the shelf sediments off Vasishta-Vainateyam Godavari river, East coast of India. Ind J Geochem 10:1–11

Rao SM, Vaidyanadhan R (1979) Morphology and evolution of Godavari delta, India. Zeitschrift fur Geomorphologie N F Bd 23:243–255

Rao CN, Anu Radha B, Reddy KSN, Dhanamjayarao EN, Dayal AM (2012) Heavy mineral distribution studies in different micro-environments of Bhimunipatnam coast, Andhra Pradesh. India Inter J Sci Res Publ 2:1–10

Rao MN, Desapati T, Subramanyam AV, Deshmukh RD, Viswanathan G, Sinha RM (2004) Natural gas at shallow depth in the placer sands of Amalapuram coast, East Godavari district, Andhra Pradesh. Curr Sci 87:144–146

Ramkumar M (1999) Sedimentary micro-environments of the Godavari delta. Technical Report submitted to ONGC (unpublished), India. Part I Distribution and morphodynamics 47 pp

Ramkumar M (2000) Recent changes in the Kakinada spit, Godavari delta. J Geol Soc India 55:183–188

Ramkumar M (2001) Sedimentary environments of the modern Godavari delta: characterization and statistical discrimination towards computer assisted environment recognition scheme. J Geol Soc India 57:49–63

Ramkumar M (2003) Progradation of the Godavari delta: a fact or empirical artifice? insights from coastal landforms. J Geol Soc India 62:290–304

Ramkumar M, Murty MV (2000) Distinction of sedimentary environments of the Godavari delta using geochemical and granulometric data through analysis of variance (ANOVA). Ind J Geochem 15:69–84

Ramkumar M, Pattabhi Ramayya M, Swamy ASR (1999a) Changing landuse/land cover pattern of coastal region: Gautami sector, Godavari delta, India. J A P Acad Sci 3:11–20

Ramkumar M, Pattabhi Ramayya M, Gandhi MS (1999b) Beach rock exposures at wave cut terraces of modern Godavari delta: Their genesis, diagenesis and indications on coastal submergence and sealevel rise. India J Mar Sci 29:219–223

Ramkumar M, Sudha Rani P, Gandhi MS, Pattabhi Ramayya M, Rajani Kumari V, Bhagavan KVS, Swamy (2000a) Textural characteristics and depositional sedimentary environments of the modern Godavari delta. J Geol Soc Ind 56:471–487
Ramkumar M, Rajani Kumari V, Pattabhi Ramyya M, Murty MV (2000b) Geochemical characteristics and depositional conditions of coastal sedimentary environments of the modern Godavari delta. Ind J Geochem. 15:31–44
Ratcliffe KT, Wright AM, HAllsworth C, Morton A, Zaitlin BA, Potocki D, Wray DS (2004) An example of alternative correlation techniques in a low accommodation setting, non-marine hydrocarbon system: the (Lower Cretaceous) Mannville Basal Quartz succession of southern Alberta. Amer Assoc Petrol Geol Bull 88:1419–1432
Reddy KSN, Deva Varma D, Dhanamjaya Rao EN, Veeranarayana B, Lakshmi Prasad T (2012) Distribution of heavy minerals in Nizampatnam-Lankavanidibba coastal sands, Andhra Pradesh, East coast of India. J Geol Soc India 79:411–418
Reineck HE, Singh IB (1980) Depositional sedimentary environments. Springer, New York 439 pp
Rengamannar V, Pradhan PK 1991 Geomorphology and evolution of Godavari delta. In: Vaidyanadhan R (ed) Quaternary deltas of India. Mem Geol Soc India 22:51–56
Sahu BK (1983) Multigroup discrimination of depositional environments using grain size statistics. India Jour Earth Sci 10:20–29
Sastry JS 1958 Some aspects of shoreline processes and physical oceanography. D.Sc. thesis submitted to the Andhra University (unpublished), Visakhapatnam
Sastry JS, Vethamony P, Swamy, GN 1991 Morphological changes at Godavari delta region due to waves, currents and the associated physical processes. In: Vaidyanadhan R (ed) Quaternary deltas of India. Mem Geol Soc India 22:139–151
Shankar R, Karbassi AR (1991) Geochemistry and magnetic susceptibility of surficial sediments of the New Mangalore port. J Geol Soc Ind 38:412–417
Singh IB, Swamy ASR 1996 Deltas of India: technical report submitted to ONGC., Dehra Dun, 247 pp
Sivaprasad G 1993 Sedimentological and geochemical studies of Krishna-Godavari prodelta, east coast of India. Ph.D. thesis submitted to the Andhra University (unpublished), Visakhapatnam
Subramanian V, Bisham G, Ramesh R (1987) Environmental geology of Peninsular river basins of India. J Geol Soc India 30:393–401
Sundararajan M, Bhat KH, Velusamy S (2010) Investigation on mineralogical and chemical characterization of Ilmenite deposits of northern Kerala coast. India World App Sci J 9:333–337
Suryanarayana A 1988 Effect of wind and freshwater discharge on hydrography and circulation of the western Bay of Bengal. Ph.D. thesis submitted to the Andhra University (Unpublished), Vishakapatnam
Swamy ASR, Rao MP, Rao BK (1990) Sediment characteristics of the Modern deltas of the East coast of India. Proceedings of the International conference on Asian Marine Geology, China, pp 251–265
Vethamony P, Gopalakrishna VV, Varkey MJ (1984) Wave spectra and statistics off Godavari during Sept–Oct 1980. Mausam 35:199–204
Vijayalakshmi CS, Trivedi D, Srinivasan P, Murthy SGN, Nair RR (2011) Subsurface reflections of prograded paleoscarps in Avis Island, North Andaman. India. Ind J Geomar Sci 40:331–337

GIS Based Quantitative Geomorphic Analysis of Fluvial System and Implications on the Effectiveness of River Basin Environmental Management

A. Venkatesan, A. Jothibasu and S. Anbazhagan

Abstract Rivers are sensitive to changes in tectonic deformation, and adjust themselves on different scales of time periods depending on the physical properties of the host rocks and climatic effects. The resultant changes are exhibited by the geomorphic indices and landform assemblages within a river basin. This paper presents the results of integrated quantitative geomorphic analysis conducted for understanding the prevalent tectonic activities in a medium sized drainage basin, the Thoppaiyar sub-basin, India. The major part of the study area is covered by gneisses and granites. The sub-basin is divided into fourteen fourth order micro basins (FOMBs) for quantitative geomorphic analysis. Prior to quantitative analysis, longitudinal river profile and channel morphology were studied. The channel morphology includes cross sections, width-to-depth ratio, entrenchment ratio, bank height ratio were measured during field investigation. Various geomorphic indices namely, the basin shape index (Bs), drainage basin asymmetry factor (Af), hypsometric integral (Hi), hypsometric curve (Hc), valley floor width-to-height ratio (Vf), transverse topographic symmetry (T) and stream length gradient index (SL) were derived using topographic maps and SRTM satellite data. The spatial distributions of these parameters were represented as thematic layers using. The results obtained from these indices were combined by ArcGIS 9.3 software to generate an index of relative active tectonics (I_{RAT}) in the sub-basin. It indicated the prevalence of differences among the FOMBs and an overall relatively low tectonic activity in the Thoppaiyar sub-basin.

Keywords GIS · Channel morphology · Geomorphic indices · Relative active tectonics

A. Venkatesan
Department of Geology, Periyar University College of Arts and Science, Mettur, India

A. Jothibasu · S. Anbazhagan (✉)
Department of Geology, Periyar University, Salem 636 011, India
e-mail: anbu02@gmail.com

Mu. Ramkumar et al. (eds.), *Environmental Management of River Basin Ecosystems*, Springer Earth System Sciences, DOI 10.1007/978-3-319-13425-3_11

1 Introduction

Geomorphology represents either qualitative or quantitative nature of landforms, landscapes and surface processes including their description, classification, origin, development and history highlighting the physical, biological, and chemical aspects (Baker 1986; Anbazhagan and Saranathan 1991; Easterbrook 1999; Keller and Pinter 2002; Morisawa 1958; Ramasamy et al. 2011). Tectonism in general has a geomorphic expression in the region where it occurs and its adjacent areas (Gerson et al. 1984). The study of tectonic affects in many fluvial systems shows that, rivers are valuable tools to understand the active and inactive tectonics prevailing in an area. Drainage system of a region records the evolution of tectonic deformation (Schumm 1986; Gloaguen 2008). Morphometric analyses in tectonic geomorphology studies basically refer to the measurement on topographic maps of quantitative parameters (Wells et al. 1988). Since the geomorphic indices were first introduced as indicators of seismic activity (Bull and McFadden 1977), the topographic and geologic maps serve as sources of elevation for determining the indices. Geomorphic indices are useful tools in the evaluation of active tectonics, because they can provide valuable insights concerning specific areas of interest (Keller 1986).

The quantitative measurement of landscape is based on the calculation of geomorphic indices using topographic maps or digital elevation models, aerial photographs or satellite imagery, and fieldwork (Keller and Pinter 2002). The development of LIDAR remote sensing technique had helped to extract high resolution topographic information and to understand the earth surface processes including river morphology and active tectonics (Tarolli 2014). Saberi et al. (2014) have adopted GIS technique for derivation of various geomorphic indices and to assess the active tectonics in Iran. Hurtgen et al. (2013) have adopted GIS technique to assess the tectonic activities in southern Spain, based on morphometric indices. Many workers have attempted morphotectonic analysis using remote sensing and GIS techniques (Jordan 2003; Korup et al. 2005; Harbor and Gunnel 2007; Peters and Van Balen 2007; Font et al. 2010; Ferraris et al. 2012; Maryam and Maryam 2013; Dutta and Sharma 2013; Bagha et al. 2014; Markose et al. 2014). Various indices including the normalized stream length gradient index (SL), valley floor width-to-height ratio (Vf), hypsometric curves (Hc), sinuosity of mountain front (Smf), asymmetry factor (Af), and elongation ratio (Bs) were demonstrated as useful geomorphic indices in evaluating relative tectonic activity classes (Seeber and Gornitz 1983; Brookfield 1998; Chen et al. 2003; Silva et al. 2003; Malik and Mohanty 2007).

The southern Indian Peninsula is considered to be a tectonically stable region consisting of ancient rocks, rivers and land surfaces. However, within this larger ensemble, certain portions of the lands show younger topographic surfaces and recent-historical seismicity and these in turn were evidenced by the investigations on the longitudinal profiles, morphotectonic indices of active tectonics and fluvial records (for example, Ramasamy et al. 2011; Kale et al. 2013). Except such generalizations on a regional scale, microscale studies on any of the river basins of south India are scarce. In this paper, we attempt demonstrating the potential use of GIS technique in the evaluation of geomorphic indices of Thoppaiyar sub-basin, Southern India. The study also attempts quantification of several geomorphic indices of relative active tectonics and topographic development to produce a single index map that can be used to characterize relative active tectonics in this region.

2 Geological Setting

Quantitative geomorphic analyses were carried out to assess the relative active tectonics in the Thoppaiyar sub-basin, bounded between northern latitudes 11°51′47″–11°59′56″ and eastern longitudes 77°53′5–78°18′2″ (Fig. 1), covering an area of about 462 km^2. Northern and southern parts of this sub-basin fall under the Dharmapuri and Salem districts respectively. The highest elevation in the sub-basin is 1,600 m above mean sea level (amsl) at Muluvi and the lowest elevation of 240 m amsl is observed at its confluence with the Cauvery River. Mean annual rainfall in the sub-basin is 707 mm, which is significantly lower than the state average (970 mm). It receives rainfall from northeast as well as southwest monsoons. The climate in the sub-basin is generally warm. The hottest period of the year spans from March to May reaching the peak of 38 °C in April. The climate becomes cool during December to February, touches minimum of 15 °C in January. The prevailing

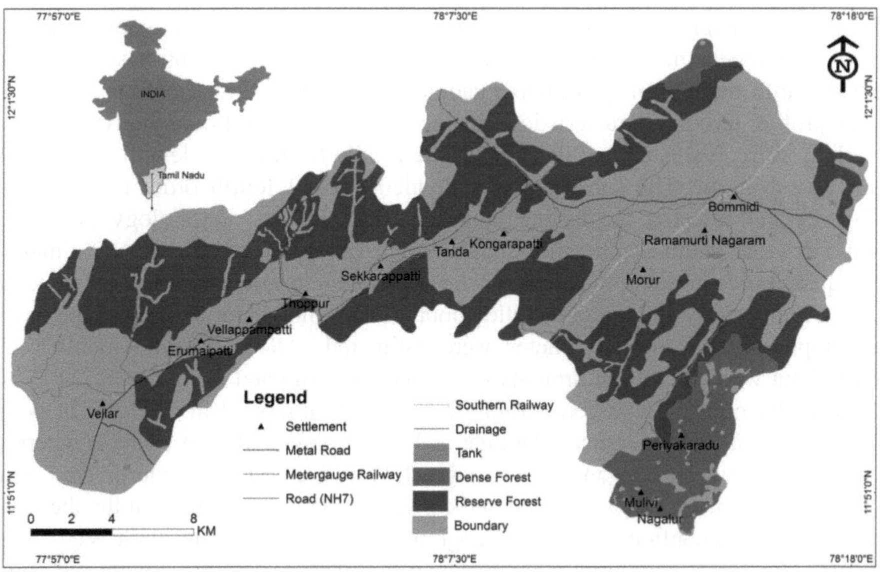

Fig. 1 Thoppaiyar sub-basin located in part of Dharmapuri and Salem districts in the state of Tamil Nadu, India selected for quantitative geomorphic analysis

hydrological, soil and climatic conditions support extensive floricultures in this sub-basin.

Structural hill system, denudational hill, fracture valleys and pediments are the major geomorphic units found to occurring in the study area. The structural hill system act as the boundary in the northern and southern part of the basin. The sub-basin is mostly covered by Precambrian crystalline rocks and recent alluvium along the river course. The major rock types are garnetferous quartzo felspathic gneiss, granites, granitoid gneiss, pink migmatite, purple conglomerate, quartz vein, syenite, sandstone, shale and shale with bands of limestone. Granites occupy about 324 km^2 in the north and central part of the sub-basin. Pink migmatite and gneissic rocks dominate in the central and eastern parts of the sub-basin. Presence of quartz veins are commonly noticed in the study area with an aerial coverage of 44 km^2. All these rock types increase the direct runoff rather than base flow contribution. However, the contribution of overland flow to the runoff is limited, because the length of overland flow is only 0.20 km. The surface infiltration is mostly restricted with weathered portion of granites. The areas covered by agricultural land and forest cover are equal—about 35 % each.

3 Materials and Methods

The Survey of India topographic maps, SRTM data, IRS P6 LISS III satellite data and data collected from field measurements were utilized in the present study. The ArcGIS 9.3 software was utilized for data generation and spatial integration. The topographic maps were utilized for the extraction of basin information such as roads, drainages and contours. Fluvial geomorphology of the sub-basin provided detailed information for quantitative geomorphic analysis. In the present study, the river profile, terrain characteristics, channel morphology and quantitative geomorphology were evaluated and discussed. The profile for 65.5 km length was plotted for the sub-basin. The sub-basin was divided into 14 fourth order micro basins (Fig. 2) and selected for further analyses. The channel morphology including, bankfull area, width-depth ratio, entrenchment ratio, height ratio were estimated. For quantitative geomorphic analysis, the stream length-gradient index, asymmetric factor, hypsometry and integral, valley floor width to height ratio, basin shape index and topographic symmetry factor were estimated. All the measurements were carried out with the help of drainages and contours extracted from the SRTM DEM in GIS environment. The geomorphic indices were imported in the form of point data along with georeferences (latitude and longitudes) and then spatially interpolated using 'Inverse Distance Weightage (IDW)' method. The interpolated geomorphic index maps were digitized and converted into 'shape file'. All the thematic maps were commonly projected to 'geographic coordinate system' with WGS 1984 datum. The fluvial geomorphic parameters were assessed for clues on the sub-basin development and relative tectonic condition of the sub-basin. Each theme was

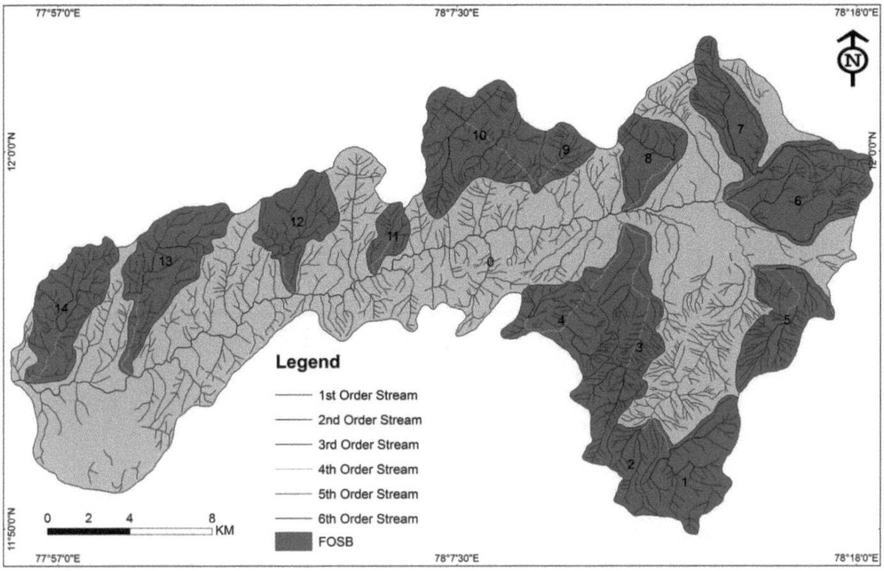

Fig. 2 Fourth order micro basins (FOMBs) in Thoppaiyar sub-basin considered for quantitative geomorphic study

divided into relative tectonic activity classes based on the range of values of individual indices (Hamdouni et al. 2008). Finally, the relative tectonic activity in the study area was divided into three classes.

4 Longitudinal River Profile

As fourth order sub-basins constitute more than 45 % of the study area, and most of the watershed management schemes elsewhere are based on fourth or lower order sub-basins (Joji and Nair 2013), the present study attempted delineating the study area, the Thoppaiyar sub-basin into 14 Fourth Order Micro Basins (FOMBs) for understanding terrain characteristics through longitudinal river profile and quantitative analysis of relative tectonic activities.

The longitudinal river profile is a curve that connects points from the source to the mouth of a river and the net effect of coarse particle inputs are presented in this profile. Individual rapids represent small-scale convexities in the longitudinal profile. They exhibit considerable changes on various time scales resulting from the frequent debris flow, deposition and river reworking (Horton 1932; Thornbury 1954 and Leopold et al. 1964). The longitudinal river profile prepared from source of Thoppaiyar to its mouth by considering the elevation and distance (Fig. 3). Survey of India (SOI) topographic maps and SRTM satellite data were utilized for preparation of river profile. The SRTM data were plotted for the 10 m contour interval.

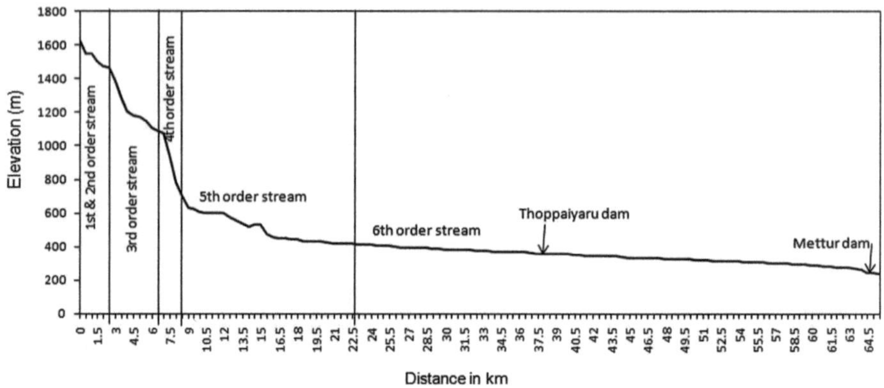

Fig. 3 Longitudinal river profile of Thoppaiyar sub-basin

In GIS environment, elevation data were collected for every 0.5 km interval from source to confluence. IRS P6 LISS III satellite data were utilized to interpret the geomorphology for all the micro basins. Other parameters including, geology, drainage density and bifurcation ratios were collected at an interval of 0.5 km for 132 points from the source to the mouth (Table 1).

5 Channel Morphology

Rosgen and Silvey (1996) developed a system for the classification of stream reaches based on their form. The system gives letter and number designations to different stream types, depending on their combination of bankfull channel characteristics such as entrenchment ratio, width to depth ratio, slope, sinuosity and bed material size. Cross-section is a line across a stream perpendicular to the flow along which measurements were taken, so that the morphological and flow characteristics of the section are described from bank to bank (Fig. 4). Stream channel morphology is often described in terms of a width/depth ratio related to the bankfull stage cross-section (Table 2; Fig. 5). The width-to-depth ratio varies primarily as a function of the channel cross-section for a given slope; the boundary roughness as a function of the stream flow and sediment regime, bank erodibility factors, including the nature of stream bank materials; degree of entrenchment; and the distribution of energy in the stream channel (Rosgen 1994).

$$\text{Width to depth ratio} = \frac{\text{Bankfull width}}{\text{Average depth}} \tag{1}$$

The direct and most reliable method of estimating channel depth is from the thickness of stratified units of point bars (Moody-Stuart 1966; Elliot 1976) and from

Table 1 Fourth order micro basins (FOMBs) along with lithology, geomorphology and morphometric parameters

FOMBs	Geology	Geomorphology	Bifurcation ratio	Drainage density (km/km^2)
1	Granitoid gneiss	Moderately weathered pediment, shallow weathered pediment, structural hills	4.03	2.83
2	Granitoid gneiss, pink migmatite, nepheline syenite, corundum syenite	Shallow weathered pediment, structural hills	3.86	2.60
3	Granite	Linear ridge/dyke, moderately weathered pediment, structural hills, shallow weathered pediment	3.03	2.27
4	Granite, pink migmatite	Moderately weathered pediment, shallow weathered pediment, structural hills	2.53	3.24
5	Garnetiferous quartz felspathic gneiss, syenite	Fracture valley, linear ridge/dyke, shallow weathered pediment, structural hills	4.20	2.60
6	Granite	Fracture valley, shallow weathered pediment, structural hills	2.88	3.57
7	Granite, pink migmatite	Moderately weathered pediment, shallow weathered pediment, structural hills	2.90	2.32
8	Granite, pink migmatite	Moderately weathered pediment, shallow weathered pediment, structural hills	3.70	2.71
9	Granite	Moderately weathered pediment, shallow weathered pediment, structural hills, residual hill	3.30	2.21
10	Granite	Moderately weathered pediment, shallow weathered pediment, structural hills, residual hill	3.83	3.51
11	Granite	Structural hills	3.51	2.35
12	Granite	structural hills, hill top weathered	2.66	2.99
13	Granite	Shallow weathered pediment, structural hills, hill top weathered, residual hill	4.53	3.08
14	Granite	Shallow weathered pediment, structural hills, hill top weathered, residual hill	4.00	3.07
Mean			3.49	2.81

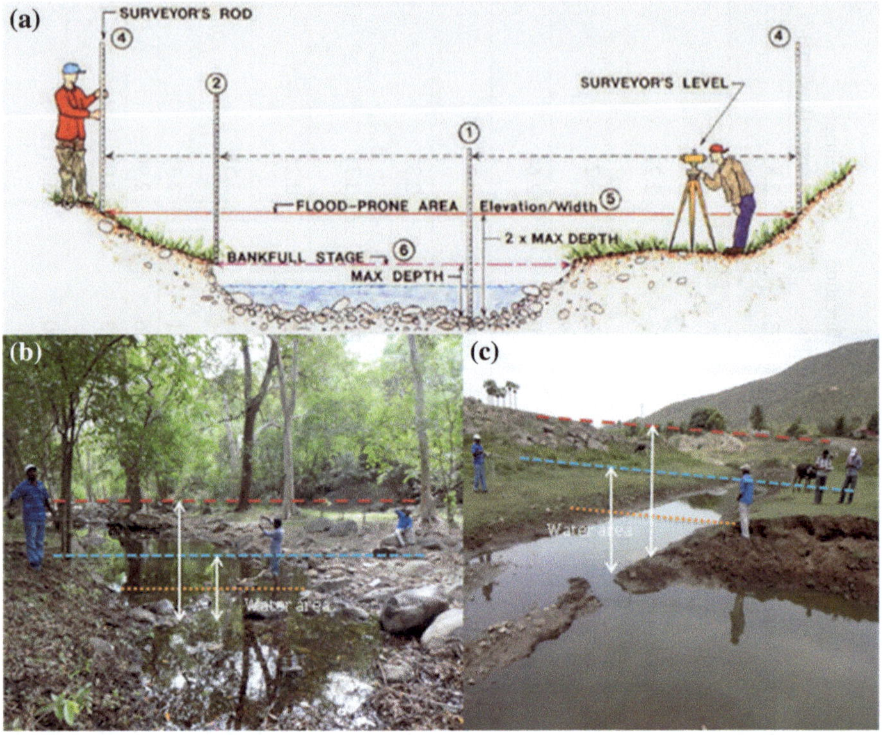

Fig. 4 Bank full estimation in Thoppaiyar river basin. **a** Bankfull model, **b** bankfull estimation at Anaimaduvu and **c** Kongarapatti

the average thickness of the coarse member in fining upward cycles (Jackson 1979). Frequently, point bar accretion surfaces cannot be recognized in ancient deposits, and it has not been possible to estimate channel depth from the thickness of the coarse sand members, because of the presence of multilevel sandstone bodies and paucity of complete cycles. Owing to this difficulty, Allen (1968) worked out an alternative relationship for estimating channel width for lower to moderate channel sinuosity streams. Alternatively, an estimate of average channel depth has been made from the thicknesses of cross-beds.

Entrenchment ratio is equal to the floodplain width at two times the bankfull depth divided by bankfull width. When a reach of stream is either straightened or narrowed, the power of the stream flow is increased. The stream may then cut down into its bed, so that flood flows are less likely to spill out into the floodplain. Through this process, the reach has incised, and that the channel has become entrenched, which can occur to varying degrees. When large flood flows are confined to the narrow channel of an incised stream, the water becomes very deep and erosive; the stream may cut down even deeper into its bed. Eventually the banks may become so high and steep that they erode away on one or both sides, widening the channel. This in turn can change previously stable areas downstream.

Table 2 Bankfull estimation in Thoppaiyar sub-basin

S. No.	Location	FPW (m)	BFW (m)	BFD (Centre) (m)	WD (m)	FPLB Alt (m)	BFLB Alt	BFD Alt	BFRB Alt	FPRB Alt	Width to depth ratio	Entrenchment ratio	BHR/Classification	
1	Aanai Maduvu	17.6	8.5	1.1	0.8	459.94	459.64	458.84	459.64	460.25	13.55	2.07	1.0	Stable
2	Annaikat	15.8	8.6	1.5	0.9	449.88	449.58	448.38	449.58	449.88	11.42	1.84	1.0	Stable
3	Bommidi	14.6	4.5	1.7	0.0	426.72	426.11	425.02	426.11	426.42	5.87	3.22	1.0	Stable
4	Karungalur	22.1	12.6	1.5	0.6	386.49	385.88	385.03	385.87	386.18	19.30	1.75	1.0	Stable
5	Thoppur dam	45.4	21.8	1.3	0.9	369.72	369.11	368.38	369.11	369.42	36.75	2.08	1.1	Moderately unstable
6	Sekkarapatti	34.7	16.2	1.8	0	340.16	339.55	338.36	339.54	339.85	19.61	2.15	1.1	Moderately unstable
7	Chinnakanavai	26.9	16.0	1.6	0	336.19	335.89	334.57	335.89	336.19	19.82	1.68	1.0	Stable
8	Vattampatti	30.2	10.7	1.1	0	335.58	335.28	334.53	335.28	335.58	20.48	2.82	1.0	Stable
9	Thoppur	36.0	13.1	2.1	0	334.06	333.76	331.96	333.75	334.06	12.46	2.75	1.1	Moderately unstable
10	Thoppur ropecar	69.5	13.2	2.3	0	332.84	331.93	330.54	331.92	333.15	10.79	5.25	1.1	Moderately unstable
11	Upparapatti	20.5	7.0	2.5	0	330.10	329.79	327.60	329.78	330.40	5.26	2.93	1.0	Stable
12	Vellappampatti	39.6	11.0	2.7	0	325.53	324.31	322.78	324.30	325.52	8.01	3.61	1.1	Moderately unstable
13	Pappireddiyur	30.2	10.9	3.1	0	317.91	317.60	314.81	317.60	318.21	6.73	2.76	1.0	Stable
14	Erumappatti	46.6	19.5	3.3	0	316.99	316.38	313.69	316.38	317.30	11.28	2.39	1.1	Moderately unstable
15	Kambampatti	31.1	12.7	3.7	0	303.28	302.97	299.58	302.97	303.27	6.89	2.44	1.1	Moderately unstable
16	Vellar	63.4	20.7	4.2	0	289.56	288.65	285.36	288.64	289.26	10.23	3.06	1.1	Moderately unstable
17	OoruVellaru	62.5	21.8	5	0	281.94	281.64	276.94	281.63	281.64	9.00	2.86	1.1	Moderately unstable
18	Soliyanur	76.5	41.1	5.2	4.8	277.06	274.02	271.88	274.01	276.15	17.42	1.86	1.2	Moderately unstable
19	Thoppaiyaru	90.5	24.0	5.5	5.1	246.58	242.01	241.10	242.01	245.36	9.86	3.76	1.2	Moderately unstable

FPW flood prone width, *BFW* bankfull width, *BFD* bankfull depth, *WD* water depth, *FPLB* flood prone left bank, *BFLB* bankfull left bank, *BFRB* bankfull right bank, *FPRB* flood prone right bank, *BHR* bank height ratio

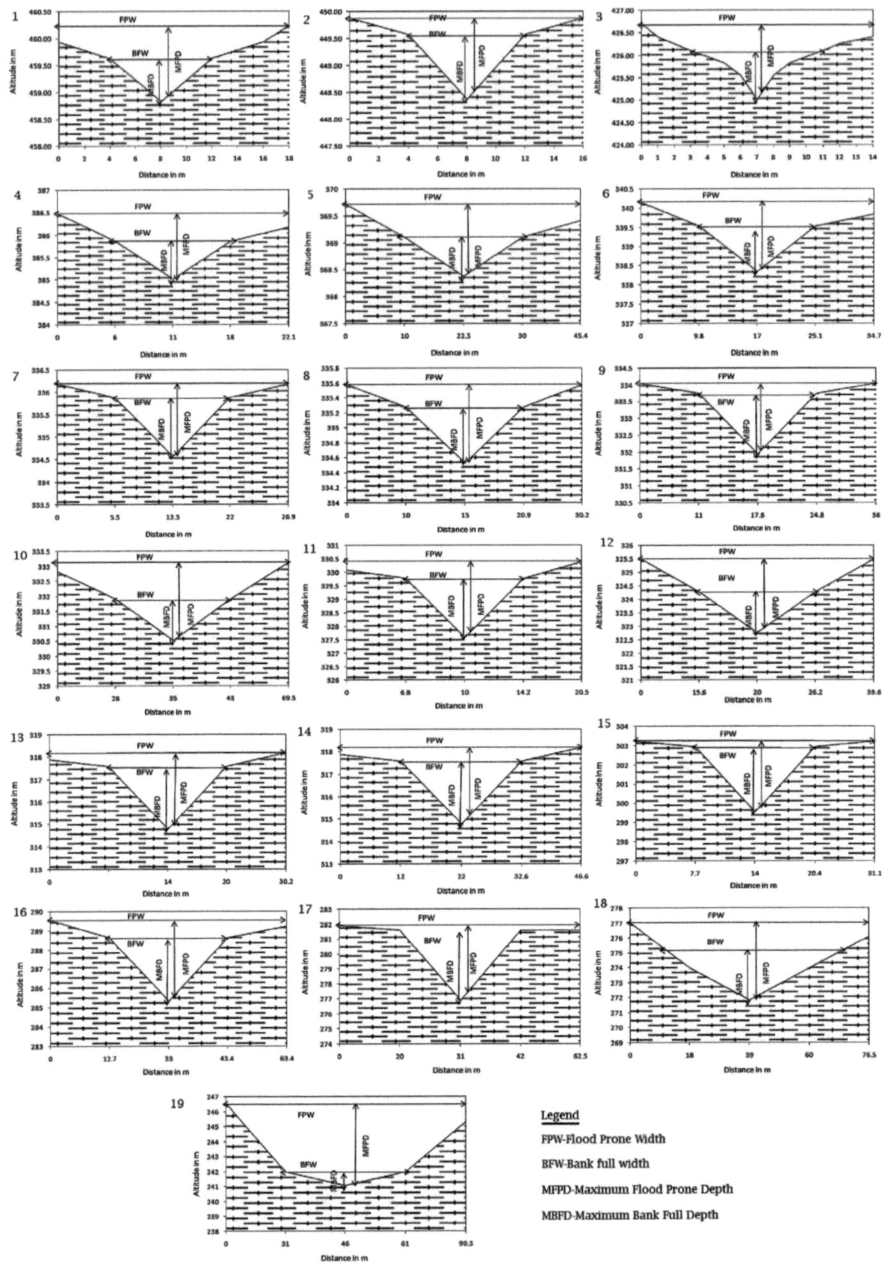

Fig. 5 Bankfull estimation at 19 locations in Thoppaiyar sub-basin

$$\text{Entrenchment ratio} = \frac{\text{Flood prone width}}{\text{Bankfull width}} \tag{2}$$

Bank height ratio (BHR) is the height of the top of the bank divided by the bankfull discharge height typically measured from the toe. The BHR is a relative measure of the floodplain connectivity to the bankfull channel.

6 Quantitative Geomorphic Analysis

The geomorphic indices are important indicators to interpret landform responses to deformation processes and together with field investigations, have been widely used to differentiate zones deformed by active tectonics (Keller and Pinter 2002; Chen et al. 2003). The tectonic quiescence or activeness in a drainage basin is reflected in the drainage pattern, lineaments, intermontane valleys, and valley incisions (Howard 1967; Cox 1994) and influences the fluvial process through the changes in slope, discharge, sediment load and bedrock erodibility. Human interventions such as gravel mining, dredging (Fig. 6) significantly alter natural conditions and can have longer impact on riparian condition. The topographic variations in a drainage basin result from adjustments between the tectonic, climatic and lithological controls as streams and rivers flow over rocks and soils of variable settings (Hack 1973). This adjustment eventually reaches a dynamic equilibrium and river systems display slightly concave longitudinal profiles. Deviation from this normal river profile may be interpreted as the result of active tectonic, lithological and/or climatic factors (Hack 1973). Many studies, including Perez-Pena et al. (2010), Mahmood and Gloaguen (2012) and Rebai et al. (2013) have demonstrated the utility of quantitative geomorphic analysis to assess the active tectonic process in a river basin.

In the present study, six geomorphic indices namely stream length gradient index (SL), asymmetrical factor (Af), hypsometric integral (Hi), valley floor width-to-height ratio (Vf), elangation ratio (Bs) and traverse topographic symmetry (T) were

Fig. 6 River bed modification due to human intervention. **a** Caving and ponding in exposure of boulders, **b** garbage dumping which obstruct the flow and contaminant surface and ground water

analyzed for 14 FOMBs of the Thoppaiyar sub-basin. The total value of each index were worked out, averaged and divided into relative tectonic classes (Hamdouni et al. 2008).

6.1 Stream Length Gradient Index (SL)

Stream length gradient index (SL) describes the morphology of a stream network using distribution of topographic gradients along rivers (Hack 1973). It is sensitive to change in channel slope, and allows evaluating the relative roles of possible tectonic activities and rock resistance (Keller and Pinter 2002; Azor et al. 2002). The topographic evolution results from an adjustment between the erosional processes as streams and river flow over rocks and soils of variable strength (Hack 1973). 'SL' was first used to reflect the stream power or differential rock erodibility and calculated using the following formula (Hack 1973),

$$\mathrm{SL} = \frac{\Delta \mathrm{H}}{\Delta \mathrm{L_r}} L_t \tag{3}$$

where SL is stream length-gradient index, $\Delta H/\Delta L_r$ is the channel slope or gradient of the reach, ΔH is change in altitude for a particular channel of the reach with respect to ΔL_r. ΔL_r is the length of a reach. L_t is the horizontal length of the watershed divide to the midpoint of the reach.

The calculated SL values for each reach of the micro basins were converted into a point shape file and the average value was estimated for each micro basin. The point data were interpolated through Inverse Distance Weighted (IDW) method available in GIS Software and a spatial map on SL index was then generated. This has been classified into three categories namely, class 1, class 2 and class 3 that contain the range values of 203–653, 98–203 and 13–98 respectively (Fig. 7a).

6.2 Asymmetry Factor (Af)

The calculations of Asymmetric Factor (AF) and Topography (T) help rapid quantitative determination of ground tilting (Cox 1994; Cox et al. 2001; Keller and Pinter 2002). Asymmetry factor (Af) is an aerial morphometric variable used in detecting the presence or absence of the regional tectonic tilt in the basin on a regional scale. 'Af' is defined as

$$\mathrm{Af} = \frac{\mathrm{A_R}}{\mathrm{A_T}} \times 100 \tag{4}$$

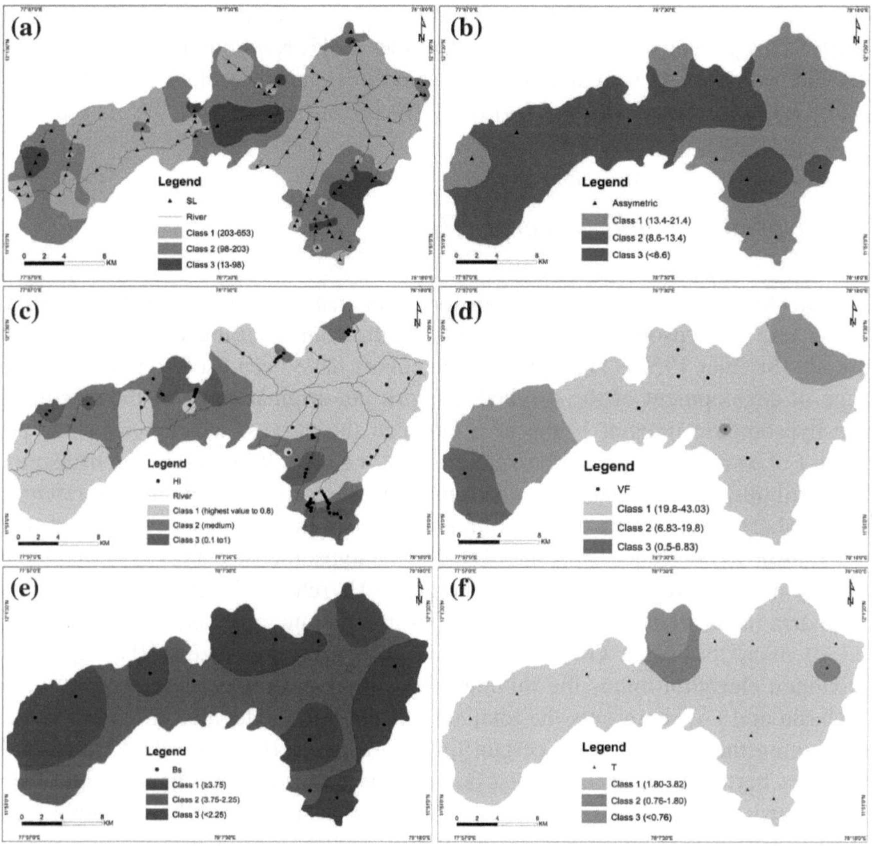

Fig. 7 Geomorphic indices. **a** SL, **b** Af, **c** Hi, **d** Vf, **e** Bs and **f** T and its classification of Thoppaiyar sub-basin

where A_R is the area of the basin to the right (facing downstream) of the trunk stream and A_T is the total area of the drainage basin. 'Af' greater or smaller than 50 indicates basin tilting, either due to active tectonics, lithological, structural control and differential erosion (Hamdouni et al. 2008). For the purpose of evaluating the relative tectonic activity (Iat), the absolute difference of Af is important. The absolute Af was calculated by subtracting the neutral value 50 from the calculated Af. The absolute Af is expressed as (Perez-Pena et al. 2010);

$$Af = \left[\frac{A_R}{A_T} \times 100\right] - 50 \quad (5)$$

Any drainage basin that was subjected to a tectonic rotation will most likely have an impact on the tributary. In case, the tectonic activity caused a tilt on the left side of the drainage basin, the tributaries located on the left of the mainstream will be shorter compared to the tributaries on the right side of the stream with an asymmetry factor greater than 50 (Hare and Gardner 1985; Keller and Pinter 2002). The absolute values of Af were are classified into three classes as class 1 (<8.6), class 2 (8.6−13.4) and class 3 (13.4−21.4) (Fig. 7b).

6.3 Hypsometry and Integral (Hi)

Hypsometry means the relative proportion of an area at different elevations within a region, and the hypsometric curve depicts the distribution of the area with respect to altitude (Strahler 1952). The hypsometric curves have been used to understand the stage of development of the original network i.e. original stage of the catchment. The hypsometric integral is the area between the curves, which relates the percentage of total relief to cumulative percent of the area. It expresses the measure of the distribution of land mass volume remaining beneath, or above a basal reference plane, otherwise it expresses the volume of a basin that has not been eroded (Pike and Wilson 1971; Keller and Pinter 2002). The hypsometric curves are obtained by plotting the proportion of total basin height (h/H_0 relative height) against the total basin area (a/A_0 relative area), and the hypsometric integrals were calculated for all the 14 micro basins in GIS environment. The maximum height (H) equals the maximum elevation minus the minimum elevation and represents the relief within the basin and (A) represents the total area of the basin. The area (a) is the surface area within the basin above a certain line of elevation (h). The relative area (a/A) measures between 1.0 at the lowest point in the basin, where relative height (h/H) equals zero, and zero at the highest point in the basin where relative height (h/H) equals 1.0 (Strahler 1952; Keller and Pinter 2002). Hypsometric integral is generally derived for a particular drainage basin and is an index that is independent of the basin area. It varies from 0 to 1. A portion of the drainage basin with hypsometric value close to 0 indicates a highly eroded nature and a value close to 1 indicate weakly eroded nature. The HI may be calculated using a following equation (Meyer 1990; Kellar and Pinter 2002).

$$HI = \frac{H_{mean} - H_{min}}{H_{max} - H_{min}} \tag{6}$$

where H_{mean} is the average height; H_{max} and H_{min} are the maximum and minimum heights of the catchments.

The topography produced by stream channel erosion and associated processes of weathering mass-movement, and sheet runoff is extremely complex, both in the geometry of the forms themselves and in the interrelations of the process which produce the forms. The formations of hypsometric curve and the value of the integral are important elements in topographic form.

The 'Hi' is similar to the 'SL' index in the rock strength as well as other factors that affect the value. Higher values generally represent that not as much of the uplands have been eroded and reflect the younger landscape, possibly produced by recent tectonics. The 'Hi' could also be interpreted as recent initiation into a geomorphic surface formed by deposition.

The 'Hi' values were grouped into three classes with respect to the convexity and concavity of the hypsometric curve. Class 1 with convex hypsometric curve (highest value to 0.8), class 2 with concave-convex hypsometric curve (medium) and class 3 with concave hypsometric curve (0.1–1). Spatial distributions of these three classes are represented in Fig. 7c.

6.4 Valley Floor Width to Height Ratio (Vf)

Valley floor width to height ratio (Vf) is a geomorphic index that discriminates the V and U shaped and flat-floored valleys. This index was adopted particularly to identify the tectonically active fronts (Pedera et al. 2009; Koukouvelas 1998; Zuchiewicz 1988; Azor et al. 2002; Silva et al. 2003) and is defined as (Bull and McFadden 1977),

$$\mathrm{Vf} = \frac{2\mathrm{Vfw}}{\mathrm{Eld} + \mathrm{Erd} - 2\mathrm{Esc}} \tag{7}$$

where 'Vfw' is the width of the valley floor, 'Eld' and 'Erd' are the elevations of the left and right valley divides respectively and 'Esc' is the elevation of the valley floor.

The valley floors tend to become progressively narrow upstream from the mountain front (Ramirez-Herrera 1998). The Vf is usually calculated at a given distance upstream from the mountain front (Silva et al. 2003). 'Vf' values were calculated for 14 micro basins where the main valleys cross the mountain fronts, using cross-sections drawn from the digital elevation model prepared from topographic map of the study area. The Vf value were categorized into three classes as class 1 (0.5−6.83), class 2 (6.83−19.8) and class 3 (19.8−43.09) and their spatial distributions are presented in the Fig. 7d.

6.5 Basin Shape Index

The elongation ratio (Bs) describes the planimetric shape of a basin. It is expressed as:

$$\mathrm{Bs} = \frac{\mathrm{Bl}}{\mathrm{Bw}} \tag{8}$$

where Bl is the length of the basin measured from its mouth to the distal point in the drainage divide, and Bw is the width of the basin measured across the short axis defined between left and right valleys divides (Ramirez-Herrera 1998).

The young drainage basins in active tectonic areas tend to be elongated in shape parallel to the topographic slope of a mountain. The index reflects the differences between the elongated basins that have high values of 'Bs' associated with relatively higher tectonic activity and circular basins that have low 'Bs' values generally associated with low tectonic activity (Bull and McFadden 1977). The 'Bs' of the 14 micro basins range from 0.58 to 5.56 and this was classified into 3 classes to assess the relative active tectonic activity in the area. The class 1 is categorized by high Bs value (≥3.75), the class 2 by moderate values (3.75−2.25) and class 3 by low values (<2.25) (Fig. 7e).

6.6 Transverse Topographic Symmetry Factor (T)

Transverse Topographic Symmetry Factor (T) is a quantitative geomorphic index that helps evaluating the basin asymmetry and is defined as:

$$T = \frac{D_a}{D_d} \tag{9}$$

where 'D_a' represents the distance from the midline of the drainage basin to the midline of the active channel or meander belt, and the 'D_d' corresponds to the distance from the basin midline to the basin divide.

For different segments of river valleys, the calculated T values indicate migration of streams perpendicular to the drainage-basin axis. For perfectly symmetric basin T = 0. As the symmetry increases, 'T' increases and approaches a value of 1.0 and indicates tilted basins (Burbank and Anderson 2000; Cox 1994; Cox et al. 2001; Keller and Pinter 2002). The T factor calculated for the 14 micro basins of the Thoppaiyar sub-basins were classified into three categories as class 1 (<0.76), class 2 (0.76–1.80), class 3 (1.80–3.82) (Fig. 7f).

7 Discussion

The longitudinal profile of the Thoppaiyar sub-basin has shown as overall concavity, which reflects a pronounced decrease in the stream gradient. The upstream condition of Thoppaiyar sub-basin is more concave than the downstream. The concave nature of the profile indicates an increase of stream discharge in the downstream direction of the Thoppaiyar sub-basin. The river profile reveals that a pronounced decrease in gradient from origin to confluence (65.5 km). The elevation difference is 1,460 m. The profile has breaks at 6th, 9th and 16.5th km. An elevation difference of 500 m has been observed in between 0 and 6th km. The profile takes sharp changes from 6th to 9th km with a difference of 400 m. This zone is probably controlled by fault. The knick point at 15th km once again indicates the structural control. The river flows along a gentle profile from 16.5th km onwards and this gentle profile forms more than 2/3rd of the river profile. The Thoppaiyar sub-basin is a sixth-order Hortonian stream and the number of streams in the first, second, third, fourth, fifth and sixth orders are 1141, 290, 60, 14, 4 and 1 respectively. The stream ordering with respect to stream profiling has revealed the variation of stream order at 20 points in the profile with respect to elevation and distance. The major portion of the stream order (up to 4th order) has been found restricted only within 16 km.

The mean bifurcation ratio (Rb) of the Thoppaiyar sub-basin is 2.49 and the mean 'Rb' value of 14 FOMBs is 3.49 and the theoretical minimum value of 2.53 is reported for FOMB-4. Most of the values range in between 2 and 3; the highest value of Rb 4.53 is reported for FOSB-13 and the marked variation in the bifurcation ratio is due to difference in geological and structural characteristics of rock, relief and stages of basin development (Schumm 1956; Bridge 1993; Pittaluga et al. 2003; Burge 2006). Mean Rb of 3.49 indicates that this value fall well within the normal range and geological structures have not distorted the drainage pattern (Horton 1945; Gopalakrishnan et al. 1997; Federici and Paola 2003). The FOMB-13 with Rb value of 4.53 indicates that this micro basin suffered structural disturbances (Nautiyal 1994). As most of the FOMBs have Rb value ranging between 2 and 5, these microbasins can be considered to posses well-developed drainage network (Strahler 1957).

The channel morphology of the sub-basin namely, the cross section, width-depth ratio, bankfull depth, channel width, entrenchment ratio and bank height ratio were studied for 19 locations. The bank height ratio indicates the prevalence of moderately unstable conditions at 11 locations and remaining 8 locations with stable condition.

The spatial map of 'SL' index has shown that major part of the sub-basin falls in class 1 followed by class 2 and 3 respectively. The class 1 with high SL values reflects hard rock or high tectonic activity, whereas the class 3 with low SL value indicates relatively low resistance or low tectonic activity (Hack 1973; Keller and Pinter 2002). In order to assess the effect of lithology, relative rock resistance map was prepared based on rock types (Fig. 8). It is inferred from this figure that the class 1 SL values are observed in high resistance rocks namely, quartz vein, garnetiferous Quartzo felspathic gneiss, granite, granite gneiss and pink migmatite. The class 3 SL index is noticed in the middle southern part of the basin comprising limestone and conglomerate. Longitudinal river profiles for the selected micro basins and Thoppaiyar sub-basin were plotted along with SL values (Fig. 9). The highest value of index was observed in the upper reach of (6,000–14,000 m) the Thoppaiyar sub-basin. This anomaly could be considered as tectonically significant zone (Hamdouni et al. 2008).

The 'Af' values for 14 micro basins show that almost all micro basins are asymmetric in nature. It is observed that micro basins located along the right bank of main channel show high 'Af' value than those located along the left bank. The absolute Af calculated for the 14 FOMBs range from 0.84 to 27.49. The FOMBs 3,

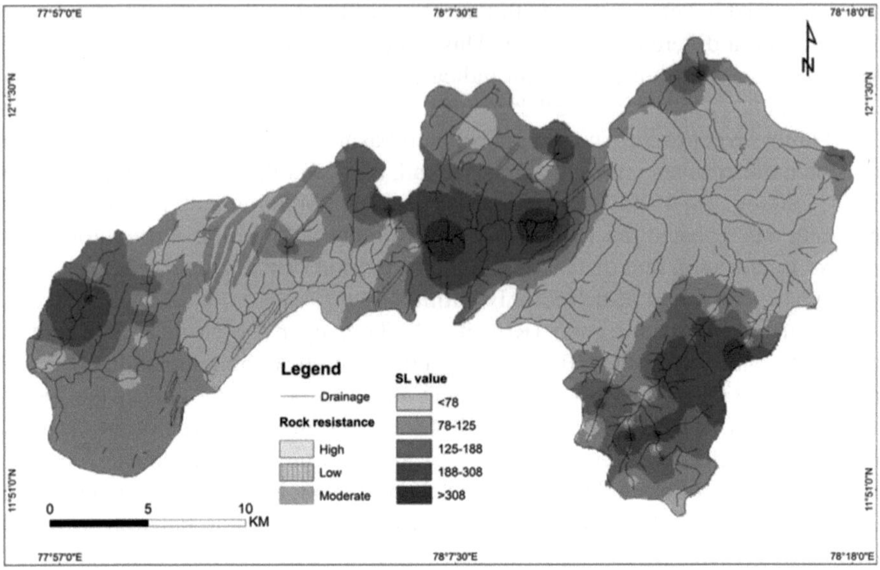

Fig. 8 Rock resistances with SL index of Thoppaiyar sub-basin

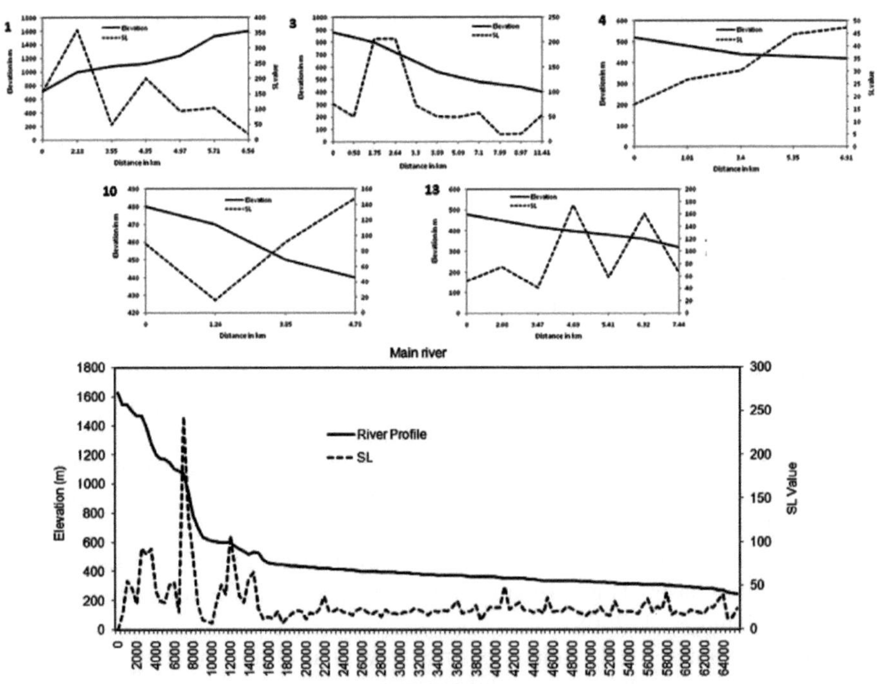

Fig. 9 Longitudinal river profiles and the measured SL index of selected micro-basins (*1*, *3*, *4*, *10*, *13*) in Thoppaiyar sub-basin

9, 11, 12 and 13 show high absolute of values and are indicative of relatively higher tectonic activity (Hamdouni et al. 2008).

The 'Hi' range from 0.11 to 0.81 and are categorized into three classes with respect to the convexity or concavity of the curves. A 'Hi' value of >0.6 indicate youthful stage, a range of 0.352–0.6 area indicative of mature stage, and 'Hi' below <0.35 characterize old stage of landscape (Strahler 1952). The higher values of 'Hi' are possibly related to young, active tectonics and the lower values of 'Hi' are indicative of the presence of older landscapes that have undergone greater erosion and less impacted by recent tectonic activities (Hamdouni et al. 2008). The hypsometric integral value of 50 % the basin area of the Thoppaiyar indicates mature stage of the watershed. The 'Hi' and hypsometric curve could be used for conceptual geomorphic models of landscape evolution. In this study, the hypsometric curves and hypsometric integral are interpreted in terms of degree of basin dissection and relative landform age. Convex up curves with high integrals are typical for youth, undissected (disequilibrium stage) landscapes; smooth, s-shaped curves crossing the center of the diagram characterize mature (equilibrium stage) landscape and concave up with low integrals typically old and deeply dissected landscapes. Accordingly, their distribution within the study area is depicted in the Fig. 10.

The "V" shaped valleys show lower 'Vf' values while higher values are associated with 'U' shaped valleys. Since the uplift is associated with incisions, were low values of 'Vf' are associated with higher rate of uplift and incisions, usually deep V shaped valleys (Vf<1) are connected with linear active down cutting streams, distinctive of zones subjected to active uplift. However, the flat and U shaped valleys (Vf>1) show an attainment of the base level of erosion, mainly in response to relative tectonic quiescence. The 'Bs' values indicate that half of the micro-basins were categorized under class 1 and 2, which indicate that the micro-basins are elongated as well as circular equally. The micro basins with elongated shape become progressively more circular with time and continued topographic evolution (Bull and McFadden 1977).

The 'T' value in the micro basin ranges from 0.01 to 0.9 which indicates that most of the micro basins are asymmetric in nature, which in turn indicates the role of regional dynamic activities. The class 1 and class 2 mostly indicate asymmetry nature of the sub-basin. In case of a negligible influence by the bedrock tilting on the relocation of the stream channels, the direction of the regional migration is an indicator of the ground tilting. The analysis of number of micro basins in a basin results in multiple spatially distributed T vectors, which, when averaged, define the irregular zones of the basin asymmetry.

Several workers (Bull and McFadden 1977; Silva et al. 2003; Hamdouni et al. 2008) have attempted combining two or more indices to extract semi-quantitative information regarding the relative tectonic activity in Active Mountain ranges. Following their precedence, in the present context, an attempt was made to evaluate the index of relative tectonic activity (Irat) in the micro-basins using various geomorphic indices namely, SL, Af, Hi, Vf, Bs and T. Based on the value of each geomorphic index and their relationship with the tectonic activity, the indices were

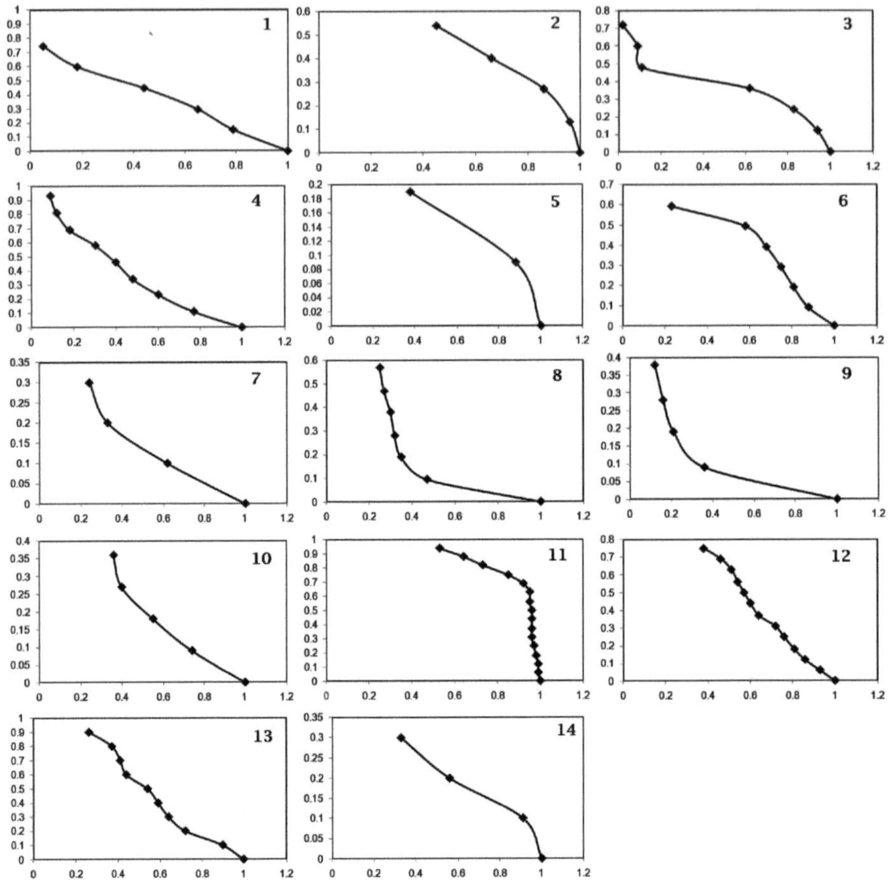

Fig. 10 Different types of hypsometric curves: downward convex curves with high Hi value and upward concave curves with low Hi values

divided into three classes; class 1 with higher geomorphic index, class 2 with moderate index, and class 3 with a lower index. The 'Irat' classes obtained by the average of different classes of geomorphic indices and again classified into three classes; class 1 high relative tectonic activity with values of Sn < 2; class 2 moderate relative tectonic activity with Sn > 2 to < 2.5; and class 3 low relative tectonic activity with values of Sn ≥ 2.5. Table 3 shows these three classes of geomorphic indices and 'Irat' classes of selected 14 micro-basins in the study area. From the table, it follows that majority of the micro basins are classified under low relative tectonic activity zones and only three micro-basins (1, 2, and 3) are in moderate category.

Table 3 Index of relative tectonic activity classes (Irat) in the selected micro-basins of the Thoppaiyar sub-basin

Micro basin	SL	Af	Hi	Vf	Bs	T	Sn	Irat	Category
1	2	3	2	1	3	1	2.2	2	Moderate
2	2	3	3	1	2	1	2.3	2	Moderate
3	3	2	2	1	1	1	2.2	2	Moderate
4	3	3	1	2	2	1	2.7	3	Low
5	2	3	3	2	3	3	3.5	3	Low
6	3	3	3	1	1	1	3.0	3	Low
7	3	3	3	1	3	1	3.5	3	Low
8	3	3	3	1	3	1	3.7	3	Low
9	2	1	3	1	3	2	3.5	3	Low
10	2	3	3	1	2	1	3.7	3	Low
11	2	2	1	2	3	1	3.7	3	Low
12	3	1	2	2	1	1	3.7	3	Low
13	3	2	1	3	1	1	4.0	3	Low
14	2	3	3	1	3	1	4.5	3	Low

SL stream length gradient index, *Af* asymmetric factor, *Hi* hypsometric integral, *Vf* valley floor width to depth ratio, *Bs* basin shape index, *T* topographic symmetry, *Sn* semi-quantitative information and Irat-Index of relative tectonic activity

8 Conclusions

The study of fluvial geomorphology provides information on the behavior of river and the influence of geology and structure on channel morphology. The analysis has shown that the majority of the micro basins currently experience low relative tectonic activity and only three micro-basins (FOMBs 1, 2, and 3) experience moderate tectonic activity. In the western, central and southern parts of the basin, SL values indicate variation of stream length gradient index, whereas in the eastern and down central part show low SL index. High SL values are associated with resistant rock formations such as quartz veins, garnetiferous quartzo felspathic gneiss, granite, granitoid gneiss and pink migmatite. The low SL values are found in limestone and conglomerate. Hypsometric integral low values in the sub-basin observed in the headward region and near the mouth, whereas high 'Hi' values are noticed in the middle part of the river basin, i.e., in the mountainous terrain. The majority of the FOMBs in the sub-basin show low 'Bs' values and indicate elongated shape. 'Vf' values of the micro basins range from 0.5 to 43 and the high 'Vf' were observed near the mouth. The class 1 and class 2 of 'T' mostly indicate the asymmetric nature of micro basins. Most of the fourth order micro-basins in the Thoppaiyar sub-basin are categorized under low index of relative tectonic activity. It is also brought out by the present study that, despite being a medium sized sub-basin located in a more or less homogenous climatic condition (within the studied basin area), prevalence of differential controls exercised by tectonic and lithological characteristics with reference to FOMBs has been brought to light with the help of remote sensing and GIS techniques. Implications of this result is that, while designing water and other natural resource management programs, the variations at micro-basin scale need to be taken into consideration, failing which, the management programs may not yield desired results.

References

Anbazhagan S, Saranathan E (1991) Hydrogeomorphological mapping in northern part of Salem dt, Tamil Nadu. J Indian Geogr Soc 66–1:65–68

Allen JRL (1968) Current ripples. North Holland, Amsterdam, p 433

Azor A, Keller EA, Yeats RS (2002) Geomorphic indicators of active fold growth: south mountain-oak ridge ventura basin, southern California. Geol Soc Am Bull 114:745–753

Baker VR (1986) Introduction: regional landforms analysis. In: Short NM, Blair RW (eds) Geomorphology from space: a global overview of regional landforms. NASA, Washington, pp 1–26

Bagha N, Arian M, Ghorashi M, Pourkermani M, Hamdouni RE, Solgi A (2014) Evaluation of relative tectonic activity in the Tehran basin, central Alborz, northern Iran. Geomorphology 213:66–87

Brookfield ME (1998) The evolution of the great river systems of southern Asia during the Cenozoic India-Asia collision: rivers draining southwards. Geomorphology 22:285–312

Bridge JS (1993) The interaction between channel geometry, water flow, sediment transport and deposition in braided rivers. In: Best JL, Bristows C (eds) Braided rivers: from process and economic applications. Geological Society Special Publication 75, Bath, pp13-71

Bull WB, McFadden LD (1977) Tectonic geomorphology north and south of the Garlock fault, California. In: Doehring, DO (ed) Geomorphology in arid regions. Proceedings of the eighth annual geomorphology symposium. State University of New York, Binghamton, pp 115−138

Burge LM (2006). Stability, morphology and surface grain size patterns of channel bifurcation in gravel–cobble bedded anabranching rivers. Earth Surf Process Land 31(10):1211–1226

Burbank DW, Anderson RS (2000) Tectonic geomorphology. Blackwell, Oxford, p 274

Chen Y, Sung Q, Cheng K (2003) Along-strike variations of Morphometric features in the western foothills of Taiwan: tectonic implications based on stream gradient and hypsometric analysis. Geomorphology 56:109–137

Cox RT, Van Arsdale RB, Harris JB (2001) Identification of possible Quaternary deformation in the northeastern Mississippi Embayment using quantitative geomorphic analysis of drainage-basin asymmetry. Geol Soc Am Bull 113:615–624

Cox RT (1994) Analysis of drainage-basin symmetry as a rapid technique to identify areas of possible quaternary tilt-block tectonics: an example from the Mississippi embayment. Geol Soc Am Bull 106:571–581

Dutta N, Sarma JN (2013) Morphotectonic analysis of Sonai river basin, Assam, NEIndia. Glob Res Anal 2(2):114–115

Elliot T (1976) The morphology, magnitude and regimen of a carboniferous fluvial distributary channel. J Sedim Petrol 46:70–76

Easterbrook DJ (1999) Surface processes and landforms, 2nd edn. Prentice Hall, New Jersey

Ferraris F, Firpo M, Pazzaglia FJ (2012) DEM analyses and morphotectonic interpretation: the Plio-quaternary evolution of the eastern Ligurian Alps, Italy. Geomorphology 149–150:27–40

Federici B, Paola C (2003) Dynamics of channel bifurcations in noncohesive sediments. Water Resour Res 39(6):271–284

Font M, Amorese D, Lagarde JL (2010) DEM and GIS analysis of the stream gradient index to evaluate effects of tectonics: the Normandy intra plate area (NW France). Geomorphology 119:172–180

Gerson R, Grossman G, Bowman D (1984) Stages in the creation of a large rift valley – geomorphic evolution along the southern Dead Sea Rift. In Morisawa M, Hack TJ (eds) Tectonic Geomorphology. State university of New York, Binghamton, pp 53–73

Gloaguen R (2008) Remote sensing analysis of crustal deformation using river networks. http://www.igarsso8.org/abstracts/pdfs/3197.pdf

Gopalakrishnan KS, Sakthivel M, Sunilkumar R (1997) Geomorphometric study of the Kodayar river basin in Western Ghat regions of Kanyakumari district, Tamil Nadu. Natl Geogr J India 43(4):295–306

Hack JT (1973) Stream-profile analysis and stream-gradient index. US Geol Sur J Res 1:421–429

Hamdouni RE, Irigaray C, Fernandez T, Chacon J, Keller EA (2008) Assessment of relative active tectonics, southwest boarder of Sierra Nevada (southern Spain). Geomorphology 96:150–173

Harbor D, Gunnell Y (2007) Along-strike escarpment heterogeneity of the Western Ghats: a synthesis of drainage and topography using digital morphometric tools. J Geol Soc India 70:411–426

Hare PW, Gardner TW (1985) Geomorphic indicators of vertical neotectonism along converging plate margins, Nicoya Peninsula, Costa Rica. In: Morisawa M, Hack JT (eds). Tectonic Geomorphology. Proceedings of the 15th annual Binghamton geomorphology symposium. Allen and Unwin, Boston, pp 123−134

Horton Robert E (1932) Drainage basin characteristics. Trans Am Geophys Union 13:350–361

Horton RE (1945) Erosional development of streams and their drainage basins. Bull Geol Soc Am 56:275–370

Hurtgen J, Rudersdorf A, Grutzner C, Reicherter K (2013) Morphotectonics of the Padul-Nigüelas Fault Zone, southern Spain. Ann Geophys, 56(6). doi:10.4401/ag-6208

Howard AD (1967) Drainage analysis in geologic interpretation: a summation, bulletin of american association of petroleum geology, 21:2246−2259. http://srtm.usgs.gov/data/

Jackson RG (1979) Preliminary evaluation of lithofacies models for meandering alluvial streams In: Miall AD (ed) Fluvial sedimentology. Canadian society of petroleum and geological memoir, 5:543−576

Jordan G (2003) Morphometric analysis and tectonic interpretation of digital terrain data: a case study. Earth Surf Proc Land 28:807–822

Joji VS, Nair ASK (2013) Terrain characteristics and longitudinal, land use and land cover profiles behavior—a case study from Vamanapuram River basin, southern kerala. Indian Arab j geosci. doi:10.1007/s12517-012-0815-z

Kale VS, Senguptab S, Achyuthanc H, Jaiswal MK (2013) Tectonic controls upon Kaveri river drainage, cratonic peninsular India: inferences from longitudinal profiles, morphotectonic indices, hanging valleys and fluvial records. Geomorphology. doi:10.1016/j.geomorph.2013.07.027

Keller EA (1986) Investigation of active tectonics: use of surficial earth processes. In: Wallace RE (ed) Active tectonics, studies in geophysics. National Academy Press, Washington DC, pp 136–147

Keller EA, Pinter N (2002) Active tectonics: earthquakes uplift and landscape. Prentice Hall, New Jersey, p 362

Korup O, Schmidt J, Mcsaveney MJ (2005) Regional relief characteristics and denudation pattern of the western southern Alps, New Zealand. Geomorphology 71:402–423

Koukouvelas IK (1998) The Egion Fault: earthquake related and long term deformation, Gulf of Corinth. Greece. J Geodyn 26(2–4):501–513

Leopold LB, Wolman MG, Miller JB (1964) Fluvial processes in geomorphology. W. H. Freeman, San Francisco

Malik JN, Mohanty C (2007) Active tectonic influence on the evolution of drainage an landscape: geomorphic signatures from frontal and hinterland areas along the northwestern Himalaya, India. J Asian Earth Sci 29:604–618

Mahmood SA, Gloaguen R (2012) Appraisal of active tectonics in HinduKush: insights from DEM derived geomorphic indices and drainage analysis. Geosci Front 3(4):407−428 (July 2012)

Meyer L (1990) Introduction to quantitative geomophology. Prentice-Hall, Englewood Cliffs, NJ,p 380

Markose VJ, Dinesh AC, Jayappa KS (2013) Quantitative analysis of morphometric parameters of Kali River basin, southern India, using bearing azimuth and drainage (bAd) calculator and GIS. Envir Earth Sci 70(2):839–848

Maryam E, Maryam AA (2013) Active tectonic analysis of Atrak river subbasin located in NE Iran (East Alborz). J Tethys 1(3):177–188

Moody-Stuart M (1966) High and low sinuosity streams deposits with examples from the Devonian of Spitsbergen. J Sedim Petrol 36:1110–1117

Morisawa Marie (1958) Measurement of drainage-basin outline form. J Geol 66:587–591

Nautiyal MD (1994) Morphometric analysis of a drainage basin using aerial photographs: a case study of Khairakulli Basin, Dehradun, Uttar Pradesh. J Indian Soc Remote Sens 22(4):251−261

Perez-Pena JV, Azor A, Azanon JM, Keller EA (2010) Active tectonics in the Sierra Nevada (Betic Cordillera, SE Spain): insights from geomorphic indexes and drainage pattern analysis. Geomorphology 119:74–87

Peters G, van Balen RT (2007) Tectonic geomorphology of the Upper Rhine Graben, Germany. Glob Planet Chang 58:310–334

Pedera A, Perez-Pena JV, Galindo-Zaldívar J, Azanon JM, Azor A (2009) Testing the sensitivity of geomorphic indices in areas of low-rate active folding (eastern Betic Cordillera, Spain). Geomorphology 105:218–231

Pike RJ, Wilson SE (1971) Elevation-relief ratio, hypsometric integral and geomorphic area altitude analysis. Geol Soc Am Bul 82:1079–1084

Pittaluga M, Repetto R, Tubino M (2003). Channel bifurcation in braided rivers: equilibrium configurations and stability. Water Resour Res 39:1046

Ramasamy SM, Kumanan CJ, Selvakumar R, Saravanavel J (2011) Remote sensing revealed drainage anomalies and related tectonics of South India. Tectonophysics 501:41–51

Ramirez-Herrera MT (1998) Geomorphic assessment of active tectonics in the Acambay graben, Mexican volcanic belt. Earth Surf Proc Land 23:317–332

Rebai N, Achour H, Chaabouni R, Bou KR, Bouaziz S (2013) DEM and GIS analysis of sub-watersheds to evaluate relative tectonic activity. a case study of the north–south axis (Central Tunisia). Earth Sci Inf 6(4):187−198. ISSN: 1865-0473 doi:10.1007/s12145-013-0121-7Springer Berlin Heidelberg, Berlin/Heidelberg (December 2013)

Rosgen DL (1994) A classification of natural rivers. Catena 22:169–199

Rosgen DL, Silvey HL (1996) Applied river morphology. Wildland Hydrology Books, Pagosa Springs, CO

Saberi MR, Pourkermani M, Nadimi A, Ghorashi M, Asadi A (2014) Geomorphic Assessment of Active Tectonics in Tozlogol Basin, Sanandaj-Sirjan Zone, Iran. Curr Trends Technol Sci 3 (3):146–154

Schumm SA (1956) Evolution of drainage systems and slopes in badlands at Perth Amboy, New Jersey. Bull Geol Soc Am 67:597–646

Schumm SA (1986) Alluvial river response to active tectonics. In: Active tectonics, studies in geophysics. National Academy Press, Washington, pp 80–94

Seeber L, Gornitz V (1983) River profiles along the Himalayan arc as indicators of active tectonics. Tectonophysics 92:335–367

Silva PG, Goy JL, Zazo C, Bardaji T (2003) Fault generated mountain fronts in Southeast Spain: geomorphologic assessment of tectonic and earthquake activity. Geomorphology 50:203–226

Strahler AN (1952) Hypsometric (area–altitude) analysis of erosional topography. Geol Soc Am Bull 63:1117–1142

Strahler AN (1957) Quantitative analysis of watershed geomorphology. Trans Am Geophys Union 38:913–920

Thornbury WD (1954) Principles of geomorphology. Wiley, London 29

Tarolli P (2014) High-resolution topography for understanding earthsurface processes: opportunities and challenges. Geomorphology 216:295–312

Wells SG, Bullard TF, Menges CM, Drake PG, Karas PA, Kelson KI, Ritter JB, Wesling JR (1988) Regional variations in tectonic geomorphology along a segmented convergent plate boundary, Pacific coast of Costa Rica. Geomorphology 1:239–365

Zuchiewicz W (1988) Quaternary tectonics of the Outer West Carpathians, Poland. Tectonophysics 297:121–132

Spatio-Temporal Variations of Erosion-Deposition in the Brahmaputra River, Majuli—Kaziranga Sector, Assam: Implications on Flood Management and Flow Mitigation

P. Kotoky, D. Bezbaruah and J.N. Sarma

Abstract Erosion-deposition activities of the Brahmaputra River channel within a stretch from Majuli to Kaziranga, Assam, India have been evaluated in the light of associated geomorphic and anthropogenic features. The results indicate that the mass failure, fluvial entrainment and sub aerial weathering and weakening play major roles over the depositional-erosional activities of the river channel. The mass failures result due to the heterogeneities of bank material. Shear failure of uneven nature and flowage (liquefaction) of the sediments are observed within the studied stretch of the Brahmaputra River channel. The transitory nature of channel configuration with time exhibited by node point generation and migration were also found to be associated with extensive lateral bank erosion processes, resulting in widening of the channel. Based on these observations, the critical areas of erosion-deposition were identified and accordingly, time-dependent flow and flood mitigation measures have to be undertaken.

Keywords Brahmaputra · Erosion · Deposition · Assam

1 Introduction

The lands adjoining the rivers courses have been the preferred regions of human settlements as these contain fertile soils and provide access to transport avenues. However, these are the flood plains that experience episodic flooding, which in turn

P. Kotoky (✉)
North East Institute of Science and Technology, Council for Scientific and Industrial Research, Jorhat 785 006, Assam, India
e-mail: probhatk@yahoo.com

D. Bezbaruah · J.N. Sarma
Department of Applied Geology, Dibrugarh University, Dibrugarh 784006, Assam, India

Mu. Ramkumar et al. (eds.), *Environmental Management of River Basin Ecosystems*, Springer Earth System Sciences, DOI 10.1007/978-3-319-13425-3_12

is a natural geomorphic process and sustains the fluvial system and ecological niches located at the downstream region. Management of flood and planning of flood plain land use are the fundamental issues where structural and non structural action have to be confined in order to achieve in a satisfactory way the social and ecological goals.

The rivers of the Indian subcontinent exhibit considerable variability in their forms, configuration and channel processes in spite of the common monsoon climate. An understanding of the geomorphic, hydrologic and hydraulic processes controlling channel shape, size, gradient and planform of different rivers is a prerequisite to a better implementation of effective management approaches. This has justified the requirement of an interdisciplinary effort involving meteorologists, hydrologists, fluvial geomorphologists, sedimentologists and engineers (Kale 1998).

Geomorphologically, the rivers of the Indian subcontinent can be broadly divided into two major fluvial systems (Rashtriya Barh Ayog 1980). The first system includes all the large rivers and their tributaries heading in the Himalayas and the second consists of rivers draining the Deccan Peninsula (Baker 1994). The dynamic Himalayan environment in relation to river flow processes have caused significant impact on the geomorphic development leading to abnormal flow parameters, channel migration, meander growth and avulsion. Rapid migration of channels or shifting of courses (Bristow 1987) may cause the flood waters to spill in unexpected areas and may alter the total area flooded by a particular discharge (Dunne 1988). Similarly, when meanders grow, rivers are lengthened and the gradient is decreased (Schumm 1985). These conditions in turn may cause the flood to rise (Jorgensen et al. 1994). Avulsion, which is common in rivers with erodible banks, can increase flood hazards by diverting the floodwaters in areas which were formerly above high flood level (Dunne 1988). In Bangladesh, the Brahmaputra (Jamuna) River had shifted some 70 km during the last half of the 18th century and is still moving at an alarming rate (Ahmed 1969; Choudhury and Kamal 1996). Natural changes in the elevation of river channels due to variations in sediment transport can also increase or decrease the flooding intensity and the extent of area affected by floods.

Several prominent earthquakes have affected the tectonically active NE-Himalayan region of India in the recent past. The 1897 earthquake for instance, raised the riverbed by several meters and caused extensive flooding. A similar disturbance after the 1950 earthquake led to the extensive increase flooding as a result of tectonically raised base level and by creating temporary artificial dams across the flow path of the Brahmaputra and other tributaries. The Brahmaputra valley experiences numerous natural hazards such as earthquakes, landslides, heavy sediments load, excessive rainfall and remarkable massive floods. The recurrent floods in the Brahmaputra River during April through October with spells of unrelenting spate, attendant deluge, sediment deposition and erosion have been ravaging upper Assam with flood hazard unabatedly since time immemorial. Explorers of ancient times—those who established *'Pragjyotishpur'* in Kamrup Province might have observed the spates of this river which is different from all other rivers and based on this behavior, named this river with a masculine term *'BrahmaPutra or the son of the Brahma'*, quite contrary to the names of almost all of the rivers of Indian subcontinent. Despite gigantic efforts and colossal expenditure (>Rs.15,000 million) in building 3,647 km of embankments, 599 km of drainage channels and 431 km^2 area of soil conservation, Brahmaputra continues to wreck havoc by uncontrollable floods year after year. Records show that catastrophic floods occurred in 1954, 1962, 1966, 1972, 1973, 1977, 1978, 1983, 1984, 1987, 1988, 1991, 1993, 1995, 1996, and 1998. Upwards of floods annually affect >9,600 km^2 land that is 12.21 % of the geographic area of Assam. In 1998, the flood which came in four frightening waves, deluged 38,200 km^2 or 48.65 % geographic area of the state, putting in peril the lives and properties of 12.5 million people (Valdiya 1999).

The Brahmaputra River system, because of its international nature and recurrent floods and attendant devastating floods has been the attention of various geoscientific studies, namely, the study of geomorphologic evolution of Majuli Island, Assam and erosion activity thereof (Sarma and Phukan 2004), aggradations and channel widening processes (Sarma 2005), assessment on the amount of bank erosion and bankline migration (Sarma and Phukan 2006) and neotectonic implication on erosion activity and geomorphology (Sarma and Acharjee 2012). Several studies (Talukdar 1995; Barman 1981; GSI 1981) have indicated that the Brahmaputra River changed its course abnormally after the Great Assam Earthquake of 8.6 magnitude in the year 1950 with the attendant historic flood. The prevalent equilibrium between sediment supply and transport was disrupted by this great earthquake which produced severe landslides within hilly tracts, and suddenly provided a large quantum of additional sediment load to the channel. Descending into the plains, the extra sediment choked the river channel gradually and initiated bank erosion causing channel widening. In addition, there has been a gradual increase in channel slope since 1920 (Bezbaruah et al. 2003; Goswami et al. 1999). The riverbed of Brahmaputra has also shoaled following heavy siltation due to the construction of flood embankments, deforestation etc. Many other towns besides “Maju—the largest river island” on the banks of the Brahmaputra River are also under threat due to abnormal changes in the river course.

2 The Brahmaputra River Basin

The river Brahmaputra originates from a glacier located at east of the Manasorovar lake in Tibet. From the headwater from snow fed catchments located at an altitude of 5,180 m, it traverses for about 2,890 km, of which approximately 1,625 km are in Tibet, 918 km in India and 337 km in Bangladesh with an average width of 8–10 km. The Brahmaputra valley in Assam is underlain by recent alluvium and represents a tectono-sedimentary province of 720 km long and 80–90 km width and elevation ranging from 120 m at Kobo in the extreme east through 50.5 m at Guwahati to 28.45 m at Dhubri in the extreme west (Goswami 1985). The geological history of the Brahmaputra River is related to the evolution of northeastern region of India. The northeastern region of the Indian subcontinent comprised of sediments deposited in two basins namely, Assam-Arakan in the south/southeast and the Himalayan basin in the north. The Brahmaputra is a comparatively young antecedent river of many names came into existence when the Himalayan Tethys dried up during Early Eocene. Sedimentation in the North Eastern region of India began with Upper Cretaceous and continued uninterrupted, excepting in parts at Oligocene-Miocene transition.

The channel of the Brahmaputra River itself occupies about one-tenth of the valley. In Assam, the river flows in a highly braided channel characterized by numerous mid-channel bars and islands (Fig. 1). The flow regime of the Brahmaputra responds to the seasonal rhythms of the monsoon and freeze-thaw cycle of the Himalayan snow. The Brahmaputra is the fourth largest river in terms of average discharge (19,830 m^3/s) at mouth and second in terms of sediment transport per unit drainage area in the world. The migration of channel towards south is a characteristic feature. Moreover, the location of the system in a highly seismic prone area with fragile geologic base warrants an indepth multidisplinary scientific evaluation keeping in view of its vast socio-economic implications (about 40 % of the valley area is under cultivation. The Brahmaputra valley in Assam is home to more than 15 million people). Salient features of the Brahmaputra River basin are presented in the Table 1.

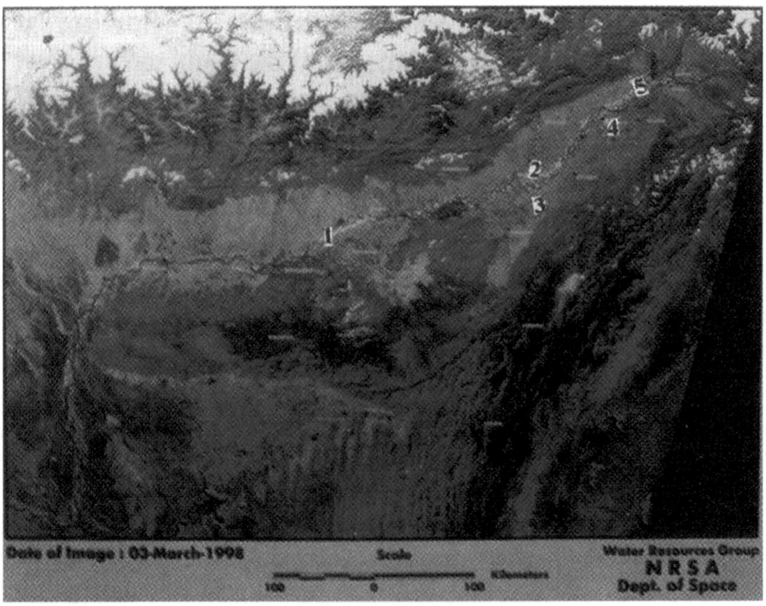

Fig. 1 The Brahmaputra river as revealed from IRS imagery

Table 1 Salient features of the Brahmaputra river basin

1.	*Total catchment area*	580,000 km^2
	(i) Catchment area within China	293,000 km^2
	(ii) Catchment area within India	195,000 km^2
	(iii) Catchment area within Bhutan	45,000 km^2
	(iv) Catchment within Bangladesh	47,000 km^2
2.	*Length from its source to confluence with ganga*	2,880 km
	(i) Length within Tibet (China)	1,625 km
	(ii) Length within India	918 km
	(iii) Length within Bangladesh up to the	337 km
3.	*Gradient*	
	(i) Reach within Tibet	0.0026
	(ii) Reach between Indo-china border and Kobo in India	0.0019
	(iii) Reach between Kobo and Dhubri	0.00014
	(iv) Reach within Bangladesh	
	First 60 km from India Border	0.00009
	Next 106 km reach	0.00008
	Next 92 km reach	0.00004
	Next 79 km reach	0.00003
4.	*Discharge characteristics*	
	(i) Maximum discharge at Pandu (Assam) on 23.08.62	72794 m^3 s^{-1}
	(ii) Minimum discharge at Pandu on 20.02.68	1,757m^3 s^{-1}
	(iii) Mean Annual flood discharge at Pandu	51,156 m^3 s^{-1}
	(iv) Mean annual dry season discharge at Pandu	4,420 m^3 s^{-1}
	(v) Mean monsoon flow (June to October)	
	Shigatse (Tibet)	507 million m^3
	Pasighat (India)	3,979 million m^3
	(i) Discharge per unit area of Basin	
	T'sela D's Zong (China)	0.01 m^3 s^{-1} km^{-2}
	Pasighat (Arunachal)	0.023 m^3 s^{-1} km^{-2}
	Pandu (Assam)	0.03 m^3 s^{-1} km^{-2}
	Bahadurabad (Bangladesh)	0.032 m^3 s^{-1} km^{-2}
5.	*Sediment load*	
	(i) Average annual suspended load during flood at Pandu	400 million metric tons
	(ii) Daily mean sediment load during flood at Pandu	2.12 million metric tons
	(iii) Sediment Yield	
	T'sela D's Zong	100 metric tons km^{-2}
	Pasighat (Arunachal)	340 metric tons km^{-2}
	Pandu (Assam)	804 metric tons km^{-2}
	Bahadurabad (Bangladesh)	1,128 metric tons km^{-2}

(continued)

Table 1 (continued)

6.	*Mean basin rainfall* (Excluding Bhutan and Tibet)	230 cm
7.	*Basin land use (in india)*	
	Total forest cover	114,992.08 km^2
	Total agricultural land	50,473.84 km^2
8.	*Basin Population in India*	30.4 million (143 persons/m^2)

3 River Bank Erosion

Riverbank erosion is a geomorphologic process exhibited by the river channels on spatial and temporal scales (Couper and Maddock 2001). It includes three primary mechanisms namely, mass failure, fluvial entrainment and sub-aerial weakening and weathering. These three mechanisms interact and exhibit themselves as observable attributes (Thorne 1982; Lawler et al. 1999). Among these, the sub-aerial process is often viewed as a 'preparatory' process, which weakens the bank surface prior to actual fluvial erosion. On a basinscale, the sub-aerial process manifests itself in the upstream regions, the fluvial erosion in the middle reaches and the mass failure in the lower reaches of the river channel, though variations occur as a result of tectonic and or anthropogenic interventions. These three mechanisms operate at different level of magnitude and frequency. Thus, the concept of process dominance has temporal, as well as spatial aspects, particularly over the short time periods that are often studied in the field. Perception of the relative efficacy of each erosive mechanism will therefore, be influenced by the temporal scale at which the bank is considered. With the advent of global climate change, both these magnitude-frequency characteristics and the consequent interactions of bank erosion mechanisms may differ. It is therefore, likely that the recognition of these temporal aspects of process dominance will become increasingly important to studies of bank erosion processes. In Assam, the river Brahmaputra flows in a highly braided channel characterized by numerous mid-channel bars and islands. Braided rivers represent high energy fluvial environments often characterized by steep valley gradients, non-cohesive banks lacking vegetation, and consequently, high rates of bank erosion and bed load transport (Lawler 1992, 1997; Thomas and Nicholas 2002).

4 Materials and Methods

The channel configuration for a distance of 260 km from Panidihing Reserve Forests to Holoukonda Bil of the Brahmaputra River (Fig. 2) was delineated from the Survey of India toposheets (1914 and 1975), and Indian Remote Sensing Satellite imagery (1998). The bank lines were superimposed upon each other and

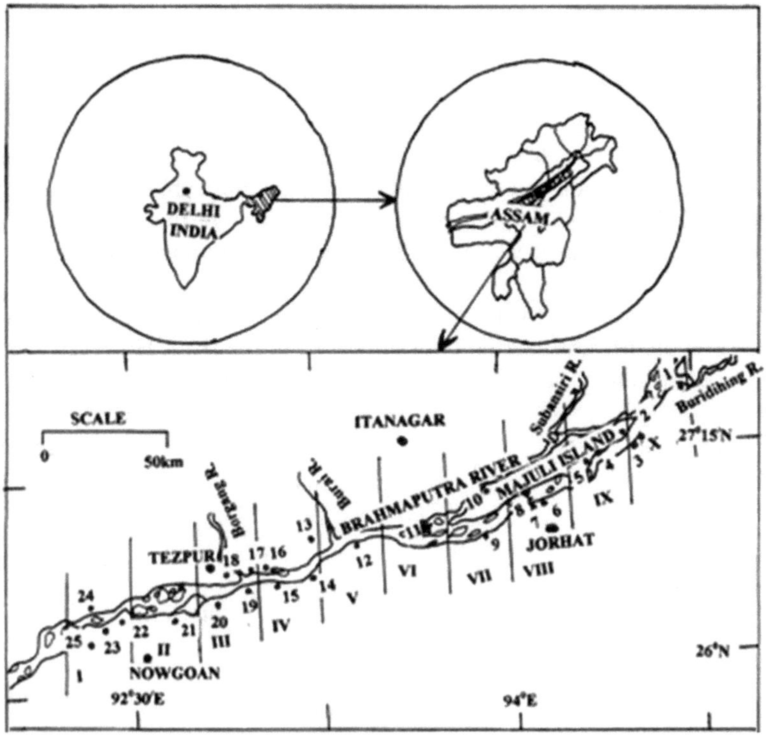

Fig. 2 Figure 1: Location map of the study area. Sectors I to X. Location of places. (1 *Panidihing Reserve Forest* 2 *Disangmukh* 3 *Dokhowmukh* 4 *Bokajan Bil* 5 *Bopkajan Bil* 6 *Nematighat* 7 *Kakila Mukh* 8 *Nayagaon* 9 *Sikarighat* 10 *Kumargaon* 11 *Pichala Chapori* 12 *Kaziranga* 13 *Behali* 14 *Dhansirimukh* 15 *Dipholumukh* 16 *Vishwanath* 17 *Kamahaya Hill* 18 *Kukurakata Hill* 19 *Bhomoraguri Hill* 20 *Jor Bil* 21 *Laokhowa Reserve Forest* 22 *Vahantola* 23 *Dhing* 24 *Orang Resrve Forest* 25 *Halaokunda Bil*)

the areas subjected to erosion and deposition were measured with the help of digital planimeter. The generated data are utilized for the evaluation and interpretation of erosional consequences exerted with respect to space and time.

The overall activity of the erosion-deposition processes spanning the period 1914–1998 were evaluated by dividing the area into 26 sectors of 10 km each (A–Z) and utilized for the present study. These information were supplemented by field mapping and examination of erosion-deposition characteristics and associated fluvial, geomorphologic and anthropogenic traits.

5 Results and Discussion

5.1 Bank Erosion

Geomorphologically, bank erosion is a central component of meander formation, lateral channel migration and movement of sediment throughout the drainage basin (Lawler et al. 1997) and a long time scale is required for these to be appreciated. Anthropogenic consequences are however, more readily apparent over just a few years.

During the period 1914–1975 the stretch under study evidenced significant erosion (Fig. 3a, b) subjected on both the banks of the river. However, the 30 km region on the southern bank of the river experienced deposition (28.15 km^2) near

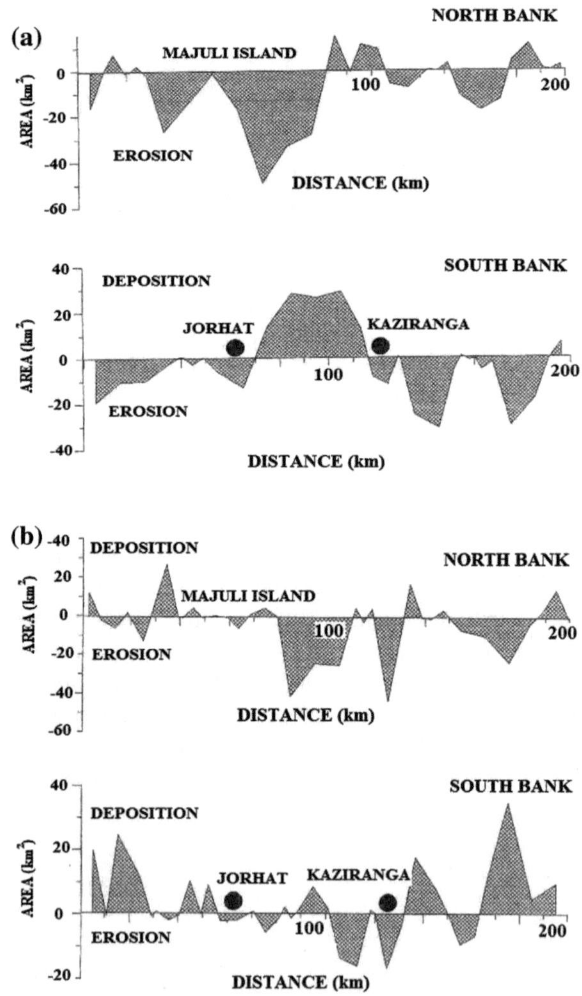

Fig. 3 Erosion and deposition within the Brahmaputra River channel during. **a** 1914–1975 and **b** 1975–1998

Sikarighat and near Bhomoraguri Hills (38.75 km^2). Erosion is much more pronounced near Jorhat–Majuli areas, and continues for 50 km up to Kumargaon. In this area, a total loss of land measuring 49.5 km^2 is observed. The area near Sohola Bil up to Dipholumukh also witnessed erosion to the tune of 83.23 km^2 until the year 1975. Interestingly the banks on both sides of the river Brahmaputra from Orang Reserve Forests to Haloukonda Bil are subjected to significant erosion. However, the erosion is much more pronounced on the northern bank of the river than the southern bank during the period from 1914 to 1975. The period during 1975–1998 is somewhat different than the earlier period under study. A slight reduction in erosion activity around Majuli area and a shift in the area of erosion are observed. The southern bank the river exhibited a depositional phase around the Neemati-Jorhat and Kaziranga National Park areas. Near the confluence of Burai River, an area of 17.65 km^2 was added to the southern side of the river Brahmaputra.

5.2 Bank Failure

Bank line stability is dependent on the behavior of the riverbed during flood and subsequent falling stages of the river. In a meandering river, the changes and migration pattern are fairly predictable, since the river cuts on one bank and deposits on the opposite. Thus, in a meander bend significant scour will take place on only one bank, whereas, the other bank will be the site of deposition. In a braided channel, on the other hand, this type of cut and fill does not occur and both the banks may experience deposition or erosion simultaneously. Several factors are responsible for the short term bank changes. These include, (a) rate of rise and fall of river level, (b) number and position of active channels during flood stage, (c) angle at which the thalwag approaches the bank line, (d) amount of scour and deposition that occur during flood, (e) formation and movement of large bed forms, (f) cohesion and variability in composition of bank materials, (g) intensity of bank slumping and (h) relationship between the bank lines of abandoned river course to present day channel. The resultant erosion-deposition is the function of interactions among these factors.

Along the Brahmaputra River, bank material is rarely homogenous in composition, and results uneven bank slumping. This causes the flow to take a different path and the orientation of the bank line to the direction of flow also changes. Field observations revealed that at some localities older alluvium protruding into the river offers significant resistance to flow regime and causes changes in the hydraulic conditions (Plate 1a, b). The finely divided bank material and the constant change in flow direction produce severe bank caving along the channel. When the flow approaches the bank at an angle, severe undercutting takes place resulting in slumping of sediments of the bank. Slumps are more common in the banks composed of clayey silt and silty-clay. Quite often, the highly saturated clayey silt liquefies and flow towards the channel. As the materials flows, the overlying less saturated bank sediments slump along the well-defined shear planes. Thus, there are two prominent types of slumping which cause bank line to recede; one operating

Plate 1 **a** and **b** Resistant older alluvium protruding into the river

during flood stage (under-cutting) and other during falling stage (flowage of highly saturated sediments). However, the intensity of slumping is more acute after the flood stage. The accumulated water level during the flood stage provides additional support to the bank material as the pore spaces of the loosely bound bank materials are occupied by the water and act as continuous system. With the recession of flood, the support diminishes abruptly and the bank materials were subjected to different degree and nature of failure. The field survey along the reach of the Brahmaputra River channel near Nematighat, Majuli and Kaziranga areas provide ample evidences on the severity of bank failure just after the flood period in comparison to the failure during flood.

In some localities, stratified fine sand, massive channel sands and silts underlie silty clay of the natural levee deposits (Kotoky et al. 2006). During high stage of the river, water is forced into these strata, and raises the pore pressure. As water level in the river falls rapidly and the pressure against channel wall is lessened abruptly, water moves from the formation back into the river. This phenomenon causes a lateral flowage of sands and silts into the channel, resulting in subaqueous failure. This normally produces a bowl-shaped shear failure in the overlying cohesive natural levee deposits (Coleman 1969). This type of bank failure is common in the studied stretch of the Brahmaputra River channel and typically observed near

Plate 2 Bowl shaped failure due to liquefaction. **a** Downstream of Nematighat and **b** Upstream of Nematighat

Nematighat (Plate 2a, b), in Majuli (Plate 3a) and also around Kaziranga. Its chordal length along the bank varies from 10 to 54 m and amplitude from the bank varies from 8 to 15 m (Plate 4a, b).

In certain locations, as observable at upstream of Nematighat in a location between spur No. 5 and 8, waterlogged areas were developed near the bank of the river, due to the construction of flood embankment. These water bodies have no direct outlets to the river. Thus, as water level in the river recedes, water from these areas move through permeable levee materials and oozes out along the bank of the river. It also causes the bank material to collapse during post flood period (Plate 5a, b).

Another type of failure related to the sub aqueous flow occurs in the upper bank and natural levee material. Because of the braided nature of the river channel and constant migration of the river, many abandoned channels intersect modern bank-lines. This give rises to a zone of well-sorted silt and fine sand, which are localized in the abandoned channel fill. During rising and flood stage, this sand and silt become highly saturated. The rapid drop in water level in channel results in rapid withdrawal of water from these sediments. The highly saturated sediments often

Plate 3 **a** Bowl shaped failure due to liquefaction near Gejeragaon, Majuli. **b** Shearing of blocks near Nematighat

liquefy and flow towards the channel. As the material flows, the overlying less saturated bank sediments tend to shear along well-defined planes. This type of failure is conspicuous in the western tip of Majuli Island and upstream of Nematighat (Plate 6a, b).

Shear failure is one of the most effective causes of bank-line recession of the Brahmaputra River (Kotoky et al. 2005). The majority of the failures occurred as a result of the current undermining the natural levee deposits, due to which large blocks of natural levee sediments shear and tilt into the river. The other major cause of shear failures is over-steepening of the bank material as thalweg of the channel hug banks (Coleman 1969). In localities where bank materials are slightly cohesive, shear failure of the bank results in rotated steps like structure leading from the top of the bank to the water edge (Plates 6 and 7a, b). Most of the shear planes diminish in slope as they penetrate into the sub-surfaces and as a result the blocks are tilted landward by rotation.

Plate 4 Shearing of bank along well defined plane. **a** Near Nematighat and **b** Spur No. 2, Near Nematighat

In some areas, as could be observed in the vicinity of Salmora in Majuli Island (Kotoky et al. 2003) and in some localities in Kaziranga National Park, bank composed of cohesive material and a slope approaching 90° or more with overhangs are observed. This type of over steepening (Plate 8a) enhances the failure of bank. Fluvial erosion, in turn is linked to mass failure processes through the concept of basal end point control (Carson and Kirby 1972; Thorne 1982). Fluvial erosion of the basal area of bank can lead to undercutting and subsequent

Plate 5 **a** and **b** Large scale flowage failure trigurred by the adjacent waterlogged areas, Downstream of Nematighat

cantilever failure. Equally, a mass failure event supplying sediments to the basal zone increases bank stability (decreasing bank angle) unless fluvial conditions results in the critical shear stress for removal of this material being exceeded (Thorne 1982). As stream power peaks in the middle reaches of a basin, fluvial erosion dominates here. Lawler (1992) suggested that with the increase in channel depth downstream, the maximum (or critical) bank height for stability with respect to mass failure is exceeded in the middle reaches of the river resulting in mass failure of banks located downstream of the location at which the critical level is crossed (Anernethy and Rutherfurd 1998; Lawler et al. 1999; Couper and Maddock 2001).

Different type of shear failure occurs during the receding stage of the river. As water level recedes in the channel, saturated levee material loose support from the channel side and result in shearing of small blocks from saturated bank due to its their own weight. However, this type of failure is always observed in small scale (Plate 3b).

A peculiar type of bank failure is observed in some localities around the Kaziranga National Park. In these localities, fine-grained overbank deposits with mud cracks are present along the banks. The processes of formation of mud cracks can directly be attributed to sub aerial processes, which include wetting and drying

Plate 6 Rotated step like shear failure of the bank **a** - at neamighat, **b** - downstream of Nematighat

of the soil (and associated desiccation). These are commonly thought as 'preparatory' rather than 'erosive' processes (Wolman 1959; Duijsings 1987; Thorne 1990; Lawler et al. 1997; Green et al. 1999). They weaken the surface of the bank prior to fluvial erosion, thus increasing the efficacy of the latter. The blocks are separated by mud cracks, detached from the bank. This type of block detachments resemble tooth of a saw (Plate-8b). The cumulative effect of block separation and fine grained sediments enhance the shearing activity, which may ultimately lead to large scale bank failure (Kotoky et al. 2005). Although sub-aerial 'weakening and weathering' of the soil can occur in a number of ways, all are associated with moisture conditions within the material (Thorne and Osman 1988; Dietrich and Gallinatti 1991; Thorne 1992) and with the physical state of this moisture (Thorne 1990). Both Wolman (1959) and Simon et al. (1999) found that the highest retreat rate occurs as a result of high flow during prolonged wet periods, rather than simply the largest storms of floods. An increase in soil moisture content decreases the magnitude of inter-particle forces within the material (Craig 1992), and reduce the 'resistance' of the bank to the shear forces associated with the flow. Alternatively, low moisture contents can also weaken the soil. As a cohesive soil mass dries, volumetric shrinkage results in the formation of a 'ped fabric' with blocks of soil

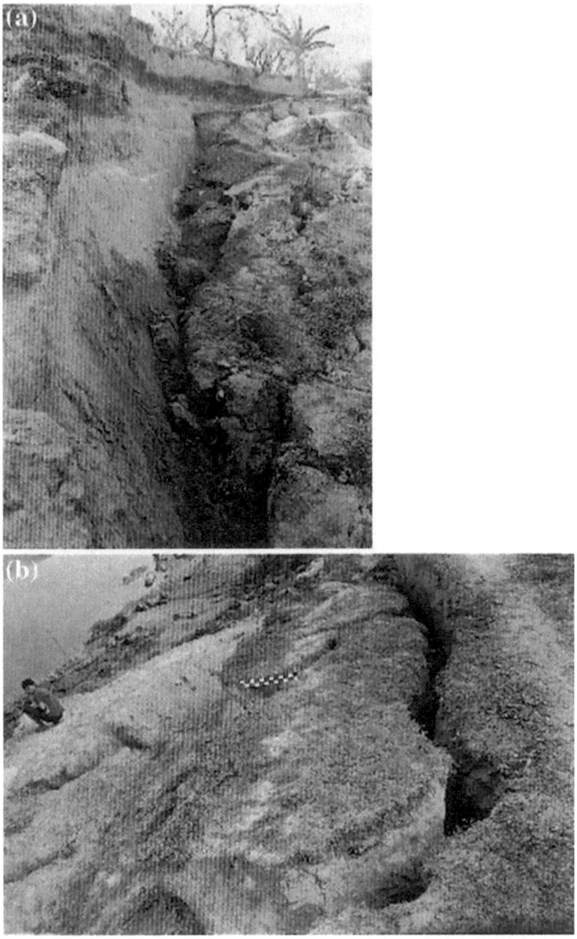

Plate 7 Rotated step like shear failure (**a**, **b**) of the bank downstream of Nematighat

separated by desiccation cracks (Thorne and Lewin 1979; Thorne and Osman 1988; Dietrich and Gallinatti 1991). This desiccation provides lines of weakness in the bank face, as cohesion is greater within the peds than between them (Thorne 1990). Green et al. (1999) found desiccation to be one of the dominant forms of bank erosion on the tributaries of the Namoi River in Australia, although they do not differentiate sub-aerial erosion rates from other processes. Slaking (detachment of aggregates from the soil surface caused by positive pore water through the soil), and leaching clay minerals, may also contribute to weakening of the bank face (Thorne and Osman 1988; Dietrich and Gallinatti 1991).

Plate 8 **a** Bankfailure due to oversteepening at Kariranga National Park. **b** Toothed replacement of bank materials related to detachment of mudcracks in blocks

5.3 Relation of Bank Erosion of the Brahmaputra River with Deposition of Bars

There exists a continuous exchange of sediments between flood plain and the channel. At any segment of the river, the sediment input in the form of suspended and bed load are sourced from the upstream reach of the river, from the tributaries and also from eroded materials from the banks and bars located within the segments. Removal of this suspended and bed load sediment from the channel results by the formation of new bars and deposition in the overbanks. In a braided river, sand bars are transient in nature and there is a constant change in shape, size and number of bars within a segment (Plate 9a, b). Any change in bars within the segment is accompanied either by erosion or deposition along the bank. To determine the quantitative relationship between the bank erosion and bar growth, the volume of sediment was calculated based on planimetric measurements of channel migration, bars and bank height.

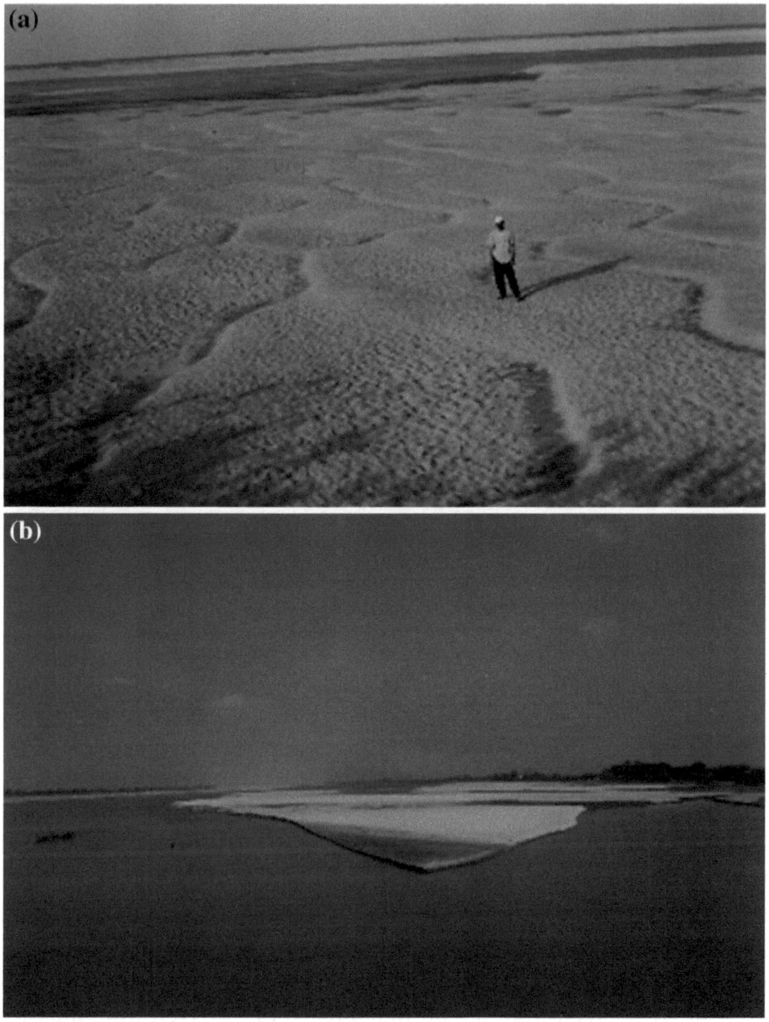

Plate 9 **a** Development of side bar with megaripples. **b** Newly immersed

In the present study, three sets of maps (Survey of India toposheets of 1914 and 1975 and Indian Remote Sensing Satellite imagery of 1998) covering the period from 1914 to1998 were used. The length of the channel was divided into ten segments at an interval of 15 min east longitude; except the first segment, which is taken at an interval of 15 min north latitude. First two sets of maps covering 1914 and 1975 are toposheet of Survey of India, while the third set, covering 1998 consists of black and white satellite imageries of IRS 1B.

The aerial measurement of erosion and/or deposition along the bank in each segment was calculated by comparing the bank line as evidenced from different base maps under use. Examination of aerial photographs or maps pertaining to river Brahmaputra reveals that it does not have simple, smooth, parallel banks. Instead, bank characteristics vary considerably along the entire length of the river. Since, most bars modify their size, shape and location from time to time, drastic changes in cross-sectional area of the channel and bars are ubiquitous. The result of this constant change in cross-sectional area is rapid cutting and filling along the channel. Thus, the bank lines of the river Brahmaputra are indistinct in many parts and utmost care is exercised while bank lines are mapped from imageries. Translation of the planimetric area to volume of sediment requires estimation of the height of the area deposited or eroded. It is assumed that each aerial measurement represented an entire column of sediment from the channel bed (Dunne et al. 1998). For each segment, net erosion or deposition along the bank, and difference in volume of sediment storage in the bar covering different periods were calculated (Table 2).

Unlike the meandering river, where erosion along one bank corresponds to deposition on the opposite bank, the braided Brahmaputra River channel shows either erosion or deposition in the corresponding segment. During the period 1914–1975, largest volume of bank sediments was eroded away from the bank near Dhing (Segment 10). Predominance of bank erosion in all the segments under study, sympathetic relationships between bar formation and net bank erosion, sympathetic relationship between decrease in sediment storage within the channel as mid channel bars and corresponding increase in deposition along the banks are all observed from these data. In addition, the volume of sediment eroded from the bank is approximately four times the increase in volume of sediment stored in the bar has also been revealed by the present study. This is in accordance to the finding of Dunne et al. (1998) that the rate of deposition in bar is about one-quarter of the rate of bank.

Table 2 Showing volume of sediment eroded or deposited in different segments

Segments	Period 1914–1975		Period 1975–1998	
	Bank net erosion/ deposition (10^6 m^3)	Bar net erosion/ deposition (10^6 m^3)	Bank net erosion/ deposition (10^6 m^3)	Bar net erosion/ deposition (10^6 m^3)
1	−399.5125	+535.6219	−62.7943	−355.8025
2	−358.6312	−5.7311	+78.3727	−123.0312
3	−568.6562	+148.025	−29.8367	+242.8656
4	+280.1625	−332.6987	−743.3967	+798.5406
5	−142.9875	+85.5406	−670.1812	−180.0906
6	−588.4187	+242.1115	+10.268	+108.5
7	−414.4312	+325.5	−275.3188	+108.6937
8	+44.95	−278.9638	+63.4537	−1.7437
9	+304.3812	−248.7673	+70.2347	−517.759
10	−749.6187	+198.4387	−273.381	+171.4532
Total	−2592.7623	+669.0767	−1525.6036	+242.6261

N.B '+' sign before numeral denotes deposition and '−' sign denotes erosion

During the later period i.e. 1975–1998, significant bank erosion occurred in the segment 4, near Kumargoan. For this period, one set of black and white satellite imagery was used for comparative evaluation of bank erosion. In panchromatic black and white satellite imageries, side or lateral bar, which is transient and part of river channel, cannot be demarcated because of poor tonal contrast. Thus, they are included with deposition of the bank. This might be probable reason that the later period manifested significant deposition along the bank.

5.4 Variation in Average Width

The Fig. 4 shows the variation of mean width at different segments from Majuli to Nowgoan district for the period from 1914 to 1998. During 1914, the river widened at upstream of Salmora in Majuli Island and narrowed near Behali, again widened upstream of Bhomoraguri Hill, reaching its maximum width of 11 km. In the later period, it is observed that river widened near Nematighat and from Kaziranga to Kukurakata Hill, while a decrease of mean width from Bhomoraguri Hills to Dhing could be observed. On the whole, it is observed that mean width of the entire studied stretch increased from 5.9 km in 1914 to 6.5 km in 1975 followed by a slight decrease (6.30 km) during 1998.

The study also revealed that there is a good correlation between the width of the river at a location and erosion. The segment downstream of the node in the Brahmaputra River usually faced problem of bank erosion. This is because, at the node due to decrease in width, the velocity of water increases which increase the sediment carrying capacity of the river. Significant sedimentation is observed downstream of this location where velocity of water dropped as a result of increase in width. Another interesting feature namely, migration of node is observed in this reach of the Brahmaputra River.

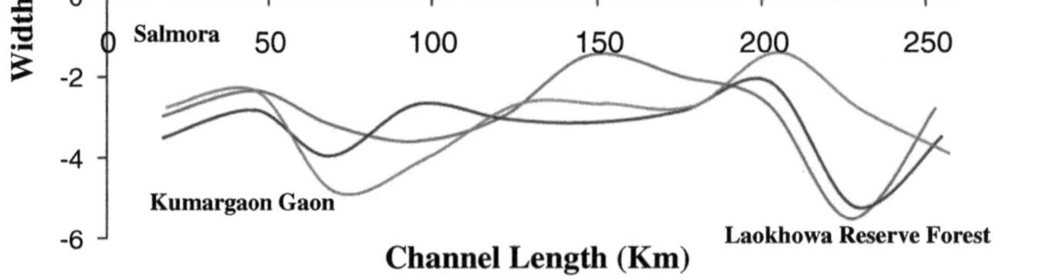

Fig. 4 Variation in average width of the Brahmaputra river channel within the studied stretch

5.5 Node Point Generation and Migration

The information available till 1914 exhibited two prominent nodal portion/points within the studied reach (Fig. 4), one in and around Salmora area and other around Behali. Just after crossing Salmora the river expands its width gradually to about 8 km and continues for 85 km till the development of another node point at Behali, where it exhibits its width as 7 km. The channel, after crossing the node point around Behali gradually widens up in downstream direction up to a maximum of 11 km near Laokhowa Reserve Forests. After crossing the Laokhowa Reserve Forests, the channel again shrinks and starts forming another node point near Dhing.

The Survey of India toposheets of 1975 revealed a significant change of width of the river from Nematighat to Kaziranga area and a node point near Kumar Gaon. After crossing Kumar Gaon, the river continues to flow with an average width of 7 km till Bhomoraguri Hills and shows the occurrence of another node point there (Kotoky et al. 2009). Thereafter, the river bulges gradually. The present day scenario, as deciphered from the 1998 IRS imagery, is somewhat different than the pre 1975 period. The scenario depicted by the 1998 IRS IB imagery, the river channel widens abruptly after the node point at Salmora and attains a maximum width of 9 km. Thereafter the channel shrinks by about 3 km and flows till Kukurakata Hill and generates a node point at Bhomoraguri and thenceforth gradually widens for another 2–3 km and reaches a maximum width (8–9 km).

The observation on the channel configuration during the period spanning from 1914 to 1998 with regard to the width of the channel revealed a significant change in its behavior (Fig. 4). Throughout the entire segments the mean width of the channel was 5.9 km in 1914, which increases to 6.5 km in 1975 and becomes 6.3 km in 1998. The slight decrease in mean width in 1998 might have a close relationship with the generation of node point in tighter fashion than the earlier periods. The geomorphologic development within the reach produced two node points in 1914 and the number of node points increased to three in the latter period and also shifted their positions. However, it is interesting to note that the position of the node around Salmora have not been changed during the entire period under study. The ground observations along the channel around Salmora area helps explain its stability due to the existence of clayey material. Since the banks are relatively stable in this area, the river scour deeper to accommodate flood discharge and the scoured materials causes a local sharp increase in transported load. Therefore, below the node point the river tends to be wider and shallower. Here the current velocities diminish, causing the increased sediment load to drop out and form islands or chars. Once the char is formed it locally decreases the cross-sectional area, and the river cuts its bank laterally to maintain a proper cross-sectional area that is in equilibrium with discharge. Thus, constriction of the channel by rigid clay materials around Salmora area has played a role in the variation of width along the Brahmaputra River channel. These type node points are relatively stable and do not move on short-term basis (50–100 years). Above and below the node points at Salmora lateral bank movement is highly exaggerated and extensive local cutting is pronounced causing significant erosion at Nematighat and Kaziranga areas. However, the width below this node point is solely dependent on discharge and local increase in sediment load in relation to variable hydraulic conditions as may be expected in a braided river like Brahmaputra. Non-uniformity in variation of width between two successive periods i.e., 1914–1975: 1975–1998 can clearly be attributed from the study within the proposed reach.

The variation in width can very well be correlated with the erosive power of the river. The segments below the node point always face acute bank erosion problem. This can be explained by the fact that due to narrowing of the channel at node points, the scouring or the undercut of the channel becomes very active to accommodate the volume of water. The generated loads of sediments were deposited just after the node point and the channel becomes shallower. To accommodate the volume, the river expands on its both sides. The growth of the sandbars also pushes the thalweg towards the bank with resultant development of severe erosion on its banks. This type of phenomena is typically observed near Nematighat and Laokhowa Reserve Forests, just after the Salmora and Bhomoraguri node points within the studied reach.

The longitudinal migration of node points within the studied reach has also been observed. Throughout the entire period of study, the position of the node around Salmora was relatively stable. However, the nodes observed near Behali with reference to 1914 SOI toposheet shifted back and forth for about 40 km towards Kaziranga–Kumargaon areas and presently located near Bhomoraguri Hills. This observation has got significant bearings on the differential hydrodynamic behavior of the channel. Within the limitation of poorly understood hydraulic conditions, it can be attributed that owing to the development of local areas of slack current, rapid deposition of bed load occur in these areas. The channel accordingly attains a tendency to widen at this point, and the islands are formed. This might be the most plausible explanation for the development of node point at Behali below the semi-permanent Salmora node point. The node points developed by this process however, are not stable and moved according to the existing hydrodynamic conditions. The movement of node point as observed during the latter period till 1998 towards Kumargaon–Kaziranga and Bhomoraguri Hills also supports is inference.

In addition, the area exhibited depositional episode near Kumargaon–Kaziranga till 1975 as observed from the Survey of India topographic maps. However, the area near Behali suffered significant erosion during both the periods. Again, just after the Kukurakata Hills the channel exhibited differential rate of deposition on both the sides of the channel as evidenced from the comparative study of the Survey of India topographic maps.

6 Conclusions

The river Brahmaputra represents a massive river system characterized by an extremely complex fluvial regime, located in a unique climatic, physiographic and tectonic setting. Bank erosion of the Brahmaputra River channel within the studied stretch is governed by the geomorphic and anthropogenic factors. Three primary mechanisms solely responsible are—mass failure, fluvial entrainment and sub aerial weathering and weakening. Interactions of these processes are also observed. Inhomogenity in composition of bank material have played a major role in bank slumping. Uneven shear failure and flowage of sediments are the principal processes as observed within the reach. The transitory nature of channel configuration exhibited through node point generation and migration also have played a dominant role in channel widening by lateral bank erosion. The results and inferences drawn from this study have major implications on designing flood control measures and flow mitigation programs.

References

Anernethy B, Rutherfurd ID (1998) where along a river's length will vegetation most effectively stabilise stream banks? Geomorphology 23:5–75

Ahmed F (1969) Maximum discharge of major rivers in large flood plains in East Pakistan-Flood and their computation. IAH-UNESCO_WHO Publication, Lenigrad: 364–369

Ayog (1980) Report of the Rashtriya Barh Ayog, Govt. of India, New Delhi, I and II

Baker VR (1994) Geomorphological understanding of floods, Geomorphology 10:139–156

Barman G (1981) Geomorphology of the Brahmaputra basin. Its floods problem and possible remedial measures. Geol Surv India Misc Pub 46:21–31

Bezbaruah D, Kotoky P, Baruah J, Sarma JN (2003) Geomorphological explanation of swamp along the Brahmaputra river channel, Assam. J Geol Soc India 62:605–613

Bristow CS (1987) Brahmaputra river: channel migration and deposition. In: Ethridge FG, Flores RM, Harvey MD (eds) Recent development in fluvial sedimentology. Soc Econ Paleontol Mineral Spec Pub 39:63–74

Carson MA, Kirkby MJ (1972) Hillslope form and process. University Press, Cambridge

Choudhury IG, kamal MM (1996) Tracking of chars of river Jamuna, GIS Asia-Pecific, December: 37–39

Coleman JM (1969) Brahmaputra river: channel processes and sedimentation. Sediment Geol 8:131–237

Couper P, Maddock IP (2001) Subaerial river bank erosion processes and their interaction with other bank erosion mechanisms on the river arrow, warwickshire, UK. Earth Surf Proc Land 26:631–646

Craig RF (1992) Soil mechanics, 5th edn. Chapman and Hall, London

Duijsings JJHM (1987) A sediment budget for a forested catchment in Luxembourg and its implication for channel development. Earth Surf Proc Land 12:173–184

Dunne T (1988) Geomorphological contribution to flood control planning In: Baker VR, Kochel RC, Patton PC (eds) Flood geomorphology, Wiley and Sons, Chichester, UK, 421–438

Dunne T, Mertes LAK, Meade RH, Richey JE, Forsberg BR (1998) Exchanges of sediments between the flood plain and channel of the Amazon river in Brazil. Geol Soc Am Bull 110 (4):450–467

Dietrich WE, Gallinatti JD (1991) Fluvial geomorphology. In field experiments and measurement programs in geomorphology. Earth Surf Proc Land 12:173–184
Geological Survey of India (1981) Proceedings of the seminar on fluvial processes and geomorphology of the Brahmaputra river basin, vol 46. Miscellaneous Publ, p 141
Goswami DC (1985) Brahmaputra river, Assam, India: physiography, basin denudation and channel aggradations. Water Resour Res 21:959–978
Goswami U, Sarma JN, Patgiri AD (1999) River channel changes of the Subansiri in Assam, India. Geomorphology 30:227–244
Green TR, Beavis SG, Dietrich CR, Jakeman JM (1999) Relating stream-bank erosion to in-stream transport of suspended sediment. Hydrol Process 13:777–787
Jorgensen DW, Harvey MD, Flamm L (1994) Morphology and dynamics of the indus river: implications for the Mohenjo daro site, In: Shroder J, kazmi A (eds) Himalaya to the Sea: Geology, Geomorphology and Quarternary, Routledge, London 288–326
Kotoky P, Bezbaruah D, Baruah J, Sarma JN (2003) Erosion activity on Majuli—the largest river island of the world. Curr Sci 84(7):929–932
Kale VS (edn) (1998) Flood studies in India, Geol. Soc. Ind. Memoir 41
Kotoky P, Bezbaruah D, Baruah J, Sarma JN (2005) Nature of bank erosion along the Brahmaputra river channel, Assam. Curr Sci 88(2):634–640
Kotoky P, Bezbaruah D, Baruah J, Sarma JN (2006) Characterisation of clay minerals along the Brahmaputra river channel, Assam. Curr Sci 91(9):1247–1250
Kotoky P, Bezbaruah D, Borah GC, Sarma JN (2009) Do node points play a role in flood proliferation? Curr Sci 96:1457–1460
Lawler, DM (1992) Process dominance in bank erosion systems In: Casling P, Petts GE (eds) Lowland floodplain rivers. Wiley, Chichester, p 117–159
Lawler DM, Thorne CR, Hook JM (1997) Bank erosion and instability. In: Thorne CR, Hey RD, Newson MD (eds) Applied fluvial Geomorphology for river engineering and management. Wiley, Chichester, pp 137–172
Lawler DM, Grove JR, Couperthwaite JS, Leeks GJL (1999) Downstream change in riverbank erosion rates in the Swale-Ouse system, northern England. Hydrol Process 13:977–992
Sarma JN (2005) Fluvial process and morphology of the Brahmaputra river in Assam, India. Geomorphology 70(3–4):226–256
Sarma JN, Acharjee S (2012) Bank erosion of the Brahmaputra river and Neotectonic activity around Rohmoria Assam, India. Comunicacoes Geologicas 99(1):33–38
Sarma JN, Phukan MK (2004) Document origin and some geomorphological changes of Majuli Island of the Brahmaputra river in Assam, India. Geomorphology 60(1–2):1–19
Sarma JN, Phukan MK (2006) Bank erosion and bankline migration of the Brahmaputra River in Assam during the twentieth century. J Geol Soc India 6:1023–1036
Schumm SA (1985) Patterns of alluvial rivers. Ann Rev Ear Sci 13:5–27
Simon A, Curini A, Darby S, Lagendoen EJ (1999) Stream bank mechanics and the role of bank and near- bank processes in Incised channels. In: Darby SE, Simon A (eds) incised river channels. John Wiley, Chichester, pp 123–152
Talukdar S (1995) Majuli in the verge of extinction. The Sentinel June 17:2 and 4
Thomas R, Nicholas AP (2002) Simulation of braided river flow using a new cellular routing scheme. Geomorphology 43:179–195
Thorne CR (1982) Processes and mechanisms of vegetation on river bank erosion. In: Hey RD, Bathurst JC, Thorne CR (eds) Gravel-bed rivers. Wiley, Chichester
Thorne CR (1990) Effects of vegetation on river on river bank erosion and stability. In: Thornes JB (ed) Vegetation and erosion. Wiley, Chichester, pp 125–144
Thorne CR, Lewin J (1979) Bank processes, bed material movement and planform development in a Meandering river. In: Rhodes DD, Williams GP (eds) Adjustments of the fluvial system. Allen and Unwin, London
Thorne CR, Osman AM (1988) The influence of bank stability on regime geometry of natural channels. In: White WR (ed) International conference on river regime. Hydraulics Research, Wallingford, pp 134–148

Thorne CR (1992) Bend scour and bank erosion on the meandering Red River, Louisiana. In: Carling MA, Petts GE (eds) Lowland floodplain rivers: geomorphological perspectives. John Wiley, Chichester, pp 95–116
Valdiya KS (1999) Why does river Brahmaputra remain untamed? Curr Sci 76(10):1301–1305
Wolman MG (1959) Factors influencing erosion of a cohesive riverbank. Am J Sci 257:204–216

Mechanisms and Spatio-temporal Variations of Meandering and Erosion-Deposition Statistics of the Dhansiri River, Assam

P. Kotoky and M.K. Dutta

Abstract The highly meandered Dhansiri River, a south bank tributary of the mighty Brahmaputra bears significant geomorphologic importance. Spanning the period 1914–2000, a stretch of the Dhansiri River channel from Dhansirimukh to Nowakota Kachari was studied with an objective to understand the erosion-deposition activities operating within the channel. Owing to its location on an alluvial plain, the river shows conspicuous migration characteristics. It has also imparted unique fluvial landscape in the study area. The river channel within the period under observation has migrated to the tune of 2.85 km towards south at Dhansirimukh in conjunction with the southward migration of the mighty Brahmaputra River channel. The study has revealed a total average annual erosion and deposition covering the entire period were 1.32 and 1.27 km^2/year respectively. The total average rate of erosion and deposition per kilometer length of the river were 0.006375 and 0.00625 km^2/km respectively. Increasing rate of erosion since the year 1914, comparatively higher erosion along the west bank than the right bank have also been observed. The areas around Butalikhowa, Golaghat and Kuruabahi have under gone severe erosion posing a threat to the population in the vicinity.

Keywords Dhansiri river · Brahmaputra · Meandering · Bankline migration · Erosion

P. Kotoky (✉)
Council of Scientific and Industrial Research, North East Institute of Science and Technology, Jorhat 785 006, Assam, India
e-mail: probhatk@yahoo.com

M.K. Dutta
Jorhat Engineering College, Jorhat 785 007, Assam, India

Mu. Ramkumar et al. (eds.), *Environmental Management of River Basin Ecosystems*, Springer Earth System Sciences, DOI 10.1007/978-3-319-13425-3_13

1 Introduction

The urge to unravel the mysteries of nature has created a strong group of workers with their own philosophy in different parts of the world since early 17th century, when attention of men was drawn to the mysteries of nature around them. Mention may be made of important contributions in this field by Hutton (1795), Lyell (1872), Gilbert (1887), Powell (1895), Geike (1905), Meinzer (1942), Grover and Harrington (1943), Horton (1932, 1940, 1941, 1945, 1954), Thornbury (1954), Morisawa (1957, 1959,1962), Brice (1964), Chow (1964), Strahler (1964), De-Weist (1965), Dury (1970), King (1970), Temple and Sandburg (1971), Allen (1971), Gregory (1977), Gregory and Walling (1973), Schumm (1977, 1985), Baker et al. (1988), Davies (1989), Goudie (1990), Allen and Allen (1990), Fielding (1993), Simon et al. (1999) and others. During the last few decades the study of river system has become one of the most important areas and gained tremendous momentum throughout the world.

Familiarity and discrete use of the concepts and methods pertaining to the flow and sediment transport in rivers has become a hallmark in geomorphological research in recent times, primarily for two reasons; first, the rivers have been providing sustenance to the ecosystem in which the human race is a part, and provide services and resources for the economic development of the humans and second, these very services and resources provided by the rivers are under threat owing to the indiscriminate use of the resources, and resultant changes in the natural fluvial system. The vastly improved *state-of-art knowledge* about many a natural phenomena of prime geomorphic significance such as hydrological, sedimentological, geomorphic, environmental and human implications of flood (Leopold and Miller 1954; White 1964; Kayastha and Yadava 1977; Ward 1978; Bevan and Carling 1987), mechanism of formation of flood plains (Wolman 1959), development of channel morphology (Schumm and Khan 1972; Leopold and Wolman 1957; Leopold et al. 1964; Yalin 1992) and aggradation and degradation of river valleys and denudation of catchments (Leopold and Miler 1954; Goswami 1985; Duijsings 1987; Dietrich and Gallinatti 1991; Baker 1994) is largely attributable to such approaches in fluvial geomorphic research. However, in most of the developing countries including India, application of ideas and principles, tools and techniques has not yet gained the desired level (Goswami 1998).

In India, the Ganga, the Brahmaputra and the Mahanadi rivers have been among the intensively studied [for example, Ahmed (1969), Coleman (1969), GSI (1977, 1981), Dutta (1980), Barman (1981), Goswami (1985, 1988), Bristow (1987), WAPCOS (1993), Hussein et al. (1993), Naik and Sing (1996), Kale (1998), Goswami et al. (1999), Sarin and Krishnaswami (1984), Sarma and Basumallick (1984), ARSAC (1990), Sarma (1986, 1993, 1995), Bezbaruah et al. (2003), Dutta and Kotoky (2006a, b), Kotoky (2011), Kotoky et al. (1997, 2003, 2005, 2006, 2009, 2011a, b, c, 2012)] fluvial systems. Geomorphologically, the rivers of the Indian subcontinent can be broadly divided into two major fluvial systems (Rashtriya Barh Ayog 1980). The first system includes all the large rivers and their tributaries heading in the Himalayas and the second consist of rivers draining the Deccan Peninsula. The striking differences between the two fluvial systems exist not only in terms of the magnitude of high flows but also in their channel morphology, plan-form and frequency and also the extent of flooding. The Himalayan rivers occupy a highly dynamic environment with profound variability in discharge and sediment load. Shifting of river courses, scouring of bed and banks and disproportionate transportation/deposition of sediments are some of the distinguishing features of the Himalayan Rivers. Conventional hydrological analysis in the Indian subcontinent is hampered by lack of sufficiently long and reliable flood record to evaluate the influence of long term climatic changes on flood magnitude and frequencies. Consequently, attempts are being made to interpret paleofloods from geological records (Baker 1994; Kale 1998). The dynamic Himalayan environment had caused significant impact on the geomorphic development leading to abnormal flow parameters, channel migration, meander growth and avulsion of the river channels draining in the area. Rapid migration of channels or shifting of courses may cause the floodwaters to spill in unexpected areas and may alter the total area flooded by a particular discharge (Dunne 1988). Similarly when meanders grow, rivers are lengthened and the gradient is decreased (Schumm 1985). These conditions in turn may cause the flood to rise (Jorgensen et al. 1994). Avulsion, which is common in rivers with erodible banks, can increase the flood hazards by diverting floodwaters in areas which were formerly above the high flood level (Dunne 1988) and considered to be safe, catching the inhabitants unawares. A case in this regard is the flooding of the Kosi River in the year 2006 that suddenly inundated hundreds of sq.km areas that were thickly populated and were considered to be safe from flooding hazard. In Bangladesh, the Brahmaputra (Jamuna) River has shifted some 70 km during the last half of the 18th century and is still moving at an alarming the rate (Ahmed 1969). Natural changes in the elevation of river channels due to variations in sediment transport can also increase or decrease the flood and/or area affected by floods.

Geomorphic analysis of a river presents unique challenges and requires a systematic and organized approach because of the spatial scale and system complexity involved. Progresses in the study of fluvial geomorphology have been aimed to develop the capability to identify, investigate and understand the continuity and connectivity of the flow processes and fluvial landforms in the river system. This prescribes the need to recognize and explore the links that bind the fluvial system in space and time (Fig. 1). Knowledge on the history and sequence of antecedent events and trends of morphologic evolution during the months, years and decades preceding the study will definitely help to understand the process-form relationships of a river.

River engineers, policy makers and managers recognize the importance of accounting for channel morphology and the dynamics of the fluvial system when dealing with alluvial rivers. Modern approaches to river management require engineers to work with rather than against the natural process-form relationships of the river, by retaining as much as possible of the natural hydraulic geometry of the

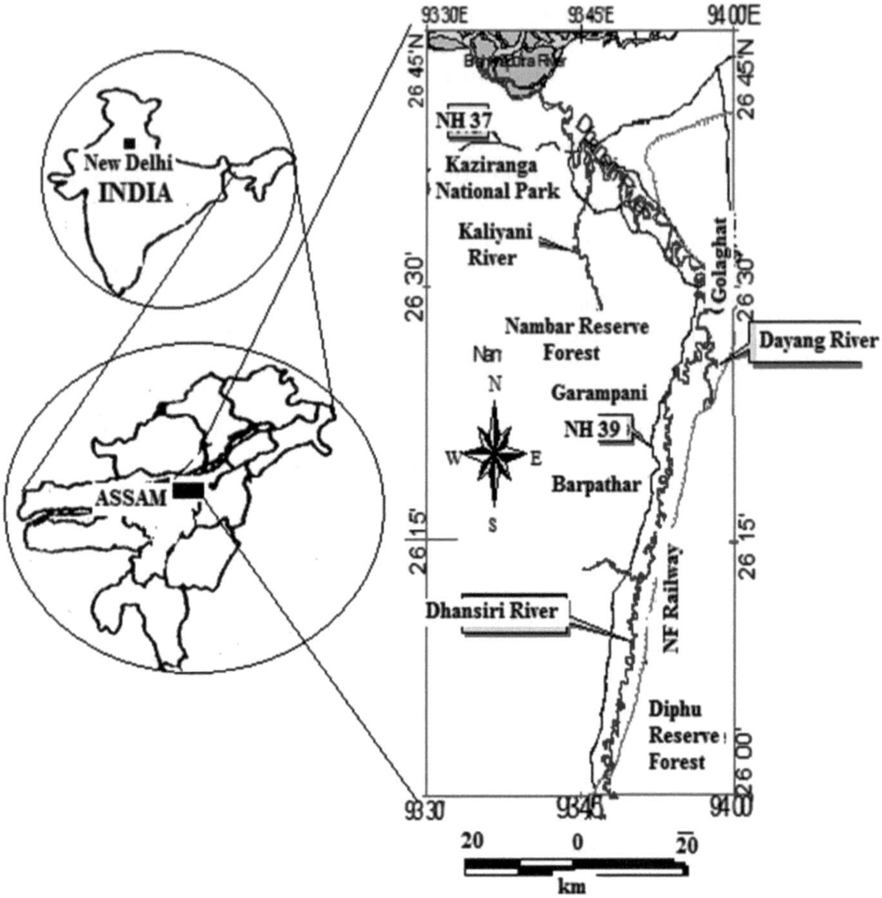

Fig. 1 Location map of the study area

self-formed channel when performing works for river regulation, channel training, navigation, flood defense, and land drainage. For this type of approach to be successful, it is essential to identify correctly the current morphological status of the river and to predict its future development with and without the proposed engineering interventions. This latter need implicitly requires the ability to predict the reaction of the channel to, for example, flow regulation (for hydropower or flood defense), or bank stabilization and training works (to control channel migration). Morphological impacts are known to spread away from the point of disturbance through process-form feedback mechanisms operating in the fluvial system. Hence, predictions of possible impacts cannot be limited to the reach directly affected by engineering works, but must extend along the mainstream and tributaries upstream, and the channel and any distributaries downstream.

Given cognizance to the nature, precepts and requirements for any geomorphic study of fluvial system as reviewed in the preceding paragraphs, this paper study has attempted understanding a unique river system in the Himalayan region, the Dhansiri River, a tributary of the mighty River Brahmaputra.

2 Location and Geological Setting of the Dhansiri River

The River Dhansiri, an important southern tributary of the mighty Brahmaputra River is originates from Thimtubum Peak of the Barail range at an altitude of 1,868 m (Sarma 1993). The important tributaries of the river are Dayang, Diphu, Sungajan, Deopani and Nambar. The tributary Dayang is much bigger than the main Dhansiri River. The total catchment's area of the river is about 12,584 km^2, the total length of the river is 352 km, out of which about 215 km falls within the plains of Assam. The river attains a maximum breadth of 132 m near Golaghat and average depth is about 6.20 m. The hydrological station on the river at Numaligarh station indicates its danger level as 77.42 m and maximum, minimum and average water discharges as 209,185, 4.88 and 513 $m^3 s^{-1}$ respectively. The annual discharge of sediment as recorded in the station was reported as 6.28 lakh ha m. The river Dhansiri was reported as one of the highly meandered river of the world (Sarma 1993).

The study area is covered by the agro-eco regions of the Golaghat and Karbianglong districts and adjoining Nagaland. It experiences nearly dry to moist sub-humid climate with a moisture index of 0.2 % and an estimated length of growing period ranging from 270 to 300 days. The average annual temperature, annual precipitation and potential evapo-transpiration are 23.9 °C, 1,223 and 1,219 mm respectively. The area encompasses major physiography and landform types of Purvanchal hills, undulating uplands and plains of Brahmaputra valley.

The soils types present in the study area include relict alluvium developed over sedimentary and metamorphic rocks. The soil profiles are deep, well drained to somewhat excessively drained, loamy to fine textured. The soils are acidic and medium in base saturation and cation exchange capacity (CEC) with appreciable exchangeable aluminium in certain places. The soils were classified as Typic/Umbic Dystrochepts, Typic Paleudalfs, Typic Hapludults in the hills and Typic/Aeric Haplaquepts, Typic aeric Haplaquents, and Typic Udorthents in the Assam plains (NBBS 1999).

The major part of the agro-eco zone is covered with mixed forests. Hilly foothills regions are presently covered by tea and coffee plantation. Major part of the plain areas are used for cultivation of paddy, sugarcane, rotational crops and oil seeds along with varieties of vegetables. The soils of the region also have tremendous potential for horticultural crops. Soil acidity and flooding/water logging in soils of lower topography and/or depressions are the main constraints in potential land use pattern.

3 Materials and Methods

Detailed geomorphologic analysis have been carried out on the Dhansiri River channel, Assam for a stretch of 215 km out of the total length of 352 km within the geographical coordinates 93° 30′E–94° 0′E longitudes and 25° 45′N–26° 45′N latitudes. However, for the study of erosion phenomena, only 95 km of the Dhansiri River channel from Oating to Dhansirimukh was taken into account. The anthropological attributes along the reach, which have got significant impact on the geomorphic behavior of the channel, were also evaluated.

The Survey of India (SOI) toposheets (1914 and 1975) and Indian Remote Sensing (IRS) satellite imagery (1990, 1995 and 2000) were utilized for investigation of spatial changes over available period of time. The IRS black and white (B/W) imagery on band four (wave length 0.77–0.86 μm) with spatial resolution 36.25 m Linear Imaging and Self Scanning II) at 1:50,000 scale were used. The area under study was digitized through the use of Arcview software and geo-referenced in a GIS environment. The meanders were then characterized, quantified and nature of movements was measured to generate database.

Encompassing a period of 87 years starting from 1914 to 2000, the study on erosion phenomena of the Dhansiri River channel from Dhansirimukh to Nowakota-Kachari with potentiality to erosion was considered. Within the logistics and limitation it is attempted to evaluate the bank erosion phenomena with ground check. The erosion intensity along the channel was determined by sequential bank line analysis and the areas of erosion/deposition were determined in a GIS environment. The river course under study has been divided into six sectors to account for the amount of erosion and other related phenomena. Each sector was further sub-divided into smaller segments of 5 km each and studied accordingly.

4 Results and Discussion

4.1 Nature of Meandering

A river is an example of an open system through which matter and energy flow, but within which are inherent tendencies of self-regulation. The interactions between discharge, load, channel shape and other variables of hydraulic geometry in a fluvial system achieve long-term self-regulation in a river channel. Although, in a regional landscape it might take a long period of time, fluvial processes might achieve mutual adjustment in a comparatively shorter time intervals. Developmental activities in the flood plains and in regions adjoining the river courses should take into account this self-regulating mechanism of the rivers. As the rivers have shorter response times than other geomorphic systems, failure to consider this mechanism leads to adverse impacts.

The river Dhansiri is an important south bank tributary of the mighty Brahmaputra River. The river was reported as one of the highly meandered channels of the world. Meandering involves inherent properties of flowing water, as well as size and shape of the channel, erodibility of the stream banks, proportion of suspended and bed loads. In turn, meandering increases the channel length between two points and thus decreases the slope of the stream. Slope influences the velocity and sediment-transporting capacity of the river water. Thus, the formation, migration and removal of meanders in river courses are indicators of the ongoing self-regulation mechanism of rivers. In order to understand this phenomenon, the objective of understanding this mechanism, documentation of spatial and temporal distribution of channels, the degree of their regularity and controls on the pattern and movement is essential.

The characteristics and controls of channel patterns have long been the topics of study in fluvial geomorphology (Jefferson 1902; Friedkin 1945; Durry 1954; Leopold and Wolman 1957). The frequency of types of change in individual bend can be analyzed by comparison with models of movement to obtain a classification. To have an idea of lateral migration, the primary elements of movements have to be identified, defined and the changes have to be established by the vectors of movement of points of inflexion (P) and the apex (A) and by change in the orientation of the apical line. The primary elements of movement are: (a) Translation, (b) Extension, (c) Rotation, (d) Change in wavelength, (e) Lateral movement and (f) Complex change. Each of the movement may have one or two directional movement i.e. up or down, increase or decrease, and to the left or right. Combination of two or three processes in a natural system is also possible (Fig. 2).

In the present study, data from Survey of India (SOI) toposheets (1914 and 1975) and Indian Remote Sensing (IRS) satellite imagery (1990, 1995 and 2000) were collected to investigate spatial changes over time. Thematic maps of different periods were prepared on 1:50,000 scale and were integrated using Arc view GIS. The meanders were then characterized, quantified from individual map and natures of movements were measured by sequential analysis.

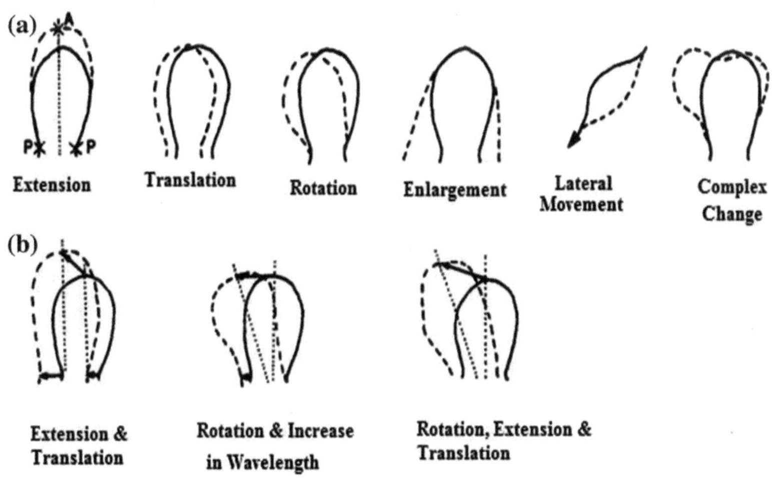

Fig. 2 Types of meandering in a river channel (After Hooke 1977)

The results indicate that out of nine different processes of meander migration, translation type was observed as dominant mechanism throughout the period under study (Table 1, Fig. 3). The order of significance can be represented as translation (49.20 %) > extension (16.04 %) > extension translation (13.37 %) > complex (12.03 %) > rotation (4.55 %) > lateral movement (2.94 %) > enlargement (1.60 %) > rotation and increase in wave length (0.27 %). The maximum number of meander bends (103) was observed in the base map of 1975. The significant rise in the number of meander bends might have a close link with the catastrophic disturbances caused by the Great Assam Earthquake of 1950 with magnitude 8.6 on

Table 1 Types of meander movement in Dhansiri river channel

Types	Periods			
	1914–1975	1975–1990	1990–1995	1995–2000
Translation	40	41	52	51
Extension	8	21	17	14
Rotation	7	5	3	2
Enlargement	1	3	1	1
Lateral movement	8	3	0	0
Complex change	13	16	8	8
Extension and translation	14	14	10	12
Rotation and increase in wave length	1	0	0	0
Total	92	103	91	88

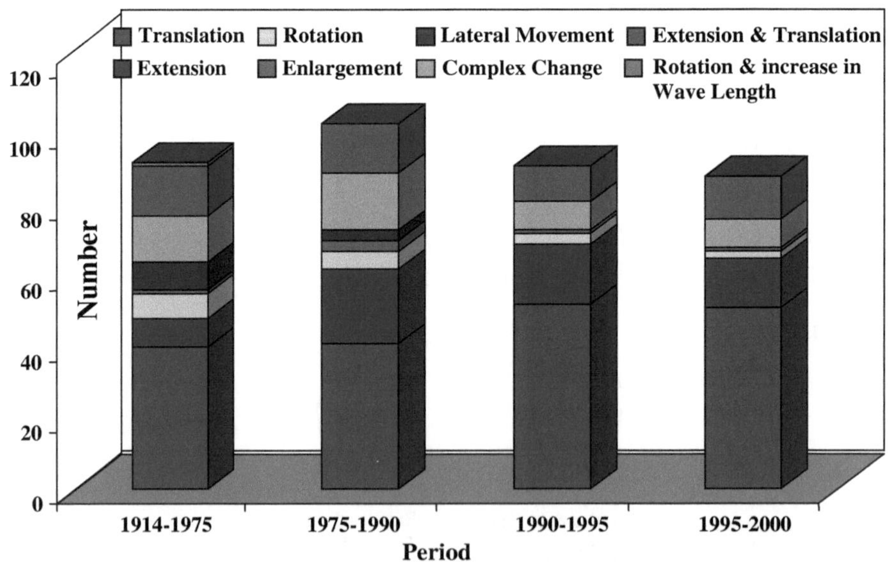

Fig. 3 Types of meander movement

Richter scale along with attendant historic flood. It is established that the earthquake has caused extensive changes to the geomorphology of the Brahmaputra Valley. Migration and abandonment of channels over a short period of time have resulted in the generation of many low-lying areas adjacent to the river system. The existence of significantly higher number of ox-bow lakes in the form of paleo-channels testifies to the transient nature of most of the former channels and the significant geomorphologic changes imparted by the earthquakes of the region. The migration of meanders at different period of time and the resultant generation of lakes (Bils) are shown in Figs. 4 and 5. Figure 6 presents the composite picture of the meander development and migration and formation of Bills for the entire study period (1914–2000).

The characteristic nature of meander migration/movement for different periods under comparative study with total number of meanders (figures in parentheses) can be represented as:

1914–1975 Translation (40) > Extension and Translation (14) > Complex (13) > Extension (8) = Lateral Movement (8) > Rotation (7) > Enlargement (1) = Rotation and increase in Wave Length (1)

1975–1990 Translation (41) > Extension (21) > Complex (16) > Extension and Translation (14) > Rotation (5) > Enlargement (3) = Lateral Movement (3)

1990–1995 Translation (52) > Extension (17) > Extension and translation (10) > Complex (8) > Rotation (3) > Enlargement (1)

1995–2000 Translation (51) > Extension (14) > Extension and translation (12) > Complex (8) > Rotation (2) > Enlargement (1)

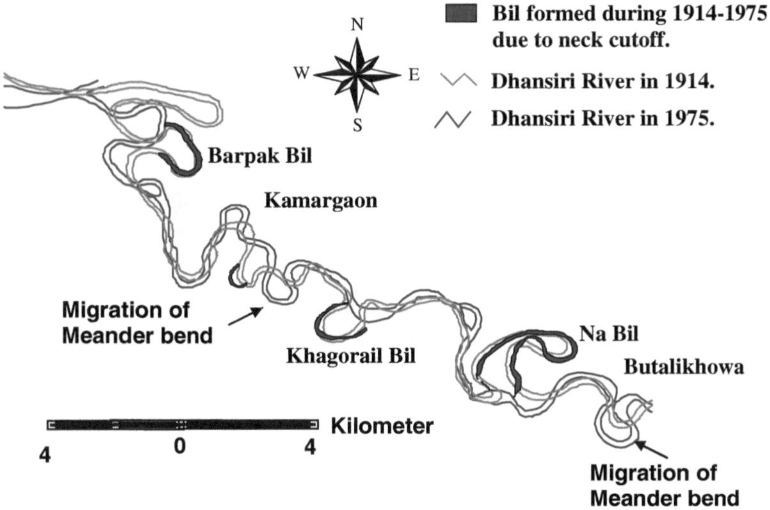

Fig. 4 Migration of meander bends and formation of Oxbow lakes (Bils) during 1914–1975

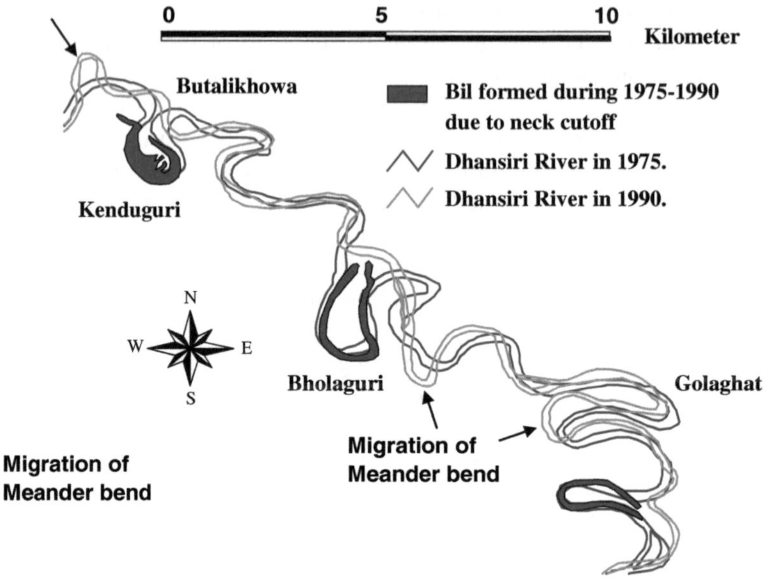

Fig. 5 Migration of meander bends and formation of Oxbow lakes (Bils) during 1975–1990

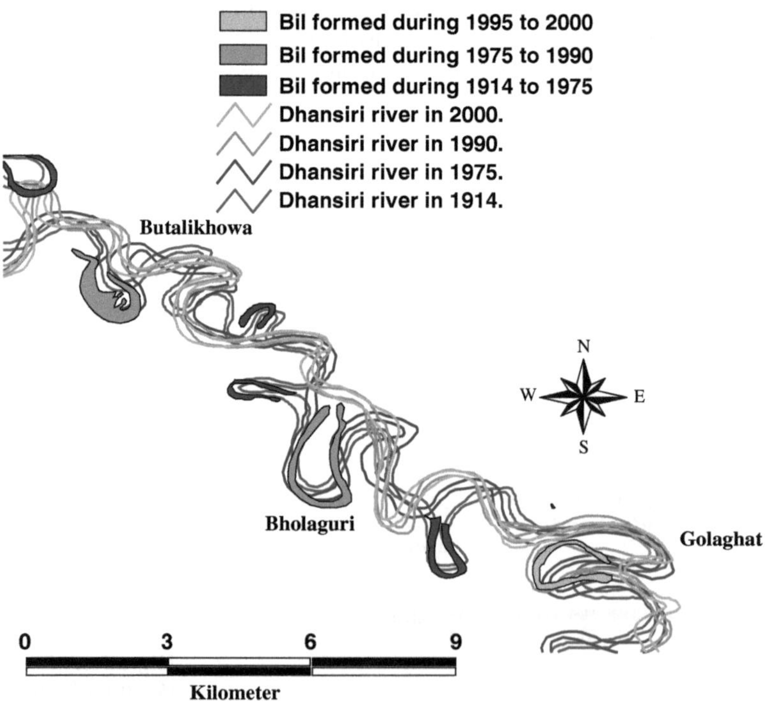

Fig. 6 Migration of meander bends and formation of Oxbow lakes (Bils) during 1914–2000

It can very well be attributed that considerable change in river channel has taken place during the last 87 years. These dynamics could be explained by the variable flow regimes of the channel, which inturn might have been influenced by natural as well as anthropogenic processes. Under constant flow conditions, the channel width stays relatively constant between stable banks, but as the current increases, a neck cutoff can occur. Otherwise, these channels are relatively stable. With mixed suspended and bed loads, however, meandering channels begin to build point bar on the inside of meanders and undercut their outer banks as the thalweg alternately impinges on opposite banks. The total load is usually larger in this type of meandering channel. Channels are wider at sharp bends than along regularly curving reaches. Such channels are quite unstable, with chute cutoffs across the back of the point bars adding to neck cutoffs as a process of shifting.

4.2 Intensity of Meanders and Sinuosity

The alluvial channel changes its position naturally with time. As it flows on erodible sediments and because the stress exerted by the flowing water often exceeds the strength of the sediments forming the bed and banks of the channel, the sediments from the bad as well as bank are mobilized towards downstream and also away from the channel, which results in changing the channel morphology into sinuous and meandering. Dynamism in river channels can very well be observed in terms of its description in plan form, cross sectional and longitudinal forms. Sinuosity represents the irregularity of the channel course and is expressed as the ratio between the channel length, measured along the center of the channel, and the valley length, measured along the valley axis (Leopold and Wolman 1957). Schumm (1963) suggested sinuosity as one of parameters to represent the intensity of meanders.

Decrease in sinuosity values of the Dhansiri River with time in almost all the sectors is recognizable from the Table 2. However, in Sector I the values after 1975 show an increasing tendency. This is also supported by the increase in channel length for different periods (Table 3). In addition, in the sectors I for the period

Table 2 Channel length and sinuosity values at different periods

Sectors	Periods				
	1914	1975	1990	1995	2000
I	2.39	1.97	2.28	2.45	2.60
II	2.98	2.45	2.13	2.17	2.19
III	3.06	3.02	2.95	2.60	2.41
IV	2.20	2.15	2.21	2.18	2.17
V	2.74	2.43	2.06	1.89	1.78
VI	2.55	2.34	1.88	1.91	1.87
Total	2.65	2.39	2.25	2.20	2.17

Table 3 Sinuosity index of the Dhansiri river channel at different periods

Sectors	Periods				
	1914	1975	1990	1995	2000
I	50.2	36	38	38.7	44.15
		(Neck cut off at Sinakangaon, Kamargaon and Barchaparigaon)			
II	43.73	36	31.27	31.84	32.21
		(Neck cut off at Bagariani near Khumtai, Bholaguri and Dhuliagaon)	(Neck cut off at Butalikhowa and Bholaguri)		
II	36.46	36	35.19	36.16	32.29
		(Neck cut off at Dachmuagaon and Thorajan Madhubangaon)	(Neck cut off at Katkatia, Golaghat)		(Neck cut off at Golaghat)
IV	36.86	36	36.95	36.46	36.38
		(Neck cut off at Dibaranigaon near Barpathar)		(Neck cut off at Silanijan)	
V	40.42	36	30.42	28.01	26.3
		(Neck cut off at Deo-pani T.G., Padamani, Jabrajangaon andDilaojan)	(Neck cut off at Sewaguri and Rongagara near Sarupathar)	(Neck cut off at Nagajuri and Kaliabil-gaon)	(Neck cut off at Jabrajan)
VI	39.2	36	29	29.48	28.74
		(Neck cut off near Mohkhuti, Bokajan and Ghorial Dubi)	(Neck cut off at Harihajan, Mara Kardaiguri and Bokajan)		(Neck cut off at Shitoyi Sema
Total length	246.87	216	200.8	200.7	200.1
Average sinuosity	2.76	2.38	2.25	2.22	2.16

1990–2000, increase in channel length was observed signifying enlargement of meander bends and migration of river channel just near the debouching point of river Dahansiri to the Brahmaputra River.

4.3 Erosion-Deposition Along the River

Effective management of rivers against bank erosion and resultant life and property loss requires data on reliable information on the effects of changes in river morphology or bank material characteristics; so that the undesirable impacts of channel changes can be avoided (Odgaard 1987). When extensive lengths of river channel become stabilized, the riverbank erosion can result in considerable riparian land loss and the delivery of large volumes of sediment downstream. The ability to predict the stability and failure geometry of eroding riverbanks is therefore an important prerequisite in estimating the rates of bank erosion and sediment yield associated with bank erosion. Several studies have contributed to the betterment of our understanding on riverbank erosion (Thorne 1982; American Society of Civil Engineers Task Committee on River Widening 1998; Darby and Thorne 1996a, b; Miller and Quick 1997; Simon et al. 1991; Osman and Thorne 1988; Rinaldi and Casagli 1999; Simon et al. 2000). The rate of channel migration (M) is likely to be dependent on stream power (essentially, the product of discharge and slope) per unit area of the bed (w), channel width (W), the force per unit area of the outer (concave) bank which resists channel migration (Y_{b}), the bank height (h), and the radius of curvature (r). Further Y_b is largely a function of the size of sediment at the base of the channel (D_{50}), such that

$$M = f\ (w, W,\ D_{50},\ h,\ r)(\text{Hickin and Nanson 1984})$$

The sediment load, particularly the bed load is known to be strongly correlated to channel migration rate (Neil 1984). Bagnold (1980), however, has shown that the bed load transport is largely a function of stream power operating on particular sediment sizes. Daniel (1971) demonstrated that channel length around a meander loop increases in proportion to the magnitude of the channel-forming discharge, whereas Hickin (1974) demonstrated that the migration operates to maintain a minimum curvature ratio (bend radius to channel width: r/W) of slightly >2. Hickin and Nanson (1975) showed that bend migration reaches a maximum value as the curvature ratio approaches 3 and declines rapidly on either side of this value. Indeed, Carey (1969), and Page and Nanson (1982) have shown that, in very tightly curving bends, deposition will occur around the outer bank and erosion will occur at the convex bank.

A different approach to the problem of channel migration has been developed by those focusing on the details of bank erosion without specific regard to channel planform. The role of frost action and ground ice has been considered by Wolman (1959), Walker and Arnborg (1966) and Outlet (1974). Knighton (1973) found that bank erosion at a cross section was largely determined by the magnitude and variability of discharge and by the degree of asymmetry in the velocity field, bank wetting being a particularly important preconditioning process. Hooke (1979, 1980) has attempted to develop predictive statistical relationships and found that erosion rate is related to catchment area (discharge) and the percentage of silt and clay in the banks.

Global estimates of erosion and sediment transport in major rivers of the world vary widely, reflecting the difficulty in obtaining reliable values for sediment concentration and discharge. Milliman and Syvitski (1992) estimate global sediment load to oceans in the mid-20th century at 20,000 million t/year, of which about 30 % comes from rivers of southern Asia (including the Yangtze and Yellow Rivers of China). Significantly, they believe that almost 50 % of the global total comes from erosion associated with high relief on islands of Oceania—a phenomenon which has been underestimated in previous estimates of global sediment production. While erosion on mountainous islands and in upland areas of continental rivers reflects natural topographic influences, Milliman and Syvitski (1992) suggested that human influences in Oceania and southern Asia cause disproportionately high sediment loads in these regions.

Erodibility and Erosivity are two important physical factors that affect the magnitude of erosion. Erodibility is a measurable characteristic, susceptible to detachment and transport by the agents of erosion. Erosivity is an expression of the ability of erosive agents. Quantification of these two factors is basic to an understanding erosional process.

4.3.1 Bank Line Migration

The sequential analyses of geomorphological maps (Fig. 7) for the period from 1914 to 2000 have shown that the total length of the channel during 1914 measured to be 246.87 km has undergone changes over the period of time and the channel configuration of 1914 was considered as the base line for evaluation of bank line migration during the period 1914–1975. The bank line migration from 1914–1975 is presented in the Fig. 8. The maximum migration was observed on both the bank lines near Dhansirimukh for a distance of about 3.0 km towards south. This information along with the field investigation revealed that, migration in this area followed the pattern of migration of the Brahmaputra River. In some other sectors the rate of migration could not be measured as the migration of meanders in the resulted development of neck cutoffs.

A significant change in the migration of bank line towards east for a distance of about 1.5 km at Dhansirimukh has been observed during the period between 1975 and 1990 (Fig. 9). Near Dighali Ati area both the banks of the river witnessed a shift towards east up to maximum of 0.6 km. The period during 1990–1995 has shown the migration of the channel towards its original configuration. The locations on the bank of the river viz., near Golaghat and Numaligarh areas, the bank lines of the river migrated towards west and east respectively (Fig. 10). Significant migration is noticed near Elengmari gaon during the 1995–2000 period. The west bank line during this period evidenced significant migration towards west near Deopani area (Fig. 11).

The river channel has migrated for a distance of about 2.5 km towards east (Fig. 12) during the entire study period from 1914 to 2000. Higher rates of migration were prevalent near Numaligarh and Butalikhowa-Bholaguri areas. Near Bholaguri, the bank line had migrated towards west and near Numaligarh it migrated towards east. From the overall migration trends of the bank line, it can stated that there has been a continuous shifting of the Dhansiri River channel towards east for a distance of about 2.5 km at Dhansirimukh. The channel course from Dhansirimukh to Golaghat has shown significant changes in the bank line whereas, beyond Golaghat till Nowakota-Kachari, the channel remained stabilized,

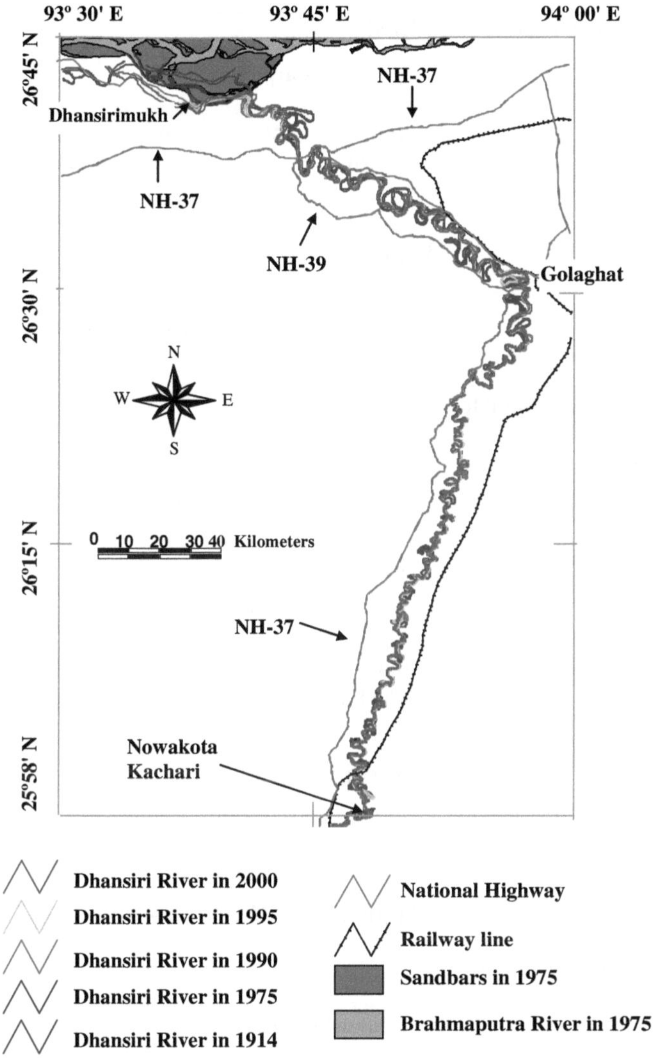

Fig. 7 Dhansiri river channel within the studied reach during 1914–2000

perhaps as a result of NS-trending fault located in the vicinity (ARSAC 1990). In other areas, the gradient of the terrain and nature of alluvium within the course might have played a crucial role in morphological adjustment of river channel.

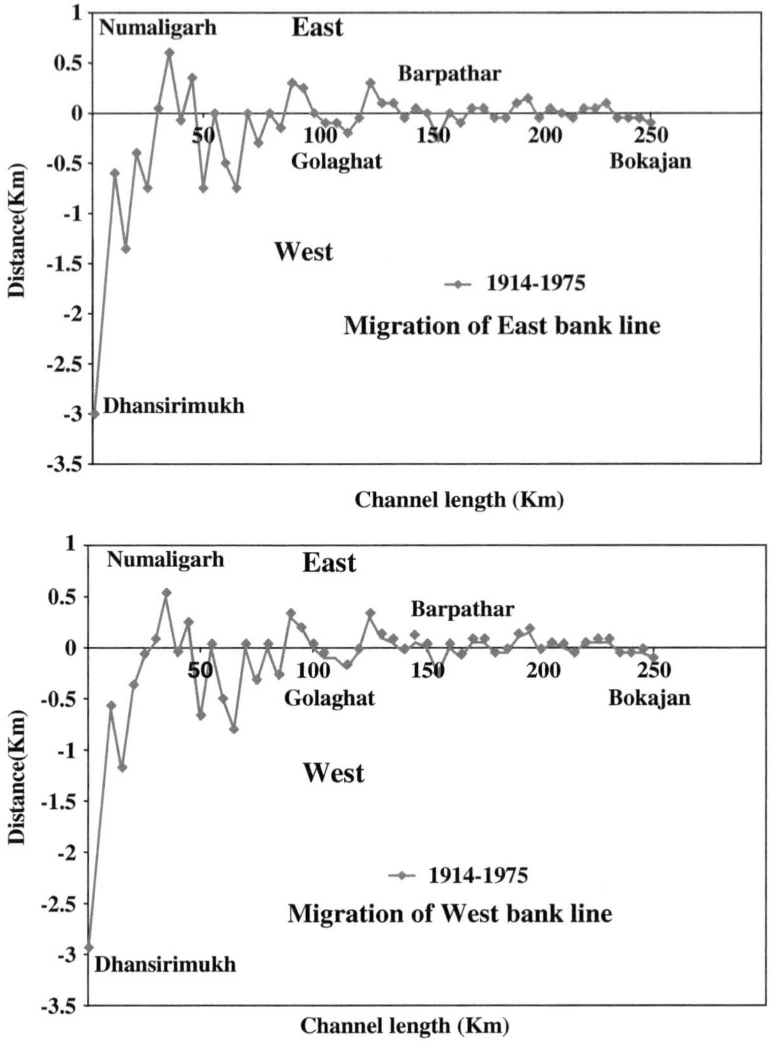

Fig. 8 Bank line migration of the Dhansiri river during 1914–1975

4.3.2 Erosion/Deposition

The erosion-deposition prevalent at different sectors during different time periods are presented in the Tables 4 and 5. From these tables, it follows that the intensity of erosion-deposition increases over time since the year 1914. From the Fig. 13, it is revealed that, the Dhansiri river experienced a higher rate of erosion along its western bank than the eastern bank. Near Butalikhowa at a distance of 60 km from Dhansirimukh the river had eroded an area measuring 1.34 km^2. However, the area

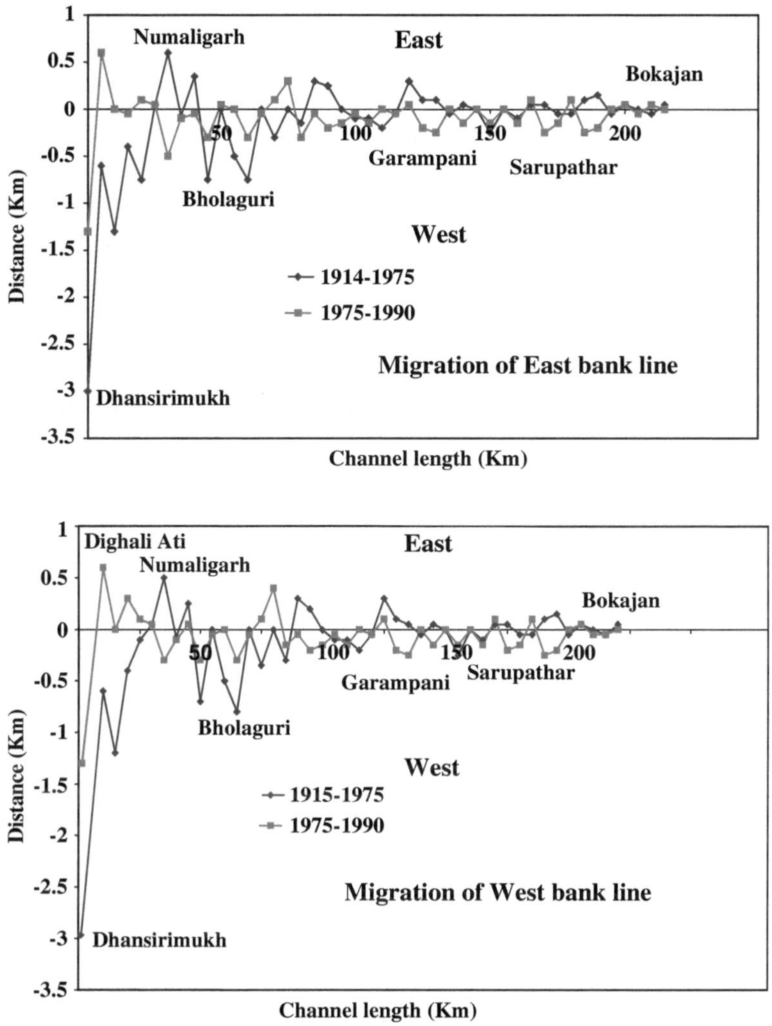

Fig. 9 Bankline migration of the Dhansiri river during 1975–1990

near Golaghat suffered significant erosion on its western bank and deposition on its eastern side. During the period 1914–1975, the total average annual erosion and deposition were estimated to be 0.37 and 0.25 km^2/year respectively, signifying a net loss of sediment/land area.

The pattern of erosion-deposition has been different during the period of 1975–1990 (Fig. 14). Fourfold increase in erosional intensity and fivefold increase in deposition are observed during this period. The region 20–30 km near Numaligarh-Kamargaon and the region 75 km near Golaghat on the eastern bank of the channel

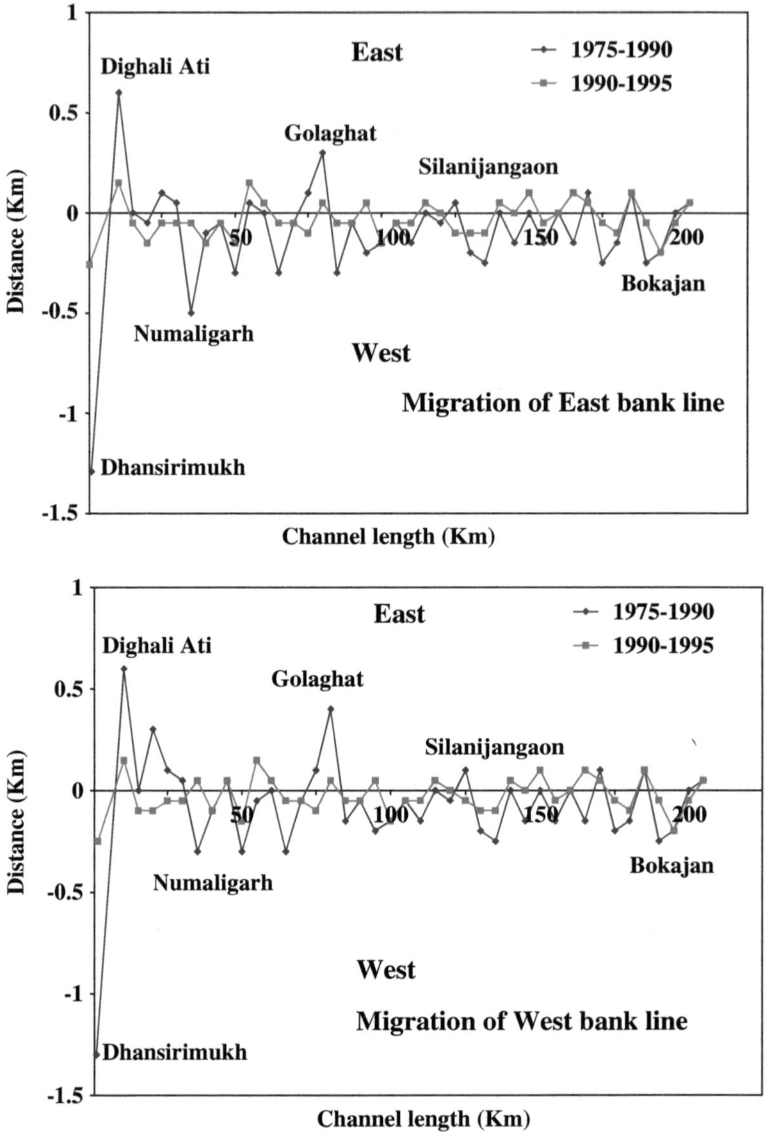

Fig. 10 Bankline migration of the Dhansiri river during 1990–1995

experienced severe erosion and lost land to the tune of 1.2 km^2. Surprisingly, the region located 65–75 km near Golaghat on the eastern bank exhibited deposition of 1.16 km^2 with concomitant erosion of 0.73 km^2 area. Net annual erosion and deposition during this period are estimated to be 1.31 and 1.25 km^2/year respectively. The period 1990 to 1995 shows average total annual erosion and deposition

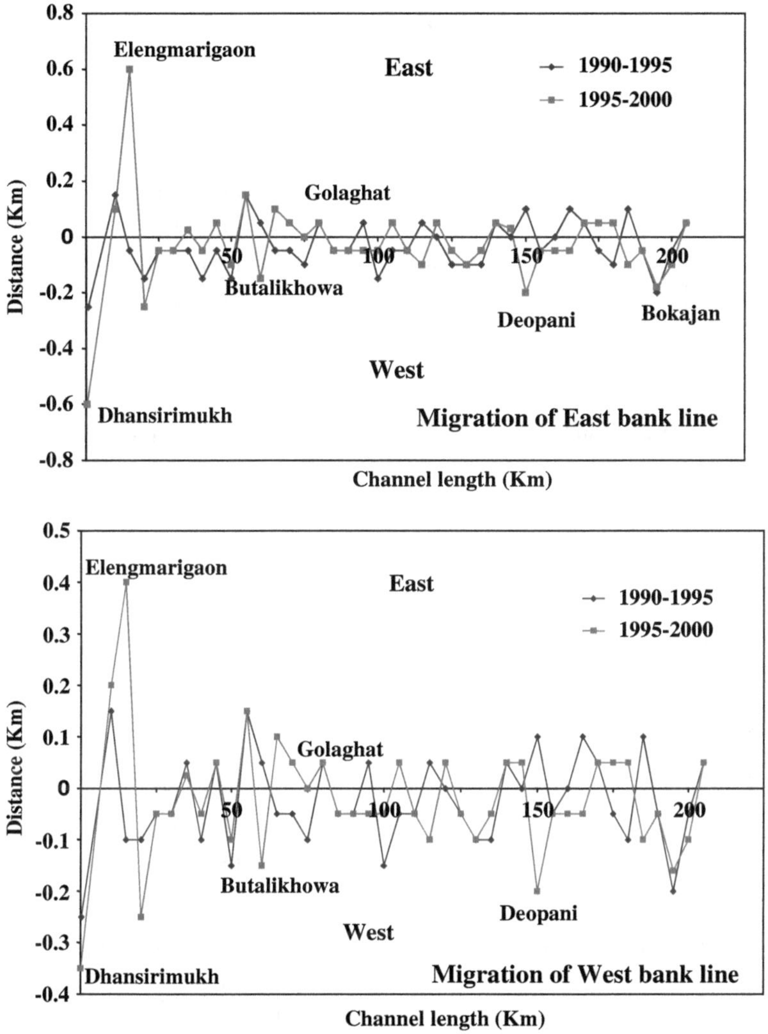

Fig. 11 Bankline migration of the Dhansiri river during 1995–2000

as 1.78 and 1.93 km^2/year respectively. The period has also represented an increase in the intensity of deposition than erosion (Fig. 15). More specifically, during this period, the eastern bank experienced consistent depositional episodes than the western bank.

The period 1995–2000 has shown phenomenal increase in erosion as well as deposition (Fig. 16). In a stretch from Kuruabahi up to Butalikhowa-Bholaguri areas, both the banks show erosional characteristics. The areas around Bholaguri on

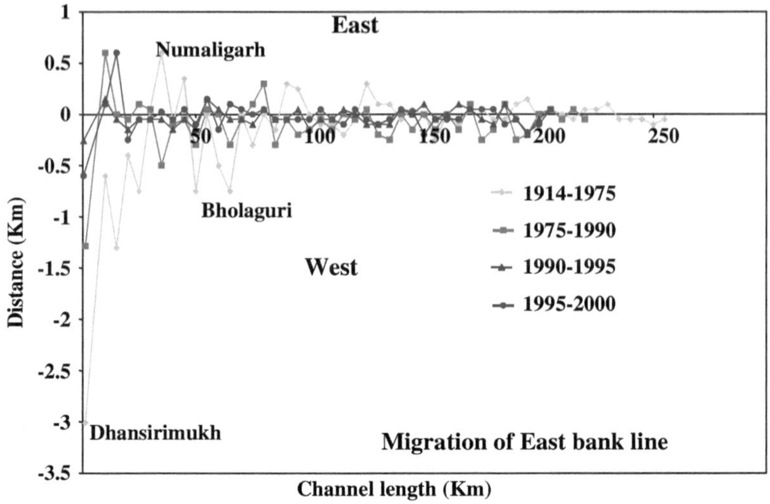

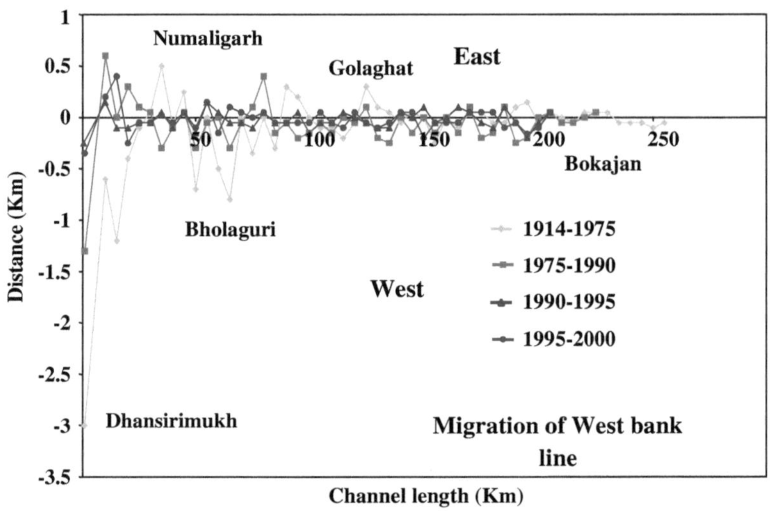

Fig. 12 Bankline migration of the Dhansiri river during 1914–2000

Table 4 Average annual erosion and deposition along with its rate within the studied reach of Dhansiri River channel

Period	Average annual erosion (km^2/year)	Average annual deposition (km^2/year)	Rate of average annual erosion (km^2/km)	Rate of average annual deposition (km^2/km)
1914–1975	0.37	0.25	0.0015	0.0010
1975–1990	1.31	1.25	0.0061	0.0058
1990–1995	1.78	1.93	0.0089	0.0096
1995–2000	1.81	1.65	0.0090	0.0082
Mean	1.32	1.27	0.006375	0.00615

Table 5 Erosion and deposition along the studied reach of Dhansiri river channel at different time and space

Sectors	Erosion (km^2)					Deposition (km^2)				
	1914–1975	1975–1990	1990–1995	1995–2000	Total	1914–1975	1975–1990	1990–1995	1995–2000	Total
I	5.59	3.92	2.59	1.99	14.09	3.65	3.44	2.46	1.67	11.22
II	7.55	5.98	1.96	2.52	18.01	5.07	5.52	2.62	2.09	15.30
III	3.44	1.96	1.34	1.23	7.97	2.53	2.22	1.46	1.01	7.22
IV	2.10	1.81	1.61	1.56	7.08	1.63	1.66	1.67	1.46	6.42
V	1.91	3.12	1.39	1.97	8.39	1.04	3.12	1.43	2.39	7.98
VI	1.90	2.92	1.81	1.58	8.21	1.24	2.73	1.95	1.56	7.48

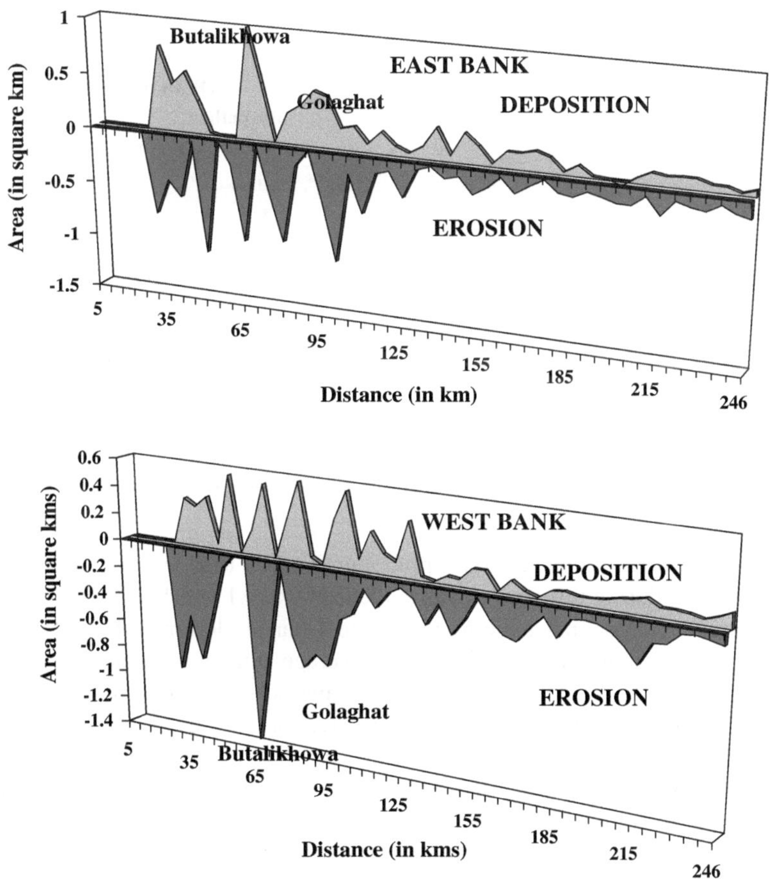

Fig. 13 Erosion and deposition in the studied reach of the Dhansiri river during 1914–1975

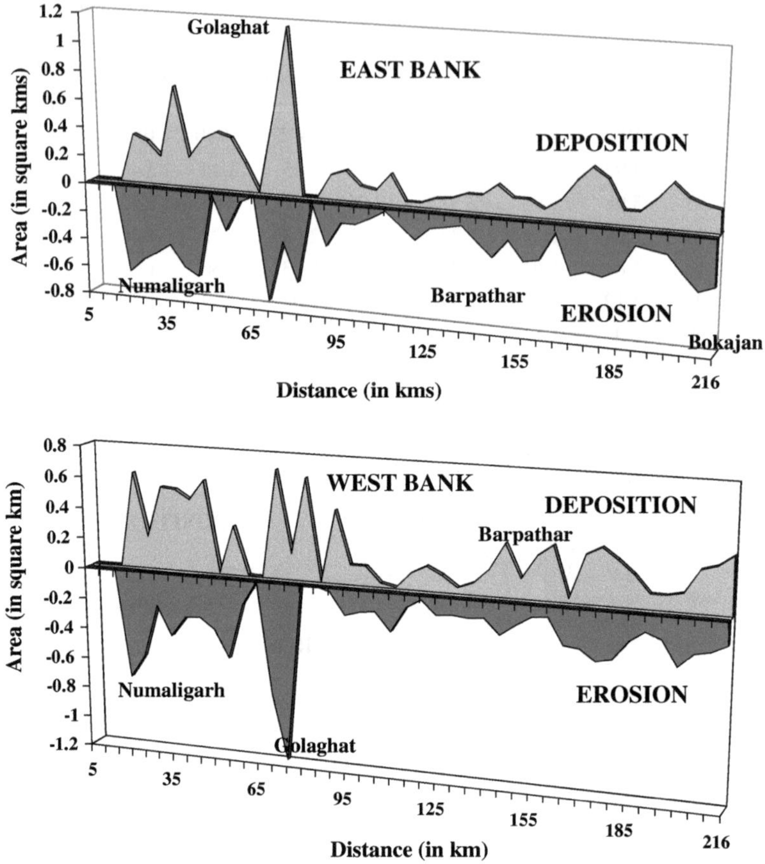

Fig. 14 Erosion and deposition in the studied reach of the Dhansiri river during 1975–1990

the east bank, depositional episodes predominate. The total average annual erosion and deposition during this period was evidenced as 1.81 and 1.65 km^2 respectively.

The erosion/deposition patterns along the Dhansiri River with average annual rate are presented in Table 4. The total average annual erosion and deposition calculated from the available data during the period from 1914 to 2000 within the studied stretch are represented as 1.32 and 1.27 km^2/year respectively. It is also observed from the available data (Table 5) that the erosion was maximum (18.01 km^2) in Sector II, whereas, it was least in the Sector IV (7.08 km^2). In addition, this sector has also evidenced the lowest amount of deposition (6.42 km^2). The total average annual erosion and deposition range from 0.37 to 1.81 km^2/year and 0.25 to 1.93 km^2/year respectively (Table 4). The nature of soft alluvial bank and contribution from the major tributaries might have played a major role in

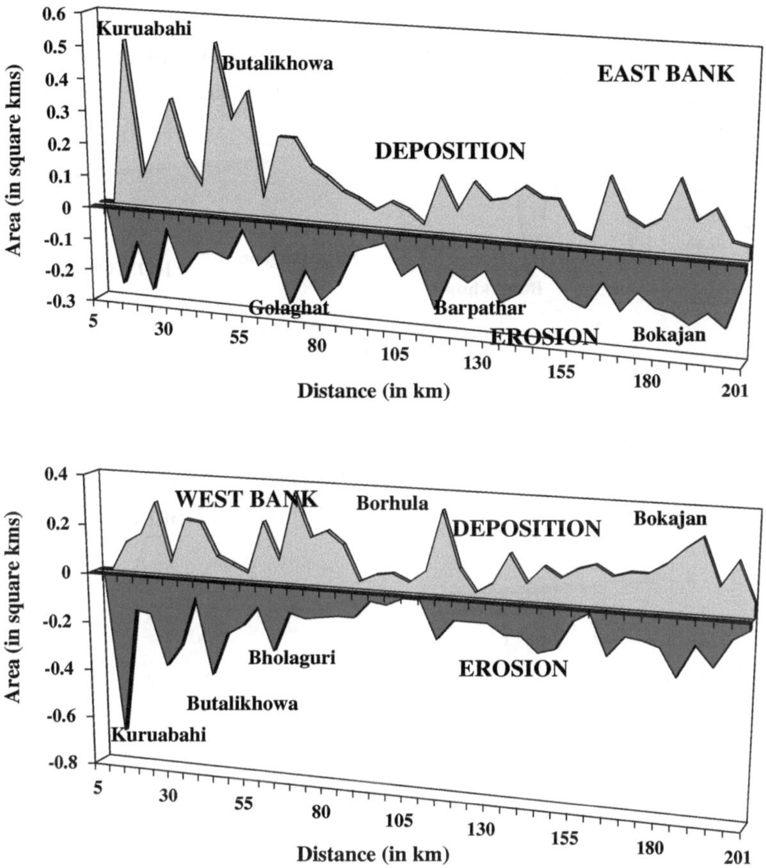

Fig. 15 Erosion and deposition in the studied reach of the Dhansiri river during 1990–1995

inhomogeneous natures of erosion/deposition activities operating within the basin. From the data of the Table 4, it is clear that the rate of bank erosion per km length of the river course ranges from 0.0015 to 0.0090 km^2/km. whereas, the rate of average annual deposition per km length of the river ranges from 0.0010 to 0.0096 km^2/km. The average rate of bank erosion and deposition per km length of the river channel, for the entire period of study, are found to be 0.006375 and 0.00615 km^2/km channel length respectively. The rate of average annual erosion and deposition for different periods are represented in the Fig. 17.

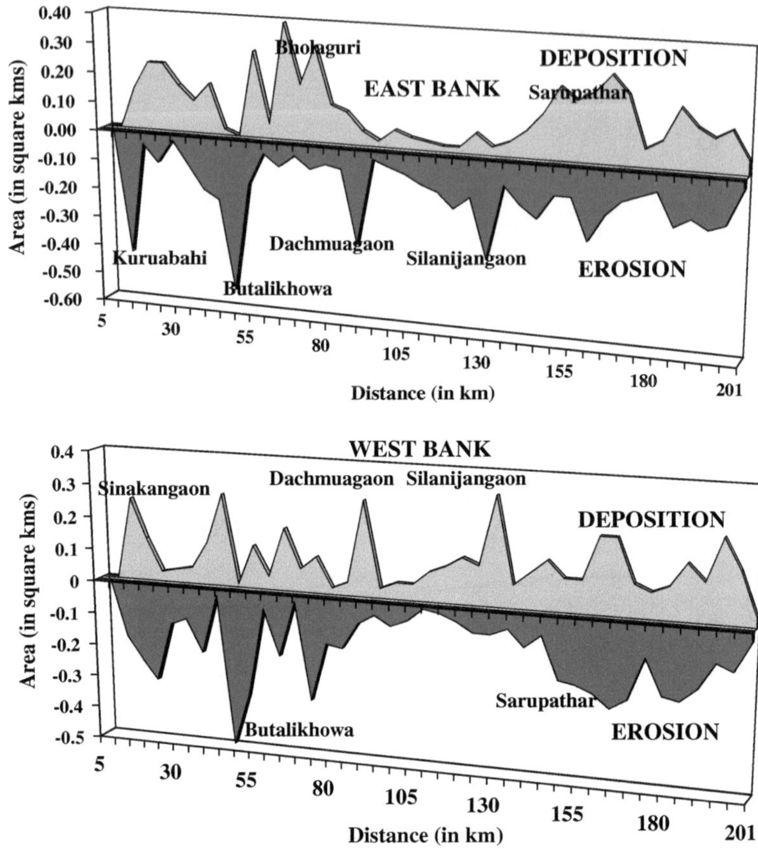

Fig. 16 Erosion and deposition in the studied reach of the Dhansiri river during 1995–2000

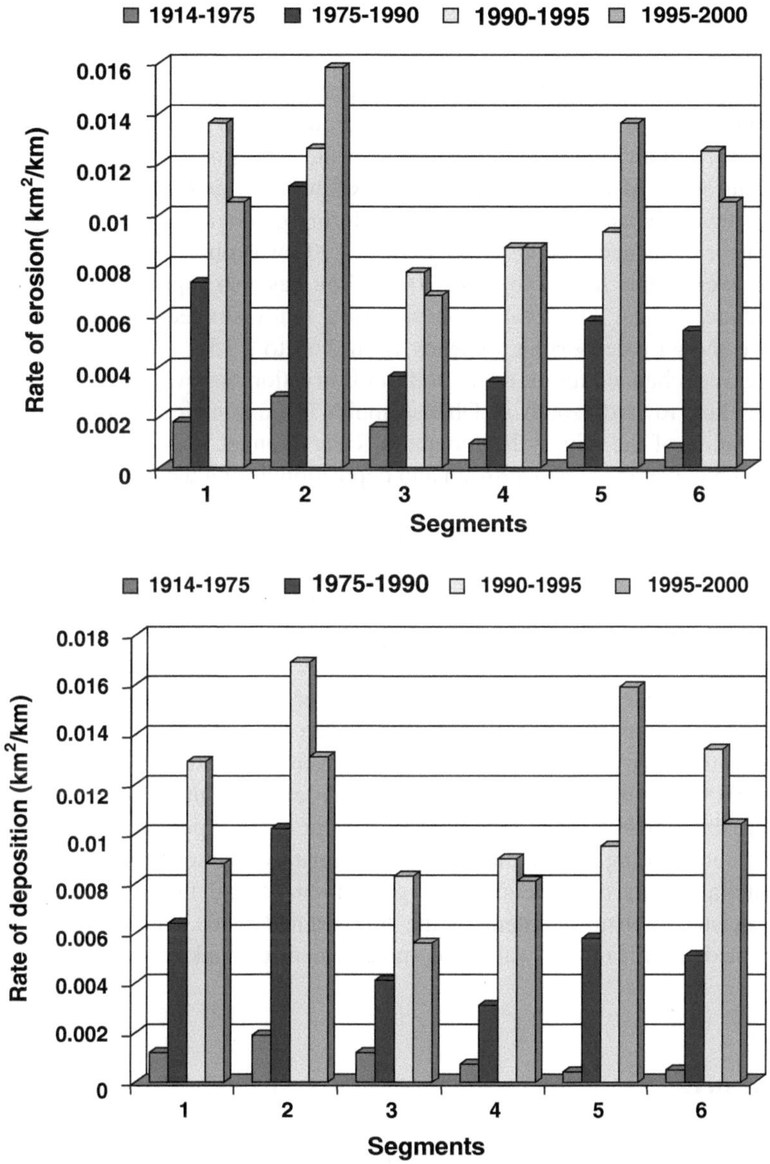

Fig. 17 Rate of average annual bank erosion and deposition along the studied reach of the Dhansiri river channel

5 Conclusions

- The highly meandered Dhansiri River channel, a south bank tributary of the Brahmaputra River has shown the predomination of translation mechanism of meander migration followed by extension and extension-translation mechanisms.
- Occurrences of abundant number of oxbow lakes and paleochannels indicate significant dynamic nature of the river during geologic past. A drop in the total number of meander bends from 1914 to 2000 was observed. The shortening of the river course from 246.87 to 200.07 km, has also supported the decrease in meander bends. The total mean sinuosity index studied for different periods have shown a decreasing tendency from 2.65 to 2.15.
- The river channel during the period of observation has evidenced a migration of 2.85 km. towards south at Dhansirimukh in conjunction with the southward migration of the mighty Brahmaputra River channel.
- The total average annual erosion and deposition during the entire period of study were 1.32 and 1.27 km^2/year respectively. Comparatively higher erosion was experienced by the west bank of the river than the east bank. The locations of the river channel in a soft alluvial plain and contribution of larger volume of water to the main channel by the larger tributaries (viz. Dayang River) might have played a major role in increasing the intensity of the erosion/deposition processes.
- The erosion/deposition processes within the studied stretch have evidenced increasing intensity from 1914 to 2000. The total average rate of erosion and deposition per kilometer length of the river were estimated to be 0.006375 and 0.00625 km^2/km respectively. The areas around Butalikhowa, Golaghat and Kuruabahi pose a threat demonstrating more erosion potentiality.
- The undercutting of soft alluvial banks ultimately lead to shearing from the bank material by its own weight. The generation of pad fabric in relation to the associated moisture content of the fine-grained bank materials has also contributed significantly towards the enhancement of bank erosion responses.

References

Ahmed F (1969) Maximum discharge of major rivers in large flood plains in East Pakistan. Floods and their computation. IAH-UNESCO-WHO Publication, Leningrad, pp 349–364

Allen JRL (1971) Changeable rivers: some aspects of their mechanics and sedimentation

Allen AP, Allen JR (1990) Basin analysis, principles and applications. Blackwell Scientific Publication, London, pp 445–470

American Society of Civil Engineers Task Committee on River Widening (1998) River width adjustment. I: Processes and mechanisms. J Hydraul Eng ASCE 124:881–902

ARSAC (1990) Landuse/landcover maps and reports of Assam. Assam Remote Sensing Application Centre, Gauhati (Unpublished Report)

Bagnold RA (1980) An empirical correlation of bedload transport rates in flumes and natural rivers. Proc Roy Soc Lond A372:453–473

Baker VR (1994) Geomorphological understanding of floods. Geomorphology 10:139–156
Baker VR, Kochel RC, Patton PC (1988) Flood geomorphology. Wiley, New York, p 503
Barman G (1981) Geomorphology of the Brahmaputra basin. Its floods problem and possible remedial measures. Geol Surv India Misc Publ 46:21–31
Bevan K, Carling P (eds) (1987) Floods: hydrological, sedimentological and geomorphological implications. Wiley, New York, p 290
Bezbaruah D, Kotoky P, Baruah J, Sarma JN (2003) Geomorphological explanation of swamps in geomorphological explanation of swamps along the Brahmaputra river channel, Assam. J Geol Soc India 62:605–613
Brice JC (1964) Channel patterns and terraces of Loupe river in Nebraska. US Geol Surv Prof Paper 422-D:1–41
Bristow CS (1987) Brahmaputra river: channel migration and deposition (In: Ethridge FG, Flores RM, Harvey MD (eds) Recent development in fluvial sedimentology). Soc Econ Paleontol Miner Spec Pub 39:63–74
Carey WC (1969) Formation of floodplain lands. Am Soc Civ Eng Proc 95(HY3):981–944
Chow VT (Ed) (1964) Handbook of applied hydrology, vol 15. McGraw Hill Inc, New York, pp 1–41
Coleman JM (1969) Brahmaputra river: channel processes and sedimentation. Sed Geol 8:131–237
Danial JF (1971) Channel movement of meandering streams. US Geol Surv Prof Pap 732-A:A1–A18
Darby SE, Thorne CR (1996a) Modelling the sensitivity of channel adjustment in destabilized sandbed rivers. Earth Surf Proc Land 21:1109–1125
Darby SE, Thorne CR (1996b) Development and testing of riverbank-stability analysis. J Hydraul Eng ASCE 1222(8):443–445
Davies WM (1989) The rivers and valleys of Pennsylvania. Nat Geog Mag 1:183–253 (also in Geographical Essays Ginn and Co, New York, pp 413–484)
De Wiest RJM (1965) Geohydrology. Wiley, New York, p 712
Dietrich WE, Gallinatti JD (1991) Fluvial geomorphology. In field experiments and measurement programs in geomorphology. Earth Surf Proc Land 12:173–184
Duijsings JJHM (1987) A sediment budget for a forested catchment in Luxembourg and its implication for channel development. Earth Surf Proc Land 12:173–184
Dunne T (1988) Geomorphological contribution to flood control planning. In: Baker VR, Kochel RC, Patton PC (eds) Flood geomorphology. Wiley, Chichester, pp 421–438
Durry G (1954) Bedwidth and wavelength in meandering valleys. Nature 176:31
Dury GH (ed) (1970) Rivers and river terraces. Macmillan Publication Co, London, p 277
Dutta SK (1980) The geomorphic history of Brahmaputra. J Palaeontol Soc India 23, 24:24–29
Dutta MK, Kotoky P (2006a) Wetlands along the Dhansiri river channel, Assam. GIS Development, Asia, pp 1–12
Dutta MK, Kotoky P (2006b) Nature of meandering in the Dhansiri river channel, Assam. National Seminar, Geological Society of India, Nagaland University, Kohima 9–11 Nov 2006
Fielding CR (1993) A review of recent research in fluvial sedimentology. Sed Geol 85:3–14
Friedkin JF (1945) A laboratory study of the meandering of alluvial rivers. United States Waterways Experiment Station, Vicksburg
Geike SA (1905) The founders of geology. Macmillan and Co Ltd, London, p 297
Geological Survey of India (1977) Contribution of geomorphology and geohydrology of the Brahmaputra valley. Geol Surv India Misc Publ 32:105–110
Geological Survey of India (1981) Proceedings of the seminar on fluvial processes and geomorphology of the Brahmaputra River basin, vol 46. Miscellaneous Publ, p 141
Gilbert GK (1887) Report on the geology of the Henry mountain. Washington, p 160
Goswami DC (1985) Brahmaputra river, Assam, India: physiography, basin denudation and channel aggradation. Water Resour Res 21(7):959–978
Goswami DC (1988) Magnitude and frequency of fluvial processes in the Brahmaputra Basin, Assam: some observations. In Singh S, Tiwari RC (eds) Geomorphology and Environment. The Allahabad Geographical Society, Allahabad, India, pp 203–211

Goswami DC (1998) Fluvial regime and flood hydrology of the Brahmaputra river, Assam. Mem Geol Surv India 41:53–75

Goswani U, Sarma JN, Patgiri AD (1999) River channel changes of the Subansiri in Assam, India. Geomorphology 30:227–244

Goudie A (1990) Geomorphological technique, 2nd edn. Unwin and Hyman, London, 570 pp

Gregory KJ (1977) River channel changes. Wiley, New York, p 443

Gregory KJ, Walling DE (1973) Drainage basin form and process, a geomorphological approach. Edward Arnold, London, p 447

Grover NC, Harrington AW (1943) Stream flow: measurements, records and their uses. Dover publication Inc. New York, p 363

Hickin EJ (1974) The development of meanders in natural river channel. Am J Sci 274:414–442

Hickin EJ, Nanson GC (1975) The character of channel migration on the Beatton river, Northeast British Columbia, Canada. Geol Soc Am Bull 86:487–494

Hickin EJ, Nanson GC (1984) Lateral migration rates of river bends. Am Soc Civ Eng, J Hydraul Eng 110, 11:1557–1567

Hooke JM (1977) The distribution and nature of changes in river channel patterns: an example from Devon. In: Gregory KJ (ed) River channel changes (British Geomorphological Research Group). Wiley, New York, pp 265–279

Hooke JM (1979) An analysis of the processes of the river bank erosion. J Hydraul 42:39–62

Hooke JM (1980) Magnitude and distribution of rates of river bank erosion. Earth Surf Process 5:143–157

Horton RE (1932) Drainage basin characteristics. Trans Am Geophys Union 13:350–361

Horton RE (1940) An approach towards a physical interpretation of infiltration capacity. Pract Soil Sci Soc Am 1:899–917

Horton RE (1941) Sheet erosion, present and past. Trans Am Geophys Union 22:299–305

Horton RE (1945) Erosional development of stream and their drainage basins, hydrophysical approach to quantitative morphology. Bull Geol Soc Am 56:275–370

Horton RE (1954) Erosional development of streams, quantitative physiographic factors. In Dury GH (ed) Rivers and rivers terraces. Misc, V 1, London

Hussein I, Choudhury JN, Ghani MU (1993) River bank erosion of Majuli Island as deciphered from the IRS imagery. In: Proceedings of national symposium on remote sensing application and resource management with special reference to NE-India region, Guwahati, pp 31–35, 25–27 Nov 1993

Hutton J (1795) Theory of the earth, Edinburgh. In: Horton RE (1945) Erosional development of streams and their drainage basins: hydrophysical approach to quantitative morphology. Bull Geol Soc Am 56:275–370

Jefferson M (1902) Limiting width of meander belts. Nat Geog Mag 13:373–384

Jorgensen DW, Harvey MD, Flamm L (1994) Morphology and dynamics of the Indian river: implication for the Mohen Jo Daro site. In: Shroder J, Kazmi (eds) Himalayas to the sea: geology, geomorphology and quaternary. Routledge, London, pp 288–326

Kale VS (1998) Flood studies in India. Geol Soc India Mem 41:27–52

Kayastha SL, Yadava RP (1977) Human perception and adjustment to environmental hazards: a case study of three villages of the Ganga flood plain. In: Robert C (ed) Man, culture and settlement. Kalyani Publishers, New Delhi

King CAM (1970) Techniques in geomorphology. Edward Arnold, London, p 342

Knighton AD (1973) Riverbank erosion in relation to stream flow conditions, River Bollin-Dean, vol 5. East Midlands Geographer, Cheshire, pp 416–426

Kotoky P (2011) Geoenvironmental perspectives. Sci Cult 77:474–479

Kotoky P, Baruah J, Baruah NK, Sarma JN (1997) Geoenvironmental studies of river Jhanji, Assam. J Hum Ecol 6:55–67

Kotoky P, Bezbaruah D, Baruah J, Sarma JN (2003) Erosion activity on Majuli- the largest river island of the world. Curr Sci 84(7):929–932

Kotoky P, Bezbaruah D, Baruah J, Sarma JN (2005) Nature of bank erosion along the Brahmaputra river channel, Assam. Curr Sci 88(2):634–640

Kotoky P, Bezbaruah D, Baruah J, Sarma JN (2006) Characterisation of clay minerals along the Brahmaputra river channel, Assam. Curr Sci 91(9):1247–1250
Kotoky P, Bezbaruah D, Borah GC, Sarma JN (2009) Do node points play a role in flood proliferation? Curr Sci 96:1457–1460
Kotoky P, Dutta MK, Goswami R, Borah GC (2011a) Geotechnical properties of the bank sediments along the Dhansiri river channel, Assam. J Geol Soc India 78:175–183
Kotoky P, Dutta MK, Borah GC (2011b) Wetland ecology along the Dhansiri river channel, Assam, India. In: Kosygin L (ed) Ecology and biodiversity of rivers and streams of NE-India. Akansa Publishing House, New Delhi, pp 185–202. ISBN: 978-81-8370-26-8
Kotoky P, Bordoloi RK, Baruah NK, Borthakur KK, Borah GC (2011c) Water chemistry of the rivers around Jorhat, Assam, India. In: Kosygin L (ed) Ecology and biodiversity of rivers and streams of NE-India. Akansa Publishing House, New Delhi, pp 203–243. ISBN: 978-81-8370-26-8
Kotoky P, Dutta MK, Borah GC (2012) Changes in landuse/landcover along the Dhansiri river channel, Assam—a remote sensing and GIS approach. J Geol Soc India 79:61–68
Leopold LB, Miller JP (1954) Apost-glacial chronology for some alluvial valleys in Wyoming. USGS water supply paper, vol 1261, p 89
Leopold LB, Wolman MG (1957) River channel patterns: braided, meandering and straight. US Geol Surv Prof Pap 282-B:39–85
Leopold LB, Wolman MG, Miller JP (1964) Fluvial processes in geomorphology. WH Freeman, San Francisco, 522 pp
Lyell SC (1872). Principles of geology, 11th edn, Chaps 2, 3 and 4D. Appleton and Co, New York
Meinzer OE (1942) Hydrology. Dover Publication Inc, New York, 712 pp
Miller RG, Quick MC (1997) Discussion on Development and testing of riverbank stability analysis by Stephen Darby and Colin Thoene. J Hydraul Eng, ASCE 123(11):1051
Milliman JD, Syvitski JPM (1992) Geomorphic/Tectonic control of sediment discharge to the ocean: the importance of small mountain rivers. J Geol 100:525–544
Morisawa ME (1957) Accuracy of determination of stream length from topographic maps. Trans Am Geogr Union 38:86–88
Morisawa ME (1959) Relation of quantitative geomorphology to stream flow in representative watersheds of the Appalachian plateau province. Project NR 389-042, Technical report 20, Columbia University, Department of Geology. ONR, Geography Branch, New York
Morisawa ME (1962) Relation of quantitative geomorphology to stream flow in representative watersheds of the Appalachian plateau province. Bull Geol Soc Am 73:1025–1046
Naik SD, Singh RKS (1996) Bank erosion at Majuli Island, Assam—a study on multitemporal satellite data. Project Report SAC/RSAM/ RSAG/ WSP/96/2, p 51
NBBS (1999) Soils of Assam for optimizing landuse. Indian Council of Agricultural Research, Nagpur Publ, pp 66, 43
Neil CR (1984) Bank erosion versus bedload transport in a gravel river: rivers 83. In: Proceedings of American society civil engineer, New Oreland, pp 204–211
Odgaard AJ (1987) Streambank erosion along two rivers in Lowa. Water Resour Res 23:1225–1236
Osman AM, Thorne CR (1988) Riverbank stability analysis-I, theory. J Hydraul Eng, ASCE 114 (2):134–150
Outlet DN (1974) Progress report on bank erosion studies in the Mackenzie river delta. Northern oil development report, Report No 74–12, pp 303–345
Page K, Nanson GC (1982) Concave bank benches and associated floodplain formation. Earth Surf Proc Land 7:529–542
Powell JW (1895) The exploration of the Colorado valley and its canyons. Dover, New York
RAB (1980) Report of the Rashtriya Barh Ayog. Government of India, New Delhi, I and II
Rinaldi M, Casagli N (1999) Stability of streambanks formed in partially saturated soils and effects of negative pore pressure: the Sieve river, Italy. Geomorphology 26:253–277
Sarin MM, Krishnaswami S (1984) Major ion chemistry of the Ganga Brahmaputra river systems, India. Nature 312(5994):538–541

Sarma JN (1986) Sediment transport in the Burhi dihing river, India. IAHS Publication, Wallingford, p 159
Sarma JN (1993) Axamar Nad Nadi, Axam Sahitya Sabha, In Assamese, 335 pp
Sarma JN (1995) A study of the hydraulic characteristics of the Noa Dihing- Burhi Dihing river sytem in Eastern India. Colloquium in floods, slopes and river beds. University of Paris, France 22–24 March 1995
Sarma JN, Basumallick S (1984) Bankline migration of the Burhi Dihing river, Assam, India. Indian J Earth Sci 11(3, 4):199–206
Schumm SA (1963) Sinuosity of alluvial rivers in Great Plains. Geol Soc Am Bull 74:1089–1100
Schumm SA (1977) The fluvial systems. Wiley, New York, 388 pp
Schumm SA (1985) Patterns of alluvial rivers. Ann Rev Earth Sci 13:5–27
Schumm SA, Khan HR (1972) Experimental study of channel pattern. Geol Soc Am Bull 83:1755–1770
Simon A, Wolfe WJ, Molinas A (1991) Mass wasting algorithms in an alluvial channel model. In: Proceedings of the 5th federal interagency sedimentation conference, Las Vegas, Nevada, p 2, 22–29 Aug 1991
Simon A, Curini A, Darby S, Lagendoen EJ (1999) Stream bank mechanics and the role of bank and near- bank processes in incised channels. In: Darby SE, Simon A (eds) Incised river channels. Wiley, Chichester, pp 123–152
Simon A, Curini A, Darby SE, Langendoen EJ (2000) Bank and near bank processes in an incised channel. Geomorphology 35:193–218
Strahler AN (1964) Quantitative geomorphology of drainage basins and channel networks. In: Chow VT (ed) Handbook of applied hydrology. McGraw Hill Book Company, New Delhi, pp 4.39–4.76
Temple H, Sundborg AKE (1971) The Rufiji river, Tanzania, hydrology and sediment transport. Geografirka Annaler 54(3–4):1972
Thornbury KD (1954) Principles of geomorphology. Wiley Eastern Ltd, New Delhi
Thorne CR (1982) Processes and mechanisms of vegetation on river bank erosion. In: Hey RD, Bathurst JC, Thorne CR (eds) Gravel bed rivers. Wiley, Chichester
Walker HJ, Arnborg L (1966) Permafrost and ice-wedge effect on riverbank erosion. In: Proceeding of Conference Permafrost, Washington DC, pp 164–171
WAPCOS (1993) Morphological study of the river Brahmaputra. Water and Power Consultancy Services (India) Ltd, New Delhi
Ward RC (1978) Floods: a geographical perspective. The Macmillan Press Ltd, London
White GF (1964) Choice of adjustment to floods. Department of Geography, University of Chicago, Research paper, p 93
Wolman MG (1959) Factors influencing erosion of a cohesive river bank. Am J Sci 257:204–216
Yalin MS (1992) River mechanics. Pergamon Press, Oxford, 23 pp

Sand Mining, Channel Bar Dynamics and Sediment Textural Properties of the Kaveri River, South India: Implications on Flooding Hazard and Sustainability of the Natural Fluvial System

Mu. Ramkumar, K. Kumaraswamy, R. Arthur James, M. Suresh, T. Sugantha, L. Jayaraj, A. Mathiyalagan, M. Saraswathi and J. Shyamala

Abstract The Kaveri River, the fourth largest river in India, undergoes the onslaught of urbanization and extensive construction activities within, along and adjoining its channel. In addition to its dwindling natural flow due to the failing monsoonal supply, and constructions of major, medium and minor dams, the extensive mining of sand from its channel bed causes severe stress on its natural fluvial processes. Reduction of carrying capacity of the channel, extensive vertical accretion of sediments within the channel, development of channel-in-channel physiography, and alteration of stream configuration and textural parameters of the stream bed sediments have contributed towards deterioration of the environmental integrity of this important river and exacerbated the flood hazard in the adjoining regions. This paper is an attempt to document the deterioration of natural fluvial dynamics due to the anthropogenic intervention and lack of required data for proper understanding for environmental management and sustenance of the fluvial system. The textural and geomorphic characteristics and the mechanism of mid-channel bar formation and stabilization documented through this study suggest that the whole of the river channel of the Kaveri River behaves like a braided bar/flood plain, which means the prevalence of slow abandonment of the fluvial processes, that could only

Mu. Ramkumar (✉)
South East Asia Carbonate Laboratory, Universiti Teknologi Petronas, Tronoh, Malaysia
e-mail: muramkumar@yahoo.co.in

Mu. Ramkumar · M. Suresh · T. Sugantha · L. Jayaraj · A. Mathiyalagan · M. Saraswathi · J. Shyamala
Department of Geology, Periyar University, Salem 636011, India

K. Kumaraswamy
Department of Geography, Bharathidasan University, Tiruchirapalli 620024, India

R.A. James
Department of Marine Science, Bharathidasan University, Tiruchirapalli 620024, India

Mu. Ramkumar et al. (eds.), *Environmental Management of River Basin Ecosystems*, Springer Earth System Sciences, DOI 10.1007/978-3-319-13425-3_14

be observed in the flood plain region of mature and or old stage of a river and/or in the event of shifting of channel course. Occurrence of such characteristics at the upper deltaic region and the observation that the channel area gets converted into mid-channel bars (in terms of textural-geomorphic traits), at a rate of 1.08 km^2/year warrant immediate measures for the restoration of natural fluvial processes.

Keywords Environmental integrity · Kaveri River · Sand mining · Channel bar · Flooding

1 Introduction

Despite being the most important life supporting systems of the nature, the river systems are being exploited by the humans for centuries, without understanding much on how the river ecosystem functions and maintains its vitality (Naiman and Bilby 1998). Rivers transport water and fertile soil from their catchment areas and distribute them in the lower reaches that form the basis for cultivation, settlements, navigation and industrial activities. In addition, river sediments are actively exploited for use in construction activities. While it is important to ensure an effective and consistent supply of sand to meet the developing economy in order to support the investment in physical infrastructure, utilities and buildings that are required to raise the living standards as well as social well-being, it is equally important to consider the sustainability of the natural fluvial processes as well. Depending on the geological and geomorphic setting, in-stream mining can cause serious environmental impacts (Fidgett 2003) and may enhance the vulnerability of the farmland, commercial and residential settlements, and the ecological habitats of the river system (Ashraf et al. 2011). The fluvial channels, the water and the sediments act as a unified system connected to each other through a delicate environmental balance. Any anthropogenic intervention (dams, bridges, settlements, mining etc.) may result in recognizable changes in the system, including adverse impacts such as flooding, erosion, and desertification, which in turn may cause loss of critical resources that provide sustenance to the human race. Though such generalizations are in the common knowledge, the environmental and geomorphic consequences are not fully understood due to the paucity of requisite data (Macfarlane and Mitchell 2003; Kale 2005; Ashraf et al. 2011; de Leeuw et al. 2010). Dramatic increase in the human population during the last few decades and associated large-scale changes in the land use and vegetation cover in the catchment areas and widespread developmental activities in the deltaic regions have adversely affected the fluvial environment (Ramkumar 2009a) including runoff (Zimmermann et al. 2006), discharge, sedimentation rates and downstream habitats (de Leeuw et al. 2010). As the fluvial systems are highly sensitive, their response to anthropogenic intervention is rapid and drastic (Kale et al. 2010). With the 70 % of its population dependent on agricultural activities and 60 % of agricultural irrigation dependent on groundwater, the role of rivers of Tamil Nadu State, India is critical in directly supplying water for irrigation and domestic consumption, in addition to recharging groundwater (Selvakumar et al. 2008) and the dependence of these rivers on monsoonal supply, that are erratic and concentrated during short duration (Kale 2003a, b, 2004). This criticality shifts towards danger-mark owing to the recent spurt in mining of river sand which affects not only the fluvial system, but also the livelihood of majority of people.

The River Kaveri, (Fig. 1) one of the holiest rivers of South India, drains through the thickly populated and industrialized regions of Indian peninsular region. Serving for the sustenance of human race since pre-historic times, the river and adjoining regions are the preferred locations for urban centers, industries, cultivation, and domestic waste disposal, etc. In addition to providing fertile soil and water, the river has been a source for sand to meet the construction activities. The mining activity within the river channel has been on the rise, at an alarming rate in the recent years. Previous studies on the Kaveri River and its flood plain regions include documentation of suspended load transport (Ramanathan et al. 1994), REE geochemistry of floodplain sediments (Sharma and Rajamani 2000a, b; Sensarma et al. 2008; Singh and Rajamani 2001a, b), phenomenon of quicksand and life loss

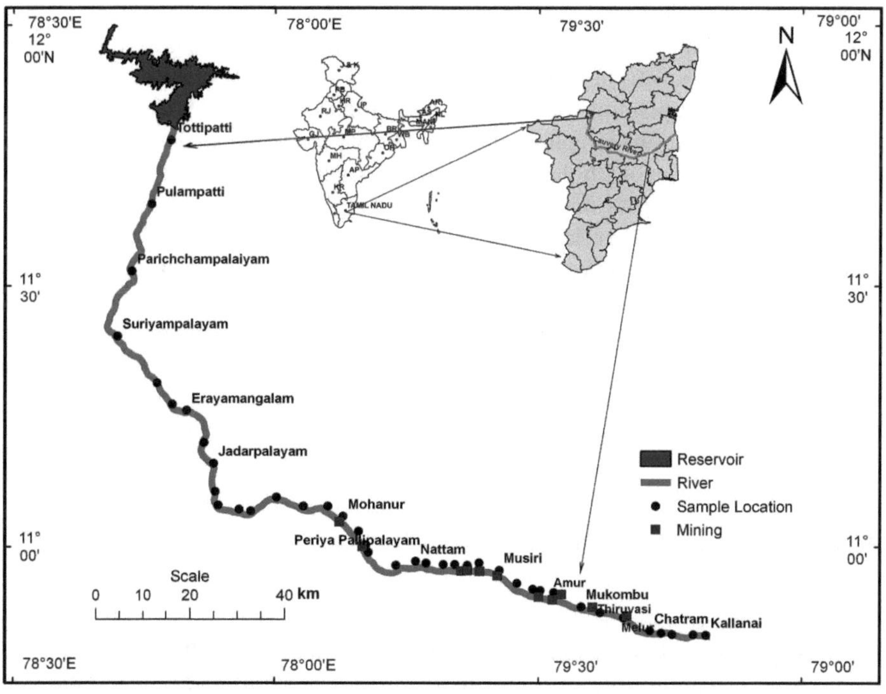

Fig. 1 Location of the Kaveri River and the sampling points from downstream of Mettur reservoir to Kallanai. At each location, three samples from the river bed surface were collected namely, right and left bank and middle of the channel. Locations of mining activity are also indicated by *Solid Squares*

(Thirukumaran and Ramkumar 2009), pattern of recurrence of floods (Ramkumar 2009b), paleoflood (Kale et al. 2010) and water quality, and heavy metal pollution, etc. (Kumarasamy et al. 2009; Mohan Raj 2010; Dhanakumar and Mohan Raj 2012; Venkatesha Raju et al. 2010, 2012; Vignesh et al. 2012). However, the information on the geomorphic and sediment textural characteristics, ongoing dynamic changes in these two important traits of a fluvial system in response to the natural and anthropogenic processes is lacking, which in turn impedeed the understanding on environmental health of the fluvial system.

2 Study Area

The Kaveri (Fig. 1) is the fourth largest river in the Indian peninsula. It originates at Talakaveri on the Brahmagiri range of the Western Ghats at an elevation of about 1341 m and traverses for about 765 km before debauching into the Bay of Bengal. Major tributaries of this river include Hemavathi, Bhavani, Noyyal and Amaravathi. Deltahead of this river is located at Mukkombu, Tiruchirapalli, from where the river branches off into Kaveri and Coleroon. At Kallanai (Grand Anicut—Tiruchirappalli), it further branches off into two distributary channels namely, the Kaveri and the Vennar. They branch further into 36 channels whose total length is 1,607 km. These in turn branch off into 2,988 channels running to a length of 18,395 km (Kandaswamy 1986). The River Kaveri has been a perennial river during the years of yore and had built a largest delta plain of south India. Being monsoon-rain fed, large floods were inherent and annual phenomenon of this river that made the ancient ruler of south India (Karikal Cholan of Chola Dynasty) to construct a major gravity dam (Kallanai or Grand Anaicut) across this river, which is dated (> 900 years) to be the oldest known dam that is in service till date. The advent of modern era of intensive-cultivation, industrialization and urbanization has necessitated construction of about a dozen more major dams for water storage and hydroelectric power generation, and about hundred small and medium dams, regulators and aqueducts for distribution of water for irrigation that literally put seize to the sediment transport to downstream regions of the delta plains, due to which, the Kaveri delta currently undergoes a stagnating phase along the confluence and adjoining coastal regions.

3 Materials and Methods

Thematic maps of lithology (Fig. 2), land use/land cover (Fig. 3) were generated based on published information and remotely sensed data (IRS IB LISS III satellite data; 24 m resolution, date of acquisition: November 2008) and limited field verification for a 10 km wide buffer zone along the course of the Kaveri River in a stretch between Mettur Reservoir and Kallanai. Using topographic sheets of Survey of India (SOI), slope for this buffer region was calculated with the analyst function of

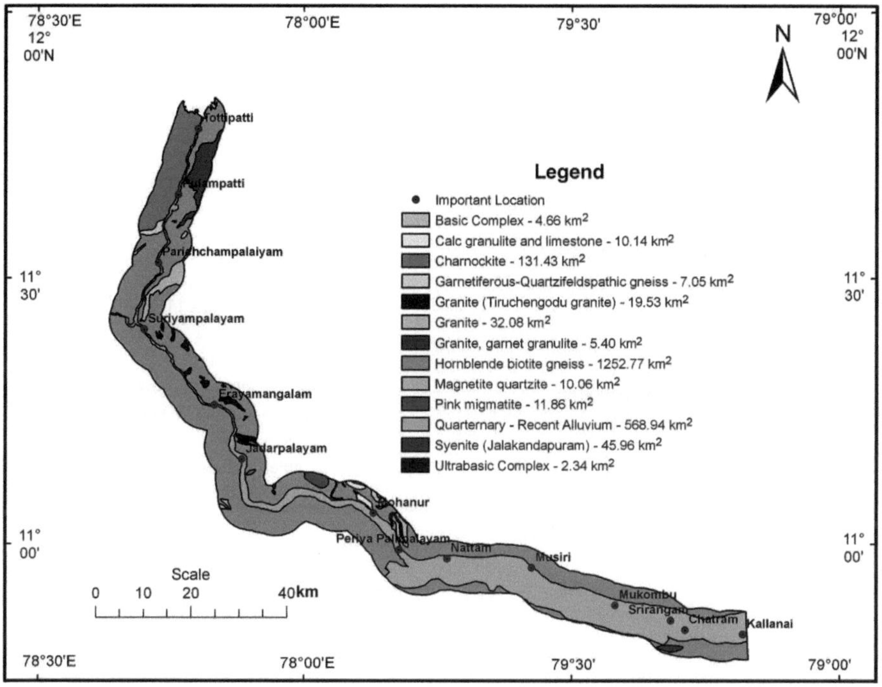

Fig. 2 Map showing the distribution of various rock types within the buffer zone. Predomination of gneissic rocks, followed by Quaternary alluvial soils, charnockite and granite in the descending order of areal distribution could be observed from the map. In addition, confining of the river channel by alluvial soils for most part of the river course could be observed. This characteristic of the Kaveri River along with the lower gradient exacerbates the possibility of erosion and breach of banks

ArcGIS software and a thematic map of slope (Fig. 4) was also generated. Another map (Fig. 5) showing the occurrences of well-developed mid-channel bars was generated with the help of visual interpretation of satellite imageries and field mapping. It was supplemented with the SOI topographic map data (published in the year 1971). Based on the land use/land cover map and field examination, 36 locations that are free from immediate contamination and or disturbance were selected for sampling stream bed sediments. The locales of sand mining were also recorded during sampling (Fig. 1). At each location, across the stream bed, three sites, namely, middle part of channel, and channel bed near right and left banks were sampled. At each site, after collecting geographic coordinates through GPS, a PVC pipe (10 cm diameter and 20 cm long) was sunk into the stream bed surface to recover sediment sample. Approximately 1 kg of sample was collected from each site, packed in air-tight PVC bag, labeled and transported to the laboratory. Laboratory analysis of these sediments and computation of textural parameters were attempted following the procedures detailed in Ramkumar et al. (2000a) and Ramkumar (2001, 2007).

In the laboratory, the samples were air dried, thoroughly mixed, homogenized and cone and quartered to obtain approximately 50 gm of sediment subsample.

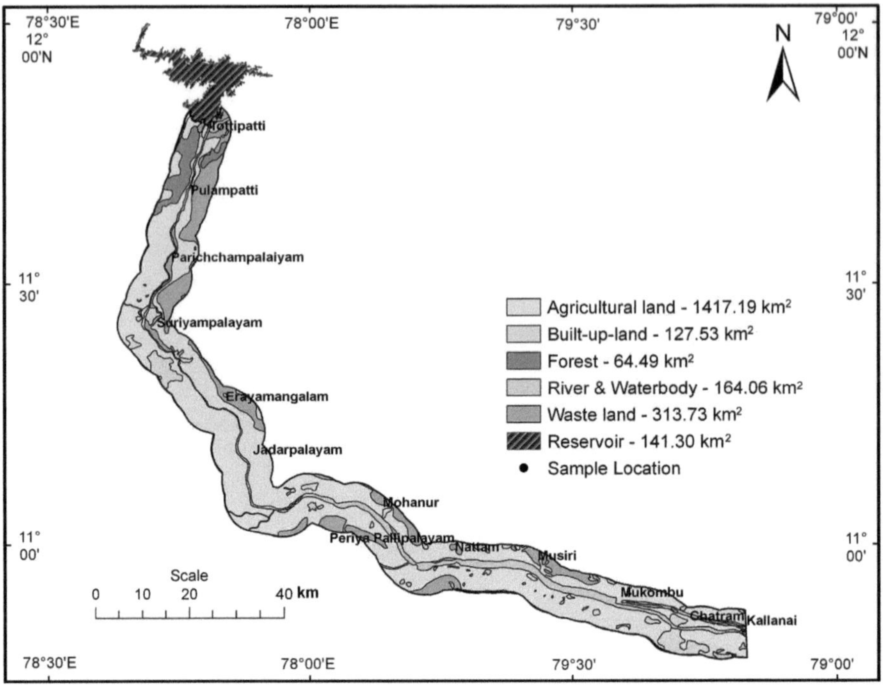

Fig. 3 Map showing the land use/land cover distribution within the buffer zone. The map also shows the predomination of irrigated cropland and built-up land. The wellbeing of these heavily depends on the safe maintenance of the Kaveri River flow within its banks. Any breach, however small it may be, could be devastating

Large pieces of shell fragments, organic material or pebble, if present, were removed. These subsamples were transferred to a pre-cleaned glass beaker. The samples were cleaned through three stages of treatments by hydrogen-per-oxide, 10 % v/v HCL and quasi-distilled water to remove the organic matter, shell calcite and other contaminants. These subsamples were then dried in an air-oven overnight at 60 °C and weighed and dry-sieved in an automated sieve-shaker with ASTM test sieve sets of ½ Ø interval (Ingram 1970) for about 20 min. The separated size classes of sediments were then weighed and tabulated. These data were converted into ¼ Ø intervals with a computer algorithm (Ramkumar et al. 2000a, Ramkumar (2001, 2007), and then cumulative percent according to Ø interval were calculated and plotted in a semi-log graph sheet for generating requisite data to compute graphic mean (Mz), standard deviation (SD), skewness (SK) and kurtosis (KG). In addition, the input data required for few bivariate and other discriminant plots (discussed in latter section) were also generated from the plots of cumulative percent of the sediment samples.

In this paper, the thematic map data, ground-truth data and textural data are discussed in terms of depositional and erosional characteristics of the river channel with reference to the geomorphic features to infer the ongoing sedimentary

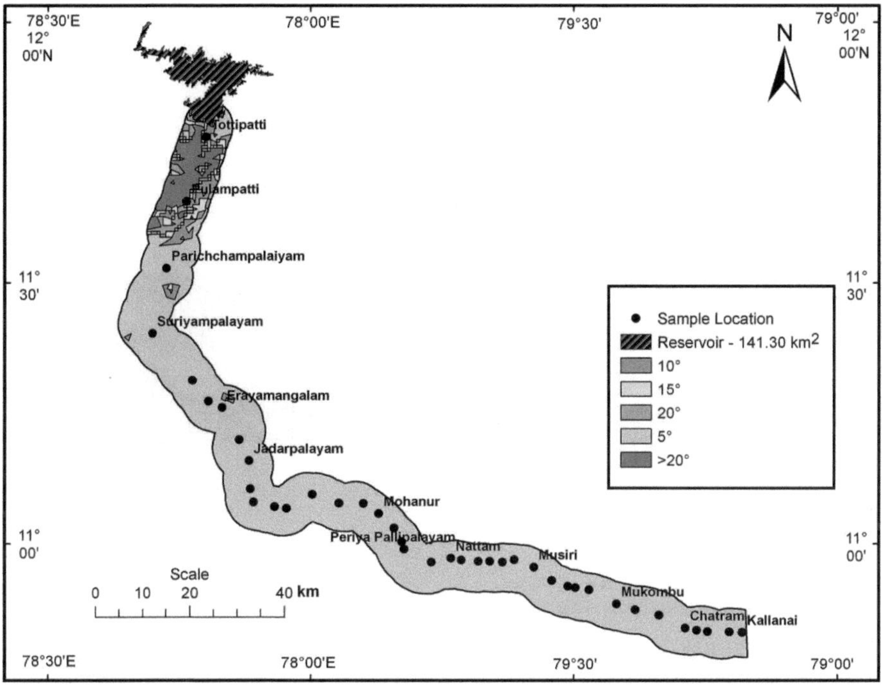

Fig. 4 Map showing the low-gradient nature of the Kaveri River channel for most of its stretch. Low gradient nature, confinement within easily erodible alluvial soils, thick population and agricultural activity immediately adjoining the river banks, together with rapid reduction of volume of the channel due to sediment accumulation are all perfect recipes for a major catastrophe in the event of copious rains

dynamics and the influence of various anthropogenic activities including mining on the fluvial processes. These inferences were utilized to constrain on the flooding hazard for the regions adjoining the river channel and the environmental integrity of the fluvial system.

4 Results and Discussion

4.1 Reduction of Downstream Transport of Sediments and Rise in the Bedlevel

Except for few tens of kilometers in the vicinity of Mettur reservoir (where steeply sloping exposures of the basement rocks within the channel and the confinement of channel itself by vertical cliff banks and isolated patches of coarse river sediments occur in the channel bed), the river traverses predominantly in a low gradient region (<5 % slope—Fig. 4) over Quaternary—Recent alluvial sediments and soils

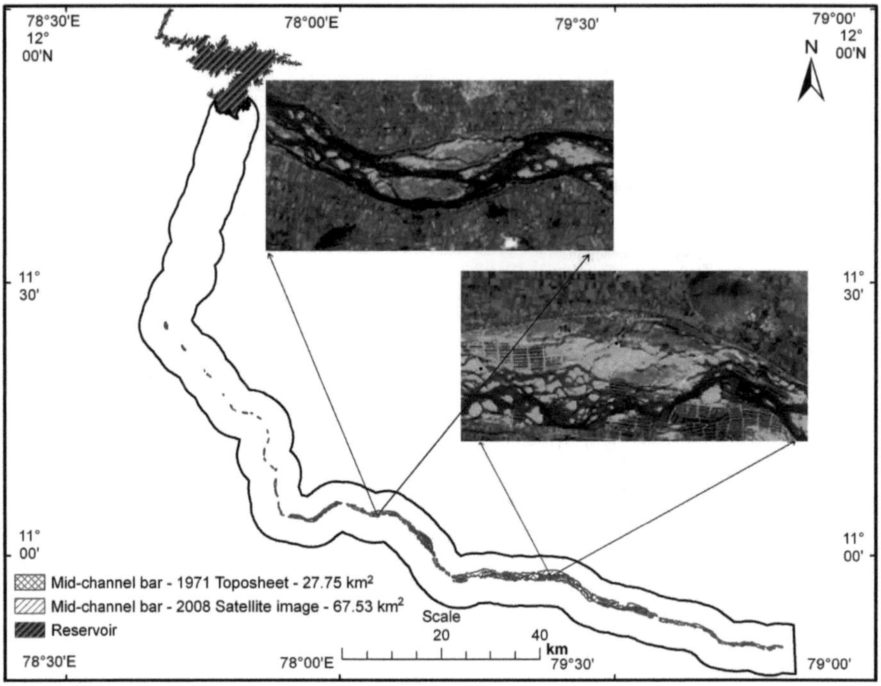

Fig. 5 Map showing the occurrences of well-developed mid-channel bars within the Kaveri River as recognized through visual interpretation of satellite imagery of IRS 1B LISS III. The mid-channel bars that existed during the year 1971 as depicted in the SOI toposheets are also shown in the map. A multifold increase of areal extent of these bars, at a rate of 1.05 km^2/year is observed. In addition to this phenomenal rate of development, it is worrisome to note the development and stabilization of these bars within the storage areas such as Upper Dam at Mukkombu and occurrence of sand sheet 1 m above the sluice level (Plate 1g). While vertical accretion of sediments all along the channel reduced the carrying capacity and channel volume, development of sand sheets and mid-channel bars within reservoir storage areas thwart the very purpose (storage of water and reduction of peak flow during flooding) for which these mid-sized dams are constructed across the Kaveri River

(Fig. 2) and is mostly confined by the banks comprising alluvial soils—a phenomenon indicating the vulnerability of the banks to erosion during bankful flow or any change in the geomorphic/environmental conditions. Owing to this very nature and the ephemeral character of this river, the sediments introduced into the river channel stay within the channel. It is also surmised that except during bankful flow, the carrying capacity of the channel is very limited and is why the channel bed level is found to be at or near the bank level (Plate 1a) in most part of the study area. These two characteristics of the Kaveri River, namely, banks of alluvial soil and occurrence of bedlevel at or near the bank level make the adjoining regions of the channel flood prone.

The river channels build land through downstream transport and distribution of sediments in the deltaic regime. With the advent of intensive-cultivation methods and dwindling water resources, dams have become an integral part of developmental activities along the river valleys and the deltaic regions. These structures considerably alter the rate and quantum of sediment and water transport to downstream regions, leading to the accumulation of sediments within the channel. However, the conditions at the catchment area in terms of weathering and sediment supply to the river channels continue unabated, making the middle part of the river system environmentally-stressed and geohazard prone. It happens in such a way that due to the reduced flow in the channel, the carrying capacity of the river also gets reduced, exacerbating the rate of accumulation of sediments within the channel. It further reduces the volume of the channel, which in turn amplifies the rate of reduction of carrying capacity of the river, even during bankful flow; all of which contribute towards rise of bed level of the channel. These characteristics are typically observed in the low-gradient Kaveri River. Thorleifson et al. (1998) recorded the occurrences of epsilon cross bedded deposits (vertical accretionary deposits of channel bed sediments) overlain by overbank deposits. All along the Kaveri River channel bed, the epsilon cross bedded deposits are observable and wherever these are at or near the bank level, breaches have either occurred or may occur in near future. Sediment deposition in the channel and continued development of meander curves would worsen the future flood damage (US Army Corps of Engineers 2012). The most significant effects on flood levels are changes in the channel bed (Williams and Swanson 1989). Large amounts of sediment and debris are introduced into tributary streams and can cause significant aggradation (filling with sediment) of the bed, particularly where the floodplain has been developed, eliminating the natural sediment storage area. This can raise flood elevations and cause flood paths to be substantially different than those predicted (Williams and Swanson 1989).

4.2 Development and Stabilization of Mid-Channel Bars

In addition to the rise in the bed level, the visible sign of reduction of carrying capacity of river channel and accumulation of sediments within the channel is the formation of mid-channel bars/braided bars. Out of 164.06 km^2 areal extent of the Kaveri River channel (Fig. 3) within the study area, it is estimated that an area of 67.53 km^2 is covered by well-developed and stabilized mid-channel bars (populated with long term plants, namely *Acacia, Casuarina*, Bamboo, and a variety of native plants including banyan, in addition to shrubs and weed grass *ipomea aquatica*—Fig. 5; Plate 1b). These well-developed mid-channel bars occur all over the studied stretch of the Kaveri River (Plate 1c–e) from south of Parichchampalayam and continue up to the Kallanai including the reservoir areas of Mukkombu (Plate 1f). Accumulation of bed sediments above the sluice level has also been observed in the Kaveri River (Plate 1g) which stands testimony to the rise of bed level, reduction of carrying capacity of the river and reduction of storage volume of the channels and reservoirs. It is observed in the field that, except the narrow channel-in-channel physiography (*sensu* Kale 2005), the entire river channel is covered by mid-channel bars (Fig. 5) and sand sheets that are vegetated by shrubs and grass. Comparison of areal extent of the mid-channel bars as per the SOI toposheet published in the year 1971 (27.75 km^2) with the present estimate based on satellite imagery of the year 2008 (67.53 km^2) shows an increase of 243.35 %, at a rate of 1.08 km^2/year. Comparison of the spread of mid-channel bars in terms of area of mid-channel bar (67.53 km^2) with reference to the area of river channel (164.06 km^2) suggests that currently 41.16 % of the channel area is occupied by the mid-channel bars. The rate of growth has been phenomenal as nearly half of the channel is covered by these bars within a span of 37 years (mid-channel bars covered only a 16.92 % of the river channel as per the toposheet of the year 1971).

Through periodic visits to selected sites since the year 2004, a set of schematic diagrams depicting various stages of sediment accumulation that lead to the rise of the stream bed level to bank level, followed by formation and stabilization of the mid-channel bars has been evolved and presented in the Fig. 6a–f. It all starts with the settling of sediments either due to obstructions to the stream flow such as dams (Fig. 6a) and or influenced by reduction in carrying capacity of the channel due to dwindling flow and excessive sediment supply (sourced from incision of stream bed due to mining and construction activities). Once the initial settling occurs at a site, the settled sediments increase the friction with the moving water, augmenting settlement of sediments exponentially. With the continued and exacerbated sediment settlement, the sediment layers rise near or above the water level, and become the preferred regions of grass development (Plate 1h). The grass (*Ipomoea aquatica)*, when grows up, arrests the sediment load being transported near channel bottom (usually larger grains owing to their density and size), adding up the quantum of sediment deposition, leading to the formation of a sand bar parallel to the stream course (Fig. 6a; Plate 2a, b). With the continued settlement and grass growth, smaller plants and finally larger plants grow over the bar, making the bar a permanent fixture within the channel.

When the vertical accretion rises above the water level, it emerges as a sand bar that results in the modification of channel cross section (Fig. 6b). This modified cross section forces the water flow to differentiate into erosional along the cut bank (Plate 2c) and depositional (Plate 1d) along the inner curve of the modified channel course and, which in turn, progressively shift the location of the channel. There are many examples of this mechanism that occur all along the Kaveri River. An example is the mid-channel bar formed in the Kaveri River from Amur up to Vathalai adjoining the southern bank. It appears that the Kaveri River channel is bifurcated along its longitudinal direction due to the development of this bar. With the accelerated vertical accretion of the stream bed, principally driven by the excessive sediment supply due to mining and other anthropogenic intervention, many mid-channel bars aligned parallel to each other are formed that move as a sand sheet (Fig. 6c; Plate 1c). The resultant cross sections of the channel across the channel (Fig. 6d) and along the channel (Fig. 6e) show the significant reduction of the channel volume and modification of the channel course. It further leads to the overbank flow even during below normal stream flow (Fig. 6f). It happens in such a way that, in addition to the change of river course, the gentle slope provided by the inner curve of the meander often promotes overbank flooding that would bypass the

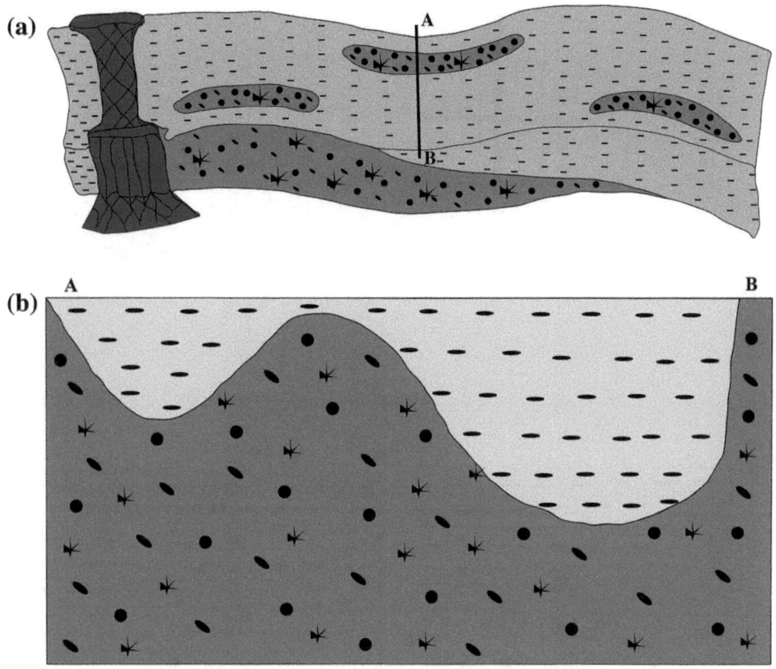

Fig. 6 Schematic diagram showing the sediment accumulation within the channel due to the construction of dams, reduction in stream flow and carrying capacity. **a** Schematic depiction of damming and resultant sediment accumulation in the channel as well as the storage area of reservoirs. In the Kaveri River, there are many examples of bar formation and stabilization within the storage area of reservoir and at places, the sedimentation has been found to be more than sluice level of reservoir (Plate 1f, g). Until recent past, these bars never existed as also appreciable from the SOI toposheets published in the year 1971. These might have grown to this size due to the extensive mechanized mining activities that were initiated a while ago. **b** Schematic depiction of the change in the cross sectional profile and areal extent of the channel due to bar formation. In addition to these, due to the vortex flow characteristics of the water, energized formation of meanders, significant vertical accretion of channel bed along inner curve and rapid shift of channel along cut bank occur. The extent of shift of channel course used to be proportional to the width of channel bars and or the point bars. As both of the banks of the Kaveri River are intensively utilized either for cultivation and or urbanization, the stream has no other go except to forcefully overflow at opportunistic places. Examples of such overflow and breach of banks are abundant, including the breaches at Melur near Srirangam (Plate 1d) and near Amur. It is noteworthy to remind here that at both the places, the channel bed level and the bank level have been equal and the breaches occurred during slightly more than normal rains. **c** With continued sedimentation on the channel bed leads to amalgamation of many channel bars into a sand sheet. **d** and **e** Pictorial depictions of sand sheets that reduce the channel volume. **f** This characteristic contributes towards significant enhancement of flooding vulnerability, as these sand sheets act as dams and are at bank level topographically

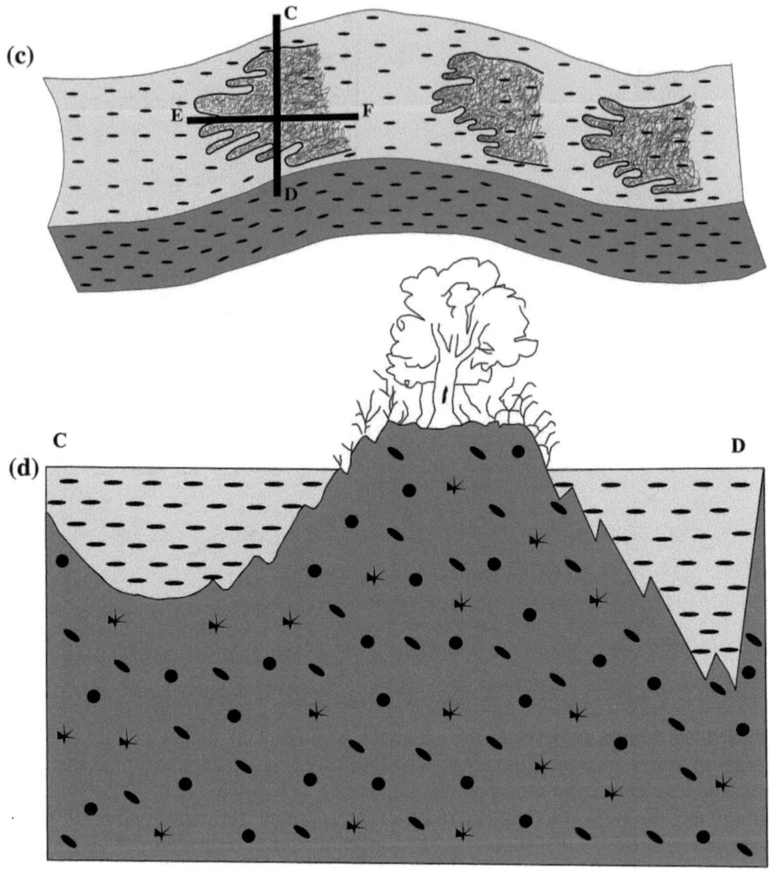

Fig. 6 (continued)

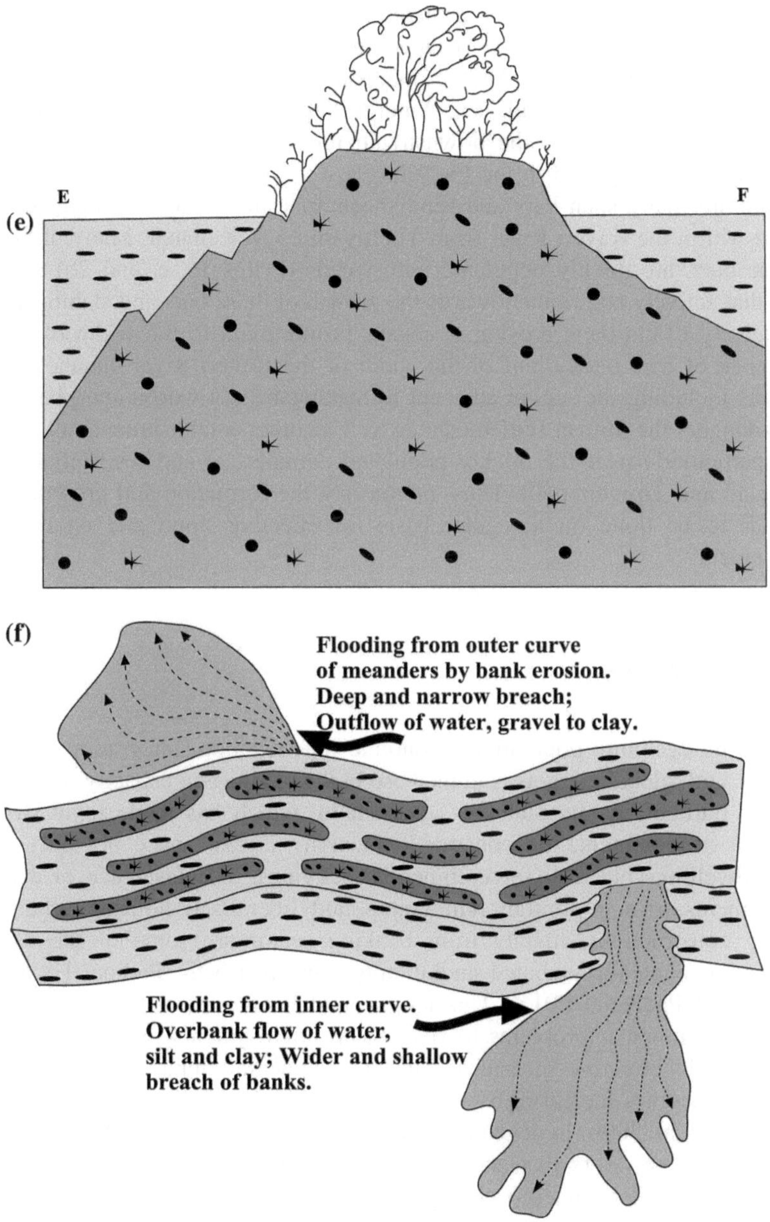

Fig. 6 (continued)

original bank of the mother channel (a natural mechanism of river to expel excessive sediment out of the channel). The breach that occurred during the year 2002 at Melur near Srirangam (Fig. 6f; Plate 2d) was due to the overbank flow adjacent to the inner curve of meander at a major mid-channel bar along the northern bank of the Kaveri River near Srirangam wherein vertical accretion of bed level reached the bank level. Another major breach of the River Kaveri during the year 2005 at its southern bank near Amur Village and inundation of thousands of hectares of land all along the downstream region occurred due to similar geomorphic features depicted in the Fig. 6f.

There are many such bars and sand sheets that are at various stages of development within the Kaveri River from Thottiyam up to Kallanai. Many, if not most of these bars, are thickly populated with weeds (Plates 1c, e, and 2b) and small plants that actively contribute towards the growth of these bars and stabilization. As the intensity of cut-bank erosion is always proportional to the width of the bars, occurrence of bars nearly half of the width of the Kaveri River channel in many locations including the region adjacent to Srirangam is a worrisome phenomenon that endangers the human settlements as well as other establishments all along the bank, particularly near the thickly populated commercial and residential areas of Srirangam and Tiruchirapalli. Thus, monitoring the formation and growth of these bars has to be done on a regular basis for effective flood and environmental management.

4.3 Sand Mining

With the bludgeoning population, economic liberalization, integration of national economy with global markets, and increase in the purchasing capacity of the people of India, particularly the State of Tamil Nadu, which has made rapid strides in economic development, the construction activity is fast apace not only for the human dwellings but also for commercial activities. The fast pace of economic developments, rise in foreign remittances and liberalized housing schemes for building constructions, mainly from banking sector are some of the causative factors responsible for unabated sand mining from river beds (Padmalal et al. 2008; Sreebha and Padmalal 2011). Unscientific and haphazard sand mining leads to severe environmental problems to river basin environments that need immediate attention and corrective measures. The environmental impact of the river sand extraction becomes increasingly well understood and is linked with the globalization (Sonak et al. 2006) in developing countries such as China and India (de Leeuw et al. 2010). The rivers that are harvested for sand at rates in excess of natural replenishments often undergo channel degradation, causing incision of the entire river system including its tributaries (Ashraf et al. 2011). Stability of the banks would also be under jeopardy as the channel banks are comprised of non-cohesive soil. Striking cases of excessive removal of river sediment removal are summarized by many researchers (Bull and Scott 1974; Sandecki 1989; Kondolf and Swanson 1993; Kondolf 1997; Macfarlane and Mitchell 2003; Hemalatha et al. 2005).

The Tamil Nadu State is endowed with construction materials (such as limestone deposits for cement manufacture), building stones and sand. While the limestone and building stones are mined from regions away from the rivers, the sand is being mined from ephemeral river channels (ironically, these were erstwhile perennial rivers!) of the Tamil Nadu State. Owing to the high quality of material available, sand from the river channels of Tamil Nadu are being mined not only for meeting the demands within Tamil Nadu, but also adjoining states such as Kerala, Karnataka and Andhra Pradesh. While mining of building stones and limestone create permanent scars in the face of earth's surface whose interaction and impact on biosphere and atmosphere are slow and thus do not result in immediate effects, the very nature of the unified system on a valley-scale (Singer et al. 2008) and quick reaction time of fluvial system (Kale et al. 2010), advocates a cause of concern. Removal of sediments from the river bed (mining) should be based on the scientific determination of availability of excessive material at a given location, and the rate of removal should never exceed the rate of natural replenishment (Padmalal et al. 2008).

While it is inevitable to exploit the sand deposits of the rivers for the economic benefits (Ashraf et al. 2011), ignoring the natural dynamics of the river systems and the haphazard placement of mining locations within the river channel would be at our own peril. Field survey by the authors has revealed that the locations (Fig. 1) of the mining activity within the river channels are randomly selected. As could be observed elsewhere (de Leeuw et al. 2010), sand mining in the study area are selected based on accessibility to the river channel, giving utter disregard to the channel configuration, sites of sediment accretion and or erosion, etc. Removing sand from the channel without understanding these characteristics, disrupts the preexisting balance between sediment supplies and transporting capacity (Padmalal et al. 2008), and creates a locally steeper gradient upon entering the pit. It typically induces incision upstream and downstream of the extraction site (Sandecki 1989). The over-steepened knick-point (with its increased stream power) commonly erodes upstream in a process known as head-cutting. As head-cuts migrate upstream, incision propagates upstream. Mining- induced incision may propagate upstream for kilometers on the Main River (Scott 1973; Stevens et al. 1990), and tributaries located at upstream (Harvey and Schumm 1987; Padmalal et al. 2008). In addition, make-shift bridges and access roads are built within the channel at every mining locality (Plate 2e) that disrupt the natural flow conditions (Plate 2f) of the stream. Factors that increase or decrease the sediment supply often destabilize beds and banks and result in dramatic channel readjustments. Activities that artificially lower the stream bed elevation cause bed instabilities that result in a net release of sediment in the local vicinity. Because the stability of sand-bed streams depends on a delicate balance among stream flow, sediment supplied from the watershed, and present channel form, mining-induced changes in the sediment supply and channel form disrupt channel and habitat development processes (Lagasse et al. 1980; Arun et al. 2006; Sreebha and Padmalal 2011). Furthermore, movement of unstable substrates above, at, and below the mine sites results in downstream sedimentation (Carling 1984; Ashraf et al. 2011).

Second major issue is the extensive use of heavy machinery for mining (Plate 2e, f) and heavy influx of large vehicles into the channel (Plate 2g), numbering thousands of sorties on a 24 h × 7 day basis. These activities result in perceptible changes in the fluvial system. Employing mechanized mining equipment and creation of deep scars in the river channels affect the flow conditions and downstream ecological niches that depend on the sediment and water drafted from upper reaches. Heavy influx of off road vehicles (ORV) over the loose sediment beds of the river channel severely increases the compaction and reduces the percolation of river water into the ground water system—a factor that seriously impede the recharge characteristics and affect the adjoining regions that depend on ground water for irrigation, domestic and industrial needs. Severe impacts on the ground water table in adjoining regions of sand mining, reduction of water storage and recharge capacity of sub-terrenes of the channel bottom, may slowly kill the self-regulating and self-repairing mechanism of the river system. Indiscriminate sand mining over the years has imposed irreparable damages to the river ecosystem of the Kaveri River. Lack of adequate information regarding the extent of environmental impacts caused by sand mining is a major lacuna that challenges the regulatory efforts and minimizing the adverse effects of in-stream sand mining.

4.4 Environmental Conditions as Deduced from Textural Parameters

Studies on the sediment size and textural parameters help correlate the sediment types and their environments of deposition (Inman 1952) and to discriminate the natural and anthropogenic processes operating within an environment. Based on the grand average of the textural parameters of 105 sediment samples collected from 36 locations, the Kaveri River sediments can be termed as moderately sorted (SD: 1.2), coarse skewed (SK: −0.25), leptokurtic (KG: 1.21) and fine sands (Mz: 2.13). While this could be a common characteristic of any comparable river at its mature stage (*for example*, the Godavari and the Krishna Rivers—Ramkumar 2001, 2007), quite oblivious to this simplified nature, the devil lies in the details as described herein.

The mean size of the sediments varies from silt (03.90) to coarse sand (0.62) with a standard deviation of 0.67. The classification of size class has an interval of 0.5 in the Ø scale and hence, the standard deviation of 0.67 could be considered as shift of more than a size class which is beyond normality in a low-gradient, ephemeral river channel with much reduced flow conditions. As the graphic mean size of the sediments indicates average size of the sediments which is influenced by the source of supply, environment of deposition and average kinetic energy of the depositional agents, the observed variation of more than a size class in a monotonous river channel is interpreted to be a result of changes in environmental and energy conditions, the reasons for which have to be recognized. Inclusive graphic standard deviation is a measure of the average dispersion of the distribution around mean. The measure of standard deviation is considered to be indicative of the level of sorting in sediments. The sorting also shows a variation from very poorly sorted (2.01) to moderately well sorted (0.57). The successive classification intervals of standard deviation are variable, but a minimum interval of 0.35 is provided (Friedman 1967). The standard deviation (0.28) of sorting (SD) is lower than the minimum interval of 0.35 and indicates not much shift in the sorting class. However, variation from very poorly sorted to moderately well sorted with an average of moderately sorted nature signifies considerable variation of sediment sorting character. This inference is also affirmed by the skewness values that vary from strongly fine skewed (0.42) to strongly coarse skewed (−0.63). The standard deviation of SK is 0.29 which is close to the minimum class interval of 0.3, indicating shift of one skewness class interval. Graphic kurtosis is a measure of peakedness of the frequency curve and is the quantitative measure used to describe the departure from normality and measures the ratio of sorting in the tails of the cumulative curve and its central portion. Similar to other textural parameters, the KG also shows a wide variation in terms of extremely leptokurtic (3.62) to very platykurtic (0.59). The grand average of KG is leptokurtic, (which means that the central portion of the sediment size within a sample is better sorted than the tails) while the minimum value of kurtosis is found to be very platykurtic (which means that the tail portion of sediment sample is better sorted than the central portion). It could be inferred from the KG that certain forces other than normal depositional conditions might have been active to create these aberrations.

As such, these observations suggest a wide variation of the studied textural parameters, which could not be expected in a monsoon-fed fluvial system draining in a low-gradient landscape and experiencing reduced flow conditions. Hence, downstream variations of these parameters, their linear and polynomial patterns were examined in the light of natural trends and perturbations with reference to the locations of anthropogenic intervention.

4.5 Influence of In-stream Mining on the Downstream Variations of Textural Parameters

River process differentiates the sediments during downstream transport in terms of progressive reduction in size, and increases the sorting, roundness and mineralogical and textural maturity (Ferrell et al. 1998; Ramkumar 2001, 2007; Sensarma et al. 2008). The downstream pattern of Mz, as indicated by the linear trend (Fig. 7a) shows a gradual shift from medium sand to silt, as could be observed elsewhere in a fluvial environment. However, there are many aberrations, as indicated by the polynomial trend as well as the absolute value curve. It follows from the Fig. 7a that immediately downstream of each mining location, there are shifts from the general linear trend of the sediment size class, perceptibly influenced by the mining activity. It also follows from the geographic locations further downstream of the mining activity that the fluvial process attempts restoration of the

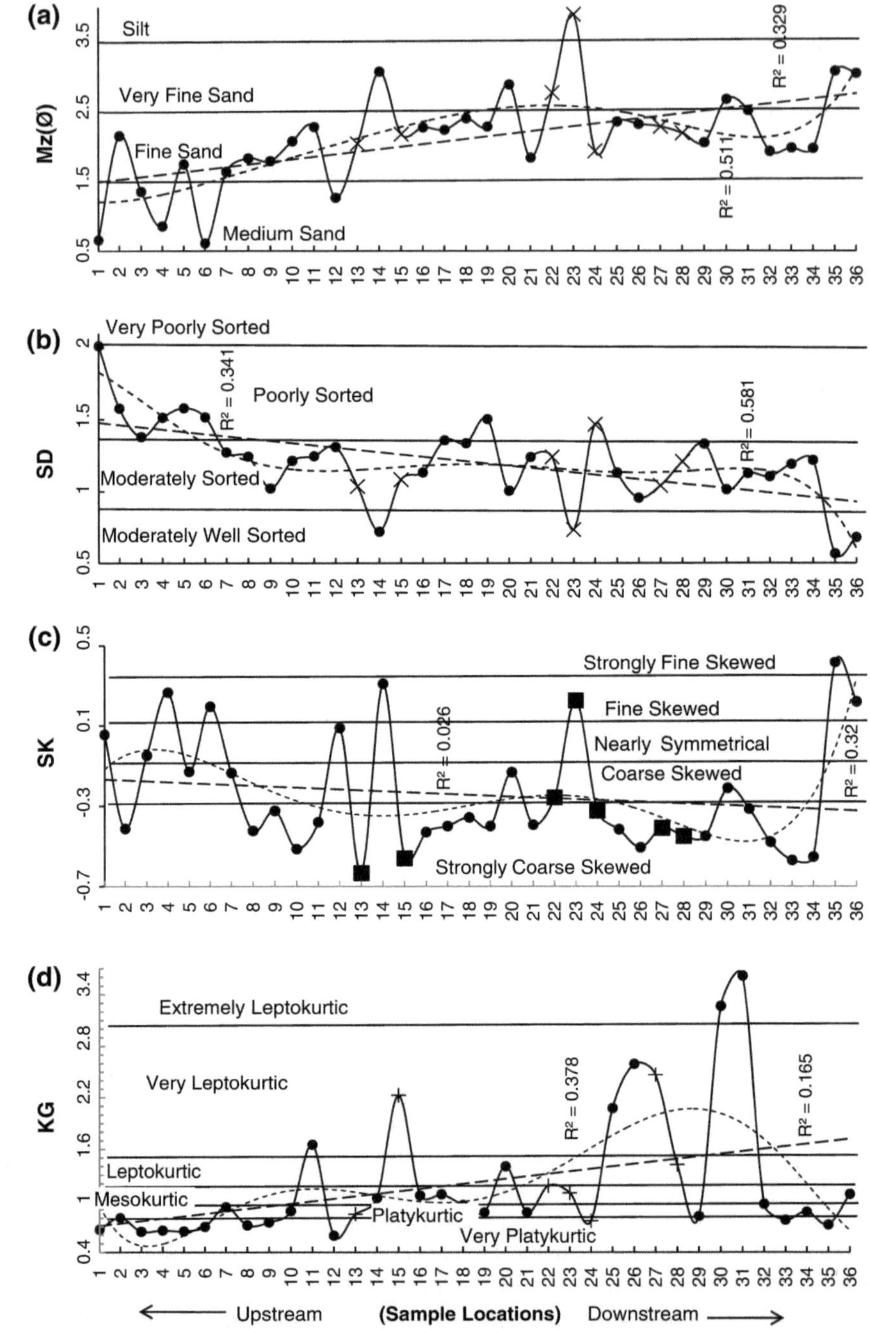

Fig. 7 Textural parameters and their downstream variations. The samples that were collected in the vicinity of mining locations are marked with solid squares. **a** Downstream variations of Mz. **b** Variation of sorting characteristics along the Kaveri River. **c** Downstream variations of SK. **d** downstream variations of KG

original trend, albeit unsuccessfully. It is to be noted that, the mining activity influences the sediment character to move both the directions, i.e., to shift towards coarser as well as finer fractions, thus disturbing the entire spectrum of the size class. It could be interpreted as the impact of mining activity over natural fluvial process in such a way that the original depositional texture of the sediments at downstream of mining locations is obliterated due to mining. It is also to be noted that the intensity of shift from one size class corresponds with the areal extent of mining activity within the channel. The locations of the mining activity and the absolute value curves also suggest the decrease of carrying capacity and quantum of sediment transport as indicated by sudden change of fine sand to either very fine sand or silt, thus providing definitive evidence for the impact of mining on mean size, carrying capacity of the stream and the selective removal of finer sediments towards downstream. The studies of Ferrell et al. (1998), Ramkumar (2001, 2007), Ramkumar et al. (2000a), have established that the mineralogy and geochemistry of deltaic sediments broadly have a size control. Selective removal of particular size class from stream bed sediments may impose constraints on the ecological niches located at downstream, that depend on the influx of specific types of sediments.

The mining activity appears to be undoing the natural process of the progressive increase in sorting towards downstream as indicated by the aberrations of sorting class from linear trend (that shows an increase from poorly sorted to moderately well sorted) in terms of shift towards sudden better sorting (locations 11–16 and 23 —see Fig. 7b) or towards poor sorting (location 24, 27 and 28 see Fig. 7b). Thus, similar to the case of mean size, mining activity disturbs the sorting character on both the directions. However, the restoration of normalcy by the stream action towards downstream could not be observed in the immediate vicinity as indicated by the absolute value curve as well as the polynomial trend line. This observation suggests the overriding and long-lasting influence of mining over the sediment sorting character and the inability of the fluvial process to return to normalcy, once disturbed.

Inclusive graphic skewness is a measure of the frequency distribution that indicates the position of the mean with respect to the median and is geometrically independent of sorting of the samples. Excess of finer parts in the sediments gives negative skewness while a dominant coarser population gives positive skewness. Thus, the energy condition of the depositing medium may be determined from the skewness values. While a general trend of progressive shift from coarse skewed nature to strongly coarse skewed nature could be observed from the linear trend line of skewness (Fig. 7c), presence of too many perturbations from this pattern in the polynomial and absolute value curves could be observed. Evidences of impact of mining over skewness in terms of shift from coarse skewed to strongly skewed to fine skewed, a conundrum of sediment textural property, principally driven by mining could be witnessed from this figure. Further, no recognizable signs of restoration of normalcy are present, which means, alteration of sediment character by mining could not all be reversed by the fluvial process, even after transporting the sediments for more than 90 km (Thottiyam to Kallanai—Fig. 1). The linear trend of the kurtosis shows a progression from very platykurtic to very leptokurtic

(Fig. 7d). Similar observations in terms of sudden shift from one class to extreme opposite class, non-return of normalcy, long term impact of mining and inability of the fluvial process to restore normalcy are all observed with reference to this sediment characteristic.

All these observations and inferences signify the impact of mining over the natural fluvial system and the inability of the natural system to cope up with the onslaught. It is also observed through the trend lines and absolute value curves that, though the river has a self-regulating mechanism, owing to the locations of mining, that are placed all along the river channel, the mechanism is not effective even at locations 90 km downstream of mining activity. At this juncture, it is important to examine the fluvial processes that are in operation within the Kaveri River channel and as explicit in the sediment textural properties and hence, the textural data were plotted in few discriminant diagrams and examined.

4.6 Influence of Fluvial Processes and Anthropogenic Intervention on Textural Parameters

Plotting the data on the bivariate discrimination diagrams of Mz and SD (Fig. 8a) and SD and SK (Fig. 8b) with discriminant lines of Friedman (1967) and Moiola and Weiser (1968) revealed that all the sediment samples fall in the field of river process except few. It is also explicit from these figures that while the sediment samples collected away from the mining and other construction activities form a cluster, the sediment samples collected in the vicinity of mining and construction activities fall either away from the cluster or in the field of beach, suggesting obliteration of the sediment textural property by the anthropogenic interventions into the natural fluvial system. The discriminant lines in these figures principally divide the sediments in terms of fluvial (unidirectional transport, multimodal sediment character, downstream fining and betterment of sorting) and beach (bidirectional transport, uni or bimodal coarse sediment with better sorting). Comparison of bivariate plots of river sediments of the Godavari River (Ramkumar et al. 2000a) and the Krishna River (Ramkumar 2007) where no sand mining of comparable scale is reported revealed the absence of river sediments in the field of beach. On the contrary, sediments collected from the beach and beach rocks of these two river deltaic regions fall on the river field and form close-clusters, signifying typical riverine and undisturbed nature, abundant fluvial sediment flux into the coastal regions, rapid burial of those sediments before inheriting "marine" character (Ramkumar and Gandhi 2000; Ramkumar et al. 2000b). Hence, incidences of Kaveri River sediments falling in the field of beach could be construed as a definitive evidence of the impact of mining and other anthropogenic intervention over the natural fluvial processes. These interventions, as detailed in the downstream trends of textural properties, introduced selective removal of specific grain sizes, and selective sorting and shifted the sediment characters on either way of

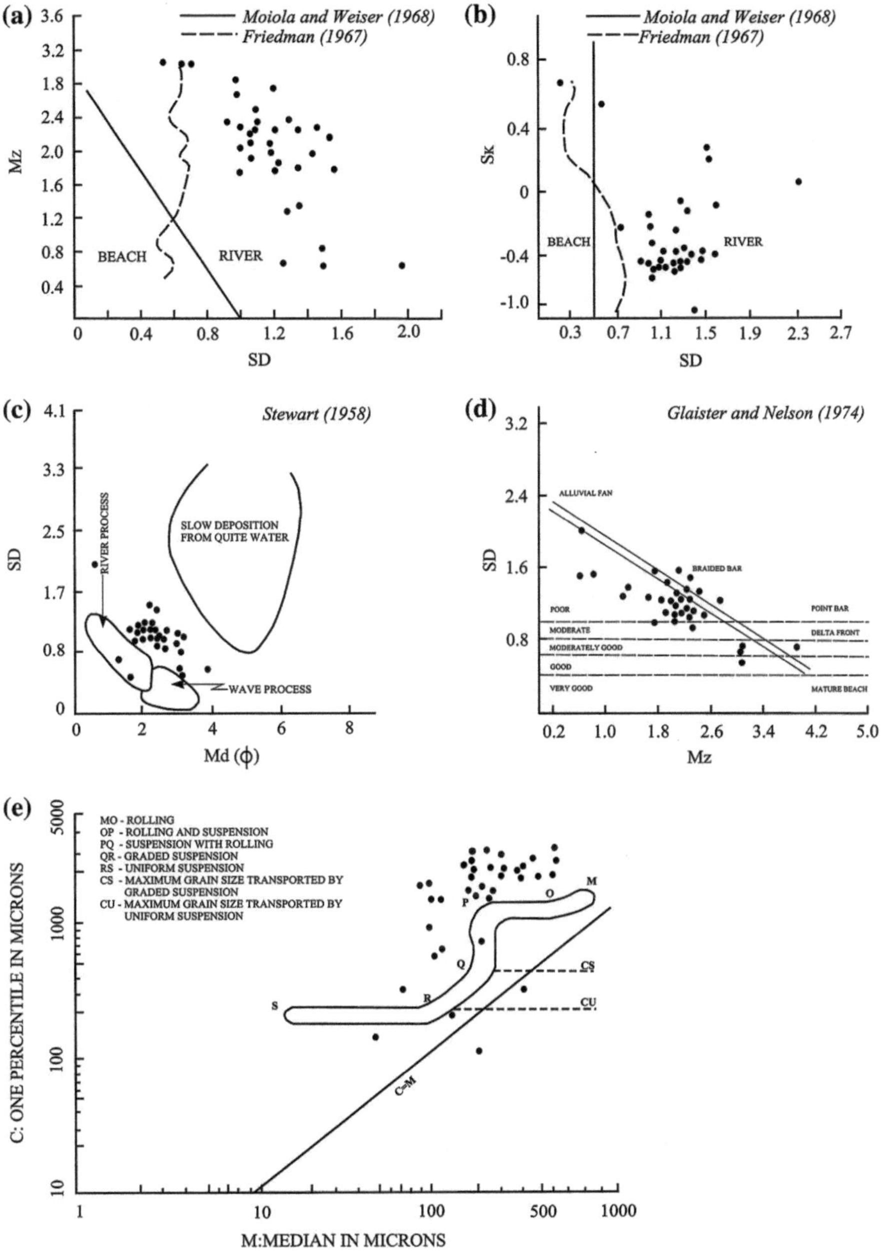

Fig. 8 Discriminant diagrams of depositional environments, depositional conditions and transportation processes. **a** Bivariate plot of Mz and SD with discriminant lines of Friedman (1967) and Moiola and Weiser (1968). **b** Bivariate diagram of SD and SK with discriminant lines of Friedman (1967) and Moiola and Weiser (1968). **c** Discriminant diagram of depositional processes (after Stewart 1958). **d** Discrimination of sedimentary environments *vis-vis* geomorphic features based on sediment textural phenomenon. **e** Discrimination of transportation processes based on textural phenomenon

(a)
(b)
(c)
(d)
(e)
(f)
(g)
(h)

◀**Plate 1** **a** Occurrence of channel bed level at or near the bank level due to accumulation of sediments within the channel and resultant vertical accretion. **b** Mid-channel bar development and stabilization by lower order seasonal grass at water level, long-term weed grass at further upper regions and shrubs at the central portion of bar. **c** Photograph showing the development of sand bars, and stabilization of these bars and sheet by vegetation (grass and shrubs). Amalgamation of these bars into a sand sheet covering major portion of the channel width could also be seen. **d** Photograph showing the development of cluster of bars along the northern bank of Kaveri and their stabilization by grass cover. Accretion of bed level in front of the bars could also be observed. **e** Well developed and stabilized mid-channel bar. Mining activity and influx of heavy vehicles (Off Road Vehicular Traffic —ORV) could also be seen in the photograph. Location of the mining activity is dependent on the accessibility of channel and availability of sand within the channel. **f** Mid-channel bar developed and stabilized within the storage area of Upper Dam located at Mukkombu. These features are unknown until the initiation of extensive mechanized mining in the upper reaches of the Kaveri River. **g** Accumulation of sediments within the storage area of reservoir above the sluice level. It means that the complete storage capacity of the dam has been exhausted and it no longer serves as a reservoir. **h** Photograph showing the stages of development of braided bar within the channel. Once settling of sediment occurs, grass and finally other higher order plants develop over the bars, leading to amalgamation of bars into big islands which, finally change the channel course

trends, principally through mixing of sediments deposited at various energy, flow and environmental conditions (layers of sub-surface river bed sediments) and hence this "beach" character could have been imparted. It is interesting to note that the sediments of clastic sediments of the Cretaceous fluvial system (which was interpreted to be ephemeral and monsoon-fed and low gradient river—Masthan 1978; Ramkumar 1996; Madhavaraju et al. 2002, 2006; Ramkumar et al. 2004, 2012; and hence comparable with the Kaveri River and also presumed to be the Proto-Kaveri River that was draining the region approximately 65–68 million years ago and had had the same catchment—Banerji 1972; Ramasamy and Banerji 1991; Ramkumar 2008; Ramkumar et al. 2004, 2005) fall exclusively in the river field and none in the beach field (Ramkumar et al. 2012). It reconfirms the inferences of obliteration of sediment textural properties by mining and other anthropogenic interventions into the natural fluvial system.

Stewart (1958) had established a bivariate diagram that classifies the sediments according to the prevalent depositional energy conditions namely, wave process (to and fro motion with high energy), river process (unidirectional flow with high to moderate energy) and slow deposition (quite water or low energy) based on the median diameter (MD in phi scale) and standard deviation of sediments. Plotting the data of the Kaveri River sediments in the discriminant diagram of Stewart (1958) shows that while a cluster of samples collected away from the vicinity of mining and anthropogenic intervention fall in the field of river process, samples collected in the vicinity of anthropogenic activities fall either away from river field and also within the wave process, but none in the quite water field (Fig. 8c) confirming the inference of mixing of sediments deposited under various environmental conditions by mining activity, selective removal and selective sorting and haphazard accumulation of sediments.

Plotting the Mz and SD data in a discriminant diagram of Glaister and Nelson (1974) shows most of the samples (samples collected away from the vicinity of

(a)
(b)
(c)
(d)
(e)
(f)
(g)
(h)

◀**Plate 2** **a** Photograph showing the close-up view of bar development in the middle portion of the Kaveri River and change of flow directions. **b** Photograph showing the development of bars near the confluence of Kudamurutti with Kaveri channel. It also obstructs the flow direction, due to which, temporary damming effects during monsoons have occurred, that led to flooding of adjoining regions of Trichy and Srirangam. **c** Photograph showing the vertical cliff formation and cut-bank development along the outer curve of the meander. While this phenomenon severely erodes the commercial and residential areas of Tiruchirapalli city, concomitant channel course shifting and vertical accretion of bed level occur along the inner curve of the meander near Srirangam which is located at northern bank of the Kaveri River. **d** Photograph showing the breach of northern bank of the Kaveri River at Melur near Srirangam after reinforcement. The onlooker shows the flood-level. At this location, the channel bed and bank levels were found to be equal, due to which, breach has occurred during normal monsoon rain. **e** Field photograph showing intensive mining activity, that disrupts the natural flow conditions of river. **f** Field photograph showing the unplanned mining and use of heavy machinery. **g** Field photograph showing the incursion of heavy vehicles. These heavy vehicles, severely compact the loose cohesionless stream bed sands and thwart recharge of underground aquifers. **h** Field photograph showing the embankment constructed along the northern bank of the Kaveri River to contain erosion along the outer curve of meander, principally driven by the mid-channel bars developed along the southern bank. Over the years, it is found that these structures shifted the sites of active erosion from one place, often at an alarming level

mining and anthropogenic activities) falling in the field of braided bar and few other samples in the mature sediment category. It is surprising to note that the samples collected from the active river channel also fall in the braided bar field (Fig. 8d), suggesting homogenization of sediments of the Kaveri River regardless of the sub-environments and the subdued influence of riverine process over the sediment characteristics. The fluvial deltaic system consists of many sub-environments that have typical sediment textural and geochemical characteristics of their own (Ramkumar and Vivekananda Murty 2000), the distinctness of which may be different to the tune of 100 % from one another (Ferrell et al. 1998; Ramkumar 2001). Ferrell et al. (1998) have demonstrated the presence of such distinction in terms of mineralogy of sediments of the Krishna delta. Ramkumar (2001) had successfully discriminated 21 sub-environments of the Godavari delta based on the textural parameters of the surface sediments. It was also possible to discriminate the differences between the depositional conditions prevalent during various seasons, geographic locations, and flood and ebb tides based on the textural data of the Krishna River (Ramkumar 2007). In this context, the homogenization of the textural property of the Kaveri River sediments with reference to the geomorphic units *vis-à-vis* sub-environments confirms the impact of in-stream mining over natural fluvial process.

Studies on the depositional characteristics and geomorphic responses of the Godavari River (Sastry et al. 1991; Ramkumar 2000, 2003) and the Krishna River (Rao and Vaidyanadhan 1979; Ramkumar and Gandhi 2000) have reported a depositional sequence that can be observed in a deltaic system, namely, active river channel → channel bar/point bar → mud-flat/natural levee → flood plain in the stratigraphic order at the places of channel shifting and or abandonment. This depositional sequence is found to be common for the East coast deltas of India

(Swamy et al. 1990; Singh and Swamy 1996). In this context, representation of the Kaveri River bed sediment samples in the field of braided bar suggests the ongoing abandonment character to the Kaveri River channel. It also suggests that due to the dwindling water flow and anthropogenic intervention, fluvial processes of this river channel are not active! Based on the observed braided bar characteristics of most of the sediment samples, ongoing vertical accretion of sediments within the Kaveri River channel is also inferred which is corroborated with the occurrences of bed level at or near the bank level. Had the sediments shown the prevalence of river channel or point bar character, it would have meant either downstream transport or lateral distribution of sediments across the delta plain. In order to further ascertain this inference, the data were plotted in a discriminant diagram of Passega (1957) to understand and to evaluate the hydrodynamic forces prevalent during the deposition of these sediments. From this figure (Fig. 8e), plot of most of the samples above the maximum grain size transported by graded suspension and plot of few samples above the uniform suspension field could be observed. In addition, predominant transportation of sediments through rolling and suspension could be observed from this figure. These information, together with the predomination of fine sand as the overall grain size class of the Kaveri River suggest that the river flow could transport sediments of fine sand category only, that too through rolling and suspension mode, an indication of reduced carrying capacity and vertical accretion of the sediments within the channel itself.

4.7 Flooding Hazard

Though monsoon-fed, the Kaveri River has been known as a perennial river since historic times. It is known to have wreaked havoc by extensive flooding since the historic times that forced the kings of yore to construct first ever gravity dam to contain the flooding and destruction by this river. There are stratigraphic and sedimentological records of short-term, high-discharge flash flood during Holocene in the upper regions or this river channel (Kale et al. 2010). Despite the efforts of the governmental agencies during the recent years, the flooding phenomenon has been recurrent and devastating. This information, along with the present observations and inferences namely, draining of the river channel over non-cohesive Quaternary-Recent alluvium, prevalence of channel abandonment characteristics and extensive sand mining and resultant negative consequences over the fluvial processes, sprawl of urban and industrial clusters all along the river channel, extensive lower deltaic plain that are put to use for cultivation, dependence of a large population of the state of Tamil Nadu over the agricultural produce from this region, necessitates evaluation of flooding hazard of this historically and inherently vulnerable river system.

The two major fluvial systems of India, namely the Himalayan river system and the peninsular river system are highly different in terms of hydrology, geomorphology, channel morphology and sediment load (Kale 2005). The Indian subcontinent is the largest monsoon-dominated region in the World. The monsoon months deliver enormous quantities of moisture. The annual flow patterns of the Indian rivers are greatly influenced by the amount of rainfall, that determines the channel morphology, sediment transport and flow conditions. However, many of the rivers including the Kaveri River remain inactive during most part of the year due to the occurrence of monsoon only during 4–5 months of a year. The Kaveri River is under the influence of southwest (June–September) and northeast (October–December) monsoons. It receives major source of water during northeast monsoon (44.44 %) followed by southwest monsoon (36.18 %). The total flow increases from the month of March to September and then it decreases. The Kaveri River with a total length of about 1,750.87 km long drainage channels, all of which receive copious rainfall almost simultaneously, and enhances the flooding vulnerability at downstream regions, especially in the low gradient regions (Jain and Sinha 2003). Added to these, occurrences of floodplain at or near the base level of the channel bed, and removal of bio-barriers along the bank and conversion of the natural barriers (natural levee) into urban centers have all but contributed towards enhancing the flooding vulnerability.

Large monsoon floods are important geomorphic events in terms of erosion and sediment transport in the Peninsular Rivers of Indian sub-continent. Flooding in the deltaic plains adjoining the river channels is caused by the inadequate capacity of the rivers (volume within the contemporary active channel banks) to contain the high flows brought down from the upper catchment due to heavy rainfall. Inadequacy of the channel volume is accentuated by the surface runoff from urban centers located along the river channel. Extraordinary floods are characterized by dramatic increase in channel boundary shear stress and power per unit area. Flood induced changes include widening of channel, erosion of bars and banks, scouring of flood plains, movement and deposition of coarse gravel, changes in channel sinuosity, avulsion and changes in channel courses, and formation of channel-in-channel physiography (Kale 2005). Few of these changes, including channel-in-channel physiography, movement of coarse sediments, and erosion of bars and banks are typical of the Kaveri River. Similar to the characteristics of the Baghmati Basin as reported by Jain and Sinha (2003), the Kaveri River also shows flooding vulnerability, due to the lower bankful capacity than the mean annual flood discharge, high sediment load and extensive sediment deposition due to the decrease in stream power.

The Kaveri River at its deltahead region displays typical characteristics of desert river system namely, wide, shallow and sandy channel, large variability of discharge and sediment load. All these are recent phenomena and may have been brought by the mining and damming activities. The problem is two-fold; the first is that, the river receives its 60–80 % of flow during the summer monsoon, thus creating a situation of heavy flow in a short span of time duration that pressurizes the channel capacity. The second is, the sediments (bedload as well as suspended sediment load) from the catchment peak during the initial phase of monsoon (Kale 2005) out of which nearly up to 75 % settle at the channel and the floodplain (Goswami 1985; Chakrapani and Subramanian 1990) and do not reach sea. Embedded within the monsoon are the cyclonic storm related copious precipitations that create flash-flood situations.

Aggravating this situation is the reduced percolation and surface runoff conditions, associated with the urban settlements that border the channel banks and the increased ORV impact due to the heavy influx of mining machinery and vehicular traffic within the channel. Hence, all these could have contributed towards the reduction of channel volume and the rate and volume of percolation (within the channel as a result of ORV and along the channel due to the changed land use from natural vegetation to irrigated lands and urban lands) that lead to the creation of conducive milieu for overbank flow or flooding during monsoon.

It is observed that, except few isolated patches, the banks of the Kaveri River and adjoining flood plains have been either converted into urban centers or croplands (Fig. 3), all of which were natural forest lands that once stood protecting the fluvial flow within the channel itself. Rapid urbanization of vegetated regions causes higher peak flows and lower low flows, especially in the river basins dominated by high-intensity rains, increasing the flood vulnerability. An increase of about 28 % stream flow as a direct result of deforestation in the Appalachian regions was observed by Swank et al. (2001). Vegetation plays several roles during extreme events. Vegetation usually mollifies the damage by dissipating the energy of the flow, and by stabilizing the banks and steep slopes against the erosive forces of the overland flow (Shroba et al. 1979). Harvesting and clearing of floodplain forests reduces the energy dissipation capacity (Jacobsen and Oberg 1993) and increases the erosivity of the riverbanks and floodplain lands. The drainage of the wetland and floodplain areas compounds the problem by speeding the conveyance of floodwaters to the main channel and increasing the flood stages. Recognition of the protective value of the vegetation during floods is shown by the construction of tree screens that stabilize the stream banks and buffer the impacts of flooding (McGuire 1989). These areas also act as floodwater retention devices, slowing flows and lowering the flood peaks (Faber 1993). Levees often bound the river at its banks and separate the channel from adjoining floodplain. However, levees, owing to their unique topography and geography (raised above from the channel and flood plain and straddle between them and contain canopies of long-term plants) are the preferred locations of conversion of natural land cover into land use for domestic, agricultural, commercial and industrial purposes. Conversion of land cover into land use and removal of this unique geomorphic unit from the natural fluvial system deprives the river channel the additional volume required (within the banks) during flooding situations. Without the natural room (provided by the levees) to expand (volume), to contain the stream power (by long-term plants), the flow velocity increases artificially until the former levees succumb and floodwaters violently reclaim their natural territories (Sinha 1996; Kale 2003a, b, 2004, 2007; Kale and Hire 2007). The separation of channel and floodplain is directly related to the alteration of the hydrologic regime and the loss of wetland areas that naturally border the river systems.

Not only the artificial conversion of levees, the anthropogenic influences are also recognized to extend beyond the realm of natural flood control structures (natural levees). In many instances, man creates hazards that rival the most dangerous and bizarre of natural flood conditions. Often man's influences on the natural systems are long-term and go unrecognized until conditions gradually deteriorate to critical levels (Hickey and Salas 1995). Agricultural crops and concrete pavements and structures have replaced the wetlands and forested floodplains that were originally capable of storing massive amounts of floodwater and decreasing flood peaks (Faber 1993). The conversion of natural floodplain into agricultural land use and urbanization reduces the timespan between precipitation and accumulation in the river channels. All of these factors alter hydrologic processes and lead to more powerful and rapid flood peaking. Flood control reservoirs alleviate potentially damaging flows by providing storage space for excess runoff. An obvious environmental effect of reservoirs is the change of long river reaches from lotic (moving water) to lentic (still water) ecosystems. By altering the hydrologic nature of the extreme events, the reservoirs mask the natural cues and depress the downstream channel alterations. By intercepting the sediment load carried by upstream floodwaters, reservoirs deprive downstream stretches of the bed material required to construct and maintain backwater areas and sand and gravel bars (Hickey and Salas 1995).

Floods have long been considered as purely hydrological phenomenon, and therefore, flood management programs have focused on hydrological variations. This has been one of many reasons for the failure of flood management efforts across the globe including India (Jain and Sinha 2003). Historical data reveal that even after continuous efforts to control the floods, flood damages and flood-affected areas in India have increased with time. High magnitude floods during the monsoon season are considered to be India's recurring and leading natural disasters (Kale et al. 1994). The country faces recurrent loss of life and damage to property due to severe floods. Purohit and Suthar (2012) stated that according to the estimate of Central Water Commission (2007) a whopping loss of 1,590 lives and damage of public utilities to the tune of Rs. 8,068 billion (USD 134.47 billion at a conversion ratio of 60 INR. per 1 USD). Despite spending large amounts of money and human effort, flooding continues to cause huge losses in terms of life and property. It indicates the failure of the flood control measures and emphasizes the need for a better understanding of the flood hazard (Jain and Sinha 2003).

Despite extensive construction of levees, bye-pass channels and other structural measures, the Sacramento River continues to flood the regions adjoining the river (Singer et al. 2008). Effective measures of flood control recognized during recent times has been channel restoration method, that includes scientific designing of channel pattern, dimension, profile and cross section of the stream, based on stream characteristics, land use and geomorphology. Unlike many others disasters, there is much that can be done to reduce comparable damage by flooding (Thorleifson et al. 1998). It requires better understanding of the stream dynamics, land use, hydrologic and geomorphic features along the river course. However, structural flood control is a deterrent to the health of natural systems, but remains a valuable tool in flood management. Thus, the challenge the policy makers may face centers on finding a balance between structural and non-structural controls (Hickey and Salas 1995). Poor understanding on the sediment transport, enhancement of flood vulnerability in regions proximal to the zones of sediment accumulation have impaired the efficiency of flood control measures (Singer et al. 2008). Studies on fluvial geomorphology of various river basins including the Yellow River, the Sacramento River, the Strickland River, and the Brahmaputra River indicate that owing to the natural characteristic of a river, it maintains Basin-scale controls on the channel alignment, conveyance of flood waters and transport and deposition of the sediments (Rogers et al. 1989; Gupta 1988, 1995; Shu and Finlayson 1993; Fischer 1994; Singer et al. 2008; Sacramento River Flood Control Project 2012). Hence, Basin-scale documentation of these characteristics and designing flood control measures with site and measure specifics is necessary.

Williams and Swanson (1989) reported that conventional flood control design methodology seeks to minimize the right-of-way required for flood control channels by increasing the flow velocities, thereby allowing a narrower channel to be built that reduces flood elevations. This is done typically by lining the channel with smooth reinforced concrete. With a suitable slope in a uniform channel, the low roughness of the concrete can allow "super-critical flow" to develop very fast-moving shallow flow. When super-critical flow occurs, the channel cross-section and right-of-way can be significantly reduced. However, many of the "managed" channels experienced failure even during modest floods, way below the super-critical levels. The primary reason for the failure of these channels in medium-sized floods was that the actual effective roughness of the channel during the flood was considerably larger due to the additional bedload sediment that was brought along with flood waters. Fluvial geomorphologic studies such as that of Limerinos (1970) have demonstrated the relation between the bedload size and the channel roughness in natural streams. As discussed in the previous sections, the Kaveri River channel has heightened bedload that could not be transported downstream and distributed in the lower deltaic regions or supplied to the coastal regions. It is followed by the settling of the sediments in the channel itself that increases the channel roughness, contributing towards the increase of the possibilities of overbank flow and enhancement of flooding vulnerability.

The breaches that occurred at Srirangam and Amur during the years 2002 and 2005 respectively have demonstrated the vulnerability of this river even during less than super-critical flows. These breaches have occurred during the years that experienced less than average monsoonal precipitation. Wherever flood flows are large, enormous quantities of sediment are mobilized and are conveyed downstream by natural watercourses (Williams and Swanson 1989). This newer sediment load from multiple sources of overland sheet flow that emanate from the urban centers located all along the river course is in addition to the channel bottom sediment that have already accumulated due to weak flow velocities of the ephemeral river create a super-critical roughness, which in turn create a situation of overbank flow or breach of cut bank. In natural streams, super-critical flow rarely occurs in the long reaches due to the size of bed material mobilized. Thus, the relationship between the heightened flood vulnerability and the accumulation of sediments within the channel in terms of mid-channel bars, sand sheets and accentuated sediment influx due to mining activity in the Kaveri River are affirmed.

The effort made by local and regional government agencies to contain the flooding vulnerability along the Kaveri Channel has been found to be construction of rip-rap walls or artificial embankments (Plate 2h). However, experience from other Indian rivers (Sinha 1998) and elsewhere state that these structures merely transferred the trouble from one place to another. These structures are also found to be interfering with the natural fluvial process (Sinha 1998). There are two types of factors that control the flooding on geomorphic response of a river channel, namely, the external and internal. The external factor includes the rainfall regime, peak discharge, contributing area, hydrograph response, basin morphometry, sediment load, vegetation, land use, soils and bed rock. The internal factor includes the channel gradient, channel morphology and the cohesion of the banks. From the data and discussion presented in previous sections, it is understood that the flood vulnerability of the study area has been exacerbated due to both of these two causes. The present observations on geomorphic and sediment textural properties of the Kaveri River warrant immediate efforts to restore the natural fluvial system to its normalcy. The stream channel restoration includes channel straightening, modification of cross section and profile and provision of pools. These are cost effective and permit more symmetric flow regime, sediment transport and reduce the erosion of banks.

5 Conclusions

- Sedimentary environments are defined by a set of physical, chemical and biological conditions in a geomorphic setup (Reineck and Singh 1982). As these conditions are intricately related, change in any of the property would be reflected in others. As the river systems act as a unified sedimentary system on a basin-scale, an ability to understand the corresponding changes at various geographic locations of river channel helps understanding the integrity of the natural fluvial system.
- Documentation of geomorphic and textural properties of the Kaveri River had revealed the ongoing phase of abandonment due to vertical accretion of stream bed, principally driven by dwindling flow conditions, damming and mining in the upper reaches and changed land use pattern all along the channel course. Due to its inherent nature such as draining in a low-gradient region and confining of its bank by non-cohesive soils, anthropogenic intervention along its channel in terms of rapid urbanization and reduction of percolation due to ORV and conversion of natural vegetation and resultant vertical accretion of stream bed, the flooding vulnerability of the river channel has been exacerbated.
- It appears that with the current rate of conversion of the river channel into mid-channel bar, the channel would be filled with sediments within a span of another 89 years.
- Systematic study of micro-geomorphic features, quantitative modeling of bed-load and hydrological properties of the river basin are needed to design and implement effective restoration measures for rejuvenating the normal fluvial processes of this river.

- Given consideration to the sustenance of river system, and to reduce the flooding vulnerability and environmental deterioration, monitoring and maintenance of sand bars within the river channel are necessary. Periodic field checking of river sand harvesting sites, recording total sand extraction, periodic environmental auditing etc., could prevent further degradation of the river channel and save the region (including the natural inhabitants, domestic, industrial, agricultural and commercial establishments) from potential geohazards.

Acknowledgments Prof. Radhakrishnan, Department of Marine Sciences, Bharathidasan University is thanked for permission to utilize the sieve-shaker. Prof. V. Rajamani, School of Environmental Sciences, Jawaharlal Nehru University, New Delhi, is thanked for enlightening the authors on the importance of conserving river and flood plain sediments for the sustenance of natural environment and mitigation of geohazards.

References

Arun PR, Sreeja R, Sreebha S, Maya K, Padmalal D (2006) River sand mining and its impact on physical and biological environments of Kerala Rivers, Southwest Coast of India. Eco-chronicle 1:1–6

Ashraf MA, Maah MJ, Yusoff I, Wajid A, Mahmood K (2011) Sand mining effects, causes and concerns: a case from the Bestari Jaya, Selangor, Peninsular Malaysia. Sci Res Essay 6:1216–1231

Banerji RK (1972) Stratigraphy and micropalaeontology of the Cauvery basin, Part I. Exposed area. J Palaeont Soc India 17:1–24

Bull WB, Scott KM (1974) Impact of mining gravel from urban stream beds in the Southwestern United States. Geology 2:171–174

Carling PA (1984) Deposition of fine and coarse sand in an open-work gravel bed. Can J Fish Aquat Sci 41:263–270

Central Water Commission (2007) Central water commission annual report. www.cwc.nic.in/main/webpages/dl_index.html

Chakrapani GJ, Subramanian V (1990) Factors controlling sediment discharge in the Mahanadi River basin, India. J Hydrol 117:169–185

de Leeuw J, Shankmann D, Wu G, de Boer FW, Burnham J, He Q, Yesou H, Xiao J (2010) Strategic assessment of the magnitude and impacts of sand mining in Poyang Lake, China. Reg Environ Change 10:95–102

Dhanakumar S, Mohanraj R (2012) Fractionation of iron in river-bed sediments: Implications for the assessment of environmental integrity of the Cauvery delta region, India. In: Ramkumar M (ed) On the sustenance of Earth's resources. Springer, Heidelberg (in press)

Faber S (1993) The mississippi flood. Environment 35:2–3

Ferrel RE, Hart GF, Swamy ASR, Bhanumurty P (1998) X-ray mineralogical discrimination of depositional environments of the Krishna delta, Peninsular India. J Sediment Petrol 54:148–154

Fidgett S (2003) River mining: planning guidelines for the management of river mining in developing countries with particular reference to Jamaica. Alliance Environment and Planning Ltd., Nottingham

Fischer KJ (1994) Fluvial geomorphology and flood control strategies: Sacramento River, California. In: Schumm SA, Winkley BR (eds) The variability of large alluvial rivers. ASCE Press, New York, pp 115–138

Friedman GM (1967) Dynamic processes and statistical parameters compared for size frequency distribution of beach and river sands. J Sediment Petrol 37:327–354
Glaister RP, Nelson HW (1974) Grain size distribution: an aid in facies classification. Bull Can Petrol Geol 22:203–240
Goswami DC (1985) Brahmaputra River, Assam, India: Physiography, basin denudation, and channel aggradation. Water Resour Res 21:959–978
Gupta A (1988) Large floods as geomorphic events in the humid tropics. In: Baker VR, Kochel RC, Patton PC (eds) Flood geomorphology. Wiley, New York, pp 151–177
Gupta A (1995) Magnitude, frequency and special factors affecting channel form and processes in the seasonal tropics. In: Costa JE, Mille AJ, Potter KW, Wilcock PR (eds) Natural and anthropogenic influences in fluvial geomorphology. American Geophysical Union, Geophys Monogr 89:125–136
Harvey MD, Schumm SA (1987) Response of dry creek, California, to land use change, gravel mining, and dam closure. In: Proceedings of the international symposium erosion and sedimentation in the pacific rim. Corvallis, pp 451–460
Hemalatha AC, Chandrakanth MG, Nagaraj N (2005) Effect of sand mining on groundwater depletion in Karnataka. In: Proceedings of the V international R&D conference of the central board of irrigation and power. Bengaluru, pp 1–15
Hickey JT, Salas JD (1995) Environmental effects of extreme floods. In: US-Italy research workshop on the hydrometeorology, impacts, and management of extreme floods. Perugia, pp 1–23
Ingram RL (1970) Sieve analysis. Procedures in sedimentary petrology. Wiley, New York
Inman DL (1952) Measures for describing size of sediments. J Sediment Petrol 19:125–145
Jacobsen RB, Oberg KA (1993) Geomorphic changes in the Mississippi River floodplain at Miller City, IL as a result of the flood of 1993, US geological survey circular 1120-J, US Government Printing Office, Washington
Jain V, Sinha R (2003) Geomorphological manifestations of the flood hazard: a remote sensing based approach. Geocarto Int 18:51–60
Kale VS, Ely LL, Enzel Y, Baker VR (1994) Geomorphic and hydrologic aspects of monsoon floods on the Narmada and Tapti Rivers in Central India. Geomorphology 10:157–168
Kale VS (2003a) Geomorphic effects of monsoon floods on Indian rivers. Nat Hazard 28:65–84
Kale VS (2003b) The spatio-temporal aspects of monsoon floods in India: implications for flood hazard management. In: Gupta HK (ed) Disaster management. University Press, Hyderabad, pp 22–47
Kale VS (2004) Floods in India: their frequency and pattern. In: Valdiya KS (ed) Coping with natural hazards: Indian context. Orient Longman, Hyderabad, pp 91–103
Kale VS (2005) Fluvial hydrology and geomorphology of monsoon-dominated Indian rivers. Rev Brasil Geomorfol 6:63–73
Kale VS (2007) Geomorphic effectiveness of extraordinary floods on three large rivers of the Indian peninsula. Geomorphology 85:306–316
Kale VS, Hire PS (2007) Temporal variations in the specific stream power and total energy expenditure of a monsoonal river: the Tapti River, India. Geomorphology 92:134–146
Kale VS, Achyuthan H, Jaiswal MK, Sengupta S (2010) Palaeoflood records from upper Kaveri River, southern India: evidence for discrete floods during Holocene. Geochronometria 37:49–55
Kandaswamy PK (1986) Irrigation development in Tamil Nadu. Bhagirath 22:67–73
Kondolf GM (1997) Hungry water: effects of dams and gravel mining on river channels. Environ Manage 21:533–551
Kondolf GM, Swanson ML (1993) Channel adjustments to reservoir construction and instream gravel mining, Stony Creek, California. Environ Geol Water Sci 21:256–269
Kumarasamy P, Vignesh S, Muthukumar K, Arthur James R, Rajendran A (2009) Enumeration and identification of pathogenic pollution indicators in Cauvery River, South India. Res J Microbiol 4:540–549

Lagasse PF, Simons DB, Winkley BR (1980) Impact of gravel mining on river system stability. J Waterw Port Coast Ocean Div ASCE 106:389–404
Limerinos JT (1970) Determination of the manning coefficient from measured bed roughness in natural channels. USGS Water Supply Paper 1898-B, 47 pp
Macfarlane M, Mitchell P (2003) Scoping and assessment of the environmental and social of river mining in Jamaica. MERN working paper, University of Warwick, 87 pp
Madhavaraju J, Ramasamy S, Ruffell A, Mohan SP (2002) Clay mineralogy of the late cretaceous and early tertiary successions of the Cauvery Basin (southeastern India): implications for sediment source and paleoclimate at the K/T boundary. Cret Res 23:153–163
Madhavaraju J, Lee Y II, Armstrong-Altrin JS, Hussain SM (2006) Microtextures on detrital quartz grains of upper Maastrichtian-Danian rocks of the Cauvery Basin, Southeastern India: implications for provenance and depositional environments. Geosci J 10:23–34
Masthan S (1978) Depositional environment of Kallamedu sandstone, Maastrichtian, Ariyalur area, south India. J Geol Mineral Met Soc India 15:61–70
McGuire WD (1989) Healing thompson bend. Mo Conserv 50:24–27
Mohan Raj R, Ravichandran A (2010) Study of soil microflora indicating pesticide contamination of Cauvery River belt in India. IndiaJ Sci Tech 3:80–82
Moiola RJ, Weiser D (1968) Textural parameters: an evaluation. J Sediment Petrol 38:45–53
Naiman RJ, Bilby RE (1998) River ecology and management in the Pacific Coastal Ecoregion. In: Naiman RJ, Bilby RE (eds) River ecology and management: lessons from the Pacific Coastal Ecoregion. Springer, New York, pp 1–22
Padmalal D, Maya K, Sreebha S, Sreeja R (2008) Environmental effects of river sand mining: a case from the river catchments of Vembanad lake, Southwest India. Environ Geol 54:879–889
Passega R (1957) Texture as characteristic of clastic deposition. Am Assoc Petrol Geol Bull 41:1952–1984
Purohit J, Suthar CR (2012) Disasters statistics in Indian scenario in the last two decade. Int J Sci Res Pub 2:1–5
Ramanathan AL, Vaithiyanathan P, Subramanian VK, Das BK (1994) Nature and transport of solute load in the Cauvery River basin, India. Water Res 28:1585–1593
Ramasamy S, Banerji RK (1991) Geology, petrography and systematic stratigraphy of pre-Ariyalur sequence in Trichirapalli district, Tamil Nadu, India. J Geol Soc India 37:577–594
Ramkumar M (1996) Evolution of Cauvery basin and tectonic stabilization of parts of South Indian shield—Insights from structural and sedimentologic data. J Geol Assoc Res Centre Misc Pub In 4:1–15
Ramkumar M (2000) Recent changes in the Kakinada spit, Godavari delta. J Geol Soc India 55:183–188
Ramkumar M (2001) Sedimentary microenvironments of modern Godavari delta: characterization and statistical discrimination—towards computer assisted environment recognition scheme. J Geol Soc India 57:49–63
Ramkumar M (2003) Progradation of the Godavari delta: a fact or empirical artifice? Insights from coastal and forms. J Geol Soc India 62:290–304
Ramkumar M (2007) Spatio-temporal variations of sediment texture and their influence on organic carbon distribution in the Krishna estuary. India J Geochem 22:143–154
Ramkumar M (2008) Cyclic fine-grained deposits with polymict boulders in Olaipadi member of the Dalmiapuram formation, Cauvery Basin, south India: plausible causes and sedimentation model. ICFAI J Earth Sci 2:7–27
Ramkumar M (2009a) Types, causes and strategies for mitigation of geological hazards. In: Ramkumar M (ed) Geological hazards: causes, consequences and methods of containment. New India Publishers, New Delhi, pp 1–22
Ramkumar M (2009b) Flooding—A manageable geohazard. In: Ramkumar M (ed) Geological hazards: causes, consequences and methods of containment. New India Publishers, New Delhi, pp 177–190
Ramkumar M, Gandhi MS (2000) Beach rocks in the modern Krishna delta. J Geol Assoc Res Centre 8:22–34

Ramkumar M, Vivekananda Murty M (2000) Distinction of sedimentary environments of the Godavari delta using geochemical and granulometric data through analysis of variance (ANOVA). India J Geochem 15:69–84
Ramkumar M, Sudha Rani P, Gandhi MS, Pattabhi Ramayya M, Rajani Kumari V, Bhagavan KVS, Swamy ASR (2000a) Textural characteristics of coastal sedimentary environments of the modern Godavari delta. J Geol Soc India 56:471–487
Ramkumar M, PattabhiRamayya M, Gandhi MS (2000b) Beach rock exposures at wave cut terraces of modern Godavari delta: their genesis, diagenesis and indications on coastal submergence and sealevel rise. India J Mar Sci 29:219–223
Ramkumar M, Stüben D, Berner Z (2004) Lithostratigraphy, depositional history and sea level changes of the Cauvery basin, South India. Ann Geol Penins Balk 65:1–27
Ramkumar M, Subramanian V, Stüben D (2005) Deltaic sedimentation during cretaceous period in the Northern Cauvery basin, South India: facies architecture, depositional history and sequence stratigraphy. J Geol Soc Ind 66:81–94
Ramkumar M, Sugantha T, Rai J (2012) Facies and textural characteristics of the Kallamedu Formation, Ariyalur Group, Cauvery basin, South India: implications on cretaceous-tertiary boundary (KTB) events. In: Ramkumar M (ed) On the sustenance of Earth's resources. Springer, Heidelberg (in press)
Rao KN, Vaidyanathan R (1979) Evolution of the coastal landforms in the Krishna delta front, India. Trans Inst India Geogr 1:25–32
Reineck HE, Singh IB (1982) Depositional sedimentary environments, 2nd edn. Springer, Berlin
Rogers P, Lydon P, Seckler D (1989) Eastern waters study: strategies to manage flood and draught in the Ganga-Brahmaputra basin. ISPAN, USAID, Washington
Sacramento River Flood Control Project (2012) Flood control and geomorphic conditions. Report accessed on 07 July 2012, 112 pp
Sandecki M (1989) Aggregate mining in River systems. Calif Geol 42:88–94
Sastry JS, Vethamony P, Swamy GN (1991) Morphological changes at Godavari delta region due to waves, currents and associated physical processes. In: Vaidyanathan R (ed) Quaternary deltas of India. Mem Geol Soc India 22:139–151
Scott KM (1973) Scour and fill in Tujunga Wash—a fan head valley in urban southern California —1969. US Geol Surv Prof Pap 732-B
Selvakumar R, Venkataraman R, Sundaravaradarajan KR (2008) Effect of sand mining on economic performance of groundwater irrigation in Cuddalore district of Tamil Nadu. Agric Econ Res Rev 21:183–190
Sensarma S, Rajamani V, Tripathi JK (2008) Petrography and geochemical characteristics of the sediments of the small River Hemavati, Southern India: implications for provenance and weathering processes. Sediment Geol 205:111–125
Sharma A, Rajamani V (2000a) Weathering of gneissic rocks in the upper reaches of the Cauvery River, south India: implications to neotectonic of the region. Chem Geol 166:203–223
Sharma A, Rajamani V (2000b) Major element, REE and other trace element behaviour in amphibolite weathering under semi-arid conditions in Southern India. J Geol 108:487–496
Sreebha S, Padmalal D (2011) Environmental impact assessment of sand mining from the small catchment river in the Southwestern Coast of India. Environ Manage 47:130–140
Singer MB, Aalto R, James LA (2008) Status of the lower Sacramento valley flood-control system within the context of its natural geomorphic setting. Nat Hazard Rev 9:104–115
Singh P, Rajamani V (2001a) REE geochemistry of recent clastic sediments from the Kaveri floodplains, southern India: implication to source area weathering and sedimentary processes. Geochim Cosmochim Acta 65:3093–3108
Singh P, Rajamani V (2001b) Geochemistry of the floodplain sediments of the Kaveri River, southern India. J Sediment Res 71:50–60
Singh IB, Swamy ASR (1996) Modern deltas. Andhra University, Visakhapatnam
Sinha R (1996) Channel avulsion and floodplain structure in the Gandak-Kosiinterfan, north Bihar plains, India. Zeit Geomorphol NF Suppl-Bd 103:249–268

Sinha R (1998) On the controls of fluvial hazards in the north Bihar Plains, eastern India. In: Maund JG, Eddleston M (ed) Geohazard Eng Geol 15:35–40

Shroba RR, Schmidt PW, Crosby EJ, Hansen WR (1979) Geologic and geomorphic effects in the Big Thompson Canyon area, Larimer County: Part B in the Storm and flood of July 31 Aug 1, 1976 in the Big Thompson River and Cache La Poudre River Basins, Larimer and Weld counties, Colorado. US Geological survey (GS Professional Paper 1115), Washington

Shu L, Finlayson B (1993) Flood management on the lower Yellow River: hydrological and geomorphological perspectives. Sediment Geol 85:285–296

Sonak S, Pangam P, Sonak M, Mayekar D (2006) Impact of sand mining on local ecology. In: Sonak S (ed) Multiple dimensions of global environmental change. Teri Press, New Delhi, pp 101–121

Stevens MA, Urbonas B, Tucker LS (1990) Public- private cooperation protects river. APWA Reporter, 25–27 Sept 1990

Stewart HB Jr (1958) Sedimentary reflections on depositional environment in San Migne lagoon, Baja California, Mexico. Am Assoc Petrol Geol Bull 42:2567–2618

Swamy ASR, Rao MP, Rao BK (1990) Sediment characteristics of the modern deltas of the East coast of India. In: Proceedings of the international conference Asian marine geology, Beijing, pp 251–265

Swank WT, Vose JM, Elliott KJ (2001) Long-term hydrological and water quality responses following commercial clear cutting of mixed hardwoods on a southern Appalachian catchment. Forest Ecol Manage 143:163–178

Thirukumaran V, Ramkumar M (2009) Quicksand: a lesser-known geohazard but not a lesser-evil. In: Ramkumar M (ed) Geological hazards: causes, consequences and methods of containment. New India Publishers, New Delhi, pp 219–223

Thorleifson H, Brooks G, Hanuta I, Kroker S, Matile G, Nielsen E, Prévost C, Rannie W (1998) Red River flooding: evolutionary geomorphic trends and evidence for major floods in recent centuries (NTS 62H/W); in manitoba energy and mines. Geol Serv Rep Activities 1998:186–195

US Army Corps of Engineers (2012) A geomorphic enhancement for flood control. Accessed 07 July 2012

Venkatesharaju K, Ravikumar P, Somashekar RK, Prakash KL (2010) Physico-chemical and bacteriological investigation on the River Cauvery of Kollegal stretch in Karnataka. Kathmandu University J Sci Eng Tech 6:50–59

Venkatesha Raju K, Somashekar RK, Prakash KL (2012) Heavy metal status of sediment in river Cauvery, Karnataka. Environ Monitor Assess 184:361–373

Vignesh S, Muthukumar K, SanthoshGokul M, Arthur James R (2012) Microbial pollution indicators in Cauvery River, Southern India. In: Ramkumar M (ed) On the sustenance of earth's resources. Springer, Heidelberg (in press)

Williams PB, Swanson ML (1989) A new approach to flood protection design and riparian management. USDA For Serv Gen Tech Rep PSW-110:40–46

Zimmermann B, Elsenbeer H, de Moraes JM (2006) The influence of land-use changes on soil hydraulic properties: implications for runoff regeneration. Forest Ecol Manage 222:29–38

Hydro-Geomorphology and Hydrogeology of the Pennar River Basin, India: Implications on Basin Scale Surface and Ground Water Resource Management

M. Sambasiva Rao and G. Rambabu

Abstract The Pennar river basin covering an area about 58,479 km^2 has been studied to delineate the landforms, structural features, and hydro-geomorphic units, ground water level variations, fluctuations and recharge. The landforms are classified into denudational, fluvio-denudational, fluvial, aeolian and coastal categories. Hydro-geomorphologically, the deltaic plains and fluvial plains are grouped under excellent ground water potential, irrigated plains other than fluvial and deltaic plains as good ground water potential, wash plains, valley fills, piedmont plains and creep built plains as fair ground water potential, shallow and moderately weathered pediment plains as poor ground water potential and slope zones of hilly terrain as run-off zones. The Pennar Basin consists of Archean unclassified granitic terrain in the western and southern parts of the basin followed by Proterozoic formations consisting of shales, quartzites and limestone in the central parts of the basin and Quaternary and Holocene sediments in the eastern side. Hydro-geologically the ground water potential is high in the weathered, fissured, fractured and faulted zones of the basin. The ground water potential in the Archean granitic terrain is very poor. The ground water potential in Quaternary and Recent sedimentary formations is good. Thus, the study of land forms in terms of their causative agents, associated lithological and hydrogeological characteristics and categorizing them into hydro-geomorphic units helps to evaluate and manage water resources efficiently.

Keywords Hydro-geomorphology · Water resource management · Pennnar delta · Remote sensing

---

M. Sambasiva Rao (✉) · G. Rambabu
Department of Geography, Sri Krishna Devaraya University, Anantapuram 515003, India
e-mail: msambasivarao2006@gmail.com

Mu. Ramkumar et al. (eds.), *Environmental Management of River Basin Ecosystems*, Springer Earth System Sciences, DOI 10.1007/978-3-319-13425-3_15

1 Introduction

Hydro-geomorphology is the study of different landforms categorized based on ground water level variations, fluctuations, recharge, porosity and specific yield. Hydro-geology is the study of hydro-geological characteristics of different geological formations based on ground water conditions, status, transmissibility, permeability and ground water yield in liters per hour. In this paper, an attempt has been made to integrate hydrogeomorphology and hydrogeology to evolve an integrated surface and ground water management scheme, following the procedures presented in Sambasiva Rao et al. (1999).

2 Study Area

The Pennar River basin covering an area of about 58,479 km^2 lies forms parts of the Karnataka and Andhra Pradesh states of India. The basin is located in between 10° and 16° North latitudes and 77° and 81° East longitudes. The river originates in northern part of Karnataka state in the Nandi hills and passes through the Anantapur and Kadapa districts of the Rayalaseema region and empties into the Bay of Bengal at 30 km ENE of Nellore. The average annual rainfall of the Pennar basin is 724 mm. The average annual potential evaporation is 1,741 mm and actual evapotranspiration is 916 mm. The average annual water deficit is 825 mm. The average moisture adequacy index is 53 % and average annual Aridity Index is 47 %. Climatologically the basin enjoys dry sub-humid climate because the average moisture index value is −28.2 %.

3 Materials and Methods

The main objectives of the study are: to study the geology of the Pennar River basin, to map and describe the landforms of the Pennar basin and to describe the hydro-geomorphic units and hydro-geological characteristics of the basin.

To this effect, the IR IB Geocoded data, LISS IV data and Survey of India topographic sheets (1:50,000 scale) were used as database to study the landforms and geology of the Pennar River basin. The annual ground water level variations and fluctuations collected from about 250 controlled wells were computed based on monthly ground water level data. The ground water recharge was computed by adopting the methods described by Seghal (1970), Krishna Rao (1970) and Radhakrishna et al. (1974). The specific yield, porosity, permeability and ground water yield were calculated for different geological formations by carrying out tests along with the State Ground Water Department, Andhra Pradesh and the details are presented in Sambasiva Rao et al. (1999).

4 Geology of the Pennar Basin

Geologically the Pennar Basin is divided into three distinct geological formations consisting of Archaeans, Proterozoic, Quaternary and Holocene sedimentary successions. The western, southern, northwestern and southwestern parts of the basin consist of Archean rocks and Archean metamorphic rocks. The Archean rocks are represented by the gnessies, schists and granites. The Archean metamorphic rocks are represented by amphibolites, peninsular chlorites, schists, biotite schists, actinolite schists, hornblende, magnetite and quartzites. The gnessies are intruded by younger granites equivalent to Closepet granites. The schists are highly folded. The older metamorphic rocks like amphibolites, quartz, mica schists and banded ferruginous quartzites occur as enclaves of different shapes and sizes. Swarms of dolerite, gabbro, quartz veins and reefs are intruded into the rocks of migmatites group. Amphibolites, hornblende, Quartz, mica schist and ferruginous quartzites occur as enclaves and also detached banded within migmatites. The schist formations are found in a linear bounded ridge from south of Kolar to north Kuderu of the Pennar basin. They are rich in gold reserves. The Archeans in Vajrakarur and Lattavaram are intruded by Kimberlite dykes and contain diamonds.

The rocks of Proterozoic formations are divided into Cuddapah Super Group and Kurnool Super Group. The Cuddapah Super Group belongs to the Middle Proterozoic period and occupies the Nallamalai hills that extend roughly for about 100 km length and 50 km width. They consist of dolomites, quartzites, conglomerates, sandstone and grit, shales, limestone and chalcedony silica. The Cuddapah Super Group has been divided into four groups namely, the Krishna, Nallamalai, Chitravathi and Papagni.

The Kurnool Super Group is mainly composed of limestone and calcareous shales. They belong to Lower Proterozoic period ranging from 600 to 1,100 million years ago. They are sub-divided into Banaganapalli formation consisting of quartzites, sandstone and conglomerates and Jammalamadugu formation overlying Banaganapalli formation consisting of Najri limestone and Owk shales. The Panyam Group of rocks comprises plateau quartzites at the bottom and pinnacle quartzites at the top. The Kunderu formation is the youngest group of rocks of Kurnool Super Group. It comprises the Koilakuntla shale beds at the bottom and Nandyal shales at the top comprising the grey, massive, flaggy limestone, and calcareous and scaly shales. The Quaternary and Recent formations are found in the form of alluvium along the river valley and in the Pennar delta of the basin. The apex of the Pennar delta is bordered by Plio-miocene sandstone lateritic uplands. The coastal sands are found along the present day coastline (Fig. 1).

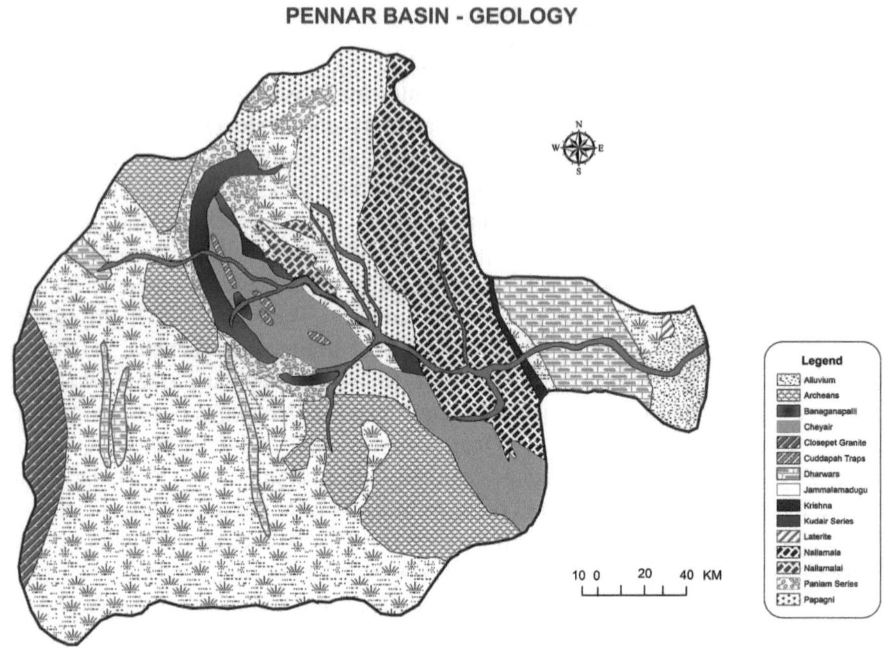

Fig. 1 Geology of the Pennar River basin

5 Landforms of the Pennar Basin

The landforms of the Pennar basin have been categorized into denudational, fluvio-denudational, fluvial, aeolian, deltaic and coastal. The denudational landforms are structural hills, ridges, escarpments, cuesta hills, mesas, buttes, structural valleys, pedimont-inselberg complex, residual hills and shallow, moderately and deeply weathered pediplains. The fluvio-denudational landforms are valley fills, bajadas, creep built plains and wash plains. The fluvial landforms consist of fluvial plains, abandoned channels, natural levees, back swamps and fluvial terraces. The deltaic landforms consist of deltaic plains, abandoned channels, natural levees and flood plains. The aeolian landforms consist of migrated sand dunes from the dry river bed of the Pennar River in some pockets and sand dunes along the coastline of the Pennar delta. The coastal landforms include the coastal beach ridges, abandoned lagoons, marshy areas, lagoons and sand bars at the confluence of the Pennar River. The distribution of the landforms has been presented in the Fig. 2.

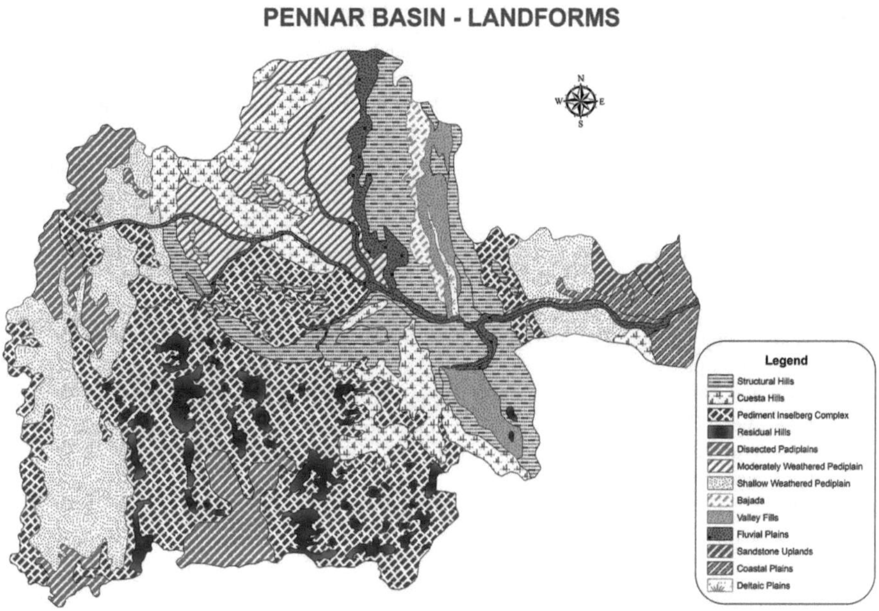

Fig. 2 Map showing the land forms and their distribution as deduced from remote sensing data and limited field verification

6 Hydro-Geomorphology of the Pennar Basin

The hydro-geomorphology of the Pennar basin has been brought out based on ground water level variations, fluctuations, specific yield, recharge and ground water potential of the individual landforms. In the fluvial and deltaic plains the annual ground water levels vary from 3 to 5 m. The porosity is high. The annual ground water fluctuations range from 0.5 to 1.0 m. The average annual specific yield is about 10 % (72.40 mm/annum). The ground water recharge is very high due to high porosity and high specific yield. The ground water potential in fluvial and deltaic plains is excellent. In the irrigated areas of other fluvial and deltaic plains, the average annual variations in ground water level ranges from 4 to 5 m. The average annual fluctuation in ground water ranges from 0.75 to 1.50 m, due to the return flow from the irrigated water. The specific yield is about 8 % (52.92 mm/ annum). The ground water recharge is good and the ground water potential is good. In the wash plains, creep built plains, valley fills, bajadas and piedmont plains the annual ground water level variations range from 4.5 to 6.5 m. The annual ground water level fluctuations range from 1.0 to 1.50 m and the average specific yield is about 5 % (36.20 mm/annum). The ground water recharge is fair and the ground water potential is fair. In the shallow moderately and deeply weathered pediplains, the annual ground water level variations range from 5 to 8 m and annual ground water fluctuations vary from 1.5 to 2.0 m. The average annual specific yield is 3 %

Table 1 Hydro-geomorphology of the Pennar River basin

Landforms	Annual ground water level variations in m	Annual ground water level fluctuation in m	Specific yield and recharge	Hydro-geomorphic unit
Fluvial plains and deltaic plains	3–5	0.5–1.0	10 % very high (72.40 mm)	Excellent
Irrigated area other than fluvial, deltaic and coastal plains	4–6	0.75–1.50	85 % Good (52.92 mm)	Good
Wash plains, creep built plains, valley fills piedmont plains and bajadas	4.5–6.5	1.0–1.50	5 % Fair (36.20 mm)	Fair
Shallow weathered, moderately weathered and deeply weathered pediplains	5–8	1.5–2.0	3 % Poor (21.72 mm)	Poor
Slope zone of hilly terrain	–	–	–	Run-off zone

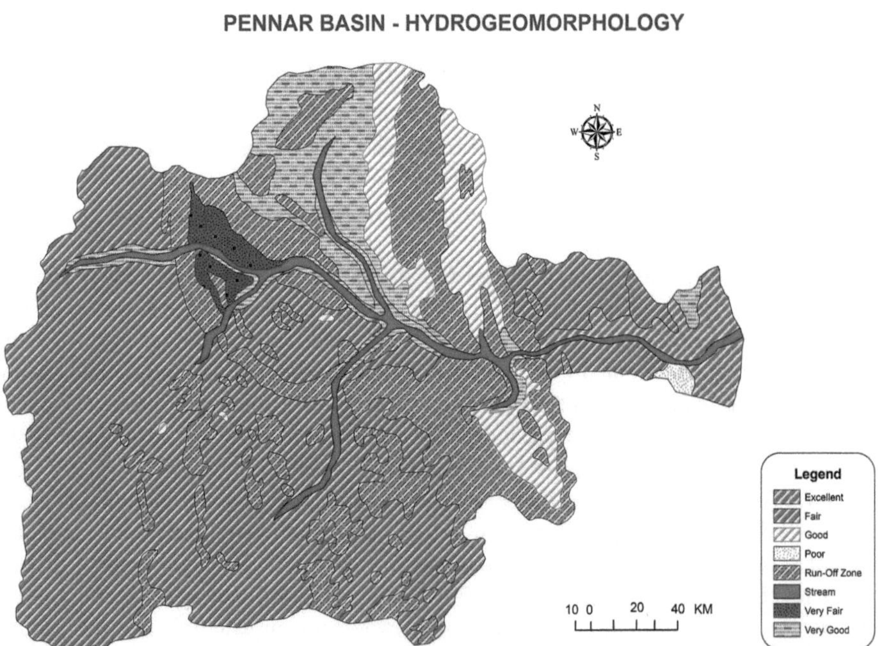

Fig. 3 Hydro-geomorphology of the Pennar River basin based on landforms, hydrogeology and other hydrological factors

(21.72 mm/annum). The ground water recharge is poor and is restricted to weathered, fissured, faulted and fractured zones. The ground water potential is poor. The slope zones of the hilly terrain act as run-off zones after soil moisture saturation (Table 1; Fig. 3).

7 Hydro-Geology of the Pennar Basin

Hydrologically the Pennar basin is divided into four major geological categories. They are: (a) Recent or sub-recent alluvial formations (b) Late Proterozoic formations consisting of Kurnool Super Group (c) Middle Proterozoic formations consisting of Cuddapah Super Group and (d) Archeans. From the study of hydrogeological characteristics, it is found that the ground water yield ranges from 25,000 to 1,00,000 Lph in Recent and Sub-recent alluvium and it is highest among all the studied geological formations. The transmissibility ranges from 1,200 to 1,750 m^3/d/m and permeability varies from 5 to 50 m/day. Ground water is found in the semi-confined aquifers in fluvial and deltaic plains (Table 2).

The ground water yield varies from 3,000 to 8,000 lph in the Late Proterozoic formations. The transmissibility ranges from 50 to 270 m^3/d/m and the permeability varies from 0.05 to 15 m/day. The transmissibility varies from 120 to 380 m^3/d/m and permeability ranges from 0.5 to 5 m/day in the Middle Proterozoic formations. The Archeans consist of young intrusive rocks, Dharwar Super Group, and Peninsular gneisses complex and old metamorphic rocks. Though the ground water is found only in the weathered, fissured, jointed, faulted and fractured zones of these rocks, there are differences in the transmissibility, permeability and ground water yield among different rock types. The young intrusive rocks consist of quartzites, basic dykes, dolerites, pegmatite, quartz and fine granites. The transmissibility varies from 35 to 280 m^3/d/m. The permeability ranges from 0.1 to 5 m/day. The ground water yield varies from 4,500 to 10,000 lph. The Dharwar Super Group consists of chlorites, hornblende, schists and granites. The transmissibility varies from 65 to 410 m^3/d/m. The permeability ranges from 0.1 to 10 m/day. The ground water yield varies from 4,500 to 20,000 lph. In the Peninsular gneiss transmissibility varies from 75 to 335 m^3/d/m and the permeability ranges from 0.5 to 20 m/day. The ground water yield varies from 6,000 to 30,000 lph. The transmissibility in the metamorphic rocks varies from 35 to 300 m^3/d/m. The permeability ranges from 0.10 to 10 m/day and the ground water yield varies from 5,000 to 12,000 lph (Table 2).

8 Surface and Ground Water Resources Management

The average annual rainfall of the Pennar basin is 724 mm. The total surface water resources are estimated to be 42,338,796,000 m^3. Out of this, 10 % is stored in about 12,000 surface tanks. About 12.57 % of surface water resources are recharged to

Table 2 Hydrogeology of the Pennar River basin

Geological formation	Ground water conditions	Transmissibility in $m^3/d/m$	Permeability m/day	Yield in lph
Recent and sub-recent	Ground water is found in semi-confined and confined aquifers in the fluvial and deltaic plains	1,200–1,750	5–50	25,000–1,00,000
Proterozoic formations Kurnool Group, quartzites (600–1,100 million years ago) shales, limestone and conglomerate	Ground water is found in weathered, jointed, caverns, bedding planes and fractures	50–270	0.05–15	3,000–8,000
Cuddapah Super Group (1,100–1,400 million years ago) quartzites, shales with basic intrusive like sills and dykes	Ground water is found in weathered, jointed, fractures faulted and fissured zones	120–380	0.5–5	12,500–45,000
Archeans young intrusive rocks (quartzites, basic dykes, dolerites, pegmatite quartz and fine granites)	Ground water is found in weathered, jointed, fractures faulted and fissured zones	35–280	0.1–5	4,500–10,000
Dharwar Super Group (chlorites, hornblende, schists and granites	Ground water is found in weathered, jointed, fractured, faulted and fissured zones	65–410	0.1–10	4,500–20,000
Peninsular gneisses complex	Ground water is found in weathered, jointed, fractured, faulted and fissured zones	75–335	0.5–20	6,000–30,000
Old metamorphic rocks	Ground water is found in weathered, jointed, fractured, fissured and faulted zones	35–300	0.1–10	5,000–12,000

ground water in different geological formations which amounts to 5,321,986,657 m^3. About 20 % of the surface water resources are lost in the form of surface run-off which amounts to 8,467,759,200 m^3. The remaining 57.43 % surface water resources are lost in the form of evaporation and evapotranspiration which amounts to 24,315,170,542 m^3. From the analysis of annual water balance estimates, it is found that the average annual potential evapotranspiration is 1,741 mm and average annual actual evapotranspiration is 916 mm. The annual water deficit is 825 mm. In view of high annual water deficit, the available water resources have to be carefully utilized for sustainable development of land use and cropping pattern. Majority of the surface tanks are silted and have reduced the storage capacity. Therefore, the tanks have to be desilted and the tank bunds have to be strengthened to store more surface water resources. It is proposed that for every 1,000 ha of land there should be one tank to store surface water resources. About 5,848 surface tanks have to be dug in the Pennar basin but only 1,200 tanks are present. About 4,648 new tanks have to be dug to store the surface water resources. The central and eastern parts of the basin receive high rainfall during northeast monsoon period due to cyclonic activity in the Bay of Bengal. However, the western part of the basin is subjected to frequent droughts and the annual rainfall is less than 600 mm. The Nallamalai, Palakonda, Seshachalam, Velikonda hills receive good annual rainfall (>800 mm). The run-off water from the slope zones during monsoon seasons have to carefully stored and used for raising crops during Kharif and Rabi seasons. As the Pennar Basin is located in a catchment area not conducive for adequate supply of runoff to the main channel, the excess flood water from Srisailam dam of the Krishna River may be diverted to the Pennar basin to fill the existing surface tanks. Interlinking of rivers may help to transport excess flood water from the Godavari and Krishna basins to Pennar Basin to meet the water domestic and irrigation needs of the Pennar Basin. Micro watershed development programs have to be implemented for a sustainable development of land and water resources of the Pennar Basin. Detailed geophysical studies have to be carried out at micro watershed level to delineate the weathered, fissured, faulted and fractured zones for excavation of ground water resources.

9 Conclusions

From the study of hydro-geomorphology of the Pennar Basin, it is concluded that fluvial and deltaic plains possess excellent ground water resources and the weathered pediplains possess poor ground water resources. Hydrogeologically the Recent and Sub-recent alluvial formations yield very high ground water resources and in old metamorphic rocks the ground water yield is very low. The annual water balance elements revealed that the Pennar Basin is located in a bad catchment area and the average annual water deficit is about 825 mm. It is concluded that about 4,648 new tanks have to be dug to store surface water resources and the existing tanks have to be desilted and tank bunds have to be strengthened to store more surface water resources.

References

Krishna Rao PR (1970) Ground water potential in hard rock areas of India. Publication of Govt of Karnataka, Bangalore

Radhakrishna BP, Duba D, Palimquist WN (1974) Ground water studies. Publication No.150, Government of Karnataka, Bangalore

Sambasiva Rao M, Narendrakumar, J, Gangadri G (1999) Water balance and water resources development of Rayalaseema region. In: Paper published in proceedings of the National seminar on water resources management for sustainable development. Water technology Centre, Department of Civil Engineering S.V. University, Tirupathi, 12–13 March 1999

Seghal SR (1970) Ground water studies. International hydrological decade News letter, India

Geochemical Perspectives on River Water of the Tropical Basins, Southwestern India

G.P. Gurumurthy and M. Tripti

Abstract The Southwestern part of Peninsular India is one among the shield terrains experiencing extreme geological, geomorphological and climatic gradients. Many small rivers originate in the western slope of Western Ghats and flow towards the Arabian Sea. The moisture from the Arabian Sea carried by the southwesterly winds forms the primary source of water for these rivers. These west flowing rivers exhibit characteristic water chemistry distinct from that of the east flowing rivers and the rivers draining the Deccan traps. The lithological heterogeneity and the intensity of weathering have significant effect on the river water as observed by the difference in water chemistry between the Deccan, East flowing and West flowing rivers. The geomorphological settings of the Western Ghats (which brings-in the rainfall over its western slope) induce higher surface runoff which has significant effect on the weathering process and thus, on the water chemistry. However, the slope of the terrain has minimal effect on the chemistry of water in this region. The weathering of sedimentary formations in the plains leads to elevated fluxes of silica and radiogenic strontium. The weathering of bedrock forms the main source of trace elements to the river water in this region. The abundance of trace elements in monsoon dominated terrain is controlled primarily by the discharge. However, the secondary processes namely, redox reactions and oxidative scavenging of surface reactive metals by the oxyhydroxides of Fe and Mn appears to have significant role in determining the geochemical abundance. Unlike the tropical river basins of Africa and South America where the organic carbon complexation plays a vital role in dissolved metal abundance, in the rivers of Southwestern India, the control of organic complexation on the dissolved metal chemistry varies widely between rivers. This could be due to variable abundance of labile fraction of organic carbon in these rivers. The secondary geochemical processes lead to the enrichment

G.P. Gurumurthy (✉)
Manipal Centre for Natural Sciences, Manipal University, Manipal 576104, India
e-mail: gurumurthy.gp@manipal.edu

M. Tripti
Department of Civil Engineering, Manipal Institute of Technology, Manipal University, Manipal 576104, India
e-mail: tripti.m@manipal.edu

Mu. Ramkumar et al. (eds.), *Environmental Management of River Basin Ecosystems*, Springer Earth System Sciences, DOI 10.1007/978-3-319-13425-3_16

of metals in sedimentary phase. However, the enrichments of these metals are within the permissible limits. From the pollution point of view, the metal contaminant studies in these river basins need to consider the rigorous fluvial geochemical redistribution of metals between particulate and dissolved phases to minimize the erroneous attribution of metal sources. The metal isotope tracing of pollution sources could be useful for accurate determination of pollution sources.

Keywords Geochemistry · Weathering · Trace elements · Tropical rivers · West flowing rivers · Peninsular india

1 Introduction

Majority of the large rivers are in the tropical regions of the world. These tropical rivers drain diversified geological and geomorphological settings which include orogenically active high mountains, cratonic shield terrains, low lands of sedimentary formations, basaltic/sedimentary formations and heterogeneous terrains (Latrubesse et al. 2005). Most of these rivers have characteristic hydrographs with peaks during rainy seasons. These tropical rivers contribute about 75 % of the global river discharge annually and about 70 % of the global sediments to the world oceans (Milliman and Meade 1983). In addition, the tropical rivers are the major sources of water for 67 % of the total human population of the earth. An increasing demand of water for domestic and industrial purposes, particularly in the rapidly developing countries like India, necessitates geochemical monitoring of water resources on a regional to global scale.

Chemical elements cycle between the various reservoirs such as lithosphere, hydrosphere and atmosphere. The elemental transfer is a continuous process and can happen at global, regional or local (between soil and soil horizons) scales. An understanding on the interactions between different compartments anticipates a thorough study on the role of continental erosion and modes of transport of chemical elements and their ultimate fate. During the continental weathering, elements may remain in situ in the form of newly formed minerals and can be an important part of the weathered soil together with primary minerals. Some elements are transferred to different reservoirs by the activity of geological agents. The fluvial activity is the most important geological agent which transfers about 90 % of the elements from the continents to the oceans (Gaillardet et al. 2003). The fluvial transfer could be in dissolved (<0.22 μm) or particulate (>0.22 μm) form. However, on a global scale, the particulate form dominates the fluvial transfer process (Viers et al. 2009), and significant part of this elemental transport is from the tropical rivers.

The weathering of rocks in the catchment supplies a major part of the chemical composition of the river water. The chemical elements carried through rainwater, both geogenic and anthropogenic, are another source of river water composition. In addition, a plethora of human activities including, but not limited to, domestic and industrial wastes, contribute to the chemical composition of river water. The chemical elements in the fluvial environment are subjected to various hydrological and geochemical processes during their journey from source to destination which determines their bioavailability, toxicity and ultimate fate. Thus, the geochemical and geobiological processes namely, organic complexation, redox processes, precipitation—co-precipitation and adsorption—desorption reactions are vital in the self-sustenance of the water bodies. These biogeochemical processes either remove or redistribute the elements between dissolved and particulate phases, from or within the aquatic systems. To understand this, a thorough study on the hydrological, geochemical and biological processes is essential. The biogeochemical perspectives on the river water are important in the monitoring and management of water resources.

In the Indian subcontinent, numerous geochemical studies pertaining to monitoring of major ion chemistry of rivers with an objective of estimating the contemporary weathering rate on a catchment scale have been carried out in recent years (reviewed in Krishnaswami and Singh 2005). Majority of these geochemical studies are motivated by the findings of Raymo and Ruddiman (1992) which suggests the link between Himalayan tectonics, chemical weathering, associated carbon dioxide consumption and global cooling during Cenozoic. Most of the geochemical studies in the 1990s are focused on the large rivers of the world. Off late, a growing consensus on the importance of small tropical rivers, draining mountainous terrains of the world, in the transport of organic matter and sediments to the Global Ocean (Milliman and Farnsworth 2011, references therein) encourage the scientific community to work on the hydrogeochemistry of small mountainous rivers. These studies either deal with the estimation of contemporary weathering rate using seasonal/monsoonal major ion composition or monitor the status of anthropogenic pollution of rivers. These tropical mountainous rivers drain, in most cases, homogenous lithology which provides a unique opportunity to understand the hydrogeochemistry of rivers and the effect of various geochemical processes on the abundance of chemical elements.

In this study, the geochemistry of small tropical rivers draining the western slope of the Western Ghats is discussed. These mountainous rivers of the Western Ghats drain a variety of lithologies including the Deccan flood basalts in the north, Peninsular Gneiss in the middle and Southern Granulites in the south. Like many tropical large rivers, these rivers have a characteristic peak discharge during the rainy season induced by the summer monsoon. Large tropical rivers and their geochemistry are discussed in detail elsewhere in the literature (Stallard and Edmond 1987; Gaillardet et al. 1995; Shiller 1997; Viers et al. 1997; Sarin et al. 1992; Pande et al. 1994; Singh et al. 1998; Krishnaswami et al. 1999; Viers et al. 2000; Dalai et al. 2002). However, lesser attention has been paid towards the geochemistry of small mountainous rivers, particularly in the Indian subcontinent. These rivers are the lifeline of many cities and industrial clusters located all along the west coast of India. Thus, the present study is aimed at documentation of geochemical traits of smaller mountainous rivers, characterizing their sources and determination of relative contribution of geogenic and anthropogenic processes.

2 Geologic, Geomorphic and Climatic Settings

Major portion of the peninsular India is occupied by the Archean Dharwar Craton which is composed of peninsular gneiss, grey gneisses, greenstone/schist belts, charnockites and younger granites. The Archean Dharwar Craton is further divided into units based on the difference in the lithological composition and age. These include, Eastern Dharwar Craton (EDC) and Western Dharwar Craton (WDC). These cratonic units are separated by Chitradurga Shear Zone located at the eastern margin of the Chitradurga Schist Belt and western margin of the Closepet Granite of age 2513 ± 5 Ma. (Ramakrishnan and Vaidyanadhan 2008; Friend and Nutman 1991). The rock types in the EDC are compositionally distinct from Granitic Gneisses of WDC and are called as 'Dharwar Batholith'. The rock types in EDC also include greenstone belts, intrusive volcanics and middle Proterozoic to more recent sedimentary formations (Ramakrishnan and Vaidyanadhan 2008). The Western Dharwar Craton (WDC) is located on the southwest India and is dominated by Archean tonalitic-trondhjemitic-granodioritic (TTG) gneisses. The regional metamorphism in the WDC increases from green schist to amphibolite facies in the north to granulite facies in the south. In the southern part of Peninsular India, high grade metamorphism of Archean and Neoproterozoic geologic era led to the formation of granulite terrains of about 2.52 Ga age. This region is commonly known as Southern Granulite Terrain. The major lithounits in this region are charnockite gneiss, mafic granulites, migmatite-gneiss- and younger granite. The northern part of Peninsular India is covered by volcanogenic basalts of Deccan formed due to a series of volcanic eruptions occurred $\sim$65 Ma ago.

Most of the mountainous rivers originate on the western slope of the Western Ghats and flow towards the Arabian Sea almost perpendicular to the coast and the Western Ghats. The river basins are divisible into three regions such as low land, midland, and high land (coastal, hinterland and ghats) depending on terrain features and their altitude (Nielsen 1972). These rivers generally flow through steep terrains at their early stages. However, the flow becomes less effective as it reaches the mid land. In the lowland region, the rivers generally take parallel route to the coast, either to the north or south, and hence two or more rivers unite before debauching into the Sea. Thus, most of the west flowing rivers form estuaries by the confluence of two rivers. These west flowing rivers are the major sources of water for domestic, agricultural and industrial purposes. The western coastline of India is a home to many megacities and the population density is more than the eastern coast of India. The combined discharge of west flowing rivers accounts for 12.5 % (200 km^3) (www.nih.ernet.in) of the entire Indian rivers. Furthermore, the average surface runoff of entire west flowing river catchment is 3.5 times higher than the Himalayan and eastern rivers.

The west flowing river basins experience four seasons viz., (a) Four wet months, June to September, with strong winds, high humidity, heavy showers and a slight lesser temperature, (b) Two warm months (October and November) of north-east monsoon with little or no rains, (c) Three relatively cool months (December to February) having dry conditions, and (d) Three hot months (March, April and May) which is the period of rising temperature. However, the climate is quite uniform throughout the coastal stretch while the inland is relatively cooler. The rainfall in the basins is mainly due to convection and orography. The convective precipitation, typical of the tropics, is caused by the heating of air at the interface with the ground and characterized by light showers or storms of extremely high intensity. The convective storms experienced by the southwest coast are usually local in nature and often result in very intense rainfall. Orographic rainfall in the southwestern region is the result of mechanical lifting of moist horizontal air currents over the Western Ghats region. The highland of the southwestern region experiences mainly orographic rainfall. The rainfall in the Ghats section ranges from 2,000 to 7,900 mm with an average more than ~2,700 mm (Francis and Sulochana 2006). The watershed is characterized by high humidity (60–75 %), and the potential evapotranspiration (PET) is estimated to be 1,100–1,400 mm (Rao and Jagannathan 1994) along the coast. The mean temperature is around 30 °C (± 5), with small variations throughout the year (India Meteorological Department, Government of India) (Fig. 1).

Vegetation reflects the environment under which it is grown. The climatic condition, rainfall, soil and water availability in the river basin favor luxurious growth of the vegetation from seashore to the Western Ghats. The basins have thick vegetation and cultivated arecanut and coconut throughout the course of the river

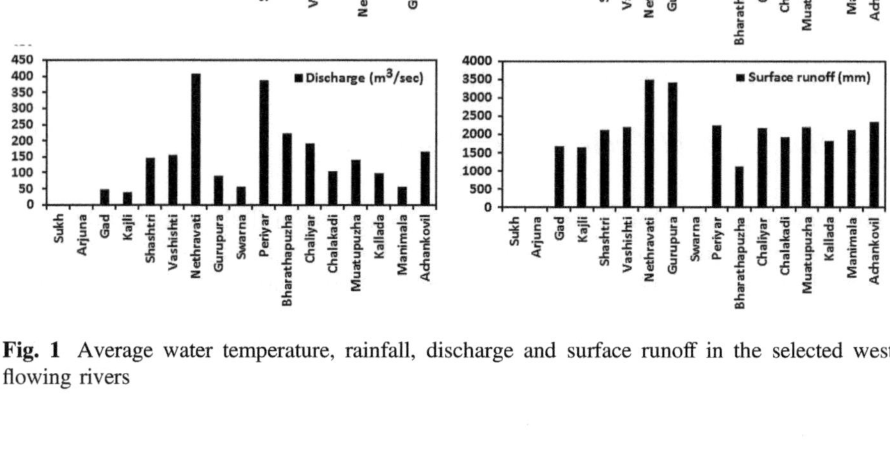

Fig. 1 Average water temperature, rainfall, discharge and surface runoff in the selected west flowing rivers

channels. Wherever there is a lateritic exposure, sparse vegetation can be noticed. Eastern part of the study area is covered with succulent dense forest whereas thick mangroves can be seen along the shoreline.

3 Materials and Methods

The data for this study are collected from different published sources as presented in Table 1. The detailed sampling methodology and analysis of water samples are explained in Khemani et al. (1994), Parashar et al. (1996), Naik et al. (2002), Das et al. (2005), Prasad and Ramanathan (2005), Pattanaik et al. (2007), Hegde (2007), Praveen et al. (2007), Prathibha et al. (2010), Gobre et al. (2010), Sathyanarayana et al. (2010) Padmalal et al. (2011), Gurumurthy et al. (2012, 2014) and Tripti et al. (2013a, b). In brief, the major anion concentrations were measured through Ion-chromatography, or following the American Public Health Association (APHA) standard procedures (as in case of Padmalal et al. 2011). The major cation concentrations were measured through Atomic Absorption Spectroscopy (Prasad and Ramanathan 2005; Pattanaik et al. 2007), Inductively Coupled Plasma-Atomic Emission Spectrometry (Das et al. 2005; Gurumurthy et al. 2012), Ion-chromatography (Satyanarayana et al. 2010) and Flame Photometer (Padmalal et al. 2011). The alkalinity measurements were performed through acid base titration and the silica measurements were done through spectrophotometry. The trace element and dissolved organic carbon (DOC) measurements were performed through Quadrupole Inductively Coupled Plasma Mass Spectrometer (ICP-MS) and Total Organic Carbon (TOC) analyser respectively (Tripti et al. 2013a; Gurumurthy et al. 2014). The stable isotopes of water (δ^2H and δ^{18}O) were analyzed through Isotope Ratio Mass Spectrometry (IRMS; Lambs et al. 2011; Tripti et al. 2013b). The analytical accuracy of the measurements was better than 85 % in all the studies. The average monsoonal composition is chosen for the study because most of the west flowing rivers discharge about 80–93 % of the total annual discharge during monsoonal period. Enrichment of ions by evaporation is assumed to be less during this period.

4 Geochemical Perspectives of River Water

4.1 Hydrological Cycle in the West Coast

The moisture source for the river basins of southwestern India is the southwestern monsoon as evidenced from a multi-year stable isotope temporal monitoring of groundwater, river water and rainwater over the Nethravati and Swarna river basins of southwest coast of India and Kozhikode area of Kerala (Yadava et al. 2007; Gupta et al. 2005; Deshpande et al. 2003; Warrier et al. 2010; Lambs et al. 2011;

Table 1 Major ion composition of west flowing and east flowing rivers of Peninsular India

River	T	pH	Runoff (mm)	Na	K	Ca	Mg	Cl	SO_4	HCO_3	SiO_2	TDS (mg L^{-1})	TZ^+	TZ^-	Reference
				μeq L^{-1}									μeq L^{-1}		
Sukh	26.2	7.0		166	6	321	262	109	12	640	337	77	756	761	Das et al. (2005)
Arjuna	26.3	6.9		181	7	334	262	115	11	633	366	80	784	759	Das et al. (2005)
Gad	24.9	7.0	1,690	179	9	306	228	108	12	604	323	74	722	724	Das et al. (2005)
Kajli	26.3	7.0	1,657	174	5	359	279	107	11	694	375	84	818	813	Das et al. (2005)
Shashtri	25.1	6.9	2,117	169	5	344	253	111	12	651	365	80	772	773	Das et al. (2005)
Vashishti	25.8	7.0	2,198	171	6	370	279	109	13	721	378	86	827	843	Das et al. (2005)
Nethravati	28.5	6.8	3,504	138	18	148	106	91	23	309	195	43	410	423	Gurumurthy et al. (2012)
Gurupura	27.9	6.9	3,425	143	22	136	90	96	21	350	135	42	392	467	Gurumurthy, (2014)
Kumaradhara river	28.3	6.9	2,594	124	30	146	116	87	22	318	188	43	416	428	Gurumurthy et al. (2012)
Nethravati @Uppinangadi	28.4	6.8	2,734	135	35	155	104	87	23	313	196	44	429	422	Gurumurthy et al. (2012)
Periyar			2,262	170	10	165	197	162	79	257	107	42	542	499	Prasad and Ramanathan, (2005)
Bharathapuzha			1,140	98	12	114	91	276	87	608	188	69	314	972	
Chaliyar			2,181	98	12	114	91	132	54	260	152	38	314	445	
Chalakadi			1,923	92	10	73	148	141	71	99	135	28	324	311	
Muatupuzha			2,200	168	13	119	156	103	204	111	118	36	456	418	
Kallada			1,820	285	32	116	181	389	85	158	182	51	614	632	
Achankovil		6.37–7.56	2,340	142	25	134	99	649	12	220	114	52	400	881	

(continued)

Table 1 (continued)

River	T	pH	Runoff (mm)	Na	K	Ca	Mg	Cl	SO_4	HCO_3	SiO_2	TDS (mg L^{-1})	TZ^+	TZ^-	Reference
				µeq L^{-1}									µeq L^{-1}		
Manimala		6.26	2,119	102	18	140	60	261	50	110	175	35	320	421	Padmalal et al. (2011)
Kaveri		7.8	259	1,471	69	2,008	1,270	1,393	688	2,954	553	388	4,818	5,035	Pattanaik et al. (2007)
Vellar		7.9		996	41	1,480	846	789	400	2,984	778	340	3,363	4,173	
Ponnaiyar		7.2		1,624	101	1,972	1,124	1,333	566	3,039	697	396	4,821	4,938	
Polar		7.5		490	28	722	360	379	418	1,258	868	194	1,600	2,055	

Tripti et al. 2013b). The stable isotope ratios of river water and groundwater in the southwestern India are similar to the isotopic ratios of southwest monsoons. It suggests that the moisture from the Arabian Sea was carried by the southwesterly winds to these river basins. However, the source isotopic signal could be altered due to the geomorphological settings of the river basins in the western part of the Western Ghats. The western part reels under warm climate and contains dense vegetation which could have supported higher evaporation and evapo-transpiration in the region. The isotopic composition and deuterium excess (d-excess) of water in the basins of Nethravati-Gurupur and Swarna-Madisal rivers suggests the dominance of evapo-transpiration over evaporation on the surface and sub-surface waters (Lambs et al. 2011; Tripti et al. 2013b). The foothills of the Western Ghats exhibits higher water vapour recycling which supports orographic rainfall in the river basins of southwestern India. In addition, the local water vapour recycling occurs in the river basins due to the widely spread thick vegetation and cultivated plantations throughout the western part of the Western Ghats.

4.2 Rainwater Chemistry Along the West Coast

The rainwater chemistry of major cities along the west coast, viz. Mumbai, Goa, Uttar Kannada and Southern Karnataka has been discussed in previous publications (Praveen et al. 2007; Prathibha et al. 2010; Naik et al. 2002; Khemani et al. 1994; Gobre et al. 2010; Parashar et al. 1996; Satyanarayana et al. 2010; Hegde 2007; Gurumurthy et al. 2012). The rainwater in the west coast of India is rich in major ions, which is consistent with the close proximity to the Arabian Sea. The concentrations of major ions in rain water of west coast range from 32 to 274 $\mu mol\ L^{-1}$ for Cl, 9–210 $\mu mol\ L^{-1}$ for SO_4, 1–66 $\mu mol\ L^{-1}$ for NO_3, 21–220 $\mu mol\ L^{-1}$ for Na, 2.5–29 $\mu mol\ L^{-1}$ for K, 17–165 $\mu mol\ L^{-1}$ for Ca and 5–64 $\mu mol\ L^{-1}$ for Mg. The higher concentrations of major ions were observed in the northern extreme of the west coast. The rain water composition shows slight enrichment of Na, Ca, Mg and SO_4 with respect to sea water which could be interpreted as presence of non-sea salt constituents/soil dust particles (Parashar et al. 2001) derived from North African and Gulf region carried through oceanic circulation (Praveen et al. 2007). However, the Na/Cl ratio (0.86) of rainwater at Arabian Sea is similar to the molar seawater ratio (0.86) whereas the Na/Cl ratio (1.42) in continental rainwater shows enrichment of Na indicating addition of Na to the rains upon arrival at the coast (Fig. 2). Therefore, the source of Na in west coast region could be local in origin while the sources of Ca and Mg in rainwater could be the dust particles carried from the Gulf and North African regions. The samples of Mumbai, Uttar Kannada and coastal regions of Southern Karnataka show lesser molar ratio, which could be explained by the influx Cl ions from industrial activity. The southwest monsoonal winds carry anthropogenically derived acidifying substances to the southwest coast causing acid rains/acidification. The concentrations of acidifying substances are observed to be significant in the rainwater of southwest coast (9–210 $\mu mol\ L^{-1}$ for SO_4 and 1–66 $\mu mol\ L^{-1}$ for NO_3).

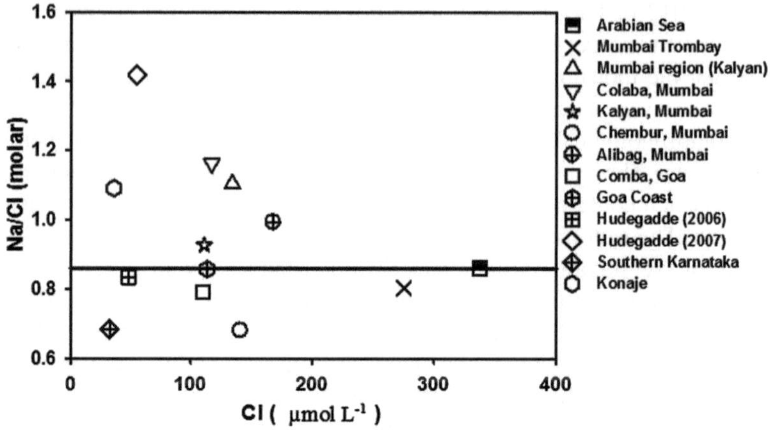

Fig. 2 Relationship between Na/Cl (molar) ratios with the chloride concentration of rainwater. The *thick line* represents the Na/Cl molar ratio in typical Seawater

The calculated pH based on the concentrations of these species varies from pH 3.3–4.84. However, the measured pH (4.8–7.1) in rainwater is higher than typical rainwater pH (5.7; Eby 2004). This means that the acidity of the rainwater is reduced by some other atmospheric process. The plausible process that can explain this reduction of acidity is dissolution of calcite dust and neutralization by abundant base cations.

4.3 Major Ion Composition of River Water of West Coast of India

The average major ion composition and physicochemical parameters of few selected rivers of the Western Ghats have been tabulated in the Table 1. The data compiled from different published sources namely, Das et al. (2005), Padmalal et al. (2011), Prasad and Ramanathan (2005), Pattanaik et al. (2007), Gurumurthy et al. (2012, 2014), Tripti et al. (2013a, 2013c) and Gurumurthy et al. (under review) are presented herein.

In general, the pH of the west flowing rivers are mildly acidic whereas the east flowing rivers are slightly alkaline. The acidic nature of the river water reflects the acidic igneous rocks over which these rivers drain. It is observed that the rivers draining basaltic rocks in the northern part of the southwest coast have slightly higher TDS compared to the rivers draining the regions predominated by granites and gneisses (located at southern part of southwest coast). Also, it is observed that the rivers flowing westerly have lesser TDS compared to the easterly flowing rivers of Karnataka and Tamil Nadu. It could be due to hydrological draining characteristics of the river i.e., dilution of concentration of major ions due to higher surface runoff. Plotting the data of east and west flowing rivers in the Gibbs plot (TDS mg L^{-1} vs Na/(Na + Ca) wt. ratio) shows that while the west flowing river water samples fall under the precipitation dominated portion, the samples of east flowing rivers fall under weathering dominated portion. This observation suggests that though waters in both of these river systems reflect lithological influence, existences of different processes—precipitation and runoff in the west and weathering in the east are also reflected in the river water chemistry.

4.3.1 Sources of Major Ion Composition of River Water and Weathering Process in the West Coast

The sources of chemical elements could be deciphered through the use of elemental ratios. Molar ratio of Na/Cl is a good proxy for deciphering the primary mineral weathering, atmospheric deposition and anthropogenic contribution. The Na/Cl ratio of sea salts/marine aerosols is close to 0.86. The weathering of primary minerals will drive the molar ratio greater than 0.86 while the anthropogenic contamination will bring down the molar ratio below 0.86. The dissolution of halites would modify these signals. However, the prevailing climatic conditions, and humid tropical climate, do not support the formation and existence of halites in the Peninsular India. The anthropogenic influence on the river basin geochemistry is significant for most of the west flowing rivers (except Periyar and Muatupuzha). The Na/Cl ratios in the studied samples are lesser than marine aerosol ratio and indicate the anthropogenic influence. Though Prasad and Ramanathan (2005) invoked the possibility of contribution from evaporite to these rivers (Achankovil river), owing to the prevailing tropical wet climate in the region the possibilities are remote. The dramatic changes in the land use-land cover over the last two decades could be the reason for changes in the water chemistry in these regions (Chattopadhyay et al. 2005; Raj and Azeez 2009).

The ternary diagram of cations of peninsular Indian rivers shows clustering of most of the samples at the centre (Fig. 3a) while the ternary anion shows the clustering of the samples in the bicarbonate apex (Fig. 3b). These suggest that the chemical composition of these river waters owe their allegiance to the silicate and carbonate weathering. It is also observed that the rivers draining the Deccan traps have higher proportion of calcium and magnesium compared to Na + K which could be due to presence of calcium and magnesium rich minerals in the basaltic rocks. Certain rivers of Kerala also have similar proportion of major cations as that of the river Nethravati and Gurupur (Karnataka rivers) and those rivers which have higher proportion of Na + K could have drawn ions from anthropogenic sources. The river waters under study also have higher proportion of Cl and SO_4 compared to bicarbonates. Hence, by analogy, suggest an anthropogenic source of major ions. The samples belonging to east flowing rivers are rich in calcium and magnesium which could be due to dominance of carbonate minerals in the catchment.

Geochemical studies pertaining to the estimation of weathering rate of rocks need to apportion different source contribution to the geochemical budget of the

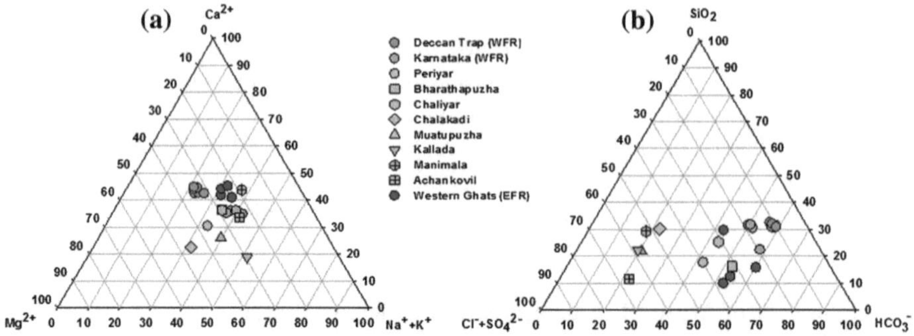

Fig. 3 Ternary cation diagram (**a**) and ternary anion diagram (**b**) for the Peninsular river water samples (*Note WFR* west flowing rivers, *EFR* east flowing rivers)

river water. The source contribution estimation would generally be done using suitable geochemical proxies. Generally, the conservative chloride ion is used for deducing the atmospheric/anthropogenic contribution, and molar ratios of major elements (for e.g., Ca*/Na* or Mg*/Na*) or isotopic ratios for correcting the carbonate rock weathering contribution. However, the suitability of the proxy needs to be evaluated before using for calculation.

The major ion compositions of the river water samples were subjected to sea salt corrections using river water chloride as a proxy to assess the atmospheric contribution of major ions to the river chemistry budget. The river water chloride is assumed to be conservative in these catchments. It is assumed that no Na and Cl are contributed from the evaporites as there is no record of presence of evaporites in these catchments. Further, the ratio of Na/Cl of river water suggests the absence of evaporites in the river catchment (discussed in earlier section). Therefore, the riverine chemical composition obtained after correcting the sea salt contribution is attributed to the composition derived from weathering. The following relation is used for the atmospheric correction:

$$X^* = X_{(\mathrm{river})} - \left((X/Cl)_{(\mathrm{rain})} \times Cl_{(\mathrm{river})}\right)$$

where X* is the element under consideration, X_{river} is the concentration of X element in the river water, $(X/Cl)_{(\mathrm{rain})}$ is the ratio of molar concentration of X element to Cl in the rain water, and $Cl_{(\mathrm{river})}$ is the chloride concentration in the river water. The weathering component of Na to the rivers draining in Kerala is not considered as these rivers have elevated concentration of Na and Cl as a result of anthropogenic activity. Non-availability of data on percentage contribution of major ions to the riverine budget forced many authors to neglect the weathering contribution of Na to the riverine budget. The literature suggests that the contribution of Ca and Mg from the anthropogenic activity could be negligible. Hence, the Ca and Mg contents, after sea water correction were attributed to the influx from weathering. Although the southwest coast of India is predominated by the granite gneiss,

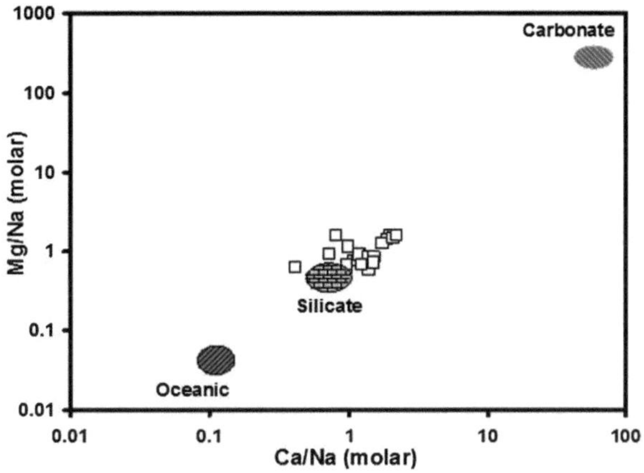

Fig. 4 Mixing diagram of molar ratio of Ca/Na plotted against Mg/Na for water samples of west flowing rivers of India

Deccan basalt, charnockite silicate rocks, the presence of carbonate minerals in minor quantities cannot be ruled out and also these minor carbonate minerals could have influenced the water chemistry. This is also evident from the mixing diagram of Ca/Na and Mg/Na (Fig. 4). In addition to that, Santosh et al. (1991) reported the presence of carbonate fluids trapped in the south Indian granulites during charnockite rock formation. The weatherability of these charnockite rocks and the relative contribution of trapped carbonate fluid on the weathering chemistry are not yet known. These charnockite rocks are the major rock types in the southern part of the Karnataka, Kerala and Tamilnadu, which compose approximately 30 % of the total drainage area. Therefore, in this study, the authors have not deciphered the silicate and carbonate weathering flux. Instead attempt was made to provide a chemical weathering flux discharging annually from the west flowing rivers. The chemical weathering rate, weathering fluxes of major elements and total chemical fluxes are estimated for the studied rivers and are presented in the Table 2. Comparison of weathering rates for different river basins of west coast and east flowing river (River Cauvery) suggests higher contribution from weathering in the Deccan rivers compared to other rivers (Fig. 5). Nevertheless, the Nethravati and Gurupur rivers have higher influx from chemical weathering compared to Kerala rivers although the draining characteristics are quite similar. This could be due to observed higher runoff in the catchment.

4.3.2 Geological and Morphoclimatic Controls on the Geochemistry of River Water

The relative intensities of chemical and physical weathering in tropical basins are dependent on the multiple co-dependent factors. The dominance of governing factors varies at different geographic locations. Lithology of the catchment is the primary governing factor which affects the weathering of rocks (Garrels and Mackenzie 1967; Gibbs 1967; Stallard and Edmond 1987; Gaillardet et al. 1997). It is obvious in

Table 2 Specific weathering fluxes of chemical elements in different rivers of South India

River	Area (km^2)	Rainfall (mm)	Runoff	Cations +Silica ($mg\ L^{-1}$)	Na	K	Ca	Mg	Cl	SO_4	HCO_3	SiO_2	TDS	CWR	Na*	K*	Ca*	Mg*
					tons km^{-2} $year^{-1}$													
Gad	981	2,600	1,690	30	6	1	19	9	1	2	57	30	114	47	3.1	0.4	18.7	8.1
Kajli	762	2,550	1,657	35	7	0	24	11	6	2	70	37	139	58	3.1	0.2	23.7	10.8
Shashtri	2,174	3,260	2,117	34	8	0	29	13	8	2	84	46	170	71	3.6	0.3	29.0	3.8
Vashishti	2,238	3,391	2,198	35	9	1	33	15	8	3	97	50	189	78	3.9	0.3	32.4	14.3
Nethravati	3,657	5,363	3,504	18	11	2	21	9	11	8	66	41	151	62	4.8	2.2	20.5	8.3
Gurupura	824	5,363	3,425	14	11	3	19	7	12	7	73	28	143	48	4.8	2.7	18.5	6.7
Kumaradhara	1,825	5,363	2,594	18	8	3	16	7	8	6	52	30	115	47	3.0	3.0	15.4	6.9
Nethravati @Uppinangadi	1,750	5,363	2,734	19	9	4	18	7	9	6	54	34	125	53	4.0	3.7	17.5	6.6
Periyar	5,398	2,919	2,262	13	9	1	15	11	13	17	36	14	94	29	1.6	0.6	14.7	10.0
Bharathapuzha	6,186	2,276	1,140	15	3	1	5	3	11	10	42	13	78	17		0.3	5.0	1.8
Chaliyar	2,788	3,201	2,181	13	5	1	10	5	10	11	35	20	84	28		0.8	9.7	4.1
Chalakadi	1,704	4,290	1,923	11	4	1	6	7	10	13	12	16	55	22		0.6	5.4	6.3
Muatupuzha	2,004	3,385	2,200	13	8	1	10	8	8	43	15	16	79	30		0.9	10.3	7.8
Kallada	1,699	2,800	1,820	16	12	2	8	8	25	15	18	20	92	29		1.8	7.9	6.3
Manimala	847	3,300	2,119	11	5	1	12	3	20	10	14	22	75	24		1.1	11.5	1.8
Achankovil	2,235	3,600	2,340	10	8	2	13	6	54	3	31	16	122	24		1.1	11.4	2.0
Kaveri	81,155		259	90	8	1	19	7	12	16	43	8	92	23	0.4	0.5	18.3	7.2

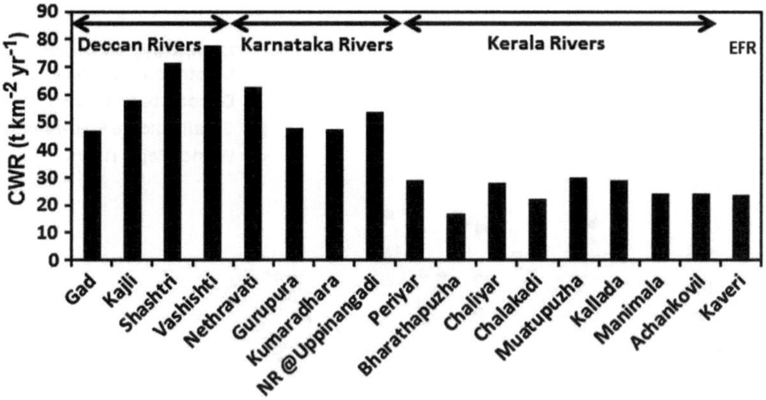

Fig. 5 Comparison between the weathering rates of the west flowing rivers and the east flowing river Cauvery (*Note NR* Nethravati river, *EFR* East flowing river)

the west coast rivers as seen in the variability in the elemental fluxes with lithology. Under similar morpho-climatic conditions, the weathering fluxes of basalts are higher than the granite gneisses and other silicate rocks (Gurumurthy et al. 2012). The basaltic rocks are composed of minerals highly susceptible to weathering and release base cations easily upon interaction with fluid. The slope of the terrain in the rivers of west coast of India has minimal effect on the weathering rate of silicates.

Runoff is thought to be the major limiting factor of chemical weathering of rocks and its role in governing the weathering of rocks has been proven at various geographical locations (Gaillardet et al. 1999; West et al. 2005). The surface runoff induces erosion of weathered rock minerals and fresh mineral surfaces to get exposed to solution which results in enhanced chemical weathering. In addition, the temperature plays an important role in the weathering of primary minerals. The weathering in the south west coast of India and many silicate terrains of the world have been reported to be controlled by the coupled effect of surface runoff and temperature (West et al. 2005; Gurumurthy et al. 2012). The basins that experience higher surface runoff and warm temperature supply higher weathering fluxes whereas, the terrains having lesser runoff and lower temperature do not exhibit higher weathering fluxes (Gurumurthy et al. 2012; Fig. 6). Thus, the silicate weathering in these river basins might be controlled by the coupled effect of temperature and runoff.

Although the weathering fluxes from the west coast rivers are higher than the east coast rivers, the significant part of the weathering fluxes is composed of silica, and the contribution of base cations to the weathering fluxes is significantly less. The lesser weathering of base cations in this region leads to lesser drawdown of carbon dioxide during silicate weathering. The recent study of strontium isotopes on the Nethravati-Gurupur River basin of southwest coast of India has revealed that the weathering of secondary minerals leads to elevated flux of dissolved silica and lesser weathering of base cations (Fig. 7). In Fig. 7, it is observed that the non-monsoonal samples contain lesser values of SiO_2/Ca ratio and slightly radiogenic Sr indicating the lithological

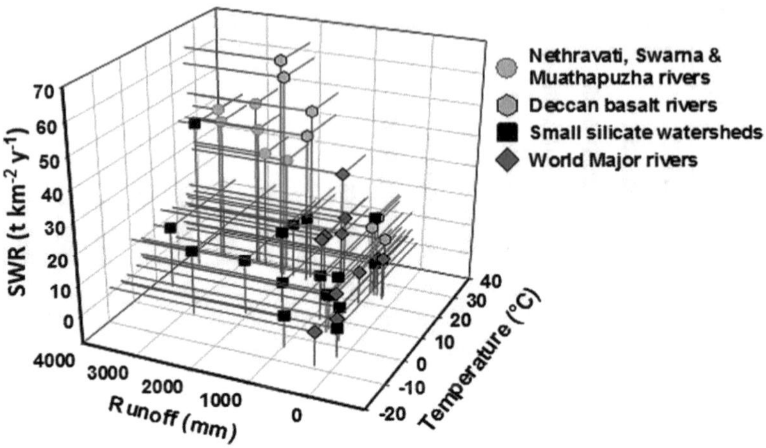

Fig. 6 3D plot of surface runoff and temperature versus silicate weathering flux to elucidate the weathering controls in the southwest coast of India (Figure modified after Gurumurthy et al. 2012; new results on the Swarna River are from Tripti et al. 2013c, AGU FM abstract volume)

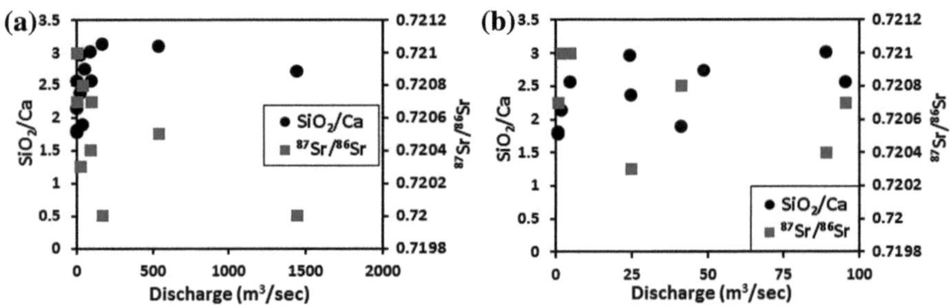

Fig. 7 Temporal variation in the molar ratio of SiO_2/Ca with respect to strontium isotope composition in the Nethravati River (a) and zoom-in view of lower discharge season (b) Data from Gurumurthy et al. (under review)

control on chemistry whereas the monsoon samples contain slightly higher SiO_2/Ca ratio and less radiogenic Sr indicating the flux from secondary soil minerals. The higher SiO_2 concentration (in spite of significant discharge dilution) in the monsoon results in the transport of significant amount of silica with radiogenic strontium to the nearby ocean. Thus, these rivers have a significant impact on the biogeochemical cycling of silica in the coastal environment.

4.4 Trace Element Composition of River Water

Anthropogenic activities introduce perturbations in many natural cycles which have impact on our food sources, health, and climate. An understanding of metal behaviour is particularly important from ecological point of view i.e., the role of metals as nutrients or toxins. The chemistry of metals is often linked to, or plays a controlling role in environmental processes including carbon cycling, ocean circulation, and weathering and transport of chemicals in nature. Most of the trace elements are nutrients to the living organisms, but become toxic when the threshold levels are crossed. The toxicity of trace elements is not dependent solely on the measure of concentrations in the aquatic system but it is a total effect of speciation, complexation, mobility and sorption behavior of elements during their cycling. Thus, an understanding on partitioning, complexation, mobility and speciation would help in better constraining the biogeochemical cycling of metals. The trace element, major ion and organic carbon data along with ancillary data namely, discharge, precipitation, alkalinity, pH, conductivity, and temperature are helpful for better interpretation of the biogeochemistry of metals in the river. The study of metal and metalloid fluxes are the need of the day as rapid increase in the industrialization led to increased fluxes of trace elements to the nearby freshwater sources (Yan et al. 2000). In addition to discharge of metal pollutants to the freshwater source, in situ biogeochemical processes affect the abundance and behavior of metals in the river water. A time series monitoring of metal chemistry and physicochemical parameters in the river would help in better understanding the metal transport from continents to the oceans (Shiller 1997).

The metal biogeochemistry of the tropical rivers has been less documented in the literature, particularly for the Indian rivers. Most of the existing data either focused on urban stretch of the river or on the estuarine regions of large rivers. Recently, Tripti et al. (2013a) and Gurumurthy et al. (2014) studied the metal geochemistry of the monsoon dominated tropical rivers discharging to the Arabian Sea with an objective to understand the biogeochemistry of the trace elements in the fluvial environment. Although there are other studies that dealt with the heavy metal pollution in the Indian estuaries (Chakraborty et al. 2014; references therein), Tripti et al. (2013a) and Gurumurthy et al. (2014) focused on the geochemical aspects of the metals in the riverine environments. Therefore, the geochemical processes and the secondary fluvial processes which redistribute the metals in the riverine environment are presented herein.

The southwestern mountainous rivers of India are fed predominantly by the summer monsoon and hence, the discharge in these river basins mainly depends on the monsoon with about 93 % of the discharge between June and October. Because of monsoon induced surface runoff, the riverine chemistry gets diluted by manifold. In order to study the geochemical processes and their effect on the abundance of metals, a time series monitoring of geochemical parameters is required. Hence, the temporal study on the metal biogeochemistry is used in the following discussion.

Some of the transition metals are biologically active and their abundance is often controlled by the biological productivity and utilization in the photosynthetic activity. The biological production in the rivers is relatively insignificant, particularly in the mountainous streams, owing to fast flow and fresh inflow of water. The geochemical abundance of the biologically active metals in the riverine environment is often controlled by the source rock weathering and their abundance in the parent rock, anthropogenic inputs and lesser known fluvial secondary geochemical processes. The tropical rivers often discharge significant amount of biologically active metals to the nearby ocean. However, the origin of these bioactive metals could be often debated as anthropogenic perturbations in these sensitive ecosystems could often obscure the natural source signals. In most of the geochemical studies, the enrichment of bioactive metals in the sedimentary phase are often considered as signals emanating from anthropogenic inputs. However, the enrichment due to geochemical redistribution processes is rarely considered.

In the west coast rivers of India, the Fe and Mn are being effectively redistributed between dissolved and particulate phases because of the redox and organic complexation reactions (Gurumurthy et al. 2014; Tripti et al. 2013a). The oxic environment prevailing in these river basins could lead to redistribution of metals through the redox reactions. Fe and Mn are among the trace elements which are sensitive to changes in the redox conditions in the fluvial environment. Mn shows linearly decreasing trend with DO (Fig. 8a) in both Nethravati and Swarna River which suggests the redox control on these rivers. Fe shows contrasting relationships between these two rivers with dissolved organic carbon (DOC). In Swarna River, the concentration of Fe increases with the increase in DOC whereas no significant correlation of Fe with DOC was found in the Nethravati River (Fig. 8b). Alternatively, Fe in Nethravati River shows linear decreasing relationship with DO. These observations suggest that Mn and Fe in Nethravati River and Mn in Swarna River are controlled by inorganic colloidal formations whereas Fe in Swarna River is controlled by small size organic colloidal formations. In spite of dominant control exercised by discharge over the variability of ionic concentrations in these river basins, the effects of redox and complexation processes are significant. However, the difference in the colloidal organic carbon—metal complexation in these two river basins could be due to relative abundance of labile fraction of dissolved organic carbon in the Swarna River and lesser abundance of labile dissolved organic carbon in the Nethravati River (Tripti et al. 2013b).

Certain heavy metals (Ni, Co, Zn and Cu) and rare earth elements are found to be controlled by sorption processes in the Nethravati River whereas lesser effect of sorption processes is found in the Swarna River. This could be due to the lesser variability of pH condition in the Swarna River and its higher variability in the Nethravati River during the sampling period. In addition, the pH was found to be in

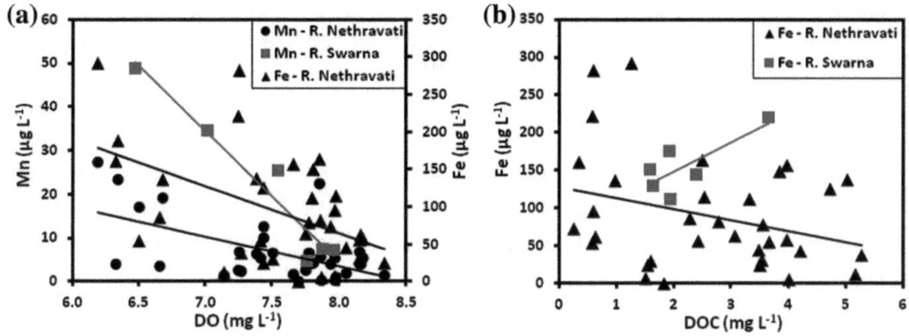

Fig. 8 Relationship of Mn and Fe with DO (**a**) and Fe with DOC (**b**) in Nethravati and Swarna rivers

the slight acidic to near neutral range with average value of 6.8 units in the Swarna River (Tripti et al. 2013a) whereas slight acidic to alkaline range with an average value of 7.2 units is observed in the Nethravati River (Gurumurthy et al. 2014). The control exercised by pH on the metal abundance in the riverine systems is well documented for major world rivers (Shiller 1997, 2002, 2010; Elbaz-Poulichet et al. 1999; Sherrel and Ross 1999; Gaillardet et al. 2003).

4.5 Comparison of Dissolved Trace Elements with World Rivers

The dissolved heavy metal concentrations in the rivers of Southwestern India is compared with world rivers draining shield terrains of tropical region (eg., Amazon, Congo, Orinocco, Niger, etc.) and major rivers of temperate region (Table 3). The concentrations of heavy metals and other transition metals in the Nethravati and Swarna rivers are below the permissible limits and there is no significant contribution from the industrial activity or human activity. Furthermore, the concentrations of heavy metals (except Zn) in the Nethravati and Gurupur river water are significantly less than the world river average, most of the tropical river basins and the temperate river basins which are heavily polluted by heavy metals because of industrial growth.

Table 3 Comparison of heavy metal concentrations in the river water of Nethravati-Gurupur (Gurumurthy et al. 2014) and Swarna (Tripti et al. 2013a) basins with other tropical and major rivers of the world (data from Gaillardet et al. 2003)

Element	Al	As	Sr	Ba	Cd	Co	Cr	Cs	Cu	Fe
	$\mu g\ L^{-1}$									
Nethravati	72	0.03	13	6	0.004	0.065	0.360	0.027	1.017	49
Gurupur	57	0.04	9	4	0.003	0.058	0.352	0.035	0.717	48
Swarna River	112	0.03	8.7	3.2	0.003	0.014	0.239	0.009	0.725	155
Ganges			90							
Brahmaputra			59							
Africa										
Oubangui	12		15	17		0.077	0.533	0.008		60
Zaire	46		21	25		0.075	0.386	0.003		202
Kasai	51		11			0.058	0.400	0.009		108
Congo	76		12	20		0.059	0.501	0.016		179
Niger	76		26	30		0.040	0.450	0.010	0.630	105
Nyong	215		10	18		0.364				241
Sanaga	29	0.2	30	27		0.059			0.952	31
Nyong	159	0.1	12	19		0.253			2.030	174
Mengong	480	0.1	18	24		0.431			1.397	614
Europe										
Seine	16	271	227	32	0.060	0.180			3.530	302
Garone										
Rhine	50									
Kalix	1,080	0.4								525
North America										
St Lawrence	15	0.9	177	23	0.011	0.063		0.005	0.936	111
Mackenzie	18	0.5	238	56	0.184	0.068	0.375	0.007	1.609	119
Fraser River	19	0.5	108	15		0.080	2.100		1.040	47
Hudson										
Mississippi Upper				73					1.850	
Mississippi at Mouth				62					1.600	
South America										
Amazon mean value	9		26	21	0.178	0.177	0.717		1.463	
Amazon (diss)	6		51	28						43
Negro	1		4	6		0.124			0.399	117
Solimoes	171		46	28		0.164				351
Orinoco	62		8	8				0.007–0.013		142
World Avg.	32	0.6	60	23	0.080	0.148	0.700	0.011	1.480	66

(continued)

Table 3 (continued)

Element	Mn	Mo	Ni	U	V	Zn	Pb	Rb	Th	Ti
	μg L^{-1}									
Element	Mn	Mo	Ni	U	V	Zn	Pb	Rb	Th	Ti
	μg L^{-1}									
Nethravati	2.2	0.014	0.307	0.007	0.435	1.646	0.059	1.2	0.006	0.284
Gurupur	2.2	0.007	0.260	0.008	0.211	1.003	0.120	1.1	0.008	0.058
Swarna River	21		0.278	0.015	0.209	0.724	0.158	1.9		
Ganges				2.000						
Brahmaputra				1.000						
Africa										
Oubangui			1.150	0.055				2.7	0.042	
Zaire			1.020	0.071				3.9	0.056	
Kasai			0.410	0.027				2.7	0.023	
Congo			0.934	0.049				3.1	0.065	
Niger	0.5		0.290	0.020	0.590	0.890		3.8	0.013	
Nyong	29.7			0.029	0.645			4.1	0.121	
Sanaga	0.4		0.700	0.028		1.020		6.1	0.012	0.231
Nyong	22.6		1.180	0.022		1.810		3.6	0.111	0.199
Mengong	20.0		5.040	0.022		3.120		0.7	0.137	5.808
Europe										
Seine	3.8		5.060	0.820	2.850	4.980	0.220	1.4	0.010	
Garone				0.750						
Rhine										
Kalix	9.4			0.090						
North America										
St Lawrence	6.3	1.292	1.330	0.373	0.439	2.580	0.233	1.0	0.004	0.509
Mackenzie	1.3	1.067	1.830	0.730	0.253	0.500	0.771	0.6	0.634	0.423
Fraser River	5.4	1.330	1.860	0.330	0.390		0.078	0.9		0.680
Hudson										
Mississippi Upper	0.4	1.114	1.660	1.285	2.055	0.210	0.008	1.2		
Mississippi at Mouth	0.7	1.630	1.120	0.620	0.820	0.180	0.001	1.1		
South America										
Amazon mean value	50.7	0.175	0.740	0.052	0.703	0.450	0.064	1.4		
Amazon (diss)	3.3			0.055		0.760		1.8	0.006	
Negro	8.2		0.210	0.034		1.210	0.170	1.1		
Solimoes	6.5			0.040		2.350	0.151	1.5	0.010	
Orinoco	6.8			0.049		1.750		1.5	0.073	
World Avg.	34.0	0.420	0.801	0.372	0.710	0.600	0.079	1.6	0.041	0.489

5 Conclusions

- The water chemistry is found to be controlled by the lithological heterogeneity and the intensity of weathering in the river basin. In addition, higher surface runoff and warm temperature triggers silicate weathering in this region.
- The weathering of sedimentary formations belonging to Quaternary and Holocene in the plain region leads to elevated fluxes of silica with radiogenic strontium.
- The trace element geochemistry in these monsoon dominated river basins is controlled primarily by the discharge dilution. However, the secondary fluvial processes namely, redox reactions, oxidative scavenging of surface reactive metals by Fe and Mn and organic carbon—metal complexation appear to have vital role in determining the geochemical abundance of metals in the river water.
- Unlike the significant uniform organic carbon—metal complexation in the tropical river basins of Africa and South America, significant variations in organic carbon—metal complexation exist between the peninsular tropical rivers of India. This could be due to the variable abundance of labile fraction of dissolved organic carbon in different rivers.
- The secondary fluvial geochemical processes lead to slight enrichment of metals in sedimentary phase. However, the enrichments of these metals are within the permissible limits. From the pollution point of view, the metal contaminant studies in these river basins need to consider the fluvial geochemical redistribution of metals to minimize the erroneous attribution of pollution sources.

Acknowledgments The authors are thankful to Dr K Balakrishna, Manipal Institute of Technology, Manipal, India, Dr Jean Riotte, Indo French Cell for Water Sciences, Indian Institute of Science, Bangalore, India, Dr Stephane Audry, and Dr Jean Jacques Braun Geosciences Environment, Toulouse, France, and Dr. Luc Lambs, ECOLAB, Toulouse, France for the support and encouragement at different stages of this work. GPG acknowledges the Director, Manipal Centre for Natural Sciences, Manipal University for the logistics. TM acknowledges the fellowship support from the Manipal University through Structured Ph. D. program.

References

Chattopadhyay S, Rani LA, Sangeetha PV (2005) Water quality variations as linked to landuse pattern: a case study in Chalakudy river basin, Kerala. Current Sci 89(12):2163–2169

Chakraborty P, Ramteke D, Chakraborty S, Nagender Nath B (2014) Changes in metal contamination levels in estuarine sediments around India—an assessment. Mar Pollut Bull 78 (1–2):15–25

Das A, Krishnaswami S, Sarin MM, Pande K (2005) Chemical weathering in the Krishna basin and Western ghats of the Deccan traps, India: rates of basalt weathering and their controls. Geochim Cosmochim Acta 69(8):2067–2084

Dalai TK, Krishnaswami S, Sarin MM (2002) Major ion chemistry in the headwaters of the Yamuna river system: chemical weathering, its temperature dependence and CO_2 consumption rates. Geochim Cosmochim Acta 66:3397–3416

Deshpande RD, Bhattacharya SK, Jani RA, Gupta SK (2003) Distribution of oxygen and hydrogen isotopes in shallow groundwater from southern India: influence of a dual monsoon system. J Hydrol 271:226–239

Eby GN (2004) Principles of Environmental Geochemistry, 1st edn. Brooks Cole, California

Elbaz-Poulichet F, Seyler P, Maurice-Bourgoin L, Guyot J-L, Dupuy C (1999) Trace element geochemistry in the upper Amazon drainage basin (Bolivia). Chem Geol 157:319–334

Francis PA, Sulochana G (2006) Intense rainfall events over the west coast of India. Meteorol Atmos Phys 94(1–4):27–42

Friend CRL, Nutman AP (1991) SHRIMP U-Pb geochronology of the closepet granite and peninsular gneiss, Karnataka, South India. J Geol Soc India 38:357–368

Gaillardet J, Dupre B, Allegre CJ (1995) A global geochemical mass budget applied to the Congo basin rivers: erosion rates and continental crust composition. Geochim Cosmochim Acta 59:3469–3485

Gaillardet J, Dupre B, Allegre CJ (1999) Global silicate weathering and CO_2 consumption rates deduced from the chemistry of large rivers. Chem Geol 159:3–30

Gaillardet J, Viers J, Dupre B (2003) Trace element in river waters. In: Drever JI (ed) Surface and groundwater, weathering, erosion and soils, vol 5. Treatise on geochemistry (Holland HD, Turekian KK eds). Pergamon, Oxford

Gaillardet J, Dupre B, Allegre CJ, Negrel P (1997) Chemical and physical denudations in the Amazon river basin. Chem Geol 142:141–173

Gobre T, Salve PR, Krupadam RJ, Bansiwal A, Shastry S, Wate SR (2010) Chemical composition of precipitation in the costal environment of India. Bull Environ Contam Toxicol 85:48–53

Gupta SK, Deshpande RD, Bhattacharya SK, Jani RA (2005) Groundwater $\delta^{18}O$ and δD from central Indian Peninsula: influence of the Arabian sea and the Bay of Bengal branches of the summer monsoon. J Hydrol 303:38–55

Gurumurthy GP, Balakrishna K, Riotte J, Audry S, Braun JJ, Udaya Shankar HN, Manjunatha BR (2012) Controls on intense silicate weathering in a tropical river, Southwestern India. Chem Geol 300–301:61–69

Gurumurthy GP, Balakrishna K, Tripti M, Riotte J, Audry S, Braun JJ, Udayashankar HN (2014) Geochemical behaviour of trace elements in a monsoon dominated humid tropical river basin, southwestern India. Environ Sci Pollut Res 21:5098–5120

Gurumurthy GP, Balakrishna K, Tripti M, Riotte J, Audry S, Braun JJ, Udaya Shankar HN (under review) Use of Sr isotopes as a tool to decipher the soil weathering processes in a tropical river catchment, Southwestern India. Applied Geochemistry. Elsevier

Garrels RM, Mackenzie FT (1967) Origin of the chemical compositions of some springs and lakes. Adv Chem Ser, Am Chem Soc 67:222–242

Gibbs RJ (1967) The geochemistry of the Amazon River system. Part I: The factors that control the salinity and composition and concentration of suspended solids. Geol Soc Am Bull 78:1203–1232

Hegde P (2007) Major ionic composition of aerosol, rainwater and its impact on surface and sub-surface waters, in and around Mangalore, west coast of India. Environ Monit Assess 133(1–3):119–125

Khemani LT, Momin GA, Rao PSP, Pillai AG, Safai PD, Mohan K, Rao MG (1994) Atmospheric pollutants and their influence on acidification of rain water at an industrial location on the west coast of India. Atmos Environ 28:3145–3154

Krishnaswami S, Singh SK (2005) Chemical weathering in the river basins of the Himalaya, India. Curr Sci 89(5):841–849

Krishnaswami S, Singh SK, Dalai TK (1999) Silicate weathering in the Himalaya: role in contributing to major ions and radiogenic Sr to the Bay of Bengal. In: Somayajulu BLK (ed) Ocean science, trends and future directions. Indian National Science Academy and Akademia International, New Delhi, pp 23–51

Lambs L, Gurumurthy GP, Balakrishna K (2011) Tracing the sources of water using stable isotopes: first results along the Mangalore-Udupi region, southwest coast of India. Rapid Commun Mass Spectrom 25:2769–2776

Latrubesse EM, Stevaux JC, Sinha R (2005) Tropical rivers. Geomorphology 70:187–206
Milliman JD, Meade R (1983) World-wide delivery of river sediments to the oceans. J Geol 91:1–22
Milliman JD, Farnsworth KL (2011) River discharge to the coastal ocean: a global synthesis. Cambridge University Press, New York
Naik MS, Momin GA, Rao PSP, Safai PD, Ali K (2002) Chemical composition of rainwater around an industrial region in Mumbai. Curr Sci 82(9):1131–1137
Neilsen E (1972) The philosophy and pedagogy of inter disciplinary coastal studies and research with particular reference to sedimentary process and to morphological and evolutionary features of the Southwest coast of India. Kobenhavn, Denmark
Padmalal D, Remya SI, Jyothi SJ, Baijulal B, Babu KN, Baiju RS (2011) Water quality and dissolved inorganic fluxes of N, P, SO4, and K of a small catchment river in the Southwestern Coast of India. Environ Monit Assess 184(3):1541–1567
Pande K, Sarin MM, Trivedi JR, Krishnaswami S, Sharma KK (1994) The Indus river system (India-Pakistan): major-ion chemistry, uranium and strontium isotopes. Chem Geol 116:245–259
Parashar DC, Granat L, Kulshretha UC, Pillai AG, Nail MS, Momin GA, Rao PSP, Safai PD, Khemani LT, Naqvi SWA, Narverkar PV, Thapa KB, Rhode H (1996) Chemical composition of precipitation in India and Nepal: a preliminary report on Indo-Swedish project on atmospheric chemistry. Report CM-90, Department of Meterology, Stockholm University
Parashar DC, Kulshrestha UC, Jain M (2001) Precipitation and aerosol studies in India. Environ Monit Assess 66(1):47–61
Pattanaik JK, Balakrishnan S, Bhutani R, Singh P (2007) Chemical and Strontium isotopic composition of Kaveri, Palar and Ponnaiyar rivers: significance to weathering of granulites and granitic gneisses of southern Peninsular India. Curr Sci 93(4):523–531
Prasad BKM, Ramanathan AL (2005) Solute Sources and Processes in the Achankovil River Basin, Western Ghats, Southern India/Sources de Solutéés et Processus Associéés Dans le Bassin du Fleuve Achankovil, Ghats Occidentaux, Inde du Sud. Hydrol Sci J 50(2):320–354
Prathibha P, Kothai P, Saradhi IV, Pandit GG, Puranik VD (2010) Chemical characterisation of precipitation at a coastal site in Trombay, Mumbai, India. Environ Monit Assess 168(1–4):45–53
Praveen PS, Rao PSP, Safai PD, Devara PCS, Chate DM, Ali K, Momin GA (2007) Study of aerosol transport through precipitation chemistry over Arabian Sea during winter and summer monsoons. Atmos Environ 41:825–836
Raj N, Azeez PA (2009) Spatial and temporal variation in surface water chemistry of a tropical river, the river Bharathapuzha, India. Curr Sci 96(2):245–251
Raymo ME, Ruddiman WF (1992) Tectonic forcing of late Cenozoic climate. Nature 359 (6391):117–122
Ramakrishnan M, Vaidyanadhan R (2008) Geology of India. Geol Soc India 1:994
Rao MKM, Jagannathan (1994). Geo Karnataka, Mysore Geological Department (MGD) Centenary volume. Hydrometeorology 388–395. Karnataka Assistant Geologists Association
Santosh M, Jackson DH, Harris NBW, Mattey DP (1991) Carbonic fluid inclusions in South Indian granulites: evidence for entrapment during charnockite formation. Contrib Mineral Petrol 108:318–330
Sarin MM, Krishnaswami S, Trivedi JR, Sharma KK (1992) Major ion chemistry of the Ganga source waters: weathering in the high altitude Himalaya. Proc Indian Acad Sci (Earth Planet Sci) 101:89–98
Satyanarayana J, Reddy LAK, Kulshrestha MJ, Rao RN, Kulshrestha UC (2010) Chemical composition of rain water and influence of airmass trajectories at a rural site in an ecological sensitive area of Western Ghats (India). J Atmos Chem 66:101–116
Sherrel RM, Ross JR (1999) Temporal variability of trace metals in New Jersey Pinelands streams: effect of discharge and pH. Geochim Cosmochim Acta 63:3321–3336
Shiller AM (1997) Dissolved trace elements in the Mississippi river: seasonal, inter-annual and decadal variability. Geochim Cosmochim Acta 61:4321–4330
Shiller AM (2002) Seasonality of dissolved rare earth elements in the lower Mississippi river. Geochem Geophys Geosyst 3(11):1–14

Shiller AM (2010) Dissolved rare earth elements in a seasonally snow-covered, alpine/subalpine watershed, Loch Vale, Colorado. Geochim Cosmochim Acta 74:2040–2052

Singh SK, Trivedi, JR, Pande K, Ramesh R, Krishnaswami S (1998) Chemical and strontium, oxygen and carbon isotopic compositions of carbonates from the lesser Himalaya: implications to the strontium isotopic composition of the source waters of the Ganga, Ghaghara and the Indus rivers. Geochim Cosmochim Acta 62(5):743–755

Stallard RF, Edmond JM (1987) Geochemistry of the Amazon. 3. Weathering chemistry and limits to dissolved inputs. J Geophys Res 92(C8):8293–8302

Tripti M, Gurumurthy GP, Balakrishna K, Chadaga M (2013a) Dissolved trace element biogeochemistry of a tropical river, Southwestern India. Environ Sci Pollut Res 20(6):4067–4077

Tripti M, Lambs L, Otto T, Gurumurthy GP, Teisserenc R, Moussa I, Balakrishna K, Probst JL (2013b) First assessment of water and carbon cycles in two tropical coastal rivers of south-west India: an isotopic approach. Rapid Commun Mass Spectrom 27:1681–1689

Tripti M, Gurumurthy GP, Balakrishna K, Audry S, Riotte J, Braun J-J, Chadaga MD, Udayashankar HN (2013c). Chemical weathering and associated carbon-dioxide consumption in a tropical river basin (Swarna River), Southwestern India. Presented at American Geophysical Union (AGU) Fall Meeting 2013, San Francisco

Viers J, Dupré B, Gaillardet J (2009) Chemical composition of suspended sediments in world rivers: new insights from a new database. Sci Total Environ 407(2):853–868

Viers J, Dupré B, Polve M, Schott J, Dandurand J-L, Braun J-J (1997) Chemical weathering in the drainage of a tropical watershed (Nsimi-Zoetele site, Cameroon): comparison between organic-poor and organic-rich waters. Chem Geol 140:181–206

Viers J, Dupré B, Braun J-J, Deberdt S, Angeletti B, NdamNgoupayou J, Michard A (2000) Major and trace element abundances and strontium isotopes in the Nyong basin rivers (Cameroon): constraints on chemical weathering processes and element transport mechanisms in humid tropical environments. Chem Geol 169:211–241

Warrier CU, Babu MP, Manjula P, Velayudhan KT, Hameed AS, Vasu K (2010) Isotopic characterization of dual monsoon precipitation evidence from Kerala, India. Curr Sci 98 (11):1487–1495

West AJ, Galy A, Bickle M (2005) Tectonic and climatic controls on silicate weathering. Earth Planet Sci Lett 235:211–228

Yadava MG, Ramesh R, Pandarinath K (2007) A positive 'amount effect' in the Sahayadri (Western Ghats) rainfall. Curr Sci 93(4):560–564

Yan XP, Kerrich R, Hendry MJ (2000) Distribution of arsenic (III), arsenic (V) and total inorganic arsenic in pore waters from a thick till and clay-rich aquitard sequence, Saskatchewan, Canada. Geochim Cosmochim Acta 64:2637–2648

Hydrogeochemical Drivers and Processes Controlling Solute Chemistry of Two Mountain River Basins of Contrasting Climates in the Southern Western Ghats, India

Jobin Thomas, Sabu Joseph and K.P. Thrivikramji

Abstract Water samples were collected from two mountain rivers of contrasting climates, viz., humid, Muthirapuzha River Basin (MRB) and semi-arid, Pambar River Basin (PRB) during monsoon (MON), post-monsoon (POM) and pre-monsoon (PRM) seasons, and were analyzed to understand the spatio-temporal variability as well as the sources and processes controlling hydrogeochemistry. In MRB and PRB, Ca^{2+} and Mg^{2+} dominate the cations, while Cl^- dominates the anions in MRB and HCO_3^- dominates the anions in PRB. PRB shows an elevated level of ionic abundance and higher degree of mineralization, due to multiple factors such as semi-aridity, discharge dominated by groundwater, lithological variations and the influences of carbonates and soil evaporites. However, K^+, Cl^- and H_4SiO_4 are relatively higher in MRB, implying significance of both anthropogenic activities and intense silicate weathering. The $Ca^{2+} + Mg^{2+}/HCO_3^-$ ratios in MRB are also relatively larger than PRB, suggesting high intensity of anthropogenic influences in MRB. Downstream variation of hydrogeochemistry implies a general decreasing trend in MRB, which is attributed to dilution due to high discharge, whereas hydrogeochemistry of PRB shows an increasing downstream trend, by which, the significance of semi-arid climate of the downstream tracts of the basin is implied. Both MRB and PRB show temporal variability in hydrogeochemical attributes implying the role of monsoon rainfall determining stream water composition. The Na^+-normalized Ca^{2+} versus

J. Thomas
Inter University Centre for Geospatial Information Science & Technology, University of Kerala, Thiruvananthapuram 695 581, Kerala, India

J. Thomas (✉) · S. Joseph
Department of Environmental Sciences, University of Kerala, Thiruvananthapuram 695 581, Kerala, India
e-mail: jobinenv@gmail.com

K.P. Thrivikramji
Department of Geology, University of Kerala, Thiruvananthapuram 695 581, Kerala, India

Mu. Ramkumar et al. (eds.), *Environmental Management of River Basin Ecosystems*, Springer Earth System Sciences, DOI 10.1007/978-3-319-13425-3_17

Na^+-normalized HCO_3^- plots suggest the control exercised by mixing between silicate and carbonate end members in both the basins. However, in PRB, dissolution of soil evaporites during MON and POM is evident by relatively lower Ca^{2+}/Na^+ ratios. The $Ca^{2+} + Mg^{2+}/Na^+ + K^+$ ratios in MRB and PRB during MON (mean = 1.96 and 2.23 in MRB and PRB respectively), POM (mean = 3.29 and 2.41) and PRM (mean = 5.74 and 4.40) also suggest sources other than silicate weathering. Relative enrichment of Cl^- (with respect to Na^+) indicates multiple sources for Cl^- (i.e., anthropogenic as well as atmospheric). Even though there are significant differences in water types between MRB and PRB, most of the waters of both the basins are considered to be "transitional". Relatively higher pCO_2 in stream waters (compared to atmosphere) is observed and the phenomenon is attributed to the influent nature of the stream discharge (i.e., contributed by groundwater which is significantly enriched in CO_2) and the slower rate of re-equilibration (i.e., solubility vs. release of CO_2) with atmosphere. Hence, evidently the hydrogeochemical composition of MRB and PRB is jointly controlled by weathering of silicate and carbonate minerals as well as anthropogenic activities and is influenced by climatic seasonality. The spatio-temporal variability of hydrogeochemical attributes of MRB and PRB is mainly due to the variations in climate, lithology, hydrologic pathways and degree of various anthropogenic activities.

Keywords Hydrogeochemistry · Tropical-mountain-rivers · Weathering · Muthirapuzha · Pambar · Western Ghats

1 Introduction

Rivers, one of the major components of the global water cycle, play a significant role in the geochemical cycling of elements (Garrels et al. 1975) by transporting continental weathered products (as solids and solutes) to the world oceans to regulate global seawater composition. The drainage basins of the world deliver roughly 37.4×10^{12} m^3 of water (Gaillardet et al. 1999) and about 20.0×10^9 tons of suspended sediment annually (Milliman and Syvitski 1992). On catchment-scale, various natural factors such as lithology, climate, tectonics, topography and vegetation influence weathering and thereby chemical composition of river water (Gibbs 1970; Stallard and Edmond 1981, 1983, 1987; Meybeck 1987; Drever 1988; Bricker and Jones 1995; White and Blum 1995; Hutchins et al. 1999; Jacobson et al. 2003). Among these factors, the significant role of lithology and climate was demonstrated by several researchers (e.g., Stallard and Edmond 1983, 1987; Bluth and Kump 1994; Johnson et al. 1994; Berner and Berner 1997). In uplifted orogenic belts, climate and tectonics regulate denudation and geomorphic development (Koons 1995; Whipple et al. 1999), and influence the patterns of river water chemistry and long-term geochemical cycles (e.g., Stallard and Edmond 1983; Jacobson et al. 2003). In addition, the role of anthropogenic activities (via point and non-point sources) on modification of chemical composition of river water has also been discussed by several workers (e.g., Carpenter et al. 1998; Zhang et al. 1999; Bennett et al. 2001).

Weathering of silicate and carbonate minerals (by various processes such as hydrolysis, dissociation, dissolution, oxidation and reduction) is an important determinant of river hydrogeochemistry (Bricker and Garrels 1967; Hem 1985; Edmond and Huh 1997; Gupta and Subramanian 1998; Gaillardet et al. 1999; White 2002). Silicate weathering has a crucial role in determining the chemical composition of river waters of the world, especially in the humid tropics (e.g., Harmon et al. 2009; Gurumurthy et al. 2012; Thomas et al. 2014) as the active orogenic belts and island arcs promote intensive chemical weathering in the humid tropics (Stallard 1988). It is evident from the fact that though the tropical ecosystems cover only 25 % of the Earth's land surface, they are responsible for 38 % of the dissolved ions and 65 % of the dissolved silica (Meybeck 1987).

Hydrogeochemical composition of rivers shows strong spatial- and temporal-variations. Spatial variation of hydrogeochemical attributes is generally influenced by lower-order tributaries (Meyer et al. 1988), land use (Townsend et al. 1983; Bucker et al. 2010), soil and/or geology (Meybeck 1987; Schultz et al. 1993; Stutter et al. 2006; Harmon et al. 2009; Leite et al. 2010) as well as groundwater contributions (Boulton et al. 1998, 1999; Banks et al. 2011). Even though temporal variability in river water chemistry is primarily controlled by discharge (Hem 1948; Smolders et al. 2004; Crosa et al. 2006; Ovalle et al. 2013), hydrologic pathways are also significant (Harriman et al. 1990; Wheater et al. 1990; Creed and Band 1998). According to Church (1997), the relative abundance of each of the four basic flow paths operating in streams (i.e., direct interception of precipitation, surface flow, subsurface flow and groundwater flow) also strongly depends on topography, geology, soil makeup and vegetation. In the semi-arid and arid rivers, temporal variation of discharge is remarkably high (due to clearly marked dry and wet seasons) and fluctuations of discharge may have enormous effects on the hydrogeochemistry (Allen 1995; Davies et al. 1996).

Since river water chemistry exhibits significant spatio-temporal variability, depending on the geologic environment, physical system and biotic responses, data collected from different sites of a river basin during different seasons can be used to model the possible relationships between chemical composition of the surface water and their controlling factors (e.g., Ahearn et al. 2004; Lindell et al. 2010). The studies on spatio-temporal patterns in river water chemistry are also important for sustainable management of river basins (Meyer et al. 1988; Petts and Calow 1996; Xie et al. 2013). The estimations of sources of dissolved load enable quantification of CO_2 utilization by the acid decomposition of continental rocks, which has critical implications in the context of climate change (e.g., Mortatti and Probst 2003; Jha et al. 2009; Moquet et al. 2011; Zhu et al. 2013).

Mean chemical composition of river and lake waters of the world has been the subject of discussion from the very early decades of 20th century (e.g., Clarke 1924; Conway 1942; Rodhe 1949; Livingstone 1963). However, after Garrels and Mackenzie (1971a), several river geochemistry studies have been carried out in two main and complimentary approaches, viz., small-scale studies of rivers draining uniform lithology under a given climate and studies on the world's largest rivers for a more global picture (Gaillardet et al. 1999). In addition, they also opined that the large-scale studies demonstrate the significance of lithology only and obscure the contributions from other parameters such as climate, topography etc. It is also very noteworthy that the world's largest rivers (e.g., Amazon, Congo-Zaire, Mississippi, Nile, Parana, Changjiang, Mackenzie, Ganges-Brahmaputra, Orinoco) have received greater attention due to their global significance towards water and sediment discharge (e.g., Gibbs 1967, 1970; Milliman and Meade 1983; Gaillardet et al. 1999; Mortatti and Probst 2003; Cai et al. 2008). In fact, the significance of small-mountain-rivers on global hydrological cycle remained underestimated until Milliman and Syvitski (1992), whose observations revealed that mountain rivers draining southern Asia and Oceania have much greater yields (two to three times) than rivers draining other mountainous areas of the world (and an order of magnitude greater than the rivers draining the high-Arctic and the non-alpine European mountains).

Documentation of hydrogeochemistry of rivers of India has also followed the global trends (e.g., Subramanian 1983; Abbas and Subramanian 1984; Biksham and Subramanian 1988; Bartarya 1993; Gupta and Subramanian 1994, 1998; Pandey et al. 1999; Das et al. 2005; Krishnaswami and Singh 2005; Jha et al. 2009; Gupta et al. 2011; Mehto and Chakrapani 2013). In addition, spatio-temporal patterns of hydrogeochemistry of several small rivers of the southern Western Ghats were reported by many researchers (e.g., Thrivikramaji 1989; Bajpayee and Verma 2001; Thrivikramaji and Joseph 2001; Prasad and Ramanathan 2005; Maya et al. 2007; Kannan 2009; Raj and Azeez 2009; Padmalal et al. 2012; Gurumurthy et al. 2012; Thomas et al. 2014). However, the rivers draining the rain shadow regions of the southern Western Ghats (in Kerala) have gone into oblivion as hardly any studies have been carried out in the east flowing rivers of Kerala. The smaller mountain river basins of varying climates of the southern Western Ghats amply qualify as candidates of study in the perspective of chemical weathering as tropical climates promote accelerated rock weathering and landscape denudation (Thomas 1994). Further, such information will be useful for a comparative understanding of tropical mountain watersheds of small to intermediate size and their contribution to global chemical weathering (White et al. 1998; Turner et al. 2003; Harmon et al. 2009). Hence, this paper examines the spatio-temporal patterns of chemical composition of surface water and processes controlling solute chemistry of two mountain river basins of contrasting climates (i.e., humid, Muthirapuzha River Basin, MRB and semi-arid, Pambar River Basin, PRB).

2 Study Area

Two contiguous and nearly equal-sized mountain river basins in the Anaimalai-Cardamom Hills of the southern Western Ghats, viz., MRB and PRB (N Lat. 10° 01′ 55″ and 10° 21′ 05″ and E Long. 76° 59′ 45″ and 77° 15′ 32″) were selected for

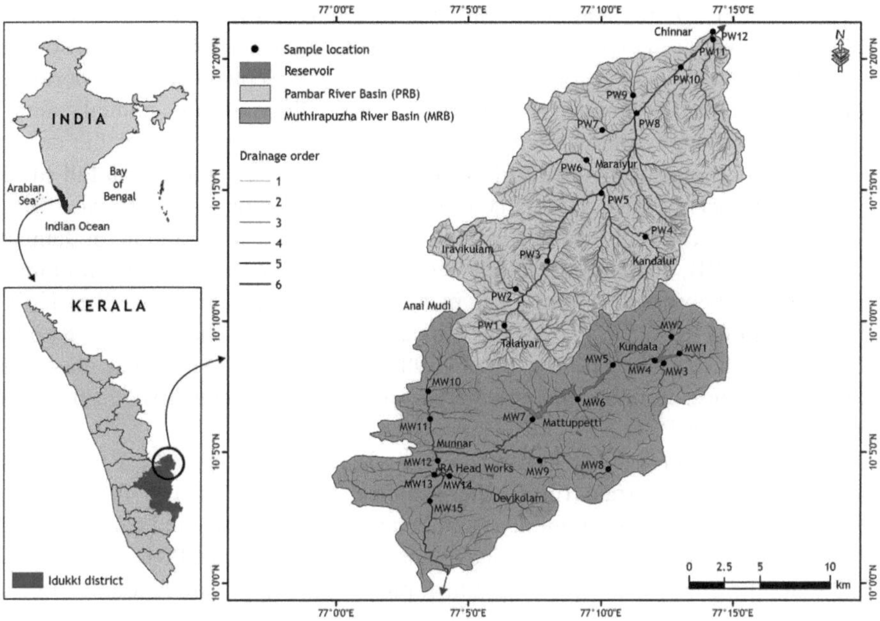

Fig. 1 Location map: stream network (from survey of India topographic maps of scale 1: 50,000) and river water sample locations of MRB and PRB

the present investigation (Fig. 1). MRB (basin order = 6; A = 271.75 km^2) is a major sub-basin of the west-flowing Periyar (the longest river of Kerala—244 km), whereas PRB (basin order = 6; A = 288.53 km^2) is a sub-basin of the east-flowing Amaravati River (a tributary of the Cauvery River). MRB is a modified hydrologic system of 14 fourth order sub-basins and three dams and reservoirs (Kundala or Setuparvatipuram, Mattuppetti and RA Head Works), whereas PRB with 18 fourth order sub-basins is devoid of dams and reservoirs with the exception of a few flow diversion structures (e.g., east of PW4 at Gokhanathapuram and immediate downstream of PW5 at Kovilkadavu). The basin elevation of MRB ranges between 2,690 and 760 m above mean sea level and that of PRB varies from 2,540 to 440 m above mean sea level. The drainage network of both MRB and PRB is influenced by the Munnar plateau, an extensive planation surface of Paleocene age (Soman 2002), which has several local planation surfaces and terrain with concordant summits at varying elevations (Thomas et al. 2012).

Due to the characteristic NNW-SSE alignment and topographic configuration (Gunnell and Radhakrishna 2001), the Western Ghats acts as a climatic barrier (Nair 2006), separating tropical humid climate of the western slopes of the Western Ghats (e.g., MRB) and semi-arid climate on the leeward side (e.g., PRB). The analysis of climatic data of this region by Thomas (2012) revealed that the mean annual rainfall (P_{ma}) of MRB was 3,700 mm (period = 1989–2009), whereas that of PRB was 1,100 mm (period = 1992–2008). Many previous studies (e.g., CESS

1984; Jose et al. 1994; Chandrashekara and Sibichan 2006) also imply that P_{ma} of the basins shows strong spatial and temporal variability. In both basins, monsoon in two different spells, i.e., southwest monsoon (June–September) and northeast monsoon (October–November), contributes major share of the annual rainfall budget. Even though southwest monsoon contributes nearly 85 % of the annual rainfall of MRB and upstream of PRB, roughly 50 % of the annual rainfall in the downstream tracts of PRB is obtained during northeast monsoon (Thomas 2012). MRB has relatively high rainfall in July (949 mm) and low rainfall in January (14 mm). In PRB, maximum rainfall occurs in July (178 mm), while minimum rainfall during February (38 mm). The mean annual temperature of MRB is 17 °C, whereas that of PRB is 26 °C. Various climate classification schemes (e.g., Koppen 1936; Trevartha 1954; Chorley et al. 1984) suggest that MRB has monsoon dominated tropical humid climate, whereas dry tropical savannah (i.e., semi-arid) climate prevails in PRB.

The basins are developed in the Madurai Granulite Block of the Precambrian Southern Granulite Terrain (in southern India). Major rock types in MRB are hornblende-biotite gneiss, granitoids (granite and granite gneiss) and calc-granulite, whereas hornblende-biotite gneiss and granite gneiss dominate PRB (Fig. 2; GSI 1992). Granite gneiss is a medium-grained, pink-colored rock and is foliated due to parallel planar arrangement of flakes of biotite, prisms of hornblende and lenticular and flattened quartz veins. Hornblende-biotite gneiss (consisting of hornblende, plagioclase, K-feldspar, quartz, clinopyroxene and biotite as major minerals and

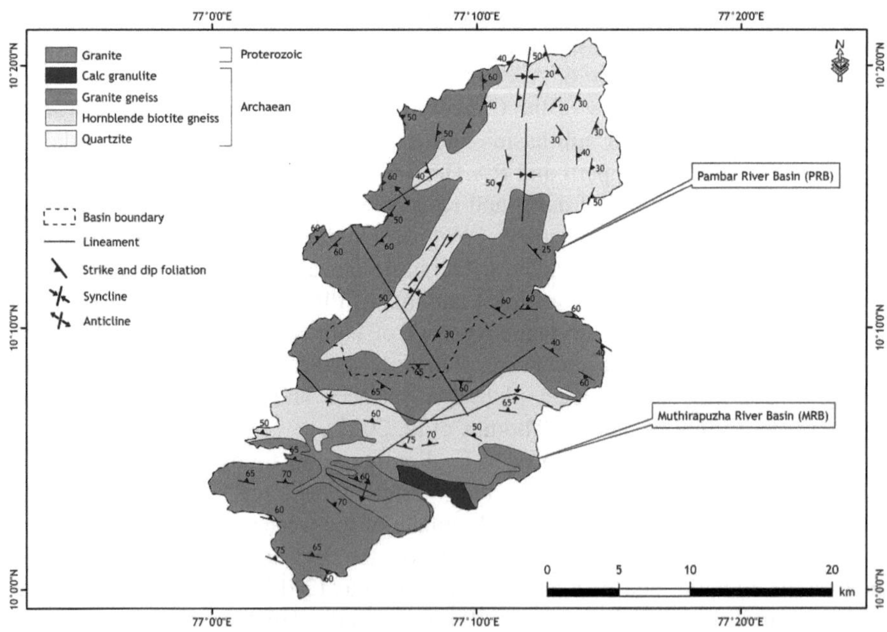

Fig. 2 Geology of MRB and PRB (after GSI 1992)

sphene, apatite, secondary calcite and opaques as the minor and accessory minerals) shows regular but alternating bands rich in quartzo-feldspathic (0.5–3.0 cm thick) and mafic (0.2–0.5 cm thick) minerals. Granite is mainly composed of K-feldspar (mostly microcline), plagioclase (albite-oligoclase) and quartz. Biotite, sphene, apatite and zircon are the accessory minerals, while magnetite is the dominant opaque (Soman 2002). Calc-granulite is a medium-grained rock, exhibiting faint layering due to the segregation of calc-silicate minerals as well as incipient development of metamorphic foliation defined by biotite and pyroxene (Thampi 1987). Pegmatite, aplite and quartzite patches as well as basic intrusives traverse the older host rocks. Nair et al. (1983) reported crystalline limestone patches and syenite-carbonatite veins in MRB. Laterite layers of varying thickness (<15 cm to ~1.0 m) are also exposed at many places in MRB, while laterite layers occur only in the upstream segments, that too sparsely in PRB.

Major soil series in MRB are the Anai Mudi, Pambadumpara, and Venmani series, whereas the Anai Mudi and Chinnar series cover PRB. The soil characteristics of each soil series in MRB and PRB are given in Table 1 (after SSO 2007). MRB is covered by several natural vegetation types such as southern montane wet temperate grassland, southern montane wet temperate forest (also known as *shola* forest), southern subtropical hill forest and southern west coast evergreen forest. On the contrary, natural vegetation of PRB covers an entire spectrum ranging from southern montane wet temperate forest to dry scrubs (characteristic of the arid plains of Tamil Nadu; Nair 1991). Dominant vegetation types in PRB are southern tropical thorn forest, southern dry mixed deciduous forest, southern moist mixed deciduous forest, tropical riparian "fringing" forest, southern montane wet temperate grassland, southern montane wet temperate forest and southern subtropical hill forest (Sankar et al. 2000). The study area is renowned for the presence of sandalwood (*Santalum album*) and *Strobilanthes* sp., especially *Strobilanthes kunthianus* (which blooms en masse once in a blue moon, i.e., roughly at an interval of 12 years).

Tea (*Camellia sinensis*) and Eucalyptus (*Eucalyptus grandis*; *E. globulus*) plantations have pervaded both basins (60 % of MRB and 10 % of PRB). Besides the plantations, intensive and ubiquitous vegetable farming is practiced in the sediment fills of the interfluves in MRB and upstream of PRB. Further, cardamom plantations are predominantly distributed in the downstream of MRB. Vast areas of farmlands (~12 % of the basin area) in the central region of PRB are under sugarcane and paddy cultivation. The regional landscape has enormous tourism potential and tourist inflow is high during non-monsoon period, i.e., December–May (Department of Tourism 2008). The river basins support a human population of ~0.1 million. Munnar and Devikolam (in MRB) and Maraiyur and Kandalur (in PRB) are the major towns, whereas settlement clusters (of smaller extent) are common and are associated with tea plantations in MRB and upstream of PRB.

Table 1 Soil characteristics of MRB and PRB (after SSO 2007)

Soil series	Anai Mudi	Pambadumpara	Venmani	Chinnar
	Both in MRB and PRB	MRB	MRB	PRB
Order	Ultisols	Ultisols	Inceptisols	Mollisols
Sub-order	Humults	Humults	Ustepts	Ustolls
Great group	Kandihumults	Kandihumults	Dystrustepts	Haplustolls
Sub-group	Typic Kandihumults	Ustic Kandihumults	Oxic Dystrustepts	Typic Haplustolls
Family	Clayey, mixed, isothermic	Clayey, mixed, isohyperthermic	Fine, mixed, isohyperthermic	Loamy skeletal, mixed, thermic
Extent	9° 15′–10° 29′N; 76° 56′–77° 25′E	9° 30′–10° 26′N; 76° 44′–77° 25′E	9° 27′–10° 22′N; 76° 48′–77° 11′E	10° 19′–10° 29′N; 77° 10′–77° 16′E
Pedogenesis	Gneissic parent; on steep to very steep slopes; above 1,200 m amsl	Gneissic parent; on steep to very steep slopes and hill tops; between 600–1,200 m amsl	Gneissic parent; on moderate to steep slopes; between 600–900 m amsl.	Gneissic parent; on gentle slopes of rain shadow region; between 400–900 m amsl
Colour	Dark reddish brown to dark brown (A horizon); dark reddish brown to reddish yellow (B horizon)	Dark reddish brown to dark brown (A horizon); yellowish red to red (B horizon)	Reddish brown to dark reddish brown (A horizon); reddish brown to red (B horizon)	Brown to dark greyish brown (A horizon); brown to very dark greyish brown (B horizon)
Texture	Silt loam to clay loam (A horizon); silty clay loam to clay (B horizon)	Silty clay to clay (A horizon); clay (B horizon)	Loam to clay (A horizon); gravelly sandy clay to gravelly clay (B horizon)	Sandy loam to sandy clay loam (A horizon); gravelly loamy sand to gravelly sandy loam (C horizon); presence of $CaCO_3$ nodules and mica flakes in the subsurface soil
Soil thickness	>150 cm	>180 cm	>150 cm	75–100 cm
Drainage	Well-drained	Moderately well-drained	Well-drained	Moderately well-drained
Permeability	Moderately rapid	Moderate	Moderately rapid	Moderately rapid
Productivity	High	Medium	Medium	Medium to high
Erodibility	Severe	Severe	Severe	Moderate
pH	Extremely to very strongly acidic	Extremely to very strongly acidic	Very strongly acidic	Neutral to mildly alkaline

3 Materials and Methods

River water samples from the main stream as well as major tributaries of MRB (n = 15; MW1–MW15) and PRB (n = 12; PW1–PW12) were collected in three different seasons, viz., monsoon (MON; July 2007), post-monsoon (POM; December 2007) and pre-monsoon (PRM; April 2008). The findings of this study are based on the analytical results of 45 water samples from MRB (n = 15 × 3) and 36 samples from PRB (n = 12 × 3). Water samples were collected at each sampling points, 20–30 cm below the water surface in prewashed (with dil. HCl) and labeled HDPE bottles (1.0 L). The sample bottles were pre-rinsed with water from the sampling sites before acquisition of the final sample. The samples for cation analysis were preserved by acidification with 2M HNO_3. Whatman No. 42 filter was used for the determination of total suspended solids (TSS) and the filtrate was analyzed following the standard procedures of APHA (Eaton et al. 2005).

Each sample was screened for various physico-chemical parameters such as pH, electrical conductivity (EC), total dissolved solids (TDS), total hardness (TH), major cations (Ca^{2+}, Mg^{2+}, Na^+, and K^+), major anions (Cl^-, SO_4^{2-} and HCO_3^-) and dissolved silica (H_4SiO_4). The data were analyzed for spatio-temporal variations, relationships with lithology, climate, land use, etc., hydrogeochemistry of other river systems, processes that controlled the observed variability, and relative contribution of natural and anthropogenic sources.

4 Results and Discussion

4.1 Spatio-temporal Variation of Hydrogeochemistry

Charge balance error (CBE) is a good measure of quality of the hydrogeochemical data of MRB and PRB, which is estimated as Eq. (1):

$$\text{CBE} = \frac{\text{TZ}^+ - \text{TZ}^-}{\text{TZ}^+ + \text{TZ}^-} \times 100 \tag{1}$$

where TZ^+ is the sum of cations and TZ^- is the sum of anions and all cations and anions are expressed in milliequivalents per liter.

According to CBE, the total dissolved cations and total dissolved anions of all the samples are well-balanced in that all the samples have normalized inorganic charge balance less than 10 %. Tables 2 and 3 provide a descriptive summary of the hydrogeochemical attributes (during the sampling seasons) of MRB and PRB. Both MRB and PRB waters are slightly acidic to slightly alkaline during all the seasons (i.e., MON: 6.99–7.52 in MRB, 6.75–7.55 in PRB; POM: 6.54–7.35 in MRB, 6.54–7.70 in PRB; PRM: 6.92–7.97 in MRB, 6.45–7.63 in PRB). Generally, river water in areas unaffected by pollution has a pH in the range between 6.50 and 8.50

Table 2 Spatio-temporal variation of hydrochemistry in MRB

Attribute	Unit	MON		POM		PRM	
		Mean ± SD	Range	Mean ± SD	Range	Mean ± SD	Range
pH		7.25 ± 0.17	6.99–7.52	6.93 ± 0.25	6.54–7.35	7.36 ± 0.33	6.92–7.97
EC	$\mu S\ cm^{-1}$	132.04 ± 30.03	87.62–215.62	259.33 ± 73.80	145.48–412.81	226.52 ± 50.48	151.63–359.89
TDS	$mg\ L^{-1}$	83.68 ± 18.97	59.11–135.88	165.71 ± 47.24	93.86–264.73	143.75 ± 32.22	95.90–228.96
TSS	$mg\ L^{-1}$	12.21 ± 2.41	9.20–18.43	13.08 ± 2.50	9.84–20.86	12.12 ± 3.75	8.55–24.69
TH	$mg\ L^{-1}$	29.60 ± 12.45	16.00–64.00	54.93 ± 14.30	36.00–80.00	55.47 ± 12.91	40.00–80.00
TA	$mg\ L^{-1}$	26.67 ± 9.03	12.00–48.00	56.00 ± 23.52	24.00–112.00	43.47 ± 15.99	24.00–92.00
Ca^{2+}	$\mu eq\ L^{-1}$	352.56 ± 174.30	167.66–839.32	570.76 ± 196.05	252.00–923.15	604.36 ± 177.28	335.83–1007.48
Mg^{2+}	$\mu eq\ L^{-1}$	240.03 ± 174.48	51.82–630.11	529.37 ± 155.84	369.35–847.28	506.23 ± 134.75	292.85–770.78
Na^{+}	$\mu eq\ L^{-1}$	261.15 ± 127.88	134.42–576.38	310.01 ± 120.89	194.88–608.13	160.98 ± 62.56	65.69–317.99
K^{+}	$\mu eq\ L^{-1}$	57.40 ± 18.37	31.71–92.56	54.21 ± 25.18	25.83–99.47	59.68 ± 20.41	29.15–92.56
Cl^{-}	$\mu eq\ L^{-1}$	549.55 ± 120.51	400.30–880.43	1339.54 ± 341.68	728.66–1840.42	1178.91 ± 242.78	800.32–1600.35
SO_4^{2-}	$\mu eq\ L^{-1}$	61.46 ± 49.38	3.33–173.64	80.16 ± 50.51	19.99–213.82	92.65 ± 48.15	33.31–213.82
HCO_3^{-}	$\mu eq\ L^{-1}$	533.22 ± 180.54	239.95–959.80	1119.76 ± 470.28	479.90–2239.53	869.15 ± 319.74	479.90–1839.61
H_4SiO_4	$\mu eq\ L^{-1}$	140.01 ± 50.10	77.00–239.73	210.40 ± 66.49	114.04–324.22	187.18 ± 55.57	116.95–302.99

MON monsoon, *POM* post-monsoon, *PRM* pre-monsoon, *SD* standard deviation, *EC* electrical conductivity, *TDS* total dissolved solids, *TSS* total suspended solids, *TH* total hardness, *TA* total alkalinity

Table 3 Spatio-temporal variation of hydrochemistry in PRB

Attribute	Unit	MON		POM		PRM	
		Mean ± SD	Range	Mean ± SD	Range	Mean ± SD	Range
pH		7.20 ± 0.23	6.75–7.55	7.11 ± 0.34	6.54–7.70	7.15 ± 0.31	6.45–7.63
EC	$\mu S\ cm^{-1}$	146.97 ± 66.38	70.09–322.48	275.28 ± 133.10	165.89–620.98	290.08 ± 192.20	136.94–855.12
TDS	$mg\ L^{-1}$	92.89 ± 42.01	46.51–204.18	175.34 ± 83.89	104.45–394.18	185.53 ± 120.98	88.35–540.07
TSS	$mg\ L^{-1}$	12.65 ± 3.11	8.04–18.48	16.25 ± 5.40	10.81–28.43	15.04 ± 5.34	9.49–26.52
TH	$mg\ L^{-1}$	36.00 ± 19.60	16.00–76.00	54.00 ± 20.07	32.00–104.00	65.67 ± 24.63	40.00–136.00
TA	$mg\ L^{-1}$	37.00 ± 19.23	16.00–88.00	68.67 ± 36.42	40.00–160.00	70.00 ± 55.75	28.00–232.00
Ca^{2+}	$\mu eq\ L^{-1}$	300.81 ± 115.70	167.66–503.49	524.53 ± 196.46	335.83–1091.31	622.46 ± 200.79	335.83–1091.31
Mg^{2+}	$\mu eq\ L^{-1}$	420.49 ± 362.41	60.05–1272.56	556.97 ± 255.52	292.85–991.23	692.77 ± 316.93	457.37–1633.68
Na^{+}	$\mu eq\ L^{-1}$	349.49 ± 325.72	54.38–1131.44	584.86 ± 571.15	90.92–2210.24	440.69 ± 533.90	62.64–2002.74
K^{+}	$\mu eq\ L^{-1}$	44.75 ± 11.09	30.94–69.55	51.10 ± 15.60	27.62–76.45	48.41 ± 21.77	15.09–100.75
Cl^{-}	$\mu eq\ L^{-1}$	373.52 ± 92.40	240.07–560.25	1073.63 ± 347.04	640.08–1840.42	1147.28 ± 494.68	720.20–2564.57
SO_4^{2-}	$\mu eq\ L^{-1}$	71.01 ± 49.32	3.33–143.66	88.52 ± 49.27	13.32–160.31	135.89 ± 63.02	40.18–237.14
HCO_3^{-}	$\mu eq\ L^{-1}$	739.84 ± 384.53	319.93–1759.63	1373.04 ± 728.15	799.83–3199.33	1399.71 ± 1114.82	559.88–4639.03
H_4SiO_4	$\mu eq\ L^{-1}$	121.43 ± 43.31	56.19–193.53	144.66 ± 65.23	64.51–282.18	149.90 ± 57.15	64.51–254.71

MON monsoon, *POM* post-monsoon, *PRM* pre-monsoon, *SD* standard deviation, *EC* electrical conductivity, *TDS* total dissolved solids, *TSS* total suspended solids, *TH* total hardness, *TA* total alkalinity

(Hem 1985) and these mountain rivers are no exception. In both basins, EC, an effective measure of dissolved ionic strength, shows relatively larger values during POM (mean = 259.33 ± 73.80 μS cm^{-1} in MRB; 275.28 ± 133.10 μS cm^{-1} in PRB) and PRM (mean = 226.52 ± 50.48 μS cm^{-1} in MRB and 290.08 ± 192.20 μS cm^{-1} in PRB), compared to MON (mean = 132.04 ± 30.03 and 146.97 ± 66.38 μS cm^{-1} respectively in MRB and PRB). Similarly, MON samples register lower values for the total dissolved ions in both MRB and PRB, whereas POM and PRM record comparatively higher values (Tables 2 and 3). Mean TDS of MRB during MON, POM and PRM is 83.68 ± 18.97 mg L^{-1} (range = 59.11–135.88 mg L^{-1}), 165.71 ± 47.24 mg L^{-1} (range = 93.86–264.73 mg L^{-1}) and 143.75 ± 32.22 mg L^{-1} (range = 95.90–228.96 mg L^{-1}) respectively. In PRB, mean TDS is 92.89 ± 42.01 mg L^{-1} (range = 46.51–204.18 mg L^{-1}) during MON, 175.34 ± 83.89 mg L^{-1} (range = 104.45–394.18 mg L^{-1}) during POM and 185.53 ± 120.98 mg L^{-1} (range = 88.35–540.07 mg L^{-1}) during PRM. According to Gaillardet et al. (1999), most of the rivers of the world have TDS less than 500 mg L^{-1}, while the exceptions are representatives of either pollution or semi-arid and arid climate. Stallard and Edmond (1983, 1987) stated that relatively high concentrations of TDS in river water imply weathering of evaporites, while waters with low TDS characterize weathering of silicates. In comparison with other rivers draining the Western Ghats (e.g., Achankovil, 54.0 mg L^{-1}, Prasad and Ramanathan 2005; Netravati, 38.0 mg L^{-1}, Gurumurthy et al. 2012; upstream of Krishna, 112.0 mg L^{-1}; west-flowing rivers of the Deccan Traps, 82.0 mg L^{-1}, Das et al. 2005), mean TDS values of MRB and PRB are relatively higher, which might be due to the differences in basin size, climate, discharge, lithology and the intensity of anthropogenic interferences.

When compared to MRB, PRB shows an elevated level of ionic abundance (Fig. 3) and hence a higher degree of mineralization, which might be due to semi-arid climate and the contribution of carbonates and evaporites occurring in the soil and shallow regolith (see Table 1). Even though lithology of MRB and PRB is similar, the mainstream of PRB (i.e., Pambar) prefers the terrain underlain by weaker hornblende biotite gneiss, sandwiched between granite gneiss to the NW and SE (Fig. 2). In addition, the areal extent of hornblende biotite gneiss is also relatively larger in PRB, compared to MRB. Hence, the higher dissolved flux of PRB might also be a result of the differences in lithology between the basins. For example, basins underlain by mafic rocks have relatively high ion flux rates compared to basins underlain by felsic rocks (Meybeck 1987). However, in contrast to the general trend, ions such as K^+, Cl^- and H_4SiO_4 show relatively higher concentrations in MRB during all the sampling seasons, implying significance of agriculture- and tourism-related activities (i.e., K^+, Cl^-) as well as intense chemical weathering of silicate minerals (i.e., H_4SiO_4) in the tropical humid climate. Unlike major cations and anions derived from multiple sources (i.e., lithologic, atmospheric, biologic and anthropogenic), H_4SiO_4 is predominantly derived from the dissolution of primary silicate minerals (Eq. 2).

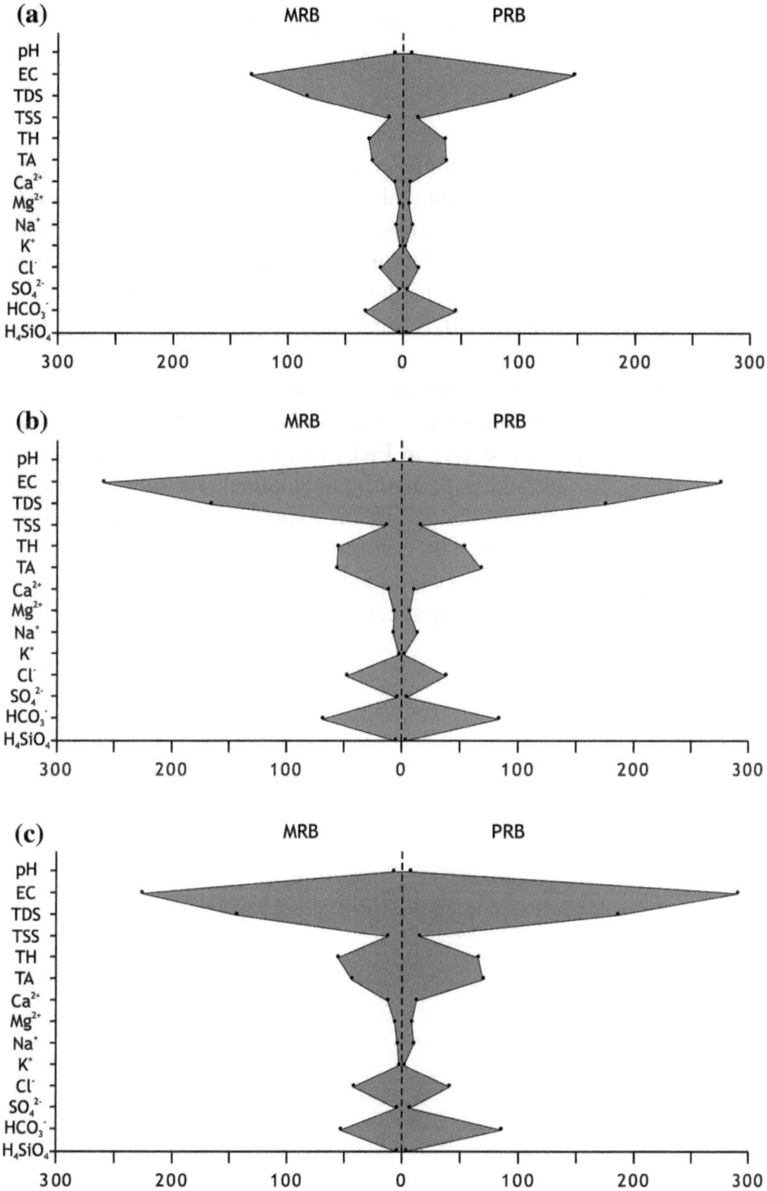

Fig. 3 Comparison of hydrogeochemistry between MRB and PRB during **a** MON, **b** POM and **c** PRM. All units in mg L^{-1} except EC ($\mu S\ cm^{-1}$) and pH

$$(\text{Na, K, Ca, Mg})\text{Silicate} + H_2CO_3 \rightarrow H_4SiO_4 + HCO_3^- + Na^+ + K^+ + Ca^{2+} + Mg^{2+} + \text{Solids} \quad (2)$$

Further, the riverine silica is mainly controlled by the natural processes which contribute approximately 80 % of annual silica load into the ocean (Treguer et al. 1995), whereas anthropogenic and atmospheric sources have only negligible contributions (Berner and Berner 1996; Nixon 2003). Hence, the spatial variability of H_4SiO_4 in PRB is mainly attributed to the difference in climate between upstream and downstream tracts, while in MRB local factors (e.g., soil moisture, reservoirs) might have decisive roles.

An inter-basin comparison of the downstream variation of hydrogeochemical attributes implies a general decrease in the attributes of water samples in MRB (Fig. 4), whereas the opposite is true in PRB, i.e., an increasing downstream trend for attributes (Fig. 5). In addition, hydrogeochemical attributes generally vary consistently within the specific zones of mainstream (i.e., upstream, midstream and downstream) of MRB and PRB. The commonly observed downstream trends are (a) a pronounced rise or fall of dissolved load towards downstream, (b) a consistent concentration gradient through the upstream zone that is considerably different and

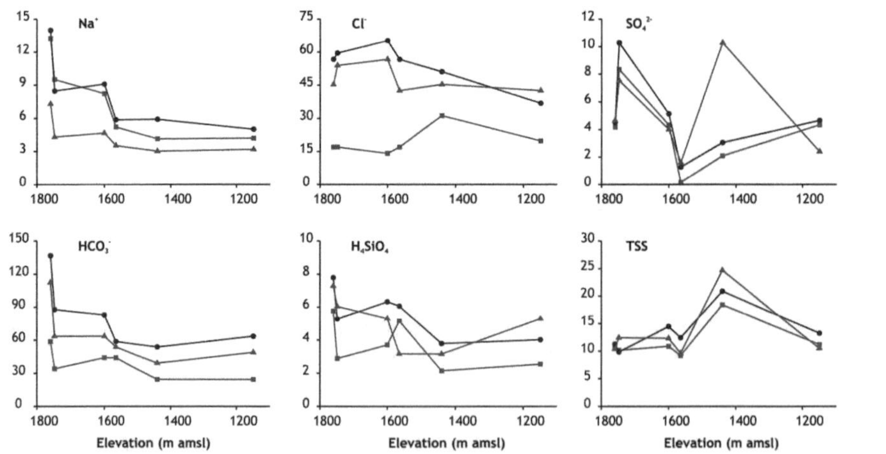

Fig. 4 Downstream variation of hydrogeochemistry, MRB. All units in mg L^{-1} except EC ($\mu S\ cm^{-1}$) and pH

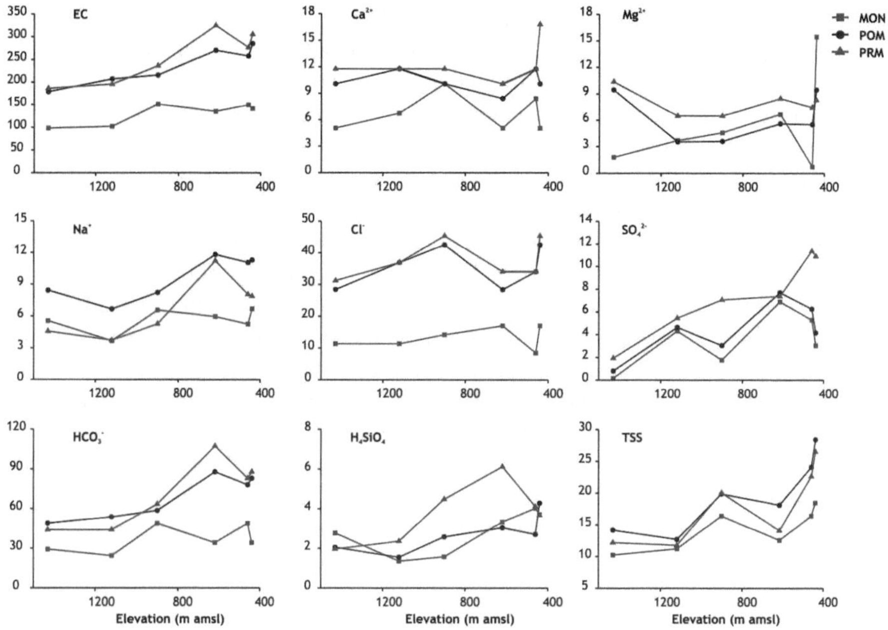

Fig. 5 Downstream variation of hydrogeochemistry, PRB. All units in mg L^{-1} except EC (μS cm^{-1}) and pH

sometimes opposite to that of downstream and (c) a trend reversal during some sampling periods, suggesting highly complex spatio-temporal relationships between hydrogeochemical attributes and their determinant factors.

In MRB, the decreasing downstream trend of hydrogeochemistry is mainly attributed to dilution due to rise in discharge. Such dilution is reported by Mortatti (1995). He had studied the relationships between the concentrations of the major dissolved ions and the river discharge in the lower Amazon and reported the decreasing patterns, similar to the theoretical dilution curves defined by Probst (1992). However, TSS shows a reversal of trend in MRB (i.e., increases toward downstream; Fig. 4) with sudden drops at MW7 (elevation = 1,564 m amsl) and MW15 (elevation = 1,149 m amsl). The increasing trend of TSS is possibly a result of logging and related activities in plantations as well as erosion from hillslopes, which add to the TSS load in the streams. However, the stations (MW7 and MW15) are located at the immediate downstream of Mattuppetti dam (i.e., MW7) and RA Head Works (i.e., MW15) respectively and the sudden drop of TSS in these stations can be a result of damming whereby the reservoirs effectively subtracted a bulk of TSS. Milliman and Mead (1983) pointed out that suspended sediment loads normally transported to the oceans by Colorado River were reduced to nearly nothing and in Mississippi River by one third due to the construction of dams.

From Fig. 4, it is also evident that only EC, Na^+ and HCO_3^- show a more or less discernible downstream decrease in MRB. But other attributes exhibit interim variations in their downstream trend (e.g., Mg^{2+}, H_4SiO_4), which might be a reflection of the contributions by the tributaries. Contrastingly, in PRB, an overall increasing trend of dissolved constituents is obvious (Fig. 5), which is chiefly due to the differing rainfall conditions between the upstream and downstream segments. In other words, the influx of highly "mineralized" water (i.e., with higher dissolved constituents) from the semi-arid downstream segment of PRB plays a significant role in enriching the dissolved load.

In PRB, samples from PW9 and PW11 in two downstream tributaries of Pambar (viz., Alampatti *odai* and Atti *odai*) do exhibit relatively higher solute levels, compared to rest of the sampling stations and in all the sampling periods. These streams drain the dry deciduous forests of Chinnar Wildlife Sanctuary, which is practically devoid of any anthropogenic activities with the exception of a few tribal settlements. Hence, these tributaries (PW9 and PW11) are considered as "hot spots" (Parkin 1987; Hill et al. 2000), which by definition is a specific form of spatial heterogeneity due to higher biogeochemical reaction rates. McClain et al. (2003) also described biogeochemical hot spots as areas (or patches) that show disproportionately high reaction rates relative to the surrounding area (or matrix).

Generally, discharge in the rivers draining the southern Western Ghats during monsoon is predominantly contributed by rainfall as well as surface runoff and recharges the groundwater reservoir (i.e., effluent rivers). But, the flow in these rivers during non-monsoon season is mainly contributed by the groundwater system (i.e., influent rivers). Hence, the relatively lower levels of various hydrogeochemical attributes measured on MON samples (Tables 2 and 3) can be directly related to dilution due to heavy monsoon rainfall (i.e., 1,146 mm during July, 2007), leading to higher water discharge, during the sampling period. However, the higher ionic content during the dry summer season is a reflection of combined contribution from aquifers and anthropogenic activities (e.g., agricultural and tourism-related). Hence, the temporal variability of hydrogeochemistry in MRB and PRB can also be attributed to the changes in hydrologic pathways between the sampling seasons. Christophersen et al. (1990), Wheater et al. (1990) and Neal et al. (1992) suggested that the rapid changes in water chemistry in upland regions are predominantly a result of changing flow paths during hydrologic events. Hence stream water may be considered to be consisting of two or more end members of differing chemistries, the proportions of which change with discharge. Mortatti and Probst (2003) documented that the seasonality in stream water chemistry is a signature of silicate weathering processes, which are highly dependent on the fluctuations in runoff. Rice and Bricker (1995) also demonstrated that there are strong seasonal cycles in the water chemistry resulting from seasonal hydrologic processes superimposed on geologically controlled groundwater compositions. Hence, the temporal variability of hydrogeochemistry can be attributed to the variation in rainfall (and thereby discharge), changes in hydrologic pathways as well as intensity of various anthropogenic activities.

Broadly, cation abundance in MRB has the following order during the sampling seasons, i.e., $Ca^{2+} > Na^{+} > Mg^{2+} > K^{+}$ (MON), $Ca^{2+} > Mg^{2+} > Na^{+} > K^{+}$ (POM and PRM). In PRB, the order of abundance of cations is $Mg^{2+} > Na^{+} > Ca^{2+} > K^{+}$,

$Na^+ > Mg^{2+} > Ca^{2+} > K^+$ and $Mg^{2+} > Ca^{2+} > Na^+ > K^+$ in MON, POM and PRM respectively. In PRB, the dominance of Mg^{2+} over Ca^{2+} might be a reflection of chemical weathering of ferromagnesian minerals of the basement rocks. Relatively lower levels of K^+, compared to Na^+, suggest conservative behavior of K^+ in river systems (Garrels and Mackenzie 1971b). Moreover, in MRB, the K^+ content is only about one-fifth of the Na^+, whereas in PRB, it is roughly one-tenth of Na^+. Such behavior of Na^+ in natural systems may be explained by its tendency to remain in solution compared to K^+, which shows strong affinity for reincorporation in the (solid) weathering products (Hem 1985). The anion abundance in MRB is $Cl^- > HCO_3^- > SO_4^{2-}$ in all the three seasons, but in PRB it is $HCO_3^- > Cl^- > SO_4^{2-}$. In MRB and PRB, the temporal variation of major cations and anions closely reflects the pattern of EC.

4.2 Hydrogeochemical Drivers

The bivariate plot of weight ratio of $Na^+/(Na^+ + Ca^{2+})$ versus TDS suggesting major natural mechanisms controlling surface water chemistry (after Gibbs 1970), provides significant information on the relative importance of climate and lithology in three different perspectives, i.e., (a) climate via atmospheric precipitation, (b) rock weathering and (c) climate via evaporation and fractional crystallization. The hydrogeochemical data of MRB and PRB on Gibbs (1970) plot (Fig. 6) clearly demonstrate the significance of rock weathering and mineral dissolution controlling chemical composition of the river waters. In comparison with other rivers of India, hydrogeochemistry of MRB is similar to the headwaters of Krishna River as well as the small west-flowing rivers draining the Deccan Traps (Das et al. 2005), whereas the PRB data show better similarity with Godavari (Jha et al. 2009), one of the east-flowing rivers of Peninsular India.

While the MRB data (during all the seasons) are clustered in the zone of rock dominance (Fig. 6), the PRB data fall along a line trending from the rock dominance zone (i.e., upstream samples) towards the zone of evaporation-crystallization dominance (i.e., downstream samples). This can be a result of the semi-arid climate prevailing in the downstream of PRB, where evaporation is significant in raising the TDS. In PRB, TDS of the downstream samples reaches up to 550 mg L^{-1} and such elevated TDS values emphasize the importance of weathering of silicates as well as dissolution of carbonates and soil evaporites. Gupta et al. (2011), in their hydrogeochemical study of Narmada River in western India, reported relatively high TDS values and inferred weathering of carbonate/saline–alkaline minerals as the reason for such a phenomenon.

The Na^+-normalized Ca^{2+} versus Na^+-normalized HCO_3^- plot (Fig. 7; after Gaillardet et al. 1999) of MRB and PRB shows that most of the samples occur outside the domain of silicate weathering, yet the sample poles align parallel to the trend line between silicate and carbonate weathering domains, which can be inferred as possible mixing between silicate and carbonate end members. Even

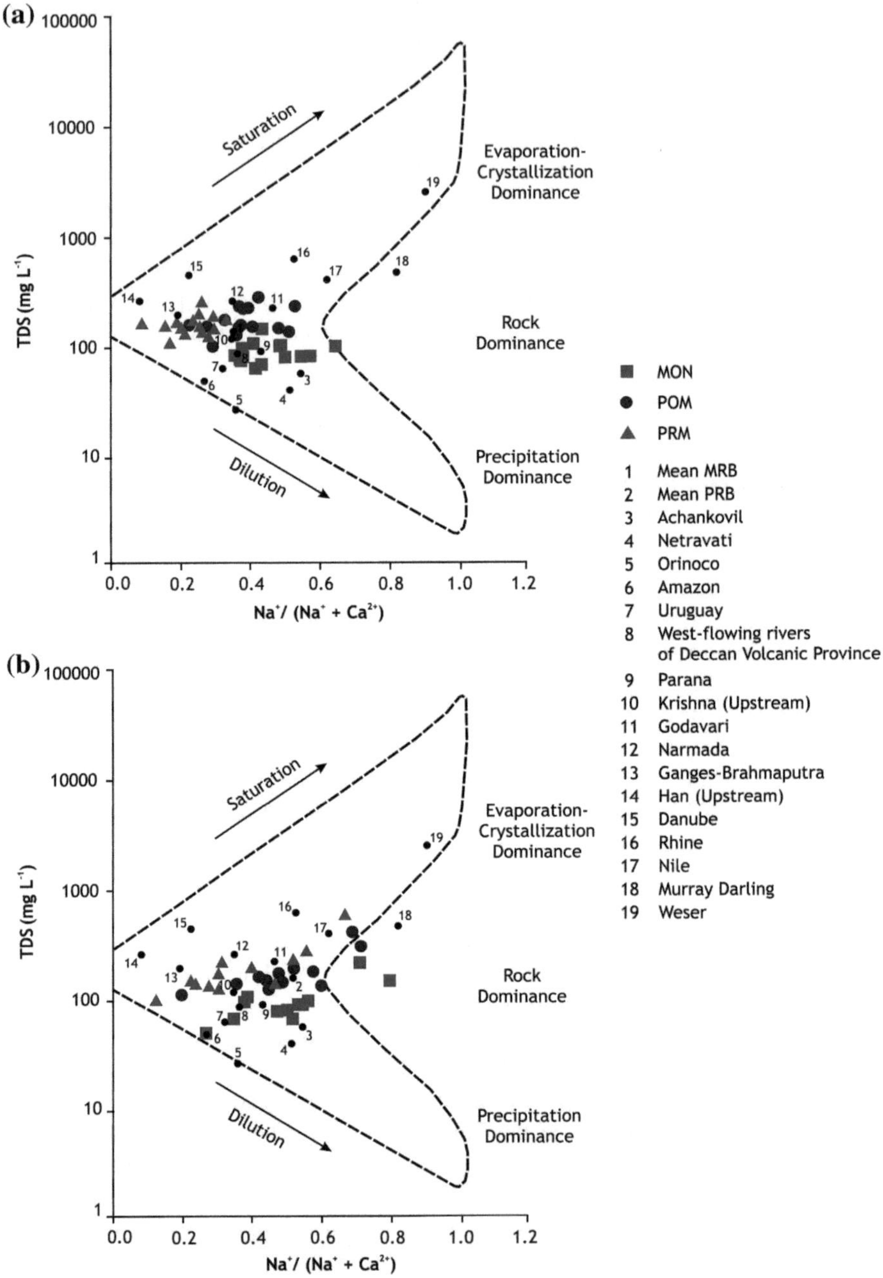

Fig. 6 Gibbs (1970) plot of **a** MRB and **b** PRB. Comparison of hydrogeochemistry data from Galy and France-Lanord (1999), Gaillardet et al. (1999), Mortatti and Probst (2003), Das et al. (2005), Prasad and Ramanathan (2005), Li and Zhang (2008), Jha et al. (2009), Gupta et al. (2011) and Gurumurthy et al. (2012)

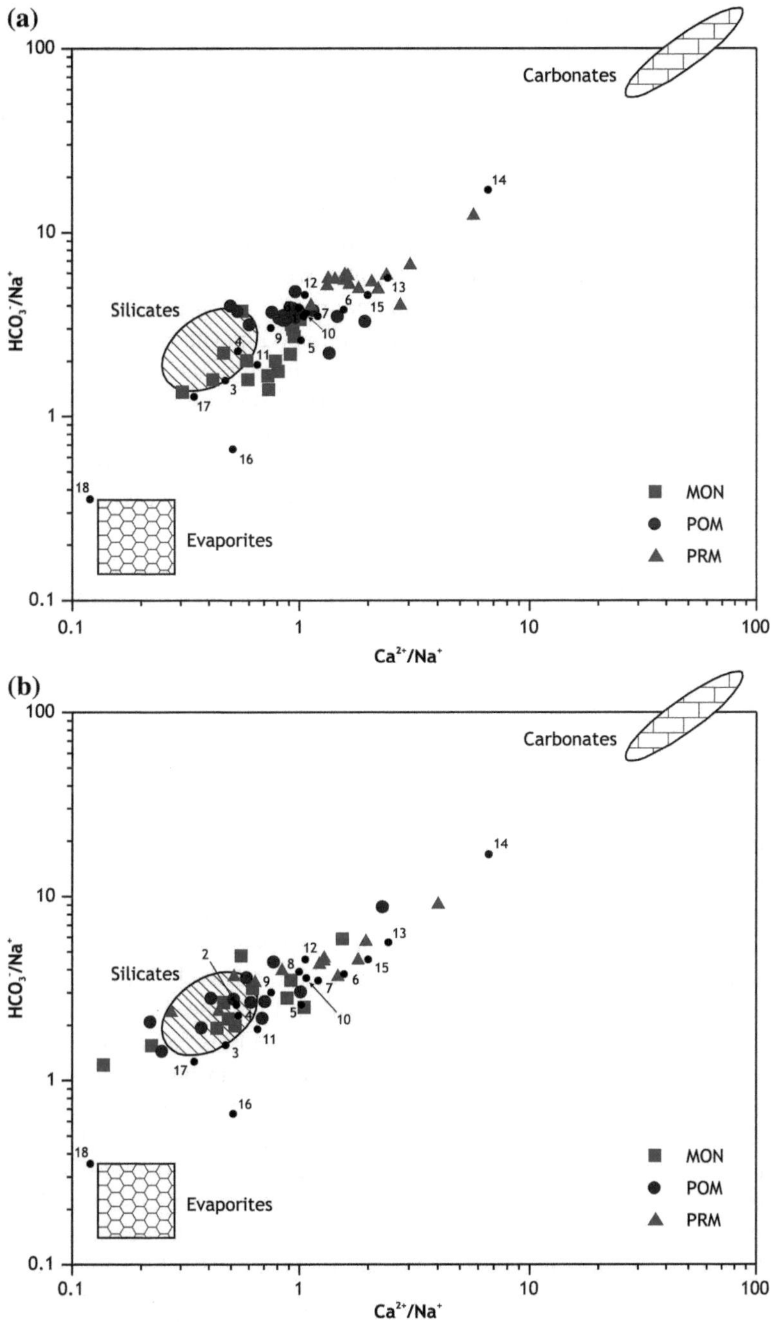

Fig. 7 Mixing diagrams (after Gaillardet et al. 1999) using Na^+-normalized Ca^{2+} versus Na^+-normalized HCO_3^- of **a** MRB and **b** PRB. The numeric river codes are same as Fig. 6

though silicate rocks dominate MRB and PRB, carbonate sources (e.g., crystalline limestone, carbonatite, kankar nodules) also co-exist in the basins (see Sect. 2). In spite of relatively minor occurrences, the carbonate minerals weather in orders of magnitude faster than Ca-Mg silicate minerals (Gaillardet et al. 1999). In addition, preferential weathering of Ca- and Mg-rich silicate minerals as well as contribution from secondary calcite can also have significance in such mixing pattern in MRB and PRB. Dessert et al. (2003) and Gupta et al. (2011) observed similar trends for the rivers draining the basaltic terrains in India. Gupta et al. (2011) opined that weathering of calcite disseminated in silicates would be limited by the degree of their exposure; nonetheless, high rate of sediment erosion coupled with monsoon climate is bound to enhance the calcite dissolution (Gupta and Chakrapani 2005). In addition, an overall linearity is observed between molar ratios of Na^+-normalized Ca^{2+} versus Na^+-normalized HCO_3^- for PRB during all the seasons (Fig. 7). However, in MRB, the sample poles are more scattered. In comparison with other rivers of India, mean of MRB plots beside upstream of Krishna River and west-flowing rivers of the Deccan Traps, while mean of PRB plots with Netravati and Godavari rivers.

In MRB and PRB, $Ca^{2+} + Mg^{2+}/Na^+ + K^+$ ratios for MON (mean = 1.96 and 2.23 in MRB and PRB respectively; Table 4), POM (mean = 3.29 and 2.41) and PRM (mean = 5.74 and 4.40), also suggest sources other than silicate weathering, e.g., carbonate dissolution, anthropogenic (domestic and farm/plantation residues) inputs etc. In both the basins, relatively lower ratios of $Ca^{2+} + Mg^{2+}/Na^+ + K^+$ in MON (in comparison with POM and PRM) can be attributed to the higher overland flow and resultant addition of Na^+ and K^+ from the terrestrial anthropogenic sources. The $H_4SiO_4/(Na^+ + K^+)$ ratio is also a proxy related to the intensity of silicate weathering (Edmond et al. 1995) and the mean ratios during MON, POM and PRM for MRB and PRB are 0.48 and 0.42, 0.59 and 0.30 and 0.92 and 0.46 respectively (Table 4), suggesting additional sources of Na^+ and K^+ other than silicate weathering. This observation affirms the inference of additional anthropogenic sources of Na^+ and K^+. In comparison with MRB, PRB has relatively smaller ratios during all the seasons, which might be due to relatively lower rate of silicate weathering (and hence lower H_4SiO_4) in semi-arid climate. In PRB, due to semi-aridity, the contribution of Na^+ and K^+ by soil evaporites can also be a factor for relatively lower $H_4SiO_4/(Na^+ + K^+)$ ratios. Further, contribution by dissolution of soil evaporites in PRB (during MON and POM) is also evidenced by relatively lower Ca^{2+}/Na^+ ratios (Fig. 6; e.g., semi-arid, downstream samples, PW9, PW11).

Relatively higher Mg^{2+}/Ca^{2+} ratios (i.e., >0.50; Table 4) in MRB and PRB suggest the contribution of Mg^{2+} from weathering of ferromagnesian minerals such as hornblende and biotite in the basement rocks (Fig. 2). When compared to MRB (i.e., 0.82, 1.02 and 0.89 in MON, POM and PRM respectively; Table 4), the PRB has relatively larger ratios during all the sampling seasons (i.e., 1.50, 1.09 and 1.12 in MON, POM and PRM respectively), which can be attributed to the relatively larger areal extent of hornblende-biotite-gneiss in PRB. Meybeck (1987) suggested that ion flux rates from basins underlain by amphibolites (mafic) are approximately five times greater than watersheds underlain by granitic (felsic) rocks and the same

Table 4 Various ionic ratios used in the present study of MRB and PRB

Ionic ratio	MRB						PRB					
	MON		POM		PRM		MON		POM		PRM	
	Mean	Range	Mean	Range	Mean	Range	Mean	Range	Mean	Range	Mean	Range
[a]$Ca^{2+} + Mg^{2+}/Na^{+} + K^{+}$	1.96	1.10–3.43	3.29	1.60–6.74	5.74	2.66–15.20	2.23	1.07–4.64	2.41	0.91–6.76	4.40	1.32–12.36
[a]$H_4SiO_4/Na^{+} + K^{+}$	0.48	0.24–0.82	0.59	0.45–0.85	0.92	0.63–2.34	0.42	0.14–0.81	0.30	0.12–0.60	0.46	0.10–0.83
[a]Mg^{2+}/Ca^{2+}	0.82	0.09–2.50	1.02	0.53–1.86	0.89	0.50–1.63	1.50	0.14–5.05	1.09	0.50–1.87	1.12	0.81–1.50
[b]Ca^{2+}/Mg^{2+}	2.67	0.40–11.33	1.13	0.54–1.90	1.25	0.61–2.01	1.43	0.20–6.99	1.06	0.54–2.01	0.94	0.67–1.23
[a]$Ca^{2+} + Mg^{2+}/HCO_3^{-}$	1.19	0.57–2.34	1.12	0.48–2.50	1.37	0.87–2.50	1.07	0.44–2.72	0.86	0.60–1.60	1.16	0.59–2.00
[a]Ca^{2+}/SO_4^{2-}	26.60	1.45–125.98	9.64	3.14–22.04	8.30	2.75–15.44	15.20	1.75–75.65	10.33	2.58–31.49	5.59	2.48–14.62
[a]$HCO_3^{-}/HCO_3^{-} + SO_4^{2-}$	0.90	0.73–1.00	0.93	0.85–0.98	0.90	0.75–0.96	0.91	0.80–0.99	0.94	0.87–0.98	0.90	0.85–0.96

MON monsoon, *POM* post-monsoon, *PRM* pre-monsoon

[a] Ratios derived from μeq values

[b] Ratios derived from μmolar concentrations

may hold true for the dominance of Mg^{2+} in PRB. In addition, Stallard and Edmond (1983), Meybeck (1987) and Bluth and Kump (1994) also underscored the well-established dependency of bedrock weathering rates on basin lithology. The mean Ca^{2+}/Mg^{2+} molar concentrations for the andesites/greenstones, diorites, gabbros and granites are 1.20, 1.18, 1.32 and 3.44 respectively (Harmon et al. 2009). Both in MRB (except during MON) and PRB, mean values of Ca^{2+}/Mg^{2+} molar ratios range from 0.94 to 1.43 (Table 4), which are far below the global average, i.e., 2.40, (Harmon et al. 2009). Even though, both MRB and PRB have granitoids, existences of relatively lower ratios might be the result of weathering of Mg-rich minerals. This is further supported by the relatively higher molar ratios of Ca^{2+}/Mg^{2+} in MRB (compared to PRB), where granitoids (granite and granite gneiss) are the main lithologic types. In addition, Ca^{2+}/Mg^{2+} ratios nearing unity (~ 1.0) may also reflect near congruent dissolution of Mg- and Ca-rich minerals (Harmon et al. 2009). The molar ratios of MRB and PRB are also comparable with several rivers draining the Western Ghats (Prasad and Ramanathan 2005; Das et al. 2005; Gurumurthy et al. 2012), e.g., Achankovil (1.37), Netravati (1.40), Bhima (1.26), upstream of Krishna (1.57), and west-flowing rivers in the Deccan traps (1.30).

The major contributions of SO_4^{2-} in surface water are from dissolution of gypsum, oxidation of pyrite, pollution, volcanism, rainout of natural biogenic emissions and cyclic salts (Berner and Berner 1996). However, in MRB and PRB, most of the SO_4^{2-} in the water samples might be derived from the fertilizers (e.g., ammonium sulphate) applied in the farmlands. $Ca^{2+} + Mg^{2+}/HCO_3^-$ ratio provides significant information on the importance of pyrite oxidation and gypsum dissolution relative to weathering reactions involving carbonic acid (Moon et al. 2007). In MRB and PRB, $Ca^{2+} + Mg^{2+}/HCO_3^-$ ratios (except POM in PRB) are greater than 1.0 (Table 4), indicating that a small proportion of the divalent cations is balanced by anions other than HCO_3^- (i.e., predominantly Cl^-). Such an enrichment of Ca^{2+} and Mg^{2+} in MRB and PRB results when weathering and anthropogenic contributions (e.g., fertilizers) co-occur. A comparison of PRB with MRB suggests that $Ca^{2+} + Mg^{2+}/HCO_3^-$ ratios of all the sampling seasons in PRB are relatively smaller (Table 4), suggesting lesser intensity of anthropogenic activities in the catchment. According to Moon et al. (2007), weathering of carbonates or Ca-Mg-silicates by sulfuric acid also results in an excess of Ca^{2+} and Mg^{2+} over HCO_3. Even though gypsum dissolution and acid hydrolysis coupled with sulphide oxidation may also increase $Ca^{2+} + Mg^{2+}/HCO_3^-$ ratio (Fairchild et al. 1994; Hodson et al. 2002; Moon et al. 2007), their contributions are comparably lesser in these basins. This is further confirmed by the higher Ca^{2+}/SO_4^{2-} ratios (Table 4) in all the three sampling seasons, indicating the supply of protons to enable chemical weathering predominantly by H_2CO_3 (Stallard and Edmond 1987). The ratio of HCO_3^- to $HCO_3^- + SO_4^{2-}$ is also used to characterize the relative importance of two major proton-producing reactions: carbonization and oxidation of sulfides (Prasad and Ramanathan 2005). The ratios of HCO_3^- to $HCO_3^- + SO_4^{2-}$ in MRB and PRB (Table 4) are also closer to unity, suggesting carbonization reaction involving dissolution and acid hydrolysis, which draws protons from atmospheric sources.

Anthropogenic signatures in hydrogeochemistry of MRB and PRB are evident in the bivariate plot of Na^+ versus Cl^- (Fig. 8), in which the data poles plot above 1:1 equiline implying an enrichment of Cl^-. Higher molar ratios between Cl^- and Na^+ (>1.0) indicate multiple sources for these ions such that Cl^- is predominantly from anthropogenic activities, while Na^+ is from weathering. However, supply of these ions from atmospheric deposition (e.g., rainfall, aerosols) as well as dissolution of soil salts might also have significance. Peters and Ratcliffe (1998) suggested supply of Cl^- from rainwater and then concentration by evaporation within the shallow soil horizon. Trace level of Cl^- (for OH^-) in amphibole minerals in the rocks is also a minor natural source (Buell and Peters 1988). Gaillardet et al. (1999) explained such Cl^- enrichment and Na^+ depletion (in Indian rivers) either as a result of pollution or presence of alkali soils (in semi-arid and arid zones) where minerals such as $NaHCO_3$, Na_2CO_3 and $CaCO_3$ precipitate. This might be true for the semi-arid zones of PRB as Chinnar soil series in semi-arid segment is slightly alkaline (Table 1), but enrichment of Cl^- in humid MRB could be from anthropogenic sources. Further, it is also evident that MRB has relatively higher ratios during all the seasons compared to PRB (Fig. 8), implying relatively larger contribution of Cl^- as a result of intense anthropogenic activities. Even though anthropogenic interferences are common to both basins, the intensity is relatively higher in MRB, compared to PRB. This is further supported by Jenkins et al. (1995), who reported comparatively higher concentrations of acid anions (e.g., Cl^- and SO_4^{2-}) in stream water draining agricultural catchments of Himalayas as a result of mineral fertilizer inputs as well as from greater water use for irrigation and, therefore, potentially increased evapotranspiration losses that would increase the concentration of Cl^-, which is considered as conservative.

4.3 Characterization of Water Types

The use of major ions as natural tracers (Back 1961) is a common method to delineate generic water types (or hydrogeochemical facies). The evolution of hydrogeochemical composition of samples of MRB and PRB can be explained by plotting the major cations (Ca^{2+}, Mg^{2+}, Na^+ and K^+) and anions (HCO_3^-, SO_4^{2-} and Cl^-) in the Piper (1944) diagram (Fig. 9). In the diagram, grouping of water types is based on the hydrogeochemical similarities and the hydrogeochemical relationships are presented in more precise terms (Walton 1970; Todd 2001), which is useful for exploring trends that provide insights into various processes (Hem 1985). Harvey et al. (2002) suggested that Piper diagrams show the effects of various factors, including major ion composition of possible source waters as well as the proportions of mixing between the source waters in samples.

In both MRB and PRB, during all sampling seasons, dominant cation facies are principally mixed type or 'no dominant type'. But, dominant anion facies of MRB is Cl^- type, while that of PRB is HCO_3^- type. In Fig. 9, sample poles of PRB are scattered on either sides of the HCO_3-Cl line, while MRB samples fall above the

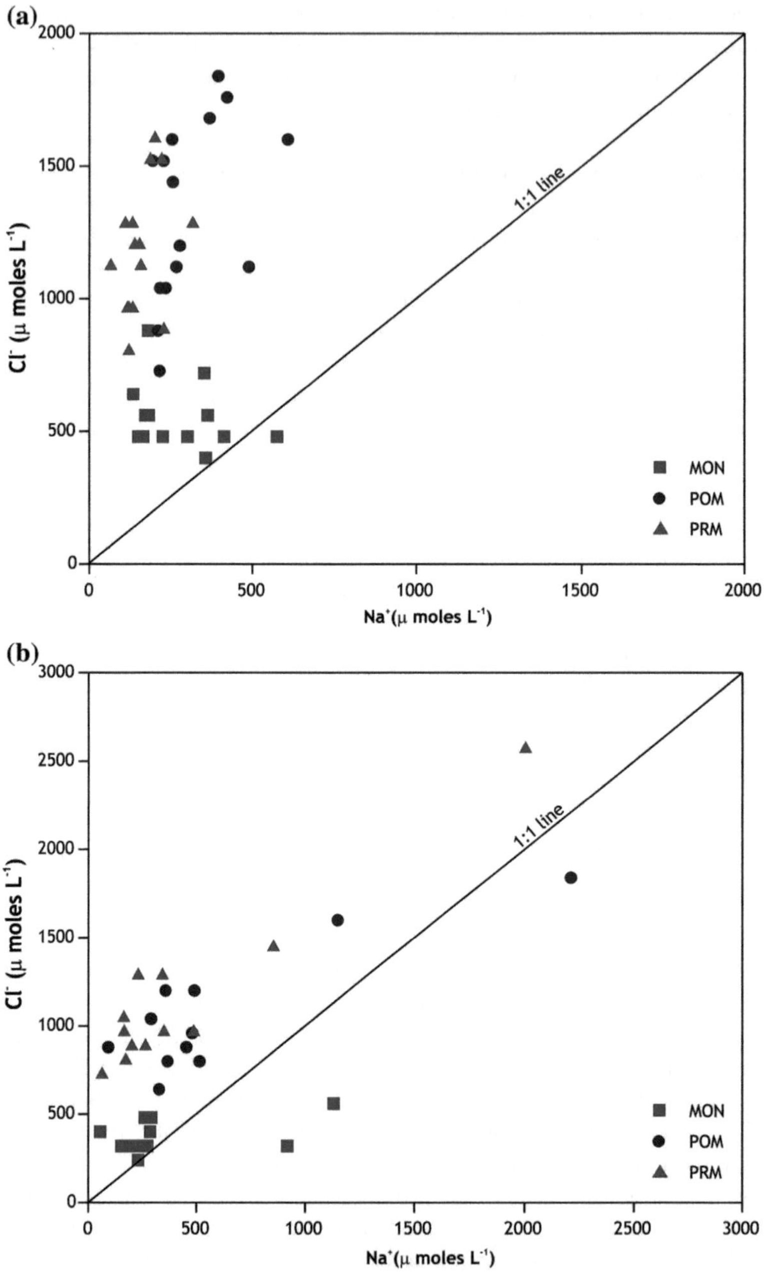

Fig. 8 Bivariate plot of Na^+ versus Cl^- of **a** MRB and **b** PRB

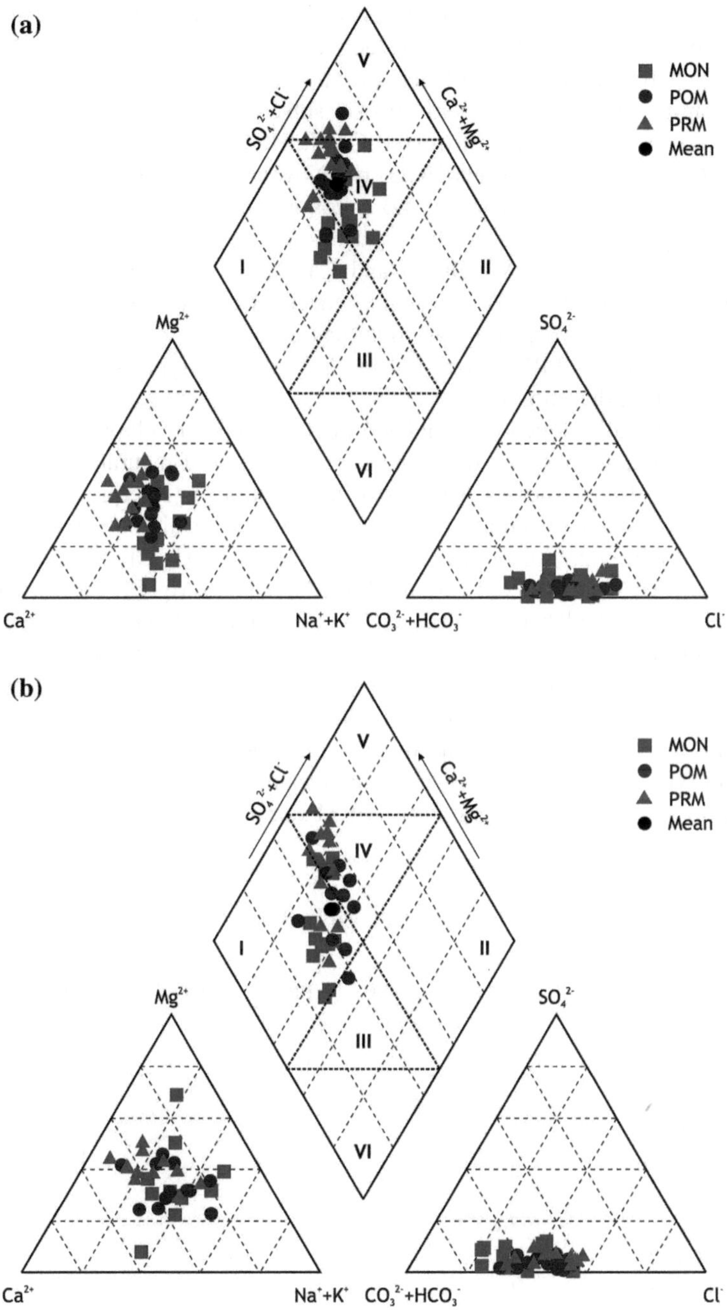

Fig. 9 Piper (1944) diagram of **a** MRB and **b** PRB

line, suggesting the dominance of Cl^- ions. Water types in PRB (semi-arid), generally dominated by HCO_3^-, indicate active groundwater flushing, whereas the waters of MRB (humid) were subjected to anthropogenic modifications.

The diagram has been classified into six subdivisions (Fig. 9), viz., I (Ca-Mg-HCO_3 type), II (Na-Cl type), III (mixed Ca-Na-HCO_3 type), IV (mixed Ca-Mg-Cl type), V (Ca-Cl type) and VI (Na-HCO_3 type), wherein III and IV are characterized by their mixed ionic content in that no cation-anion pair exceeds 50 %. In MRB, Ca-Mg-Cl type dominates all the three seasons followed by Ca-Mg-HCO_3 type. On the contrary, in PRB, Ca-Mg-HCO_3 and Ca-Mg-Cl types are roughly in equal proportions. In general, MRB waters (i.e., mean hydrogeochemical concentration) belong to Ca-Mg-Cl type, implying modification of water chemistry by anthropogenic inputs, whereas PRB waters fall in Ca-Mg-HCO_3 type (suggesting groundwater-dominating discharge).

Frazee (1982) developed a specialized interpretive water classification and later Upchurch (1992) suggested a descriptive classification of natural waters based on Piper diagram. Both schemes were used in this study to interpret the hydrogeochemical composition of samples, which enabled identification of 11 water types in MRB and PRB, viz., mixed cation-HCO_3, Ca-Na-HCO_3-Cl, Mg-Na-HCO_3-Cl, mixed cation-HCO_3-Cl, mixed cation-Cl, Ca-Na-Cl, Ca-Mg-Cl, Ca-Mg-HCO_3-Cl, Mg-Na-HCO_3, Ca-Na-HCO_3 and Mg-HCO_3-Cl (Table 5). Interestingly, these water types do show significant spatio-temporal variations (Table 5). MON samples of MRB are dominantly mixed cation-HCO_3-Cl (6 of 15 samples) followed by Ca-Na-HCO_3-Cl type (4 of 15 samples), while dominant water types of PRB are mixed cation-HCO_3 (5 of 12 samples) and mixed cation-HCO_3-Cl type (4 of 12 samples). During POM, mixed cation-HCO_3-Cl dominates in MRB (9 of 15 samples) as well as in PRB (8 of 12 samples). Similarly, PRM samples in MRB (12 of 15 samples)

Table 5 Classification of water types in MRB and PRB based on Piper (1944) diagram

	MRB			PRB		
Water type	MON	POM	PRM	MON	POM	PRM
Ca-Mg-Cl		2				
Ca-Mg-HCO_3-Cl		2	12		1	7
Mg-HCO_3-Cl				1		
Mg-Na-HCO_3				1		
Mg-Na-HCO_3-Cl	1				1	
Ca-Na-HCO_3				1		
Ca-Na-Cl	1					
Ca-Na-HCO_3-Cl	4					
Mixed cation-HCO_3	1			5	2	2
Mixed cation-Cl	2	2	1			
Mixed cation-HCO_3-Cl	6	9	2	4	8	3
Total	15	15	15	12	12	12

MON monsoon, *POM* post-monsoon, *PRM* pre-monsoon

and PRB (7 of 12 samples) are dominantly Ca-Mg-HCO_3-Cl type. In summary, the temporal variability of dominant water types in MRB is as follows: mixed cation-HCO_3-Cl and Ca-Na-HCO_3-Cl (MON) → mixed cation-HCO_3-Cl (POM) → Ca-Mg-HCO_3-Cl (PRM). Similarly, the pattern in PRB is as: mixed cation-HCO_3 and mixed cation-HCO_3-Cl (MON) → mixed cation-HCO_3-Cl (POM) → Ca-Mg-HCO_3-Cl (PRM). Such spatio-temporal variations might be reflective of variability in climate, hydrologic pathways as well as degree of anthropogenic actions (e.g., farming, tourism etc.).

The interpretations of various water types as detailed herein are based on Frazee (1982), Upchurch (1992) and Harvey et al. (2002). Ca-Na-HCO_3 water types are generally "fresh recharge waters" derived from rainfall and its interaction with soil and bedrock. Further these water types have relatively higher ionic strength (compared to Ca-HCO_3 type) and have a significant role for Na due to the occurrence of Na-rich minerals in the basement rocks and greater reaction times (due to more prolonged contact time or exposure). Mixed cation-HCO_3-Cl and Ca-Na-HCO_3-Cl types are considered "transitional", where these water types refer to waters that are evolving by geochemical reactions with bedrock and soil-matrix or waters that changed their geochemical character by mixing with other geochemically distinct waters (Frazee 1982). These waters can also be a product of mixing among two or more end members, e.g., Ca-Na-HCO_3-Cl waters might have been derived from mixing between "fresh recharge waters" and waters with anthropogenic signatures. In addition, leaching of soil evaporites (in semi-arid areas) by "fresh recharge waters" can also yield Ca-Na-HCO_3-Cl water types. However, the occurrence of Ca-Na-HCO_3-Cl waters uniquely in MRB suggests mixing of "fresh recharge waters" and waters with anthropogenic signatures. Similarly, mixed cation-HCO_3-Cl type is also comparable with Ca-Na-HCO_3-Cl type, but the former type has additional inputs of Mg, which can be derived from geogenic (i.e., ferromagnesian minerals) as well as anthropogenic sources (e.g., fertilizers). The significance of Mg^{2+} is clearly evident in the water types of MRB and PRB in that 8 of the 11 water types have Mg^{2+} either with dominant or equal controls on the hydrogeochemical composition (Table 5).

In addition, Ca-Mg-Cl and mixed cation-Cl types in MRB strongly imply the significance of anthropogenically-derived ions. Similar to the mixed cation-HCO_3-Cl type, the Ca-Mg-HCO_3-Cl type is also "transitional" water, but hardly any significance for Na. This type of water is common to both MRB and PRB, but in both basins it is prominent only during PRM. Mixed cation-HCO_3 waters, showing dominance in PRB (compared to MRB) might be derived from the interaction of "fresh recharge waters" with Mg-rich minerals in the basin lithology. In PRB, mixed cation-HCO_3 water samples were collected from locations viz., PW4, PW5, PW6, PW7 and PW11, which is a hornblende-biotite gneiss terrain (Fig. 2).

4.4 Partial Pressure of CO_2 (pCO_2)

Partial pressure of CO_2 (pCO_2) of surface waters reflects both productivity and the dynamic state of the rivers (Stallard and Edmond 1987; Zhu et al. 2013). pCO_2 levels of the water samples of MON, POM and PRM were estimated from pH and HCO_3^- content. The results show that pCO_2 during MON, POM and PRM is significantly higher than the atmospheric value of $10^{-3.5}$ atm (Table 6). Similar observations were made by Anshumali and Ramanathan (2007) for lentic and Prasad and Ramanathan (2005) for lotic systems. Earlier, Garrels and Mackenzie (1971a) and Raymahashay (1986) reported that the global trend of slightly higher pCO_2 indicates the existence of disequilibrium in natural waterbodies vis-à-vis the atmosphere. Kempe (1982) also reported such supersaturation in most of the World Rivers with respect to the atmosphere.

In MRB, mean pCO_2 of MON water samples is six times that of atmospheric pCO_2, while during POM it is more than 25 times and in PRM it is nearly ten times the atmospheric pCO_2 (Table 6). In PRB, mean pCO_2 of water samples during MON, POM and PRM is roughly 10, 22 and 23 times the atmospheric pCO_2. Relatively higher pCO_2 in stream waters is mainly due to influent (groundwater-dominating) stream discharge, which is significantly enriched in CO_2 and the slower rate of re-equilibration (i.e., solubility vs. release of CO_2) with the atmosphere (Stumm and Morgan 1970; Holland 1978). This inference is further confirmed by the relatively higher pCO_2 during POM and PRM (compared to MON) since stream discharge of the rivers draining the Western Ghats during POM and PRM is chiefly derived from aquifers. Various studies (e.g., Huh et al. 1998; Wu et al. 2005) also showed that most rivers are nearly 10 times supersaturated, while the tributaries of the Amazon are roughly 40 times supersaturated due to respiration by roots and decomposition of organic matter (Stallard and Edmond 1987; Richey et al. 2002). An inter-basin comparison of pCO_2 in water samples demonstrates that PRB (with the exception of POM) has relatively higher pCO_2, which is attributable to semi-arid climate of the basin. In both MRB and PRB, relatively lower pCO_2 during MON might be the result of dilution during monsoon. Nevertheless, lower pCO_2 can also result from relatively high photosynthetic activity of the riverine (biotic) system (Moon et al. 2007). Albeit monsoon is common to both MRB and PRB, the quantum of rainfall is significantly different, i.e., >3,000 mm in MRB versus <1,000 mm in PRB (see Sect. 2), which could be the reason for relatively higher pCO_2 of PRB during MON.

The excess CO_2 in the water samples may also have originated from the labile organic matter released during monsoon by soil erosion (Gao and Kempe 1987). In spite of higher rate of soil erosion during MON, the lower pCO_2 (in both the basins) suggests that the dilution effect due to monsoon rainfall is not compensated by the increased rate of soil erosion. Further, the plots of TSS versus log pCO_2 of MRB and PRB (Fig. 10) also do not illustrate any causal relationships between pCO_2 and soil erosion during the sampling seasons (with the exception of MON in PRB).

Table 6 Log $mHCO_3^-$ and log pCO_2 of water samples of MRB and PRB

MRB							PRB						
Sample-ID	MON		POM		PRM		Sample-ID	MON		POM		PRM	
	Log $mHCO_3^-$	Log pCO_2	Log $mHCO_3^-$	Log pCO_2	Log $mHCO_3^-$	Log pCO_2		Log $mHCO_3^-$	Log pCO_2	Log $mHCO_3^-$	Log pCO_2	Log $mHCO_3^-$	Log pCO_2
MW1	−3.02	−2.54	−2.65	−2.08	−2.74	−1.89	PW1	−3.32	−2.42	−3.10	−1.96	−3.14	−1.80
MW2	−3.25	−2.50	−2.77	−1.88	−3.06	−2.18	PW2	−3.49	−3.12	−3.10	−2.20	−3.25	−2.89
MW3	−3.19	−2.87	−2.77	−2.26	−2.95	−2.26	PW3	−3.40	−2.70	−3.06	−2.40	−3.14	−2.61
MW4	−3.25	−2.89	−2.84	−2.01	−2.98	−2.71	PW4	−3.19	−2.52	−2.84	−2.21	−3.02	−2.43
MW5	−3.14	−2.64	−2.87	−2.21	−2.98	−2.54	PW5	−3.10	−2.60	−3.02	−2.00	−2.98	−2.46
MW6	−3.32	−2.57	−3.06	−2.47	−3.19	−2.66	PW6	−3.14	−2.80	−2.98	−2.33	−3.14	−2.53
MW7	−3.14	−2.60	−3.02	−1.76	−3.06	−2.29	PW7	−3.14	−2.68	−3.06	−1.80	−3.06	−2.36
MW8	−3.62	−2.82	−3.19	−2.07	−3.10	−3.22	PW8	−3.25	−2.61	−2.84	−2.15	−2.75	−1.67
MW9	−3.25	−2.59	−3.10	−2.16	−3.14	−3.32	PW9	−2.95	−2.28	−2.62	−2.23	−2.70	−1.95
MW10	−3.49	−2.93	−3.32	−2.26	−3.32	−2.79	PW10	−3.10	−2.05	−2.89	−2.34	−2.87	−2.22
MW11	−3.25	−2.56	−3.10	−2.01	−3.19	−2.90	PW11	−2.75	−2.51	−2.49	−2.40	−2.33	−1.89
MW12	−3.40	−3.12	−3.06	−1.94	−3.19	−2.80	PW12	−3.25	−2.61	−2.87	−2.58	−2.84	−2.68
MW13	−3.40	−3.00	−3.06	−2.09	−3.14	−2.63							
MW14	−3.32	−2.90	−2.98	−2.11	−3.10	−2.45							
MW15	−3.40	−2.73	−2.98	−2.54	−3.10	−3.13							

MON monsoon, *POM* post-monsoon, *PRM* pre-monsoon

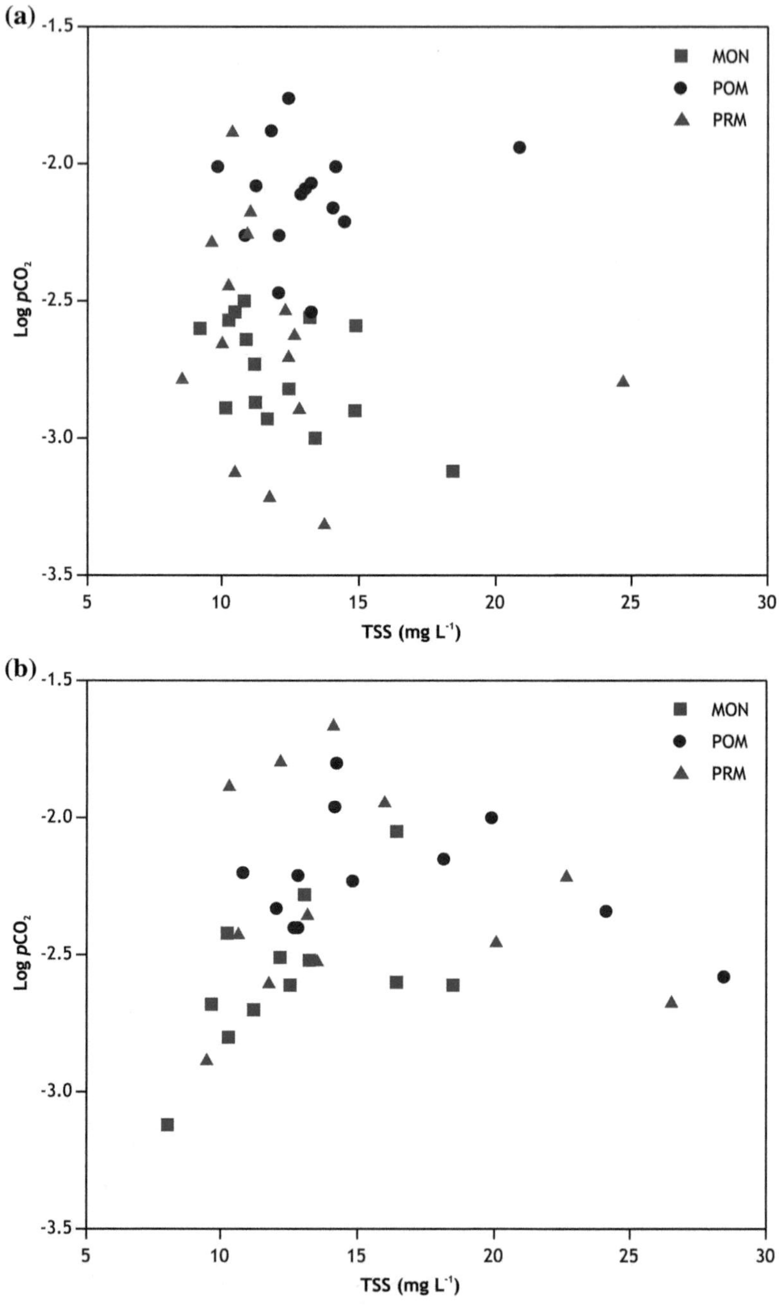

Fig. 10 Bivariate plot of TSS versus log pCO_2 of **a** MRB and **b** PRB

4.5 Suitability for Domestic and Irrigation Purposes

The chemical composition of water samples of MRB and PRB determines its suitability for domestic and irrigation purposes, which can be assessed by evaluating certain parameters such as corrosion coefficient (Cc), sodium adsorption ratio (SAR) and percent sodium (Na%). The results are summarized in the Table 7.

4.5.1 Corrosion Coefficient (Cc)

Larson and Scold (1958) proposed Cc, an important parameter in water quality evaluation, which is estimated as Eq. (3):

$$Cc = \frac{Cl^{-} + SO_4{}^{2-}}{HCO_3{}^{-}} \tag{3}$$

In MRB, mean Cc is 1.31 ± 0.59 (range = 0.59–2.71), 1.39 ± 0.47 (range = 0.74–2.56) and 1.56 ± 0.32 (range = 0.87–2.33) during MON, POM and PRM respectively (Table 7), whereas for PRB, mean Cc is 0.71 ± 0.31 (range = 0.35–1.26), 0.94 ± 0.27 (range = 0.47–1.31) and 1.09 ± 0.34 (range = 0.59–1.60) in MON, POM and PRM respectively. Balasubramanian (1986) suggested a ratio >1.0 for Cc as a safe limit for delivery of water through metal pipes. Relatively higher Cc values in MRB reflect non-corrosive nature of waters, in contrast to lower values in PRB suggesting higher probability for corrosion.

4.5.2 Sodium Adsorption Ratio (SAR)

SAR, an important parameter determining the suitability of waters for irrigation, is a measure of alkali/sodium hazard to crops. Richards (1954) defined SAR (Eq. 4) as a measure of cation exchange of irrigation water with the soil.

$$SAR = \frac{Na^{+}}{\sqrt{\frac{Ca^{2+} + Mg^{2+}}{2}}} \tag{4}$$

where concentrations are in meq L^{-1}.

In MRB, mean SAR is 0.47 ± 0.16 (range = 0.28–0.73) in MON, 0.42 ± 0.14 (range = 0.21–0.70) in POM and 0.22 ± 0.08 (range = 0.08–0.36) in PRM. SAR in PRB is 0.55 ± 0.39 (range = 0.13–1.41), 0.74 ± 0.54 (range = 0.14–2.17) and 0.49 ± 0.45 (range = 0.09–1.71) in MON, POM and PRM respectively (Table 7). Richards (1954) suggested that low to medium SAR of river water makes it suitable for irrigating most of the agricultural crops. In addition, based on the Bouwer's (1978) classification, the water samples belong to 'no problem category' of irrigation water quality (i.e., SAR < 6.0).

Table 7 Estimated parameters of suitability for domestic and irrigation purposes in MRB and PRB

Parameter	MON		POM		PRM	
	Mean ± SD	Range	Mean ± SD	Range	Mean ± SD	Range
MRB						
Cc	1.31 ± 0.59	0.59–2.71	1.39 ± 0.47	0.74–2.56	1.56 ± 0.32	0.87–2.33
SAR	0.47 ± 0.16	0.28–0.73	0.42 ± 0.14	0.21–0.70	0.22 ± 0.08	0.08–0.36
Na%	34.98 ± 6.68	22.12-47.25	24.87 ± 6.05	12.50–38.46	16.67 ± 5.32	6.45–27.27
PRB						
Cc	0.71 ± 0.31	0.35–1.26	0.94 ± 0.27	0.47–1.31	1.09 ± 0.34	0.59–1.60
SAR	0.55 ± 0.39	0.13–1.41	0.74 ± 0.54	0.14–2.17	0.49 ± 0.45	0.09–1.71
Na%	33.65 ± 9.75	17.84–48.51	33.29 ± 10.55	13.04–52.40	23.02 ± 10.53	7.69–43.10

MON monsoon, *POM* post-monsoon, *PRM* pre-monsoon, *SD* standard deviation, *Cc* corrosion coefficient, *SAR* sodium adsorption ratio

4.5.3 Percent Sodium (Na%)

When concentration of Na^+ is high, the ions tend to be absorbed by clay particles displacing Mg^{2+} and Ca^{2+}. Exchange of Na^+ in water with Mg^{2+} and Ca^{2+} in soil reduces permeability, eventually resulting in poor internal drainage (Collins and Jenkins 1996; Saleh et al. 1999). Excess Na^+ combining with CO_3^{2-} will lead to the formation of alkaline soils, while with Cl^- saline soils are formed and both these soils are unsuitable substrates for crops (Wilcox 1948). Hence, Na% (Eq. 5), a parameter of suitability for irrigation, is calculated as:

$$\text{Na\%} = \frac{\text{Na}^+ + \text{K}^+}{\text{Ca}^{2+} + \text{Mg}^{2+} + \text{Na}^+ + \text{K}^+} \times 100 \quad (5)$$

where all the concentrations are in meq L^{-1}.

Na% values of MRB range from 22.12 to 47.25 (mean = 34.98 ± 6.68) in MON, 12.50 to 38.46 (mean = 24.87 ± 6.05) in POM and 6.45 to 27.27 (mean = 16.67 ± 5.32) in PRM. In PRB, it spans between 17.84 and 48.51 (mean = 33.65 ± 9.75) in MON, 13.04 and 52.40 (mean = 33.29 ± 10.55) in POM and 7.69 and 43.10 (mean = 23.02 ± 10.53) in PRM (Table 7). The lower Na% suggests that the water samples of MRB and PRB during all the seasons are "excellent to good" for irrigation (after Wilcox 1955).

5 Conclusions

- In comparison with MRB, PRB has elevated levels of ionic abundance and higher degree of mineralization, possibly a result of multiple factors such as semi-arid climate, discharge dominated by groundwater, lithological variations and influence of carbonates and evaporites in the soil and shallow regolith.
- Higher K^+, Cl^- and H_4SiO_4 in MRB waters during all the seasons, implying significant contributions from farms, plantations and tourism sectors (i.e., K^+, Cl^-) as well as intense chemical weathering of silicate minerals (i.e., H_4SiO_4) in tropical humid climate.
- Downstream variation of hydrogeochemistry exhibits a general decreasing trend in MRB, which is attributed to dilution due to higher discharge, whereas hydrogeochemistry in PRB shows an increasing downstream trend, implying the significance of gradually changing rainfall (and climate), i.e., humid upstream versus semi-arid downstream.
- In MRB and PRB, hydrogeochemistry shows considerable temporal variability, which is a result of the variation in rainfall, changing hydrologic pathways controlling the stream discharge as well as intensity of various anthropogenic activities.
- MRB data (during all the seasons) on Gibbs plot are more or less clustered in the zone of rock dominance, whereas PRB data fall along a line trending from the

rock dominance zone (upstream samples) towards the zone of evaporation-crystallization dominance (downstream samples). Such a trend implies the influence of semi-arid climate prevailing in the downstream of PRB, where evaporation tends to increase the TDS.
- The Na^+-normalized Ca^{2+} versus Na^+-normalized HCO_3^- plots show that most of the sample poles occur outside the silicate weathering domain, instead fall parallel to the trend line between silicate and carbonate end members, which can be inferred as due to possible mixing between the end members.
- The Ca^{2+}+Mg^{2+}/Na^++K^+ ratios in MRB and PRB during MON (mean = 1.96 and 2.23 in MRB and PRB respectively), POM (mean = 3.29 and 2.41) and PRM (mean = 5.74 and 4.40) also reflect sources other than silicate weathering (e.g., weathering of carbonate minerals, anthropogenic inputs). Mean H_4SiO_4/(Na^++K^+) ratios during MON, POM and PRM in MRB and PRB are 0.48 and 0.42, 0.59 and 0.30 and 0.92 and 0.46 respectively, which also suggest additional sources of Na^+ and K^+ other than silicate weathering.
- According to classification schemes of Frazee (1982) and Upchurch (1992), 11 types waters have been delineated in MRB and PRB, viz., mixed cation-HCO_3, Ca-Na-HCO_3-Cl, Mg-Na-HCO_3-Cl, mixed cation-HCO_3-Cl, mixed cation-Cl, Ca-Na-Cl, Ca-Mg-Cl, Ca-Mg-HCO_3-Cl, Mg-Na-HCO_3, Ca-Na-HCO_3 and Mg-HCO_3-Cl. These water types show significant variability between the basins (MRB vs. PRB) as well as between the seasons (predominantly MON and POM vs. PRM). Most of the waters are considered "transitional", where these water types refer to waters that are evolving by geochemical reactions with bedrock as well as soil-matrix or waters that changed their geochemical character by mixing with other geochemically distinct waters.
- In both MRB and PRB, pCO_2 during MON, POM and PRM is significantly higher than the atmospheric value of $10^{-3.5}$ atm. In MRB, mean pCO_2 of stream water is 6 times the atmospheric pCO_2, while during POM it is roughly 25 times and during PRM, it is nearly 10 times that of the atmospheric pCO_2. In PRB, mean pCO_2 of water during MON, POM and PRM is roughly 10, 22 and 23 times of the atmospheric pCO_2. An inter-basin comparison of pCO_2 in stream waters demonstrates that PRB (except during POM) has relatively higher pCO_2, which might also be a result of the semi-arid climate of the basin.

In summary, the chemical composition of water samples of MRB and PRB is jointly controlled by weathering of silicate and carbonate minerals as well as anthropogenic activities and is influenced by climatic seasonality. However, the spatio-temporal variability of hydrogeochemical attributes is mainly due to the variations in climate, lithology, hydrologic pathways and degree of various anthropogenic activities.

Acknowledgments First author (JT) is indebted to (late) Dr. R. Satheesh (SES, Mahatma Gandhi University, Kerala) for moral support and mentoring during the early stages of research career. JT also acknowledges the helps rendered during chemical analyses by Ms. Manjusree, T.M. (Department of Environmental Sciences), the HOD (Department of Geology), University of Kerala and the Director, Central Ground Water Board, Thiruvananthapuram. Financial support from

Kerala State Council for Science, Technology, and Environment, Thiruvananthapuram and permission and logistics for the field studies in the protected areas by Kerala Forest Department are also acknowledged.

References

Abbas N, Subramanian V (1984) Erosion and sediment transport in the Ganges river basin (India). J Hydrol 69(1–4):173–182. doi:10.1016/0022-1694(84)90162-8

Ahearn DS, Sheibley RW, Dahlgren RA, Keller KE (2004) Temporal dynamics of stream water chemistry in the last free-flowing river draining the western Sierra Nevada, California. J Hydrol 295(1–4):47–63. doi:10.1016/j.jhydrol.2004.02.016

Allen JD (1995) Stream ecology: structure and functioning of running waters. Chapman & Hall, London

Anshumali, Ramanathan AL (2007) Seasonal variation in the major ion chemistry of Pandoh Lake, Mandi district, Himachal Pradesh, India. Appl Geochem 22(8):1736–1747. doi:10.1016/j.apgeochem.2007.03.045

Back W (1961) Techniques for mapping of hydrogeochemical facies. USGS Professional Paper 424-D:380–382

Bajpayee SK, Verma A (2001) Water quality of rivers of Kerala, southwestern India. In: Subramanian V, Ramanathan AL (eds) Proceedings of the international workshop on ecohydrology, School of Environmental Sciences, Jawaharlal Nehru University, New Delhi, 26–29 November 2001. Capital Publishing Company, New Delhi, pp 305–316

Balasubramanian A (1986) Hydrogeological investigations in the Tambraparni river basin, Tamil Nadu. PhD dissertation, University of Mysore, Mysore

Banks EW, Simmons CT, Love AJ, Shand P (2011) Assessing spatial and temporal connectivity between surface water and groundwater in a regional catchment: implications for regional scale water quantity and quality. J Hydrol 404(1–2):30–49. doi:10.1016/j.jhydrol.2011.04.017

Bartarya SK (1993) Hydrogeochemistry and rock weathering in a sub-tropical lesser Himalayan river basin in Kumaun, India. J Hydrol 146:149–174. doi:10.1016/0022-1694(93)90274-D

Bennett EM, Carpenter SR, Caraco NF (2001) Human impact on erodable phosphorus and eutrophication: a global perspective increasing accumulation of phosphorus in soil threatens rivers, lakes, and coastal oceans with eutrophication. BioScience 51(3):227–234. doi:10.1641/0006-3568(2001)051[0227:HIOEPA]2.0.CO;2

Berner EK, Berner RA (1996) Global environment: water, air and geochemical cycles. Prentice-Hall, Englewood Cliffs

Berner RA, Berner EK (1997) Silicate weathering and climate. In: Ruddiman WF, Prell W (eds) Tectonic uplift and climate change. Plenum Press, New York, pp 353–365

Biksham G, Subramanian V (1988) Nature of solute transport in Godavari basin, India. J Hydrol 103(3–4):375–392. doi:10.1016/0022-1694(88)90145-X

Bluth GJS, Kump LR (1994) Lithological and climatologic control of river chemistry. Geochim Cosmochim Acta 58(10):2341–2359. doi:10.1016/0016-7037(94)90015-9

Boulton AJ, Findlay S, Marmonier P, Stanley EH, Valett HM (1998) The functional significance of the hyporheic zone in streams and rivers. Annu Rev Ecol Syst 29:59–81. doi:10.1146/annurev.ecolsys.29.1.59

Boulton AJ, Marmonier R, Davis JA (1999) Hydrological exchange and subsurface water chemistry in streams varying in salinity in south-western Australia. Int J Salt Lake Res 8(4):361–382. doi:10.1007/BF02442121

Bouwer H (1978) Groundwater hydrology. McGraw Hill, New York

Bricker OP, Garrels RM (1967) Mineralogical factors in natural water equilibria. In: Faust SD, Hunter JV (eds) Principles and applications of water chemistry. Wiley, New York, pp 449–469

Bricker OP, Jones BF (1995) Main factors affecting the composition of natural waters. In: Salbu B, Steinnes E (eds) Trace elements in natural waters. CRC Press, Boca Raton, pp 1–5

Bucker A, Crespo P, Frede HG, Vache K, Cisneros F, Breuer L (2010) Identifying controls on water chemistry of tropical cloud forest catchments: combining descriptive approaches and multivariate analysis. Aquat Geochem 16(1):127–149. doi:10.1007/s10498-009-9073-4

Buell GR, Peters NE (1988) Atmospheric deposition effects on the chemistry of a stream in northeastern Georgia. Water Air Soil Pollut 39(3–4):275–291. doi:10.1007/BF00279474

Cai WJ, Guo X, Chen CTA, Dai M, Zhang L, Zhai W, Lohrenz SE, Yin K, Harrison PJ, Wang Y (2008) A comparative overview of weathering intensity and HCO_3^- flux in the world's major rivers with emphasis on the Changjiang, Huanghe, Zhujiang (Pearl) and Mississippi rivers. Cont Shelf Res 28(12):1538–1549. doi:10.1016/j.csr.2007.10.014

Carpenter SR, Caraco NF, Correll DL, Howarth RW, Sharpley AN, Smith VH (1998) Non-point pollution of surface waters with phosphorous and nitrogen. Ecol Appl 8(3):559–568. doi:10.1890/1051-0761(1998)008[0559:NPOSWW]2.0.CO;2

CESS (1984) Resource atlas of Kerala. Centre for Earth Sciences Studies, Thiruvananthapuram

Chandrashekara UM, Sibichan V (2006) Logs and snags in a shola forest of Kerala, India. J Mt Sci 3(2):131–138. doi:10.1007/s11629-006-0131-8

Chorley RJ, Schumm SA, Sugden DE (1984) Geomorphology. Methuen, London

Christophersen N, Neal C, Hooper RP, Vogt RD, Anderson S (1990) Modelling stream water chemistry as a mixture of soil water end members: a step towards second generation acidification models. J Hydrol 116(1–4):307–320. doi:10.1016/0022-1694(90)90130-P

Church MR (1997) Hydrogeochemistry of forested catchments. Annu Rev Earth Planet Sci 25:23–59. doi:10.1146/annurev.earth.25.1.23

Clarke FW (1924) The data of geochemistry, 5th edn. US Geological Survey Bulletin 770, Government Printing Office, Washington DC

Collins R, Jenkins A (1996) The impact of agricultural landuse on stream chemistry in the middle hills of Himalayas, Nepal. J Hydrol 185(1–4):71–86. doi:10.1016/0022-1694(95)03008-5

Conway EJ (1942) Mean geochemical data in relation to oceanic evolution. Proc R Irish Acad Sect B: Biol Geol Chem Sci 48:119–159

Creed IF, Band LE (1998) Export of nitrogen from catchments within a temperate forest: evidence for a unifying mechanism regulated by variable source area dynamics. Water Resour Res 34 (11):3105–3120. doi:10.1029/98WR01924

Crosa G, Froebrich J, Nikolayenko V, Stefani F, Galli P, Calamari D (2006) Spatial and seasonal variations in the water quality of the Amu Darya river (Central Asia). Water Res 40(11):2237–2245. doi:10.1016/j.watres.2006.04.004

Das A, Krishnaswami S, Sarin MM, Pande K (2005) Chemical weathering in the Krishna basin and Western Ghats of the Deccan Traps: rates of weathering and their controls. Geochim Cosmochim Acta 69(8):2067–2084. doi:10.1016/j.gca.2004.10.014

Davies BR, Thoms MC, Walker KF, O'Keefe JH, Gore JA (1996) Dryland rivers: their ecology conservation and management. In: Calow P, Petts GE (eds) The rivers handbook: hydrological and ecological principles. Blackwell, Cambridge

Department of Tourism (2008) Tourist statistics-2008. Department of Tourism, Government of Kerala. https://www.keralatourism.org/tourismstatistics/statistics2008.html

Dessert C, Dupre B, Gaillardet J, Francois LM, Allegre CJ (2003) Basalt weathering laws and the impact of basalt weathering on the global carbon cycle. Chem Geol 202(3–4):257–273. doi:10.1016/j.chemgeo.2002.10.001

Drever JI (1988) The geochemistry of natural waters, 2nd edn. Prentice-Hall, New Jersey

Eaton AD, Clesceri LS, Rice EW, Greenberg AE, Franson MAH (2005) Standard methods for the examination of water and wastewater, 21st edn. American Public Health Association (APHA), the American Water Works Association (AWWA), and the Water Environment Federation (WEF)

Edmond JM, Huh Y (1997) Chemical weathering yields from basement and orogenic terrains in hot and cold climates. In: Ruddiman WF, Prell W (eds) Tectonic uplift and climate change. Plenum Press, New York, pp 329–351

Edmond JM, Palmer MR, Measures CI, Grant B, Stallard RF (1995) The fluvial geochemistry and denudation rate of the Guayana Shield in Venezuela, Colombia and Brazil. Geochim Cosmochim Acta 59(16):3301–3325. doi:10.1016/0016-7037(95)00128-M

Fairchild IJ, Bradby L, Sharp M, Tison JL (1994) Hydrogeochemistry of carbonate terrains in alpine glacial settings. Earth Surf Proc Land 19(1):33–54. doi:10.1002/esp.3290190104

Frazee JM Jr (1982) Geochemical pattern analysis-method of describing the southeastern limestone regional aquifer system. In: Beck BF (ed) Studies of the hydrogeology of the southeastern United States. Georgia Southwestern College, Americus, special publications no 1, pp 46–58

Gaillardet J, Dupre B, Louvat P, Allegre CJ (1999) Global silicate weathering and CO_2 consumption rates deduced from the chemistry of large rivers. Chem Geol 159(1–4):3–30. doi:10.1016/S0009-2541(99)00031-5

Galy A, France-Lanord C (1999) Weathering processes in the Ganges–Brahmaputra basin and the riverine alkalinity budget. Chem Geol 159(1–4):31–60. doi:10.1016/S0009-2541(99)00033-9

Gao W, Kempe S (1987) The Changjiang: its long-term changes in pCO_2 and carbonate mineral saturation. In: Degens ET, Kempe S, Soliman H (eds) Transport of carbon and minerals in major world rivers, part 4, vol 64. SCOPE/UNEP Sonderband Heft, Hamburg, pp 207–215

Garrels RM, Mackenzie FT (1971a) Evolution of sedimentary rocks. Norton, New York

Garrels RM, Mackenzie FT (1971b) Gregor's denudation of the continents. Nature 231:382–383. doi:10.1038/231382a0

Garrels RM, Mackenzie FT, Hunt C (1975) Chemical cycle and the global environment. William Kaufman, New York

Gibbs RJ (1967) The geochemistry of the Amazon river system: Part I. The factors that controls the salinity and the composition and concentration of the suspended solids. Geol Soc Am Bull 78(10):1203–1232. doi:10.1130/0016-7606(1967)78[1203:TGOTAR]2.0.CO;2

Gibbs RJ (1970) Mechanisms controlling world water chemistry. Science 170(3962):1088–1090. doi:10.1126/science.170.3962.1088

GSI (1992) District resource map, Idukki district Kerala. Part-I. Geology and minerals. Geological Survey of India, Kolkata

Gunnell Y, Radhakrishna BP (eds) (2001) *Sahyadri*: the great escarpment of the Indian Subcontinent. Memoir 47(1):1054 (Geological Society of India, Bangalore)

Gupta H, Chakrapani GJ (2005) Temporal and spatial variations in water flow and sediment load in Narmada river basin, India: natural and man-made factors. Environ Geol 48(4–5):579–589. doi:10.1007/s00254-005-1314-2

Gupta LP, Subramanian V (1994) Environmental geochemistry of the river Gomti: a tributary of the Ganges river. Environ Geol 24(4):235–243. doi:10.1007/BF00767084

Gupta LP, Subramanian V (1998) Geochemical factors controlling the chemical nature of water and sediments in the Gomti river, India. Environ Geol 36(1–2):102–108. doi:10.1007/s002540050325

Gupta H, Chakrapani GJ, Selvaraj K, Kao S-J (2011) The fluvial geochemistry, contributions of silicate, carbonate and saline-alkaline components to chemical weathering flux and controlling parameters: Narmada river (Deccan Traps), India. Geochim Cosmochim Acta 75(3):800–824. doi:10.1016/j.gca.2010.11.010

Gurumurthy GP, Balakrishna K, Riotte J, Braun JJ, Audry S, Udaya Shankar HN, Manjunatha BR (2012) Controls on intense silicate weathering in a tropical river, southwestern India. Chem Geol 300–301:61–69. doi:10.1016/j.chemgeo.2012.01.016

Harmon RS, Lyons WB, Long DT, Ogden FL, Mitasova H, Gardner CB, Welch KA, Witherow RA (2009) Geochemistry of four tropical montane watersheds, Central Panama. Appl Geochem 24(4):624–640. doi:10.1016/j.apgeochem.2008.12.014

Harriman R, Gillespie E, King D, Watt AW, Christie AEG, Cowan AA, Edwards T (1990) Short-term ionic responses as indicators of hydrogeochemical processes in the Allt a'Mharcaidh Catchment, Western Cairngorms, Scotland. J Hydrol 116(1–4):267–285. doi:10.1016/0022-1694(90)90127-J

Harvey JW, Krupa SL, Gefvert C, Mooney RH, Choi J, King SA, Giddings JB (2002) Interactions between surface water and groundwater and effects on mercury transport in the North-central Everglades. Water resources investigation report 02-4050. US Department of the Interior, USGS, Reston, Virginia

Hem JD (1948) Fluctuations in the concentration of dissolved solids of some southwestern streams. Trans Am Geophys Union 29(1):80–84

Hem JD (1985) Study and interpretation of the chemical characteristics of natural water, 3rd edn. USGS Water Supply Paper 2254

Hill AR, Devito KJ, Campagnolo S, Sanmugadas K (2000) Subsurface denitrification in a forest riparian zone: interactions between hydrology and supplies of nitrate and organic carbon. Biogeochemistry 51(2):193–223. doi:10.1023/A:1006476514038

Hodson A, Porter P, Lowe A, Mumford P (2002) Chemical denudation and silicate weathering in Himalayan glacier basins: Batura Glacier, Pakistan. J Hydrol 262(1–4):193–208. doi:10.1016/S0022-1694(02)00036-7

Holland HD (1978) The chemistry of the atmosphere and oceans. Wiley, New York

Huh Y, Panteleyev G, Babich D, Zaitsev A, Edmond JM (1998) The fluvial geochemistry of the rivers of Eastern Siberia: II. Tributaries of the Lena, Omoloy, Yana, Indigirka, Kolyma, and Anadyr draining the collisional/accretionary zone of the Verkhoyansk and Cherskiy ranges. Geochim Cosmochim Acta 62(12):2053–2075. doi:10.1016/S0016-7037(98)00127-6

Hutchins MB, Smith B, Rawlins BG, Lister TR (1999) Temporal and spatial variability of stream waters in Wales, the Welsh borders and part of the west midlands, UK—1. Major ion concentrations. Water Res 33(16):3479–3491. doi:10.1016/S0043-1354(99)00057-3

Jacobson AD, Blum JD, Chamberlain CP, Craw D, Koons PO (2003) Climatic and tectonic controls on chemical weathering in the New Zealand Southern Alps. Geochim Cosmochim Acta 67(1):29–46. doi:10.1016/S0016-7037(02)01053-0

Jenkins A, Sloan WT, Cosby BJ (1995) Stream chemistry in the middle hills and high mountains of the Himalayas, Nepal. J Hydrol 166(1–2):61–79. doi:10.1016/0022-1694(94)02600-G

Jha PK, Tiwari J, Singh UK, Kumar M, Subramanian V (2009) Chemical weathering and associated CO_2 consumption in the Godavari river basin, India. Chem Geol 264(1–4):364–374. doi:10.1016/j.chemgeo.2009.03.025

Johnson CE, Litaor MI, Billett MF, Bricker OP (1994) Chemical weathering in small catchments: climatic and anthropogenic influences. In: Moldan B, Cerny J (eds) Biogeochemistry of small catchments: a tool for environmental research. Wiley, New York

Jose S, Sreepathy A, Kumar BM, Venugopal VK (1994) Structural, floristic and edaphic attributes of the grassland-shola forests of Eravikulam in Peninsular India. For Ecol Manage 65(2–3):279–291. doi:10.1016/0378-1127(94)90176-7

Kannan N (2009) Studies on the surface and groundwater, soil and sediments of Palakkad and Chittur thaluks of Bharathapuzha basin, Kerala. PhD dissertation, University of Kerala, Kerala, India

Kempe S (1982) Long-term records of CO_2 pressure fluctuations in fresh waters. In: Degens ET, Kempe S, Soliman H (eds) Transport of carbon and minerals in major world rivers, Part 1, vol 52. SCOPE/UNEP Sonderband Heft, Hamburg, pp 91–332

Koons PO (1995) Modeling the topographic evolution of collisional belts. Annu Rev Earth Planet Sci 23:375–408. doi:10.1146/annurev.ea.23.050195.002111

Koppen W (1936) Das geographische system der klimate. In: Koppen W, Geiger R (eds) Handbuch der klimatologie, Part C, vol 1. Verlag von Gerbruder Borntraeger, Berlin

Krishnaswami S, Singh SK (2005) Chemical weathering in the river basins of the Himalaya, India. Curr Sci 89(5):841–849

Larson TS, Scold RW (1958) Laboratory studies relating mineral quality of water to corrosion of steel and cast iron. Corrosion 14(6):285t–288t

Leite MGP, Fujaco MAG, Nalini HA Jr, Castro PTA (2010) Influence of geology in the geochemistry signature of Itacolomi State Park waters, Minas Gerais-Brazil. Environ Earth Sci 60(8):1723–1730. doi:10.1007/s12665-009-0306-z

Li S, Zhang Q (2008) Geochemistry of the upper Han river basin, China, 1: Spatial distribution of major ion compositions and their controlling factors. Appl Geochem 23(12): 3535–3544. doi:10.1016/j.apgeochem.2008.08.012

Lindell L, Astrom M, Oberg T (2010) Land-use change versus natural controls on stream water chemistry in the Subandean Amazon, Peru. Appl Geochem 25:485–495. doi:10.1016/j.apgeochem.2009.12.013

Livingstone DA (1963) Chemical composition of rivers and lakes. USGS Professional Paper 440-G

Maya K, Babu KN, Padmalal D, Seralathan P (2007) Hydrogeochemistry and dissolved nutrient flux of two small catchment rivers, south-western India. Chem Ecol 23(1):13–27. doi:10.1080/02757540601084029

McClain ME, Boyer EW, Dent CL, Gergel SE, Grimm NB, Groffman PM, Hart SC, Harvey JW, Johnston CA, Mayorga E, McDowell WH, Pinay G (2003) Biogeochemical hot spots and hot moments at the interface of terrestrial and aquatic ecosystems. Ecosystems 6(4):301–312. doi:10.1007/s10021-003-0161-9

Mehto A, Chakrapani GJ (2013) Spatio-temporal variation in the hydrogeochemistry of Tawa river, Central India: effect of natural and anthropogenic factors. Environ Monit Assess 185 (12):9789–9802. doi:10.1007/s10661-013-3291-3

Meybeck M (1987) Global chemical weathering of surficial rocks estimated from river dissolved loads. Am J Sci 287(5):401–428. doi:10.2475/ajs.287.5.401

Meyer JL, McDowell WH, Bott TL, Elwood JW, Ishizaki C, Melack JM, Peckarsky BL, Peterson BJ, Rublee PA (1988) Elemental dynamics in streams. J N Am Benth Soc 7(4):410–432

Milliman JD, Meade RH (1983) World-wide delivery of river sediment to the oceans. J Geol 91 (1):1–21

Milliman JD, Syvitski JPM (1992) Geomorphic/tectonic control of sediment discharge to the ocean: the importance of small mountainous rivers. J Geol 100(5):525–544

Moon S, Huh Y, Qin J, van Pho N (2007) Chemical weathering in the Hong (Red) river basin: rates of silicate weathering and their controlling factors. Geochim Cosmochim Acta 71 (6):1411–1430. doi:10.1016/j.gca.2006.12.004

Moquet J-S, Crave A, Viers J, Seyler P, Armijos E, Bourrel L, Chavarri E, Lagane C, Laraque A, Casimiro WSL, Pombosa R, Noriega L, Vera A, Guyot J-L (2011) Chemical weathering and atmospheric/soil CO_2 uptake in the Andean and Foreland Amazon basins. Chem Geol 287:1–26. doi:10.1016/j.chemgeo.2011.01.005

Mortatti J (1995) Erosao na Amazonia: Processos, Modelos e Balanco. Tese de Livre-Docencia, Escola Superior de Agricultura Luiz de Queiroz. Universidade de Sao Paulo, pp 150

Mortatti J, Probst JL (2003) Silicate rock weathering and atmospheric/soil CO_2 uptake in the Amazon basin estimated from river water geochemistry: seasonal and spatial variations. Chem Geol 197(1–4):177–196. doi:10.1016/S0009-2541(02)00349-2

Nair NGK, Santosh M, Thampi PK (1983) Geochemistry and petrogenesis of the alkali granite of Munnar, Kerala (India) and its bearing on rift tectonics. Neues Jahrb Mineral Abh 148:223–232

Nair SC (1991) The southern Western Ghats: a biodiversity conservation plan. Indian National Trust for Arts and Cultural Heritage, New Delhi

Nair VG (2006) Impact of Western Ghats orography on the weather and climate over southern Peninsular India: a mesoscale modelling study. PhD dissertation, Cochin University of Science and Technology, Kerala

Neal C, Robson A, Reynolds B, Jenkins A (1992) Prediction of future short term stream chemistry: a modeling approach. J Hydrol 130(1–4):87–103. doi:10.1016/0022-1694(92)90105-5

Nixon SW (2003) Replacing the Nile: are anthropogenic nutrients providing the fertility once brought to the Mediterranean by Great River? Ambio 32(1):30–39. doi:10.1579/0044-7447-32.1.30

Ovalle ARC, Silva CF, Rezende CE, Gatts CEN, Suzuki MS, Figueiredo RO (2013) Long-term trends in hydrogeochemistry in the Paraiba do Sul river, southeastern Brazil. J Hydrol 481:191–203. doi:10.1016/j.jhydrol.2012.12.036

Padmalal D, Remya SI, Jissy Jyothi S, Baijulal B, Babu KN, Baiju RS (2012) Water quality and dissolved inorganic fluxes of N, P, SO_4 and K of a small catchment river in the southwestern Coast of India. Environ Monit Assess 184(3):1541–1557. doi:10.1007/s10661-011-2059-x

Pandey SK, Singh AK, Hasnain SI (1999) Weathering and geochemical processes controlling solute acquisition in Ganga Headwater-Bhagirathi river, Garhwal Himalaya, India. Aquat Geochem 5(4):357–379. doi:10.1023/A:1009698016548

Parkin TB (1987) Soil microsites as a source of denitrification variability. Soil Sci Soc Am J 51 (5):1194–1199. doi:10.2136/sssaj1987.03615995005100050019x

Peters NE, Ratcliffe EB (1998) Tracing hydrologic pathways using chloride at the Panola Mountain research watershed, Georgia, USA. Water Air Soil Pollut 105(1–2):263–275. doi:10.1023/A:1005082332332

Petts G, Calow P (1996) River restoration. Blackwell, Oxford

Piper AM (1944) A graphic procedure in the geochemical interpretation of water-analyses. Trans Am Geophys Union 25(6):914–923

Prasad MBK, Ramanathan AL (2005) Solute sources and processes in the Achankovil river basin, Western Ghats, southern India. Hydrol Sci J 50(2):341–354. doi:10.1623/hysj.50.2.341.61798

Probst JL (1992) Geochimie et Hydrologie de l'Erosion Continentale. Mecanisms, Bilan Global Actuel et Fluctuations au Cours des 500 Derniers millions d'annes. Sci Geol Mem 94 (Strasbourg, 161 pp)

Raj N, Azeez PA (2009) Spatial and temporal variation in surface water chemistry of a tropical river, the river Bharathapuzha, India. Curr Sci 96(2):245–251

Raymahashay BC (1986) Geochemistry of bicarbonate in river water. J Geol Soc India 27(1):114–118

Rice KC, Bricker OP (1995) Seasonal cycles of dissolved constituents in streamwater in two forested catchments in the mid-Atlantic region of the eastern USA. J Hydrol 170(1–4):137–158. doi:10.1016/0022-1694(95)92713-N

Richards LA (1954) Diagnosis and improvement of saline and alkali soils. USDA Handbook No 60

Richey JE, Melack JM, Aufdenkampe AK, Ballester VM, Hess LL (2002) Outgassing from Amazonian rivers and wetlands as a large tropical source of atmospheric CO_2. Nature 416 (6881):617–620. doi:10.1038/416617a

Rodhe W (1949) The ionic composition of lake waters. Verhandlungen der Internationalen Vereinigung fur Theoretische und Angewandte Limnologie 10:377–386

Saleh A, Al-Ruwaih F, Shehata M (1999) Hydrogeochemical processes operating within the main aquifers of Kuwait. J Arid Environ 42(3):195–209. doi:10.1006/jare.1999.0511

Sankar S, Easa PS, Nair KKN (2000) Chinnar wildlife sanctuary: an overview. In: Ramakrishnan PS, Chandrashekhara UM, Elouard C, Guilmoto CZ, Maikhuri RK, Rao KS, Sankar S, Saxena KG (eds) Mountain biodiversity, landuse dynamics and traditional ecological knowledge, MAB Programme UNESCO. Oxford & IBH Publishing, New Delhi, pp 157–176

Schultz AM, Begemann MH, Schmidt DA, Weathers KC (1993) Longitudinal trends in pH and aluminium chemistry of the Coxing Kill, Ulster County, New York. Water Air Soil Pollut 69 (1–2):113–125. doi:10.1007/BF00478352

Smolders AJP, Hudson-Edwards KA, Van der Velde G, Roelofs JGM (2004) Controls on water chemistry of the Pilcomayo river (Bolivia, South-America). Appl Geochem 19(11):1745–1758. doi:10.1016/j.apgeochem.2004.05.001

Soman K (2002) Geology of Kerala. Geological Society of India, Bangalore

SSO (2007) Benchmark soils of Kerala. Soil survey organization. Department of Agriculture, Government of Kerala, Kerala

Stallard RF, Edmond JM (1981) Geochemistry of the Amazon: 1. Precipitation chemistry and the marine contribution to the dissolved load at the time of peak discharge. J Geophys Res: Oceans 86(C10):9844–9858. doi:10.1029/JC086iC10p09844

Stallard RF, Edmond JM (1983) Geochemistry of the Amazon: 2. The influence of geology and weathering environment on the dissolved load. J Geophys Res: Oceans 88(C14):9671–9688. doi:10.1029/JC088iC14p09671

Stallard RF, Edmond JM (1987) Geochemistry of the Amazon: 3. Weathering chemistry and limits to dissolved inputs. J Geophys Res: Oceans 92(C8):8293–8302. doi:10.1029/JC092iC08p08293
Stallard RF (1988) Weathering and erosion in the humid tropics. In: Lerman A, Meybeck M (eds) Physical and chemical weathering in geochemical cycles. Kluwer Academic Publishers, Dordrecht, pp 225–246
Stumm W, Morgan JJ (1970) Aquatic chemistry: an introduction emphasizing chemical equilibria in natural waters. Wiley, New York
Stutter MI, Deeks LK, Low D, Billett MF (2006) Impact of soil and groundwater heterogeneity on surface water chemistry in an upland catchment. J Hydrol 318(1–4):103–120. doi:10.1016/j.jhydrol.2005.06.007
Subramanian V (1983) Factors controlling the chemical composition of river waters of India. Proc Int Assoc Hydrol Sci Symp Hamburg 141:145–151
Thampi PK (1987) Geology of Munnar granite, Idukki district, Kerala, India. PhD dissertation, University of Kerala, Kerala
Thomas J (2012) Channel characteristics of two upland river basins of contrasting climate: a study from Kerala. PhD dissertation, University of Kerala, Thiruvananthapuram, Kerala
Thomas J, Joseph S, Thrivikramji KP, Manjusree TM, Arunkumar KS (2014) Seasonal variation in major ion chemistry of a tropical mountain river, the southern Western Ghats, Kerala, India. Environ Earth Sci 71(5):2333–2351. doi:10.1007/s12665-013-2634-2
Thomas J, Joseph S, Thrivikramji KP, Abe G, Kannan N (2012) Morphometrical analysis of two tropical mountain river basins of contrasting environmental settings, the southern Western Ghats, India. Environ Earth Sci 66(8):2353–2366. doi:10.1007/s12665-011-1457-2
Thomas MF (1994) Geomorphology in the tropics: a study of weathering and denudation in low latitudes. Wiley, Chichester
Thrivikramaji KP (1989) River geochemistry. Final report, Department of Science, Technology and Environment, Government of Kerala, Thiruvananthapuram, Kerala
Thrivikramaji KP, Joseph S (2001) Solute variation in some minor watersheds, Kerala, India. In: Subramanian V, Ramanathan AL (eds) Proceedings of the international workshop on ecohydrology. School of Environmental Sciences, Jawaharlal Nehru University, New Delhi, 26–29 November 2001. Capital Publishing Company, New Delhi, pp 295–304
Todd DK (2001) Groundwater hydrology. Wiley, New York
Townsend CR, Hildrew AG, Francis J (1983) Community structure in some southern English streams: the influence of physicochemical factors. Freshw Biol 13(6):521–544. doi:10.1111/j.1365-2427.1983.tb00011.x
Treguer P, Nelson DM, van Bennekom AJ, DeMaster DJ, Leynaert A, Queguiner B (1995) The silica balance in the world ocean: a reestimate. Science 268(5209):375–379. doi:10.1126/science.268.5209.375
Trevartha GT (1954) An introduction to climate. In: Fairbridge RW (ed) The encyclopedia of atmospheric sciences and astrogeology. Reihold Publishing Corporation, New York
Turner BF, Stallard RF, Brantley SL (2003) Investigation of in situ weathering of quartz diorite bedrock in the Rio Icacos basin, Luquillo Experimental Forest, Puerto Rico. Chem Geol 202 (3–4):313–341. doi:10.1016/j.chemgeo.2003.05.001
Upchurch SB (1992) Quality of water in Florida's aquifer systems. In: Maddox GL, Lloyd JM, Scott TM, Upchurch SB, Copeland R (eds) Florida's groundwater quality monitoring programme: background hydrogeochemistry. Florida Geological Society, Special publication no 34, Tallahassee, pp 12–51
Walton WC (1970) Groundwater resources evaluation. McGraw Hill, New York
Wheater HS, Kleissen FM, Beck MB, Tuck S, Jenkins A, Harriman R (1990) Modeling short-term flow and chemical response in the Allt a' Mharcaidh catchment. In: Mason, BJ (ed) The surface waters acidification programme. Cambridge University Press, Cambridge, pp 477–483
Whipple KX, Kirby E, Brocklehurts SH (1999) Geomorphic limits to climate-induced increases in topographic relief. Nature 401(6748):39–43. doi:10.1038/43375

White AF (2002) Determining mineral weathering rates based on solid and solute weathering gradients and velocities: application to biotite weathering in saprolites. Chem Geol 190(1–4): 69–89. doi:10.1016/S0009-2541(02)00111-0

White AF, Blum AE (1995) Effects of climate on chemical weathering in watersheds. Geochim Cosmochim Acta 59(9):1729–1747. doi:10.1016/0016-7037(95)00078-E

White AF, Blum AE, Schulz MS, Vivit DV, Stonestrom DA, Larsen M, Murphy SF, Eberl D (1998) Chemical weathering in a tropical watershed, Luquillo Mountains, Puerto Rico: I. Long-term versus short-term weathering fluxes. Geochim Cosmochim Acta 62(2):209–226. doi:10.1016/S0016-7037(97)00335-9

Wilcox LV (1948) The quality of water for irrigation use. USDA Technical Bulletin 962, Washington DC

Wilcox LV (1955) Classification and use of irrigation water. USDA Circular 969

Wu L, Huh Y, Qin J, Du G, van Der Lee S (2005) Chemical weathering in the Upper Huang He (Yellow River) draining the eastern Qinghai-Tibet Plateau. Geochim Cosmochim Acta 69 (22):5279–5294. doi:10.1016/j.gca.2005.07.001

Xie Y, Li X, Wang H, Li W (2013) Spatio-temporal variation analysis of hydrogeochemical characteristics in the Luanhe river basin, China. Water Sci Technol 67(6):1332–1338. doi:10. 2166/wst.2013.007

Zhang J, Zhang ZF, Liu SM, Wu Y, Xiong H, Chen HT (1999) Human impacts on the large world rivers: would the Changjiang (Yangtze river) be an illustration? Global Biogeochem Cycles 13 (4):1099–1105. doi:10.1029/1999GB900044

Zhu B, Yu J, Qin X, Rioual P, Liu Z, Zhang Y, Jiang F, Mu Y, Li H, Ren X, Xiong H (2013) The significance of mid-latitude rivers for weathering rates and chemical fluxes: evidence from northern Xinjiang rivers. J Hydrol 486:151–174. doi:10.1016/j.jhydrol.2013.01.016

Assessment of Water Availability in Chennai Basin under Present and Future Climate Scenarios

J. Anushiya and A. Ramachandran

Abstract Climate change may significantly impact the hydrological processes of a watershed system and lead to water scarcity or increased flooding. It may also cause serious problems to humans including loss of biodiversity and risks to the ecosystem. Quantifying and understanding the hydrological response to a changing climate are necessary for water resource management and formulation of adaptive strategies. In this study, changes in water balance components of the Chennai Basin under present and future climate scenarios had been assessed using Soil and Water Assessment Tool (SWAT). High resolution climate outputs ($0.25° \times 0.25°$) from PRECIS regional climate model for present (1961–1990 BL), mid-century (2041–2070 MC) and end-century (2071–2098 EC) under the IPCC SRES A1B emission scenario were used to assess the hydrological changes in the Chennai Basin. The study had determined the present and future water availability in space and time without incorporating any man-made changes like dams, diversions etc. The results indicated a decrease of precipitation in future scenarios as a result decrease of total water yield and ground water flow component in mid-century and end-century. Though both of these scenarios showed decreases in water balance components, the decrease in end-century would be lesser than the mid-century. In the season-wise analysis, the ET would be increased in winter and post monsoon seasons. Water yield had shown decrease in all the seasons of the mid-century scenario and increase during the EC winter and summer seasons.

Keywords Hydrological response · Chennai Basin · SRES · PRECIS · SWAT

J. Anushiya · A. Ramachandran (✉)
Centre for Climate Change and Adaptation Research, Anna University,
Chennai 600025, India
e-mail: ram7@annauniv.edu

Mu. Ramkumar et al. (eds.), *Environmental Management of River Basin Ecosystems*,
Springer Earth System Sciences, DOI 10.1007/978-3-319-13425-3_18

1 Introduction

According to the Fourth Assessment Report of the IPCC, the climate change will affect water resources through its impact on the quantity, variability, timing, form and intensity of precipitation. The report has also highlighted the imminent intensification of the global hydrological cycle affecting both ground and surface waters. The IPCC report concluded that it is highly likely that "the negative impacts of climate change on freshwater systems outweigh its benefits' with runoff declining in most streams and river (IPCC 2007). In recent times, several studies show that the climatic change is likely to impact significantly freshwater resource availability. It is expected to have several impacts on water resources, including the diminishing of snow pack and increase in evaporation, which affect the seasonal availability of water (Field et al. 2007). A warmer climate will accelerate the hydrological cycle, altering rainfall, magnitude and timing of runoff. Changes in precipitation intensity and duration will probably be the main factors altering the hydrological cycle leading to more floods and droughts (Gleick 2010). Climate-change-induced changes of the seasonal runoff regime and inter annual runoff variability can be as important for water availability as changes in the long-term average annual runoff amount if water is not withdrawn from large groundwater bodies or reservoirs (US Global Change Research Program 2000).

Availability or scarcity of water will vary greatly depending on the region. A number of global-scale (Alcamo and Henrichs 2002; Arnell 2004), national-scale (Thomson et al. 2005), and basin-scale assessments (Barnett et al. 2004) show that the semi-arid and arid basins are most vulnerable. Among these two types of basins, the semi-arid river basins located in the developing countries are more vulnerable to climate change than those basins located in the developed countries, as population, and thus water demand, is expected to grow rapidly in the future and the coping capacity is low (Millennium Ecosystem Assessment 2005).

It is a complex task to assess the impacts of climate change on future availability of fresh water resources, given the uncertainty in predicting climate change (Boorman and Sefton 1997). The uncertainties in climate projection can be examined with the range of available SRES scenarios. These SRES scenarios are developed by the Intergovernmental Panel on Climate Change (IPCC) to capture a wide range of potential changes in future emissions due to demographic, socio-economic, and technological changes (IPCC SRES 2000). These scenarios are plausible representations of the future that are consistent with assumptions about future emissions of GHGs and their effects on global climate and defined as A1, A2, B1, and B2 scenarios (Fig. 1).

A1 scenario describes a future world of very rapid economic growth, global population that peaks in mid-century and declines thereafter, and rapid introduction of new and more efficient technologies. A2 scenario describes a very heterogeneous world with continually increasing global population and regionally oriented economic growth that is more fragmented and slower. B1 scenario describes a convergent world with the same global population as in the A1, but with rapid changes

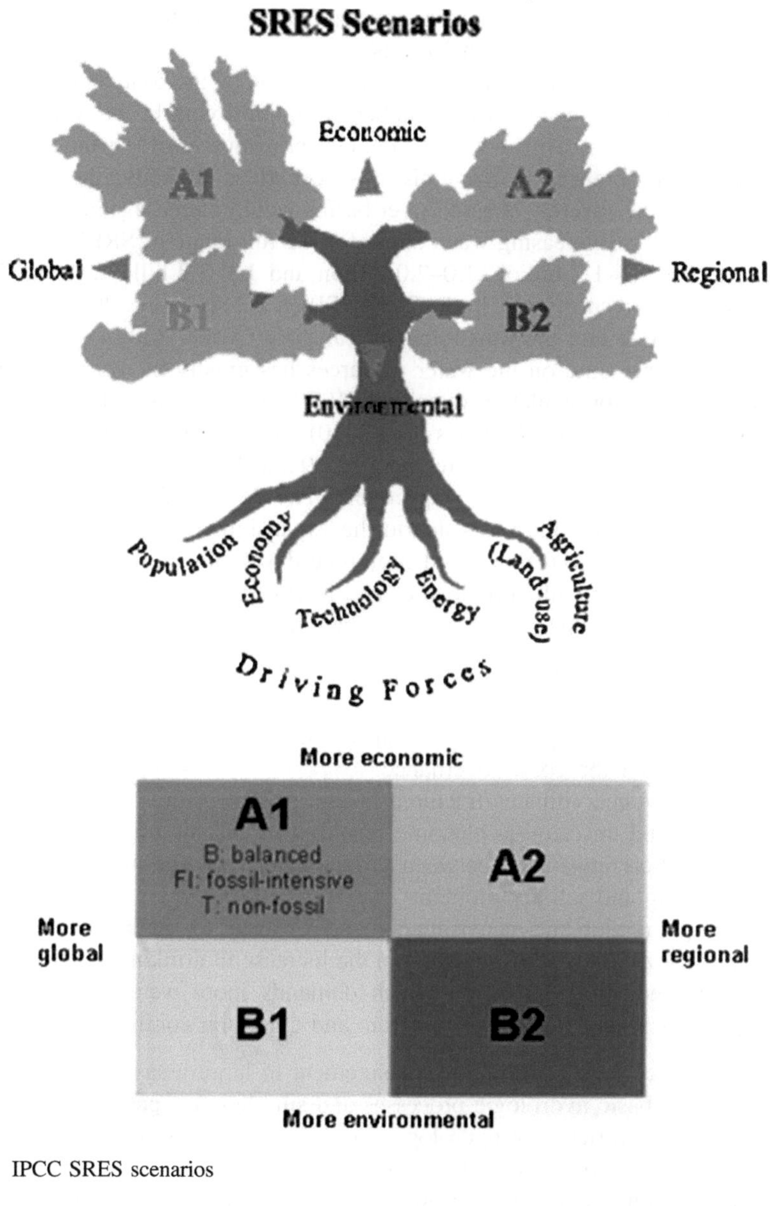

Fig. 1 IPCC SRES scenarios

in economic structures towards a service and information economy with reductions in material intensity and the introduction of clean and resource-efficient technologies. B2 scenario describes a world in which the emphasis is on local solutions to economic, social and environmental sustainability, with continuously increasing population (lower than A2) and intermediate levels of economic development and less rapid and more diverse technological change than in the B1 and A1 Scenarios.

This scenario is also oriented towards environmental protection and social equity and it focuses on local and regional levels.

In 2050s, differences in the population projections of the four SRES scenarios would have a greater impact on the number of people living in the water-stressed river basins (defined as basins with per capita water resources of less than 1,000 m^3/year) than the differences in the emissions scenarios. Globally, the number of people living in the severely stressed river basins would increase significantly. The population at risk of increasing water stress for the full range of SRES scenarios is projected to be: 0.4–1.7 billion, 1.0–2.0 billion and 1.1–3.2 billion, in the 2020s, 2050s, and 2080s, respectively. In the 2050s (SRES A2 scenario), 262–983 million people would move into the water-stressed category (Arnell 2004).

In India, the pressure on the water resources has greatly increased, leading to severe competition for available water since 1970s (Gulati et al. 2009). According to the Government of India Statistics (GoI 2010), the per capita surface availability (based on census 1991 and 2001) are about 2,309 and 1,902 m^3 for the populations of 1991 census and 2001 census respectively. The demand for water has already increased manifold over the years due to the expansions of urbanization, agriculture, population, industrialization and economic development. The river basins in south Asia are marred with complexities and fractured; the problems identified by Janakarajan (2006) in this regard are listed herein:

- Lack of information flow
- Lack of scientific data generation
- Myopic policies, competitive populism of successive government and lack of political will for the good governance
- Disintegrated/uncoordinated/fractured institutional structures
- Inadequate and unscientific planning resulting in chronic upstream and downstream conflicts, mismatch between groundwater recharge and extraction and water logging and salinity problems
- Growing population and increasing demand for water for attaining food security
- Rapid urbanization process resulting in the increase in drinking water needs and sanitation, industrial expansion which demands more water, and competing demands for scarce water across sectors and emerging conflicts.

Water resource development and management in large areas require an understanding on the basic hydrologic processes and simulation capabilities at the river basin scale. Wide varieties of hydrological models as well as applications of them have been developed over the past decades. Important among them include, but not limited to, the continuous stream flow simulation model, tank model, Hydrologic Simulation Program-Fortran (HSPF) and European Hydrologic system (SHE) model (Maidment 1993). While few models describe the processes by differential equations based on simplified hydraulic laws while other few attempt describing the processes by empirical algebraic equations (Arnold et al. 1998). Two types of hydrological models are in most of the applications, namely, the lumped conceptual models and physically based models. A lumped model is generally applied in a single point or a region for the simulation of various hydrological processes.

The parameters used in the lumped model represent spatially averaged characteristics in a hydrological system and are often unable to be directly compared with the field measurements (Yu 2002). The major drawback of lumped models is the incapability to account for spatial variability (Dhar and Majumdhar 2009).

The Physically based hydrological models represent the hydrological processes and spatially distributed data. The need for the research on the better representation of physical processes in space and time is evident given the availability of digital products (e.g., distributions of elevation, soil, vegetation) and remotely sensed data (e.g., soil moisture, vegetation), along with new technologies for measuring temporal and spatial variability in precipitation (Yu 2002). The common feature of these models is that they can incorporate the spatial distribution of various inputs and boundary conditions, such as topography, vegetation, land use, soil characteristics, rainfall and evaporation, and produce spatially detailed outputs such as soil moisture fields, water table positions, groundwater fluxes and surface saturation patterns (Troch et al. 2003). In these models, difference is approximated by spatial variation of precipitation, catchment parameters and hydrologic responses. Temporal variations of hydrologic responses are modeled by introducing threshold values for different processes. Representation of the catchments by individual sub-basins or grids of individual elements are used to integrate the spatial variability of the parameters with the model. One of such model is SWAT.

1.1 Description of SWAT (Soil and Water Assessment Tool) Model

The Soil and Water Assessment Tool (SWAT) model (Arnold et al. 1998; Neitsch et al. 2005) is a distributed parameter and continuous simulation model. SWAT is a public domain model actively supported by the Grassland, Soil and Water Research laboratory of the USDA Agriculture Research Service (http://swat.tamu.edu/). It is physically based and computationally efficient. Unlike the other conventional conceptual simulation model, it does not require much calibration. Therefore it can be used on ungauged watersheds

In SWAT, a watershed is divided into multiple sub watersheds which are then further subdivided into unique soil/land use characteristics called Hydrologic Response Unit (HRU). Flow generation, sediment yield, and non point—source loadings from each HRU in a sub watershed are summed and the resulting loads are routed through channels, ponds, and/or reservoirs to the watershed outlets. Hydrological processes are estimated with the following water balance equation:

$$SW_t = SW + \sum (R_{it} - Q_i - ET_i - P_i - QR_i)$$

where SW is the soil water content, minus the wilting point water content and R, Q, ET, P and QR are the daily amounts (in mm) of precipitation, runoff, evapotranspiration, percolation and ground water flow respectively. The soil profile is subdivided into multiple layers that support soil water processes, including infiltration, evaporation, plant uptake, lateral flow and percolation to the lateral layers. The soil percolation component of SWAT uses a storage routing technique to predict the flow through each soil layer in root zone. Downward flow occurs when the field capacity of the soil is exceeded, and the layer below is not saturated. Percolation from the bottom of the soil profile recharges the shallow aquifer. The percolation is not allowed from the layer, if the temperature in a particular layer is 0 °C or below. Lateral subsurface flow in the soil profile is calculated simultaneously with percolation. The contribution of ground water flow to the total stream flow is simulated by routing a shallow aquifer storage component to the stream (Gosain and Sandhya 2012).

The SWAT model has been used extensively as a useful tool to evaluate the hydrologic process impacts, with various soil, land use, agricultural management, and weather conditions over long periods (Neitsch et al. 2005; Gassman et al. 2007; Srinivasan et al. 2010). The SWAT is physically based model and uses readily available inputs facilitated by the GIS interface and it is suitable for the study of watersheds from small to very large sizes. Advantages of SWAT model include, ungauged watersheds with no monitoring data (e.g. stream gauge data) can be successfully modeled, the relative impact of alternative input data (e.g. changes in management practices, climate, vegetation etc.,) on water quantity, quality or other variables of interest can be quantified, the model uses readily available inputs and computationally efficient, simulation of large basins or a variety of management strategies can be performed without excessive investment of time and money, the model is also capable of incorporating climate change conditions to quantify the impacts of climate change. Based on these traits, it has gained wide global acceptability.

2 Study Area

The Chennai basin is situated between latitudes 12° 3′ 00″–13° 35′ 00″ N and longitudes 79°15′00″–80°22′30″E and is located in the northern part of Tamil Nadu State (Fig. 2). It comprises 8 sub basins and covers Chennai, Kancheepuram, Thiruvallur and Vellore districts. The Chennai basin is bordered in the north by Andhra Pradesh and Pulicat lake, southern and western sides by the Palar river basin and eastern side by the Bay of Bengal.

The total geographical area of the Chennai basin within Tamil Nadu State is 6118.34 km^2. Four major rivers namely, Araniar, Kosasthalaiyar, Cooum and Adyar are draining in this basin. In addition to numerous small-scale freshwater ponds and tanks, the Chennai Basin is home to the major freshwater supply tanks namely, the Poondi reservoir, Cholavaram lake, Redhills lake and Chembarambakkam tank and brackish waterbodies namely, the Pulicat lake, Ennore estuary, Cooum estuary, Adayar estuary and Covelong estuary. Typically, the ground water

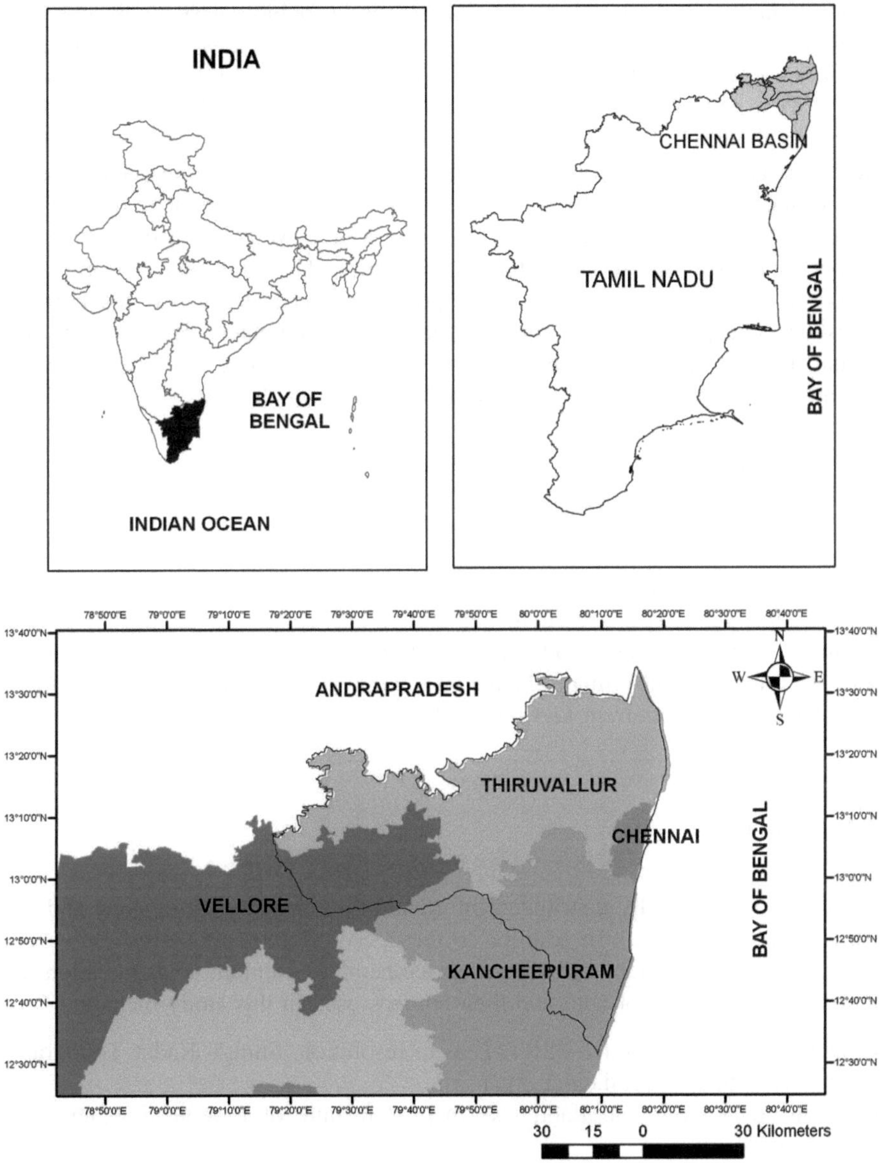

Fig. 2 Administrative map of Chennai Basin

distribution in this basin is not uniform and is subjected to wide spatio-temporal variations, depending on the underlying rock formation, their structural fabric, geometry, surface expression, etc. Thus, hydro-geomorphology controls the groundwater potential of the Chennai Basin.

Major land use categories of the Chennai basin include, agriculture and waste land which accounts nearly 79 % of the total rainfall. The area that other land use categories cover are: Build up 291.13 km^2, Forest 397.14 km^2, Water bodies 430.75 km^2 and Mining and industries 112.78 km^2 (Micro Level Study 2007). There are about 652 large, medium industries and 92,381 small scale industries in the basin. The major crops grown in the basin are rice, food grains, and oilseed.

Chennai basin lies in a tropical monsoon zone and experiences four seasons namely, winter (January–February), summer (March–May), monsoon (June-September) and post monsoon (October–December). The monsoon season is also known as southwest monsoon and the post-monsoon is known as the northeast monsoon. The post-monsoon months account approximately 50 % of the total annual rainfall. Monsoon precipitation helps to improve the recharging of groundwater as well as storage of surface water.

3 Materials and Methods

A distributed hydrological model SWAT (ArcSWAT 2009.93.7b compatible with Arc GIS 9.3 version) has used in this study. The daily weather data from high resolution regional climate model are used to carry out the future simulations. The potential impacts of the climate change on water availability are quantified using the SWAT outputs with current and future climates.

3.1 Data Used

SWAT requires spatially distributed information on elevation, soil, slope and land use (Arnold et al. 1998). In addition to these, SWAT requires daily weather data which includes maximum and minimum temperature, rainfall, solar radiation, relative humidity. Types of data and their sources used in this study include;

- Digital Elevation Model—SRTM 90 m resolution, Shuttle Radar Topography mission (http://srtm.csi.cgiar.org/)
- Drainage network and Land Use maps—Institute of Remote Sensing, Anna University
- Soil map and associated soil characteristics map—Tamil Nadu Agriculture University
- Meteorological data—daily rainfall, maximum and minimum temperature, solar radiation, relative humidity and wind speed
  - High resolution IMD gridded data (1971–2000)—IMD Pune
  - Baseline (1961–1990), Mid-century (2041–2070) and End-century (2071–2098) climate scenario data—PRECIS

3.2 PRECIS (Providing Regional Climate Scenarios for Impact Studies)

PRECIS is an atmospheric and land surface model developed by Hadley Centre, UK Met office (http://www.metoffice.gov.uk/precis/). It runs on Linux based PC in the area of interest. The Centre for Climate Change and Adaptation Research, Anna University is the licensed user of PRECIS. For the first time in India, Centre for Climate Change and Adaptation Research, Anna University has developed high resolution (25 km resolution) future climate projections with PRECIS (http://www.annauniv.edu/CCAR/precis.html). UK Met office has provided the boundary data (HADCM3Q0-Q16) for 17 member perturbed physics ensemble 'QUMP' (McSweeney and Jones 2010). The selection of subsets of the 17 available QUMP members for India have been done based on their performance ability to project reasonable present day climate. The selected members were simulated under A1B scenario (a mid range emission scenario) for a continuous run till 2100. A1B scenario is characterized by a future world of very rapid economic growth, global population that peaks in mid-century and declines thereafter, and rapid introduction of new and more efficient technologies with the development of balanced across energy sources.

The outputs have been post processed and used for impact studies. The PRECIS grid points which cover the Chennai basin have been considered and processed for SWAT hydrological assessment (Fig. 3). PRECIS model outputs for three periods namely Base line (1961–1990 BL), mid-century (2040–2070, MC) and end-century (2071–2098, EC) were further processed with WGN Excel Macro to create weather input files readable by SWAT (http://swat.tamu.edu/media/41586/wgen-excel.pdf).

3.3 Hydrological Assessment and Simulation of Chennai Basin

Automatic delineation of the basin was done using SRTM 90 m DEM (Fig. 4). The sub basin threshold area was set to 532,745 h. Table 1 shows the land use categories and the area covered under each category.

The agriculture and settlements cover approximately 11 % of the total area (Fig. 5). Predomination of sandy loam, followed by lesser coverage by sandy clay; sandy clay loam, clay, sand and clay loam are observable in the basin. The Chennai Basin has moderate to gentle slope and nearly 88 % of the basin exhibit slope of 1 %. Nearly 10 % of the area has slope of 1–5 %.

The basin has been subdivided into 60 sub-basins for spatial aggregation (Fig. 6). Each sub-basin was further divided into hydrological response units (HRUs) which have unique combination of soil, slope, and land use. Then, the model was simulated to run a for control period to assess the present water availability in space and time without incorporating any man-made changes like dams,

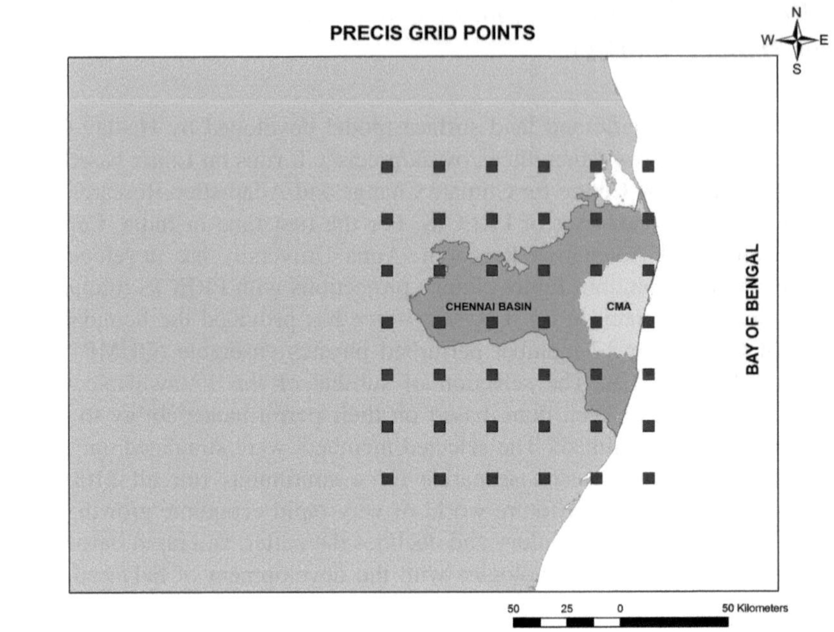

Fig. 3 PRECIS grid points covering Chennai Basin

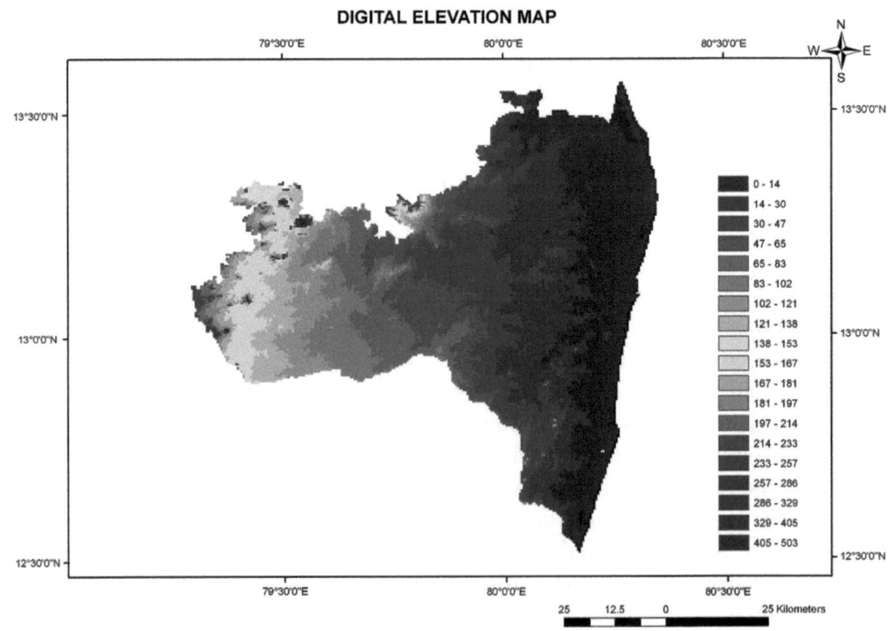

Fig. 4 Digital Elevation Map of Chennai Basin

Table 1 Land use categories of Chennai Basin

Land use	% on Total area	SWAT LU
Agricultural land-close-grown	34.27	AGRC
Agriculture land-generic	21.6	AGRL
Water	12.64	WATR
Pasture	8.89	PAST
Residential	6.81	URBN
Range-grasses	4.12	RNGE
Residential-low density	2.69	URLD
Forest-mixed	2.5	FRST
Forest-deciduous	1.51	FRSD
Agricultural land-row crops	1.26	AGRR
Wetlands-mixed	1.26	WETL
Institutional	0.96	UINS
Commercial	0.57	UCOM
Industrial	0.56	UIDU
Forest-evergreen	0.31	FRSE
Transportation	0.06	UTRN

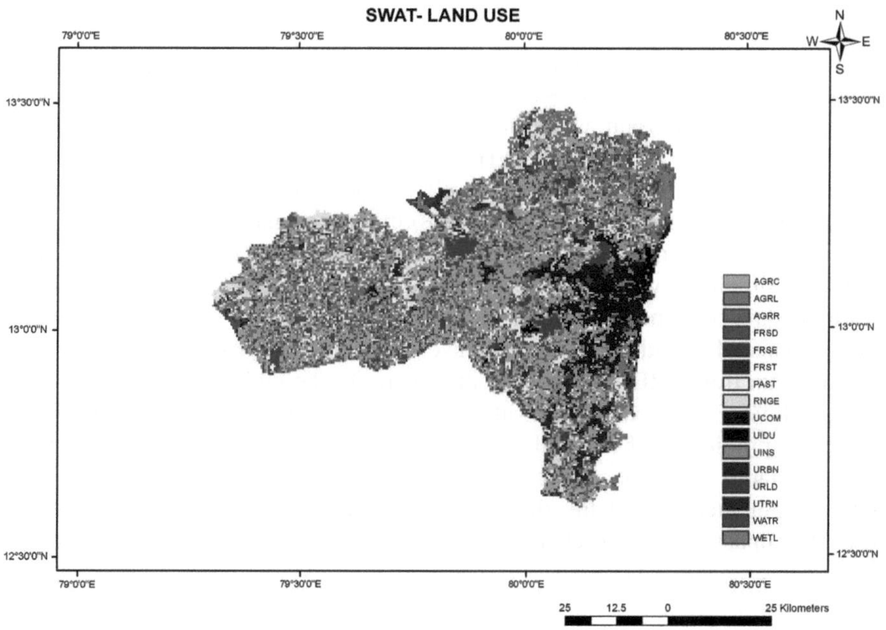

Fig. 5 Land use map of Chennai Basin

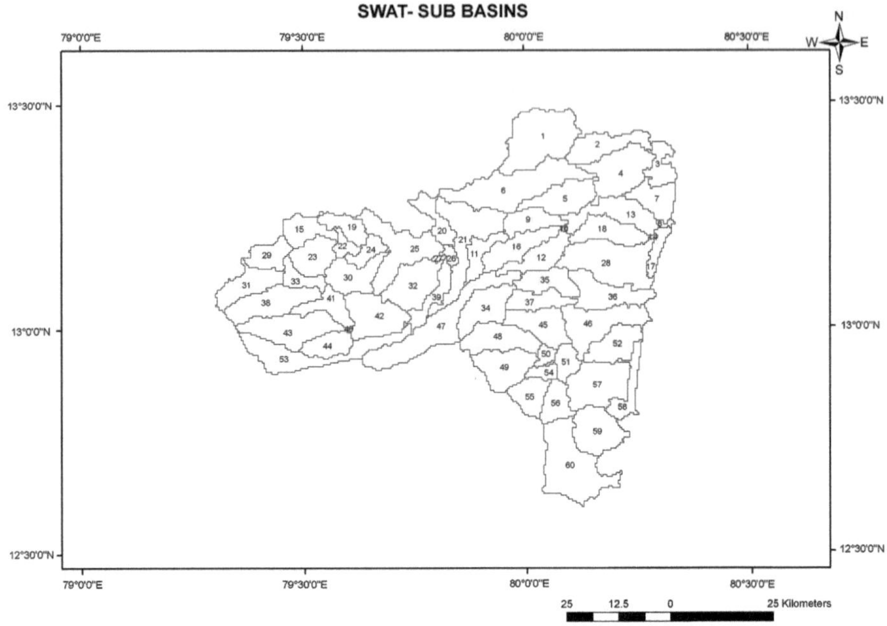

Fig. 6 Sub basin delineation of the Chennai Basin with SWAT using DEM

diversions, etc. The same framework was then used to predict the impact of climate change on the water resources with the assumption that the land use shall not change over time (Gosain and Sandhya 2012). Out of the 88 years of simulation, 30 years belong to IPCC SREA A1B baseline (1961–1990), 30 years belong to IPCC SRES A1B mid-century (2041–2070) and remaining 28 years belong to IPCC SRES A1B end-century (2071–2098) climate scenarios.

The model does not require elaborate calibration. The calibration is not meaningful, if simulated weather data is used for a control period, which is not the historical data corresponding to the recorded observed runoff (Gosain and Sandhya 2012). Hence, this study does not include calibration practice since it uses simulated weather data from PRECIS for the control period. The hydrological model SWAT generated very detailed outputs at daily intervals for each sub basin. Actual evapotranspiration, outflow, soil moisture, surface and subsurface runoff, ground water recharge are some of the main outputs available at daily intervals. These outputs simulated for all the three scenarios have been used for analyzing the possible impact of climate change on water balance components of the Chennai basin.

4 Results and Discussion

The impacts of climate change on the Chennai basin's water availability are calculated as changes from baseline in percentages. Table 2 shows the changes in water balance components from the baseline scenario (BL) to mid-century (MC) and end-century (EC) scenarios. From this table it is perceptible that there is a decrease in rainfall predicted over the basin due to climate change in MC as well as EC periods. In MC, there is about −14.3 % decrease of annual precipitation and in EC the decrease of annual precipitation is −9.4 %. Though both scenarios show decrease in rainfall, the decrease is less in EC when compared to MC and as a consequence; there is a decrease in the total water yield in MC, and EC scenarios and the ground water flow component in the Chennai Basin.

The annual evapotranspiration shows a slight decrease from BL scenario under increased GHG scenarios. But seasonal analysis shows an increase in rainfall during winter seasons till the end of the century. Post-monsoon season in end century also shows 11.7 % increase of rainfall. Though the summer and monsoon seasons show decreasing trends, the decreases are severe in end century.

Figure 7 shows the changes in seasonal water balance components under IPCC SRES BL, MC and EC climate scenarios. In the future scenarios, ET will be increased in winter and post-monsoon season. When compared to the baseline, the ET has an increase of 12.3 % in mid-century winter season and 16.7 % in end-century winter season. In post-monsoon season it has increased to 9.5 % in MC and 39.1 % in EC climate scenarios (Table 3). The increase in ET will decrease soil moisture, which leads to severe agricultural drought. More frequent and severe droughts arising from climate change will have serious management implications for water resource users. Agricultural producers and urban areas are particularly vulnerable, as evidenced by recent prolonged droughts in the western and southern United States, which are estimated to have caused over $6 billion in damages to the agricultural and municipal sectors. If the runoff season occurs primarily, water availability for seasonal crops will decline and water shortages will occur earlier in the growing season, particularly in watersheds that lack large reservoirs (Richard and Dannele 2008). ET will be decreased in summer and monsoon seasons. In MC, it shows a slight decrease to −8.2 % in summer season and −5.0 % in monsoon season. While in EC climate scenario, the decrease is more. In the summer season, it decreased to as low as −23.8 % and in monsoon months to −17.9 %.

Mean monthly value of water balance components for all the three climate scenarios are presented in Fig. 8. The ET values are higher during the monsoon months (August, July and September). The month of May shows lower ET. In mid century scenario, increases of 4.9, 7.3, 9.3, 6.5, 2.7, 5.8, and 0.9 % ET for January, February, March, September, October and November months respectively. While April, May, June, July and August months show decreasing ET. The percentage decrease is −2.5, −15, −0.2, −7.0, and −4 respectively. In the end century ET increase is much in December (Fig. 9). October and November months also show nearly 14 % increase from baseline. During the month of August, not much change

Table 2 Chennai Basin's water balance for IPCC SRES A1B baseline, Mid-century, end century scenarios and the changes with respect to baseline scenario

Scenario	Precipitation (mm)	Change in % w. r.t BL	Water yield (mm)	Change in % w. r.t BL	ET (mm)	Change in % w. r.t BL	Shallow GW (mm)	Change in % w. r.t BL
BL	1081.89		561.14		493.05		339.52	
MC	927.48	−14.3	415.81	−25.9	491	−0.5	249.1	−26.7
EC	981.4	−9.4	464.33	−17.3	492.3	−0.2	262.82	−22.6

Negative change indicates decrease from baseline, positive change indicates increase from baseline scenario

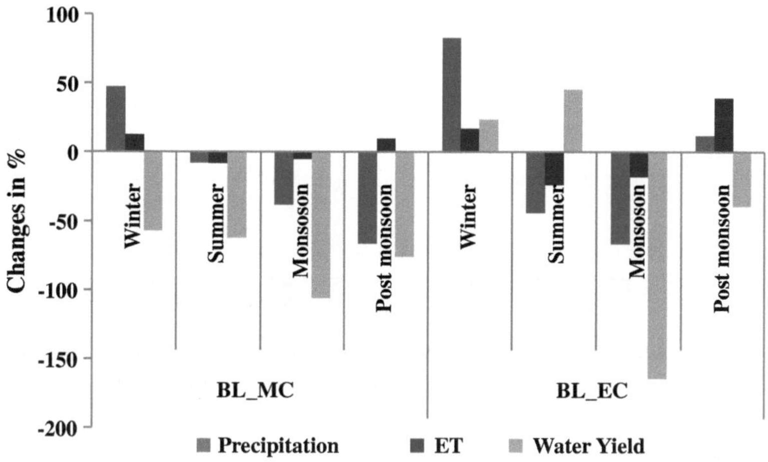

Fig. 7 Seasonal water balance components

Table 3 Season wise changes in water balance components

Changes from baseline	Seasons	Precipitation	ET	Water yield
		Changes in %		
BL_MC	Winter	47.3	12.3	−57.0
	Summer	−8.0	−8.2	−62.1
	Monsoon	−38.3	−5.0	−105.9
	Post monsoon	−66.4	9.5	−76.0
BL_EC	Winter	82.7	16.7	23.6
	Summer	−43.9	−23.8	45.3
	Monsoon	−66.6	−17.9	−164.2
	Post monsoon	11.7	39.1	−39.4

Negative change indicates decrease from baseline, positive change indicates increase from baseline scenario

in both mid and end centuries could be observed while decrease is more in the month of May, viz., −15 % in MC to −35 % in EC scenarios.

Normally, there is a decreasing tendency of water yield in both mid and end centuries. About −25.9 % reductions in MC and −17.3 % reductions in EC period are observed. There is a decrease in water yield in all the seasons of mid century scenario. But in the end century, water yield increases in winter (23.6 %) and summer seasons (45.3 %). This may be due to the increase of rainfall in winter season and decrease of ET in summer season. Water yield is very low in monsoon seasons of both MC and EC scenarios. The reduction of water yield is lowest in end century post monsoon period (Table 3). Although all the three major water

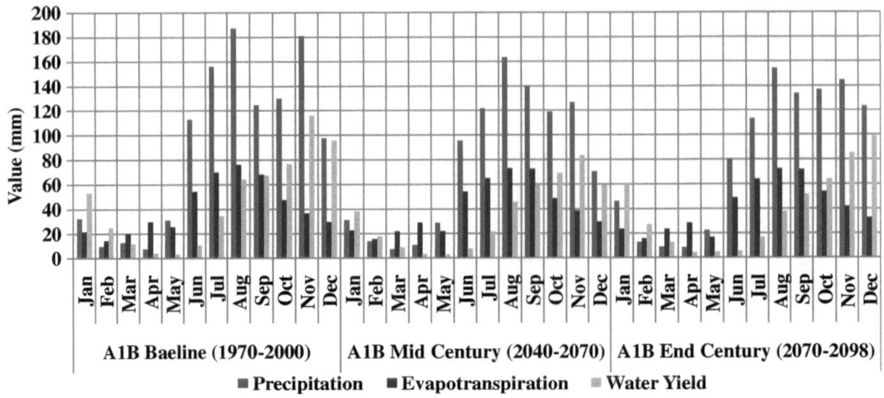

Fig. 8 Monthly water balance components

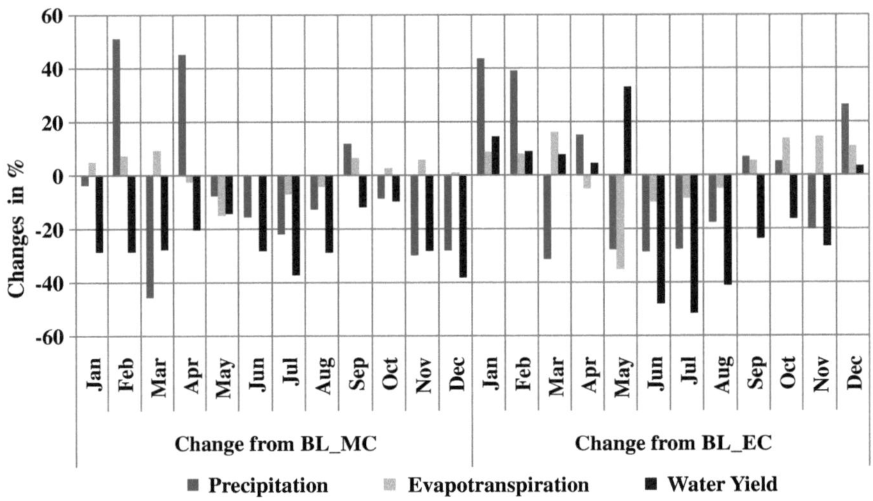

Fig. 9 Percentage change in mean monthly water balance components in MC and EC periods

components show decrease over MC and EC scenarios, the decrease is higher in mid-century when compared to end-century.

Effect of climate change on water balance components has also been analyzed spatially. The spatial distributions of precipitation, water yield, and evapotranspiration along with ground water recharge are analyzed for IPCC SRES A1B BL, MC and EC periods and are shown in the Fig. 10. The spatial variability of the components will help to identify the hot spots and to frame suitable adaptation strategies.

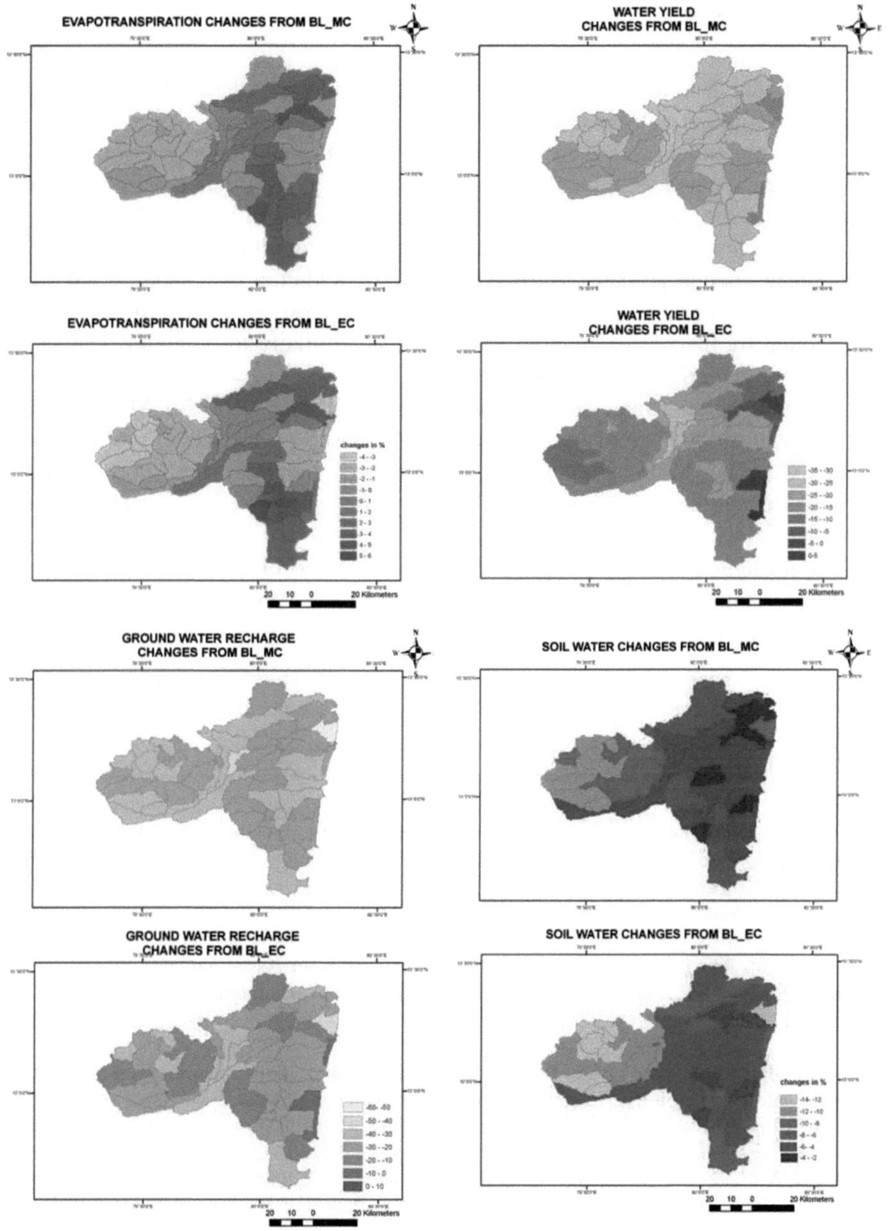

Fig. 10 Chennai basin's changes of water balance components under IPCC SRES AIB scenarios of MC and EC periods

5 Conclusions

- The model used in the present study has generated very detailed outputs at daily intervals and at sub-basin level.
- The results indicate an increase in evaporation rates and reduction of water yield that are expected to reduce water supplies. The greatest deficits are expected to occur in the post-monsoon season, leading to significantly decreased soil moisture levels, and more frequent and severe agricultural drought.
- Climate change is likely to exacerbate the degradation of resources and socio-economic pressures. The decline and degradation of the natural resources, resulted by the unsustainable rates of usage are likely to be aggravated due to climate change in the next 50 years.
- In Tamil Nadu, water supply-demand gap will be 14,100 MCM (504 TMC) (29.7 %) by 2025 (Palanisamy et al. 2011). Based on the model predictions for the Chennai basin, the climate change will increase this supply demand gap. Hence, the development and implementation of climate change adaptation strategies are essential.
- The present study also identifies the hotspots based on the spatial variability of water resource components, wherein site and measure specific water resource management practices have to be implemented.

References

Alcamo J, Henrichs T (2002) Critical regions: a model-based estimation of world water resources sensitive to global changes. Aquat Sci 64:1–11

Arnell NW (2004) Climate change and global water resources: SRES scenarios and socio-economic scenarios. Global Environ Change 14:31–52

Arnold JG, Srinivasan R, Muttiah RR, Williams JR (1998) Large area hydrologic modeling and assessment. Part I: Model development. J Am Water Resour Assoc 34(1):73–89

Barnett TP, Malone R, Pennell W, Stammer D, Semtner B, Washington W (2004) The effects of climate change on water resources in the West: introduction and overview. Clim Change 62:1–11

Boorman DB, Sefton CEM (1997) Recognizing the uncertainty in the quantification of the effects of climate change on hydrological response. Clim Change 35(4):415–434

Dhar S, Mazumdar A (2009) Hydrological modelling of the Kangsabati river under changed climate scenario: case study in India. Hydrol Process 23:2394–2406. doi:10.1002/hyp.7351

Field CB, Mortsch LD, Brklacich M, Forbes DL, Kovacs P, Patz JA Running SW, Scott MJ (2007) North America. In: Parry ML, Canziani OF, Palutikof JP, van der Linden PJ, Hanson CE (eds) IPCC. 2007. Climate change 2007: Impacts, adaptation and vulnerability. Contribution of working group II to the fourth assessment report of the Intergovernmental panel on climate change. Cambridge University Press, Cambridge http://www.ipcc.ch/pdf/assessment-report/ar4/wg2/ar4-wg2-chapter14.pdf

Gassman PW, Reyes MR, Green CH, Arnold JG (2007) The soil and water assessment tool: historical development, applications, and future research. Trans ASABE 50(4):1211–1250

Gleick P (2010) The world's water 2008–2009. The Biennial report on Freshwater Resources. Island press, Washington DC

GoI (2010) Ministry of water resources. Government of India. http://mowr.gov.in/ index1.asp? linkid = 1407langid = 1

Gosain AK, Sandhya Rao (2012) Climate change Impacts assessment on water resources of the Godavari river basin Chap 5 in book Water and Climate change. In: Udaya Sekhar Nagothu, Gosain AK, Palanisamy K (eds) Macmillan publishers India Ltd ISBN: 978-935-059-059-1

Gulati A, Shah T, Sreedhar G (2009) Agricultural performance in Gujarat since 2000. IWMI-IFPRI Joint Report, New Delhi

IPCC (2007) Contribution of working group II to the fourth assessment report of the intergovernmental panel on climate change. In: Parry ML, Canziani OF, Palutikof JP, van der Linden PJ, Hanson CE (eds) Cambridge University Press, Cambridge

IPCC SRES (2000) Emission scenarios, In: Summary for Policymakers. A Special Report of IPCC Working Group III, Available at https://www.ipcc.ch/pdf/special-reports/spm/sres-en.pdf

Janakarajan S (2006) Approaching IWRM through multi-stakeholders' dialogue: some experiences from South India. Integrated water resources management-Global theory, emerging practice and local needs: In: Peter P, Molinga, Ajaya D, Kusum A (eds) sage publications: New Delhi

Maidment DR (1993) Handbook of hydrology. McGraw Hill Publications

McSweeney C, Jones (2010) Selecting members of the 'QUMP' perturbed—physics ensemble for use with PRECIS, UK Met office

Micro level study Chennai basin (2007) Vol I Govt. of Tamil Nadu, PWD, Water Resources Organization, IWS, Taramani, Chennai

Millennium Ecosystem Assessment (2005) Ecosystems and human well-being: synthesis. Island Press, Washington, District of Columbia, p 155

Neitsch SG, Arnold JG, Kiniry JR, Wiliams JR (2005) Soil and water assessment tool theoretical documentation version 2005. Temple, Tex.: grassland, soil and water research laboratory, agricultural research service, blackland research center, texas agricultural experiment station

Palanisami K, Ranganathan CR, Vidhyavathi A, Rajkumar M, Ajjan N (2011) Performance of agriculture in river basins of Tamil Nadu In the last three decades—A Total Factor Productivity Approach. A Project Sponsored by Planning Commission, Government of India. Centre for Agricultural and Rural Development Studies, Tamil Nadu Agricultural University pp 1–171

Richard M Adams, Dannele E Peck (2008) Effects of climate change on water resources. Choices 1st Quarter 23(1):12–14

Srinivasan R, Zhang X, Arnold J (2010) SWAT ungauged: hydrological budget and crop yield predictions in the upper Mississippi river basin. Tran ASABE 53(5):1533–1546

Thomson AM, Brown RA, Rosenberg NJ, Srinivasan R, Izaurralde RC (2005) Climate change impacts for the conterminous USA: an integrated assessment. Part 4. Water resources. cimatic change 69:67–88

Troch PA, Paniconi C, McLaughlin D (2003) Catchment-scale hydrological modeling and data assimilation, Preface. Adv Water Resour 26:131–135

US Global Change Research Program (2000) Water: the potential consequences of climate variability and change. national waterassessment group, us global change research program, us geological survey and pacific Institute, Washington, District of Columbia, pp 160

Yu Z (2002) Modeling and prediction. Hydrology. In: Shankar M (ed) pp 1–8 rwas.2002.0172

A Review on the Riverine Carbon Sources, Fluxes and Perturbations

Sumi Handique

Abstract Carbon is transported from the land to the oceans via rivers and groundwater. The transfer of organic matter from the land to the oceans via fluvial systems is a key link in the global carbon cycle. Rivers also provide a key link in the geological scale carbon cycle. Nevertheless, an appreciation of their roles is yet to be made. Even when their roles are included, data are drawn only from selected large rivers, often neglecting the small mountainous rivers. Previous studies have demonstrated that, the tropic rivers, especially located in Asian region play crucial role in regulating the global carbon budgets. Superimposed on the natural sources and fluxes, the anthropogenically-induced fluxes, primarily emanating from reduced sediment and discharge (as a result of constructions of dams and reservoirs), and enhanced detrital organic matter (as a result of increased surface flow due to land use change) introduce perturbations.

Keywords Carbon cycle · River transport · Sources and fluxes of carbon · Perturbations

1 Introduction

Global biogeochemical cycles have shaped the Earth's climate and surface environment since the earliest days of the planet. Carbon in the biosphere is unevenly distributed among three major reservoirs: terrestrial, oceanic and atmospheric. Simplified depictions of the global carbon cycle have generally consisted of two biologically active boxes (oceans and land) connected through gas exchanges with a third box, the atmosphere (Bolin 1981; Siegenthaler and Sarmiento 1993; IPCC 2007). As models developed further, more sub-compartments and processes have been added in an attempt to unravel the more intricate interactions among them

S. Handique (✉)
Department of Environmental Sciences, Tezpur University, Tezpur 784028, Assam, India
e-mail: sumihan@tezu.ernet.in; sumihandique@gmail.com

Mu. Ramkumar et al. (eds.), *Environmental Management of River Basin Ecosystems*, Springer Earth System Sciences, DOI 10.1007/978-3-319-13425-3_19

(for example, Parton et al. 1994; Foley et al. 1996; Canadell et al. 2000; Cramer et al. 2001). Work during the 1970s and 1980s demonstrated that rivers deliver significant amounts of terrestrially-derived organic and inorganic C from land to the sea (Degens et al. 1991; Schlesinger and Melack 1981). This riverine "pipe" transports C from land to the ocean. When inland aquatic systems are included in global models, it is usually only for the transport of C through the riverine pipe. This delivery of terrestrial C through the riverine drainage network is in fact the end result of a number of transformations and losses in aquatic systems en route.

In the global carbon cycle, rivers have a critical role in connecting terrestrial, oceanic and atmospheric carbon reservoirs. Atmospheric Carbon is transported by rivers from terrestrial ecosystems as soil dissolved and particulate organic C (Dissolved Organic Carbon or DOC, Particulate Organic Carbon or POC) and Dissolved Inorganic C (DIC) and supplied to the oceans. Terrestrial organic matter thus represents approximately one third of the organic matter buried in all marine sediments and is stored over geological timescales leading to atmospheric carbon dioxide sequestration (Berner 1989). Of the portion of riverine that originates from the atmosphere, organic carbon is formed by the photosynthesis reaction; the fraction of atmospheric carbon, that is, dissolved inorganic carbon (DIC) comes from soil CO_2, fixed from the atmosphere by the weathering of rocks and air-water exchange. Of the terrestrial portion, DIC and particulate inorganic carbon (PIC) are associated with the weathering of rock.

Given cognizance to the importance of the fluvial system and the dynamics within drainage basins on the global carbon cycle, this paper reviews the importance of rivers in collecting, sequestering, transporting and delivering carbon to other realms of the Earth system.

2 Sources and Transport of Riverine Carbon

Carbon is transported from the land to the oceans via rivers and groundwater. The transfer of organic matter from the land to the oceans via fluvial systems is a key link in the global carbon cycle. Rivers also provide a key link in the geological scale carbon cycle by moving weathering products to the ocean. But organic carbon does not passively through the river systems; even very old and presumably recalcitrant soil carbon may be partially remineralized in aquatic systems. Remineralization of organic carbon during transport leads to elevated levels of dissolved CO_2 in rivers, lakes and estuaries worldwide. These high concentrations subsequently lead to outgassing to the atmosphere on the order of 1 Pg C year^{-1} with the majority in the humid tropics.

The most obvious source of export of C from the continental margins occurs through riverine flux. These fluxes are large, fairly well quantified, and have been derived from estimates of water discharge (for example, Dai and Trenberth 2002) and measurements of aqueous carbon concentrations. River export of organic carbon to the sea has thus been estimated as ranging from 0.38 (Degens et al. 1991) to 0.53 Pg C year^{-1} (Stallard 1998) with several other estimates falling within this range (for example, Schlesinger and Melack 1981; Meybeck 1982; Ludwig et al. 1996, b; Aitkenhead and McDowell 2000). Riverine export of dissolved inorganic carbon resulting from the fixation of atmospheric carbon through rock weathering is likely to be between 0.21 and 0.3 Pg C year^{-1} (Stallard 1998). Globally about half of the bicarbonate transported by rivers originates from silicate weathering (in which case 100 % of the bicarbonate came from CO_2 sequestration) and half from carbonate weathering (in which case only half the bicarbonate came from CO_2 sequestration (Stallard 1998; Meybeck 1993).

Groundwater export to the sea has not been considered as yet in global C budgets. Groundwater comprises 97 % of the world's liquid freshwater (van der Leeden et al. 1990) and can contain substantial quantities of organic and inorganic carbon (Cai 2003; Hem 1985). Some groundwater discharges as the base flow of rivers and is included in river carbon export. However, estimates of submarine groundwater discharge (SGD) span a broad range (Church 1996; Cai 2003). Imbalances in the world water budget (van der Leeden et al. 1990; Dai and Trenberth 2002; Shiklomanov and Rodda 2003) and groundwater residence times from 3 to 25 ka suggest SGD equal to 1.4–12 % of river influx, with the most accepted values between 5 and 10 % (Taniguchi et al. 2002; Slomp and Van Cappellen 2004). Estimates of groundwater alkalinity of around 60 mg^{-1} (Cai 2003) and a minimum DOC concentration of 1 mg^{-1} (Simpkins and Parkin 1993) suggest SGD of carbon of 0.13–0.25 Pg C year^{-1}. Collectively, using mid-range values for the river and groundwater components, inland waters thus deliver about 0.9 Pg C year^{-1} to the oceans, roughly equally as inorganic and organic carbon.

3 Types of Fluxes

Large rivers tend to integrate the biogeochemical activities within the drainage basin and the total carbon observed in river water is a mixed component that originates from different sources. In a pristine environment, the basic nature of riverine carbon consists of three categories: (a) Dissolved inorganic carbon (DIC) derived from chemical weathering of rocks which is largely transported as HCO_3^- ion, (b) Particulate organic carbon (POC) derived from soil organics, litter fall and autochthonous production; and (c) Dissolved organic carbon (DOC) arising from leaching of top-soil, peat and regulated by in situ pH. (Sarin et al. 2002).

This input from rivers is composed of four fluxes. The first and largest one is soil-derived C that is released to inland waters, mainly in organic form (particulate and dissolved), but also as free dissolved CO_2 from soil respiration (Sarmiento and Sundquis 1992). The flux is estimated to be 1.9 Pg C year^{-1}, by subtracting, from a total median estimate of 2.8 Pg C year^{-1}, the smaller contributions from the other three fluxes: chemical weathering, sewage and net C fixation. The soil-derived C flux is part of the terrestrial ecosystem C cycle and represents about 5 % of soil heterotrophic respiration. Current soil respiration estimates neglect the C released to inland waters. A downward revision of the estimate of soil heterotrophic respiration to account for the soil C channeled to inland freshwater systems would nevertheless remain within the uncertainty of this flux (Meybeck 1982). The second flux involves the chemical weathering of continental surfaces (carbonate and silicate rocks). It is part of the inorganic (often called 'geological') C cycle and causes an additional ~0.5 Pg C $year^{-1}$ input to upstream rivers (Beusen et al. 2005; Schlesinger and Melack 1981; Borges and Abril 2012; Laruelle et al. 2010). About two-thirds of this C flux is due to removal of atmospheric CO_2 in weathering reactions and the remaining fraction originates from chemical weathering of C contained in rocks. The pathway for chemical weathering is nevertheless largely indirect with most of the CO_2 removed from the atmosphere being soil CO_2, having passed through photosynthetic fixation. Weathering releases C to the aquatic continuum in the form of dissolved inorganic C, mainly bicarbonate, given that the average pH is in the range of 6–8 for freshwater aquatic systems (Chen and Borges 2009).

In contrast to soil-derived organic C, it is assumed that C derived from rock weathering will not degas to the atmosphere during its transfer through inland waters (Borges et al. 2005). Over geological timescales, silicate weathering coupled with carbonate precipitation in the Ocean is responsible for a large fraction of atmospheric CO_2 sequestration that balances the mantle and metamorphic CO_2 inputs into the atmosphere and therefore regulates the global climate (e.g. Walker et al. 1981; Berner et al. 1983). While modern weathering rates are often derived from river solute fluxes (e.g. Meybeck 1987; Gaillardet et al. 1999a; West et al. 2005) their solid counterparts have received far less attention (e.g. Gaillardet et al. 1999b; France-Lanord and Derry 1997; Gislason et al. 2006) probably because of the difficulty of integrating the variability of detrital sediments over space and time (Lupker et al. 2011; Bouchez et al. 2010, 2011a, b). Sediment records are however one of the rare archives that can be reliably used to trace past erosion fluxes at regional scales.

The third flux represents the C dissolved in sewage water originating from biomass consumption by humans and domestic animals, which releases an additional ~0.1 Pg C $year^{-1}$ as an input to freshwaters (Borges 2005). The fourth flux involves photosynthetic C fixation within inland waters, potentially high on an areal basis. A substantial fraction of this C is returned to the atmosphere owing to decomposition within inland waters (Duarte et al. 2005) but a percentage remains for export and burial (Breithaupt et al. 2012) and priming of terrestrial organic matter decomposition (Liu et al. 2010). Thus, although aquatic systems can emit CO_2 to the atmosphere, they still can be autotrophic.

4 Role of Chemical Weathering in Regulating the Carbon Cycle

Chemical weathering of silicate rocks consumes significant quantities of CO_2 that has regulated the global carbon cycle and in so doing Earth's climate over several eons (Arvidson et al. 2006; Berner 2004; Kempe and Degens 1985; Walker et al. 1981).

The carbon dioxide in the atmosphere dissolves in rainwater forming carbonic acid, which, once in contact with rocks, slowly dissolves them. This atmospheric carbon is then transported by rivers into the oceans, where it is trapped for several 1,000 years, before returning to the atmosphere or alternatively being stored in marine sediments or in organisms secreting aragonite/calcite shells and tests. During the weathering of silicate rocks, the totality of the HCO_3^- ions released in solution comes from atmospheric/soil CO_2 as it can be noted for example in the albite hydrolysis:

$$2NaAlSi_3O8 + 2CO_2 + 11H_2O \; --- > A_{12}SiO_5(OH)_4 + 2HCO_3^- + Na^+ + 4H_4SiO_4$$

Considering the weathering of carbonate rocks, only half of the HCO_3^- ions released in solution come from the atmospheric/soil CO_2 as it can be seen in the calcite dissolution:

$$CaCO_3 + CO_2 + H_2O \rightarrow Ca^{++} + 2HCO_3^-$$

Chemical weathering is central in surface biogeochemical cycles because it redistributes the chemical elements between Earth's surface reservoirs such as continental crust and the Ocean. Over geological timescales, silicate weathering coupled with carbonate precipitation in the Ocean is responsible for a large fraction of atmospheric CO_2 sequestration that balances the mantle and metamorphic CO_2 inputs into the atmosphere and therefore regulates the global climate (e.g. Walker et al. 1981; Berner et al. 1983). Continental weathering consumes about 0.3 Gt of atmospheric CO_2 each year (Gaillardet et al. 1999a, b).

Chemical weathering of rock helps regulate the supply of nutrients and solutes to soils, streams and the ocean, and is also the long-term sink for atmospheric CO_2, thus modulating Earth's climatic evolution via the greenhouse effect. Thus, to the extent that chemical weathering rates increase with temperature, weathering feedbacks should, over millions of years, buffer Earth's climate against large temperature shifts (Riebe et al. 2004). This chemical weathering process stores around 0.3 billion tons of atmospheric carbon in rivers and in the oceans every year: although this is considerably less than human-induced CO_2 production (around 8 billion tons per year), it is roughly equivalent to the net exchange flux between the atmosphere and the terrestrial biosphere (vegetation, soil, humus, etc.) under preindustrial conditions (0.4 billion tons). The long-term cooling effect is due to the higher weatherability of basalts compared to granite. In the same climatic conditions, basaltic surfaces consumes between 5 and 10 times more atmospheric CO_2 than granitic surfaces. Despite this, chemical weathering of the continents has never been taken into account until now in models of future climate change.

5 Current Understanding on the Riverine Carbon Transport

Transport of material via rivers is under study for over 100 years in geochemical budgets. It gives essential information both on processes affecting the continental surface (weathering, plant production, pollution etc.) and on the amount and nature of material carried to water bodies such as lakes, seas and oceans. Many world-wide budgets have been published based on ever increasing studies of river dissolved and particulate material (Martin and Meybeck 1979; Meybeck 1979). Two approaches are available for estimating global fluvial carbon fluxes. One uses carbon data for large rivers in various regions. For instance, Meybeck (1979) estimated global DIC and PIC fluxes (0.38 and 0.17 Pg C/year) based on data for 60 large rivers or groups of rivers that together are responsible for 63 % of global river discharge and considered runoff and average watershed temperature to obtain information regarding the other 37 %. Ludwig et al. (1996) utilized a database of mean annual dissolved organic carbon (DOC) and particulate organic carbon (POC) fluxes of 29 and 19 rivers respectively and other ecological factors to calculate DOC and POC fluxes (0.21 and 0.17 Pg C/year, respectively). The main determinants of DOC fluxes are the drainage intensity, basin slope and amount of organic soil carbon. The main factors that govern POC fluxes are the total mass of suspended matter (TSM) and sediment load. The other approach considers the mass balance. For example, Mackenzie et al. (1998) evaluated fluvial inorganic and organic carbon fluxes (0.72 and 0.61 Pg C/year) using a conceptual model. Notably, published results considered only the total quantity of inorganic or organic carbon (in the case of Mackenzie et al. 1998) or only some of the four carbon components (in the case of Meybeck 1979 and Ludwig et al. 1996). Importantly, the IPCC (2007) report considered only DIC and DOC fluxes. As reported by various studies worldwide, the amount of terrigenous carbon that enters the river systems is on the order of 1.5 Pg C a^{-1} (range: 0.8–2 Pg C a^{-1}). The present-day bulk C input (natural plus anthropogenic) to freshwaters was recently estimated at 2.7–2.9 Pg C $year^{-1}$, based on upscaling of local C budgets (Raymond et al. 2008).

Studies generally focus on large river systems like the Mississippi, the Ganga–Brahmaputra, Amazon or large Arctic rivers which integrate differences in lithology, vegetation, soils and climate. Small mountainous rivers directly connected to the oceans are less studied than large rivers, although they play an important role in transport of organic matter, their yields and runoff being inversely proportional to the watershed area (Milliman and Meade 1983). Recent works have demonstrated that these small rivers are major sources of POC, DOC and dissolved major elements to the oceans. Due to their location in the tropical zone, numerous small mountainous rivers are affected by periodic intense precipitation events such as cyclones or tropical storms that can play an important role on soil erosion and can potentially increase total organic carbon fluxes released by these systems. Gaps exist in the understanding of spatial and temporal variations of organic carbon and their fluxes from the tropical rivers, which are less studied because of their location in the developing countries. This is in spite of their high water discharge (>60 %) and 34 % of total suspended load supply into the global oceans (Martin and Meybeck 1979; Meybeck 1988; Ludwig et al. 1996; Ludwig and Probst 1998; Balakrishna and Probst 2005).

Particulate inorganic carbon concentrations in tropical rivers are negligible because of the dissolution of PIC into DIC during the weathering of carbonate rocks (Sarin et al. 2002). Tropical rivers provide 0.53 Pg C/year of riverine carbon to the oceans, of which 39.8 % is DIC, 25.7 % is DOC, 9.7 % is PIC and 24.8 % is POC. The largest DIC flux within the tropical region is found in Asia, because the DIC concentration is highest there and the discharge is second highest. The Americas have the highest DOC flux in the tropical area, owing to the sheer volume of discharge. The highest PIC flux in tropical regions is also found in the Americas because they have the highest PIC/TSM ratio and the second highest sediment load. Asia has the highest specific carbon yields in the tropical region, because of the high ratios of discharge to surface area, and sediment load to surface area. Anthropogenic activities, however, such as reducing sediment load and increasing the amount of detrital organic matter in rivers, may continue to change the fluvial carbon fluxes of tropical rivers. In India scattered studies have been made on the organic carbon. For example, Sarin et al. (2002) studied the DOC, POC and DIC concentrations on Godavari, the largest peninsular Indian tropical river.

6 Anthropogenic Perturbation of the Carbon Fluxes

During the past two centuries, human activities have greatly modified the exchange of carbon and nutrients between the land, atmosphere, freshwater bodies, coastal zones and the open ocean (Likens et al. 1981; Mulholland. and Elwood 1982; Wollast and Mackenzie 1989; Degens et al. 1991; Smith and Hollibaugh 1993; Stallard 1998; Ver et al. 1999; Richey 2004; Raymond et al. 2008). Together, land-use changes, soil erosion, liming, fertilizer and pesticide application, sewage-water production, damming of water courses, water withdrawal and human-induced climatic change have modified the delivery of these elements through the aquatic continuum that connects soil water to the open ocean through rivers, streams, lakes, reservoirs, estuaries and coastal zones, with major impacts on global biogeochemical cycles (Mackenzie et al. 2005; Cotrim da Cunha et al. 2007; Quinton et al. 2010). Carbon is transferred through the aquatic continuum laterally across ecosystems and regional geographic boundaries and exchanged vertically with the atmosphere, often as greenhouse gases.

Although the importance of the aquatic continuum from land to ocean in terms of its impact on lateral C fluxes has been known for more than two decades, the magnitude of its anthropogenic perturbation has only recently become apparent. The lateral transport of C from land to sea has long been regarded as a natural loop in the global C cycle unaffected by anthropogenic perturbations. Thus, this flux is at present neglected in assessments of the budget of anthropogenic CO_2 reported, for instance, by the Intergovernmental Panel on Climate Change (IPCC) or the Global Carbon Project. Quantifying lateral C fluxes between land and ocean and their implications for CO_2 exchange with the atmosphere is important to further our understanding of the mechanisms driving the natural C cycle along the aquatic continuum, as well as for closing the C budget of the ongoing anthropogenic perturbation. Superimposed on these natural forms, the present-day increased amounts of industrial effluents, fertilizers, sewage and other human wastes are modulating the riverine concentrations of carbon. Anthropogenic changes and river eutrophication is an important factor in future for algal POC, which can create near-anoxic conditions when reaching coastal waters.

Land use change is currently extremely rapid and its consequences are more evident in the tropical regions, in part because of the disproportionate share of human population growth that is taking place in the tropics. Land clearing and conversion causes substantial loss of carbon and nitrogen and a lesser loss of sulphur and phosphorus from cleared sites in most regions. Climate change and increased nutrient deposition from the atmosphere will affect soils, plant productivity and biogeochemical cycles. The overall emphasis of the biogeochemistry is the terrestrial regulation of element pools, transformation gains and losses as they are altered by components of global change. In addition, there are a number of regions in which land-use and atmospheric composition and anticipated climate change are likely to alter the biogeochemistry of terrestrial ecosystems significantly and consequently to cause significant change on the riverine biogeochemistry.

Model simulations suggest that the transport of riverine C has increased by about 20 % since 1750, from ~0.75 Pg C year^{-1} in 1750 to 0.9–0.95 Pg C year^{-1} at present. The existence of such an enhanced riverine delivery of C is supported by the available published data (Milliman and Meade 1983; Meybeck 1982; Wollast and Mackenzie 1989; Richey 2004; Richey et al. 1991) and has been attributed to deforestation and more intensive cultivation practices that have increased soil degradation and erosion. This leads to an increase in the export of organic and inorganic C to aquatic system (Raymond et al. 2008). For example, erosion of particulate organic C in the range 0.4–1.2 Pg C year^{-1} has been reported for agricultural land alone (Stallard 1998; Quinton et al. 2010). However, only a percentage of this flux represents a lateral transfer of anthropogenic CO_2 fixed by photosynthesis (Stallard 1998; Smith et al. 2001; Billings et al. 2010).

7 Conclusions

- The carbon cycle plays an important role in regulating interactions among the lithosphere, hydrosphere, atmosphere, and biosphere. An understanding on the transfer of C between these spheres is essential from the environmental point of view. Rivers act as conduits in transporting, sequestering, and delivering C from lithosphere to ocean basins.

- At present, there is substantial lacuna in understanding of the sources, transport pathways and rates of C transfer to different spheres which inhibits our ability to predict the present and future contribution of the riverine fluxes to the global C budget involving anthropogenic CO_2.
- The extremely complex nature of riverine systems, due to variations in climate, land use, soil composition, hydrology and man's impact, complicates interpretation of relative roles and contributions of geogenic and anthropogenic sources and fluxes of carbon. Therefore, adequate characterization of such complex systems requires measurement of a number of parameters over extended periods of time.

Acknowledgments The review could not have been possible without the publications listed in this paper, for which, I express my sincere thanks to all those authors. Thanks are also due to the authors of these publications, for having enlightened me through their publications on the importance of understanding carbon transfer among earth's components.

References

Aitkenhead JA, McDowell WH (2000) Soil C/N ratio as a predictor of annual riverine DOC flux at local and global scales. Global Biogeochem Cycles 14:127–138

Arvidson RS, Mackenzie FT, Guidry M (2006) MAGic: a Phanerozoic model for the geochemical cycling of major rock-forming components. Am J Sci 306(3):135–190

Balakrishna K, Probst JL (2005) Organic carbon transport and C/N ratio variations in a large tropical river: Godavari as a case study, India. Biogeochemistry 73:457–473

Berner RA (1989) Biogeochemical cycles of carbon and sulfur and their effect on atmospheric oxygen over Phanerozoic time. Global Planet Change 75:97–122

Berner RA (2004) The phanerozoic carbon cycle: CO_2 and O_2. Oxford University Press, Oxford, p 150

Berner R, Lasaga A, Garrels R (1983) The carbonate–silicate geochemical cycle and its effect on atmospheric carbon dioxide over the past 100 million years. Am J Sci 283:641–683

Beusen AHW, Dekkers ALM, Bouwman AF, Ludwig W, Harrison J (2005) Estimation of global river transport of sediments and associated particulate C, N, and P. Global Biogeochem Cycles 19

Billings SA, Buddemeier RW, Richter DdeB, Van Oost K, Bohling G (2010) A simple method for estimating the influence of eroding soil profiles on atmospheric CO_2. Global Biogeochem Cycles 24:GB2001

Bolin B (ed) (1981) Carbon cycle modelling. Wiley, New York

Borges AV, Abril G (2012). In: Wolanski E, McLusky DS (eds) Treatise on estuarine and coastal science, vol 5. Academic Press, pp 119–161

Borges AV, Delille B, Frankignoulle M (2005) Budgeting sinks and sources of CO_2 in the coastal ocean: diversity of ecosystem counts. Geophys Res Lett 32:1–4

Bouchez J, Metivier F, Lupker M, Gaillardet J, France-Lanord C, Perez M, Maurice L (2010) Prediction of depth-integrated sedimentary fluxes in large rivers: particle aggregation as a complicating factor. doi:10.1002/hyp.7868

Bouchez J, Gaillardet J, France-Lanord C, Dutra-Maia P, Maurice L (2011a) Grain size control of river suspended sediment geochemistry: clues from amazon river depth

Bouchez J, Lupker M, Gaillardet J, France-Lanord C, Maurice L (2011b) How important is it to integrate riverine suspended sediment chemical composition with depth? Clues from amazon river depth-profiles. Geochim Cosmochim Acta 75:6955–6970

Breithaupt JL, Smoak JM, Smith TJ, Sanders CJ, Hoare (2012) A organic carbon burial rates in mangrove sediments: strengthening the global budget. Glob Biogeochem Cycles 26

Cai W-J (2003) Riverine inorganic carbon flux and rate of biological uptake in the Mississippi river plume. Geophys Res Lett 30:1032

Canadell et al (2000) Carbon metabolism of the terrestrial biosphere: a multitechnique approach for improved understanding. Ecosystems 3:115–130

Chen CTA, Borges AV (2009) Reconciling opposing views on carbon cycling in the coastal ocean: continental shelves as sinks and near-shore ecosystems as sources of atmospheric CO_2. Deep-Sea Res II 56:578–590

Church TM (1996) An underground route for the water cycle. Science 380:579–580

Cotrim da Cunha L, Buitenhuis ET, Le Quéré C, Giraud X, Ludwig W (2007) Potential impact of changes in river nutrient supply on global ocean biogeochemistry. Glob Biogeochem Cycles 21:GB4007

Cramer W, Bondeau A, Woodward FI, Prentice IC, Betts RA, Brovkin V, Cox PM, Fisher V, Foley JA, Friend AD, Kucharik C, Lomas MR, Ramankutty N, Sitch S, Smith B, White A, Young-Molling C (2001) Global response of terrestrial ecosystem structure and function to CO_2 and climate change: results from six dynamic global vegetation models. Global Change Biol 7:357–373

Dai, Trenberth KE (2002) Estimates of freshwater discharge from continents: latitudinal and seasonal variations. J Hydrometeor 3:660–687

Degens ET, Kempe S and Richey JE (1991) Summary: biogeochemistry of the major world rivers. In: Degens ET et al (eds) Biogeochemistry of major world rivers, SCOPE 42. Wiley, New York, pp 323–347

Duarte CM, Middelburg JJ, Caraco N (2005) Major role of marine vegetation on the oceanic carbon cycle. Biogeosciences 2:1–8

Foley JA, Prenticen IC, Ramunkutty S, Levis D, Pollard S, Sitch and Haxeltine A (1996) An integrated biosphere model of land surface processes, terrestrial carbon balance and vegetation dynamics. Global Biogeochem Cycles 10(4):603−628

France-Lanord C, Derry LA (1997) Organic carbon burial forcing of the carbon cycle from Himalayan erosion. Nature 390:65–75

Gaillardet J, Dupre B, Allegre CJ (1999a) Geochemistry of large river suspended sediments: silicate weathering or recycling tracer? Geochim Cosmochim Acta 63(23–24):4037–4051

Gaillardet J, Dupre B, Louvat P, Allegre CJ (1999b) Global silicate weathering and CO_2 consumption rates deduced from the chemistry of large rivers. Chem Geol 159(1–4):3–30

Gislason SR, Oelkers EH, Snorrason A (2006) Role of river-suspended material in the global carbon cycle. Geology 34:49–52

Hem, John D (1985) Study and interpretation of the chemical characteristics of natural water. 3rd edn. US geological survey water supply paper 2254. Alexandria, VA, 263

IPCC (2007) IPCC: fourth assessment report climate change 2007. Geneva: I intergovernmental panel on climate change

Kempe S, Degens ET (1985) An early soda ocean. Chem Geol 53(1–2):95–108

Laruelle GG, Dürr HH, Slomp CP, Borges AV (2010) Evaluation of sinks and sources of CO_2 in the global coastal ocean using a spatially-explicit typology of estuaries and continental shelves. Geophys Res Lett 37

Likens GE, Mackenzie FT, Richey JE, Sedwell JR, Turekian KK (1981) Flux of organic carbon from the major rivers of the world to the oceans (National technical information service, US department of commerce)

Liu KK, Atkinson L, Quiñones R, Talaue-McManus (2010) L. carbon and nutrient fluxes in continental margins: a global synthesis, Springer

Ludwig W, Amiotte-Suchet P, Probst JL (1996) River discharges of carbon to the world's oceans: determining local inputs of alkalinity and of dissolved and particulate organic carbon. CR Acad Sci Paris 323:1007−1014

Ludwig W, Probst JL (1998) River sediment discharge to the oceans: present-day controls and global budgets. Am J Sci 298(4):265–295

Ludwig W, Probst J-L, Kempe S (1996) Predicting the oceanic input of organic carbon by continental erosion. Global Biogeochem Cycles 10:23–41

Lupker M, France-Lanord C, Lave J, Bouchez J, Galy V, Metivier F, Gaillardet J, Lartiges B, Mugnier JL (2011) A Rouse-based method to integrate the chemical composition of river sediments: application to the Ganga basin. J Geophys Res [Solid Earth]. doi:10.1029/2010JF001947

Mackenzie FT, Lerman A, Ver LMB (1998) Role of the continental margin in the global carbon balance during the past three centuries. Geology 26:423–426

Mackenzie FT, Andersson AJ, Lerman A, Ver LM (2005) In: Robinson AR, Brink KH (eds) The sea. vol 13. Harvard University Press, pp 193–225

Martin JM, Maybeck M (1979) Elemental mass of material balance carried by major world rivers. Mar Chem 7:173–206

Meybeck M (1979) Concentrations des eaux fluviales en elements majeurs et apports en solution aux oceans. Rev Geol Dyn Geogr Phys 21:215–246

Meybeck M (1982) Carbon, nitrogen, and phosphorus transport by world rivers. Am J Sci 282:401–450

Meybeck M (1987) Global chemical weathering from surficial rocks estimated from river dissolved loads. Am J Sci 287:401–428

Meybeck M (1988) How to establish and use world budgets of riverine materials. In: Lerman A, Meybeck M (eds) Physical and chemical weathering in geochemical cycles. Kluwer Academic Publishers, pp 247−272

Meybeck M (1993) Riverine transport of atmospheric carbon—sources, global typology and budget. Water Air Soil Pollut 70:443–463

Milliman J, Meade R (1983) World-wide delivery of river sediment to the oceans. J Geol 91:1–21

Mulholland PJ, Elwood JW (1982) The role of lake and reservoir sediments as sinks in the perturbed global carbon cycle. Tellus 34:490–499

Parton WJ, Ojima DS, Cole DV, Schimel DS (1994) A general model for soil organic matter dynamics: sensitivity to litter chemistry, texture and management. In: Quantitative modeling of soil forming processes. SSSA Special Publication 39. Soil Science Society of America

Quinton JN, Govers G, Van Oost K, Bardgett RD (2010) The impact of agricultural soil erosion on biogeochemical cycling. Nature Geosci 3:311–314

Raymond PA, Oh NH, Turner RE, Broussard W (2008) Anthropogenically enhanced fluxes of water and carbon from the Mississippi River. Nature 451:449–452

Richey JE (2004) In: Field CB, Raupach MR (eds) The global carbon cycle, integrating humans, climate, and the natural world, vol 17. Island Press, pp 329–340

Richey JE, Victoria RL, Salati E (1991) The biogeochemistry of a major river system: the amazon case study. In: biogeochemistry of major world rivers, SCOPE/UNEP 42, Wiley, New York, pp 57−74

Riebe CS, Kirchner JW, Finkel RC (2004) Erosional and climatic effects on long-term chemical weathering rates in granitic landscapes spanning diverse climate regimes. Earth Planet Sci Lett 224:547–562. doi:10.1016/j.epsl.2004.05.019

Sarin MM, Sudheer AK, Balakrishna K (2002) Significance of riverine transport: a case study of a large tropical river, Godavari (India). Sci China Ser C Life Sci 45:97−108

Sarmiento JL, Sundquist ET (1992) Revised budget of the oceanc uptake of anthropogenic uptake of anthropogenic carbon dioxide. Nature 356:589–593

Schlesinger WH, Melack JM (1981) Transport of organic carbon in the world's rivers. Tellus 33:172−187

Shiklomanov IA, Rodda JC (eds) (2003) World water resources at the beginning of the 21st century. UNESCO and Cambridge University Press, Cambridge, UK

Siegenthaler U, Sarmiento JL (1993) Atmospheric carbon dioxide and the ocean. Nature 365:119–125

Simpkins WW, Parkin TB (1993) Hydrogeology and redox geochemistry of CH_4 in a late Wisconsinan till and loess sequence in central Iowa. Water Resour Res 29:0043–1397

Slomp CP, Van Cappellen P (2004) Nutrient inputs to the coastal ocean through submarine groundwater discharge: controls and potential impact. J Hydrol 295:64–86

Smith SV, Hollibaugh JT (1993) Coastal metabolism and the oceanic organic carbon balance. Rev Geophys 31:75–89

Smith SV, Renwick WH, Buddemeier RW, Crossland CJ (2001) Budgets of soil erosion and deposition for sediments and sedimentary organic carbon across the conterminous United States. Glob Biogeochem Cycles 15:697–707

Stallard RF (1998) Terrestrial sedimentation and the carbon cycle: coupling weathering and erosion to carbon burial. Glob Biogeochem Cycles 12:231–257

Taniguchi M, Burnett WC, Cable JE, Turner JV (2002) Investigation of submarine groundwater discharge. Hydrol Processes 16:2115–2129

Van der Leeden F, Troise FL, Todd DK (eds) (1990) The water encyclopedia, 2nd edn. Lewis Publishers, Chelsea, Mich, p 808

Ver LMB, Mackenzie FT, Lerman A (1999) Biogeochemical responses of the carbon cycle to natural and human perturbations: past, present, and future. Am J Sci 299:762–801

Walker JCG, Hays PB, Hastings JF (1981) A negative feedback mechanism for the long term stabilization of earth's surface temperature. J Geophy Res 86:9776–9782

West JB, HilleRisLambers J, Lee TD, Hobbie SE, Reich PB (2005) Legume species identity and soil nitrogen supply determine symbiotic nitrogen fixation responses to elevated atmospheric CO_2. New Phytol 167:523–530

Wollast R, Mackenzie FT (1989) In: Berger A, Schneider S, Duplessy JCl (eds) Climate and geosciences, vol 285. Academic Publishers, pp 453–473

GIS-Based Modified SINTACS Model for Assessing Groundwater Vulnerability to Pollution in Vellore District (Part of Palar River Basin), Tamil Nadu, India

K. Rutharvel Murthy, S. Dhanakumar, P. Sundararaj, R. Mohanraj and K. Kumaraswamy

Abstract Though traditionally considered to be less vulnerable to pollution than surface water, the groundwater resources face multiple sources of contamination during the recent times. In this paper, we employed modified SINTACS (Normal and Severe) model over a region known for thick clusters of leather processing industries, and dependence of domestic and industrial water supply on groundwater resources. Several parameters including, depth to groundwater, effective infiltration, unsaturated zone attenuation capacity, soil attenuation capacity, hydrogeological characteristics of the aquifer, hydraulic conductivity and topographical slope were spatially evaluated and subjected to overlay analysis after assigning appropriate ratings and weights to identify the different vulnerability levels in the study area. The results show that, the groundwater vulnerability to pollution is very high on the eastern part of the study area, under the influences of gentle slope, alluvial soil and higher permeability. In addition, scattered occurrences of vulnerable zones aligned along the river course are also identified.

Keywords Contamination · Pollution · Vulnerability · Infiltration · Subsurface · Hydrogeology

K. Rutharvel Murthy
Department of Geography, Arignar Anna Government Arts College,
Namakkal 637002, India

S. Dhanakumar
Department of Environmental Science, PSG College of Arts and Science,
Coimbatore 641024, India

P. Sundararaj
Department of Geography, Government Arts College (Autonomous),
Karur 639005, India

R. Mohanraj
Department of Environmental Management, Bharathidasan University,
Tiruchirappalli 620024, India

K. Kumaraswamy (✉)
Department of Geography, Bharathidasan University, Tiruchirappalli 620024, India
e-mail: kkumargeo@gmail.com

Mu. Ramkumar et al. (eds.), *Environmental Management of River Basin Ecosystems*,
Springer Earth System Sciences, DOI 10.1007/978-3-319-13425-3_20

1 Introduction

Groundwater contamination is a growing environmental concern, especially in urban and industrial areas where dependency on groundwater to meet the demands is more. The problem gets aggravated in developing countries, where indiscriminate disposal of municipal wastes, industrial effluents and agriculture is common (Mato 2002). The concept of 'vulnerability of groundwater to contamination' was first introduced in France in the late 1960s (Albinet and Margat 1970; Chilton 2006). The fundamental principle of groundwater vulnerability is that some land areas are more vulnerable to pollution than others, and the goal of preparing a vulnerability map is to subdivide an area into micro units for planning and management of available groundwater resources and for mitigating the effects of contamination. The first vulnerability map of groundwater contamination at a scale of 1:1 million was prepared in France by Margat (1968). Vrba and Zaporozec (1994a) recognized that there could be more than one type of vulnerability: (i) intrinsic (natural) is defined purely as a function of hydrogeological factors, and (ii) specific pollutants (agricultural nitrate, pesticides, atmospheric deposition, industrial or municipal wastes, etc.).

The assessment of groundwater quality through field investigations on a regional scale is often expensive and time-consuming process. To circumvent this problem, process-based methods, statistical methods, and overlay and index methods (Vrba and Zaporozec 1994b; Tesoriero et al. 1998; Gogu and Dassargues 2000a) are employed using the available attributes namely, precipitation, soil groups and properties, depth to groundwater, aquifer properties, etc. The process-based methods are used to predict the contaminant transport in both space and time, and require numerical equations and simulation models. Statistical methods are based on statistical correlations between spatial variables and actual occurrence of pollutants in the groundwater. These methods are limited by shortage of water quality observations, data accuracy and choices in spatial variables (Babiker et al. 2005). As application of simulation and statistical methods in regional scale are relatively difficult, overlay and index methods are mostly preferred (Gogu and Dassargues 2000b). Overlay and index methods rely mainly on the quantitative or semi-quantitative compilation and interpretation of mapped data. These methods combine the results (maps) of the factors that control the movement of contaminants from the ground surface to the saturated zone of subsurface. The results are in the form of vulnerability levels (indices) at different locations. The most commonly used overlay and index methods are DRASTIC (Aller et al. 1987) and SINTACS (Civita 1990, 1993, 1994; Civita and De Maio 2000).

The SINTACS was created and developed by Civita (Civita 1993, 1994) and this method is originally derived from DRASTIC method (Civita and De Maio 1997). The SINTACS is a tool with a remarkable degree of flexibility for optimization of data and analysis. It can operate within GIS environment, and thus extensively being used in the vulnerability studies (Civita and De Maio 1998; Marsico et al. 2004; Al Kuisi et al. 2006; Draoui et al. 2008) by exploiting the

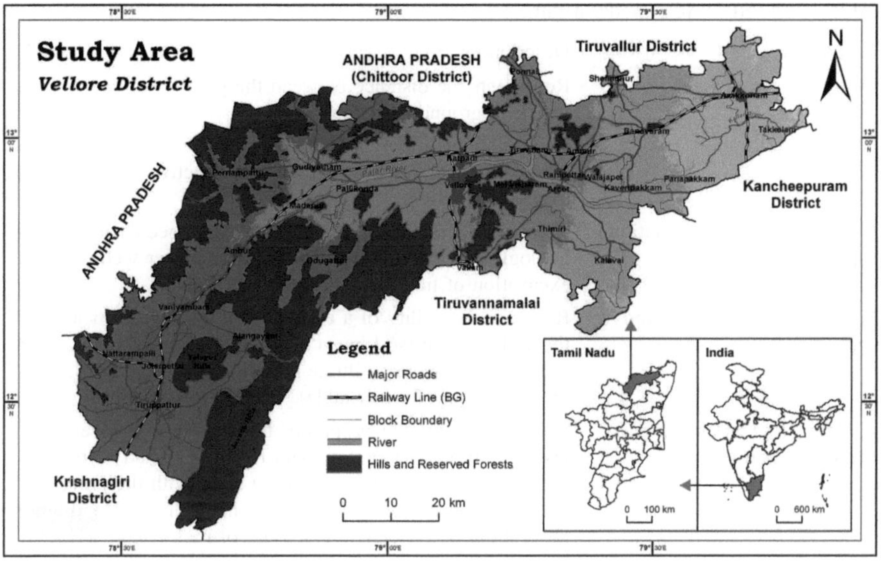

Fig. 1 Study area (Vellore district)

capabilities of a decision support system involving the integration of spatially referenced data (Cowen 1988). This paper analyzed the intrinsic vulnerability of groundwater resources of the Vellore district, Tamil Nadu state, India (Fig. 1), by employing geospatial technology and regional scale groundwater vulnerability model of SINTACS (Normal and Severe versions).

2 Materials and Methods

The SINTACS scheme of aquifer pollution vulnerability mapping was established for hydrogeological, climatic and impacts settings (Civita 1990). This assessment procedure incorporates seven parameters, relevant for the contaminant attenuation and vertical flow capacity: **S**—water table depth, **I**—effective infiltration, **N**—unsaturated conditions, **T**—soil media, **A**—aquifer hydrogeologic characteristics, **C**—hydraulic conductivity, and **S**—topographic slope (Table 1).

This model yields numerical indices that are derived from ratings and weights assigned to all the parameters. The important media types or classes of each parameter represent the ranges, which are rated from 1 to 10 based on their relative effect on the aquifer vulnerability. These parameters are then assigned weights ranging from 1 to 5 reflecting their relative importance. The SINTACS index is then computed by applying a linear combination of all the factors by adopting the following equation:

Table 1 The SINTACS model parameters

Factor	Description
Depth to groundwater	Represents the distance between the ground surface and the top level groundwater table
Effective infiltration	Represents the difference between total precipitation and the cumulative loss by direct runoff and effective evapotranspiration
Unsaturated zone attenuation capacity	Refers to the impact of vadose zone examined by interpreting geological maps, drilling logs of groundwater wells and excavation of trenches
Soil/overburden attenuation capacity	Refers to the ability of a contaminant to move vertically into the vadose zone (soil has a significant impact on the amount of recharge that can infiltrate into the ground and the quantum of upper limit of water table)
Hydrogeological characteristics of the aquifer	Refers to the aquifer properties to describe the process that takes place when a contaminant becomes mixed with groundwater (the aquifer typology deals with the processes which take place below water level that is dispersion, dilution, absorption, and chemical reactivity of rocks: they are deeply influenced by the lithological characters as well as by the permeability of the aquifer)
Hydraulic conductivity	Refers to the ability of the aquifer materials to transmit water, which in turn controls the rate at which groundwater will flow under a given hydraulic gradient under conditions
Topographic slope	Represents the slope that determines the extent of runoff of the pollutant and the degree of settling sufficient for infiltration

Source: Civita (1990)

$$I_v = \sum_{i=1}^{7} P_i \times W_i$$

where,

I_v Vulnerability Index by SINTACS method,
P_i Rating for parameter i,
W_i Weight factor for parameter i.

Groundwater level data (pre- and post-monsoon seasons) from 57 monitoring wells for 20 years (1989–2008) are used in this study. The net recharge was estimated by using the norms framed by the Groundwater Estimation Committee (GEC), Ministry of Water Resources, Government of India. The data of groundwater status in 20 locations were calculated based on basin and hydrological attributes collected from the Ground Water Division, Public Works Department, Vellore (Government of Tamil Nadu) to estimate the spatial distribution of net recharge over the study area.

Soil samples were collected up to 60 cm depth from 40 locations to examine their textural characteristics in the study area. These results were cross checked and supplemented with the results of Soil Atlas Report, procured from the Directorate of Agriculture, Vellore. The interpolated raster surfaces of subsurface soil for every 15 cm depth BGL up to 1.8 m thickness (mean depth of soil in the study area—Soil Atlas Report, Directorate of Agriculture, Vellore) were generated for different type of soil textures through converting the qualitative data into quantitative representations by assigning code to the data with simple numerical and logical approach. The information related to borehole lithology (from 134 boreholes) has been collected from the Ground Water Division, Public Works Department, Vellore (Government of Tamil Nadu) and processed to understand the subsurface lithological characteristics at every 2 m depth BGL, up to the extent of 24 m depth BGL. The interpolated raster surfaces of subsurface lithological characteristics for every 2 m depth BGL (up to 24 m depth) were generated through the conversion process of qualitative data into quantitative representations by assigning code to the data with numerical and logical approach.

The subsurface details about soil and lithological sections were further analyzed to identify the vadose zone characteristics and infiltration capacity of the study area. These parameters were broadly classified with reference to their infiltration capacities. Data on hydraulic conductivity were collected from the results of pumping tests. In addition to the parameters of SINTACS model, land use/land cover details of the study area were also considered as an additional parameter to identify the vulnerability zones. For this, the IRS 1C satellite imagery (2006 data) were used to identify the land use/land cover status and their impact over the groundwater quality in the study area.

The layers or themes used to evaluate the groundwater vulnerability to contamination were converted from vector (polygon feature themes) to raster format. The seven critical parameters of SINTACS (Normal and Severe versions) model, namely, depth to water table, net recharge, aquifer media, soil media, topography, impact of the vadose zone, hydraulic conductivity and infiltration rates were spatially analyzed using reclassification techniques and by assigning weights and ratings. In this study, each data layer was reclassified into a common scale of its potential to contaminate the groundwater quality. This common scale consists of five classes for each data layer with a value ranging from 5 to 1, denoting high to low level of pollution potential. The ratings and weights were assigned based on the importance of the parameters with reference to their theme and local conditions.

The SINTACS model is not originally designed for use in GIS, even though it has been revealed that such a kind of attempts can provide various substantial benefits (Merchant 1994). The spatial analyst tool available in ArcGIS 9.3 was used to develop the data layers for the components of SINTACS model. The performance of ArcGIS on raster overlay analysis and indexing techniques were applied to evaluate the groundwater vulnerability to pollution over the study area. The groundwater vulnerability zones were then cross validated with the results of groundwater quality for drinking and agricultural standards to find out their relevance.

3 SINTACS Model

3.1 Depth to Groundwater

Depth to groundwater indicates the depth to which a contaminant must travel before reaching the aquifer. It is an important factor since agricultural and other chemicals most often affect the nearby surface or uppermost aquifers (Leonard and Knisel 1988). The average groundwater level in the study area varies from 1.29 (Arcot) to 17.56 m (Kallapadi). During pre-monsoon season, it varies from 2.09 (Arcot) to 19.21 m (Kallapadi) and during post-monsoon season, it varies from 0.5 (Arcot) to 15.91 m (Kallapadi) BGL. The groundwater fluctuation ranges from 0.99 (Ambur) to 5.13 m (Asanampattu) BGL. About 65 % (37 Nos.) of the monitoring wells indicate groundwater fluctuation between 2 and 4 m BGL. A high range of groundwater fluctuation during both the seasons (pre- and post-monsoon seasons) impacts the groundwater at different depths. Hence, separate layers of depth to groundwater for pre- and post-monsoon seasons and interpolated raster surface of depth to water table were created from point data. Geologic formation and aquifer media data from well logs were also considered to finalize the groundwater table and its classes (Fig. 2a, b). The ranges of interpolated raster surfaces were classified into five types from ≤4 m (shallow) to >16 m (very deep) BGL. The highest ratings were assigned to depth to water levels that are nearer to the surface with the depth of ≤4 m (shallow), and the lowest ratings are assigned to the surfaces with the depth of >16 m (very deep) BGL during pre- and post-monsoon seasons. The ratings range from 10 (shallow groundwater table with ≤4 m depth BGL) to 1 (very deep groundwater table with >16 m depth BGL) for different depths of groundwater table (Table 2).

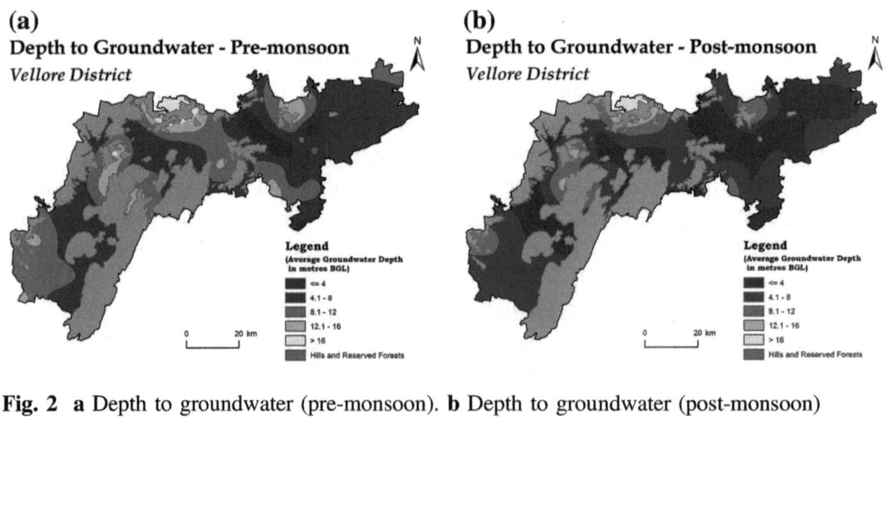

Fig. 2 **a** Depth to groundwater (pre-monsoon). **b** Depth to groundwater (post-monsoon)

Table 2 Ratings and weights assigned in SINTACS model (normal and severe versions)

SI. No.	Parameters	Class	Normal			Severe		
			Rating	Weight	Index	Rating	Weight	Index
1	Depth to groundwater (S)	≤4 m	10	5	50	10	5	50
		4.01–8 m	8		40	8		40
		8.01–12 m	3		15	3		15
		12.01–16 m	2		10	2		10
		>16 m	1		5	1		5
2	Effective infiltration (I)	*Lithology*						
		Gravel	10	4	40	10	5	50
		Sand	8		32	8		40
		Pegmatite	6		24	6		30
		Kankar	6		24	6		30
		Weathered gneiss	5		20	5		25
		Jointed gneiss	4		16	4		20
		Fissured gneiss	4		16	4		20
		Charnockite	2		8	2		10
		Fresh rock	1		4	1		5
		Soil texture						
		Loamy sand	7	4	28	7	5	35
		Sandy loam	7		28	7		35
		Sandy clay loam	6		24	6		30
		Clay loam	5		20	5		25
		Silty clay loam	4		16	4		20
		Sandy clay	4		16	4		20
		Silty sand	3		12	3		15
		Clay	1		4	1		5
		Net recharge						
		≤15 mcm	1	4	4	1	5	5
		15.01–25 mcm	2		8	2		10
		25.01–35 mcm	4		16	4		20
		35.01–45 mcm	6		24	6		30
		>45 mcm	8		32	8		40

(continued)

Table 2 (continued)

SI. No.	Parameters	Class	Normal			Severe		
			Rating	Weight	Index	Rating	Weight	Index
3	Unsaturated zone attenuation capacity (N)	*Soil texture*						
		Loamy sand	10	5	50	10	4	40
		Sandy loam	9		45	9		36
		Sandy clay loam	8		40	8		32
		Clay loam	7		35	7		28
		Silty clay loam	4		20	4		16
		Sandy clay	3		15	3		12
		Silty sand	2		10	2		8
		Clay	1		5	1		4
		Lithology						
		Gravel	10	5	50	10	4	40
		Sand	9		45	9		36
		Pegmatite	8		40	8		32
		Kankar	7		35	7		28
		Weathered gneiss	6		30	6		24
		Jointed gneiss	5		25	5		20
		Fissured gneiss	5		25	5		20
		Charnockite	2		10	2		8
		Fresh rock	1		5	1		4
4	Soil attenuation capacity (T)	Loamy sand	10	3	30	10	5	50
		Sandy loam	9		27	9		45
		Sandy clay loam	8		24	8		40
		Clay loam	7		21	7		35
		Silty clay loam	4		12	4		20
		Sandy clay	3		9	3		15
		Silty sand	2		6	2		10
		Clay	1		3	1		5

(continued)

Table 2 (continued)

SI. No.	Parameters	Class	Normal			Severe		
			Rating	Weight	Index	Rating	Weight	Index
5	Hydrogeological characteristics of the aquifer (A)	Gravel	10	3	30	10	3	30
		Sand	9		27	9		27
		Pegmatite	8		24	8		24
		Kankar	7		21	7		21
		Weathered gneiss	6		18	6		18
		Jointed gneiss	5		15	5		15
		Fissured gneiss	5		15	5		15
		Charnockite	2		6	2		6
		Fresh rock	1		3	1		3
6	Hydraulic conductivity (C)	≤10 m/day	2	3	6	2	2	4
		10.01–20 m/day	3		9	3		6
		20.01–30 m/day	4		12	4		8
		30.01–40 m/day	5		15	5		10
		>40 m/day	6		18	6		12
7	Topographical slope (S)	≤2 %	5	3	15	5	2	10
		2–6 %	4		12	4		8
		6–12 %	3		9	3		6
		12–18 %	2		6	2		4
		>18 %	1		3	1		2

3.2 Effective Infiltration

The effective infiltration plays an important role in aquifer vulnerability assessment because of its dragging effect on pollutants into groundwater through the processes like dilution and percolation from unsaturated zone to saturated zone. Infiltration is the only component of the net recharge in all the areas, where there is no interflow between aquifers, surficial water bodies or irrigation practices using large volumes of water.

The calculation of the effective infiltration is based on net recharge, hydrogeological characteristics of the aquifer and soil texture. The estimated net recharge ranges between 8.56 and 56.75 mcm. The interpolated raster surface shows the occurrences of high net recharge in the eastern portion of the study area and low net recharge in isolated, small pockets distributed in the central and western portions (Fig. 3). Considering the soil types of the study area, higher range of infiltration is

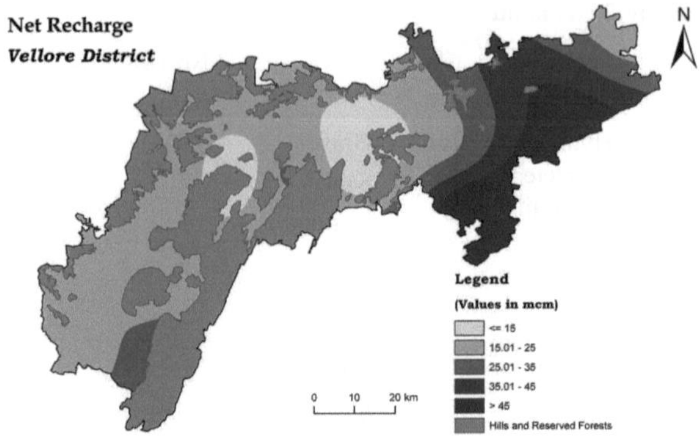

Fig. 3 Net recharge

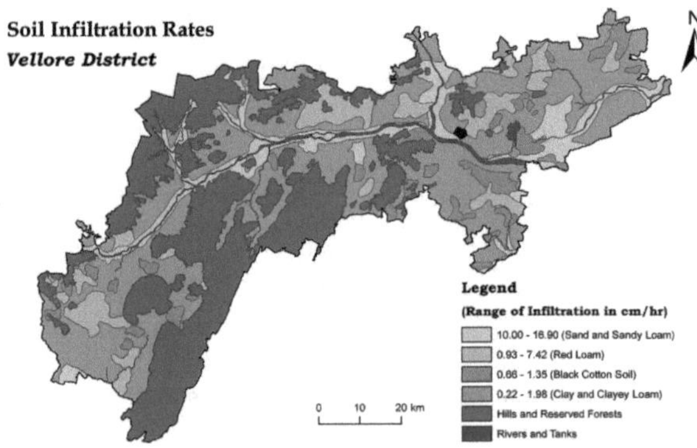

Fig. 4 Soil infiltration rates

observed in sand and sandy loam with 10.00–16.90 cm/h, moderate infiltration in red loam with the range of 0.93–7.42 cm/h, low infiltration in black cotton soil with the range of 0.66–1.35 cm/h, and least or nil infiltration in clay and clayey loam with the range of 0.22–1.98 cm/h which are reported by the Public Works Department (Fig. 4). Besides these, the subsurface soil and lithological characteristics were also considered based on their textural properties. The soil media was classified into eight types from loamy sand to clay (as per the textural properties of the soil). The aquifer media (subsurface lithology) was classified into nine types from gravel to fresh rock (as per the textural properties and age of the rocks). Based on these information and spatial variation of the attributes, the infiltration rates (high to low) at different regions have been estimated.

3.3 Unsaturated Zone Attenuation Capacity

The vadose zone is the portion of the subsurface in which the available pore spaces are unsaturated or partially saturated. A key element in pollution attenuation is the behavior of contaminants in the vadose zone, as the media is the home to many natural organisms which break down many polluting substances into secondary by-products both harmful and harmless. Various attenuation processes may occur between the soil horizon and the water table; namely: biodegradation, neutralization, mechanical filtration, chemical reaction, volatilization and dispersion (Aller et al. 1987). Two attenuation factors namely, soil texture and lithology, in the vadose zone at two different depths for pre-monsoon (10 m BGL) and post-monsoon (6 m BGL) were considered. It was made due to the greater variation of groundwater level between seasons (5.45 m BGL during post-monsoon and 8.18 m BGL during pre-monsoon).

The typical soil textures: loamy sand, sandy loam, sandy clay loam, clay loam, silty clay loam, sandy clay, silty clay and clay were rated with 10, 9, 8, 7, 4, 3, 2 and 1 for the versions of SINTACS model (Table 2). The subsurface lithological types: gravel, sand, pegmatite, kankar, weathered gneiss, jointed gneiss, fissured gneiss, charnockite and fresh rocks are rated with 10, 9, 8, 7, 6, 5, 5, 2 and 1 for SINTACS model (Table 2). The ratings were assigned as per the infiltration and attenuation capacities of soil and lithological features.

3.4 Soil Attenuation Capacity

The soil parameters that can influence vulnerability include the texture and thickness of soil profile under study and the presence/absence of organic matter in the soil. The 1.8 m thick soil profile, constituting the vadose zone supports significant biological activity. Though soil profile is a continuous medium, there exists spatial variability in its physical, chemical, and biological properties which introduce attenuation function (Zaporozec 1985). Soil permeability and contaminant migration are directly linked to soil type, shrink and swell potential, and grain size of the soil (Aller et al. 1987).

The qualitative features of the subsurface soil have been mapped through converting qualitative data into quantitative classes by assigning numerical code to the data with logical approach. Then, the textural characteristics of the subsurface soil were mapped by using interpolation technique (spline method) from the point features of subsurface soil texture data. Finally, the layers of different depths (Ground level—1.8 m BGL with the results of every 15 cm) were subjected to overlay to find out the distribution of subsurface soil textural characteristics in the study area. The study area is characterized by eight types of soil namely, loamy sand, sandy loam, sandy clay loam, clay loam, silty clay loam, sandy clay, silty clay and clay which correspond with the ratings of 10, 9, 8, 7, 4, 3, 2 and 1 (Table 2; Fig. 5a–l) according to their quality, and transportation and dispersion of surface water.

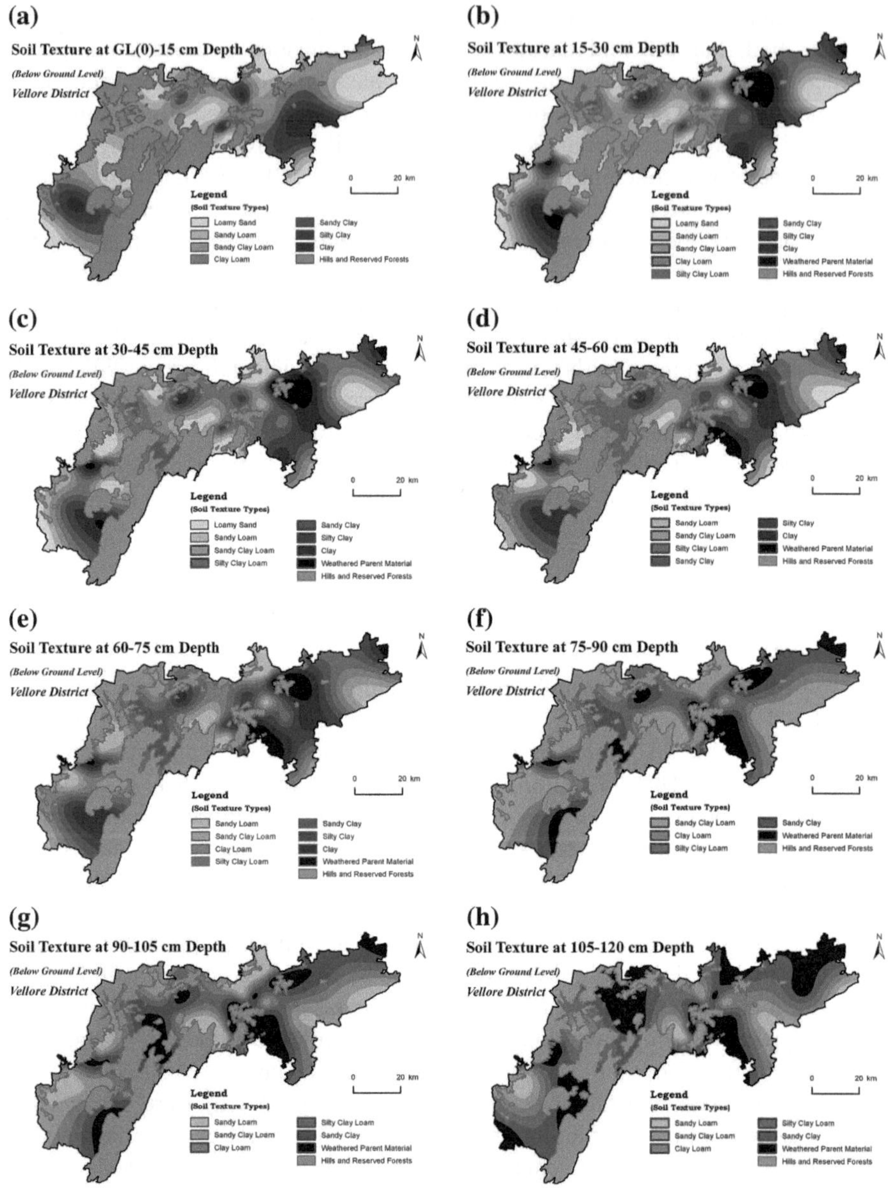

Fig. 5 **a** Soil texture at 0–15 cm depth. **b** Soil texture at 15–30 cm depth. **c** Soil texture at 30–45 cm depth. **d** Soil texture at 45–60 cm depth. **e** Soil texture at 60–75 cm depth. **f** Soil texture at 75–90 cm depth. **g** Soil texture at 90–105 cm depth. **h** Soil texture at 105–120 cm depth. **i** Soil texture at 120–135 cm depth. **j** Soil texture at 135–150 cm depth. **k** Soil texture at 150–165 cm depth. **l** Soil texture at 165–180 cm depth

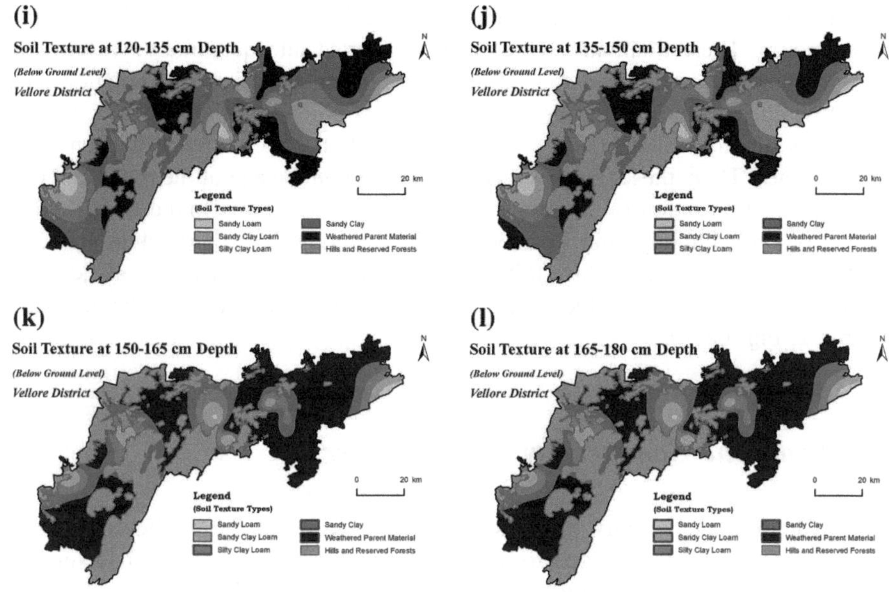

Fig. 5 (continued)

3.5 Hydrogeological Characteristics of the Aquifer

An aquifer is a subsurface rock unit which yields water for use. Aquifer media describes the consolidated and unconsolidated rock where water is contained in the pore spaces and fractures of the media. The aquifer media therefore affects the flow within the aquifer. This flow path controls the rate of contaminant contact within the aquifer (Aller et al. 1987) and plays a significant role in the attenuation process. The pathways of groundwater flow are strongly influenced by the grain size of the medium, fractures or openings within the aquifer. Larger grain size and more fractures or openings imply a higher permeability and thus, a attenuation capacity. Alternatively, the presence of clay materials in the aquifer lowers the rate of passage of the pollution potential.

From the results of groundwater level (pre- and post-monsoon seasons), aquifer and borehole lithological characteristics, the aquifer boundary has been approximately identified with the thickness of 7.31–24.05 m BGL. The spatial characteristics of subsurface lithology using data of one hundred and thirty-four (134) locations were applied for generating raster layers for every 1 m depth up to 4 m and then for every 2 m depth interval up to 24 m depth BGL. The different types of subsurface lithological features were broadly classified into nine major groups namely, gravel, sand (coarse/medium/fine/clay), pegmatite (jointed/weathered/fissured), weathered gneiss (amphibolite/hornblende/biotite/mica), jointed gneiss (granitic gneiss/gneiss/charnockite), fissured gneiss (hornblende/biotite/quartz), charnockite (jointed/weathered/fissured) and fresh rock (charnockite/epidote/granitic gneiss). These qualitative types were converted into quantitative values with simple numerical codes and logical approach. Then, by using interpolation technique (spline method), from the point features of borehole lithology data, the raster layers of different depths were generated to find structurally weak zones and strong zones. Then, these features were rated based on the age of the rocks and their capacity for infiltration on a scale of 10–1. The infiltration is found to be high in gravel type of lithology and low (Nil) in the materials like charnockite and fresh rocks. The layers of subsurface lithology (from 1 to 24 m BGL) were subjected to overlay to find out the suitable raster surface of aquifer media and its characteristics (Table 2; Fig. 6a–h).

3.6 Hydraulic Conductivity of the Aquifer

The hydraulic conductivity of an aquifer is a measure of the aquifer's ability to transmit water when subjected to a hydraulic gradient. It is a critical factor as it controls the velocity of groundwater flow; which in turn influences the velocity of contaminant flow within the aquifer. An aquifer with high conductivity is vulnerable to substantial contamination as a plume of contamination can move easily through the aquifer (Rahman 2008). Hence, it is assumed that the areas with high hydraulic conductivity values are prone to contamination. Thus, the hydraulic conductivity maps were generated using two components of conductivity: transmissivity and saturated thickness based on the formula $K = T/h$ where K is the hydraulic conductivity, T represents the transmissivity, and h the thickness of the aquifer.

The hydraulic conductivity (Permeability) values range between 0.94 (Lalapet) and 157.74 m/day (Mattrapalli). Hydraulic conductivity of an area has direct relationship with groundwater potential. The interpolated raster has been classified into five classes. 47.6 % of the test wells show hydraulic conductivity (Permeability) with ≤10 and 9.5 % of the test wells a range of hydraulic conductivity from 10.1 to 20 m/day. Zones with very high and high levels of hydraulic conductivity of >40 and 30.1–40 m/day were observed in with 28.6 % of the test wells. The remaining portions fall under medium level of hydraulic conductivity with 20.1–30 m/day in 14.3 % of the test wells in the study area (Fig. 7). The following classes of hydraulic conductivity: ≤10, 10.1–20, 20.1–30, 30.1–40, >40 m/day are represented with the ratings of 2, 3, 4, 5 and 6 respectively (Table 2).

3.7 Topographic Slope

In terms of slope and slope variability, topography is a controlling factor for pollutant runoff or infiltration. As the infiltration probability of contaminant is lowered,

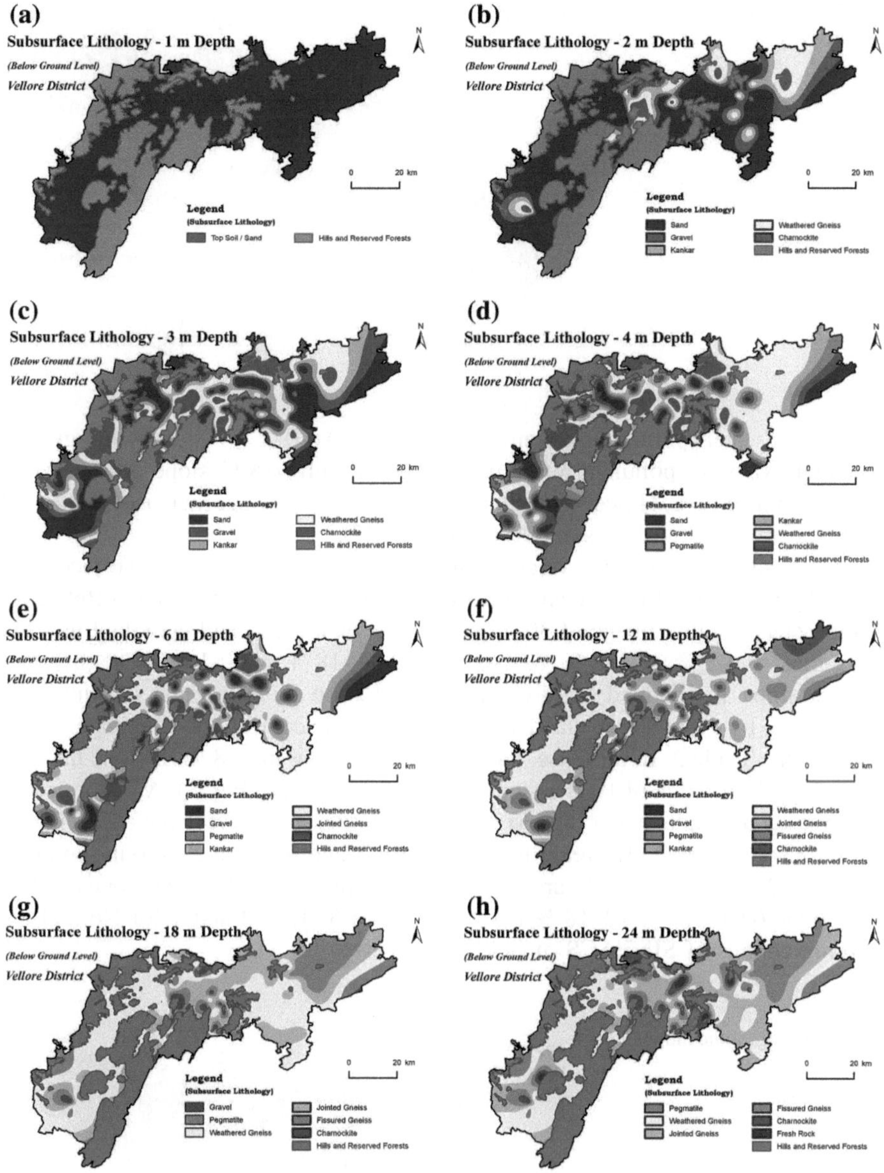

Fig. 6 **a** Subsurface lithology at 1 m depth. **b** Subsurface lithology at 2 m depth. **c** Subsurface lithology at 3 m depth. **d** Subsurface lithology at 4 m depth. **e** Subsurface lithology at 6 m depth. **f** Subsurface lithology at 12 m depth. **g** Subsurface lithology at 18 m depth. **h** Subsurface lithology at 24 m depth

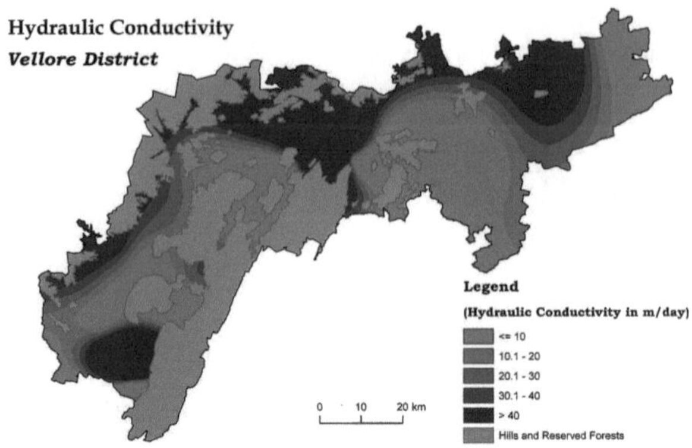

Fig. 7 Hydraulic conductivity

the groundwater pollution potential decreases. At 0–2 % slope, the greatest potential exists for pollutant infiltration, whereas with >18 % slope little potential exists for infiltration. However, contamination to surface water increases along with a greater probability of erosion (Aller et al. 1987).

In this study, topography refers to the percent slope of the land surface, generated by using ASTER (Advanced Space borne Thermal Emission and Reflection Radiometer) satellite image of 30 m gridded elevation data and classified into five categories as ≤2 %, 2.01–6 %, 6.01–12 %, 12.01–18 %, and >18 %. The (average) percent slope is obtained by dividing vertical drop by horizontal distance (and multiplying by 100). These classes were reclassified according to the range criteria given in the Table 2. In this case, ratings corresponding to >18 % of slope have the value 1, and the value of 10 is assigned for 0–2 % of slope (Fig. 8). A major portion of the study area falls under highest ratings, as topography is very flat (≤2 %) and it covers 32 % of the study area. Slopes between 2 and 6 % are also found in larger tracts (34.22 %) of the study area. The assigned ratings of different slopes (≤2 %, 2.01–6 %, 6.01–12 %, 12–18 % and >18 %) are 5, 4, 3, 2 and 1 for Normal and Severe versions of SINTACS model respectively.

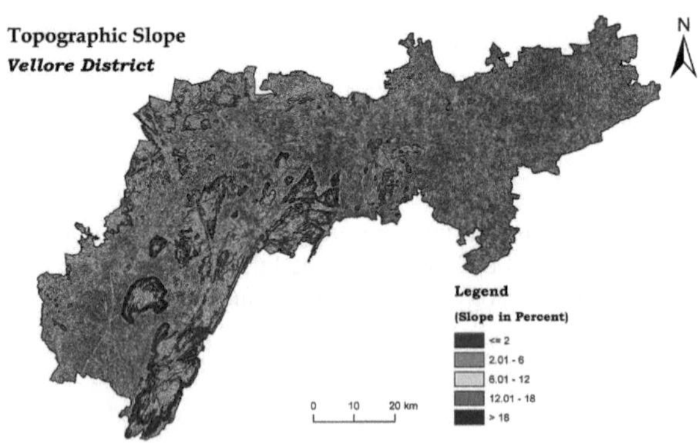

Fig. 8 Topographic slope

4 Land Use/Land Cover for Modified SINTACS Model

In order to evaluate the land use/land cover parameter of Modified SINTACS model, the land use/land cover layer was prepared from IRS 1C LISS III and PAN merged satellite data procured from the Institute of Remote Sensing (IRS), Anna University, Chennai. The National Remote Sensing Centre's (NRSC's) classification techniques were used to classify the land use/land cover categories of the study area by manual and digital interpretations.

Among all categories, agricultural land covers 56.18 % of the total area (kharif—14.41 %, rabi—1.71 %, double/triple crops—37.28 %, current fallow—2.78 and 0.03 % under plantations/orchards), followed by forest lands to the tune of 29.65 % (evergreen forest—5.44 %, deciduous forest—20.33 % and shrub/degraded forests—3.88 %). The wastelands and shrub lands together occupy 12.01 % and water bodies occupy 1.4 % of the study area. The built-up land covers around 0.72 % of the study area (Fig. 9).

The significant sources of contamination in the study area are agricultural practices, including the heavy application of fertilizers and pesticides and leakages/discharges from Effluent Treatment Plants (ETPs) and Common Effluent Treatment Plants (CETPs) of leather tanneries in the study area. The discharge of sewages from urban area may also considerably contaminate the groundwater quality of the study area.

With respect to non-agricultural land cover, built-up lands (discontinuous urban areas) were assigned with highest ratings of 10, followed by evergreen forest—6, deciduous forest—5 and degraded forest—5. The scrub lands and wastelands were rated 3 each and water bodies with 1. While looking at agricultural land cover, the kharif and rabi crop lands were assigned with highest rating of 9, whereas double/triple crop lands, current fallow lands and plantations/orchards (semi-natural areas) were rated with 6 (Table 3).

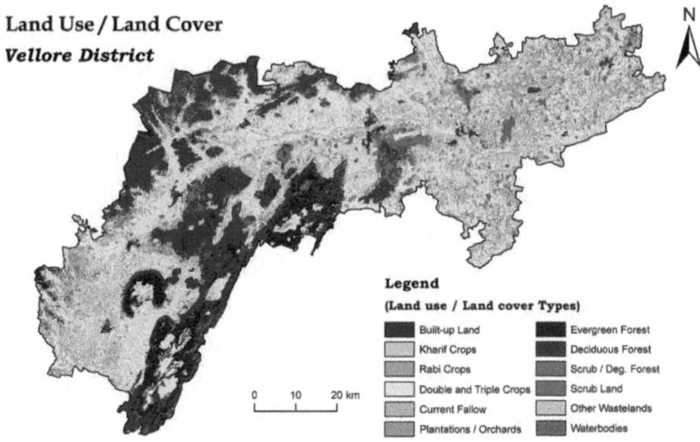

Fig. 9 Land use/land cover

Table 3 Ratings and weights assigned to land use/land cover in modified SINTACS model

Sl. No.	Parameters	Class	Rating	Weight	Index
1	Land use/land cover	Built-up land	10	3	30
		Kharif crops	9		27
		Rabi crops	9		27
		Double/triple crops	8		24
		Current fallow land	6		18
		Plantation/orchard	6		18
		Evergreen forest	6		18
		Deciduous forest	5		15
		Scrub/degraded forest	5		15
		Scrub land	3		9
		Other wastelands	3		9
		Water bodies	1		3

5 Reclassification of Data Layers

The reclassification techniques are used to replace the input cell values with new output cell values. This technique is employed essentially to replace values based on new information, to group certain values together, and to reclassify values to a common scale (McCoy and Johnston 2001). In this study, each data layer was reclassified to a common scale of its potential to cause pollution of groundwater by chemical fertilizers, pesticides and domestic and industrial wastes. This scale consists of many classes for each data layer with a value ranging from 10 to 1, meaning high to low contamination potential. The reclassification of all data layers were performed by using raster calculator available in ArcGIS spatial analyst tool.

The ratings of depth to groundwater (1–10) are weighted with a value of 5 indicating the relative importance of this model element. The values of ratings were used to reclassify each data layer of SINTACS (Normal and Severe versions) model. Later, the ratings of each layer were multiplied with assigned weights for Normal and Severe versions of SINTACS model (Table 2). The effective infiltrations of soil media and aquifer media (pre- and post-monsoon) are reclassified based on their infiltration rates and textural properties. The assigned ratings for net recharge range from 1 (for low recharging areas) to 8 (for high recharging areas). These reclassified data layers were multiplied by 4 and 5 for Normal and Severe versions of SINTACS model (Table 2).

According to Aller et al. (1987), the vadose zone of an unconfined aquifer system is the same as the aquifer media. The soil data layer (top layer of vadose zone) was reclassified according to its textural characteristics and rated with 1 for fine textured soil (clay) and 10 for coarse textured soil (loamy sand). The subsurface lithological characteristics of vadose zone were analyzed based on the thickness of saturated zone (due to groundwater fluctuation) during pre- and post-monsoon seasons. These lithological features were reclassified with their rates ranging from 1 (fresh rock) to 10 (gravel) (Table 2). The variations in the thickness of vadose zone (unsaturated zone attenuation capacity) were dependent on the fluctuations in the groundwater levels during pre- and post-monsoon seasons. During pre-monsoon season, it extends up to the depth of 10 m BGL (Average groundwater level of pre-monsoon season—8.18 m BGL) and in post-monsoon season, it is restricted within 6 m BGL (Average groundwater level of post-monsoon season—5.45 m). The reclassified data layers of rated Normal and Severe versions (SINTACS model) were multiplied with weights of 5 and 4 respectively.

The soil texture classes (loamy sand to clay) of different depths were reclassified with a range of values 1 (clay) to 10 (loamy sand). The reclassified layers (soil attenuation capacity) were subjected to overlay analysis to generate a single layer of soil media (Table 2). Then, this reclassified layer was weighted with values 3 and 5 for Normal and Severe versions of SINTACS model. The data layers of hydrogeological characteristics of the aquifer classes (gravel to fresh rock) at different depths were reclassified with different rates and subjected to overlay analysis to generate a single layer for aquifer media. The ratings of reclassified layer were multiplied by 3 for Normal and 5 for Severe versions of SINTACS model (Table 2). Aquifer with high hydraulic conductivity is more vulnerable to potential contamination than an aquifer with low hydraulic conductivity and hence, ratings were assigned accordingly—a rating of 6 for zones with high hydraulic conductivity and 2 for zones with low hydraulic conductivity. Then this layer was reclassified with the ratings and multiplied with weights of 3 and 2 for Normal and Severe versions of SINTACS model respectively (Table 2).

Topographic slope layer was reclassified with a range from 1 to 5. The minimum rate of 1 was assigned to the slope with >18 % and the maximum rate of value 5 was assigned to the slope with ≤2 % (Table 2). The reclassified layer was weighted with 3 and 2 for the Normal and Severe versions of SINTACS model.

The land use/land cover data layer was reclassified as per the rates assigned to available land use/land cover types and their potential impact on groundwater contamination. The ratings range from 10 (built-up land) to 1 (water bodies). The built-up lands were rated 10 due to their role on groundwater pollution by the discharge of urban sewage and effluents from industries (particularly from leather tanneries). Agricultural lands were rated with 9–6 (kharif—9; rabi—9; double/triple crops—8; current fallow land—6; and plantation/orchard—6), based on the relative amount of pesticides applied for crop cultivation. Lands with vegetative cover were assigned with ratings 6–3 (evergreen forest—6; deciduous forest—5; scrub/degraded land—5; scrub land—3). Other wastelands were rated with 3 and the water bodies with 1. The weighting criterion was assigned with 3 for the version of modified SINTACS model (Table 3).

6 Analysis of Data Layers Using GIS

The data layers or themes used to evaluate groundwater vulnerability to pollution were manipulated by converting the vector to raster as the raster data structures are comparatively easier to conceptualize (DeMers 2000). In this study, raster overlay was performed using arithmetic operation (available in ArcGIS 9.0—Raster Calculator tool). The arithmetic overlay means that the values of two or more input themes are combined arithmetically to generate an output grid (ESRI 2000). With reference to the Tables 2 and 3, the reclassified data layers with ratings were first multiplied by assigned weights and then added together to generate an output grid of vulnerable zones as per the Normal and Severe versions of SINTACS model. This kind of arithmetic overlay is called 'Additive Overlay' (Ormsby and Alvi 1999).

7 SINTACS Vulnerability Zones

The resultant vulnerable zones were classified into five types (very low–low–moderate–high–very high). The vulnerability layer contains grid cells with unique values showing the level of vulnerability to contamination (called as vulnerability index).

The total area of the study area (Vellore district) is 6,077 km^2. Excluding the (approximately) 2,000 km^2 of Hills and Reserved Forests area, the present investigation was carried out in the remaining 4,077 km^2 of area. The results of vulnerability assessments based on Normal and Severe versions of SINTACS during pre- and post-monsoon seasons are presented in Tables 4 and 5. Their spatial distributions are presented in the Figs. 10, 11, 12 and 13. While comparing the

Table 4 Groundwater vulnerability levels of pollution in Vellore district (pre-monsoon)

Methods	Level of vulnerability in Vellore district—pre-monsoon (area in square kilometer and percentage)									
	Very low	In %	Low	In %	Moderate	In %	High	In %	Very high	In %
SINTACS (Normal)	213.2	5.2	1167.4	28.6	2115.2	51.9	480.5	11.8	100.7	2.5
SINTACS (Severe)	253.9	6.2	1323.2	32.5	1999.9	49.1	408.0	10.0	92.0	2.3
Modified SINTACS (Normal)	137.2	3.4	1063.1	26.1	2203.1	54.0	567.2	13.9	106.4	2.6
Modified SINTACS (Severe)	98.6	2.4	977.5	24.0	2295.3	56.3	609.0	14.9	96.7	2.4

Table 5 Groundwater vulnerability levels of pollution in Vellore district (post-monsoon)

Methods	Level of vulnerability in Vellore district—post-monsoon (area in square kilometer and percentage)									
	Very low	In %	Low	In %	Moderate	In %	High	In %	Very high	In %
SINTACS (Normal)	104.4	2.6	898.9	22.1	2241.0	55.0	718.8	17.6	113.8	2.8
SINTACS (Severe)	160.2	3.9	1023.8	25.1	2220.0	54.5	574.6	14.1	98.4	2.4
Modified SINTACS (Normal)	32.6	0.8	694.8	17.0	2336.4	57.3	890.6	21.9	122.5	3.0
Modified SINTACS (Severe)	37.6	0.9	688.2	16.9	2375.8	58.3	875.5	21.5	99.9	2.5

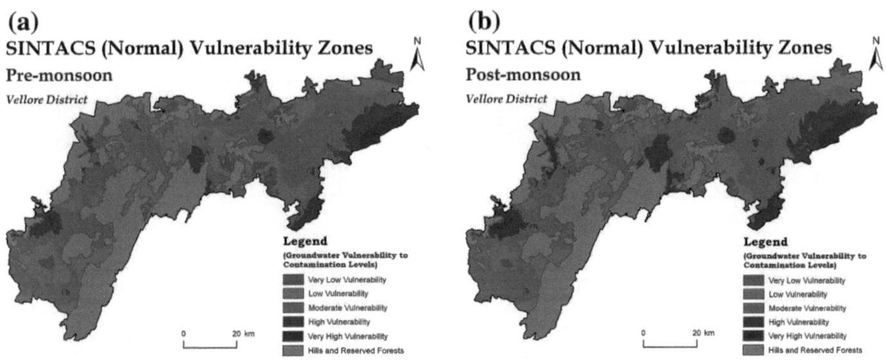

Fig. 10 **a** Vulnerability zones of SINTACS (Normal) in pre-monsoon. **b** Vulnerability zones of SINTACS (Normal) in post-monsoon

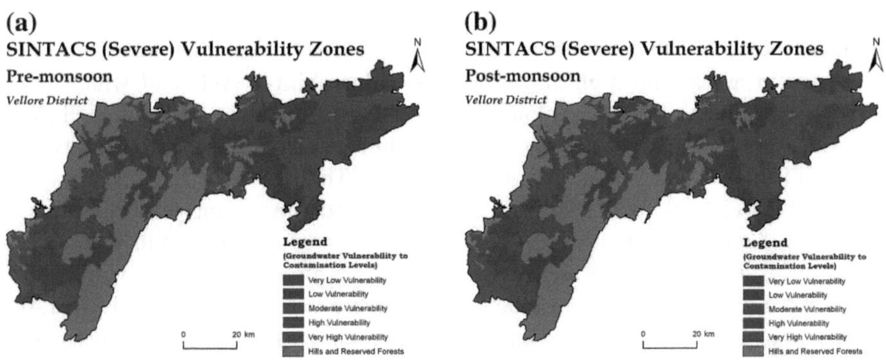

Fig. 11 **a** Vulnerability zones of SINTACS (Severe) in pre-monsoon. **b** Vulnerability zones of SINTACS (Severe) in post-monsoon

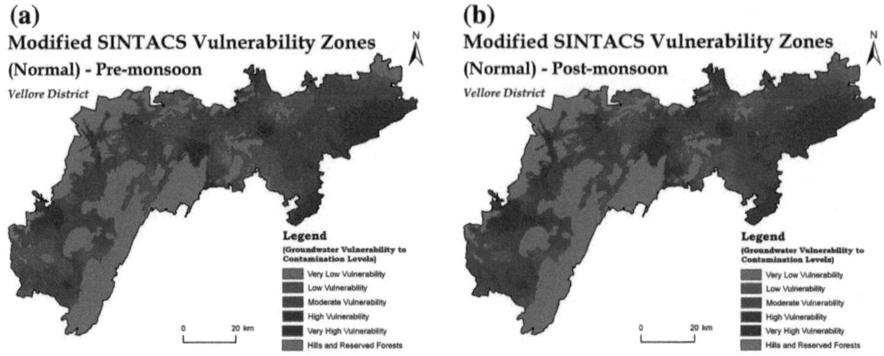

Fig. 12 **a** Vulnerability zones of modified SINTACS (Normal) in pre-monsoon. **b** Vulnerability zones of modified SINTACS (Normal) in post-monsoon

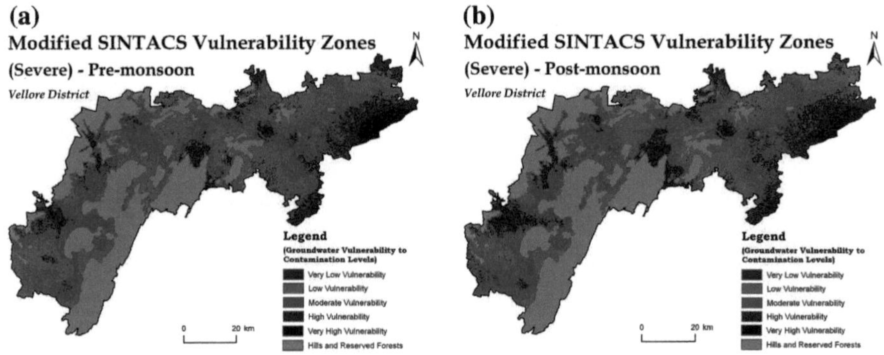

Fig. 13 **a** Vulnerability zones of modified SINTACS (Severe) in pre-monsoon. **b** Vulnerability zones of modified SINTACS (Severe) in post-monsoon

results of Normal and Severe versions of SINTACS model during pre- and post-monsoon seasons, the post-monsoon season results reveal that there are some increases in areas with high and very high vulnerability levels and which might have been influenced by the monsoon rainfall, and resultant percolation through favorable zones, as supported by the soil, lithology, slope, and permeability conditions (Tables 4 and 5; Fig. 10a, b and 11a, b). The results of Normal version of modified SINTACS model suggest that during post-monsoon season there is a decrease of areal extents of areas with very low and low vulnerability levels and increase of areal extents of areas with moderate, high and very high vulnerability (Tables 4 and 5; Fig. 12a, b). In the case of Severe version of modified SINTACS model, the results during pre- and post-monsoon season show that there is no appreciable difference in all the levels of vulnerability classes (Tables 4 and 5; Fig. 13a, b).

8 Conclusions

- The vulnerability maps of SINTACS (Normal and Severe versions) model serve as tools for land use management as they show the areas of potential groundwater contamination and also help to identify the areas of high natural protection against pollution. These vulnerability maps were then coupled with land use to create the modified version of outputs during pre- and post-monsoon seasons. As the vulnerability map illustrates intrinsic weaknesses of the hydrogeological system and the land use map represents potential sources of contamination, combining these two maps produces a better and reliable spatial decision support tool.
- The results of SINTACS model (Normal and Severe versions) show that 2.3–2.6 % and 2.4–3.0 % of the investigated area falls under very high vulnerability levels during pre- and post-monsoon seasons respectively. High level of vulnerability levels are found with the range of 10–14.9 % and 14.1–21.9 % of the investigated area during pre- and post-monsoon seasons respectively.
- The very high vulnerability levels are observed in the eastern portion of the study area and high vulnerability levels are found in the areas adjacent to very high vulnerable regions all over the study area.

Acknowledgments The authors are grateful to the organizations, State Ground Water & Surface Water Resources Data Centre, Water Resources Organisation, Public Works Department, Chennai for providing groundwater quality and level data; Ground Water Division, Public Works Department, Vellore for giving geophysical, pumping test and net recharge data; and the Directorate of Agriculture, Vellore for providing the Soil Atlas of Vellore District.

References

Al Kuisi M, El-Naqa A, Hammouri N (2006) Vulnerability mapping of shallow groundwater aquifer using SINTACS model in the Jordan Valley area, Jordan. Environ Geol 50(5):651–667

Albinet M, Margat J (1970) Cartographie de la vulnerabilite a la pollution des nappes d'eau souterraine. Bulletin BRGM Second Series, Orleans, Section 3(4):13–22

Aller L, Bennett T, Lehr JH, Petty RJ, Hackett G (1987) DRASTIC: a standardized system for evaluating groundwater pollution potential using hydrogeologic settings. United States Environmental Protection Agency, Robert S. Kerr Environmental Research Laboratory, Ada. EPA-600/2-87-035

Babiker IS, Mohamed MAA, Hiyama T, Kato K (2005) A GIS-based DRASTIC model for assessing aquifer vulnerability in Kakamigahara Heights, Gifu Prefecture, central Japan. Sci Total Environ 345(1–3):127–140

Chilton J (2006) Assessment of aquifer pollution vulnerability and susceptibility to the impacts of abstraction. In: Schmoll O, Howard G, Chilton J, Chorus I (eds) Protecting groundwater for health: managing the quality of drinking-water sources, World Health Organization. International Water Association (IWA) Publishing, London, pp 199–239

Civita MV (1990) Legenda unificata per le Carte della vulnerabilita dei corpi idrici sotterranei. Studi sulla Vulnerabilità degli Acquiferi, 1 (Annex), Pitagora Editrice, Bologna, p 13

Civita MV (1993) Groundwater vulnerability maps: a review. In: Lucca B (ed) Proceedings of the IX symposium on pesticide chemistry, mobility and degradation of xenobiotics. Piacenza, pp 587–631

Civita MV (1994) Le carte della vulnerabilità degli acquiferi all'inquinamento: Teoria & pratica. Pitagora Editrice, Bologna, p 325

Civita MV, De Maio M (1997) SINTACS Un sistema parametrico per la valutazione e la cartografia della vulnerabilita degli acquiferi all'inquinamento. Metodologia and Automatizzazione, vol 60. Pitagora Editrice, Bologna

Civita MV, De Maio M (1998) Mapping groundwater vulnerability by the point count system model SINTACS. In: Managing hydrogeological disasters in a vulnerable environment (IHP-UNESCO), GNDCI 1900, pp 243–273

Civita MV, De Maio M (2000) Valutazione e cartografia automatica della vulnerabilita degli acquiferi all'inquinamento con il sistema parametrico SINTACS R5. Pitagora Editrice, Bologna, p 226

Cowen DJ (1988) GIS versus CAD versus DBMS: what are the differences? Photogramm Eng Remote Sens 54(11):1551–1555

DeMers MN (2000) Fundamentals of geographic information systems, 2nd edn. Wiley, New York, p 498

Draoui M, Vias J, Andreo B, Targuisti K, El Messari JS (2008) A comparative study of four vulnerability mapping methods in a detritic aquifer under mediterranean climatic conditions. Environ Geol 54(3):455–463

ESRI (2000) Model builder for arcview spatial analyst 2: an ESRI white paper. Environmental Systems Research Institute (ESRI), Redlands, California

Gogu RC, Dassargues A (2000a) Current trends and future challenges in groundwater vulnerability assessment using overlay and index methods. Environ Geol 39(6):549–559

Gogu RC, Dassargues A (2000b) Sensitivity analysis for the EPIK method of vulnerability assessment in a small karstic aquifer, southern Belgium. Hydrogeol J 8(3):337–345

Leonard RA, Knisel WG (1988) Evaluating groundwater contamination potential from herbicide use. Weed Technol 2(2):207–216

Margat J (1968) Vulnerabilite des nappes d'eau souterraine a la pollution. Bases de la cartographie, Doc. BRGM, 68 SGL 198 HYD, Orleans

Marsico A, Giuliano G, Pennetta L, Vurro M (2004) Intrinsic vulnerability assessment of the south-eastern Murge (Apulia, Southern Italy). Nat Hazards Earth Syst Sci 4(5–6):769–774

Mato RRAM (2002) Groundwater pollution in urban Dar es Salaam, Tanzania: assessing vulnerability and protection priorities. Ph.D. thesis, Eindhoven University of Technology, Eindhoven, pp 108–113

McCoy J, Johnston K (2001) Using ArcGIS spatial analyst. Environmental Systems Research Institute (ESRI), Redlands, p 230

Merchant JW (1994) GIS-based groundwater pollution hazard assessment: a critical review of the DRASTIC model. Photogramm Eng Remote Sens 60(9):1117–1127

Ormsby T, Alvi J (1999) Extending ArcView GIS: teach yourself to use ArcView GIS extensions. Environmental Systems Research Institute (ESRI), Redlands, p 527

Rahman A (2008) A GIS based DRASTIC model for assessing groundwater vulnerability in shallow aquifer in Aligarh, India. Appl Geogr 28(1):32–53

Tesoriero AJ, Inkpen EL, Voss FD (1998) Assessing ground-water vulnerability using logistic regression. In: Proceedings of the source water assessment and protection 98 conference. Dallas, pp 157–165

Vrba J, Zaporozec A (1994a) Guidebook on mapping groundwater vulnerability. International contributions to hydrogeology, vol 16. International Association of Hydrologists, Verlag Heinz Heise, Hannover, p 7

Vrba J, Zaporozec A (1994b) Guidebook on mapping groundwater vulnerability. International contributions to hydrogeology, vol 16. International Association of Hydrologists, Verlag Heinz Heise, Hannover

Zaporozec A (1985) Groundwater protection principles and alternatives for Rock County. Wisconsin geological and natural history survey, special report 8, Madison, p 73

Natural and Anthropogenic Determinants of Freshwater Ecosystem Deterioration: An Environmental Forensic Study of the Langat River Basin, Malaysia

Ahmad Zaharin Aris, Wan Ying Lim and Ley Juen Looi

Abstract Freshwater ecosystems face numerous threats that challenge the local authorities' ability on tackling down the water security (quantity and quality) issues and their management. The quality of surface water is an essential component of the natural environment and is considered as the main factor for controlling ecosystem health and potential hazard to the surrounding environment. The Langat River Basin in Selangor, Malaysia is exposed to natural and anthropogenic activities. A forensic investigation via the use of geostatistical and geochemical approaches and different standard criteria revealed two sources controlling the evolution of Langat River Basin water chemistry: (i) anthropogenic (agricultural and industrial activities) and (ii) natural processes (seawater intrusion and geological weathering). In addition, the suitability of river water for various purposes was determined based on the application of selected indicators and indices. The findings serve as an essential platform for the protection of water resources.

Keywords Environmental forensics · Geostatistics · Geochemistry · Heavy metals · Pollution indices · Indicators

1 Introduction

The freshwater ecosystem contributes to both human welfare and aquatic ecosystem by providing water for various purposes throughout the world. Therefore, the river water quality is an essential component of the natural environment and is considered as the main factor for controlling environmental health and potential hazard to the surrounding ecosystem. However, rivers may also carry a significant load of pollutants from different sources and affect not only the regions in the vicinity, but

A.Z. Aris (✉) · W.Y. Lim · L.J. Looi
Environmental Forensics Research Centre, Faculty of Environmental Studies, Universiti Putra Malaysia, 43400 Serdang, Selangor, Malaysia
e-mail: zaharin@upm.edu.my

Mu. Ramkumar et al. (eds.), *Environmental Management of River Basin Ecosystems*, Springer Earth System Sciences, DOI 10.1007/978-3-319-13425-3_21

also the downstream reaches and possibly, the coastal and marine regions to which the river debauches (Shrestha and Kazama 2007; Alkarkhi et al. 2009). The land area within the basin, especially from the middle to lower reaches of the river always have been the preferred locales of housing, industrial, and agricultural activities. Since the waste discharges emanating from these activities are commonly not regulated and effectively controlled, it is more likely that levels of pollutants will increase in line with the scale of development (Suratman et al. 2009). These contaminants eventually end up into the runoff, land based area, and mangroves (Mokhtar et al. 2009a; Yu et al. 2010) and pollute the aquatic ecosystems. Therefore, the river quality has become a critical issue in many countries due to the concern that freshwater will be a scarce resource in the future.

Several workers attempted identification of metal pollution status in the river waters (Mokhtar et al. 1994, 2009a; Lim and Kiu 1995; Ahmad et al. 2009; Naji et al. 2010; Nasrabadi et al. 2010; Sultan et al. 2011; Eneji et al. 2012; Sultan 2012). Elevated concentrations of As, Pb, Cu, Cd, Cr, Ni, and Zn have been reported in various environmental media, especially in water, biota, and sediments. These metals were the most common pollutants found in the Strait of Malacca (Sarmani 1989; Gadgil 1998; Abdullah et al. 1999; Shazili et al. 1999, 2006; Yap et al. 2002). The Pb, Cd, Cr, Hg, and As are notorious contaminants, and are listed as the most hazardous inorganic contaminants in the EPA Hazardous Substance Priority List (EPA 2012). These elements are dissolved in water, highly toxic, and can cause significant health effects towards human and organism (Awofolu et al. 2005). They could exhibit extreme toxicity even at trace amounts and cause deadly diseases such as like edema of eyelids, tumor, neurological and genetic malfunctions (Hem 1970; Tsuji and Karagatzides 2001; WHO 2011). The Department of Environment recorded that areas with high anthropogenic pressure, particularly in Johor and Selangor states, have the highest number of water pollution sources. Sources of these pollutions are mostly attributed to the manufacturing industries, sewage treatment plants, agro-based industries, and animal husbandries (Ismail et al. 1993; Shazili et al. 2006; DOE 2010; Naji et al. 2010; Zulkifli et al. 2010b). Rahman and Surif (1993) and Zulkifli et al. (2010b) found that metal processing and petroleum-related industries produce e-waste which contain elevated levels of metals, including Zn, Cu, Ni, Fe, Al, Pb, Mn, Cr, and Sn. The by-product derived from nutrient supplements in animal feeds (such as copper sulfate pentahydrate) contributed elevated levels of Cu and has been correlated to contamination of sediments and molluscs (Sarmani et al. 1992; Ismail and Ramli 1997; UPUM 2002; JICA and MGDM 2002; Lee et al. 2006). Port and shipping activities are the complementary pollution sources contributing tributyltin, Pb, Cu, and As (Abdullah et al. 1999; Zulkifli et al. 2010b). Mining of Cu, Sn, Fe, and Au also increase the prevalence and occurrence of metal contamination in the aquatic ecosystems (Yusuf 2001; Ali et al. 2004).

Considerable efforts have been made in the past 2 decades to understand the environmental conditions of the river ecosystems in Malaysia. Notable among them are: river basin management (Mokhtar et al. 2011), hydrological properties and water quality (Alkarkhi et al. 2009; Suratman et al. 2009; Fulazzaky et al. 2010), the distributions of heavy metals in soil, sediment and surface water (Sarmani 1989; Lim and Kiu 1995; Ali et al. 2004; Mokhtar et al. 2009a; Suratman et al. 2009; Sultan et al. 2011), the distribution and biodiversity of benthic macroinvertebrates (Azrina et al. 2006; Al-Shami et al. 2010), and the compositional patterns and occurrence of organic pollutants (Osman et al. 2012; Zainuddin et al. 2012). A common inference from all these diverse studies is that, the freshwater ecosystems are increasingly facing multiple pressures from a variety of contaminants.

The Langat River Basin is one of the most important freshwater ecosystems in the tropical west coast of Peninsular Malaysia. This tropical catchment area is now experiencing rapid urban expansion (Azrina et al. 2006; Amini et al. 2009; Ali et al. 2012; Zainuddin et al. 2012). The rapid development of this catchment area poses a threat to the river quality from manufacturing and agro-based industries (palm oil mills and rubber processing plants), sewage treatment plants, wastewater from animal farms (pig farms), ship waste and domestic activities (JICA and MGDM 2002; Lee et al. 2006; Bahaa-Eldin et al. 2008; Mokhtar et al. 2009b; Juahir et al. 2010; Zainuddin et al. 2012). The inadequate water management, uncontrolled contaminant discharge from industrial, domestic and economic activities bear a direct effect on the river ecosystem (Mokhtar et al. 2009b). Besides, the tropical climates, with warm temperature and high rainfall intensity enhance both physical and chemical weathering of rocks (Sultan et al. 2011), thus enhancing the suspended and solute fluxes.

Sarmani (1989), Yap et al. (2003), Bahaa-Eldin et al. (2008) and Mokhtar et al. (2009b) studied the heavy metal contamination in the Langat River. However, these studies have not attempted documenting the potential toxicity and suitability of the river waters for domestic and agricultural purposes, as is being studied elsewhere (Sultan et al. 2011). The present study attempts identification of the determinant factors that influence the hydrochemistry of the Langat River, Malaysia.

2 Regional Setting

The Langat River Basin is a trans-state river basin of the Peninsular Malaysia and has been recognized by the United Nations Educational Scientific Organization (UNESCO) as an Evolving HELP basin (Mokhtar et al. 2011; Fig. 1). The river flows from the western slope of Titiwangsa Mountains and draining into the Strait of Malacca. The Langat River Basin is located close to the equator, and thus is under the influences of northeast (November–March) and southwest (May–September) monsoons. There are two inter-monsoonal periods in between these monsoons, which are characterized by variable winds and thunderstorms (JICA and MGDM 2002). The spatial extent of the Langat River Basin and its hydrological characteristics are given in the Table 1. The topography of this basin can be separated into three distinct zones, namely, mountainous area, hilly area and lowland area (DOA 1995; Lim et al. 2012). The bedrock in mountainous area includes Permian igneous rocks, Pre-Devonian schist, and phyllite of the Howthornden Formation

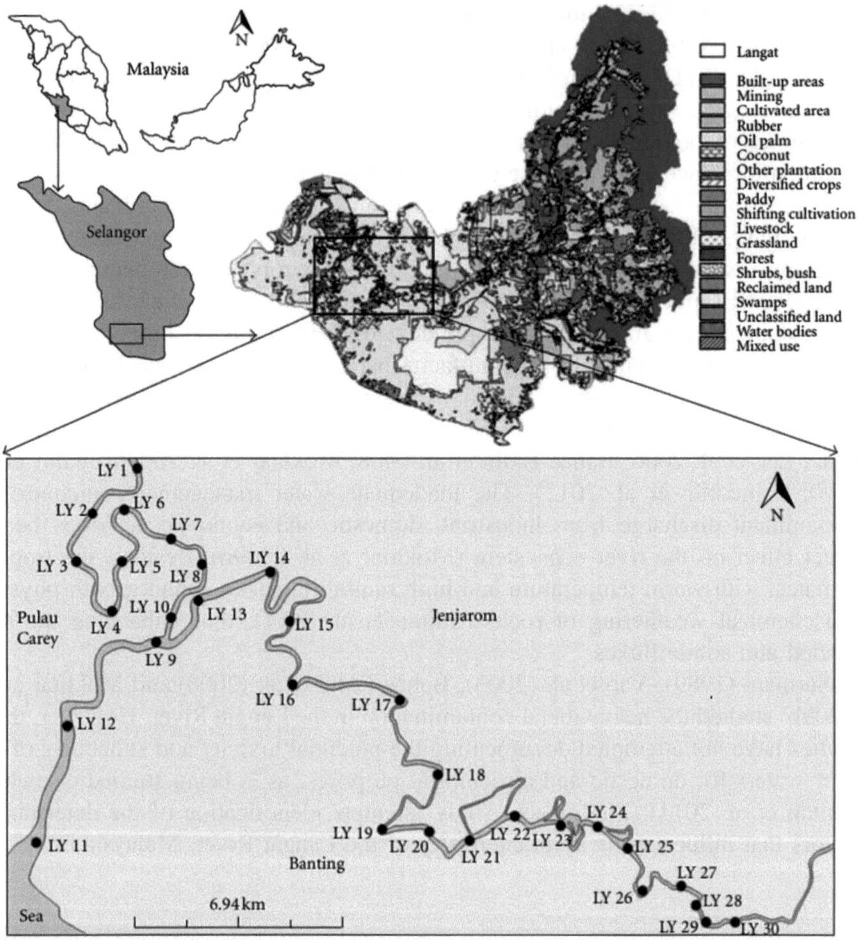

Fig. 1 Map of sampling points in Langat River

(Gobbett and Hutchison 1973). The Permo-Carboniferous meta-sandstone, Kajang Formation, and Kenny Hill Formation are predominated in the hilly area while the lowland area is occupied by the Quaternary deposits of Beruas, Gula, and Simpang Formations (Gobbett and Hutchison 1973; Taha 2003). A downstream trend of change of river bed sediments from gravel in the catchments to clayey in the lowland area is conspicuous (JICA and MGDM 2002) as could be observed elsewhere.

There are three constructed dams, namely the Semenyih Dam, the Langat Dam, and the Putrajaya Dam, which supply water to the entire basin. The basin, especially its lower region has been experiencing massive development since the past few decades (Ali et al. 2012). On a regional scale, the upper region of the basin is predominated by natural forests, the middle region by built-up area and mining activities and the lower region by agricultural activities including oil palm

Table 1 Spatial extent and hydrological characteristics of the Langat River Basin

Parameters	Characteristics
Coordinate	Latitudes 2° 40′ 152″ N–3° 16′ 15″ N Longitudes 101° 19′ 20″ E–102° 1′ 10″ E
Basin area	1,815 km^2
Main river course	141 km
Total length	120 km
Total number of tributaries	39
Annual average rain fall	2,469 mm
Minimum and maximum range of rainfall	1,521 and 2,883 mm, respectively
Average rainy day per year	190 days/year
Basin level water production capacity	1,052 mL/day
Average annual flow	35 m^3/s
Annual flood	300 m^3/s
Water abstraction range	3.41–5.86 m^3/s
Average temperature	32 °C
Humidity	80 %

Sources Modified from DOA (1995), Yusuf (2001), DID (2010), Juahir et al. (2011), Mokhtar et al. (2011), Lim et al. (2012)

plantation, rubber, mixed horticulture and livestock (Fig. 1; DOA 1995; JICA and MGDM 2002; Ali et al. 2012). It is quite understandable from these that the demand for water for all these activities has been on the rise concomitant with deterioration of quality of the available water.

3 Materials and Methods

3.1 Sample Collection and Preservation

During July 2011, a total of 90 water samples (3 replicates of 30 samples) were collected at 30 different sampling stations, which were labeled as LY 1–LY 30 (Fig. 1). All the variables were measured in triplicates to estimate the variability resulting from the sampling and analytical procedure. The sampling, preservation, transportation and storage of samples were based on the Standard Method for Water and Wastewater Analysis (APHA 2005). An appropriate quality control and quality assurance were practiced to provide and maintain a degree of confidence in data integrity and accuracy. In addition, all the chemical and reagents used were of analytical grade or equivalent and free from any contaminants. The collected samples were filtered with 0.45 μm cellulose acetate membrane filter (Whatman Millipore, Clifton, NJ, USA) to prevent clogging during analysis and to obtain the dissolved ions for metal analysis (APHA 2005). Each sample was then separated into two pre-cleaned polyethylene bottles. The first bottle was utilized for determination of sulfate (SO_4^{2-}) and nitrate (NO_3^-), and the second bottle was for the determination of cations and heavy metals. Samples taken for metal analysis were acidified with concentrated HNO_3^- (pH < 2) to prevent adsorption onto the walls of the bottles, to prevent precipitation and to retard any biological activities (APHA 2005).

3.2 In Situ Parameters Measurement

The in situ variables (temperature, dissolved oxygen (DO), pH, redox potential (*Eh*), electrical conductivity (EC), total dissolved solids (TDS), and salinity) were measured immediately at the sampling locations. This was to acquire representative value of water quality and to avoid biochemical changes in the samples (APHA 2005; Radojević and Bashkin 2007). The DO and temperature were measured using YSI 52-dissolved oxygen meter (YSI incorporated, Yellow Spring OH, USA). pH and Eh were determined using the SevenGo Pro-SG78 probe meter while EC, TDS and salinity were determined using the SevenGo Pro-SG7 probe meter (Mettler Toledo AG, Schwerzenbach, Switzerland). Before measurements, each meter was calibrated with freshly prepared buffer solutions to ensure flawless functioning and accurate reading.

3.3 Major Ions and Heavy Metals Analyses

The HCO_3^- and Cl^- concentrations were measured on site based on titration and Argentometric methods, respectively, by using raw water samples collected (APHA 2005). Filtered un-acidified samples were used for determination of SO_4^{2-} (Surf-aVer 4 method) and NO_3^- (NitraVer 5 method) through UV-spectrophotometry (DR/2500 Spectrophotometer, HACH Odyssey, Loveland, Colorado, USA). The cations (Ca^{2+}, Na^+, Mg^{2+}, and K^+) and heavy metals (^{27}Al, ^{75}As, ^{138}Ba, ^{111}Cd, ^{59}Co, ^{63}Cu, ^{52}Cr, ^{57}Fe, ^{55}Mn, ^{60}Ni, ^{208}Pb, and ^{66}Zn) were measured from filtered and acidified samples by flame atomic absorption spectrometry (FAAS, AA6800, Shimadzu Scientific Instruments, Kyoto, Japan) and by inductively coupled plasma mass spectrometry (ICP-MS, ELAN DRC-e, Perkin Elmer, Massachusetts, USA), respectively. The operating conditions of ICP-MS are given in Table 2. Several isobaric interferences were programmed (Table 3), and the interference corrections were automatically applied (if used) for standard-mode analysis by the ELAN software. The cations concentrations are expressed as milligrams per liter (mg/L) while the heavy metals concentrations are expressed as micrograms per liter (μg/L). The accuracy of metal analysis was assessed by external standards, which were prepared by diluting the ICP Multi-Element Mixed Standard III (Perkin Elmer, Massachusetts, USA) into series of concentrations with the same acid mixture used for sample dissolution. Recovery rates of 97–103 % (±5 %) were achieved.

Table 2 The instrument setting of ICP-MS

	Condition used
Instrument	Perkin Elmer SCIEX, ELAN DRC-e
RF power	1,100 W
Gas	Argon gas-flow
Nebulizer gas flow rate	0.72 L/min
Auxiliary gas flow rate	1.2 L/min
Nebulizer	Cross-flow Gem-tip
Spray chamber	Cyclonic
Detector mode	Dual (electron multiplier operating in both pulse counting and analog modes)
Sampler/skimmer cones	Nickel
Dwell time per amu	50 ms
Points per peak	1
Sweeps per reading	10
Readings per replicate	3
Replicates	3
Sample flush	40 s (48 rpm)
Sample reading delay	20 s (20 rpm)
Sample wash	45 s (48 rpm)
Isotopes monitored	^{27}Al, ^{75}As, ^{138}Ba, ^{111}Cd, ^{59}Co, ^{63}Cu, ^{52}Cr, ^{57}Fe, ^{55}Mn, ^{60}Ni, ^{208}Pb and ^{66}Zn

Table 3 Isotopes monitored and applicable parameters in standard mode

Analyte	Mass	Potential interferences	Correction equation
Al	26.9815	BO, CN, BeO	–
As	74.9216	ArCl, Sm^{2+}, Nd^{2+}, Eu^{2+}	−3.127 * [ArCl77 − (0.815 * Se82)]
Ba	137.9050	La, Ce	−0.000901 * La139 − 0.002838 * Ce140
Cd	110.9040	MnO	–
Co	58.9332	CaO	–
Cu	62.9298	PO_2, TiO	–
Cr	51.9405	ArN, ClO, ArO, SO, ArC, HClO	–
Fe	56.9354	CaO, ArO	–
Mn	54.9381	ArN, HClO, ClO	–
Ni	59.9332	CaO	–
Pb	207.9770	–	–
Zn	65.9260	TiO, VO, SO_2, Ba^{2+}	–

3.4 Data Analyses

Descriptive analysis and skewness were performed using the PASW Statistics 18 (formerly known as SPSS Statistics 18, or SPSS Base). The descriptive statistic was performed to calculate maximum, minimum, mean, standard deviation (SD), and coefficient of variance (CV). SD was used as an indication of the precision of each parameter while CV was calculated based on the sum value of standard deviation from each parameter divided by its mean value. One-way ANOVA was applied to test the significant difference for all water quality variables among stations. Principal component analysis (PCA) was applied to apportion pollution sources and the contribution of each variable to the study area. The ionic ratios including (Cl^-/HCO_3^-) versus Cl^-, Na^+ versus Cl^-, Ca^{2+} versus ($Ca^{2+} + SO_4^{2-}$) and ($SO_4^{2-} + HCO_3^-$) versus (Ca^{2+} + Mg^{2+}) were computed to delineate the possible mechanisms, which contribute to the river constituents. Meanwhile, the suitability of river water for irrigation purpose is evaluated based upon the estimation of the parameters like sodium percentage, magnesium hazard, Kelly's ratio and Wilcox diagram. The results were also compared with the different drinking water quality standards stipulated by the Malaysian Ministry of Health (MOH 2004) and World Health Organization (WHO 2011).

4 Results and Discussion

4.1 Source Apportionment of River Pollution in Relation to Natural and Anthropogenic Activities

Source apportionment approach improves the knowledge on the natural and human impacts on the aquatic environment. This explanatory analysis estimates the contribution of each variable and affords data reduction with a minimum loss of the original information (Shrestha and Kazama 2007; Mustapha et al. 2012). The PCA was applied on the water quality data set to identify the spatial sources of pollution in the Langat River. To ensure that there is no violation, preliminary analysis on the assumptions of Kaiser–Meyer–Olkin (KMO) measure of the sampling adequacy and Bartlett's test of sphericity were conducted. The KMO result was 0.916, and the Bartlett's test of sphericity was significant ($p < 0.001$). They proved that PCA can be considered as appropriate to provide significant reduction in the data dimensionality (Mustapha et al. 2012). Figure 2 and Table 4 display the component loadings after varimax rotation. Four components with eigenvalues greater than 1.0 were extracted that explained 89 % of total variance (Table 4). More than 70 % of the data variance was explained by the first two components (Fig. 2; Table 4). The first PC explains the highest variation among the variables; while the remaining PCs explain the variations among the variables in order (Civan et al. 2011).

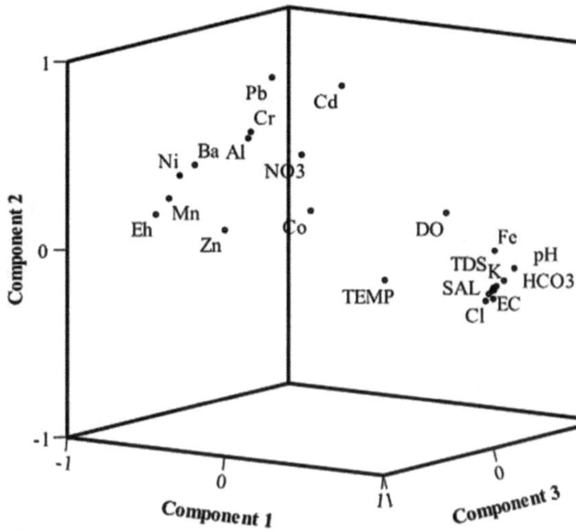

Fig. 2 Component plot of PCA in rotated space

The components expose the potential factors responsible for variation in river water quality and eventually lead to sources identification of river pollution.

PC 1 with total variance of 64 % has a strong loadings of SO_4^{2-}, salinity, EC, TDS, Ca^{2+}, K^+, Mg^{2+}, Cu, As, Na^+, HCO_3^-, Fe, Mn, Cl^-, pH, Eh, Ba, temperature, Cr, Ni, and NO_3^-. The presence of dissolved ions including major ions (SO_4^{2-}, Ca^{2+}, K^+, Mg^{2+}, Na^+, HCO_3^-, Cl^-, NO_3^-) and heavy metals (Cu, As, Fe, Mn, Ba, Cr, Ni) may trigger the values of EC, salinity and TDS (Radojević and Bashkin 2007; Lim et al. 2012; Shafie et al. 2013a). In addition, the strong positive loadings on EC, salinity, TDS, and major ions explained that the ion-exchange reactions between freshwater (river) and seawater corresponded to seawater intrusion into the Langat River (Aris et al. 2012; Lim et al. 2012). Salinity portrays a vital role in the adsorption and desorption mechanism as it gives strong correlation with all the exchangeable cations (Shafie et al. 2013a). Therefore, the mixing of seawater and river water caused the competition between cations and heavy metals for binding sites in particulates (Connell and Miller 1984; Elder 1988; Lim et al. 2012). The cations being more prominent than other ions, especially the heavy metals and thus caused these ions desorbed from sediment. Thus, it will increase the concentration of heavy metals in river water (Connell and Miller 1984; Elder 1988). The presence of these metals also indicated that the quality of river water was affected by the inorganic compounds perhaps from anthropogenic activities in the study area. Heavy metals including Cu, As, Fe, Mn, Ba, Cr, and Ni were prominent in PC 1. These elements could be characterized by their natural availability in the river basin. Their presence is likely due to the oxisols and ultisols, which are the dominant soil in Peninsular Malaysia (Nieuwolt et al. 1982; Tessens and Jusop 1983; JICA and MGDM 2002). In addition, the oxides of Fe and Mn (goethite and hematite) are commonly found in tropical soils (Nieuwolt et al. 1982; Tessens and Jusop 1983) and have a great effect on the chemical behaviour of metals in sediment

Table 4 Component loadings of river water quality variables on varimax rotated matrix

	PC 1	PC 2	PC 3	PC 4
SO_4^{2-}	**0.98**	−0.15	−0.03	0.09
Salinity	**0.98**	−0.15	−0.04	0.11
EC	**0.98**	−0.16	−0.03	0.11
TDS	**0.98**	−0.16	−0.03	0.11
Ca^{2+}	**0.98**	−0.17	−0.04	0.11
K^+	**0.98**	−0.15	−0.06	0.12
Mg^{2+}	**0.97**	−0.17	−0.02	0.11
Cu	**0.97**	−0.18	0.00	0.10
As	**0.97**	−0.21	−0.04	0.05
Na^+	**0.97**	−0.16	−0.05	0.15
HCO_3^-	**0.96**	−0.13	−0.15	−0.07
Fe	**0.95**	0.03	−0.09	−0.16
Mn	**−0.94**	0.16	0.16	0.19
Cl^-	**0.93**	−0.23	−0.03	0.21
pH	**0.89**	−0.10	−0.35	−0.14
Eh	**−0.89**	0.10	0.35	0.14
Ba	**−0.88**	0.32	0.02	0.13
Temperature	**0.61**	−0.08	0.43	0.48
Cr	**−0.61**	0.51	−0.10	−0.39
Ni	**−0.60**	0.37	0.55	0.22
NO_3^-	**−0.56**	0.33	−0.49	0.04
Pb	−0.28	**0.87**	0.18	0.08
Cd	0.00	**0.82**	−0.05	0.15
Al	−0.48	**0.51**	0.10	−0.40
Zn	−0.12	0.17	**0.82**	0.18
DO	0.14	0.05	**−0.79**	0.03
Co	−0.07	0.17	0.13	**0.76**
Initial eigenvalue	17.21	2.77	2.44	1.51
Percent of variance	63.73	10.27	9.02	5.60
Cumulative percent	63.73	74.01	83.03	88.62

EC Electrical conductivity; *Eh* Redox potential; *DO* Dissolved oxygen; *TDS* Total dissolved solids. The values in bold are factor loadings above 0.50 that were taken after varimax rotation was performed

(Alloway 1995). However, Cu, As, Cr and Ni may be attributed to anthropogenic pollution. Agriculture is the dominant activity in the study area. The use of inorganic fertilizers such as arsenical herbicides or ammonium nitrate is frequent and widespread. Therefore, high loadings of As and NO_3^- could have been caused by agricultural activities. Besides, another source of contamination might come from pig farming activities. The by-products from pig food (copper sulfate) can contribute to Cu concentration via effluent discharged into the river (Sarmani et al. 1992;

UPUM 2002; Lee et al. 2006). This inference is also supported by several previous studies, which explained the effects of anthropogenic activity toward this river (Sarmani 1989; Sarmani et al. 1992; Lee et al. 2006; Mokhtar et al. 2009b; Juahir et al. 2011). Thus, PC 1 is believed to be interpreted by two contributions, the first is related to seawater intrusion; and the latter, related to the agricultural and farming activities.

PC 2 with total variance of 10 % consists of Pb, Cd, and Al. Relatively high loadings of these elements indicate that the river waters suffer from metal pollution, possibly by the heavy shipping traffic (Shazili et al. 2006; Mokhtar et al. 2009b; Berandah et al. 2010). The steel manufacturing and metal finishing factories located in the upstream (Dengkil) of the sampling location (Mokhtar et al. 2009b) could also have contributed. Yet another reason, namely the ongoing intensive dredging, reclamation, and construction activities in the river channel and in the adjoining regions might resuspend the stream-bed sediments and reintroduce the trace metals bound in the sediments into the river water (Nayar et al. 2004; Zulkifli et al. 2010a). Similar inferences were reported in previous studies, which explained the effects of anthropogenic inputs (Sarmani 1989; Shafie et al. 2013a) into this river. PC 3 and PC 4 have strong loadings of Zn, DO, and Co which illustrate that the DO plays a role in controlling the behavior and distribution of metals in the river water. The Co was possibly derived from lithogenic or anthropogenic sources, especially from agricultural activities (Hooda 2010). However, Co was observed to be present in low concentration compared with other metals. Hence, lithogenic source to Co is attributed.

4.2 Ionic Ratio

As discussed in the PCA section, it was deduced that the mixing of seawater is an influential hydrochemical process in the study area. To ascertain the origin and chemical behavior of the river water, several ionic ratios were attempted. The relationships of (Cl^-/HCO_3^-) versus Cl^-, Na^+ versus Cl^-, Ca^{2+} versus ($Ca^{2+} + SO_4^{2-}$) and ($SO_4^{2-} + HCO_3^-$) versus ($Ca^{2+} + Mg^{2+}$) were computed to delineate the possible mechanisms that may contribute to the river water constituents (Hounslow 1995; Moujabber et al. 2006; Aris et al. 2012; Isa et al. 2012; Lim et al. 2012). Typically, Ca^{2+} and HCO_3^- are the dominant constituents found in freshwater whereas Na^+ and Cl^- are the most abundant constituents in seawater and the enrichment of these constituents could be an indicator of saline water intrusion to river water (Aris et al. 2012). From the Fig. 3a, it follows that sampling stations located near coastal area (LY 1–LY 15), are saline than other stations, resulting in a higher (Cl^-/HCO_3^-) versus Cl^- ionic ratio. The result represents a greater contribution of saline water to the river water relative to the amount of Cl^- that might be present. In addition, 50 % of the samples with low ratio of (Cl^-/HCO_3^-) versus Cl^- indicate freshening of river water. A Na^+/Cl^- ratio that is greater than 1 describes silicate weathering and about 44 % of the samples fall within this range (Fig. 3b). Around 50 % of the samples were below 1, and represent the occurrence of ionic exchange process while 6 % of the samples were equal to 1, representing halite dissolution. The Na^+/Cl^- ratio of the

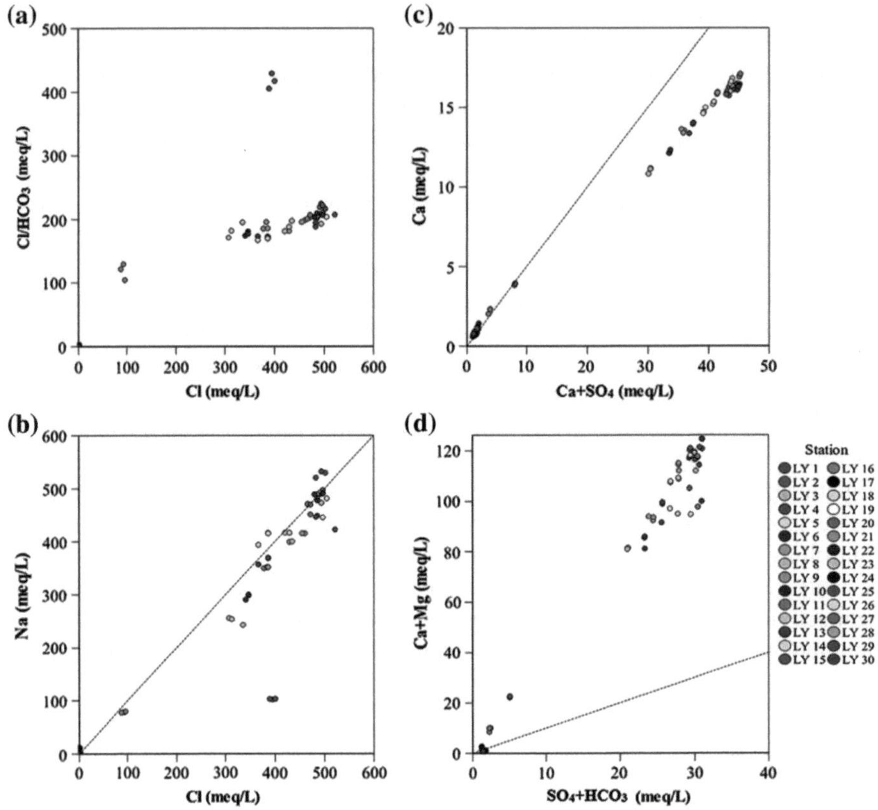

Fig. 3 Distribution of ionic ratio for major ions describing mechanism involve

samples collected from stations located near coastal area (LY 15–LY 30) indicate salinization due to the result of mixing with seawater. The stations LY 16–LY 30, show heightened levels of the Na^+ and the same may have been resulted by irrigation return flow or due to evaporation (Jeevanandam et al. 2007). The scatter plot $Ca^{2+}/Ca^{2+} + SO_4^{2-}$ (Fig. 3c) is another example to illustrate the ionic exchange process. It was deduced that 52 % of samples were below 0.5, which indicates the occurrence of Ca depletion, either through ion exchange or calcite precipitation. The order of cation affinity tends to be $Na^+ > K^+ > Ca^{2+} > Mg^{2+}$ if there is mixing with seawater, where Ca being displaced from the exchanger in the first order and Na eventually dominating the water and the exchanger (Aris et al. 2007, 2012). 48 % of the samples contain Ca > 0.5, and indicate that the Ca may have been drawn from carbonate ($CaCO_3$) or silicate (Ca_2SiO_4) sources. This inference is affirmed by ($Ca^{2+} + Mg^{2+}$) versus ($SO_4^{2-} + HCO_3^-$) ratio (Fig. 3d), as 43 % of the samples fall below the equiline which indicates weathering of minerals. The rest of samples lie above equiline, and suggest that carbonate weathering might have contributed ions to the samples under study.

4.3 Suitability for Drinking and Domestic Purposes

The analytical results of in situ variables, major ions and heavy metals are presented in the Table 5. Significant differences of all the studied variables among sampling stations ($p < 0.05$) are observed. All the variables were compared with the drinking water quality standards stipulated by the Malaysian Ministry of Health (MOH 2004) and the World Health Organization (WHO 2011). As depicted in the table,

Table 5 Descriptive analysis of in situ variables, major ions and heavy metals (n = 90)

	Unit	Range	Mean	SD	CV	MOH	WHO
In situ							
pH	–	6.20–7.61	6.82	0.34	4.94	6.5–9.0	NA
Eh	mV	−25.90–61.90	23.30	21.04	90.27	NA	NA
Temp	°C	28.10–30.20	29.40	0.68	2.31	NA	NA
EC	μS/cm	173.10–39500.00	17024.69	17665.66	103.76	NA	NA
Salinity	ppt	0.09–25.10	10.68	11.21	104.94	NA	NA
DO	mg/L	0.72–3.17	1.55	0.51	32.88	NA	NA
TDS	mg/L	86.50–19740.00	8511.26	8831.83	103.77	1000	NA
Major ions							
HCO_3^-	mg/L	26.84–156.16	95.02	41.36	43.53	NA	NA
Cl^-	mg/L	16.00–18494.27	7802.55	7773.31	99.63	NA	NA
NO_3^-	mg/L	0.06–3.50	1.32	0.79	60.05	NA	50
SO_4^{2-}	mg/L	15.00–1375.00	590.18	600.65	101.77	250	NA
Na^+	mg/L	11.63–12240.00	4645.45	4857.57	104.57	200	NA
Ca^{2+}	mg/L	11.92–343.09	153.40	142.34	92.79	NA	NA
K^+	mg/L	5.53–363.41	135.22	133.96	99.07	NA	NA
Mg^{2+}	mg/L	0.98–1319.30	528.28	553.90	104.85	150	NA
Heavy metal							
Al	μg/L	11.72–122.90	46.28	32.71	70.67	200	NA
As	μg/L	1.79–21.48	11.18	8.29	74.13	10	10
Ba	μg/L	3.42–27.40	12.18	8.14	66.81	700	700
Cd	μg/L	<0.01–0.38	0.07	0.09	126.84	3	3
Co	μg/L	0.10–0.56	0.21	0.07	33.13	NA	NA
Cu	μg/L	0.40–58.82	26.02	25.45	97.79	1,000	2,000
Cr	μg/L	<0.005–3.09	0.67	0.90	133.22	50	50
Fe	μg/L	61.90–162.53	116.23	28.68	24.67	300	NA
Mn	μg/L	<0.0005–504.46	200.54	190.71	95.10	100	NA
Ni	μg/L	<0.005–3.07	0.46	0.66	142.80	20	70
Pb	μg/L	<0.005–1.34	0.16	0.23	145.91	10	10
Zn	μg/L	0.34–19.21	3.10	3.29	106.26	3,000	NA

SD Standard deviation; *CV* Coefficient of variance; *Eh* Redox potential; *Temp* Temperature; *EC* Electrical conductivity; *DO* Dissolved oxygen; *TDS* Total dissolved solids

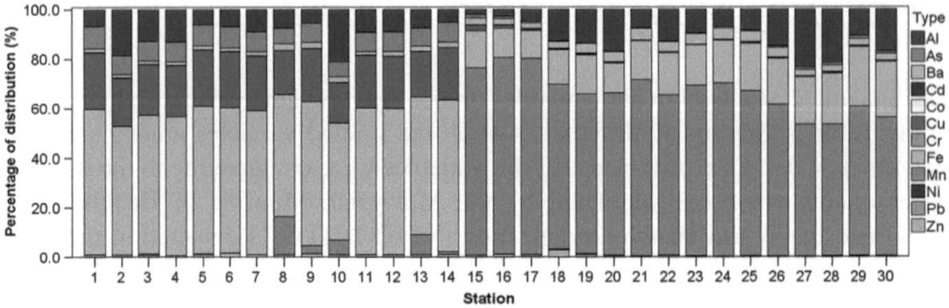

Fig. 4 Distribution of heavy metals according to its sampling stations

the pH value of the water samples is weakly acidic to circum-neutral ranging from 6.20 to 7.61. Most of the stations were well within the limit prescribed by MOH (6.50–9.0). The TDS values ranged from 86.50 to 19740.00 mg/L and 53 % of them exhibited a value greater than 1,000 mg/L recommended by MOH. For major ions, it was found that 57, 50, and 50 % of samples exceeded the permissible limits of Na, Mg and SO_4 respectively (MOH 2004; WHO 2011; Table 5). The Fig. 4 depicts the spatial distribution of metals.

The mean concentration of heavy metals follows the order of: Mn > Fe > Al > Cu > Ba > As > Zn > Cr > Ni > Co > Pb > Cd. Except As and Mn at certain locations, all the heavy metals in the studied water samples are within the permissible limits (MOH 2004; WHO 2011; Table 5). 46 % of the samples exceed the permissible As content (10 μg/L; MOH 2004; WHO 2011) For As, 46 % of samples were found to exceeded the MOH and WHO permissible limits. Several studies stated that the As pollution in the Langat River basin result by agricultural application (arsenical herbicides) and tin mining (Sarmani 1989; Shafie et al. 2013a, b). 53 % of stations exhibited Mn concentration higher than the MOH standard (100 μg/L; MOH 2004). The direct reduction of particulate manganese oxides in aerobic environments by organic matters, the natural weathering of Mn-bearing minerals and acid drainage are the possible factors that control Mn mobility and bioavailability besides anthropogenic inputs (Heal 2001; Howe et al. 2005; WHO 2011). Mn does not occur naturally as a base metal but is a component of various minerals (WHO 2011). In the aquatic environment, Mn exists in two main forms: Mn^{2+} and Mn^{4+}, while the solubility of Mn^{2+} is higher than Mn^{4+}. Mn tends to become more bioavailable with decreasing pH and redox potential (Heal 2001; Eq. 1).

$$Mn^{2+} + 2HCO_3^- + H_2O + \frac{1}{2}O_2 \rightarrow Mn(OH)_4 + 2CO_2 \quad (1)$$

Considering the pH value of the Langat River, it ranged between 6.20 and 7.61. The acidic conditions observed at certain stations might have caused the release of Mn from Mn-bearing minerals or Fe–Mn oxide minerals. The guideline values for variables such as *Eh*, temperature, EC, salinity, DO, K, Na^+, Cl^-, and HCO_3^-, have not been established since they are not of health concern at levels found in drinking water. However, some substances may affect the color, taste, odour or appearance of drinking water, which indirectly affect the acceptability of water for drinking and domestic purposes (WHO 2011). The application of these standards has shown that most of samples collected from the Langat River are suitable for domestic use, whereas only few samples were found to be unsuitable based on elevated contents of certain elements.

4.4 Suitability for Irrigation Uses

The suitability of irrigation water depends upon the chemical constituents of water. Major ions such as Ca^{2+}, Mg^{2+}, Na^{+}, Cl^{-}, HCO_3^{-} and SO_4^{2-} are the main dissolved constituents which determine the suitability of water for irrigation purposes. Several indicators and indices including sodium percentage, magnesium hazard, Kelly's ratio and Wilcox diagram were employed to evaluate the usefulness of river water for irrigation (US Salinity Laboratory Staff 1954; Wilcox 1955).

The sodium percentage was calculated and the quantities of all cations are expressed in meq/L (Eq. 2). Based on the sodium percentage present in the samples under study, the water can be categorized as excellent (<2 %), good (2–40 %), permissible (40–60 %), doubtful (60–80 %) and unsuitable (>80 %). As higher Na^{+} increases the hardness and reduces its hydraulic conductivity of soil or permeability to water (Glover 1996; Jeevanandam et al. 2007), waters with high concentration of Na^{+} are considered to be undesirable for irrigation purposes (Glover 1996). In the study area, the sodium percentage varied from 27 to 90 % and their classification varied from 'good' to 'unsuitable' classes (Fig. 5a). The magnesium hazard (MH) is used to evaluate the potential of Mg^{2+} hazard to irrigation water where the value of magnesium hazard above 50 is considered as harmful and unsuitable for irrigation purposes (Szabolcs and Darab 1964; Eq. 3). From the calculated MH values, 57 % of the water samples (LY 1–LY 17) are classified as unsuitable for irrigation use (Fig. 5b). The Kelly's ratio is another index employed to identify the suitability of water for irrigation use. The Kelly's ratio is calculated by the level of Na^{+} measured against Ca^{2+} and Mg^{2+} (Kelly 1951; Eq. 4). From the calculated value, 65 % of water samples are classified as unsuitable for irrigation due to the excess level of Na^{+} in the water, where the Kelly's ratio were above unity (Fig. 5c). The rest of samples were classified as good quality for irrigation as these fall along the equiline or below the unity.

$$\text{Sodium percentage} = \frac{[\mathrm{Na^+}]}{([\mathrm{Ca^{2+}}] + [\mathrm{Mg^{2+}}] + [\mathrm{K^+}] + [\mathrm{Na^+}])} \times 100 \quad (2)$$

$$\text{Magnesium hazard} = \frac{[\mathrm{Mg^{2+}}]}{[\mathrm{Ca^{2+}}] + [\mathrm{Mg^{2+}}]} \times 100 \quad (3)$$

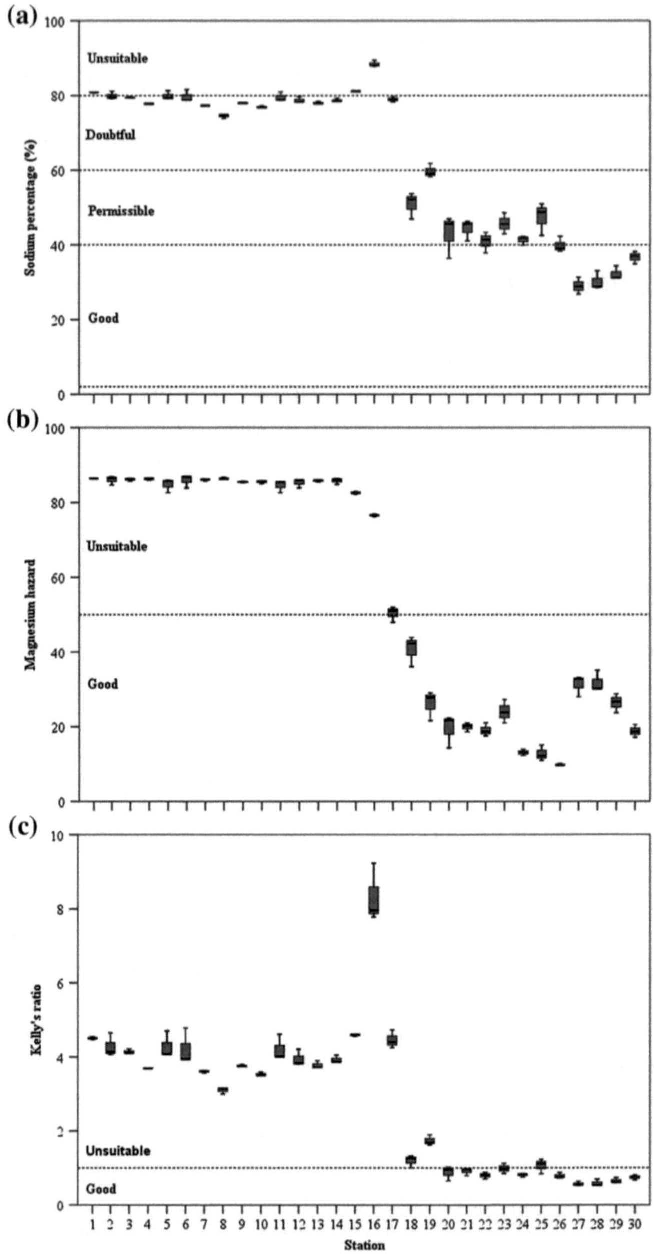

Fig. 5 **a** Sodium percentage, **b** magnesium hazard and **c** Kelly's ratio for overall study location

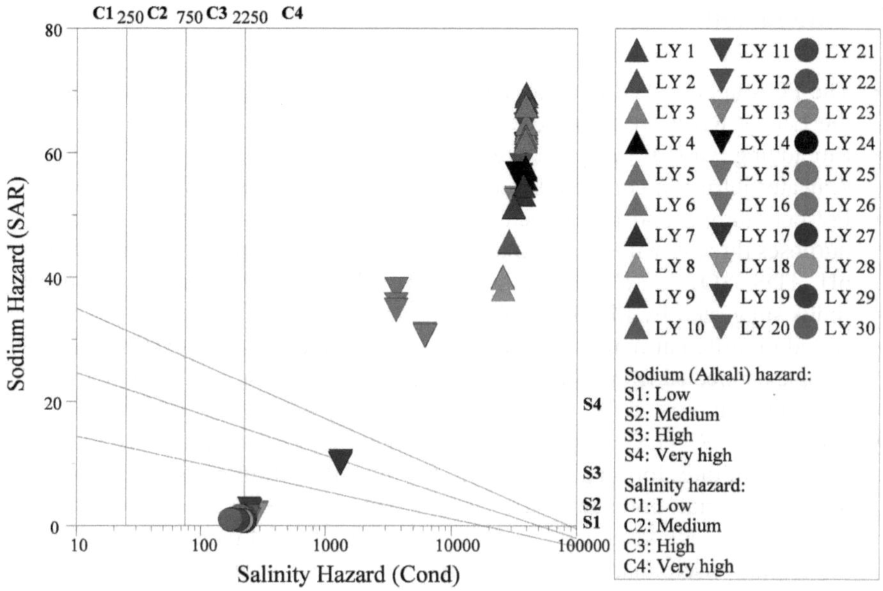

Fig. 6 Wilcox diagram of irrigation water accordingly to its sampling stations

$$\text{Kelly's ratio} = \frac{[\text{Na}^+]}{[\text{Ca}^{2+}] + [\text{Mg}^{2+}]} \tag{4}$$

$$\text{SAR} = \frac{[\text{Na}^+]}{\sqrt{\frac{1}{2}\left([\text{Ca}^{2+}] + [\text{Mg}^{2+}]\right)}} \tag{5}$$

Wilcox's diagram comprises the sodium adsorption ratio (SAR, or known as sodium hazard) and salinity hazard (with respect to EC). The SAR is the ratio of the Na^+ to the combination of Ca^{2+} and Mg^{2+} in relation to the recognized effects on soil dispersibility (Eq. 5). It measures the degree to which cation exchange between Na^+ with Ca^{2+} or Mg^{2+} can occur in the soil, which influence the soil structure (Glover 1996; Radojević and Bashkin 2007). Based on the SAR value, the water can be categorized into four classes as S1 (<10), S2 (10–18), S3 (18–26) and S4 (>26). For salinity hazard, water with EC value less than 250 μS/cm is considered as low salinity water (C1), 250–750 μS/cm as medium salinity water (C2), 750–2,250 μS/cm as high salinity water (C3) and above 2,250 μS/cm as very high salinity water (C4). The Wilcox diagram (Fig. 6) shows that 53 % of samples (LY 1–LY 16) were unsuitable for irrigation purposes, while the rest of the samples can be used for irrigation on most crops in most soils. High SAR and salinity hazard for water is inappropriate for irrigation purpose as this will reduce the osmotic activity of plants and restrict the roots of plants from absorbing water from the soil (Hiscock 2005).

5 Conclusions

- The PCA apportioned that the major sources controlling the hydrochemistry of the Langat River are from both natural and anthropogenic activities.
- Employing various ratios suggested that weathering in the catchments and seawater intrusion contribute to the hydrochemistry of the river.
- Present study found that selected water samples were unsuitable for drinking purposes directly without treatment because their water quality variables such as pH, TDS, SO_4^{2-}, Na^+, Mg^{2+}, As, or Mn values were beyond the permissible limits suggested by MOH and WHO.
- Based on different indicators and indices calculated, it was found that the water samples collected from sampling stations close to coastal area were not suitable for irrigation.
- The Langat River faces pollution from multiple sources, primarily associated with agriculture, forming, ship traffic and manufacturing units. These activities also reintroduce pollutants such as Mn into the river water by resuspending channel-bed sediments and accelerated erosion of flood-plain regions. A basin-scale multi-disciplinary study based restoration scheme has to be undertaken in this basin.

Acknowledgments This research was funded by Research University Grant Scheme (RUGS) vot no. 9199751, no. project 03-01-11-1142RU from Universiti Putra Malaysia and the Academy of Sciences for the Developing World (TWAS) project number 09-09 RG/EAS/AS_C/UNESCO FR: 3240231216.

References

Abdullah AR, Tahir NM, Loong TS, Hoque TM, Sulaiman AH (1999) The GEF/UNDP/IMO Malacca Straits demonstration project: sources of pollution. Mar Pollut Bull 39(1–12):229–233

Ahmad AK, Mushrifah I, Shuhaimi-Othman M (2009) Water quality and heavy metal concentrations in sediment of Sungai Kelantan, Kelantan, Malaysia: a baseline study. Sains Malaysiana 38(4):435–442

Al-Shami SA, Rawi CSM, HassanAhmad A, Nor SAM (2010) Distribution of Chironomidae (Insecta: Diptera) in polluted rivers of the Juru River Basin, Penang, Malaysia. J Environ Sci 22(11):1718–1727. doi:http://dx.doi.org/10.1016/S1001-0742(09)60311-9

Ali MF, Heng LY, Ratnam W, Nais J, Ripin R (2004) Metal distribution and contamination of the Mamut River, Malaysia, caused by copper mine discharge. Bull Environ Contam Toxicol 73 (3):535–542. doi:10.1007/s00128-004-0462-5

Ali ZM, Ibrahim NA, Mengersen K, Shitan M, Juahir H, Ahmad Shahabuddin FA (2012) Temporal water quality assessment of Langat River from 1995–2006. In: Voudouris K, Voutsa D (eds) Water quality monitoring and assessment. InTech. doi:10.5772/32959

Alkarkhi AM, Ahmad A, Easa A (2009) Assessment of surface water quality of selected estuaries of Malaysia: multivariate statistical techniques. Environmentalist 29(3):255–262. doi:10.1007/s10669-008-9190-4

Alloway BJ (1995) Heavy metals in soils, 2nd edn. Springer, Blackie Academic and Professional, London

Amini A, Ali TM, Ghazali AH, Huat BK (2009) Adjustment of peak streamflows of a tropical river for urbanization. Am J Environ Sci 5:285–294

APHA (2005) Standard methods for the examination of water and wastewater. American Water Works Association, Water Environment Federation, Washington

Aris AZ, Abdullah MH, Ahmed A, Woong KK (2007) Controlling factors of groundwater hydrochemistry in a small island's aquifer. Int J Environ Sci Technol 4(4):441–450

Aris AZ, Praveena SM, Abdullah MH (2012) The influence of seawater on the chemical composition of groundwater in a small island: The example of Manukan Island, East Malaysia. J Coastal Res 28(1):64–75. doi:10.2112/jcoastres-d-10-00020.1

Awofolu OR, Mbolekwa Z, Mtshemla V, Fatoki OS (2005) Levels of trace metals in water and sediment from Tyume River and its effects on an irrigated farmland. Water SA31(1):87–94

Azrina MZ, Yap CK, Rahim Ismail A, Ismail A, Tan SG (2006) Anthropogenic impacts on the distribution and biodiversity of benthic macroinvertebrates and water quality of the Langat River, Peninsular Malaysia. Ecotoxicol Environ Saf 64(3):337–347. doi:http://dx.doi.org/10.1016/j.ecoenv.2005.04.003

Bahaa-Eldin EAR, Yusoff I, Rahim SA, Wan Zuhairi WY, Abdul Ghani MR (2008) Heavy metal contamination of soil beneath a waste disposal site at Dengkil, Selangor, Malaysia. Soil Sediment Contam: Int J 17(5):449–466. doi:10.1080/15320380802304342

Berandah FE, Yap CK, Ismail A (2010) Bioaccumulation and distribution of heavy metals (Cd, Cu, Fe, Ni, Pb and Zn) in the different tissues of *Chicoreus capucinus Lamarck* (Mollusca: Muricidae) collected from Sungai Janggut, Kuala Langat, Malaysia. EnvironmentAsia 3(1):65–71

Civan MY, Kuntasal ÖO, Tuncel G (2011) Source apportionment of ambient volatile organic compounds in Bursa, a heavily industrialized city in Turkey. Environ Forensics 12(4):357–370. doi:10.1080/15275922.2011.622345

Connell DW, Miller GJ (1984) Chemistry and ecotoxicology of pollution. Wiley, New York

DID (2010) Monthly rainfall data (January–December) from 2000 to 2009. Department of Irrigation and Drainage, Kuala Lumpur

DOA (1995) Landuse of Selangor and Negeri Sembilan. Department of Agriculture, Kuala Lumpur

DOE (2010) Malaysia environmental quality report 2010. Department of Environment Ministry of Natural Resources and Environment, Kuala Lumpur

Elder JF (1988) Metal biogeochemistry in surface-water systems: a review of principles and concepts. In: U.S. Geological Survey circular, Issue 1013. Department of the Interior, U.S. Geological Survey, Washington, D.C., p 194

Eneji I, Sha"Ato R, Annune P (2012) An assessment of heavy metals loading in River Benue in the Makurdi Metropolitan Area in Central Nigeria. Environ Monit Assess 184(1):201–207

EPA (2012) Ecological risk assessment. Environmental Protection Agency. http://www.epa.gov/reg3hwmd/risk/eco/index.htm

Fulazzaky MA, Seong TW, Masirin MIM (2010) Assessment of water quality status for the Selangor River in Malaysia. Water Air Soil Pollut 205(1–4):63–77. doi:10.1007/s11270-009-0056-2

Gadgil A (1998) Drinking water in developing countries. Annu Rev Energy Environ 23(1):253–286

Gobbett DJ, Hutchison CS (1973) Geology of the Malay Peninsula: West Malaysia and Singapore. Wiley-Interscience, New York

Glover CR (1996) Irrigation water classification systems. Department of Agriculture Cooperating, New Maxico, pp 1–4

Heal KaV (2001) Manganese and land-use in upland catchments in Scotland. Sci Total Environ 265(1–3):169–179. doi:http://dx.doi.org/10.1016/S0048-9697(00)00657-4

Hem JD (1970) Study and interpretation of the chemical characteristics of natural water, 3rd edn. Geological survey water-supply paper, U.S. Government Printing Office, Washington

Howe PD, Malcolm HM, Dobson S (2005) Manganese and its compounds: environmental aspects. World Health Organization, Geneva

Hiscock KM (2005) Hydrogeology: principles and practice. Blackwell Publishing, Wiley, New York, p 389
Hooda P (2010) Trace elements in soils. Wiley, Chichester
Hounslow A (1995) Water quality data: analysis and interpretation. CRC Press, Lewis Publishers, Boca Raton
Ismail A, Ramli R (1997) Trace metals in sediments and molluscs from an estuary receiving pig farms effluent. Environ Technol 18(5):509–515
Ismail A, Badri MA, Noor Ramlan M (1993) The background levels of heavy metal concentration in sediments of the west coast of Peninsular Malaysia. Sci Total Environ 134(Supplement 1):315–323
Isa NM, Aris AZ, Sulaiman WNAW (2012) Extent and severity of groundwater contamination based on hydrochemistry mechanism of sandy tropical coastal aquifer. Sci Total Environ 438:414–425. doi:http://dx.doi.org/10.1016/j.scitotenv.2012.08.069
Jeevanandam M, Kannan R, Srinivasalu S, Rammohan V (2007) Hydrogeochemistry and groundwater quality assessment of lower part of the Ponnaiyar River Basin, Cuddalore District, South India. Environ Monit Assess 132(1–3):263–274. doi:10.1007/s10661-006-9532-y
JICA and MGDM (2002) The study on the sustainable groundwater resources and environmental management for the Langat Basin in Malaysia. Final report, Kuala Lumpur, Malaysia
Juahir H, Zain SM, Aris AZ, Yusoff MK, Mokhtar MB (2010) Spatial assessment of Langat river water quality using chemometrics. J Environ Monit 12(1):287–295
Juahir H, Zain SM, Yusoff MK, Hanidza TIT, Armi ASM, Toriman ME, Mokhtar M (2011) Spatial water quality assessment of Langat River Basin (Malaysia) using environmetric techniques. Environ Monit Assess 173(1–4):625–641. doi:10.1007/s10661-010-1411-x
Kelly WP (1951) Alkali soils—their formation, properties and reclamation. New York, Reinhold
Lee YH, Abdullah MP, Chai SY, Mokhtar MB, Ahmad R (2006) Development of possible indicators for sewage pollution for the assessment of Langat River ecosystem health. Malays J Anal Sci 10(1):15–26
Lim PE, Kiu MY (1995) Determination and speciation of heavy metals in sediments of the Juru River, Penang, Malaysia. Environ Monit Assess 35(2):85–95. doi:10.1007/bf00633708
Lim WY, Aris AZ, Praveena SM (2012) Application of the chemometric approach to evaluate the spatial variation of water chemistry and the identification of the sources of pollution in Langat River, Malaysia. Arab J Geosci 1–11. doi:10.1007/s12517-012-0756-6
MOH (2004) National standard for drinking water quality (NSDWQ). Engineering Services Division, Ministry of Health Malaysia, Kuala Lumpur
Mokhtar MB, Awaluddin A, Guan L (1994) Water quality of Inanam river estuary and the Ko-Nelayan tiger prawn aquaculture ponds in Sabah, Malaysia. Hydrobiologia 285(1):227–235
Mokhtar MB, Aris AZ, Abdullah MH, Yusoff MK, Abdullah MP, Idris AR, Raja Uzir RI (2009a) A pristine environment and water quality in perspective: Maliau Basin, Borneo's mysterious world. Water Environ J 23:219–228. doi:10.1111/j.1747-6593.2008.00139.x
Mokhtar MB, Aris AZ, Munusamy V, Praveena SM (2009b) Assessment level of heavy metals in *Penaeus monodon* and *Oreochromis spp.* in selected aquaculture ponds of high densities development area. Eur J Sci Res 30:348–360
Mokhtar MB, Toriman MEH, Hossain MAA, Tan KW (2011) Institutional challenges for integrated river basin management in Langat River Basin, Malaysia. Water Environ J 25 (4):495–503. doi:10.1111/j.1747-6593.2010.00245.x
Moujabber ME, Samra BB, Darwish T, Atallah T (2006) Comparison of different indicators for groundwater contamination by seawater intrusion on the Lebanese coast. Water Resour Manage 20(2):161–180. doi:10.1007/s11269-006-7376-4
Mustapha A, Aris A, Juahir H, Ramli M (2012) Surface water quality contamination source apportionment and physicochemical characterization at the upper section of the Jakara Basin, Nigeria. Arab J Geosci 1–13. doi:10.1007/s12517-012-0731-2
Naji A, Ismail A, Ismail AR (2010) Chemical speciation and contamination assessment of Zn and Cd by sequential extraction in surface sediment of Klang River, Malaysia. Microchem J 95 (2):285–292

Nasrabadi T, Nabi Bidhendi G, Karbassi A, Mehrdadi N (2010) Evaluating the efficiency of sediment metal pollution indices in interpreting the pollution of Haraz River sediments, southern Caspian Sea basin. Environ Monit Assess 171(1):395–410

Nayar S, Goh BPL, Chou LM (2004) Environmental impact of heavy metals from dredged and resuspended sediments on phytoplankton and bacteria assessed in in situ mesocosms. Ecotoxicol Environ Saf 59(3):349–369

Nieuwolt S, Zaki MG, Gopinathan B (1982) Agro-ecological regions in Peninsular Malaysia. Malaysian Agricultural Research and Development Institute (MARDI), Kuala Lumpur

Osman R, Saim N, Juahir H, Abdullah M (2012) Chemometric application in identifying sources of organic contaminants in Langat river basin. Environ Monit Assess 184(2):1001–1014. doi:10.1007/s10661-011-2016-8

Radojević M, Bashkin VN (2007) Practical environmental analysis. Royal Society of Chemistry, United Kingdom

Rahman RA, Surif S (1993) Metal finishing wastewater: characteristics and minimization. In: Yeoh BG, Chee KS, Phang SM, Isa Z, Idris A, Mohamed M (eds) Waste management in Malaysia: current status and prospects for bioremediation. Ministry of Science, Technology and the Environment, Malaysia

Sarmani S (1989) The determination of heavy metals in water, suspended materials and sediments from Langat River, Malaysia. Hydrobiologia 176–177(1):233–238. doi:10.1007/bf00026558

Sarmani S, Abdullah MP, Baba I, Majid A (1992) Inventory of heavy metals and organic micropollutants in an urban water catchment drainage basin. Hydrobiologia 235–236(1):669–674. doi:10.1007/bf00026255

Shafie N, Aris A, Puad N (2013a) Influential factors on the levels of cation exchange capacity in sediment at Langat river. Arab J Geosci 6(8):3049–3058. doi:10.1007/s12517-012-0563-0

Shafie NA, Aris AZ, Zakaria MP, Haris H, Lim WY, Isa NM (2013b) Application of geoaccumulation index and enrichment factors on the assessment of heavy metal pollution in the sediments. J Environ Sci Health, Part A 48(2):182–190. doi:10.1080/10934529.2012.717810

Shazili NAM, Rashid MKA, Husain ML, Nordin A, Ali S (1999) Trace metals in the surface sediments of the South China Sea, Area I. In: Proceedings of the first technical seminar on marine fishery resources survey in the South China Sea Area I: Gulf of Thailand and East Coast of Peninsular Malaysia 1997 November 24–26. Southeast Asian Fisheries Development Center, Samutprakan, Thailand, pp 73–85

Shazili NAM, Yunus K, Ahmad AS, Abdullah N, Rashid MKA (2006) Heavy metal pollution status in the Malaysian aquatic environment. Aquat Ecosyst Health Manage 9(2):137–145. doi:10.1080/14634980600724023

Shrestha S, Kazama F (2007) Assessment of surface water quality using multivariate statistical techniques: a case study of the Fuji river basin, Japan. Environ Model Softw 22(4):464–475. doi:http://dx.doi.org/10.1016/j.envsoft.2006.02.001

Sultan K (2012) Hydrochemistry and baseline values of major and trace elements in tropical surface waters of the Terengganu River (Malaysia). Water Int 37(1):1–15

Sultan K, Shazili NA, Peiffer S (2011) Distribution of Pb, As, Cd, Sn and Hg in soil, sediment and surface water of the tropical river watershed, Terengganu (Malaysia). J Hydro-Environ Res 5 (3):169–176. doi:http://dx.doi.org/10.1016/j.jher.2011.03.001

Suratman S, Awang M, Ling LA, Tahir NM (2009) Water quality index study in Paka River Basin, Terengganu (Kajian indeks kualiti air di Lembangan Sungai Paka, Terengganu). Sains Malaysiana 38(2):125–131

Szabolcs I, Darab C (1964) The influence of irrigation water of high sodium carbonate content of soils. In: Proceedings of 8th international congress of ISSS, Transaction II, pp 803–81

Taha M (2003) Groundwater and geoenvironmental quality issues in the Langat Basin, Malaysia. In: Kono I, Nishigaki M, Komatsu M (eds) Proceedings of the international symposium on groundwater engineering-recent advances, Okayama, Japan. Taylor & Francis

Tessens E, Jusop S (1983) Quantitative relationships between mineralogy and properties of tropical soils. Penerbit Universiti Pertanian Malaysia, Serdang (UPM Press), Malaysia

Tsuji LJS, Karagatzides JD (2001) Chronic lead exposure, body condition, and testis mass in wild mallard ducks. Bull Environ Contam Toxicol 67(4):489–495

UPUM (2002) Program Pencegahan dan Peningkatan Kualiti Air Sungai Langat. Final draft report, Universiti Malaya Consultancy Unit, Kuala Lumpur, Malaysia

US Salinity Laboratory Staff (1954) Diagnosis and improvement of saline and alkali soils. Handbook no 60. US Department of Agriculture, Washington, DC, p 160

WHO (2011) Guidelines for drinking-water quality, 4th edn. World Health Organization, Geneva

Wilcox LV (1955) Classification and use of irrigation waters. U.S. Department of Agriculture Circular 969. U.S. Department of Agriculture, Washington, DC, p 19

Yap CK, Ismail A, Tan SG, Omar H (2002) Correlations between speciation of Cd, Cu, Pb and Zn in sediment and their concentrations in total soft tissue of green-lipped mussel Perna viridis from the west coast of Peninsular Malaysia. Environ Int 28(1–2):117–126

Yap CK, Ismail A, Tan SG (2003) Cd and Zn concentrations in the straits of Malacca and intertidal sediments of the west coast of Peninsular Malaysia. Mar Pollut Bull 46(10):1349–1353

Yu R, Hu G, Wang L (2010) Speciation and ecological risk of heavy metals in intertidal sediments of Quanzhou Bay, China. Environ Monit Assess 163(1):241–252

Yusuf MA (2001) River water quality and ecosystem health in Langat Basin. Universiti Kebangsaan Malaysia, Bangi

Zainuddin K, Zakaria MP, Al-Odaini NA, Bakhtiari AR, Latif PA (2012) Perfluorooctanoic acid (PFOA) and perfluorooctane sulfonate (PFOS) in surface water from the Langat River, Peninsular Malaysia. Environ Forensics 13(1):82–92. doi:10.1080/15275922.2011.643335

Zulkifli S, Ismail A, Mohamat-Yusuff F, Arai T, Miyazaki N (2010a) Johor Strait as a hotspot for trace elements contamination in Peninsular Malaysia. Bull Environ Contam Toxicol 84 (5):568–573

Zulkifli S, Mohamat-Yusuff F, Arai T, Ismail A, Miyazaki N (2010b) An assessment of selected trace elements in intertidal surface sediments collected from the Peninsular Malaysia. Environ Monit Assess 169(1):457–472

Phosphorous Fractionation in Surface Sediments of the Cauvery Delta Region, Southeast India

S. Dhanakumar, K. Rutharvel Murthy, R. Mohanraj, K. Kumaraswamy and S. Pattabhi

Abstract Phosphorus is both a nutrient and a key factor responsible for eutrophication of freshwater ecosystem. Knowledge on geochemical forms of phosphorous is one among the proxies to monitor the quality of an aquatic ecosystem. Channel-bed sediment samples were collected from the Cauvery River in delta region, fractionated into five namely, exchangeable P, Fe-P, Al-P, Ca-P and Residual-P and were studied. Total phosphorous concentration of ranged between 360 and 1,070 mg/kg. The mean values of P fractions are: HCl-P (68 %), NaOH-P (12 %) and NH_4Cl-P (8 %). Among the phosphorous fractions studied, HCl extractable P recorded as a dominating chemical form. It implies P had a preferential association with Ca. The pH of the river sediments varied from 7.37 to 8.69 implying alkaline nature. Total Organic Carbon (TOC) and Organic Matter (OM) of the sediments showed ranges of 0.04–1.0 % and 0.14–3.45 %, respectively. Total Nitrogen (TN) of the sediment samples was in the range between 0.08 and 0.20 %. The C/P and N/P ratios ranged 1.65–78.67 and 1.33–6.69, respectively.

Keywords Cauvery river · Sediment grain size · Sequential extraction · Phosphorus fractions · Mobility · Nutrient ratio

---

S. Dhanakumar (✉) · S. Pattabhi
Department of Environmental Science, PSG College of Arts and Science, Coimbatore 641014, India
e-mail: ecodhanan.phd@gmail.com

K. Rutharvel Murthy
Department of Geography, Arignar Anna Government Arts College, Namakkal 637002, India

R. Mohanraj
Department of Environmental Management, Bharathidasan University, Tiruchirappalli 620024, India

K. Kumaraswamy
Department of Geography, Bharathidasan University, Tiruchirappalli 620024, India

Mu. Ramkumar et al. (eds.), *Environmental Management of River Basin Ecosystems*, Springer Earth System Sciences, DOI 10.1007/978-3-319-13425-3_22

1 Introduction

Freshwater systems provide essential ecosystem services, both for human populations and for the diverse fauna and flora on the Earth. These ecosystems are complex entities that consist of groups of species at various trophic levels, the hydrological and physical environment that makes up their habitat, the chemical properties of that environment, and the multiple physical, biogeochemical, and ecological processes that act on and within the system. Unscientific exploitation of natural resources by industrial, urban and commercial activities, lack of stringent legislation and awareness ultimately leads to the deterioration of the freshwater systems. In particular, freshwater systems such as river and lake are vulnerable to pollution due to their proximity to populated centers, sensitivity to land use as well as land cover change (Walsh et al. 2005; Withers and Jarvie 2008).

Nutrient loading into rivers is one among the major concerns in river water quality management. Overloading of nutrients in freshwater sources ultimately causes eutrophication. It also affects the water quality and aquatic lives of the rivers and increases the costs of water treatment (Dubey et al. 2012). Phosphorus (P) plays a key role in the eutrophication of surface waters, especially as a result of excessive sewage loads from urbanized areas. Unregulated population growth and changes in food habits have greatly increased the total amount of P in wastewater discharged to the environment (Caraco 1995). According to Van Drecht et al. (2009), global sewage associated P emissions are predicted to increase from 0.042 Tmol-P/yr in 2000 to 0.077–0.10 Tmol-P/yr in 2050. Atmospheric deposition, leaf fall, urban litter, residential activities and livestock excreta, domestic and industrial fertilizers, soil particles deposited by vehicles and eroded from roadside verges, detergents and lubricants are major sources of P in water bodies located along the urban centers (Duncan 2005; Dunk et al. 2007).

The stability and chemical form of particulate P in association with environmental conditions regulate its retention and release from the sediment phase and determine the level of dissolved P in the water column (House 2003). Howarth et al. (1988) found the phenomenon of secondary eutrophication as a result of release of primary compounds of P and N from the sediments to the water phase. The nutrient load that travels from the river toward lakes and seas gives an estimate of the degree of eutrophication of the ecosystems (Diaz and Rosenberg 2008; Zhigang et al. 2009). A recent study found that there are more than 400 dead zones in the world's coastal areas, most of them formed during the past half century. These dead zones, which cover a quarter of a million square km, are found in coastal areas where rivers discharge large amounts of fertilizers and sewage into relatively enclosed ocean areas (Diaz and Rosenberg 2008). Several studies have focused on the status and extent of P fractionation in sediments of lake, estuaries and rivers (Chao et al. 2008; Zhu et al. 2010; Wang et al. 2013). Anshumali and Ramanathan (2007) revealed the roles of F, Ca, organic matter and silt and clay contents of sediments in regulating the seasonal P budget. Chao et al. (2008) stated that the physicochemical properties of sediments influence the distribution of P fractions in lake sediments.

Zhu et al. (2010) studied the influence of sediment particle size, P fractions and bioavailability. Wang et al. (2013) found the higher P-buffering capacity of marsh sediments than the mudflat sediments.

In riverine environment, a major proportion of the total P is associated with sediment in the form of particulate P. Other occurrence of P is in the form bio-available P in the dissolved form, and it forms only a small fraction of total P in the system. The particulate P acts as a long-term source of available P to the river system (Sharpley et al. 1992). However, not all of the P fraction can be released from sediments into the water column and thus knowledge on the total P is not sufficient to assess the risk of eutrophication (Wang et al. 2005, 2006). The behavior of P can be evaluated by fractionation study.

Fractionation studies can be carried out through sequential extractions of operationally defined P fractions using specific reagents at various conditions. These intend classification of P according to physical (e.g. size, solubility) or biological (e.g. bio-availability) or more frequently chemical properties that determine involvement of P in geological, chemical and biological processes (Katsaounos et al. 2007). Since accumulation of P in sediments also acts as a potential nutrient source for the overlaying water during eutrophication process, it is important to evaluate the availability of P to the aquatic environment through chemical and reactive forms in relation to its total content in sediments. Among the fractions of phosphorus, the loosely adsorbed P (exchangeable P or exch-P), Al, Fe oxides and hydroxides bounded P (Fe/Al–P), Ca-associated P (Ca–P), inorganic P (IP), and organic P (OP) are widely reported from the freshwater sediments (Hupfter et al. 1995).

River Cauvery and its tributaries are the major sources of water for drinking, agriculture, and industrial needs of two states (Karnataka and Tamil Nadu) and an Union Territory (Pondicherry) of India. Water pollution due to anthropogenic activities in Cauvery River has been reported by several researchers (Senthilnathan 2004; Jameel and Hussain 2007; Solaraj et al. 2010; Dhanakumar et al. 2011). Industrial effluent discharge, sewage, solid waste dumping and agrochemical runoff are the major contributing sources of pollution in the River Cauvery. Most of the studies were focused on water quality and heavy metal pollution, but there is no detailed study was made on particulate P fractionation and the influencing factors. In this background, the present study was attempted to investigate the phosphorous fractionation in the surface sediments of Cauvery delta region and to assess the influence of sediment texture on phosphorous fractionation and organic carbon distribution.

2 Regional Setting

The Cauvery River is one of the major interstate rivers of the country flowing eastward through the States of Karnataka and Tamil Nadu and Union Territory (Pondicherry) of India, and draining into the Bay of Bengal. It has a drainage area of about 90,000 km^2 and traverses a distance of approximately 800 km from the Brahmagiri hills in the Sahyadri Range in the Western Ghats before debauching at the Bay of Bengal. In terms of catchment area, it is the eighth largest among the fourteen major river basins in India. The basin receives substantial rain during the northeast monsoon (October – December), and the water flow during summer (April to June) is normally minimal. Downstream of the Grand Anicut (located at the deltahead), the Cauvery River branches into two, namely, the Cauvery and Vennar, which get further sub-divided into 36 distributary channels to feed the delta through a network of thousands of channels. The basin comprises of Precambrian rocks, principally the Dharwars, Peninsular granitic Gneiss, Charnockites and the Closepet Granite. Physiographically, the basin can be divided into three parts: the Western Ghats area, the Plateau of Mysore and the Delta. The delta area is the most fertile tract in the basin. The principal soil types found in the basin are black soils, red soils, laterites, alluvial soils, and mixed soils. Red soils occupy large areas in the basin. Alluvial soils are found in the delta areas. The cultivable area of the basin is about 58,000 km^2 which is about 3 % of the cultivable area of the country.

3 Materials and Methods

Twelve freshly deposited channel-bed sediment samples were collected from Cauvery River in January 2011. The sampling locations are shown in Fig. 1. These locations, namely, Thirumukudallur (S1), Kulithalai (S2), Sathanur (S3), Anaikarai (S4), Kollidam (S5), Pichavaram (S6), Kumbakonam (S7), Aduthurai (S8), Mayavaram (S9), Poompuhar (S10), Nagapattinam (S11) and Muthupet (S12) were selected based on the occurrences of various point and non-point source pollutants emission in the vicinity and possible run off into the river, and urban development in the river basin. The surface sediments were collected by scooping with a plastic spade from the upper 5 cm beneath a water depth of about 20 cm. The collected sediment samples were packed and sealed in pre-washed polyethylene bags and transferred to the laboratory immediately, where they were dried at room temperature. For determining the relationship between grain size and phosphorous fractionation as well as organic carbon distribution, the sediment samples were fractionated into seven sizes (53–2,000 μm) by mechanical sieve shaker. Details of sediment analysis methods are given in the Table 1.

In order to characterize various P-fractions in the sediments, a sequential extraction scheme (Fig. 2) was performed by following Psenner et al. (1984) and the modifications proposed by Hupfter et al. (1995). This sequential scheme extracts different chemical forms of P in five steps namely, F1—loosely sorbed P (NH_4Cl-P), F2—redox-sensitive P (BD-P), F3—metal oxide bound P (NaOH-P), F4—calcium bound P (HCl-P), and F5—residual P (Res-P) (organic and refractory P). The residual phosphorous (F5) was calculated from the difference between total P concentration

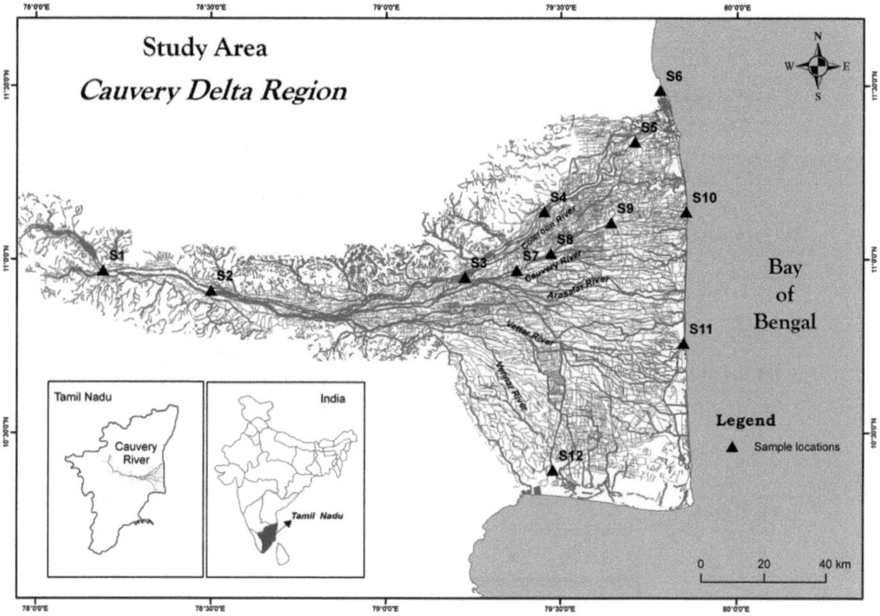

Fig. 1 Study area of Cauvery delta region

Table 1 Sediment analysis methods

Parameter	Method	Instrument/Apparatus
Soil texture	Standard sieve method	Mechanical sieves
pH (1:5 soil suspension)	Potentiometry	Water analyser model: systronics-371
Oxidation reduction potential	Potentiometry	
Total organic carbon (Walkley and Black 1934)	Wet digestion method	Titration assembly
Organic matter	Thermal combustion method	Muffle furnace method
Calcium carbonates (Allen 1989)	Rapid titration method	Titration assembly
Total phosphorous	Perchloric acid digestion	Spectrophotometric method
Fractionation of phosphorous (Psenner et al. 1984; Hupfter et al. 1995)	Sequential extraction	Spectrophotometric method

and the sum of the first four fractions. After each stage of extraction, the samples were centrifuged and phosphorus concentration in the supernatant was measured by the molybdate method with ascorbic acid as a reducer.

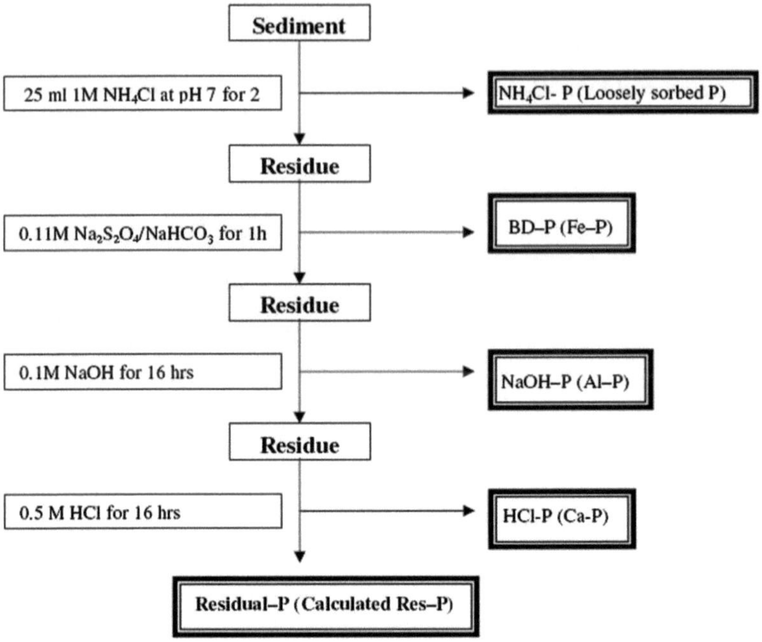

Fig. 2 The sequential extraction method for phosphorous fractionation

4 Results and Discussion

4.1 Sediment Characteristics

Descriptive statistics of physico-chemical characteristics of the sediments are given in the Table 2.

Grain size distribution in the Cauvery River suggests that this river is a low gradient and coarse-grained system. The sediment texture appeared to be different spatially in most of the sampling sites, which might have been caused by the spatial variations of prevalent sediment transport, deposition and assimilation patterns. The sand fraction is the dominant fraction, accounting for about 89.36–99.75 % of the sediments, whereas silt and clay fractions contribute only 0.25–10.64 % in the total sediment size classes.

The pH of river sediments varied from 7.37 to 8.69 and implies alkaline nature. Total Organic Carbon (TOC) and Organic Matter (OM) in the sediments show the ranges of 0.04–1.0 % and 0.14–3.45 %, respectively. Higher organic matter content recorded in few sampling sites such as Nagapattinam (S11) and Muthupet (S12) suggests active organic decomposition and mineralization of organic matter deposited in the sediment (Sharma 1994). Total Nitrogen (TN) in the sediment samples was observed to be in the range between 0.08–0.20 % with an average of 0.11 %.

Table 2 Sediment characteristics

Parameters	Min	Max	Mean	Std. dev
pH	7.37	8.69	7.97	0.48
ORP (mv)	−141.00	−40.00	−84.92	29.69
TOC (%)	0.04	1.00	0.32	0.30
OM (%)	0.14	3.45	1.12	1.03
TN (%)	0.08	0.20	0.11	0.03
C:N ratio	1.15	20.21	6.47	5.41
N:P ratio	1.33	6.69	2.59	1.43
C:P ratio	1.65	78.67	17.93	21.64
$CaCO_3$ (%)	0.25	5.75	1.84	1.88
Coarse texture (%)	89.36	99.75	98.08	2.91
Fine texture (%)	0.25	10.64	1.92	2.91

The C/P and N/P ratios were observed in the ranges of 1.65–78.67 and 1.33–6.69, respectively. Higher C/P and N/P ratios recorded in the sediment samples indicate substantial anthropogenic inputs of nitrate and phosphate. Nasnolkar et al. (1996) also found similar trend of nutrient ratios in the Mandovi River, India.

4.2 Effect of Grain Size on Total Organic Carbon Distribution

Although, the relationship between organic carbon distribution and sediments texture were well understood, studies on organic carbon distribution in various size fractions are important as the partition and transport of river sediments may influence the redistribution and accumulation of sediment bound organic carbon (Magni et al. 2008). In the study area, total organic carbon in sediments ranged between 0.04 and 2 %. Organic carbon in various size fractions of sediments shows that the highest concentration was recorded in the <53 μm fraction followed by 100–53 μm fraction. High level of organic matter and fine grains in sediments are generally associated with reduced heavy metal bioavailability and toxicity (Ankley et al. 1996). Higher concentration of organic carbon may also lead to adverse effects on benthic communities (Hyland et al. 2005) that feed on fine detritus.

4.3 Phosphorus Fractionation

Evaluation of contributions of individual P fractions to total P can provide vital information about durability of its accumulation in sediments, and possibility of its

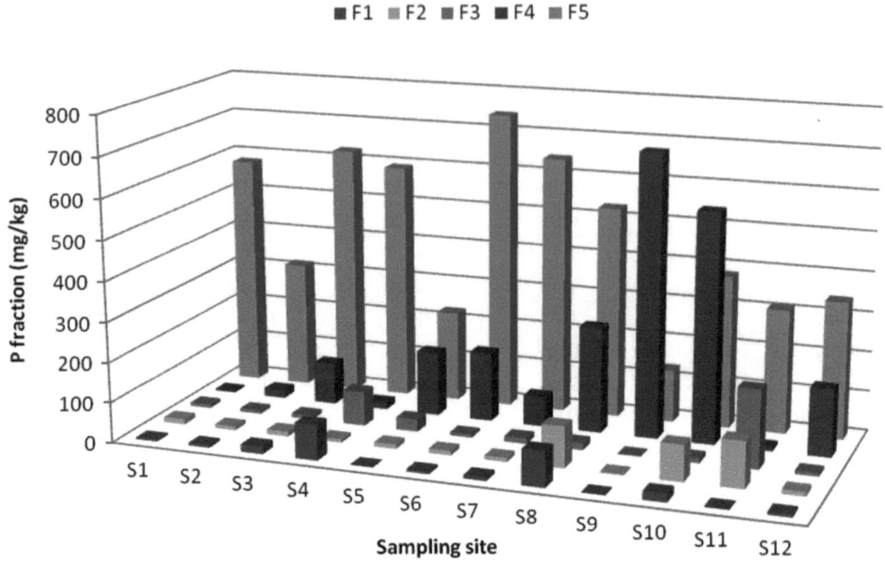

Fig. 3 Spatial variation of phosphorous fractionation

release into the water column (Kentzer 2001). In the present study, the sediment phosphorus was fractionated into five fractions namely, the exchangeable P (F1), Fe–P (F2), Al-P (F3), Ca-P (F4) and Residual–P (F5) (Fig. 3). Total phosphorous concentrations of the sediment samples ranged between 360 and 1,070 mg/kg.

The NH_4Cl extractable P (F1) represents loosely sorbed P, which is the most reactive form of sediment P. This fraction may include dissolved P in the pore water (Kaiserli et al. 2002). In the present study, exchangeable P fractions were recorded in the range between 0.85 and 87.75 mg/kg. Maximum concentration 87.75 mg/kg was recorded at Aduthurai (S8) followed by Anaikarai (87.70 mg/kg). This fraction represents on average 6.7 % of the total P in the sediments. Extensive application of di-ammonium phosphate (DAP) fertilizers and sewage discharges are some of the major anthropogenic sources of phosphate in the River Cauvery. The Cauvery delta support extensive agricultural activities. An estimate shows that use of chemical fertilizers has been on the rise during the past 15 years, which in turn could have contributed towards the elevated levels of NH_4Cl extractable P. The percentages of usages of nitrogenous, phosphatic and pottasic fertilizers were recorded as 52.84, 22.47 and 24.69 %, respectively (Department of Environment 2001). Apart from these agrochemicals runoff into the river, massive quantities of untreated municipal sewages and industrials effluents are discharged into the river from urban settlements and industries which are located all along the banks of the River Cauvery (Dhanakumar et al. 2011).

BD-P (F2) fraction is a potentially mobile and redox-sensitive form of P (Kozerski and Kleeberg 1998; Rydin 2000). The BD-P concentration ranged between 2.79 and 114.5 mg/kg with a mean of 32.8 mg/kg. This fraction accounts for about 5 % of the total P content. In response to dissolved oxygen depletion in overlaying water column, this P fraction can be released from anaerobic sediments and it also acts as an internal P source in the aquatic environment (Kleeberg and Kozerski 1998). The BD-P is known for strong upward diffusion tendency from sediments to the overlying water during summer (Shang et al. 2011).

The NaOH-P fraction (F3) represents P associated with amorphous oxyhydroxide surfaces and crystalline Fe and Al oxides (Kozerski and Kleeberg 1998; Kaiserli et al. 2002). The studied samples record a maximum NaOH-P of 194.3 mg/kg at S11 (Nagapattinam) and a mean of 33.1 mg/kg. The average concentration and relative contribution of NaOH-P in the downstream is higher than the upper reaches of the delta region. This fraction represents 12 % of the total P in sediments. This fraction may be unstable in fluctuating redox conditions. Thus there may be considerable movements of P into and out of this fraction, depending on the prevailing environmental conditions (Moore and Reddy 1994). F3 fraction of P can be used to estimate both short-term and long-term available phosphorus in the sediments (Wang et al. 2005).

According to Pardo et al. (2004), HCl extractable P (F4) is commonly associated with inorganic P. In the sediments under study, HCl-P content was recorded between 16 and 703 mg/kg with an average of 199 mg/kg. HCl-P fraction usually represents higher proportion in the total P depending on the sediment characteristics along with anthropogenic inputs. The average distributions obtained for all sediments were as follows: HCl-P (68 %), NaOH-P (12 %), NH_4Cl-P (8 %). The dominating chemical form of P was HCl-P, suggesting that P had a preferential association with Ca at the expense of other P fractions. This fraction is also considered as mobile and exchangeable in the presence of biological activities in both anoxic and aerobic conditions (Jiang et al. 2008). The HCl-P fraction had been reported to be the largest fraction for P in sediments (Kaiserli et al. 2002). For example, Vaithiyanathan et al. (1993) recorded about 70 % of P associated with HCl-P in the Hooghly Estuary, India. McDowell and Sharpley (2003) reported HCl-P representing 48–85 % of TP in an agricultural stream environment.

Residual-P (Res–P) stands as the main P fraction and it is also considered to be permanently combined with minerals to be non-bioavailable in the sediments (Wang et al. 2006). Contents of Res-P ranged between 132 and 742 mg/kg with a mean value of 452 mg/kg. Maximum of Res-P (742 mg/kg) was recorded at Thenperambur (S6). This fraction recorded a range of 16–95 % with a mean of 64 % in total P.

4.4 Effect of Particle Size on P-fractionation

Particle size is one among the factors which influence the release and adsorption of P in sediments (Zhu et al. 2010). In the present study, two sampling sites (Aduthurai S8 and Muthupet S12) were selected to find the role of particle size on P fraction (Figs. 4 and 5). The sediment texture appeared to be different spatially in both

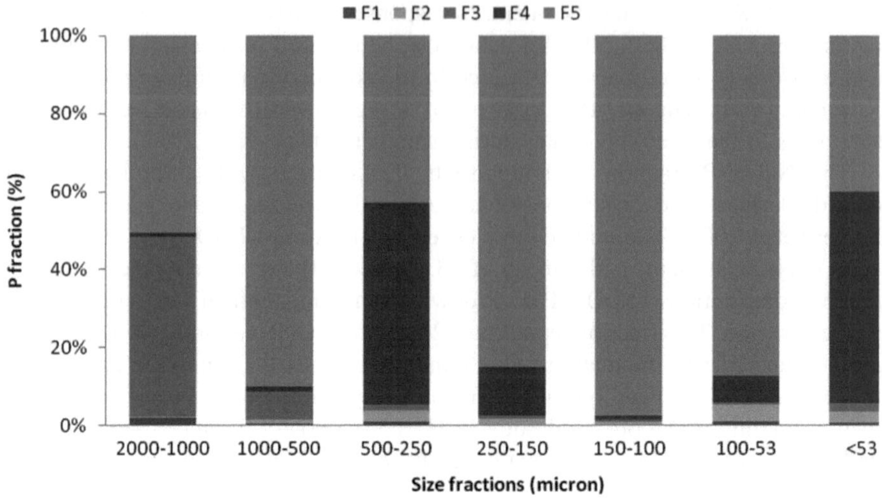

Fig. 4 Effect of particle size on P-fractionation in Aduthurai (S8)

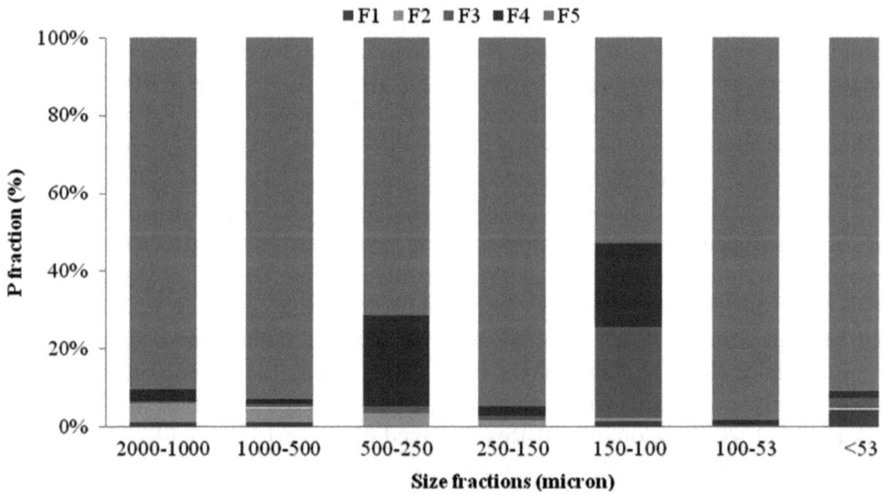

Fig. 5 Effect of particle size on P-fractionation in Muthupet (S12)

sample sites, which might have been caused by the variations in sediment transport, deposition and assimilation patterns and are under the influences of varying sewage discharge and agricultural runoff.

The results of statistical correlation analysis showed a negative relationship between P fraction and sand proportion and positive correlations of silt and clay with P fractions. Owing to the availability of higher surface-volume ratio and other surface properties namely higher cation exchange capacity of the finer fractions, particularly clays exhibit significant (50 % or more) P enrichments than the coarser fractions of the sediments (Forstner and Salomons 1980; Vervier et al. 2009). Evidently, sediments with smaller grain size (higher portion of clay and silt) have a greater capacity to adsorb P (Moore and Reddy 1994; Jin et al. 2006). Sand has typically a very low P sorption potential because of its low amorphous Fe and Al concentrations (Bridgham et al. 2001).

5 Conclusions

- Phosphorous fractionation results reveal that the residual phosphorus forms major part of the total P in the sediments. However, periodical monitoring is essential because changes in the physico-chemical characteristics of the water column may change the fractionation characteristics and release P into the system.
- Chemistry of the size-fractioned sediments showed the enrichment of P in the fine fractions. However, due to contributions from anthropogenic effluents, P was found enriched in coarse fractions at certain polluted pockets.
- Unregulated use of fertilizers shall be streamlined to control nutrient enrichments in the river system.

References

Allen SE (1989) Analysis of ecological materials. Blackwell Scientific Publications, Oxford p 368

Ankley GT, Ditoro DM, Hansen DJ, Beny WJ (1996) Technical basis and proposal for deriving sediment quality criteria for metals. Environ Toxicol Chem 15:2056–2060

Anshumali, Ramanathan AL (2007) Phosphorus fractionation in surficial sediments of Pandoh Lake, Lesser Himalaya, Himachal Pradesh, India. Appl Geochem 22:1860–1871

Bridgham SD, Johnston CA, Schubauer-Berigan JP, Weishampel P (2001) Spatial controls over phosphorus sorption in soils and coupling with surface water in two riverine wetlands. Soil Sci Soc Am J 65:577–588

Caraco NF (1995) Influence of human populations on P transfers to aquatic systems: a regional scale study using large rivers. In: Tiessen H (ed) Phosphorus in the global environment—Transfers, Cycles and Management. SCOPE 54. Wiley, New York, pp 235–244

Chao W, Jin Q, Guo Z, Li Z, chen LX (2008) Vertical distributions of phosphorus fractions in sediments of three typical shallow urban lakes in P.R. China. Pol J Environ Stud 17(1):155–162

Department of Environment (2001) Department of Environment, Government of Tamil Nadu, India. http://www.tnenvis.nic.in/DtProfiles/trichy. Accessed Mar 19 2009

Dhanakumar S, Mani U, Murthy RC, Veeramani M, Mohanraj R (2011) Heavy metals and their fractionation profile in surface sediments of upper reaches in the Cauvery River Delta, India. Int J Geol Earth Environ Sci 1(1):38–47

Diaz RJ, Rosenberg R (2008) Spreading dead zones and consequences for marine ecosystems. Science 321(5891):926–929

Dubey VK, Srivastav AL, Singh PK, Sharma YC (2012) The nutrients level in middle Ganga Basin, India. J Appl Technol Environ Sanit 2(2):121–128

Duncan H (2005). Urban storm water pollutant concentrations and loads. In: Australian runoff quality: a guide to water sensitive urban design. National Committee on Water Engineering, Australia

Dunk MJ, McMath SM, Arikans J (2007) A new management approach for the remediation of polluted surface water outfalls to improve water quality. Water Environ J 22:32–41

Forstner U, Salomons W (1980) Trace metal analysis on polluted sediments. Assessment of sources and intensities. Environ Technol Lett 1:494–505

House W (2003) Geochemical cycling of phosphorus in rivers. Appl Geochem 18:739–748

Howarth RW, Marino R, Cole JJ (1988) Nitrogen fixation in freshwater, estuarine, and marine ecosystems: 2. biogeochemical controls. Limnol Oceanogr 33:688–701

Hupfter M, Gachter R, Giovanoli R (1995) Transformation of phosphorus species in setting seston and during early sediment diagenesis. Aquat Sci 57:305–324

Hyland J, Balthis L, Karakassis I, Magni P, Petrov A, Shine J, Vestergaard O, Warwick R (2005) Organic carbon content of sediments as an indicator of stress in the marine benthos. Mar Ecol Prog Ser 295:91–103

Jameel AA, Hussain AZ (2007) Assessment of ground water on banks of Uyyakondan channel of River Cauvery at Tiruchirappalli. Indian J Environ Prot 27(8):713–716

Jiang X, Jin XC, Yao Y, Li LH, Wu FC (2008) Effects of biological activity, light, temperature and oxygen on phosphorus release processes at the sediment and water interface of Taihu Lake, China. Water Res 42:2251–2259

Jin X, Wang SH, Pang Y, Wu FC (2006) Phosphorus fractions and the effect of pH on the phosphorus release of the sediments from different trophic areas in Taihu Lake, China. Environ Pollut 139:288–295

Kaiserli A, Voutsa D, Samara C (2002) Phosphorus fractionation in lake sediments—Lakes Volvi and Koronia, N. Greece. Chemosphere 46:1147–1155

Katsaounos CZ, Giokas DL, Leonardos ID, Karayannis MI (2007) Speciation of phosphorus fractionation in river sediments by explanatory data analysis. Water Res 41:406–418

Kentzer A (2001) Phosphorus and its biologically available fractions in sediments of lakes of different trophy, habilitation dissertation (in Polish). Wydawnictwo UMK, Torun

Kozerski H, Kleeberg A (1998) The sediments and the benthic pelagic exchange in the shallow lake Muggelsee. Int Rev Hydrobiol 83(8):77–112

Magni P, De Falco G, Como S, Casu D, Floris A, Petrov AN, Castelli A, Perilli A (2008) Distribution and ecological relevance of fine sediments in organic-enriched lagoons: the case study of the Cabras lagoon (Sardinia, Italy). Mar Pollut Bull 56:549–564

McDowell RW, Sharpley AN (2003) Uptake and release of phosphorus from overland flow in a stream environment. J Environ Qual 32:937–948

Moore AJ, Reddy KR (1994) Role of Eh and pH on phosphorus geochemistry in sediments of Lake Okeechobee, Florida. J Environ Qual 23:955–964

Nasnolkar MC, Shirodkar PV, Singbal SYS (1996) Studies on organic carbon, nitrogen and phosphorous in the sediments of Mandovi estuary, Goa. Indian J Mar Sci 25:120–124

Pardo P, Rauret G, López-Sánchez JF (2004) Shortened screening method for phosphorus fractionation in sediments: a complementary approach to the standards, measurements and testing harmonized protocol. Anal Chim Acta 508:201–206

Psenner R, Puesko R, Sager M (1984) Die Fractionierung Organischer and Anorganischer Phosphorverbindungen von Sedimenten Versuch einer Definition Okologisch Wichtiger Fractionen. Arch Hydrobiol Beih 10:115–155

Rydin E (2000) Potentially mobile phosphorus in Lake Erken sediment. Wat Res 24(7):2037–2042

Senthilnathan S (2004) Micro level environmental status report of River Noyyal basin. Environmental Cell Division, Public Works Department, Coimbatore, India

Shang LH, Li QH, Qiu HB, Qiu GL, Li GH, Feng XB (2011) Chlorophyll-a distribution and phosphorus cycle in water body of Hongfeng Reservoir Guizhou. Chin J Ecol 30(5):1023–1030

Sharma S (1994) Productivity of some submerged macrophytes in relations to the physico-chemical characteristics of water and bottom sediments of lakes Mansar ans Surinsar. In: Jammu. PhD thesis, University of Jammu, India, p 81

Sharpley AN, Smith SJ, Jones OR, Berg WA, Coleman GA (1992) The transport of bioavailable phosphorus in agricultural runoff. J Environ Qual 21:30–35

Solaraj G, Dhanakumar S, Rutharvel Murthy K, Mohanraj R (2010) Water quality in select regions of Cauvery delta River basin, southern India, with emphasis on monsoonal variation. Environ Monit Assess 166:435–444

Vaithiyanathan P, Jha PK, Subramanian V (1993) Phosphorus distribution in the sediments of the Hooghly (Ganges) Estuary, India. Estuar Coast Shelf Sci 37(6):603–614

Van Drecht G, Bouwman AF, Harrison J, Knoop JM, (2009) Global nitrogen and phosphate in urban wastewater for the period 1970 to 2050. Global Biogeochemical Cycles GB0A03. doi:10.1029/2009GB003458

Vervier P, Bonvallet-Garay S, Sauvage S, Valett HM, Sanchez- Perez JM (2009) Influence of the hyporheic zone on the phosphorus dynamics of a large gravel-bed River, Garonne River, France. Hydrol Processes 23:1801–1812

Walkley A, Black IA (1934) An examination of the Degljareff method for determining soil organic matter and a proposed modification of the chromic acid titration method. Soil Sci 37:29–38

Walsh CJ, Roy AH, Feminella JW, Cottingham PD, Groffman PM, Morgan RP II (2005) The urban stream syndrome: current knowledge and the search for a cure. J North Am Benthol Soc 24(3):706–723

Wang SR, Jin XC, Pang Y, Zhao HC, Zhou XN, Wu FC (2005) Phosphorus fractions and phosphate sorption characteristics in relation to the sediment compositions of shallow lakes in the middle and lower reaches of Yangtze River region, China. J Colloid Interf Sci 289:339–346

Wang SR, Jin XC, Zhao HC, Wu FC (2006) Phosphorus fractions and its release in the sediments from the shallow lakes in the middle and lower reaches of Yangtze River area in China. Colloids Surf A Physicochem Eng Aspects 273:109–116

Wang C, Zhang Y, Li H, Morrison RJ (2013) Sequential extraction procedures for the determination of phosphorus forms in sediment. Limnology 14:147–157

Withers PJA, Jarvie HP (2008) Delivery and cycling of phosphorus in rivers: a review. Sci Total Environ 400:379–395

Zhigang Y, Zhengyu B, Lifa Z, Pu G (2009) A statistical approach for determining the environment impact of surface sediments from the Dongting Lake area, central China. Chin J Geochem 28:97–104

Zhu YR, Zhang RY, Wu FC, Fu PQ (2010) Distribution of bioavailable phosphorus and their relationship with particle size in sediments of Lake Hongfeng Guizhou Province. J Lake Sci 22 (4):513–520

[illegible]

Water Pollution in the Vicinity of Stanley Reservoir by Point and Non-point Sources, Cauvery Basin, India

R. Jayakumar, S. Dhanakumar, K. Kalaiselvi and M. Palanivel

Abstract Surface and groundwater samples were collected from the Stanley reservoir and its vicinity for two seasons (pre-monsoon and monsoon) and analyzed for physico-chemical characteristics. Variations between surface and groundwater and between pre monsoon and monsoon samples could be discerned through the studied parameters, signifying the roles of rainfall, and solute processes. Superimposed on these natural phenomena, elevated levels of chemical oxygen demand and biological oxygen demand indicative of high organic load in surface water and higher Cl and F contents in surface and groundwater as a result of anthropogenic inputs were observed. These results signify that, despite the dilutions from rainfall and potential percolation from large reservoir, the anthropogenic perturbations, mainly emanating from point and non-point sources contaminate the surface and groundwater resources.

Keywords Groundwater · Surface water · Industrial pollution · River Cauvery · Point and non-point source

1 Introduction

In addition to being a commodity, the water resources provide important services and environmental benefits to the human society in particular and to the ecosystem at large. The unprecedented increase in population demands increased allocations of surface and groundwater for the domestic, agriculture and industrial sectors, leading to conflicts among users, and excessive pressure on the environment worldwide. India is facing a serious crisis of fresh water scarcity, especially in the light of explosive population growth and accelerated economic development. India is the

R. Jayakumar · S. Dhanakumar · K. Kalaiselvi · M. Palanivel (✉)
Department of Environmental Science, PSG College of Arts and Science,
Coimbatore 641 014, India
e-mail: empe2k2@gmail.com

Mu. Ramkumar et al. (eds.), *Environmental Management of River Basin Ecosystems*,
Springer Earth System Sciences, DOI 10.1007/978-3-319-13425-3_23

largest consumer of groundwater in the world with an estimated usage of 230 km^3/ year (World Bank 2010). Approximately 60 % of the demand for agriculture and irrigation, and about 80 % of the domestic water demand are met through groundwater. Water pollution is adding to India's water woes with almost 70 % of surface water and an increasing percentage of groundwater being contaminated by biological as well as chemical, organic, inorganic and toxic pollutants.

Pollutants can enter into the aquatic environment from point and non-point sources. As the name indicates, in point-source pollution, a contaminant or nutrient enters the water at an identifiable point. Point source pollution usually originates from industrial or municipal waste outfalls, leachate from municipal and hazardous waste dump sites, hazardous spills, underground storage tanks, storage piles of chemicals, mine-waste ponds and septic tanks (Loague and Corwin 2005). According to UN-WWAP (2003), about 2 million tons of sewage, industrial and agricultural waste is discharged into the world's water per day, the equivalent of the weight of the entire human population of 6.8 billion people. The UN estimates that the amount of wastewater produced annually is about 1,500 km^3, six times more water than that exists in all the rivers of the world (UN-WWAP 2003).

Discharge of untreated sewage is the most critical water pollution source for surface and groundwater in India. Estimated sewage generation from Class I and Class II cities is about 38,000 million l/day (80 % of the water supply) and only 31.5 % of the generated sewage can be treated per day on the basis of the installed capacity. It shows a large gap between generation and treatment of wastewater in India (CPCB 2010). Thus, most of the Indian rivers and their tributaries are also polluted due to the discharge of untreated sewage and industrial effluents directly into the rivers.

Industries are a large promoter of economic growth and as India strives to maintain and accelerate its economic growth, the demand for water to the industrial sector is also bound to increase. India is the tenth most industrialized country in the World with about 88 industrial clusters scattered across the country (CPCB 2009). Although the industrial sector accounts only a paltry 3 % of the annual water withdrawals in India, it causes water pollution in most of the river basins (Prakasan and Joseph 2000; Jayaprakash et al. 2005; Tiwari et al. 2005; Panda et al. 2006; Alam et al. 2007; Kannel et al. 2007; Akan et al. 2009; Solaraj et al. 2010; Varunprasath and Daniel 2010; Shraddha et al. 2011). According to the Ministry of Water Resources, India (2003), wastewater generation from industrial sector is about 55,000 million m^3/day, of which 68.5 million m^3 are dumped directly into rivers and streams without prior treatment. Disposal of treated and untreated industrial effluents from different industries, e.g., tannery, steel plants, battery industries, thermal power plants, etc. may lead to serious environmental problems posing threat to human beings (Akhtar et al. 2009).

Effluents from tanneries and textile manufacturing/processing units increase Chemical Oxygen Demand (COD) and total organic carbon (Akan et al. 2009). Massive accumulations of various organic and inorganic pollutants in River Rapti due to the impact of the sugar industry and distillery effluents were reported by Chaurasia and Tiwari (2011). Biochemical Oxygen Demand (BOD), one of the key pollution indicators was recorded at very high levels in few of the river basins as a result of industrial and domestic wastewater discharges (Suthar et al. 2010; Chaurasia and Tiwari 2011; Sundaray et al. 2011; Yadav and Rajesh 2011). Utilization of untreated or partially treated industrial effluents for irrigation activity may lead to ground water and soil degradation through the accumulation of pollutants. In India, in addition to the disposal of industrial effluents on land and/or surface water bodies, untreated effluents are also disposed into subsurface reservoirs through dedicated pipelines and wells in some industrial sites in order to avoid pollution abatement costs (Behera and Reddy 2002; Ghosh 2005). As an outcome, groundwater resources in the vicinity become unsuitable for agriculture and/or drinking purposes. Further, long term application of polluted surface and ground water for irrigation can also increase the soil salinity or alkalinity problems in farmlands (Mukherjee and Nelliyat 2006). There are many laws as well as emission standards to minimize the impact of industrial discharge, but it is non-compliance and poor monitoring that has aggravated the situation in the most of the industrial clusters of India. Evidently, the health implications of poor water quality are enormous, and water and sanitation related diseases are responsible for 60 % of the environmental health burden in India (Planning Commission 2008). A study conducted by the World Bank has estimated that the total cost of environmental damage amounts to US$9.7 billion annually, or 4.5 % of the gross domestic product of India. Of this, 59 % results from the health impacts of water pollution (World Bank 1999).

Stanley Reservoir is located in the banks of Mettur industrial town in Salem district, Tamil Nadu, India. Three largest industries namely Chemplast, Madras Aluminum Company and the Mettur Thermal Power Plant are located in Mettur town. Besides these industries, several chemical industries are situated on the banks of the River Cauvery in Mettur as part of Small Industries Development Corporation (SIDCO) industrial estate. According to a report by the Department of Environment of the State Government of Tamil Nadu (2005), the estimated waste water discharge from Salem district into the Cauvery River is 640 lakh liter/day. Although numerous reports on surface water pollution in the River Cauvery are available, no records are available pertaining to the water quality of surface and groundwater resources in the vicinity of the Stanley reservoir from where River Cauvery flows into downstream region. This region is considered to be critical and environmentally sensitive, as any impact on the environmental quality in this region may affect the riparian regions as well.

2 Materials and Methods

Surface water (10 samples) and groundwater samples (49 samples) were collected in acid washed polythene bottles from the Stanley Reservoir and its vicinity (Figs. 1 and 2) during pre-monsoon (May 2013) and monsoon season (September 2013). Water samples were analyzed for pH, Electrical Conductivity (EC), Total Dissolved

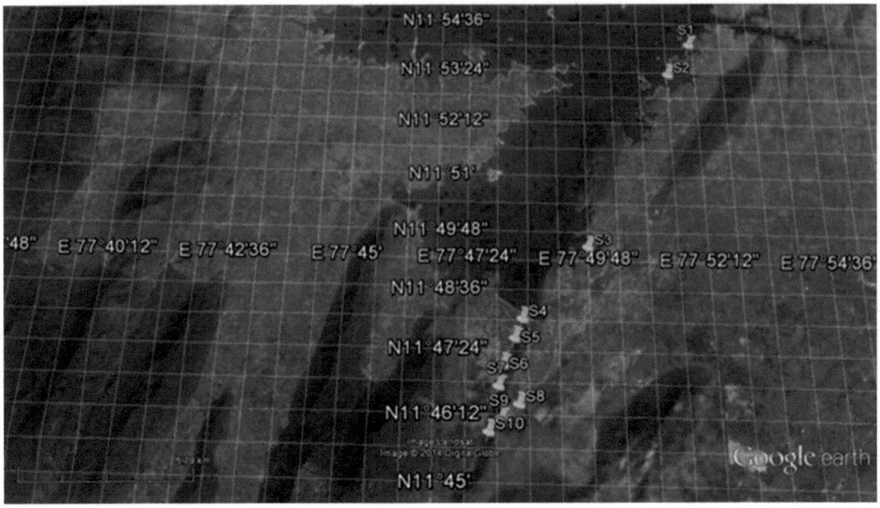

Fig. 1 Surface water sampling sites- Stanley reservoir (*Source* Google Earth)

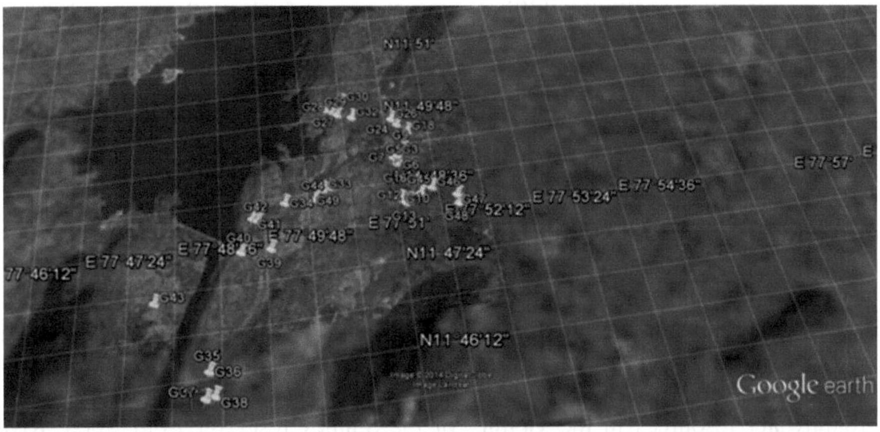

Fig. 2 Groundwater sampling sites- Stanley reservoir and its vicinity (*Source* Google Earth)

Solids (TDS), Total Suspended Solids (TSS), Dissolved Oxygen (DO), Chemical Oxygen Demand (COD), Biological Oxygen Demand (BOD), total hardness, chloride (Cl), alkalinity, calcium (Ca), magnesium (Mg), sodium (Na), potassium (K), nitrates (NO_3), sulfate (SO_4), phosphate (PO_4) and fluoride (F) using standard methods (APHA 1995) and are presented briefly herein.

The pH, EC and TDS were measured using Water analyzer (Make: Systronics) in the field immediately after sampling. DO was measured as per Winkler's method. COD was determined by open reflux method. Total hardness, Ca and Mg were estimated titrimetrically using standard ethylenediamine tetraacetic acid. Alkalinity and chloride were estimated by standard sulfuric acid titration method and silver nitrate titration method, respectively. Sodium and potassium were measured using flame photometer. Sulphate was measured spectrophotometrically by precipitating as $BaSO_4$ using $BaCl_2$. PO_4 and NO_3 were determined by stannous chloride method and brucine sulphate method with spectrophotometer respectively. Fluoride was estimated using SPADNS method. Results obtained were subjected to statistical analysis using SPSS (16th version).

3 Results and Discussion

3.1 Surface Water Characteristics

A summary on the physico-chemical characteristics of the surface water samples is provided in the Table 1 against the permissible limits suggested by WHO and BIS. The pH level of surface water samples ranged between 7.67 and 9.15 indicating alkaline nature. In many sampling stations, pH ranged above 8.5 exceeding the WHO (1997) permissible limit of 6.5–8.5. Electrical conductivity (EC) has recorded in the range of 142–1,320 µS/cm. A significant disparity between pre monsoon and monsoon samples in terms of mean total dissolved solids (TDS) was observed (390.5 and 211.6 mg/l during pre-monsoon and monsoon respectively). Mean total suspended solids (TSS) in pre-monsoon samples (0.67 mg/l) is many orders higher than the monsoon samples (0.04 mg/l). Higher contents of TDS and TSS recorded in the pre-monsoon season were probably due to the very low flow or no flow in river which ultimately led to non-dilution of pollutants. Similar results are also reported from Kshipra River, India (Gupta et al. 2013). Total hardness varied from 52.5 to 346.5 mg/l and 40.4 to 111.1 mg/l in pre-monsoon and monsoon, respectively. According to Russian epidemiological study, high hardness of drinking water may lead to risk of urinary and salivary stone formation (Mudryi 1999). In general, mean cations (Na, K, Ca and Mg) contents were found to be high in pre-monsoon than the monsoon samples.

Dissolved Oxygen (DO) is considered as a vital indicator of industrial and municipal effluent contamination on freshwater system (Rudolf et al. 2002). DO content in the pre-monsoon samples were quite normal (mean level 5.55 mg/l), but an alarming condition was observed in few of the sampling stations during monsoon. It implies river flow carries the wastewater generated from industries and municipal sewage from the vicinity into the river during the high water flow. More than half the numbers of surface water samples recorded DO less than the permissible limit of Bureau of Indian standards (6 mg/l). A decreasing trend of DO from upstream to downstream in the Varuna River, India was reported by Aggarwal et al. (2000) and these authors have interpreted the trend to be the result of point source pollution.

Table 1 Physico-chemical characteristics of surface water samples

Parameters	Unit	Pre-monsoon			Monsoon			WHO (1993)	BIS (1991)
		Min	Max	Mean	Min	Max	Mean		
pH	–	7.67	8.21	7.94	7.82	9.15	8.10	6.5–9.2	6.5–8.5
EC	μS/cm	0.52	1.32	0.72	0.16	0.41	0.22	–	–
TDS	mg/l	300.00	770.00	423.50	90.80	740.00	276.32	1,000	500
TSS	mg/l	0.50	1.00	0.67	0.01	0.08	0.04	–	–
TS	mg/l	300.50	771.00	424.17	90.81	740.08	276.35	–	–
TH	mg/l	52.50	346.50	150.58	40.40	111.10	74.74	100	300
Ca	mg/l	2.02	21.48	9.34	2.30	14.60	5.04	75	75
Mg	mg/l	9.03	71.46	31.03	8.25	25.58	16.91	30	30
Na	mg/l	46.44	167.40	97.57	5.10	23.70	10.36	200	–
K	mg/l	2.90	4.94	3.68	1.41	2.15	1.58	–	–
Chloride	mg/l	46.26	61.68	50.54	9.25	101.77	35.15	250	250
Carbonate	mg/l	25.65	85.50	51.30	0.00	23.70	4.74	–	–
Bicarbonate	mg/l	920.00	1,200.00	1,033.33	114.00	171.00	139.65	–	500
DO	mg/l	3.33	7.78	5.55	2.22	6.67	4.89	–	6
BOD	mg/l	9.00	34	16.1	5	15	8.33	5	–
COD	mg/l	160.00	640.00	360.00	40.00	560.00	200.00	10	–
PO_4	mg/l	0.11	1.03	0.71	1.60	2.50	1.91	–	–
NO_3	mg/l	0.48	1.45	0.88	1.00	3.70	2.20	50	45
NO_2	mg/l	0.19	0.35	0.23	0.10	0.21	0.11	–	–
SO_4	mg/l	18.55	161.28	45.55	8.50	58.00	22.50	250	200
F	mg/l	0.65	1.1	0.85	0.86	1.18	1.00	1.5	1.5
Salinity	mg/l	250.00	600.00	338.33	50	180	90	–	–

Organic load, Chemical oxygen demand (COD) and Biological Oxygen Demand (BOD) range of 40–640 and 5–34 mg/l, respectively. These depleted levels of oxygen might have resulted by high organic load. The slight drop in COD monsoon samples might have resulted by rainfall and higher waterflow. All the samples have COD higher than the permissible limit of 10 mg/l. Mukherjee et al. (1993) opined that about 47 % of COD are contributed by industrial sources in Indian River systems. The BOD/COD ratio is an indicator of biodegradation capacity (Metcalf and Eddy 1985). According to Contreras et al. (2003), a value of >0.5 BOD/COD ratio denotes rapid biodegradation, and a range of 0.2–0.4 indicates biodegradation only in favorable thermal condition. In the present study, the BOD/COD ratio range between 0.009 and 0.125, which indicate the presence of a considerable amount of non-degradable organic materials.

In general, nitrite (NO_2) concentration is three fold higher than the nitrate (NO_3) contents. In all the sampling sites, phosphate (PO_4) content in surface water samples (0.11–2.5 mg/l) exceeded the permissible limit (0.1 mg/l) of US Public Health Standards (De 2002) in both seasons. Higher concentration of PO_4 in few sample sites might be attributed to the mixing of municipal sewage and industrial effluents into River Cauvery in the study area. Existences of significant differences in PO_4 contents of samples collected during pre monsoon and monsoon are revealed by the *Paired T-test* and may indicate differences in the phosphorus inputs during these seasons. Similar observations on NO_3 and PO_4 inputs and differences in inputs from domestic sewage and industrial effluents into the Vaigai River, India were made by Jesu et al. (2013). Although geogenic phosphates occur naturally in water, significant enrichments are made only by anthropogenic inputs (Girija et al. 2007). Mean Fluoride (F) contents of the studied samples are 0.85 and 1.0 mg/l during pre-monsoon and monsoon, respectively. Slightly higher F content in the monsoon sample could have been drawn from higher runoff that can be expected during monsoon. Solaraj et al. (2010) reported similar trend in the mid regions of the River Cauvery. However, agricultural return flow or leachetes from irrigated lands (as excessive use of phosphatic fertilizers is a common practice (Jameel and Sirajudeen 2006) and enhanced sewage discharge (a common phenomenon associated with monsoons in ephemeral rivers of India) could also have contributed towards the slight increase in F content.

3.2 Groundwater Characteristics

A summary of the physico-chemical characteristics of the groundwater samples is provided in the Table 2 along with the permissible limits suggested by WHO and BIS. Mean pH values of the groundwater samples are 7.23 and 7.66 in pre-monsoon and monsoon samples, respectively. The electrical conductivity values have shown the ranges of 0.53–13.9 mS/cm during pre-monsoon and 0.32–20.4 mS/cm during monsoon. Relatively, higher total dissolved solids (TDS) was recorded in the pre-monsoon samples (2,895.8 mg/l) than the monsoon samples (2,433.8 mg/l).

Table 2 Physico-chemical characteristics of groundwater samples

Parameters	Unit	Pre-monsoon			Monsoon			WHO (1993)	BIS (1991)
		Min	Max	Mean	Min	Max	Mean		
pH	–	6.67	8.27	7.23	6.67	8.69	7.66	6.5–9.2	6.5–8.5
EC	μS/cm	0.53	13.90	4.67	0.32	20.40	4.54	–	–
TDS	mg/l	306	8,110	2,895.82	190.00	7,840.00	2,433.84	1,000	500
TH	mg/l	115.50	5040	1,461.02	90.90	4,848.00	1,506.34	100	300
Ca	mg/l	4.20	759.90	120.11	2.10	1,123.30	127.34	75	75
Mg	mg/l	23.70	812.08	283.80	20.64	942.25	289.51	30	30
Na	mg/l	25.30	3,125	541.13	9.70	3,033.00	525.77	200	–
K	mg/l	2.02	30.86	8.35	1.58	31.95	8.46	–	–
Chloride	mg/l	42.12	2,878.54	852.15	27.75	6,014.09	956.59	250	250
Carbonate	mg/l	0	85.65	18.15	0.00	76.95	22.30	–	–
Bicarbonate	mg/l	719	2,240	1,439.97	171.00	954.75	472.29	–	500
PO_4	mg/l	0.36	5.92	1.01	1.28	5.90	2.52	–	–
NO_3	mg/l	0.3	4.31	1.85	0.30	9.90	2.71	50	45
NO_2	mg/l	0.13	54.40	1.47	0.09	3.05	0.43		
SO_4	mg/l	17.85	3,090	368.12	20.00	3,923.96	432.38	250	200
F	mg/l	0.70	1.35	0.92	0.82	1.42	1.10	1.5	1.5
Salinity	mg/l	270	5,040	1,961.84	70.00	8,780.00	1,898.98	–	–

Significant pollutant flux into the groundwater from the surficial contaminants (for example Motoor—G19 wherein partially or untreated industrial pollutants are discharged into the open surface) are inferred from two observations: (a) water samples collected from shallow wells contain higher TDS than the water samples collected from deep wells and deep bore wells; (b) water samples collected from topographically low-lying areas contain higher TDS than those samples collected from relatively topographically higher regions. Though such generalizations on pollutant movement from topographically higher regions to subsurface (Rao et al. 2007) could be made, these data provide an affirmative evidence for a phenomenon of migration of pollutants from surface contaminants through percolation and diffusion into sub-surface aquifers.

Maximum values of total hardness recorded during pre-monsoon and monsoon were 5,040 and 4,848 mg/l respectively. The *paired-T* test has not yielded any significant difference between these two seasons and may indicate insignificant difference between the geogenic and anthropogenic sources. Nevertheless, about 80 % of the groundwater samples exceeded the desirable limit of 300 mg/l (WHO 1997). Alarmingly higher levels of hardness could have been from the improper disposal of partially or untreated industrial effluents into the vicinity of industries, which ultimately diffuse into the subsurface aquifers. Similar observations were also made by Rao et al. (2007) based on the data from an industrialized area located in Visakhapatnam, India.

Mean concentrations of cations were observed in the order Na > Mg > Ca > K. The calcium and magnesium levels vary from 2.1 to 1,123.3 and 20.6 to 942.2 mg/l, respectively against the desirable limits of 75 mg/l (Ca) and 30 mg/l (Mg), (BIS 1991) for potable use. Ca and Mg are highly mobile elements that are released extensively from natural geological sources by weathering and these also form significant portions of effluents which explain the high-variation as well as exorbitant levels. Among the cations studied, sodium (Na) is most dominant in groundwater samples. Mean value of Na was slightly higher in pre-monsoon samples (541.1 mg/l) than the monsoon samples (525.7 mg/l) and indicates monsoon influenced dilution in sodium concentration. Maximum sodium content was recorded at Kunjandiyur (G1) (3,125 mg/l) in pre-monsoon season. Although Na plays a vital role in electrolyte regulation, at elevated levels it adversely affects the cardiac, renal and circulatory functions (Srivastava 2007). The potassium contents of all the groundwater samples were within the prescribed limit (30 mg/l), except at Krupurediyur (48) (31.95 mg/l) during monsoon season.

Mean values of anion were observed in the order HCO_3 > Cl > SO_4 > CO_3 > NO_3 > PO_4 > NO_2. Bi-carbonate and carbonate values fluctuated from 719 to 2,240 and 17.1 to 85.65 mg/l during pre-monsoon season and from 171 to 954.75 and 8.55 to 76.95 mg/l during monsoon season, respectively. Except few, all the samples exceeded the permissible limit (500 mg/l) of World Health Organization. Irrespective of seasons studied, chloride concentration in groundwater samples varied between 27.75 and 6,014.09 mg/l. The maximum chloride content was recorded at Motoor (G19) (2,878.54 mg/l) during pre-monsoon and at Motoor Katuvalavu (G22) (6,014.09 mg/l) during monsoon season. Higher chloride contents are probably from the industrial activities in the vicinity, as large industries (Chemplast, Malco and other subsidiary industries) are located in the vicinity of the sampling sites. Geologically, the region contains rocks devoid of Cl, and hence, the elevated levels of Cl might have been supplied by the anthropogenic sources. Srinivasamoorthy et al. (2008) also reported elevated Cl contents in the groundwater of this area.

In order understand the relationships between the physico-chemical parameters studied, statistical correlation analysis was performed and the results are provided in the Table 3. The strongly positive correlation observed between EC, TDS and Na, K, Ca, Mg, Cl, SO_4 in both the seasons indicates major contribution of these parameters in the ionic load of groundwater. This relationship also suggests the contributions from weathering and anthropogenic sources (Shrestha et al. 2008).

Higher concentration of phosphate (PO_4) is an indicator of pollution. Mean PO_4 of monsoon samples (2.5 mg/l) has been found to be higher than the pre-monsoon season (1.0 mg/l). Throughout the study period, the PO_4 values exceeded the permissible limit (0.1 mg/l) of US Public Health Standards (De 2002). The major sources of anthropogenic PO_4 are leaching of sewage, industrial effluents and agrochemicals (Sinha et al. 2000). The nitrate (NO_3) varied between 0.3 and 9.9 mg/l. Maximum nitrite (NO_2) content recorded is 54.4 and 3.05 mg/l in pre-monsoon and monsoon season, respectively. Maximum (54.4 mg/l) was recorded at Karupurettiyur (G12) in pre-monsoon season. Nitrites are carcinogenic (Ensafi et al. 2004). Abnormally concentrations of PO_4 and NO_2 in monsoon samples may have been resulted by leaching of agrochemicals under the influence of rainfall and shallow water table. Sulphate (SO_4) values in most of the sampling stations exceeded the prescribed limit of Bureau of Indian standards (400 mg/l). Since the study area is not known for sulphate minerals, anthropogenic activities such as agrochemical application and industrial effluent disposal (Central Ground Water Board 2008) are presumed to be the reasons for the higher sulphate contents. An investigation conducted by Rao et al. (2013) also recorded high concentration of SO_4 in groundwater due to the disposal of tannery sludge in Ranipet Industrial area, India.

Although the fluoride contents of the studied samples are lower than the maximum permissible limit of 1.5 mg/l (WHO 1993), they are slightly higher than the safe limit (0.6 mg/l) and can cause dental caries (BIS 2003). The mean Fluoride (F) contents are 0.92 and 1.10 mg/l in pre-monsoon and monsoon samples, respectively. Slightly higher F recorded in the monsoon samples implies significant input from water softeners and agrochemical runoff in addition to natural leaching from fluoride-bearing minerals (Srinivasamoorthy et al. 2012). It is worthy to mention that the excessive utilization of fertilisers and pesticides for agricultural activities in the study area has resulted in a localized enrichment of nitrate, phosphate, sulphate and fluoride in the groundwater zone (Central Ground Water Board 2008).

Table 3 Correlation matrix for groundwater quality

Pre-monsoon season

	pH	EC	TDS	Alk	TH	Ca	Mg	Cl	Na	K	NO_3	NO_2	PO_4	SO_4	Sal	F
pH	1															
EC	−0.445[b]	1														
TDS	−0.459[b]	0.962[b]	1													
Alk	−0.410[b]	0.536[b]	0.495[b]	1												
TH	−0.425[b]	0.774[b]	0.812[b]	0.487[b]	1											
Ca	−0.373[b]	0.760[b]	0.750[b]	0.231	0.680[b]	1										
Mg	−0.366[b]	0.632[b]	0.684[b]	0.506[b]	0.948[b]	0.411[b]	1									
Cl	−0.469[b]	0.969[b]	0.939[b]	0.523[b]	0.774[b]	0.775[b]	0.626[b]	1								
Na	−0.265	0.600[b]	0.590[b]	0.243	0.333[a]	0.538[b]	0.182	0.608[b]	1							
K	−0.163	0.256	0.236	0.185	0.292[a]	0.193	0.279	0.182	0.032	1						
NO_3	−0.053	0.410[b]	0.416[b]	0.262	0.469[b]	0.266	0.466[b]	0.306[a]	0.195	0.239	1					
NO_2	−0.123	0.278	0.258	0.136	0.027	0.267	−0.084	0.313[a]	0.115	0.141	−0.016	1				
PO_4	0.096	−0.077	−0.115	−0.025	−0.093	−0.103	−0.070	−0.052	−0.086	−0.083	−0.017	−0.011	1			
SO_4	−0.186	0.497[b]	0.456[b]	0.390[b]	0.427[b]	0.211	0.438[b]	0.417[b]	0.319[a]	0.413[b]	0.424[b]	0.253	−0.152	1		
Sal	−0.417[b]	0.984[b]	0.944[b]	0.537[b]	0.769[b]	0.734[b]	0.638[b]	0.961[b]	0.593[b]	0.269	0.406[b]	0.291[a]	−0.062	0.501[b]	1	
F	0.135	0.122	0.064	0.136	−0.059	0.011	−0.078	0.052	0.254	−0.038	−0.028	0.090	−0.128	0.408[b]	0.110	1

Table 3 (continued)

Monsoon season

	pH	EC	TDS	Alk	TH	Ca	Mg	Cl	Na	K	NO_3	NO_2	PO_4	SO_4	Sal	F
pH	1															
EC	−0.186	1														
TDS	−0.129	0.691[b]	1													
Alk	−0.125	−0.069	−0.112	1												
TH	−0.163	0.583[b]	0.800[b]	−0.044	1											
Ca	−0.061	0.466[b]	0.574[b]	−0.083	0.637[b]	1										
Mg	−0.201	0.479[b]	0.693[b]	0.030	0.867[b]	0.206	1									
Cl	−0.172	0.956[b]	0.613[b]	−0.079	0.454[b]	0.349[a]	0.384[b]	1								
Na	−0.203	0.609[b]	0.322[a]	0.152	0.383[b]	0.388[b]	0.268	0.665[b]	1							
K	−0.019	0.376[b]	0.271	−0.058	0.260	0.115	0.263	0.302[a]	0.116	1						
NO_3	−0.157	0.053	−0.030	−0.125	−0.125	−0.133	−0.061	0.097	0.027	−0.220	1					
NO_2	−0.058	0.116	0.142	−0.085	−0.132	−0.071	−0.115	0.139	−0.079	−0.109	0.042	1				
PO_4	−0.221	0.125	0.168	0.095	0.208	−0.016	0.306[a]	0.081	−0.098	0.064	0.027	−0.062	1			
SO_4	0.071	0.345[a]	0.437[b]	−0.108	0.495[b]	0.401[b]	0.363[a]	0.206	0.107	0.664[b]	−0.272	−0.110	−0.021	1		
Sal	−0.155	0.978[b]	0.716[b]	−0.093	0.586[b]	0.464[b]	0.483[b]	0.949[b]	0.596[b]	0.383[b]	0.095	0.117	0.102	0.363[a]	1	
F	0.028	0.118	0.055	0.163	0.051	0.158	−0.079	0.133	0.207	−0.125	0.080	−0.231	0.054	−0.105	0.137	1

[a] Correlation is significant at the 0.05 level (2-tailed)
[b] Correlation is significant at the 0.01 level (2-tailed)

4 Conclusions

- The present investigation reveals that the surface water and groundwater resources of the study area are polluted due to various point and non-point source pollution (seepage of industrial effluents, municipal sewage, agrochemical sewage and return flow from irrigated lands).
- Taking into account of the high loads of TDS, Na, total hardness, NO_2, PO_4, SO_4 and F in the groundwater samples, the present situation implies potential risks to human health.
- Measures to reduce non-point source pollution should be implemented as a long term approach.
- The gap between industrial effluent and sludge generation and treatment, uncontrolled and unmonitored discharges into the river, less ground water recharge and encroachments in the river bed severely affect the water quality.
- Given cognizance to the existences of large industries in the vicinity of large reservoir, the observed seepage of polluted surface water into the groundwater and the dependence of domestic, agricultural and industrial demands over available surface and ground water resources, a real time monitoring system for keeping track of water pollution is essential for the study area.

References

Aggarwal TR, Singh KN, Gupta AK (2000) Impact of sewage containing domestic wastes and heavy metals on the chemistry of Varuna River. Pollut Res 19(13):491–494

Akan JC, Abdulrahman FI, Ayodele JT, Ogugbuaja VO (2009) Impact of tannery and textile effluent on the chemical characteristics of Challawa River, Kano State, Nigeria. Electron J Environ Agric Food Chem 8(10):1008–1032

Akhtar M, Ahman N, Booij MJ (2009) Use of regional climate model simulations as input for hydrological models for the Hindukush-Karakorum-Himalaya region. Hydrol Earth Syst Sci 13:1075–1089

Alam MJB, Islam MR, Muyen Z, Mamun M, Islam S (2007) Water quality parameters along rivers. Int J Environ Sci Technol 4(1):159–167

APHA (American Public Health Association) (1995) Standard methods for the examination of water and wastewater, 19th edn. American Public Health Association, Washington

Behera B, Ratna Reddy V (2002) Environment and accountability: impact of industrial pollution on rural communities. Econ Polit Weekly, 19 Jan 2002. pp 257–265

BIS (Bureau of Indian Standard) (1991) Indian standards drinking water specification, Indian Standard: 10500 New Delhi India

BIS (2003) Drinking water specifications, Bureau of Indian Standards, IS: 10500

Central GroundWater Board (2008) District Groundwater Brochure Salem District Central Ground Water Board, Government of India, Tamil Nadu

Chaurasia NK, Tiwari RK (2011) Effect of industrial effluents and wastes on physico-chemical parameters of river Rapti. Adv Appl Sci Res 2(5):207–211

Contreras S, Rodriguez M, Al Momani F, Sans C, Esplugas S (2003) Contribution of the ozonation pre-treatment to the biodegradation of aqueous solutions of 2,4 dichlorophenol. Water Res 37:3164–3171

CPCB (Central Pollution Control Board) (2009) Comprehensive environmental assessment of industrial clusters. The Energy and Resources Institute Press, New Delhi
CPCB (Central Pollution Control Board) (2010) Status of water supply, wastewater generation and treatment in class-I cities & class-II towns of India. Ministry of Environment and Forests, Govt. of India, New Delhi
De AK (2002) Environmental chemistry, 4th edn. New Age International, New Delhi India
Ensafi AA, Rezaei B, Nouroozi S (2004) Simultaneous spectrophotometric determination of nitrite and nitrate by flow injection analysis. Anal Sci 20:1749–1753
Ghosh Padmaparna (2005) Drug abuse: Ranbaxy, Dutch pharma put paid to groundwater. Down Earth 14(17):7–8
Girija TR, Mahanta C, Chandramouli V (2007) Water quality assessment of an untreated effluent impacted urban stream: the Bharalu tributary of the Brahmaputra River, India. Environ Monit Assess 130:221–236
Gupta RC, Gupta AK, Shrivastava RK (2013) Assessment and management of water quality of Kshipra River in Ujjain City (Madhya Pradesh), India. J Environ Sci Eng 55(2):189–196
Jameel AA, Sirajudeen J (2006) Risk assessment of physico-chemical contaminants in ground water of Pettavaithalai, Tiruchirappalli, Tamilnadu, India. Environ Monit Assess 123:299–312
Jayaprakash M, Srinivasalu S, Jonathan MP, Ram Mohan V (2005) A baseline study of physico-chemical parameters and trace metals in waters of Ennore Creek, Chennai, India. Mar Pollut Bull 50:583–608
Jesu A, Kumar LP, Kandasamy K, Dheenadayalan MS (2013) Environmental impact of industrial effluent in Vaigai River and the ground water in and around the River at Anaipatti of Dindigul District, Tamil Nadu, India. Int Res J Environ Sci 2(4):34–38
Kannel PR, Lee S, Kanel SR, Khan SP, Lee Y (2007) Spatial-temporal variation and comparative assessment of water qualities of urban river system: a case study of the river Bagmati (Nepal). Environ Monit Assess 129:433–459
Loague K, Corwin DL (2005) Point and nonpoint source pollution. In: Anderson MG (ed) Encyclopedia of hydrological sciences. Wiley, New York, pp 84–118
Metcalf and Eddy Inc (1985) Wastewater engineering treatment, disposal and reuse, 3rd edn. McGraw Hill, New York
Ministry of Water Resources (2003) Fresh water for all ministry of water resources. Government of India, New Delhi
Mudryi IV (1999) Effects of the mineral composition of drinking water on the population's health. Gig I Sanit. 1:15–18
Mukherjee D, Chattopadhyay M, Lahiri SC (1993) Water quality of river Ganga (The Ganges) and some of its physico-chemical properties. Environmentalist 13:199–210
Mukherjee S , Nelliyat P (2006) Ground water pollution and emerging environmental challenges of industrial effluent irrigation In: a case study of Mettupalayam Taluk, Tamil Nadu. Working Paper 7/2006, Madras School of Economics, Chennai
Panda UC, Sundaray SK, Rath P, Nayak BB, Bhatta D (2006) Application of factor and cluster analysis for characterization of river and estuarine water systems—A case study: Mahanadi River (India). J Hydrol 331:434–445
Planning Commission (2008) Eleventh five year plan 2007–2012 (Vol II Chapter 5) Oxford University Press New Delhi
Prakasan VR, Joseph ML (2000) Water quality of Sasthamcotta lake in relation to primary productivity and pollution from anthropogenic sources. J Environ Biol 21:305–307
Rao GT, Rao VVSG, Ranganathan K (2013) Hydrogeochemistry and groundwater quality assessment of Ranipet industrial area, Tamil Nadu, India. J Earth Syst Sci 122(3):855–867
Rao KVS, Rao NS, Krishna BM, Rao PS, Subaramaniyam A, Devadas DJ, Rao BT (2007) Temporal changes in groundwater quality in an industrial area of Andhra Pradesh, India. Curr Sci 93(11):1616–1619
Rudolf A, Ahumada R, Perez C (2002) Dissolved oxygen content as an index of water quality in San Vicente Bay, Chile. Environ Monit Assess 78:89–100

Shraddha S, Rakesh V, Savita D, Praveen J (2011) Evaluation of water quality of Narmada River with reference to physicochemical parameters at Hoshangabad City, MP, India. Res J Chem Sci 1(3):40–48

Shrestha S, Kazama F, Nakamura T (2008) Use of principal component analysis, factor analysis and discriminate analysis to evaluate spatial and temporal variations in water quality of the Mekong River. J Hydro inf 10:43–56

Sinha AK, Srivastava KP, Sexena J (2000) Impact of urbanization on groundwater of Jaipur, Rajasthan. In: Sinha AK, Shrivastava PK (eds) Earth resources and environmental issues. ABD Publishers, Jaipur, pp 173–179

Solaraj G, Dhanakumar S, Rutharvel Murthy K, Mohanraj R (2010) Water quality in select regions of Cauvery delta River basin, southern India, with emphasis on monsoonal variation. Environ Monit Assess 166:435–444

Srinivasamoorthy K, Chidambaram S, Prasanna MV, Vasanthavihar M, Peter J, Anandhan P (2008) Identification of major sources controlling groundwater chemistry from a hard rock terrain— a case study from Mettur taluk, Salem District, Tamil Nadu, India. J Earth Syst Sci 117:49–58

Srinivasamoorthy K, Vijayaraghavan K, Vasanthavigar M, Chidambaram S, Anandhan P, Manivannan R (2012) Assessment of groundwater quality with special emphasis on fluoride contamination in crystalline bed rock aquifers of Mettur region, Tamil Nadu, India. Arab J Geosci 5:83–94

Srivastava SK (2007) Groundwater quality in parts of Uttarakhand Groundwater In: Proceedings of National Seminar on agricultural development and rural drinking water, Bhopal, pp 305–312

State of Environment of Tamil Nadu (2005) Department of Environment, Government of Tamil Nadu

Sundaray SK, Nayak BB, Lin S, Bhatta D (2011) Geochemical speciation and risk assessment of heavy metals in the river estuarine sediments– a case study: Mahanadi Basin India. J Hazard Mater 186(2–3):1837–1846

Suthar S, Sharma J, Chabukdhara M, Nema AK (2010) Water quality assessment of river Hindon at Ghaziabad, India: impact of industrial and urban wastewater. Environ Monit Assess 165(1–4):103–112

Tiwari RK, Rajak GP, Mondal MR (2005) Water quality assessment of Ganga River in Bihar region, India. J Environ Sci Eng 47(4):326–355

UN-WWAP (2003) United Nations World Water Assessment Programme. The World water development report 1: water for people. Water for Life, Paris

Varunprasath K, Daniel NA (2010) Physico-chemical parameters of river bhavani in three stations, Tamil Nadu, India. Iranica J Energy Environ 1(4):321–325

WHO (1993) World Health Organization. Water, sanitation and health: Guidelines for drinking water quality: vol 1 recommendations, 2nd edn. World Health Organization, Geneva, p 47

WHO (World Health Organization) (1997) Guidelines for drinking water quality (vol 1) recommendations. World Health Organization, Geneva

World Bank (1999) World development indicators. World Bank, Washington DC, p 420

World Bank (2010) Deep well and prudence: towards pragmatic action for addressing groundwater overexploitation in India. The World Bank, Washington

Yadav SS, Rajesh K (2011) Monitoring water quality of Kosi River in Rampur District, Uttar Pradesh. India. Adv Appl Sci Res 2(2):197–201

Environmental Integrity of the Tamiraparani River Basin, South India

R. Arthur James, R. Purvaja and R. Ramesh

Abstract Monitoring the surface run-off and physicochemical parameters of a river on a regular basis provides valuable information on the eco-hydrologic conditions of the river basin. The resultant data provide valuable insights into spatial and temporal variation on water quality, considered as a measure of the health of a river. The physicochemical characteristics of any aquatic ecosystem and the nature and distribution of its biota are directly related to and influenced by each other and controlled by a multiplicity of natural regulatory mechanisms. The River Tamiraparani, a perennial river in southern India was studied for comprehensive environmental issues. The major ion chemistry and nutrients of the studied samples show seasonal and spatial variations, under the predominant influence of geogenic (natural weathering) and to certain extent, lesser influence of anthropogenic sources. Application of various geochemical indices (SAR, RSC, PI, PS and Na %) and plots revealed that, a majority of the samples (90 %) are within the permissible limits of the domestic and agricultural usages. The occurrences of organochlorine compounds in the studied samples suggest influx from agricultural activities.

Keywords Biogeochemistry · GIS · Organochlorine pesticides · Tamiraparani river · Water quality index

1 Introduction

Fresh water resources are unevenly distributed; in terms of precipitation: there is a range from almost no rainfall in deserts to several meters per year in the most humid regions; most of the flow is in a limited number of rivers: the Amazon River carries

R. Arthur James (✉)
Department of Marine Science, Bharathidasan University, Tiruchirappalli 620 024, India
e-mail: james.msbdu@gmail.com

R. Purvaja · R. Ramesh
Institute for Ocean Management, Anna University, Chennai 600 025, India

Mu. Ramkumar et al. (eds.), *Environmental Management of River Basin Ecosystems*, Springer Earth System Sciences, DOI 10.1007/978-3-319-13425-3_24

16 % of global runoff, while the Congo-Zaire River basin carries one third of the river flow in all of Africa; the arid and semi-arid zones of the world, constituting 40 % of the landmass, have only 2 % of global runoff (Subramanian 2000). Livingstone (1963) was the first to initiate the studies on chemical composition of the inland waters of the world. Since then, a slew of publications emerged on the world's biggest rivers: for example, Amazon (Gibbs 1970), Chinese rivers, Ganges-Brahmaputra (Raymahasay 1970; Subramanian 1987), and Congo (Gaillardet et al. 1995). High quality freshwater is limited in quantity and hence there is a need for comprehensive water management involving representatives of all those who use water (Kumarasamy et al. 2013). Effective management must ensure that the best use is made of available supplies, including protection from pollution, and conflict over freshwater access (Chen et al. 2013). River ecosystems are often extensively modified by man, at a rate higher than marine or terrestrial processes combined, and appear to be the most perilous (Berner and Berner 1987). Physical alteration, habitat loss, water withdrawal, overexploitation and pollution are main threats to these ecosystems and their associated biological resources. 41 % of the world's population lives in river basins under water stress (Madavan and Subramanian 2001). Industrial revolution, rapid economic development and population growth, have brought about transformations of these ecosystems and biodiversity loss on an unprecedented scale.

For effective maintenance of water quality through appropriate control measures, continuous monitoring of large number of quality parameters is essential. Variation in the quality and quantity of river water is widely studied in the case of several world rivers (Riedel et al. 2000; Solairaj et al. 2010). Continental weathering and erosion are major components of the exogenic cycles of elements on the earth. Weathering breaks down rocks and the resultant dissolved and particulate materials are transported by rivers to the sea. Chemical weathering of rocks and minerals determines the flux of the dissolved materials carried by the rivers whereas physical weathering regulates the particulate transport (Kumarasamy et al. 2014). Among the Indian rivers, those flowing through the Indo-Gangetic plains are the most studied. Subramanian (1993) documented inconsistent down-stream variations in river water chemistry. Mukherjee et al. (1993) and Singh and Singh (2007) assessed the physical, chemical and biological aspects of the Ganga River. There are several studies that documented the physicochemical parameters of the tributaries of the Ganges, including the rivers Yamuna (Saxena et al. 2001; Dalai et al. 2004; Singh et al. 2007), Gomti (Gaur et al. 2005; Singh et al. 2005) and Hindon (Jain and Sharma 2002).

Multivariate statistical analysis is used as a tool to reduce and organize large hydro-geochemical datasets into groups with similar characteristics (Kumarasamy et al. 2014). Statistical correlation and factor analysis are widely used in aqueous chemistry, to interpret the collected river water quality data and relating them to specific hydrogeological processes. The basic purpose of such an analysis is to study the biogeochemistry of the surface water ecosystem to find a set of factors with few numbers in which one can explain large amount of the variance of the analytical data (Ruiz et al. 1990).

The Tamiraparani River has been studied with a variety of tools for monitoring and interpreting its water quality. Previous studies have focused on biogeochemistry, heavy metals, morphometric evaluation and trace organic pollutants (Ravichandran et al. 1996; Ramesh et al. 2002; Magesh and Chandrasekar 2012; Kumarasamy et al. 2012, 2013). The present study attempts compiling the water quality data to assess the status of water retrieved from this river for domestic and agricultural uses, and to determine spatial and temporal differences of various hydrochemical parameters related to water quality issues.

2 Study Area

The Tamiraparani is a perennial river in southern India (Fig. 1) and is categorized under the minor river basin category. The basin is located between 8° 30′ and 9° 15′N latitudes and 77° 10′ and 78° 10′E longitudes. It originates on the eastern slopes of the Western Ghats at an altitude of 2,000 m above mean sea level (MSL), and runs for about 125 km and drains an area of 5,869 km^2 (Ravichandran et al. 1996; Kumarasamy et al. 2012). It traverses the hilly slope for a distance of 30 km and then flows on relatively low lands for 95 km and debauches into the Gulf of Mannar (Bay of Bengal). According to Balasubramanian et al. (1985), the entire basin is occupied by three distinct geological formations: (a) Archean, (b) Tertiary and (c) Recent to subrecent. The Archean formations comprise 90 % of the aerial distribution of rock types in the catchments, which consist of charnockites, granitic gneiss, calc gneisses, calc granulites, crystalline limestones and quartzites. The coastal belt is characterized

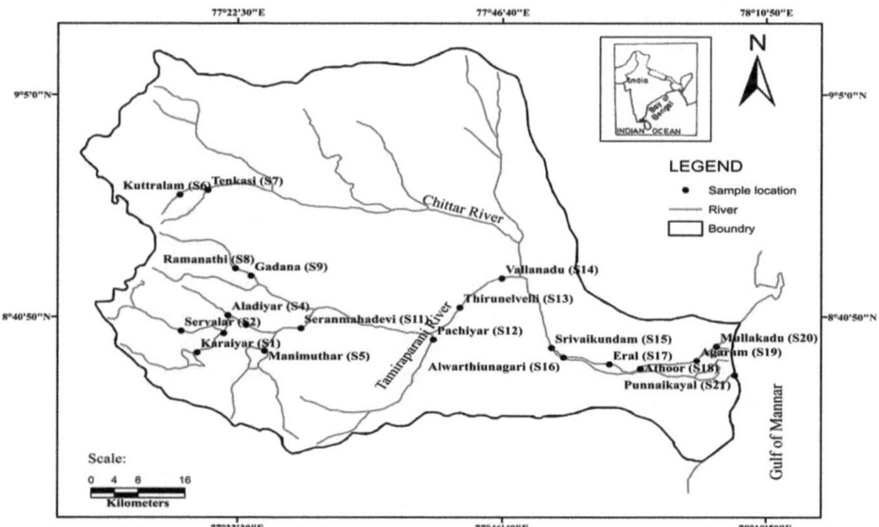

Fig. 1 Study area and sampling details of the river basin

Table 1 Geology of the Tamiraparani river basin

Age	Lithology	Distribution in the basin
Recent to sub-recent	Soils, coastal sands, red "Teri" sands. Kanker, Tufa and laterites, calcareous sandstorne and shell limestorne	Eastern region adjacent to the coastline
Tertiary	Hard and compact calcareous sandstones and shell limestone	Underlying the beach sands extending up to Srivaikundam
	Unconformity	
	Younger granitic and pegmatitic intrusions Composite gneiss, granitic mica gneiss	Northern part, central and northern parts
Archaen	Garentiferous mica gneiss amphibolites, epidiorites and pyroxene granutites	Southern and south western parts
	Charnockites	Western and northern parts of Tenkasi
	Crystalline limestornes and calc granulites	Central parts around Tirunelveli
	Quartzites	Ambasamudram

After Balasubramanian et al. (1985)

by Tertiary formations and Recent sediments, which occur in the form of a narrow belt parallel to the coastline with increased thickness towards the shore (Table 1).

The Tamiraparani basin receives an annual rainfall of 1,100 mm during the southwest (pre-monsoon: June–September) and northeast monsoons (monsoon: October–December) alone. The river has six major tributaries and their yields are: upper Tamiraparani (368 million cubic meter (Mm^3)), Servalar (289 Mm^3), Manimuthar (243 Mm^3), Chithar (119 Mm^3), Pachaiyar (100 Mm^3) and Gadana (55 Mm^3). The Tamiraparani River is being intensively used (>90 %) for agricultural purposes. The Tamiraparani irrigation system is one of the oldest systems in Tamil Nadu. The irrigation development dates back several centuries, and eight dams have been constructed across the main river. These eight dams serve a total of eleven channels with an authorized command area of about 34,443 ha, which represents about 47 % within the direct command area and 53 % indirectly fed through canals. The average withdrawal of irrigation water by the eleven channels of the irrigation project is 844 million m^3/year (Ravichandran et al. 1996). The irrigated land is supplied with water by direct irrigation to 48 % (34,934 ha) and the rest of 52 % by indirect methods through numerous tanks located within the river basin (IWS 1988). The main cropping plants include paddy, banana, groundnut, coconut. Additionally, cotton, sorghum, ragi, pulses and ginger are the other crops varieties planted in the rain-fed lands (Table 1).

The Tamiraparani River provides most of the water supplies to two major corporations (Tirunelveli and Tuticorin) as well as to small towns and several villages of southern Tamilnadu (James 2000). In addition, about 20 million l/day is supplied to the Tuticorin industrial complex and township (Ravichandran 2003). The river receives large quantities of land materials, urban and rural sewage effluents, industrial and agricultural wastes throughout the year, especially during September to December, the months of intensive rainfalls (Kumarasamy et al. 2012).

3 Materials and Methods

Water samples were collected from 20 locations covering the entire stretch of the river. Based on the locations of sampling points, the study area was divided into upper (S1–S9), middle (S10–S16) and lower (S17–S22) stretches. Sampling locations were chosen to represent a river gradient in land drainages and other effluents. Water samples were collected in two liter pre-sterilized bottles from each location for all the four seasons (pre-monsoon, monsoon, post-monsoon and summer). The samples were stored in an ice box, brought to the laboratory, and stored at 4° C until the physiochemical parameters were analyzed. The parameters namely, pH, temperature (°C) and electrical conductivity were measured in the field immediately after sampling by using a field kit. In the laboratory, the samples were filtered though 0.45 μm size fiber glass filter to remove suspended particles and then analyzed for total dissolved solids (TDS), carbonate (CO_3^-), bicarbonate (HCO_3^-), chloride (Cl^-), dissolved silica (H_4SiO_4), sulfate (SO_4^{2-}), nitrite (NO_2-N), orthophosphate (PO_4-P), hardness, alkalinity, sodium (Na^+), potassium (K^+), calcium (Ca^{2+}) and magnesium (Mg^{2+}) by using standard methods (APHA 1999; Ramesh and Anbu 1996). Each analysis was repeated for three times and only the mean of this triplicate analysis was used for further calculations namely, sodium adsorption ratio (SAR), potential Index (PI), potential salinity (PS), percent sodium (Na %) and residual sodium carbonate (RSC) and statistical analyses. Precision sampling and analyses were ensured through careful standardization, procedural blank measurements, and duplicate sampling.

4 Results and Discussion

4.1 Major Anions

The alkalinity of the fresh water is determined by the direct measure of bicarbonate (HCO_3) concentration. It is generally accepted that the river water is an imprint of a complex atmosphere-hydrosphere-lithosphere interactions (Raymahasay 1987). The bicarbonate concentration varies between 15 and 110 mg/l and 15 to 256 mg/l during monsoon and summer season, indicating that intense chemical weathering is taking place in the river basin particularly during summer season. Though intense chemical weathering in the catchment and resultant solutes to the river water may not have been possible, absence of monsoon-dilution effect and resultant enhanced

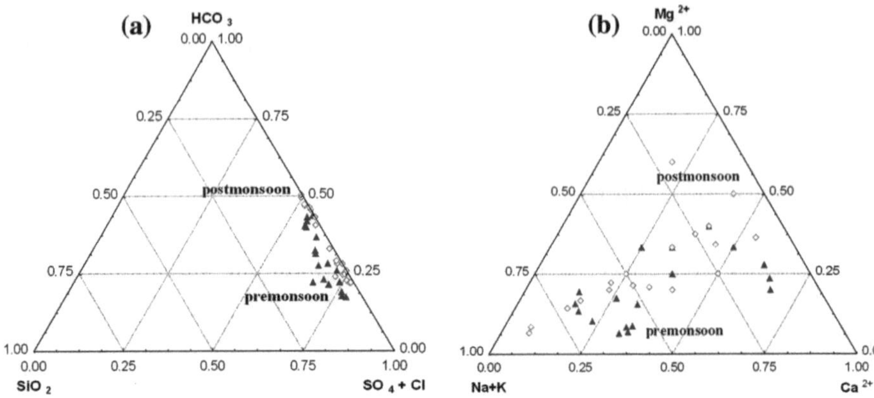

Fig. 2 **a** Ternary plots of anion chemistry in the water quality. **b** Ternary plots of cation chemistry in the water quality

levels of bicarbonates in the river water could be inferred as an alternative interpretation of significant flux during summer season. Sulphate contributes >36 % of the average chemical composition followed by bicarbonate (27 %); chloride (18 %) and H_4SiO_4 (1.5 %) in the samples under study. These anions account 73 % of the compositions. An increasing trend of anion composition from lower order channels towards higher order channel is discernible (Fig. 2a). Downstream variations of major ions have been reported for a number of other rivers (Subramanian 1983; Gupta and Subramanian 1994; Zhang 1995; Datta and Subramanian 1998). Though an abundance order of $H_4SO_4 > HCO_3 + Cl + SO_4$ anions is reported for rivers that drain areas where intense silicate weathering is prevalent, (Zhang et al. 1990), the Tamiraparani River do not show such a trend. On the other hand it has a characteristic feature of predominance in Ca^{2+} and Mg^{2+} cations and high (Ca + Mg): (Na + K) ratio, typical of river basins that drain regions where carbonate weathering prevail.

Chloride is the third major anion accounting about 20 % in the river water. An annual average of 43 mg/l has been reported in this river, which is comparable with other South Asian rivers and World rivers (Table 2). Berner and Berner (1987) estimated that 55 % of the Cl in the river is derived from rock formation and the remaining 45 % is recycled through the atmosphere. In general, a very high positive correlation of Cl^- has been observed with Na during all the seasons. The seasonal variations in EC, TDS and HCO_3^- were maximum in postmonsoon and minimum in premonsoon. This strongly suggests that the flow of river during postmonsoon and summer seasons are drastically reduced resulting in enrichments of EC, TDS and HCO^{3-}. As the summer months are always associated with warmer (average summer temperature is 38° C) weather conditions, role of evaporation could also be envisaged. Together, the seasonal variations of EC, TDS and HCO_3^-, particularly the maximum values observed during post monsoon and summer have not been influenced by changes in the rates of fluxes. This inference is further affirmed by the

Table 2 Average chemical composition of Tamiraparani river with other world rivers

River	HCO_3	Cl	SO_4	Ca	Mg	Na	K	H_4SiO_4	TDS	References
Tamiraparani	67	43	67	11	4	20	8	4	248	Kumarasamy et al. (2014)
Kerala rivers	12	7	4	3	2	5	1	9	39	Subramanian (2004)
Cauvery	135	20	13	21	9	43	4	23	272	Subramanian (2004)
Gomti	274	9	15	30	19	27	5	15	394	Subramanian (2004)
Krishna	178	38	49	29	8	30	2	24	360	Subramanian (2004)
Godavari	105	17	8	22	5	12	3	10	181	Subramanian (2004)
Mahanadi	122	23	3	24	13	14	8	17	224	Subramanian (2004)
Narmada	225	20	5	14	20	27	2	9	322	Subramanian (2004)
Tapti	150	65	1	19	22	48	3	16	322	Subramanian (2004)
Indus	64	5	23	54	12	10	0.3	5	173	Subramanian (2004)
Brahmaputra	56	11	4	14	5	7	3	7	107	Subramanian (2004)
Ganges	128	10	11	25	8	11	3	18	241	Subramanian (2004)
Indian average	74	15	13	30	7	12	3	7	159	Subramanian (2004)
Chinese rivers	113	7	14	33	5.5	5	1.5	3	181	
World average	62.2	3.7	9.2	16	4	4.4	1.5	12.4	115	Sarin et al. (1989)
Zaire	11.2	3	3	2.4	1.3	1.7	1.1	10	33	Meybeck (1987)
Amazon	23	4	3	6.5	1	3.1	1	11.2	53	Gibbs (1970)

fact that the total TDS values of the surface water were at reduced levels in pre-monsoon and monsoon seasons as a result of heavy rains during the months of June-September, October-December in the catchment area. Heavy rainfall during monsoon acts as a dilution factor which results in the dilution of contaminants intensity. The main source of the HCO_3^- in surface water is due to the rock weathering processes (Gupta and Subramanian 1994). Together, absence of seasonal variations of influx, and enrichment/dilution as a function of rainfall-river flow volume are interpreted.

The concentration of dissolved silica fluctuates widely throughout the river. The possible source of dissolved silica is kaolinization of silicates such as feldspar, augite and biotite. The temperature of the river basin exceeds 38° C during summer and immediately fresh rainfall follows during the southwest monsoon in the catchments areas, which accelerates the weathering of the basement rocks such as charnockites, granitic gneiss, calc gneisses, and calc granulites resulting in high concentration of silica. The alkaline nature of the river further enhances the solubility of amorphous silica (Gupta and Subramanian 1994). At the earth surface temperature, quartz has a solubility of 6.5 mg/l whereas other forms of silica have solubility up to 115 mg/l (Garrels and Mackenzie 1971). Meybeck (1979) suggested that the dissolved silica content of river water is controlled by the average temperature and geology of the basin. In summary, the wide variation and higher average values of silica are attributable to the weathering conditions promoted by climatic fluctuations.

4.2 Major Cations

Calcium and potassium in river water are normally drawn from weathered products in the catchments. Calcium constitutes about 5 % of the average chemical composition and Mg about 1.8 % of the total dissolved solids of the river basin. This is probably derived from the dissolution of kankar, limestones, plagioclase feldspar and other ferromagnesian minerals. Magnesium may have been derived from the hydrolysis of olivine bearing rocks in the sources area. In summer, the Ca^{2+} concentration ranged from 2 to 38 mg/l in the upstream and 12–23 mg/l in the downstream. During premonsoon season, the calcium concentration ranged from 4 to 38 mg/l and magnesium concentration ranged of 1–16 mg/l. The ratio of (Ca + Mg):(Na + K) indicates that the carbonate weathering could be the primary source of the major ions to these waters (Fig. 2b). Nevertheless, their fluctuations may have been attributable towards evaporation and dwindling flow induced enrichment during summer and dilution by rainfall and higher river flow during monsoon. Seasonal disparity and wide spatial variations are observed in the total dissolved solids (TDS) contents. The (Ca + Mg):(Na + K) ratio in the Tamiraparani River suggests that the contribution of these ions by silicate weathering is less significant. Gaillardet et al. (1997) computed Ca/Na; HCO_3/Na and Mg/Na ratios to differentiate between the carbonate and silicate weathering in the Amazon River basin. In the upstream region of river basin, high values of Ca/Na (0.61–7.3) has been observed when compared to the downstream region (0.07–4.48) indicating the predominance of carbonate weathering in the catchment region.

Sodium is the dominant cation in the river system and accounts for 9 % of the average TDS. The silicate weathering in the catchment and atmospheric inputs are the major sources of the Na^+ in river water. The sodium concentration is lower than chloride values from upstream to downstream for all the seasons. This indicates that the atmospheric recycling is the main source for sodium and is less influenced by physical weathering (Subramanian et al. 1987). Average sodium content during premonsoon season is higher (590 mg/l) than the monsoon season (300 mg/l). Potassium concentration is lower by about two times when compared to Na in all the four seasons possibly due to the low mobility of potassium than sodium. Based on the environmental factors that influence river water chemistry, Gibbs (1970) suggested a threefold classification of rivers namely, Low total dissolved solids (TDS) and high Na/(Na + Ca) water controlled by rainfall, Intermediate TDS and low Na/(Na + Ca) water controlled by rock weathering, and High TDS and high Na/(Na + Ca) water controlled by evaporation.

The Gibbs diagram has three distinct fields namely precipitation, evaporation and rock dominance areas. The Gibbs ratio I-Cl/(Cl + HCO_3) for anion and ratio II-Na + K/(Na + K+Ca) for cations in water samples were plotted separately against the respective values of total dissolved solids. This gives a characteristic boomerang shaped plot, depicting the major factors operating behind the water chemistry. From the Fig. 3, it can be observed that all the fall in the region of rock dominance, consistent with the inference (presented in previous paragraphs and previous section

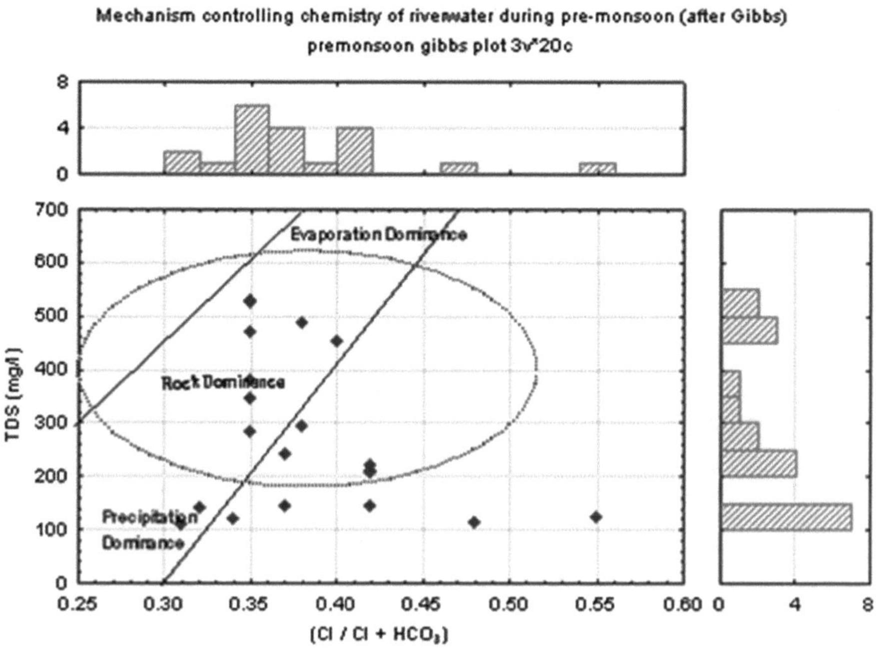

Fig. 3 Gibbs plot predicting factors controlling water chemistry

that the river water chemistry owes its allegiance predominantly to the weathering in the catchment areas, particularly carbonate weathering and also that spatial and temporal variations if any, were the results of enrichments during summer by evaporation and reduced river flow and depletions during monsoon by rainfall dilution and heightened river flow) of geogenic flux controlled nature of the water chemistry of the Tamiraparani basin. It is further ascertained by the plot of (Ca + Mg) versus Total cations and (Ca + Mg) versus HCO_3 (Fig. 4a and b) which shows that the river basin is dominated by rock weathering. One of the unifying factors is the excellent relationship between the major parameters like calcium and magnesium. All the rivers, in general, show high carbonate alkalinity independent of local lithology; rock weathering involving atmospheric CO_2 and minerals in different lithology uniformly releases Ca^{2+}, Mg^{2+} and HCO_3^- to the river water (Berner and Berner 1996).

In the water chemistry, Ca^{2+} and Mg^{2+} together constitute 36 % of the total cations and HCO_3^- contributes about 32 % to the total anionic balance. Carbonate and silicate weathering and evaporite dissolution can supply Ca^{2+} and Mg^{2+} to the river water, whereas the Na^+ and K^+ are drawn from many sources including atmospheric deposition, evaporite dissolution and silicate weathering. On an average (Na + K) contributes 63 % of the total cationic balance. Relatively high contents of dissolved silica and higher value of (Na + K)/TZ^+ ratio (1.6) suggest that silicate weathering could be the major source of alkalis. However, this is an

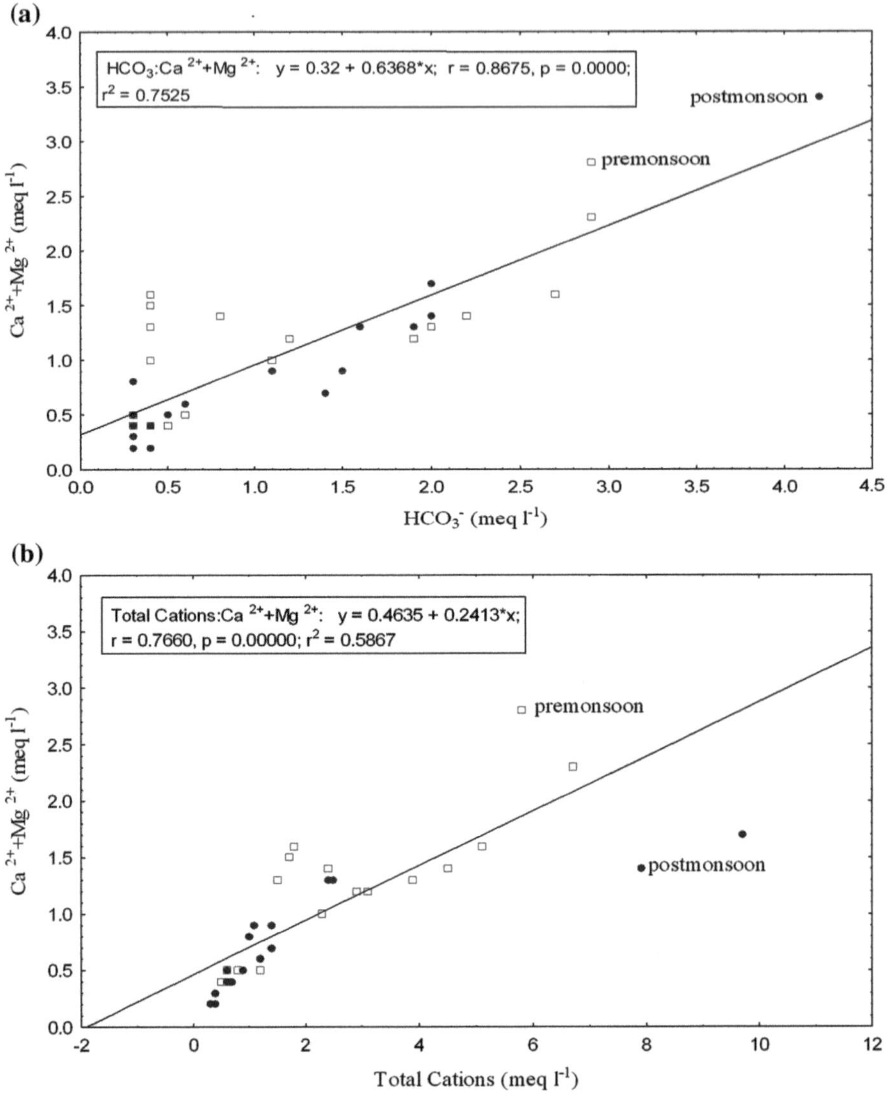

Fig. 4 **a** Cation chemistry plot against HCO_3. **b** Cation chemistry plot against Ca and Mg

upper limit of silicate weathering, as there could be other sources namely, evaporite, sodium/potassium minerals and alkaline soil and groundwater. Contributions from evaporite are considered to be highly feasible than the contribution from silicate weathering as the occurrences of evaporite encrustations are reported in many parts of the drainage basin. These encrustations develop due to cyclic wetting and drying during high and low flow period. This aids the formation of alkaline/saline soils, which may also serve as a source of sodium and potassium (Sarin et al. 1989).

4.3 Nutrients

The occurrence of nitrite in surface waters is closely connected with the nitrogen cycle and the main nitrite source is the oxidation of ammonia and organic nitrogen. Microorganisms play an important role in these processes, although nitrite can also be produced on photo-degradation of dissolved organic matter (DOM) (Kieber et al. 1999). The fast oxidation of nitrite to nitrate by different pathways limits the concentration of nitrite in surface waters. Accordingly, nitrite concentration is usually considerably lower than that of nitrate. The higher absorption rate of solar UV radiation by nitrite compared to nitrate and the higher photolysis quantum yield makes nitrite a competitive photo reactant (Boule et al. 1999). Nitrate pollution usually originates from diffused sources, like intensive agriculture or point sources, such as irrigation of land by sewage effluent (Eckhardt and Stackelberg 1995; McLay et al. 2001). It may also originate from industrial effluents, including paper and munitions manufacturing, septic tanks and human and animal wastes, due to biochemical activity of nitrifying bacteria.

In living organisms, nitrogen cycle is considered to be most important after carbon cycle. Nitrogen occurs in the form of nitrogen gas, oxides (N_2O, NO and NO_2) and reduced nitrogen (NH_4). The major forms of nitrogen occurring in river water are nitrite (NO_2), nitrate (NO_3^- most oxidized), ammonium (NH_4-most reduced) and organic nitrogen. The occurrence and distribution of these forms of nitrogen are governed by the chemistry of the subsoil, land use pattern, geology, vegetation, river water quality etc. The formation of nitrates assumes great importance in an aquatic ecosystem, since it is utilized by phytoplankton or lost from the environment through denitrification. Many agricultural practices affect soil aeration and enhance mineralization of organic nitrogen in soil. In the upstream region of the Tamiraparani River basin, the landscape is dominated by forested area, resulting in significant amount of nitrogen fixation (i.e. the conversion from N_2 gas to organic nitrogen) from the atmosphere. Due to death and decay of leaves and other organic matter in the environment, and through microbial activity, the NH_3 is first converted to the assimilatable form of NO_3 and NO_2 termed 'nitrification'. In the downstream region, because of the high organic input from natural forest and agricultural runoff, the NO_3 is converted to N_2 by the process of 'denitrification' resulting in oxygen stress in the environment.

During summer season, high concentration of NO_2-N was observed at upstream [Tenkasi (47 μg/l) and Papanasam (10 μg/l)] and low at downstream region [Atoor (4 μg/l)]. The upstream region is dominated by dense evergreen forests of the Western Ghats. The death and decay of leaves, twigs, plankton, etc., results in the degradation of organic matter and an enrichment of nitrogenous nutrients particularly NO_2-N are supplied to the river water (James 2000) which results in a consequent increase in primary productivity. By contrast, in the midstream region, the NO_2-N concentration was very low, due to the fact that in the midstream region, the river water is diverted through eight weirs and channels, for agricultural and other uses, resulting in the trapping of nutrients. However, a distinct loading of

NO_2-N has been observed in the downstream region during summer, suggesting that anthropogenic input contributes significantly to nutrient enrichment in the surface water.

Phosphorus is a very important element in biogeochemical cycling and mass transfer between the continent-ocean, because of its involvement in the biological and inorganic processes respectively. The present day flux of phosphorus provides a clue regarding the sources of phosphorus for the formation of phosphorites (Subramanian 1984). The amount of phosphorus transported by the rivers depends on the natural processes (climate and rock type), man's influence (deforestation, agriculture and population) and biological removal processes. Orthophosphate levels in natural waters are normally low, range between 1 and 24 μg/l (Meybeck 1982) with a mean of 12 μg/l in the tropical rivers. Of the inorganic forms, the complex commonly referred as PO_4-P is important; being the only form of phosphorus that is capable of being taken up by autotrophic organisms (Vaithiyanathan et al. 1989). Eutrophication is one of the most serious problems facing the ecology of freshwaters. The nutrient status of many lakes and rivers has increased dramatically over the past 30 years in response to increased collection and discharge of domestic wastes, increased loadings to collection systems (including the use of phosphate detergents) and widespread agricultural intensification. Phosphorus is delivered to the river system from a range of sources. Soluble phosphorus can be incorporated into inorganic phosphate minerals by precipitation, particularly in association with calcium (in hard water rivers), iron and aluminum (in soft water rivers). The precipitation of soluble phosphorus with calcium is particularly likely to occur below sewage treatment works in rivers with calcareous waters (House and Denison 1997) where both calcium and soluble phosphorus concentrations are very high. Colloids of calcium phosphate minerals can be generated in the water column, whilst algal biofilms are thought to be involved in the co-precipitation of calcite and phosphorus onto bed sediments and plants (Hartley et al. 1997). The orthophosphate concentration ranges from 0.6 to 16 μg/l during monsoon, 2–12 μg/l during summer, 0.6–17 μg/l during premonsoon and 0.6–14 μg/l during postmonsoon. Higher contents of orthophosphate were observed at Agaram (17 μg/l) and at Tirunelveli (16 μg/l) during monsoon.

4.4 Organochlorine Pesticides

The use of the pesticides in India started only after the independence and the first plant to produce a pesticide (benzene hexachloride-BHC) on a commercial basis was set up in 1952, while the first public-sector DDT plant came into existence in 1955. In 1958, India was producing over 5,000 metric tons of pesticides. Currently, there are approximately 145 registered pesticides for use, and the production has increased to approximately 94,000 metric tons (Ministry of Chemicals and Fertilizers (MoCF) 2005–06). The agricultural sector consumes about 67 % and the remainder is as domestic programs (Lal 2007). The pesticide pollution and their impact on Indian rivers are reported by several researchers (for example, Ramesh et al. 1990; Rajendran and Subramanian 1997, 1999; Kumarasamy et al. 2012). The present study seeks to identify and quantify OCPs concentrations in both water and sediment samples of the Tamiraparani River to assess the impacts of agricultural inputs on the river system.

Figure 5 shows the distribution pattern of total OCPs residues for three seasons in the river basin. It clearly reveals that the OCPs residues in post-monsoon and summer were much higher than pre-monsoon season. This could be due to the release of some OCPs from non point sources and subsequent leaching processes during monsoon period throughout the basin and intense agriculture practices in summer season followed by southwest monsoon in the upper region. Higher concentrations of OCPs in upstream region are probably resulted during monsoon season from farmland runoff (Tenkasi) and by the discharge of untreated municipal waste into the river. This is affirmed by the low levels of OCPs in Servalar region which is associated with lower areal extents of agricultural lands and predomination of dense natural forests. In the midstream regions, removal of OCPs from water by sedimentation and accumulation by soils under the influence of restricted flow (dam) could explain lower contents of OCPs in river water. The highest concentrations of 58 ng/l was observed at Papanasam lower dam during monsoon season. This could have been due to the extensive sprawl of rubber plantation, known for the use of large quantities of organochlorine insecticides (dieldrex and octalox). Higher OCPs are observed in the downstream areas (Eral and Punnakayal) in all the seasons as a result of collective down flow loads. The concentrations of OCPs are comparable with other Indian and World river.

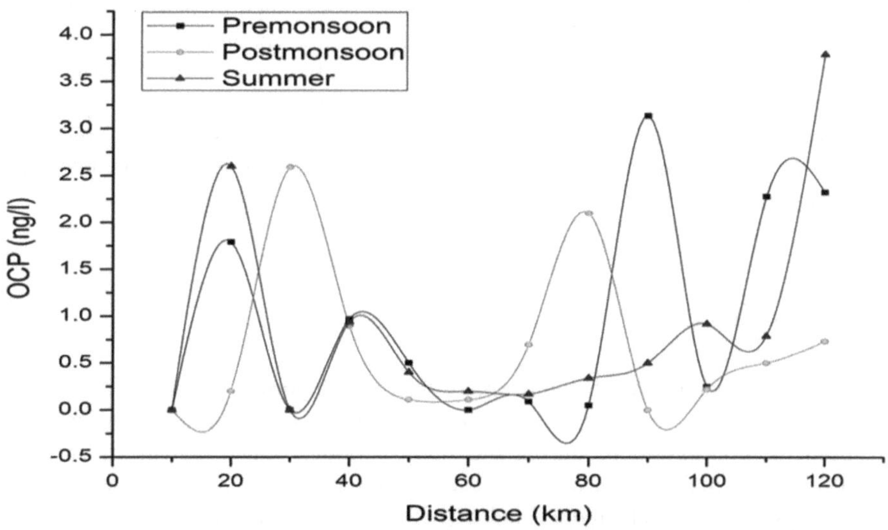

Fig. 5 Spatial and seasonal variation of organochlorine pesticides surface water

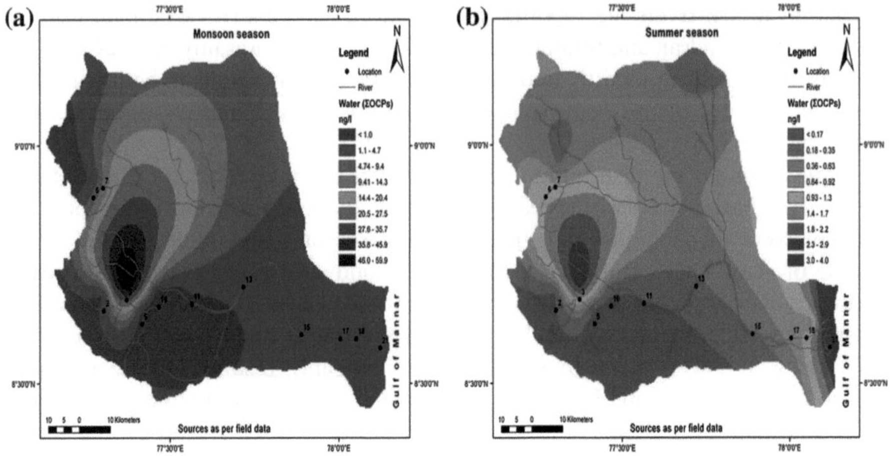

Fig. 6 **a**, **b** Spatial prediction on OCP in the catchment area

Other OCPs including aldrin, endrin, dieldrin, heptachlor, trans-chlordane, cis-chlordane and mirex were found in small quantities in all the sampling sites and seasons (Fig. 6a and b). Specifically, endrin and dieldrin were found in most of the locations in water samples. Dieldrin was widely used as an effective insecticide around the world until the middle 1970s, mainly for the control of soil pests, like termites, grasshoppers, locusts, beetles, textile pests, and for the treatment of seeds. Additionally it has also been used to control vectors of tropical diseases, including malaria, yellow fever. It has uses in industries to protect electric and telephone cables and to preserve timber materials etc. In the study area, dieldrin is effectively used as soil insecticide, especially for perennial crops, such as palm, and banana. Aldrin and dieldrin are used for controlling termites in houses by direct soil injection in the past (ATSDR 2002). Aldrin is rapidly converted to dieldrin in the environment. Investigations have shown that aldrin is found pervasively in the study area with a range of <0.02–4.0 ng/l in different seasons and the same trend was observed for dieldrin in the range of 1.4–12.7 ng/l. These compounds were probably originated from the industrial and public health practices.

5 Conclusions

- The Tamiraparani river basin is one of the most important perennial river basins in Tamil Nadu, South India. The dissolved major ion, inorganic nutrients and pesticide concentrations during the study period revealed that the river water receives its major share of major anions and cations from natural weathering processes in the catchments and also that, anthropogenic contamination are minimal pertaining to these major anions and cations.

- Contaminants emanating from agricultural leachets and return flows and municipal wastes have been polluting the river water and the effects are felt up to the downstream regions such as Eral and Punnakayal.
- Most of the pollution indicators and indices suggest that the Tamiraparani River water is suitable for domestic and irrigation uses. However, the ubiquitous presence of OCPs almost in all of the basin and during all the seasons is a cause of worry.

References

APHA (2001) Supplement to standard methods for the examination of water and wastewater, 20th edn. APHA, American water works association, and water pollution control federation, Washington

Balasubramanian A, Sastri JCV (1987) Studies.on the quality of groundwater of Tamiraparani river basin, Tamil Nadu, India. J Assoc Explor Geophys 8:41–51

Balasubramanian A, Sharma KK, Sastri JCV (1985) Geological and hydrogeochemical evolution of coastal aquifers of Tamiraparani basin, Tamil Nadu. Geophys Res Bull 23:203–206

Berner EK, RA Berner (1996) Global environment: water, Air and geochemical cycles. Prentice Hall, New Jersey, pp.376

Berner EK, Berner RA (1987) The global water cycle: geochemistry and environment. Prentice Hall, Englewood Cliffs, p 397

Boule P, Bolte M, Richard C (1999) Phototransformations induced in aquatic media by NO_3−/ NO_2−Fe-III and humic substances. In: Boule P (ed.) The Handbook of Environmental Chemistry, vol. 2.L (Environmental Photochemistry). Springer, Berlin, pp.181–215

Chen L, Yang L, Wei W, Wang Z, Baoru M, Cai G (2013) Towards sustainable integrated watershed ecosystem management: a case study in Dingxi on the Loess Plateau, China. Environ Manage 51:126–137

Dalai TK, Rengarajan R, Patel PP (2004) Sediment geochemistry of the Yamuna River System in the Himalaya: implications to weathering and transport, Geochem J 38:441–453

Datta D, Subramanian V (1997) Nature of solute load in the rivers of Bengal Basin, Bangladesh. J Hydrol 198:196–208

Datta DK, Subramanian V (1998) Distribution and fractionation of heavy metals in the surface sediments of the Ganges-Brahmaputra-Meghna river system in the Bengal basin. Environ Geol 36:93–102

Eckhardt DAV, Stackelberg PE (1995) Relation of ground-water quality to land use on Long Island, New York. Groundwater 33:1019–1033

Gaillardet J, Dupré B, Allègre CJ (1995) A global geochemical mass budget applied to the Congo basin river: erosion rates and continental crust composition. Geochim Cosmochim Acta 59 (17):3469–3485

Gaillardet J, Dupré B, Allègre CJ, Négrel P (1997) Chemical and physical denudation in the Amazon river basin. Chem Geol 142:141–173

Garrels RM Mackenzie FT (1971) Gregor's denudation of the continents. Nature 231(5302):382–383

Garrels RM, Mackenzie FT, Hunt C (1975) Chemical cycle and the global environment. William Kaufmann, New York, p 260

Gaur VK, Gupta SK, Pandey SD, Gopal K, Misra V (2005) Distribution of heavy metals in sediment and water of river Gomti, Environ Monit Assess 102:1–3

Gibbs RJ (1970) Mechanisms controlling world water chemistry. Science 170:1088

Gupta LP, Subramanian V (1994) Environmental geochemistry of Gomti river: a tributary of the Ganges river. Environ Geol 24:235–243

Gupta LP, Subramanian V (1998) Geochemical factors controlling the chemical nature of water and sediments in the Gomti River, India. Environ Geol 30:1–21

Hartley AM, House WA, Callow ME, Leadbetter BSC (1997) Co-precipitation of phosphorus with calcite in the presence of photosynthesizing green algae. Water Res 31:2261–2268

House WA, Denison FH (1997) Nutrient dynamics in a lowland stream impacted by sewage effluent—Great Ouse, England. Sci Total Environ 205:25–49

IWS (1988) Water research management studies in Tamil Nadu Final Project Report, UND-CTD. Technical Report IWSIB5I01 Institute for water studies, Public Works Department, Government of Tamil Nadu, Chennai

Jain CK, Sharma MK (2002) Heavy metal transport in the Hindon river basin, India. Water Air Soil Pollut 137:1–19

James RA (2000) Environmental Biogeochemistry of Tamiraparani river basin, South India. Ph.D. Thesis submitted to Anna University, Chennai

Kieber RJ, Li A, Seaton PJ (1999) Production of nitrite from the photodegradation of dissolved organic matter in natural waters. Environ Sci Technol 33:993–998

Kumarasamy P, Govindaraj S, Vignesh S, Babu Rajendran R, James RA (2012) Anthropogenic nexus on organochlorine pesticide pollution: a case study with Tamiraparani river basin, South India. Environ Monit Assess 184:3861–3873

Kumarasamy P, Dahms HU, Jeon HJ, Rajendran A, James RA (2013) Irrigation water quality assessment: an example from the Tamiraparani river, Southern India. Arab J Geosci. doi:10.1007/s12517-013-1146-4

Kumarasamy P, James RA, Dahms HU, Buyon CW, Ramesh R (2014) Multivariate water quality assessment from the Tamiraparani river basin, Southern India. Environ Earth Sci. doi:10.1007/s12665-013-2644-0

Livingston DA (1963) Chemical composition of rivers and lakes date on geochemistry, U.S Geological survey, 440G, G1-G64, pp 223

Madavan N, Subramanian V (2001) Fluoride concentration in river waters of south Asia. Curr Sci 80:1312–1319

Magesh NS, Chandrasekar N (2012) GIS model-based morphometric evaluation of Tamiraparani sub basin, Tirunelveli district, Tamil Nadu, India. Arab J Geosci 7:131–141

McLay CDA, Dragten R, Sparling G, Selvarajah N (2001) Predicting groundwater nitrate concentrations in a region of mixed agricultural land use: a comparison of three approaches. Environ Pollut 115:191–204

Meybeck M (1979) Dissolved load of world rivers. Rev Geol Dyn Geogr phys 21:215–246

Meybeck M (1982) Carbon, nitrogen and phosphorus transport by world rivers. Am J Sci 282:401–450

Meybeck M (1987) Global chemical weathering of surficial rocks estimated from river dissolved load. Am J Sci 287:401–428

Mukherjee D, Chattopadhaya M, Lahiri SC (1993) Water quality of the River Ganga (The Ganges) and some of its physicochemical properties, Environmentalist13:199–210

Narayansawamy S, Purnalakshmi N (1967) Charnockiitc rocks of Tirunelveli district, Madras. Geol Soc India 8:38–50

Narayansawamy S, Purnalakshmi N (1993) Charnockiitc rocks of Tirunelveli district, Madras. In: Continental crust of South India, Memoir of Geological Society of India, vol 25. pp 135–153

Rajendran RB, Subramanian AN (1997) Pesticide residues in water from the river Kaveri, South India. Chem Ecol 13:223–236

Ramesh R, Purvaja R, Ramesh S, James RA (2002) Historical pollution trends in coastal environments of India. Environ Monit Assess 79:151–176

Ramesh R, Anbu M, (1996) Chemical methods for environmental analysis. Macmillan India Ltd, Chennai, pp 210

Ramesh A, Tanabe S, Iwata H, Tatsukawa R, Subramanian AN, Mohan D, Venugopalan VK (1990) Seasonal variation of persistent organochlorine insecticide residues in Vellar River waters Tamil Nadu, South India. Environ Pollut 67(4):289–301

Ravichandran S (2003) Hydrological influences on the water quality trends in Tamiraparani basin, South India. Environ Monit Assess 87:293–309

Ravichandran S, Ramanibai R, Pundarikanthan NV (1996) Ecoregions for describing water quality patterns in Tamiraparani basin, South India. J Hydrol 178:257–276

Ray SB, Mohanti M, Somayajulu BLK (1984) Suspended matter, major cations and dissolved silicon in the estuarine waters of the Mahanadi river, India. J Hydrol 69:183–196

Raymahashay BC (1970) Characteristic of stream erosion in the Himalayan region of India. Hydrogeochemistry Biogeochemistry 1:81–89

Raymahashay BC (1986) Geochemistry of bicarbonate in river water. J Geol Soc India 27:114–118

Raymahasay BC (1987) A Comparative study of clay minerals for pollution control. J Geol Soc India 30:408–413

Riedel GF, Tvvilliams SA, Riedel GS, Oilmour CC, Sanders JG (2000) Temporal and spatial patterns of trace elements in the Patuxent river: a whole watershed approach. Estuaries 23:521–535

Sarin MM, Krishnaswamy S, Dilli K, Somayajulu BLK. Moore WS. (1989) Major ion chemistry of the Ganges-Brahmaputra river system, weathering process, fluxes to the Bay of Bengal. Geochim Et Cosmochim Acta 53:997-1009

Saxena DP, Joos P, Grieken V, Subramanian V (2001) Sedimentation rate of the floodplain sediments of the Yamuna river basin (tributary of the river Ganges, India) by using 210Pb and 137Cs techniques. J Radio Anal Nucl Chem 251:399–408

Singh KP, Malik A, Sinha S, Singh VK, Murthy RC (2005) Estimation of source of heavy metal contamination in sediments of Gomti river (India) using principal component analysis. Water Air Soil Pollut 166:321–341

Singh AP, Ghosh SK, Sharma P (2007) Water quality management of a stretch of river Yamuna: an interactive fuzzy multiobjective approach. Water Resour Manage 21:515–532

Solaraj S, Dhanakumar S, Murthy KR, Mohanraj R (2010) Water quality in select regions of Cauvery Delta River basin, southern India, with emphasis on monsoonal variation. Environ Monit Assess 166(1–4):435–444

Stallard RF, Edmond LM (1983) Geochemistry of the Amazon, 2: the influence of geology and weathering environment on the dissolved load. J Geophys Res 88:9671–9688

Subramanian V (1983) Factors controlling the chemical composition of river waters of India. In: Proceedings of the Hamburg Symposium, vol 141, pp 145–151

Subramanian V (1984) River transport of phosphorus and genesis of ancient phosphorites. Spec Publ Geol Surv India 17:11–15

Subramanian V (1987) Environmental geochemistry of Indian river basins-a review. J Geol Soc India 29:205–220

Subramanian V (1993) Sediments load of the Indian rivers. Curr Sci 64:928–930

Subramanian V (2000) Transfer of phosphorus from the Indian sub-continent to the adjacent oceans. Soc Sediment Geol SEPM Spec Publ 66:77–88

Subramanian V (2004) Water quality in South Asia. Asian journal of water. Environ Pollut 1(1 and 2): 41–54

U.S. Environmental Protection Agency (1999) Draft guidance for water quality-based decisions: the TMDL process. D.C.7 U.S. Environmental Protection Agency, Washington

U.S. Environmental Protection Agency (2001) Protocol for developing1 pathogen TMDLs EPA 841-R-00-002. Office of water (4503F). DC7 United States Environmental Protection Agency, Washington

Vaithiyanathan P, Subramanaian V, Ramanathan AL (1989) Transport an distribution of phosphorus by the rivers of India. Geol Soc India Mem 13:127–137

Zhang J (1995) Geochemistry of trace metals from Chinese river/estuary systems: an overview. Estuar Coast Shelf Sci 41:631–658

Zhang J, Huang WW, Li L, Zhou Q (1990) Drainage basin weathering and major element transport in two large Chinese rivers, (Huanghe and Chanjiang). J Geophysic Res 95:13277–13288

Changes in Water Quality Characteristics and Pollutant Sources Along a Major River Basin in Canada

Jianxun He, M. Cathryn Ryan and Caterina Valeo

Abstract Temporal and spatial variations of water quality along the Bow River (Alberta, Canada) were investigated using monthly water quality data (chloride, sulphate, nitrate, sodium, and conductivity) collected from 2004 to 2011. Non-point and point (notably three wastewater treatment plants) pollutant loads were characterized along the river. The river was divided into three reaches, namely, the Upper river reach, the Calgary reach, and the Downstream river reach, based on the distribution of point pollutant sources and geographic conditions. A mass balance approach and statistical analyses were employed to analyze water quality. The results demonstrated that the point sources, Calgary's three wastewater treatment plants (WWTPs), are largely responsible for the observed spatial and temporal trends in the investigated quality parameters. However, the contribution of non-point sources appears to vary along the river, which might be related to the flow pathways taken by non-point pollutants discharging into the river and the geochemical characteristics of the groundwater within the alluvial aquifer that is hydraulically connected to the river. Apart from the identified point and non-point sources, the effects of other processes such as biological reactions need to be further ascertained and quantified for a better assessment of pollutant loads, in particular nutrients. Further understanding of these issues will allow a more accurate quantification of pollutant loads and consequently, better knowledge for formulating reliable water quality management strategies.

J. He
Civil Engineering, University of Calgary, 2500 University Dr. NW, Calgary, AB T2N 1N4, Canada

M.C. Ryan
Geoscience, University of Calgary, 2500 University Dr. NW, Calgary, AB T2N 1N4, Canada

C. Valeo (✉)
Mechanical Engineering, University of Victoria, 3055 Stn CSC, EOW Bldg, Victoria, BC V8W 3P6, Canada
e-mail: valeo@uvic.ca

Mu. Ramkumar et al. (eds.), *Environmental Management of River Basin Ecosystems*, Springer Earth System Sciences, DOI 10.1007/978-3-319-13425-3_25

Keywords River water quality variations · Surface non-point pollutants · Subsurface pollutants · Groundwater pollutant loads · Anthropogenic impacts · Wastewater treatment impacts · Alluvial aquifer

1 Introduction

Both natural factors (including geological factors) and anthropogenic factors can contribute to changes in the water quality of natural aquatic environments. In the last several decades, anthropogenic activities, which generate human sewage, industrial effluents, agricultural nutrients, pesticides and other pollutants, have been continually blamed for degrading water quality in natural water systems. For example, urbanization has been identified to be the cause of increases in nutrient inputs and consequent degradation of water quality in receiving water bodies (Busse et al. 2006).

To understand water quality in aquatic environments, monitoring systems have been developed to collect water quality data *in a consistent manner*. For example, in Canada on some major rivers, long term water quality monitoring programs provide adequate information leading to better knowledge of river hydrochemistry and pollution levels for management purposes. But it also creates challenges for interpreting that water quality data for insight into ecosystem processes that are functions of many variables. Conventionally, exploratory data analyses such as box plots, and univariate analyses such as analysis of variance, have been employed to examine water quality variations in hopes of understanding ecosystem processes. However, such analyses have been considered inadequate to statistically characterize water quality data, which normally consist of more than one variable. A large number of studies (e.g., Alberto et al. 2001; Singh et al. 2004; Shrestha and Kazama 2007) have been conducted to evaluate spatial and temporal variations in water quality of rivers using multivariate statistical techniques including factor analysis, principle component analysis, discriminant analysis, and cluster analysis. These works have focused on identifying spatial and temporal differences in water quality and factors significantly affecting water quality. However, the conventional methods have continued to be employed in water quality characterization and assessment.

It has been well acknowledged that surface water quality responds to population growth and changes in land use at a variety of spatial and temporal scales (Rhodes et al. 2001). These changes lead to the deterioration of water quality in rivers due to increases in both point and non-point pollution sources. For water quality management that aims to reduce pollutant loads to receiving water bodies and consequently, to protect the health of river ecosystems, the challenge of identifying pollutant sources and quantifying pollutant loads is an initial but crucial step. Pollutant loading can originate from both point sources and non-point sources, and the latter can exist ubiquitously on the land surface and underground. Thus, they are not always easily quantified nor managed due to this characteristic. Point sources, such as wastewater treatment plants (WWTPs) effluent, have often been identified as one of the major causes of water quality degradation in aquatic environments across Canada (Environment Canada 2001). On the other hand, non-point sources, such as surface stormwater runoff and discharge from groundwater, could significantly affect water quality in receiving water bodies. Urban stormwater runoff is a primary transport route of non-point pollutants to receiving water bodies (He et al. 2010); however, stormwater runoff could flow through routes other than over the land surface prior to discharging into the receiving body. Precipitation falling within a catchment could discharge into surface water bodies after flowing through the subsurface (Grasby et al. 1999) and therefore, transport contaminants along the groundwater flow pathway of the alluvial aquifer (Winter et al. 1998). Thus, groundwater pollutant load contribution, which has often been ignored in water management considerations due to the difficulty in observing and measuring this source, could be a significant concern in surface water contamination (Winter et al. 1998). In recognition of the complexity in quantifying pollutant loads, in particular non-point pollutant loads, various water quality models, from simple/empirical models to process based models have been employed. However, the model complexity and incomplete understanding of water quality limits the application of process based models in practice (Obropta and Kardos 2007). The export coefficient model, which calculates the pollutant loads using the estimated export rate from catchments given the characteristics of land use type, has also been applied to estimate non-point pollutant loads at the catchment scale (Malver et al. 2012; Robinson and Melack 2013). McLeod et al. (2006) quantified pollutant loads from a city according to water quality characteristics and the relationship between rainfall and runoff pollutant loads. However, to implement these methods for quantifying pollutant loads, export coefficients or water quality characteristics need to be estimated using large data sets.

The Bow River begins in the Rocky Mountains and flows through Alberta, Canada. It is one of major tributaries of the South Saskatchewan River. Calgary (approximately one million inhabitants) is the largest urban centre along the river. The water quality of the Bow River has been intensely affected by Calgary's WWTPs, which has been identified to significantly contribute to pollutant load and alter the geochemistry of the river downstream of the city (Grasby et al. 1999; Iwanyshyn et al. 2008). In contrast, three Bow River tributaries contribute smaller mass loads to the Bow River in Calgary (Iwanyshyn et al. 2008). In the Bow River watershed, flow in the river, particularly close to the headwaters, is primarily generated through the replacement of groundwater via recharging rainwater and meltwater into the alluvial aquifer; rather than from surface runoff, even during high flow events (Grasby et al. 1999; Katvala 2008). Most recently, Cantafio and Ryan (2014) demonstrated through a field investigation in short river reaches (3–5 km long) within Calgary in 2010–2011 that the mass flux of chloride loading into the Bow River from groundwater is significant, although the volumetric discharge contribution is negligible. Therefore, the role of surface and subsurface non-point sources cannot be ignored when quantifying and characterizing pollutant loads into the Bow River. These sources needs to be further investigated in order to develop an appropriate surface water quality management framework for this watershed.

It has been a challenge to reduce pollutant loads into the Bow River, which supports a blue ribbon fishery and provides water supply to over half of Calgary's population and many downstream communities-all of which require high water quality. In past decades, the City of Calgary has made tremendous efforts to reduce the pollutant loads by enhancing wastewater treatment at its plants with the aim of improving water quality levels in the Bow River. To better manage water quality in the river, it is necessary to understand the roles of different pollutant sources and the possibility of control. As aforementioned, non-point pollution cannot be ignored and needs to be well understood. In addition, water quality characterization needs to be conducted given the site-specific observations, since water quality is affected by various factors including geographic, climatological, and hydrologic conditions. To the authors' best knowledge, little to no research has been conducted to characterize spatial and temporal variations of water quality along the Bow River, nor to accurately quantify non-point pollutant loads discharged into the river. Thus, the twofold objectives of this research are to understand: (a) the spatial and temporal variations of water quality and the dependence between quality parameters using statistical analyses; and (b) the pollutant load from non-point sources along the Bow River using water quality observations collected from the long term monitoring stations on the river. Improved understanding of these issues, particularly the pollutant contribution from non-point sources would better inform water quality management decisions.

2 Study Area

The Bow River originates in the eastern slopes of the Rocky Mountains in Banff National Park in Alberta, Canada and flows eastward through the foothills. The Bow River has a total length of 587 km flowing through three major physiographic regions: the mountains, the foothills, and the plains. The drainage area is approximately 25,000 km^2. Average annual discharge is about 90 m^3/s. Following spring freshet, the discharge characteristically peaks in mid-June due to rainfall and mountain snowmelt. The flows then decrease slowly from July until October, when baseflow conditions are usually reached (Alberta Environment River Basins 2010). In addition to several flow regulated hydraulic structures (e.g., hydroelectric reservoirs), there are flow diversions for irrigation and municipal WWTPs withdrawing from/discharging to, the river.

Four small communities (population <15,000, but with significant summer tourism) discharge treated wastewater into the Bow River in the Upper river reach. Three wastewater treatment plants serve Calgary and environs. There are no other licensed wastewater treatment plants for large communities (>15,000) in the reach studied here.

The *Upper river reach* from the headwater to the upstream limit of Calgary flows along roughly 200 km through largely undeveloped and low intensity agricultural land before entering the City of Calgary. The water quality in this river reach is generally considered to be near pristine. In the river reach within Calgary (henceforth referred to as the *Calgary reach*), there are three WWTPs releasing effluents and three tributaries (Nose Creek, Elbow River, and Fish Creek) flowing into the river. The land uses of the watershed in the *Downstream river reach* from the Calgary's downstream limit to the confluence with the South Saskatchewan River are predominantly agricultural. There are also a number of irrigation diversions and irrigation return flows within this reach.

3 Materials and Methods

3.1 Water Quality Parameters

Several water quality parameters including chloride (Cl), sulphate (SO_4), sodium (Na), nitrate (NO_3), and electrical conductivity were selected for the analysis. Cl is a conservative solute and affected little by chemical and biological reactions in natural rivers (Vandenberg et al. 2005). Cl has been frequently used to aid in the investigation of pollutant loading, since it usually exists in wastewater at high concentrations while occurring naturally at low concentrations. Iwanyshyn et al. (2008) employed Cl mass balance to separate the physical loadings and loadings due to photosynthesis/respiration processes in the Bow River within the City of Calgary. Jasechko et al. (2012) used Cl mass balance to derive saline groundwater seepage to surface waters in the Athabasca oil sands region.

A regional trend of SO_4 in groundwater showing higher concentration at those areas east of the Laurentide glacial limit in Alberta was identified (Alberta Environment 2004; Grasby et al. 2010). Furthermore a strong linear relationship between SO_4 and Na concentrations was detected in Paskapoo groundwater (Grasby et al. 2010). It is speculated that non-point pollutants from groundwater in the alluvial aquifer may largely contribute to pollutant loads into the Bow River. Thus, SO_4 and Na were also selected for this study.

In addition, NO_3 and electrical conductivity were included in the analysis. Nutrient loads have always been of interest because of the significant effect of nutrients on natural aquatic environments, such as eutrophication in freshwater bodies. Electrical conductivity is an indirect and inexpensive measure of dissolved ions.

3.2 Hydrologic, Meteorological and Water Quality Data

Alberta Environment runs five long-term water quality monitoring stations on the Bow River. In addition, the City of Calgary has several regular monitoring stations within the City limits on the river. The locations of the selected long-term monitoring stations and several monitoring stations within Calgary for this study are shown in Fig. 1. The effluent from the Bonnybrook WWTP, the largest point-source contributor on the Bow River, is fully mixed within 12–14 km downstream before

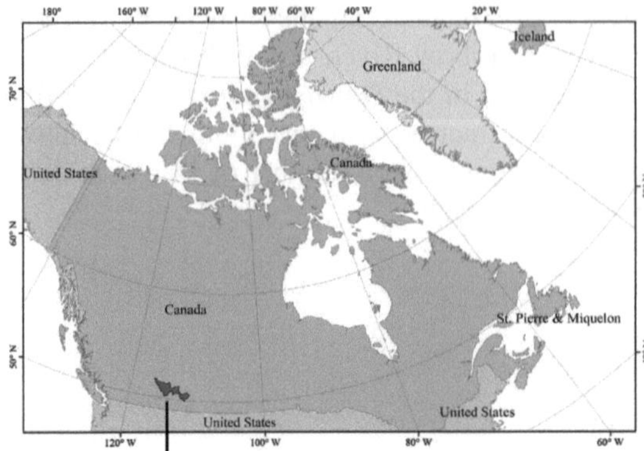

Station	M_LAT	M_LONG
Above Canmore	51.123	-115.387
Cochrane	51.174	-114.467
Below Bearspaw	51.101	-114.280
Cushing Bridge	51.038	-114.013
Policeman's Flats	50.848	-113.944
Above Highwood	50.812	-113.737
Below Carseland	50.831	-113.417
Near Ronalane	50.046	-111.591
Bonnybrook WWTP	51.045	-114.010
Fish Creek WWTP	50.992	-113.989
Pine Creek WWTP	50.899	-113.993

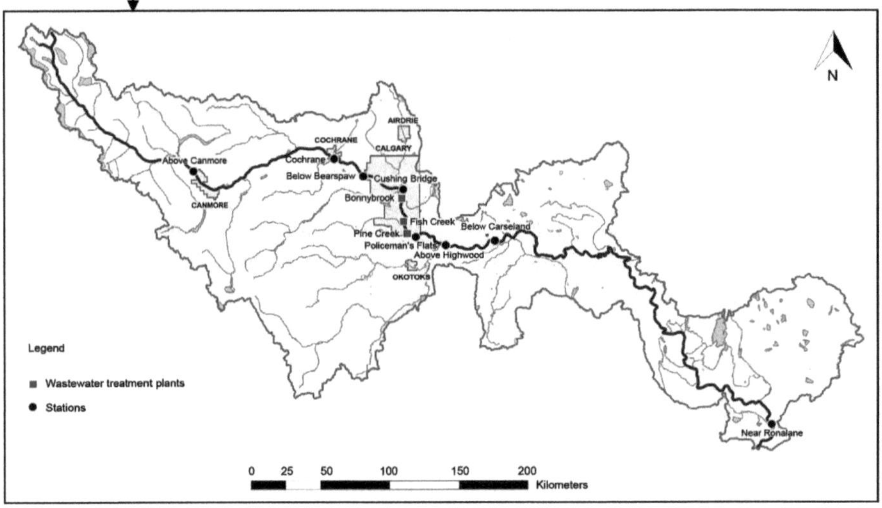

Fig. 1 Map showing the Bow River basin, major population centres, Calgary's wastewater treatment plants, river gauging and water quality monitoring locations. The mainstem of the Bow River is indicated by the *dark blue line*

Policeman's Flats (Vandenberg et al. 2005), which is also close to the City's downstream limit. Monthly water quality data has been collected at these selected water quality monitoring stations by Alberta Environment and the City of Calgary. The water quality data collected from WWTP effluents, generally on a weekly basis, were obtained from the City of Calgary. Daily hydrometric (flow) data were collected by Water Survey of Canada (WSC) at several stations along the Bow River. Annual meteorological data were obtained from Environment Canada collected at the Calgary International Airport. All water quality, hydrometric and meteorological data collected in the time period from 2004 to 2011 were used in this work. There are longer data available at the long-term monitoring stations of Alberta Environment; while the city's data on the river are available from 2004 to 2011.

3.3 Study Methodology

Water quality data are usually not normally distributed. In order to relax the restriction of a normal distribution in parametric statistical analysis, non-parametric approaches were selected, in general, for this study. Seasonality was demonstrated in all investigated water quality parameters, thus, the nonparametric seasonal Mann-Kendall (MK) test was employed to detect persistent temporal changes in the water quality parameters. The developed analysis tool for MK analysis by the USGS (2005) was applied. The non-parametric rank sum test is an approach for identifying whether one of two independent data sets tends to be larger than the other. This test was used to compare water quality levels at one station with its next downstream station. In addition, the analysis of covariance (ANCOVA) was adopted to investigate the relationships between Cl and other water quality parameters. The multiple comparisons of regression slopes obtained from ANCOVA was also conducted at all monitoring stations. ANCOVA was applied to the ranks of data instead of the original data sets (Conover and Iman 1981; Quade 1967). In other words, this approach was applied to identify the differences in regression slopes between Cl and other water quality parameters along the river. If there is no obvious difference in the regression slopes between the upstream and downstream sites, it might suggest that pollutants are likely to have originated from the same sources in the river reach; otherwise, they may have originated from different sources or affected by different factors. All the statistical analyses except the MK test were conducted using Matlab.

This paper also employed a mass balance approach to quantify the contribution of pollutant loads from non-point sources plus unknown sources given that point-source contributions are known or insignificant on an annual basis. For non-conservative constituents (e.g., nutrients), the changes in pollutant mass from upstream to downstream in a reach might result from chemical and biological processes along with potential non-point sources contribution. In this paper, pollutant load in a time interval (here approximately a month) was quantified by taking the average of pollutant loading rates, which are a product of concentration and flow rate on sampling days, of the two next samples and then multiplying by the time interval. The annual pollutant load is then estimated by summing the monthly load over a year.

4 Results

4.1 Hydrometric Variation

Before conducting the analysis on the water quality parameters, variations in flow were examined in order to determine low flow versus high flow periods along the river. The results for daily flow from analysis of variance (ANOVA) and multiple comparisons at three hydrometric stations, Banff (unregulated), Below Bearspaw in

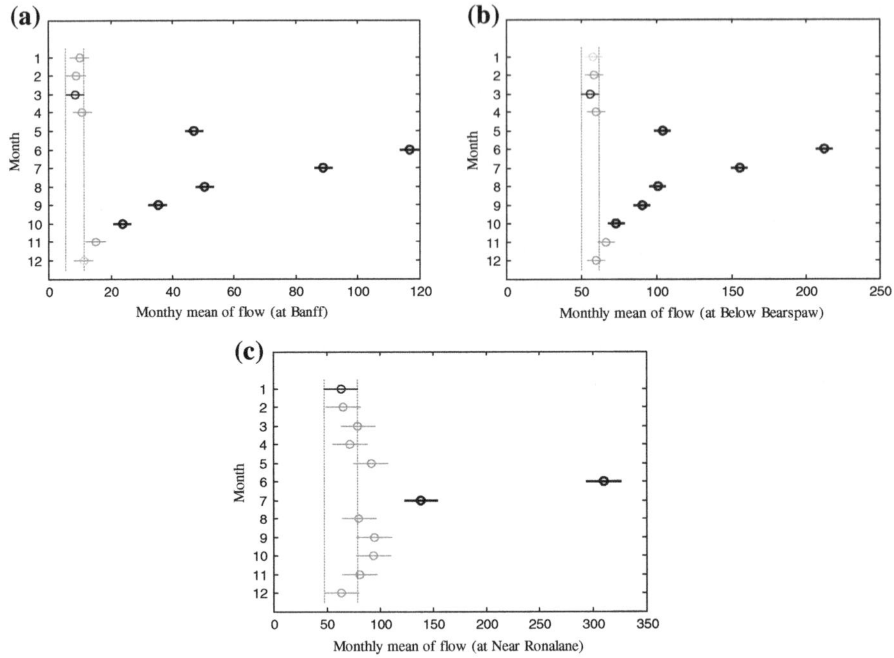

Fig. 2 Comparison of monthly flow (ANOVA followed by multiple comparison) at three sites. The *dashed lines* indicate the range (from the minimum to the maximum of daily flow) of the selected month (in *black*) that is of the lowest minimum daily flow in 12 months. The *thicker black lines* show that the months are significantly different from the selected months; while the *grey lines* indicate that the months are not significantly different from the selected months. **a** At Banff (unregulated). **b** At Below Bearspaw (regulated). **c** At Near Ronalane (regulated)

Calgary (regulated), and Near Ronalane closed to the confluence with the South Saskatchewan River (regulated), are presented in Fig. 2. Flow on the river shows variation from upstream to downstream. Similar annual variation patterns of flow were found from the Upper river reach to Calgary; however different annual flow patterns were observed at the most downstream hydrometric station. In the river reaches above and within Calgary, there are no significant differences in the monthly mean flows from November to next April; while mean flows in other months are significantly higher than mean flows from November to next April. Below Calgary at Near Ronalane, monthly mean flows in May and from August to October are not significantly higher than those in low flow period compared to two upstream stations. In general, the months from November to next April are classified as the low/base flow period; and the months from May to October are in high flow period.

4.2 Temporal and Spatial Variation in Water Quality Parameters

Seasonality was observed in all these water quality parameters (results not shown) discussed in this paper; however, the seasonal division is usually arbitrary such as using four seasons (winter, spring, summer, and autumn), or two seasons (ice-covered period and open period), or 12 months. In this work, water quality data were grouped in two different ways for seasonal MK analysis: (a) two seasons (from November to the next April and from May to October) based on results from Sect. 3.1; and (b) 12 seasons (each month). Seasonal MK analysis was applied to monthly grouped data and two-season grouped data of Cl. The results from these two ways of data grouping were not obviously different; thus, the MK test was applied to monthly group data for all the parameters over the time period from 2004 to 2011. The calculated Kendall τ correlation coefficients are represented in the Table 1. Overall, in the Downstream river reach from Policeman's Flats (downstream of Calgary's WWTPs) to the most downstream station Near Ronalane, significantly positive temporal trends were detected, except in NO_3 between Above Highwood and Below Carseland. Significant temporal trends were not found in the river reach upstream of Calgary's WWTPs (above Cushing Bridge), except in SO_4, at all stations upstream of WWTPs, and in Cl and conductivity at Above Canmore, the most upstream monitoring station.

Box-Whisker plots of concentrations of all the water quality parameters along the river are presented in the Fig. 3. It follows from the figure that all the water quality parameters are at their elevated levels in the region stretching from Cushing Bridge (just upstream of Calgary's WWTPs) or Policeman's Flats (after Calgary's WWTPs). Table 2 presents the results of the rank sum test for comparing all stations to their next downstream station. As shown by these results, significant positive spatial trends were confirmed in the river reach within Calgary, including the river reach from below Bearspaw to Cushing Bridge within Calgary where it is not subjected to the influence of Calgary's WWTPs. In the Downstream river reach, no

Table 1 Kendall tau correlation coefficients based on monthly grouped data

River reach	Station	Cl	SO_4	NO_3	Na	Conductivity
Upper river reach	Above Canmore	**−0.41**	**0.47**	NaN*	−0.17	**0.35**
	Cochrane	0	**0.51**	−0.02	−0.04	0.17
Calgary reach	Below Bearspaw	−0.01	**0.43**	0.01	−0.12	0.06
	Cushing Bridge	0.08	**0.44**	0.04	0.04	0.06
	Policeman's Flats	**0.33**	**0.36**	**0.18**	0.08	0.08
Downstream river reach	Above Highwood	**0.38**	**0.46**	0.07	**0.29**	**0.33**
	Below Carseland	**0.39**	**0.27**	0.05	**0.19**	**0.29**
	Near Ronalane	**0.42**	**0.37**	**0.26**	**0.26**	**0.24**

bold font numbers indicate significant trends at significance level of 10 %

* *NaN* Not a number

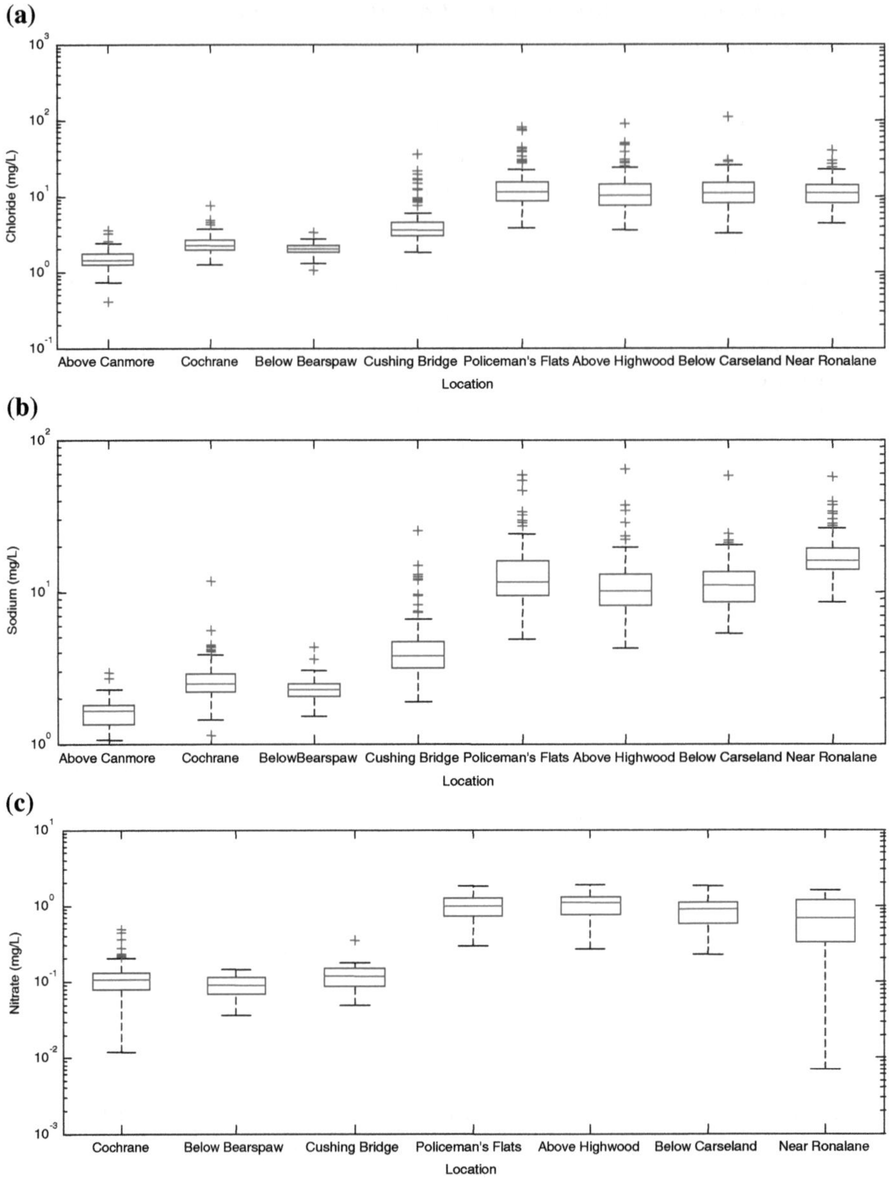

Fig. 3 Box-Whisker plots of water quality parameters along the Bow River for **a** chloride; **b** sodium; **c** nitrate; **d** sulphate; and **e** conductivity. Outliers are indicated by "+" (defined as data points that are larger than $q_3 + 1.5(q_3 - q_1)$ or smaller than $q_1 - 1.5(q_3 - q_1)$, where q_1 and q_3 are the 25th and 75th percentiles.)

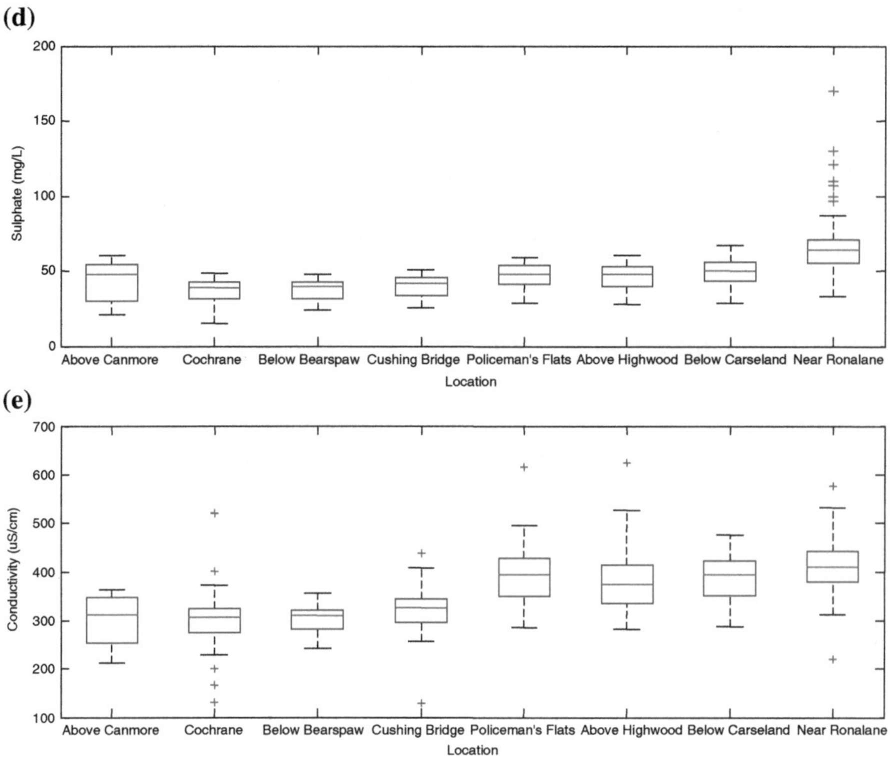

Fig. 3 (continued)

Table 2 The results of non-parametric rank sum test for significant parameter concentration changes in upstream versus downstream stations in each reach

River reach	Stations	Cl	SO_4	NO_3	Na	Conductivity
Upper river reach	Above Canmore versus Cochrane	↑	↑	–	↑	→
	Cochrane versus Below Bearspaw	↓	→	→	↓	→
Calgary reach	Below Bearspaw versus Cushing Bridge	↑	↑	↑	↑	↑
	Cushing Bridge versus Policeman's Flats	↑	↑	↑	↑	↑
Downstream river reach	Policeman's Flats versus Above Highwood	→	→	→	↓	→
	Above Highwood versus Below Carseland	→	↑	↓	→	→
	Below Carseland versus Near Ronalane	→	↑	↓	↑	↑

↑: significant increase; ↓: significant decrease; → : no significant change significant trends were identified in the water quality parameters except for Na between Policeman's Flats and Above Highwood; while significant spatial trends were found in NO_3 and SO_4 after Above Highwood and in Na and conductivity between Below Carseland and Near Ronalane. The spatial trends in the Upper river reach and the Downstream river reach appear to be variable for different water quality parameters. In particular, overall negative spatial trends of NO_3 were found after Policeman's Flats. The absence of trend in NO_3 between Policeman's Flats and Above Highwood might be due to their proximity in distance.

4.3 Relationships Between Chloride and Other Parameters

The relationships between Cl and other water quality parameters, which may imply whether these pollutants originate from the same sources as Cl, were investigated. The regression slopes between Cl and other water quality parameters at all the locations were derived using non-parametric ANCOVA, and then the regression slopes were compared non-parametrically by using a multiple comparison test. In the analysis, Cl was the independent variable; the other parameters were dependent variables. The results are presented in the Fig. 4a–d. These figures depict some results obtained from multiple comparison tests as examples. The location with the lowest maximum parameter value was selected for comparison to results from other

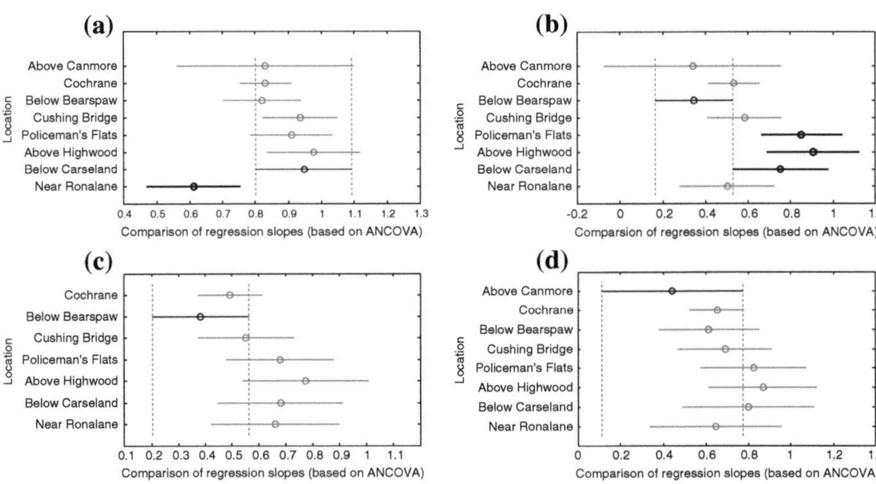

Fig. 4 Comparison of regression slopes between chloride and other water quality parameters. *Black lines* denote the selected locations of lowest maximum values; the *dashed lines* show the ranges of the selected locations; the *thicker black lines* indicate that the locations are significantly different from the selected locations; while *grey lines* show that the results at the locations are not significantly different from the selected locations. **a** Chloride and sodium. **b** Chloride and sulphate. **c** Chloride and nitrate. **d** Chloride and conductivity

locations in each figure. As shown in the Fig. 4a, the associations between Cl and Na at the most downstream location appear to largely deviate from their relationships at the other seven sites along the river. Significant differences between Near Ronalane and its next four upstream sites were found. Their association at sites within the City and Below Carseland appear to be different from the sites from Above Canmore to Below Bearspaw, although the differences are not significant. The regression slopes of SO_4 and Cl from Policeman's Flats (downstream of WWTPs) to Below Carseland are relatively larger than those at all other sites (Fig. 4b); while significant differences between these three sites and one out of five other sites at Below Bearspaw were found. Between NO_3 and Cl, the regression slope in the Upper river reach is evident apart from all other sites, although the differences are not significant (Fig. 4c). Insignificant differences in regression slopes of Cl and conductivity were found (Fig. 4d); while the pattern of variation in the slopes appears to be similar to that of NO_3 and Cl. Overall, relatively greater slopes of Cl and these investigated quality parameters are presented downstream of Below Bearspaw after the river enters Calgary.

4.4 Annual Pollutant Load

Pollutant loads into the Bow River were quantified and loading rates from non-point sources were estimated for each river reach: the Upper river reach, the Calgary reach, and the Downstream river reach. The Calgary reach was further divided into two sub-reaches: *Calgary upstream* and *Calgary downstream*. The Calgary upstream reach is from Below Bearspaw to Cushing Bridge (upstream of Calgary's WWTPs) where it is not subjected to effects from identified point source pollution; while the Calgary downstream reach is from Cushing Bridge to Policeman's Flats, and receives effluent from three Calgary's WWTPs. In the Calgary upstream reach, the contribution of pollutant loads is attributed to non-point sources due to the absence of identified point sources of pollution. Both point and non-point sources contribute to pollutant loads into the Calgary downstream reach.

4.4.1 Upper River Reach

In the Upper river reach where no identified point sources exist, annual pollutant loads of Cl, SO_4, and Na were assessed between Above Canmore and Below Bearspaw. NO_3 data are not available at Above Canmore. The normalized results are shown in the Fig. 5. For each pollutant, the normalized data were calculated by dividing each annual pollutant load by the root of the sum of the squares of the loads. Similar patterns in the annual loads can be seen. The annual loads for all the pollutants peaked in 2005, which was a high flow year. Precipitation and rainfall data obtained from the weather station at Calgary International Airport were used in this analysis. The meteorological data at Banff located within the watershed of this

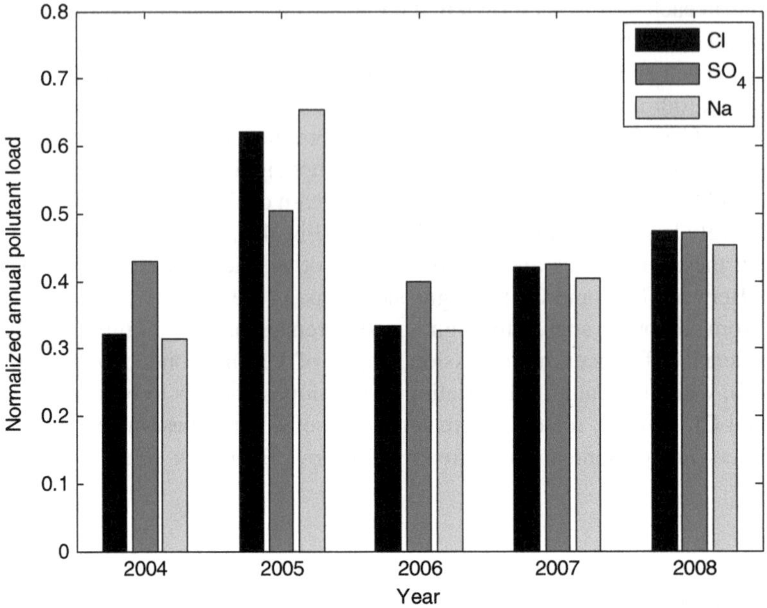

Fig. 5 Normalized annual pollutant loads in the Upper river reach

river reach are not available for the time period from 2007 to 2009. Precipitation usually varies over space and time as a result of topography. In this watershed, significant correlations in annual rainfall and precipitation were identified between Banff and the Calgary International Airport from 1997 to 2006. Meteorological data collected at the Calgary International Airport were thus, used for this river reach and all other river reaches herein. Correlations between all the annual pollutant loads and meteorological variables were attempted for the study time period. The correlation coefficients between precipitation and annual pollutant loads (Cl, SO_4, and Na) are 0.89, 0.71, and 0.85, respectively; while the correlation coefficients with rainfall are 0.94, 0.72, and 0.94, respectively. The estimated annual average Cl loading rates, which are calculated by dividing annual Cl load by the time interval (a year), for these years were in the range of 106–206 g/s.

4.4.2 Calgary Reach

In the Calgary reach, annual pollutant loads were quantified for the Calgary upstream and Calgary downstream reaches and normalized annual pollutant loads are presented in the Figs. 6 and 7, respectively. The Calgary downstream reach is heavily affected by the WWTPs' effluent, and thus the pollutant loads in this reach are contributed from both point and non-point sources. Pollutant loads from point sources including effluents from three WWTPs (Bonnybrook, Fish Creek and Pine Creek) were estimated. The annual pollutant loads from three WWTPs combined

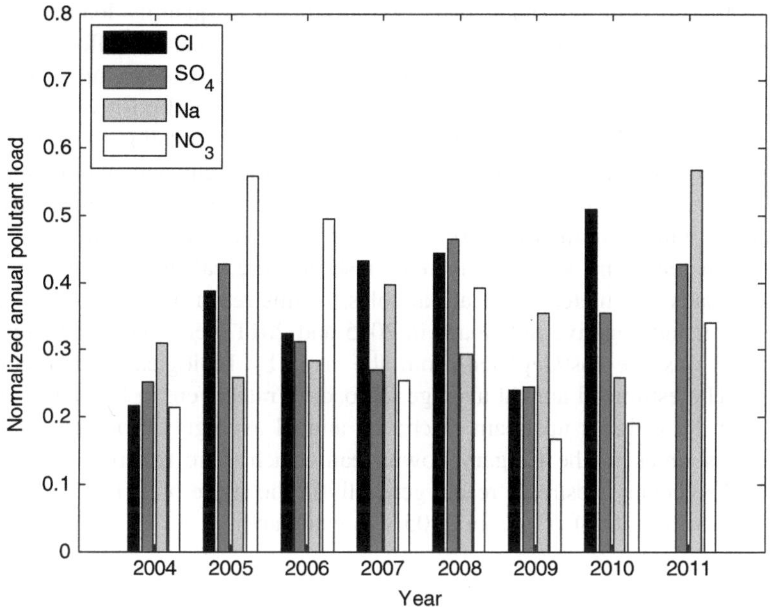

Fig. 6 Normalized annual pollutant loads in the Calgary upstream reach

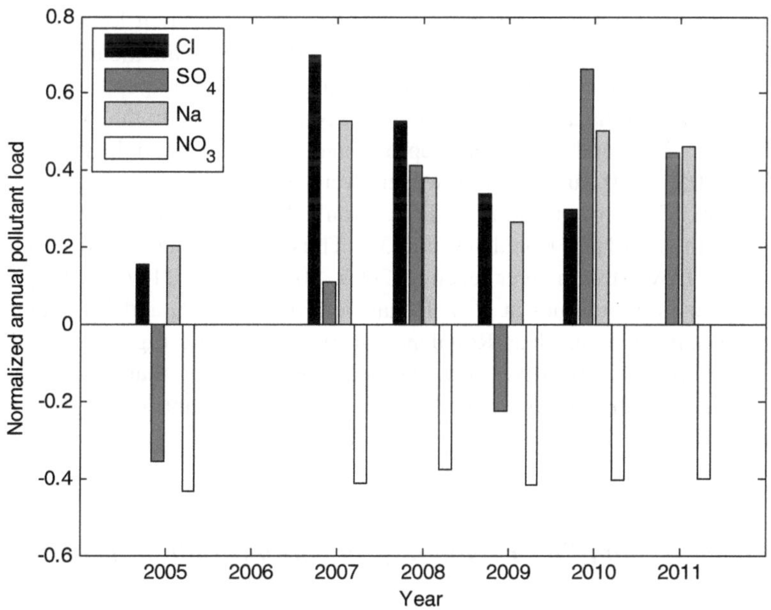

Fig. 7 Normalized annual pollutant loads in the Calgary downstream reach

were subtracted from pollutant loads into this reach to quantify loads from non-point sources. Various patterns of annual pollutant loads that are different from the Upper river reach are demonstrated in these figures. In the Calgary upstream reach, annual pollutant loads of Cl, SO_4, and NO_3 are strongly and positively associated with meteorological variables; while weak and negative correlations were computed for Na. The correlation coefficients between precipitation and annual pollutant loads (Cl, SO_4, and NO_3) are 0.74, 0.76, and 0.56, respectively; while the correlation coefficients with rainfall are 0.60, 0.69, and 0.73, respectively. In the Calgary downstream reach, no such strong relationships were calculated between all the pollutant loads and meteorological variables. In this reach, negative NO_3 loads in all the years and negative SO_4 loads in 2005 and 2011 were calculated suggesting that these ions are lost/depleted from the river by biological and/or chemical reaction. The estimated annual average Cl loading rates were in the range of 109–225 g/s in the Calgary upstream reach; the annual average Cl loading rates from non-point sources in the Calgary downstream reach were relatively higher than those in the Calgary upstream reach, generally in the range of 141–298 g/s, except that low loading rate of 67 g/s in 2005 was estimated.

4.4.3 Downstream River Reach

Annual pollutant loads in this reach were estimated between Policeman's Flats and Near Ronalane and normalized results are presented in the Fig. 8. Positive pollutant loads of SO_4 and Na were assessed in all the years. Negative loads of Cl were calculated except for the load in 2005 and for 2010–2011; while the annual Cl loads calculated in the Upper river reach and the Calgary reach are positive in all these years. The ratios of the annual flow volume at Near Ronalane to that at Policeman's Flats between 2006 and 2008 were approximately equal to or slightly over 1.0; the ratios in 2005, 2010, and 2011 however, were above 1.0 (1.38, 1.19, and 1.21, respectively). Cl concentrations at Policeman's Flat were equivalent or slightly higher than those at Near Ronalane (Fig. 3a). Thus, the estimated positive annual Cl loads in the Downstream river reach in 2005, 2010, and 2011 might be primarily due to higher flow volumes at Near Ronalane. For both SO_4 and Na, significantly higher concentrations at Near Ronalane can be seen in the Fig. 3b, d and in the Table 2. Thus, their positive pollutant loads were lower than those expected. Moderate and positive correlations between annual pollutant loads (SO_4 and Na) and meteorological variables were calculated.

4.5 Effect of Sampling Frequency on Annual Pollutant Loads

Investigations into the effect of sampling schemes on the estimation of pollutant loads is constrained by data availability, and thus, the authors conducted this investigation only for continuously monitored conductivity data (at 15 min intervals) and monthly

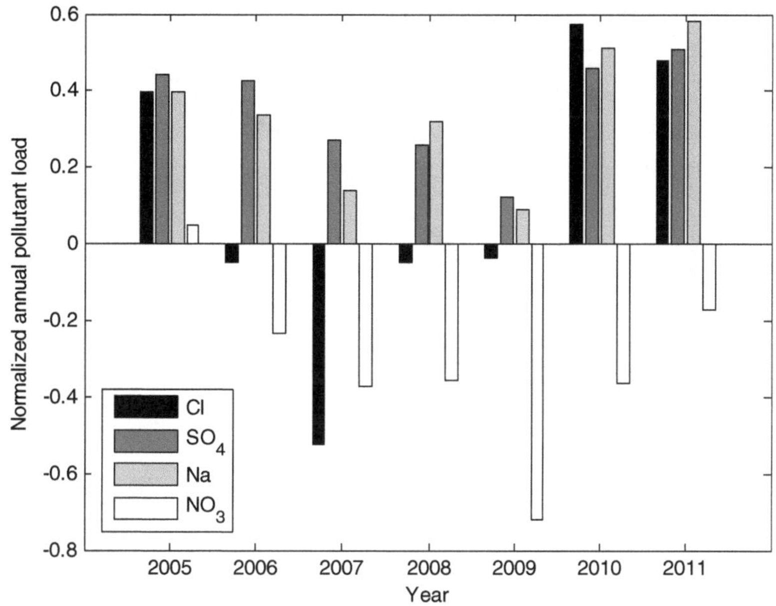

Fig. 8 Normalized annual pollutant loads in the Downstream river reach

observed data at Below Bearspaw from May 1 to December 31 in 2008 (collected by the City of Calgary). Here conductivity was treated the same as pollutants that are usually measured in units of mg/L; and thus, conductivity load was calculated in the same way as the other pollutants. Normalized cumulative equivalent load of conductivity versus number of days are plotted in the Fig. 9.

5 Discussion

5.1 Point Source Contributions

In the Upper river reach, there are several point sources (WWTPs) from small communities such as Canmore, Deadman's Flats, and Exshaw. But the water quality in this reach is generally considered the be fairly good and thus, the point source contributions from these communities is considered to be insignificant. In the Downstream river reach, there is one WWTP effluent from the Town of Strathmore, and a number of irrigation diversions and return flows. In addition, a tributary—the Highwood River, flows into the Bow River. Annual flow volumes in the Highwood River were in the range of 16–27 % of annual flow volumes in the main stream. The annual pollutant loads from the Highwood River, another point source, were not included in the mass balance of the Downstream river reach. The ratios of annual

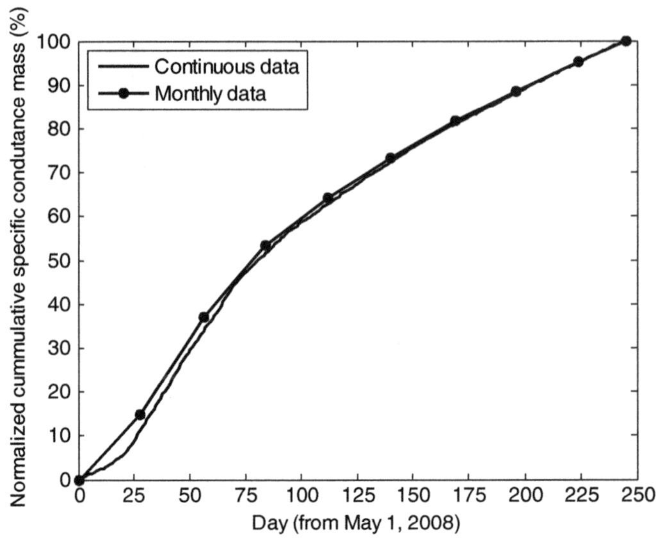

Fig. 9 Normalized cumulative load of conductivity

flow volumes at Calgary to at Near Ronalane are often slightly below 1.0. All these might lead to the negative annual Cl loads in four out of seven years in the Downstream river reach; the negative loads in three out of four years however, are negligible (Fig. 8). For a more accurate assessment of pollutant loads for this river reach, all the point sources from Highwood River should be included.

The effluents from Calgary's three WWTPs are significant sources of all the investigated pollutants. Elevated concentrations in all the water quality parameters occur after Calgary's WWTPs (Iwanyshyn et al. 2008) and are clearly shown in the results. These results, however, are under expectation since WWTPs have been ascribed to contribute significant pollutant loads to the Bow River within Calgary. The WWTPs' effluent also contributes to the variability in the associations between Cl and other water quality parameters within the City. The effect can be observed in all these parameters (Fig. 4); although the effects in NO_3 and conductivity are not as significant as that in Na and SO_4.

5.2 Non-point Sources Contribution

In the Upper river reach, variations in the concentrations of all the water quality parameters were observed. However, they were not consistent. In addition, the changes in this reach were not as significant as those detected within the City of Calgary, where both point and non-point sources contribute to pollutant loads into the Bow River.

In the Calgary reach, pollutant loads are largely contributed from the non-point sources in the Calgary upstream before Calgary's WWTPs. In Calgary downstream both point and non-point sources result in increases in pollution levels in the Bow River, except for NO_3 in all the years and SO_4 in two years, which saw a decrease. These two sub-reaches within the City are approximately equal in length. As for Cl, annual average loading rates into the Calgary upstream are similar to those into the Calgary downstream. The estimated annual average Cl loading rate is also equal to the rate obtained by Cantafio and Ryan (2014) in the Calgary upstream reach. Cantafio and Ryan (2014) investigated non-point source contributions to Cl load using a reach of a short distance located in this reach. These results suggest that the contribution of Cl from non-point sources is equal within Calgary. The contribution of non-point pollution also can be inferred from the associations between Cl and other parameters in the Calgary reach. Although significant differences in the associations between Cl and other parameters were not detected in Calgary upstream compared to the Upper river reach in our study, obvious variations in the associations are demonstrated after entering into the City. More prominent increases in Cl and Na can be observed between Below Bearspaw and Cushing Bridge. This reflects the effect of urban non-point pollution in the absence of significant point pollution between these two locations.

Surface runoff generated from stormwater is one mechanism for bringing non-point pollutants into receiving waters, but it is not the only possible pathway for non-point pollutant transport. In the upstream watershed portion of the Bow River, the major flow contribution is from the groundwater replaced by rain and snow melt water instead of from direct surface runoff, even during high flow (Sklash and Farvolden 1979; Grasby et al. 1999; Katvala 2008). In other words, precipitation falling within the watershed typically flows through the subsurface prior to discharging into the surface water. The significant contribution of Cl loads from non-point sources through the alluvial aquifer was identified in a one year (2010–2011) study of short river reaches within Calgary before Calgary's WWTPs (Cantafio and Ryan 2014). The calculated annual average Cl loading rate from this study is equal to that of the estimates provided by Cantafio and Ryan (2014). The hydrological characteristics along with the results on Cl loading might suggest the contribution of Cl from groundwater in the upstream watershed. Compared to surface water, the more or less constant pollutant concentrations in groundwater are expected. These may explain the strong and significant associations between pollutant loads and precipitation and/or rainfall amount on an annual basis in the Upper river reach and the Calgary upstream reach. Therefore, it might lead to the conclusion that pollutants from non-point sources are mainly being transported into the river through the subsurface in both the Upper river reach and the Calgary upstream reach. However, strong relationships were not identified in the Calgary downstream reach in all the parameters nor in the Downstream river reach in Cl and NO_3. The lack of relationships between pollutant loads from non-point sources and rainfall/precipitation in the Calgary downstream reach may imply that non-point pollutants discharge into the surface water through different flow pathways governed by different mechanisms. However, further investigation on flow pathways of pollutants is recommended. In the Downstream river reach, the transport of non-point sources is not clear through the mass balance investigation, since there are irrigation diversions, irrigation return flows and a tributary (the Highwood River). In order to accurately estimate non-point pollutant, accurate accounting of all the point sources into the Downstream river reach is necessary. While the pollutant load from the three major WWTPs is accurately known, little data are available on other point sources. The similar hydrometric volumes at the upstream and downstream ends of this reach suggest that the effect of the hydraulic structures on the reach and the tributary, is negligible. The negative loads of Cl and NO_3 might be due to the slight reduction in water volume at the downstream end of this river reach, although their concentrations are approximately similar; while the negative annual NO_3 load might also be partly due to biological uptake. However, significant upward trends were detected in Na and SO_4 in this reach, which resulted in positive loads of Na and SO_4 even with no obvious increase in hydrometric volume. In some watersheds, the groundwater recharge zone is mainly located in the Upper reach like the Wasi Tharad watershed in western Saudi Arabia (Subyani 2004). In the Downstream river reach of the Bow River in the Western Alberta Plains, the direct recharge to the aquifer from precipitation may decrease markedly. All of these may explain the differences in pollutant loads in the Upper river reach and the Downstream river reach of the Bow River. As a result, the contribution and transport route of non-point pollutants could vary along the river due to the geological characteristics. A more detailed field investigation, particularly on the alluvial aquifer along the river would help to draw more definitive conclusions.

Novotny et al. (2009) found a linear relationship between Cl applied per watershed area and average annual chloride concentration in the major rivers in an urban watershed. However in this study, the potential relationship between the amount of de-icing salts applied and Cl concentration in the river within Calgary were not calculated. The result might be consistent with the findings of aforementioned. Alluvial groundwater is known to be a long-term and year-round source of Cl to surface waters due to road salt application (Howard and Haynes 1993). The increasing concentration of Cl in the groundwater over a long period of time may be a concern. Thus, it is no surprise that no direct relationship between applied salt and Cl concentration exists.

5.3 Temporal and Spatial Variations of Water Quality

Water quality in the Bow River degrades along the river and with time. Monotonic spatial trends were not detected in *all* the quality parameters but in general, upward spatial trends from upstream to downstream were demonstrated (Fig. 3 and Table 2). The contribution of the point sources, WWTPs, is very significant in all the water quality parameters. As shown in the Table 1, the upward temporal trends in the water quality parameters including Cl and NO_3 after Calgary's WWTPs and moreover, spatial trends in all the water quality parameters within the Calgary reach were detected. Significant upward temporal trends in Cl, Na, and NO_3 in Bonnybrook WWTP (the biggest WWTP in Calgary) were also observed at a significance level of 10 % in the time period of this study; while the upward temporal trend in SO_4 was calculated, but it was not significant (results not shown). In addition, the associations between nutrient pollutant loads by WWTPs have been related to macrophyte and periphyton biomass downstream of WWTPs in Calgary (Sosiak 2002). The detected general downward spatial trends in NO_3 (Table 2) in the Downstream river reach, in which significant point sources are absent, might suggest a sink of nutrients through processes such as biological uptake and/or other mechanisms (for example, sedimentation). The estimated negative annual loads of NO_3 in the Calgary downstream reach and the Downstream river reach are therefore, not unexpected. The growth of macrophyte and periphyton is generally more significant downstream of Calgary's WWTPs (Robinson et al. 2009). Even with the significant loadings from the WWTPs, the contribution from the non-point sources is apparently not negligible, in particular within Calgary. As aforementioned, the mode of transport taken by non-point pollutants in different reaches needs further investigation because different transport routes/mechanisms along with geochemical characteristics of the alluvial aquifer is likely the cause of the temporal and spatial variation in water quality. In the Bow River watershed, spatial changes in SO_4 and Na in groundwater have been demonstrated (Grasby et al. 2010).

5.4 Sensitivity of the Results (Sampling Frequency)

The sampling scheme is usually recognized as a factor that affects the estimation of pollutant loads. But this effect may not be altogether clear (Littlewood 1995). As shown in the Fig. 9, for comparing the loads of conductivity calculated from monthly sampling and continuous sampling, the monthly sampling frequency appears to be sufficient for accurate quantification of the load. In this study, spikes in the concentrations of Cl and Na (outliers in the box-whisker plots of the Fig. 3a, b were frequently observed during snowmelt and very evident within the reach in Calgary, in particular at Policeman's Flats. Novotny et al. (2009) demonstrated that a biweekly sampling frequency could capture the chloride dynamics in snowmelt or storm events of short duration and was adequate to estimate the annual load of Cl entering or exiting the study watershed. In the Bow River watershed, it is expected that pollutants from non-point sources may take a subsurface route into the river in the upstream portion of the watershed; while a different route, such as surface pathway, may be taken at the downstream end of the watershed. This may be related to the obvious spikes in Cl concentrations observed in the Calgary downstream reach after the WWTPs, while no such large spikes were found at other sites both before and after Policeman's Flats. Pollutants transported through a surface flow pathway into receiving waters may be subject to other processes such as pollutant deposition on the land surface. When quantifying the pollutant loads, such factors should be taken into account for accurate estimation. In this study, the temporarily elevated Cl concentrations were smoothed for estimating the Cl load. If pollutants such as Cl are sensitive to events like snowmelt and rainfall, a more frequent sampling scheme might be needed to capture the dynamics of temporal variations at a finer scale.

6 Conclusions

- This paper investigated the spatial and temporal variations in several selected water quality parameters including Cl, Na, SO_4, NO_3, and conductivity. A mass balance approach was employed to quantify and characterize non-point pollutant loads along the Bow River. In general, both increasing temporal and spatial trends were demonstrated in the investigated water quality parameters, in particular in the river reach after Calgary's WWTPs. The results show that anthropogenic activities contribute to the degradation of water quality in this river. Moreover, the contribution of pollutant loads from non-point sources is not negligible, and their contribution within the City of Calgary is more significant.
- The majority of pollutant loads from non-point sources is thought to enter into the river through flow pathways under the shallow subsurface upstream of Calgary's WWTPs; while different path ways (such as surface flow pathways) might be contributing non-point pollutants into the river downstream of Calgary's WWTPs. These may result in different relationships between non-point pollutant loadings and meteorological variables along the river.
- Other processes or pollutant sources need to be further investigated to more accurately characterize pollutant loads into the river. For example, understanding the role of biological uptake and/or sedimentation of nutrients is needed to better quantify nutrient non-point loads. Further comprehension of these issues will allow identification of pollutant sources and more accurate quantification of pollutant loads, which will definitely help formulate reliable water quality management plans.

Acknowledgments The authors thank the City of Calgary and Alberta Environment and Sustainable Resource Development for providing data.

References

Alberta Environment River Basins (2010) Bow River at Calgary. Alberta Environment Webpage http://environment.alberta.ca/apps/basins/DisplayData.aspx?Type=Figure&BsinID=8&DataType=1&StationID=RBOWCALG. Accessed 16 Aug 2010

Alberta Environment (2004) Alberta groundwater data, Final edn. Groundwater information centre two CD set, West of 4M and West of 5/6M, Data Current April 2003

Alberto WD, Maria del Pilar D, Maria Valeria A, Fabiana PS, Cecilia HA, Maria de los Angeles B (2001) Pattern recognition techniques for the evaluation of spatial and temporal variations in water quality. A case study: Suquia River basin (Cordoba-Argentina). Water Res 35(12):2881–2894

Busse LB, Simpson JC, Cooper SD (2006) Relationship among nutrients, algae, and land use in urbanized southern California streams. Can J Fish Aquat Sci 63:2621–2638

Cantafio LJ, Ryan MC (2014) Quantifying baseflow and water-quality impacts from a gravel-dominated alluvial aquifer in an urban reach of a large Canadian river. Hydrogeol J 22:957–970

Conover WJ, Iman RL (1981) Rank transformations as a bridge between parametric and nonparametric statistics. Am Stat 35(3):124–129

Environment Canada (2001) The state of municipal wastewater effluents in Canada (state of the environment report). Cat. No. En1-11/96E, ISBN 0-662-29972-8, Ottawa, ON, 74 pp

Grasby SE, Hutcheon I, McFarland L (1999) Surface-water–groundwater interaction and the influence of ion exchange reactions on river chemistry. Geology 27(3):223–226

Grasby SE, Osborn J, Chen Z, Wozniak PRJ (2010) Influence of till provenance on regional groundwater geochemistry. Chem Geol 273:225–237

He J, Valeo C, Chu A, Neumann N (2010) Characterizing physicochemical quality of stormwater runoff from an urban area in Calgary. Alta J Environ Eng 136(11):1206–1217

Howard KWF, Haynes J (1993) Groundwater contamination due to road De-icing chemicals—salt balance implications. Geosci Can 20(1):1–8

Iwanyshyn M, Ryan MC, Chu A (2008) Separation of physical loading from photosynthesis/respiration processes in rivers by mass balance. Sci Total Environ 390:205–214

Jasechko S, Gibson JJ, Birks J, Yi Y (2012) Quantifying saline groundwater seepage to surface waters in the Athabasca oil sands regions. App Geochem 27:2068–2076

Katvala SM (2008) Isotope hydrology of the Upper Bow River Basin, Alberta, Canada. Masters Thesis, University of Calgary

Littlewood I (1995) Hydrological regimes, sampling strategies, and assessment of errors in mass load estimates for United Kingdom Rivers. Environ Int 21:211–220

Malve O, Tattari S, Riihimaki J, Jaakkola E, Vob A, Williams R, Barlund I (2012) Estimation of diffuse pollution loads in Europe for continental scale modelling of loads and in-stream river water quality. Hydrol Process 26:2385–2394

McLeod SM, Kells JA, Putz GJ (2006) Urban runoff quality characterization and load estimation in Saskatoon, Canada. J Environ Eng 132:1470–1481

Novotny EV, Sander AR, Mohseni O, Stefan HG (2009) Chloride ion transport and mass balance in a metropolitan are using road salt. Water Resour Res 45:W12410. doi:10.1029/2009WR008141

Obropta CC, Kardos JS (2007) Review of urban stormwater quality models: deterministic, stochastic, and hybrid approaches. J Am Water Resour Assoc 43(6):1508–1523

Quade D (1967) Rank analysis of covariance. J Am Stat Assoc 62:1187–1200

Rhodes AL, Newton RM, Pufall A (2001) Influences of land use on water quality of a diverse New England watershed. Environ Sci Technol 35:3640–3645

Robinson K, Valeo C, Ryan MC, Chu A, Iwanyshyn M (2009) Modelling aquatic vegetation and dissolved oxygen after a flood event in the Bow River Alberta, Canada. Can J Civil Eng 36:492–503

Robinson TH, Melack JM (2013) Modeling nutrient export from coastal California watersheds. J Am Water Resour Assoc 49(4):793–809

Shrestha S, Kazama F (2007) Assessment of surface water quality using multivariate statistical techniques: a case study of the Fuji river basin, Japan. Environ Modell Softw 22:464–475

Singh KP, Malik A, Mohan D, Sinha S (2004) Multivariate statistical techniques for the evaluation of spatial and temporal variations in water quality of Gomti River (India)—a case study. Water Res 38:3980–3992

Sklash M, Farvolden R (1979) The role of groundwater in storm runoff. J Hydrol 43(1–4):45–65

Sosiak A (2002) Long-term response of periphyton and macrophytes to reduced municipal nutrient loading to the Bow River (Alberta, Canada). Can J Fish Aquat Sci 59(6):987–1001

Subyani AM (2004) Use of chloride-mass balance and environmental isotopes for evaluation of groundwater recharge in the alluvial aquifer, Wadi Tharad, western Saudi Arabia. Environ Geol 46:741–749

USGS (2005) Computer Program for the Kendall Family of trend tests. USGS report 2005–5275

Vandenberg JA, Ryan MC, Chu A (2005) Field evaluation of mixing length and attenuation of nutrients and fecal coliform in a wastewater effluent plume. Env Monit Assess 107:45–57

Winter T, Harvey J, Franke O, Alley W (1998) Groundwater and surface water a single resource. U.S. Geological Survey Circular 1139, United States Geological Survey, Denver, CO http://pubs.usgs.gov/circ/circ1139/#pdf. Accessed 16 Aug 2010

Assessment of Groundwater Quality in the Amaravathi River Basin, South India

K. Narmada, G. Bhaskaran and K. Gobinath

Abstract Sustainable water management in a river basin requires knowledge on the water availability in the basin and current and future demands. The problems of water quality have become more important than the quantity and any sustainable management program must take into account the availability and suitability of water resources for various purposes. Being a solvent and depleted in major and trace ions, water is susceptible to contamination by a variety of sources including, but not limited to, rocks, soil, effluents, and sewage with which it comes into contact. In this study, attempt is made to evaluate the status of ground water quality and its effect on human health, irrigation and environment and domestic purposes. Water samples were collected from 53 locations distributed in Karur, Erode and Thiruppur districts in the year 2011 during both pre-monsoon and post-monsoon seasons. Water quality parameters namely, pH, Total Dissolved Solids (TDS), Total Hardness (TH) and Chlorides were analyzed and were interpreted using geospatial techniques. The results were evaluated in the context of human health, irrigation, domestic effects. It is revealed by the study that the groundwater in the Amaravathi River basin is affected by effluents from textile industries, irrigation return flows and domestic sewage, in addition to the natural geogenic contaminants.

Keywords Watershed · Effluent · Geo-informatics · Qualitative analysis · Environmental degradation · Toxicity · Industrialization

1 Introduction

The safe portable water is absolutely essential for healthy living. Ground water is ultimate and most suitable fresh water resource for human consumption in both urban as well as rural areas. The importance of ground water for existence of human

---

K. Narmada · G. Bhaskaran (✉) · K. Gobinath
Department of Geography, University of Madras, Chepauk, Chennai 600 005, India
e-mail: grbhaskaran@gmail.com

Mu. Ramkumar et al. (eds.), *Environmental Management of River Basin Ecosystems*, Springer Earth System Sciences, DOI 10.1007/978-3-319-13425-3_26

society cannot be overemphasized. Various textile industries are present in Gobichettipalayam in Erode. From the water quality analysis that was carried out by Palanisamy it was found that the water quality was suitable for domestic purposes. (Palanisamy et al. 2007). A systematic study has been carried out to assess the ground water quality of the Noyyal River Basin in and around Thiruppur Town. (Geetha et al. 2008). It was found that there was constant fluctuation of all the parameters in different periods. So it was concluded that periodic monitoring is required for future sustainability. During the last three decades Karur has emerged as a major center for textile manufacture, the partially treated effluent is released into the streams. Hence the dyeing units are either provided with Individual Effluent Treatment Plant or connected to the common effluent treatment plants to manage the effluent quality. (Rajamanikam 2010). Ground water quality modelling of Amaravathi water shed using MODFLOW was also carried out by Rajamanickam. The validated model is used for simulation of ground water quality in the next 15 years. The simulation shows that there will be no improvement in water quality even the effluent meet the discharge standards for the next 10 years. (Rajamanickam and Nagan 2010). A semi distributed water balance model for Amaravathi River Basin was developed to estimate the average spatial distribution of water balance components (Jenifa Latha et al. 2010). Here spatially semi-distributes water balance model was developed to simulate mean monthly hydrological processes using landuse, soil texture, topography and hydro meteorological data as input parameters in the Amaravathi River basin, a semi-arid region of Tamil Nadu in India. Shivakumar carried out the assessment of the physiochemical parameters of the Amaravathi water shed (Sivakumar et al. 2011). Ground water quality for two seasons was studied. The water quality was found to be better during the monsoon period than during the post monsoon period. Landuse and landcover categories of Karur district was analyzed using remote sensing and from this study it was found that agricultural land was predominant in that area. (Balachander et al. 2011). Ground water potential zones were demarcated using remote sensing and GIS techniques in Trichy and Karur districts in Tamil Nadu. (Pandian and Kumanan 2013). This study helped to delineate the potentiality as moderate potential, and some part have been classified as high potential zones, low and very low potential zones and only few areas have been classified as very high groundwater potential zones. Ground water pollution was assessed on the banks of the Amaravathi River at Karur district (Zahir Hussain and Rajadurai 2013). The ground water quality assessment was carried out at 28 sampling stations in Amaravathi water basin all the physiochemical parameters were found to be at elevated levels at all stations. The physico-chemical parameters such as pH, electrical conductivity, total dissolved solids, total hardness, carbonate, bicarbonate, chloride, calcium, magnesium, nitrate, sulphate and phosphate, BOD, COD, and DO were analyzed. Most of the parameters were found to be higher than the permissible limit. Water for irrigation and food production constitutes one of the greatest pressures on fresh water resources. Globally we use 70 % of our water resources for agriculture and irrigation and only 10 % for domestic use.

- Every day, 2 million tons of sewage and industrial and agricultural waste are discharged into the world's water (UN WWAP 2003), the equivalent of the weight of the entire human population of 6.8 billion people.
- The UN estimates that the amount of wastewater produced annually is about 1,500 km^3, six times more water than exists in all the rivers of the world. (UN WWAP 2003).
- With the Millennium Development Goals, the international community committed to have the proportion of people without access to safe water and sanitation by 2015.
- 70 % of industrial wastes in developing countries are disposed of untreated into waters where they contaminate existing water supplies. (UN-Water 2009).
- In Chennai, India, over-extraction of groundwater has resulted in saline groundwater nearly 10 km inland of the sea and similar problems can be found in populated coastal areas around the world. (UNEP 1996).

There are several states in India where more than 90 % population is dependent on groundwater for drinking and other purposes. The water resources potential of the country which occurs as natural runoff in the rivers is about 1,869 BCM as per the estimates of Central Water Commission (CWC), considering both surface and ground water into account. The estimated per capita availability of water works out to be 1,588 m^3 (cu.m) as on March 2010. Due to the various constraints of topography, uneven distribution of resource over space and time, it has been estimated that only about 1,123 BCM of total potential of 1,869 BCM can be put to beneficial use. Globally, 70 % of the available water resources are used for agriculture and irrigation and 10 % is available for domestic use. Total replenishable ground water potential of the Indian subcontinent has been estimated as 433 BCM per year. In Tamil Nadu the annual replenishable water resources accounts to 23.07 BCM.

During the last few decades, the Coimbatore, Erode, and Tiruppur districts which are located within the Amaravathi River basin, Tamil Nadu State, India have emerged major hubs of textile, and tannery industries and poultry and intensive agricultural practices Brief industrial profile of Karur District (2012–2013). Associated with these are the unplanned developments of urban centers and ancillary industries and resultant drastic changes in the land use and unscientific disposal of sewage and effluents into the surface and underground water bodies District ground water brochure, Karur District (2008). Though several workers have attempted characterizing the ground water quality in this region (for example, Palanisamy et al. 2007; Geetha et al. 2008; Rajamanikam and Nagan 2010; Sivakumar et al. 2011; Balachander et al. 2011; Pandian and Kumanan 2013; Zahir Hussain and Rajadurai 2013), they focused on areas limited in areal extent and were confined to fewer parameters. Hence, the present study attempts documenting the environmental impact on the Amaravathi River in terms of the ground water quality and the probable effects of the studied parameters on the landscape, soil quality, agriculture, human settlements and human health.

2 Study Area

The Amaravathi River basin is located between the north latitudes 11° 00′ N and 10° 00′, and east longitudes 77° 00′ and 78° 15′ (Figs. 1 and 2). The Amaravathi River is a major tributary of the Cauvery River. It originates from Thirumurthimalai in the Coimbatore district, Tamil Nadu State. Throughout its course of 256 km, before joining the Cauvery River, it receives a number of small streams. Shanmuganadhi, Nankanchi and Kodaganar rivers are the tributaries of Amaravathi. It has a catchment area of 8,280 km^2. The flow in the river is seasonal and is contributed by the northeast and southwest monsoon seasons. Many small, medium and large scale textile industries are situated in the basin. The major rock types present in the basin are: Hornblend Biotite Gneiss (occupying around 77 % of the basin) and charnockite (covering about 18 % of the basin). Soil types present in the basin include alfisols, entisols and inceptisolsis. About 76.3 % of the basin area is under agricultural practice and 13 % is covered by forest cover.

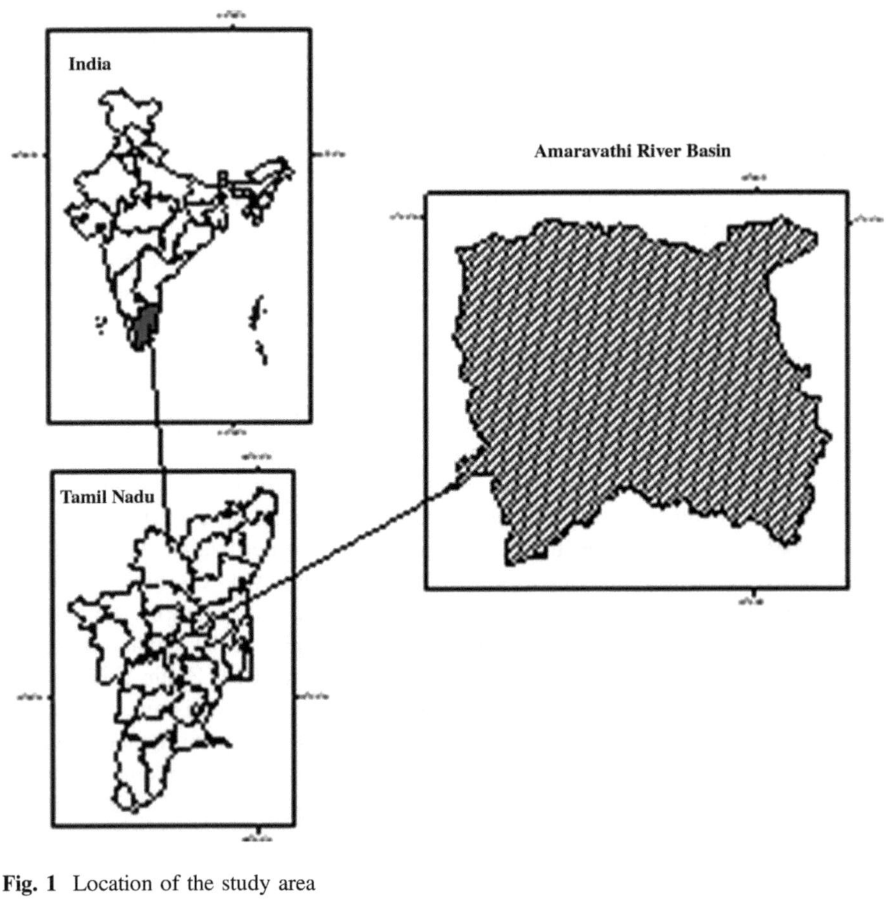

Fig. 1 Location of the study area

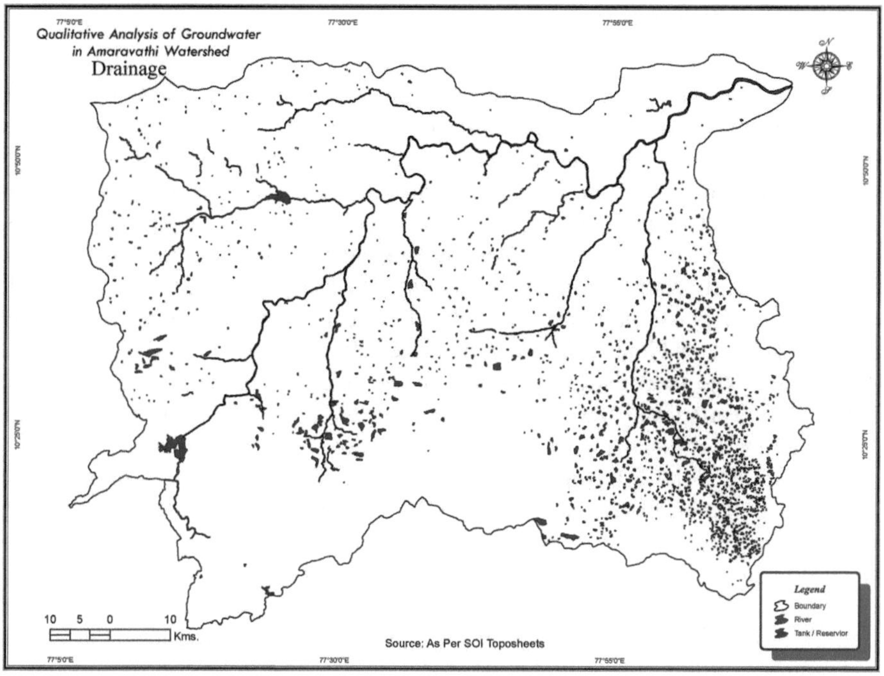

Fig. 2 Drainage map of the Amaravathi River basin

During the last three decades, the region emerged as a major textile center with more than 1,000 power loom and handloom. The main raw material is cotton yarn. This cotton yarn and the finished product (i.e.) grey cloth are bleached and dyed by the industries located in this basin. At present around 487 bleaching and dyeing units are in operation. These bleaching and dyeing units are located on either side Amaravathi River within 2 km from the river. By average one unit generates effluent of 30 kl/day (KLD). After treatment (treatement plants are either located within the industry or clusters of industries avail the services of common treatment plant), the effluent is discharged into the river Amaravathi. Approximately around 14,600 KLD of effluent is discharged into the river every day. On an average about 43 tons of TDS and 101 tons of chlorides are released per day. During the summer, when the river receives no rainfall, flow of only the effluents in the river can be noticed, suggesting the extensive damage being inflicted on this river. The ground water quality of the study area is thus, adversely affected by the industrialization. Increased population and improper drainage system have potential to influence the ground water quality.

3 Need for the Study

During the last three decades the town emerged as a major textile center with more than 1,000 power loom and handloom producing bed sheets, towels and furnishings. The main raw material is cotton yarn. This cotton yarn and the finished product (i.e.) grey cloth are bleached and dyed by the industries located in this town. At present around 487 bleaching and dyeing units are in operation. These bleaching and dyeing units are located on either side Amaravathi River within 2 km from the river. By average one unit generates 30 KLD of trade effluent. These units have provided either individual Effluent Treatment Plant or joined in common Effluent Treatment Plant. After treatment the effluent is discharged into the river Amaravathi. The total dissolved Solids in the river discharge is in the range of 5,000–10,000 mg/l and the chloride is in the range of 200–1,500 mg/l. Around 14,600 KLD of effluent is discharged into the river every day. On an average about 43 tons of TDS and 101 tons of chlorides are released per day. This salt accumulates and in the soil and increase the TDS level in the groundwater. The effluent reaching the river is dark brown in color. During the summer period there is no water flow in the river, only effluent flow can be noticed. In the monsoon period the color in river water can be noticed up to Kattali where the river confluence with Cauvery. Hence a detailed study on the impact of industrial effluent on ground water quality in this area is highly needed.

4 Objectives

Given cognizance to the basin characteristics, population grown, industrial and agricultural developments and the status of domestic, agricultural and industrial effluent discharge, following objectives were chosen for the present study.

- To study the physical and geographical characteristics of the Amaravathi River basin.
- To study the environmental impact on the Amaravathi River in terms of the ground water quality.
- To conduct a detailed qualitative analysis of the ground water samples taken from different locations on either side of the Amaravathi River basin.
- To study the effect of the deteriorating water quality on the landscape, human settlements, soil quality, agriculture and human health.
- Integration of the entire hydrological data to analyze the extent of deterioration of water resource and land cover.

5 Methodology

5.1 Sample Collection

- Random sampling was done by collecting water samples from different locations on either side of the Amaravathi River basin in the three districts Karur, Thiruppur and Erode (Fig. 3).

Fig. 3 Map showing sampling locations

- Samples were collected in plastic container which was cleaned and dried in sunlight to avoid unpredictable changes in characteristic as per standard procedure.
- The collected samples were analyzed for different physicochemical parameters such as pH, Electrical Conductivity, TDS, Total hardness, Chlorides, etc., as per the standard methods and the results were compared with the Indian Standards (IS:10500) for potable water.

6 Mapping

- Various thematic maps for the study area like geology, soil types, landuse/landcover and watershed maps were prepared using supervised classification with the help of ARC GIS software. (Figures 4, 5, 6, and 7 respectively).
- Land use, land cover maps were prepared using ERDAS Imagine 10.0 image processing software.
- Interpolation of the data was done to study the extent of variation of different sets of parameters in different localities by preparing thematic maps using ARC GIS to show its variations in the study area.

7 Results and Discussion

7.1 Water Quality Parameters

Table 1 presents the studied parameters listed according to sampling locations and seasons. The pH values were found to be between 6.65 and 8.6 during the pre-monsoon period and 6.5–7.8 during the post-monsoon period signifying the seasonality and dilution-enrichment by monsoon rainfall and evaporation-enrichment by summer weather respectively. The results also show that the parameter is within the permissible limit (according to World Health Organization or WHO, the permissible limit for pH is 7.0–8.5) except for a few places and a maximum of >8.5 in Sellipalayam. The TDS values vary between 699 in Bolikavandan-palayam and 5,968 mg/l in Veluswamypuram and are far beyond the permissible limits (500–1,500 mg/l for drinking water—WHO). The hardness varies from 368 mg/l in Vellagavai to a maximum of 1,523 mg/l in Veluswamypuram as against the permissible limit of 200–600 mg/l for drinking water (WHO). Lowest level of Chlorides (226 mg/l) is recorded in Bolikavandanpalayam, and maximum is reported in Sellipalayam and Kodaiyar (1,275 mg/l) (Fig. 3).

The spatial distribution of the studied parameters during pre-monsoon and post-monsoon are presented in Figs. 8, 9, 10, 11, 12, 13, 14 and 15. These figures show the high spatial and seasonal variability of the studied parameters. Incidences of high TDS, Chloride and Hardness and varying pH have been reported mostly in the

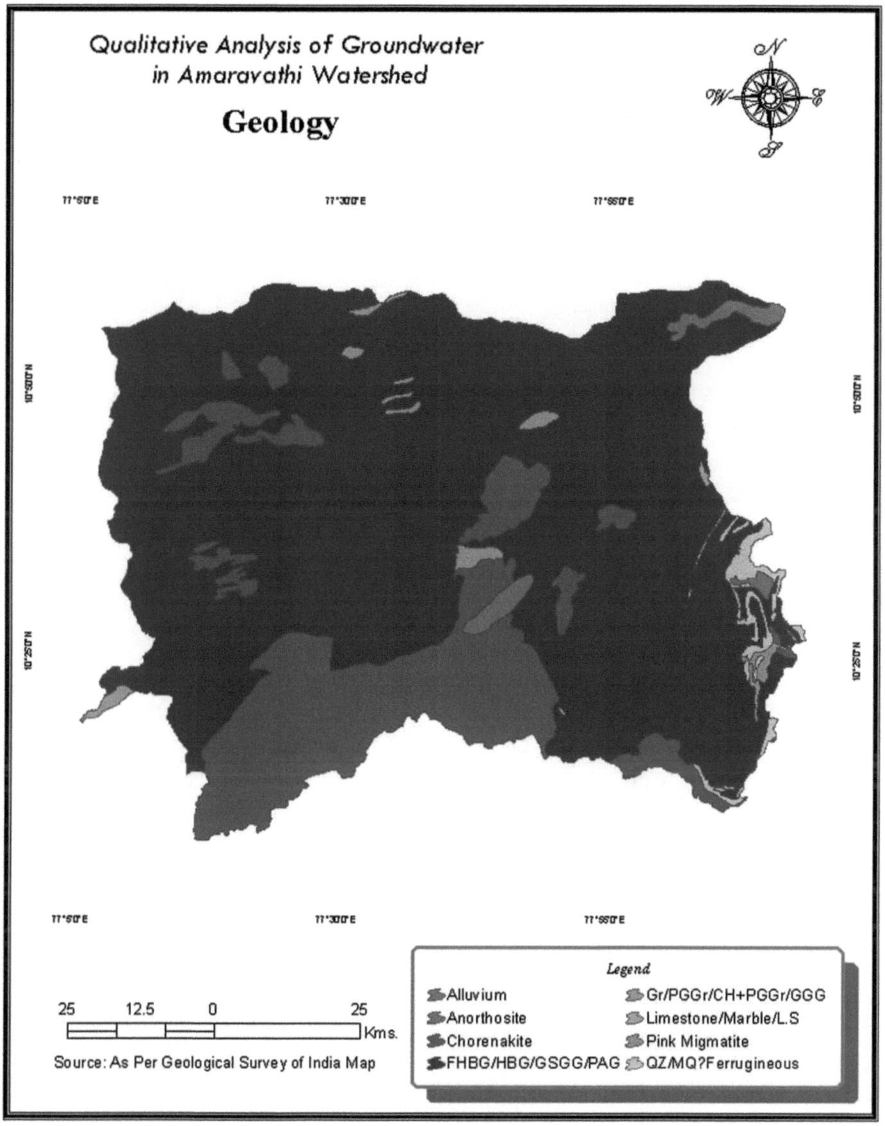

Fig. 4 Distribution of rock types within the Amaravathi River basin

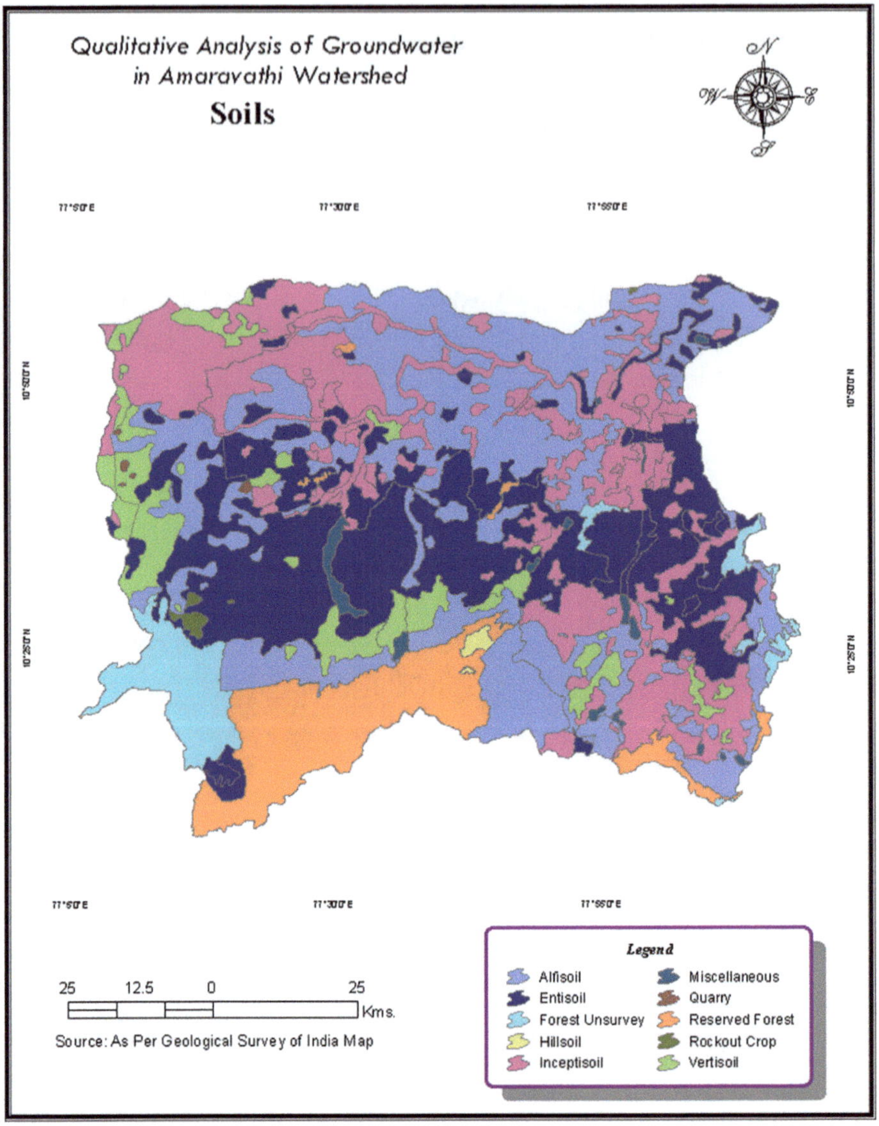

Fig. 5 Distribution of soil types

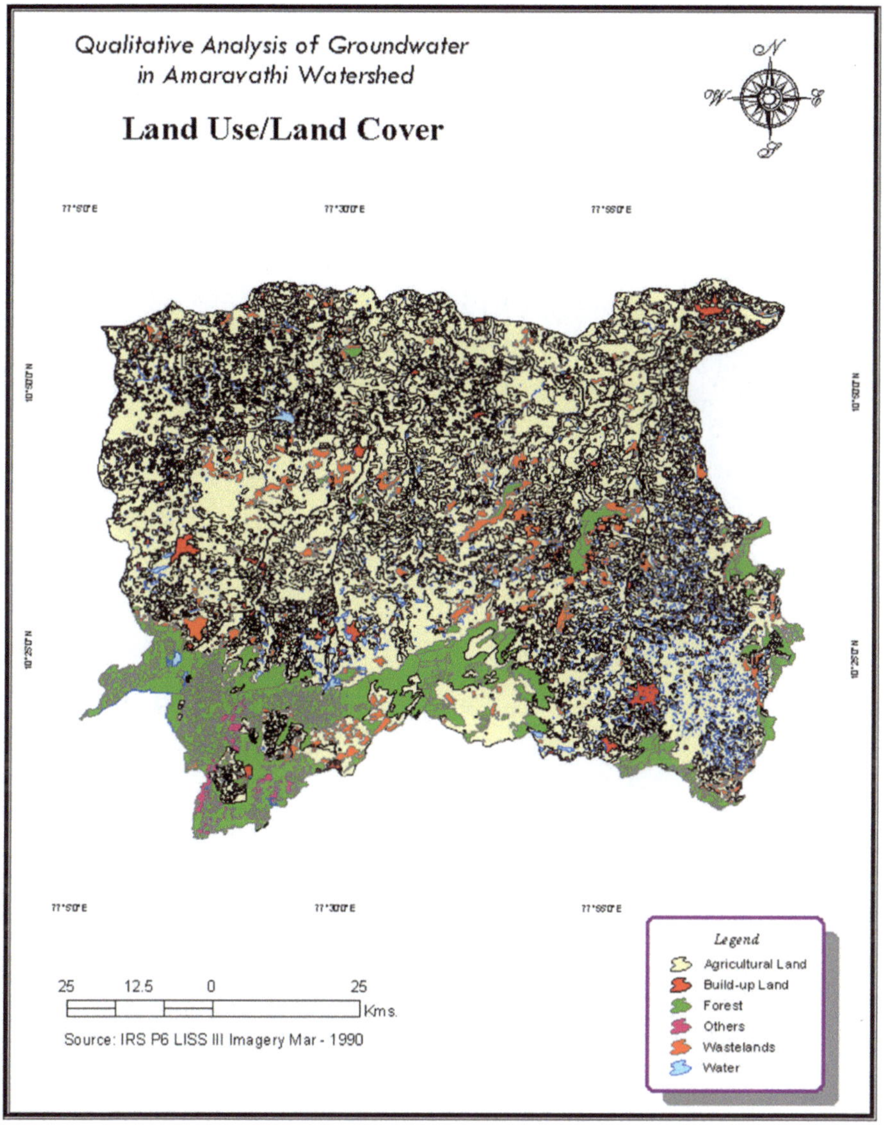

Fig. 6 Landuse/land cover

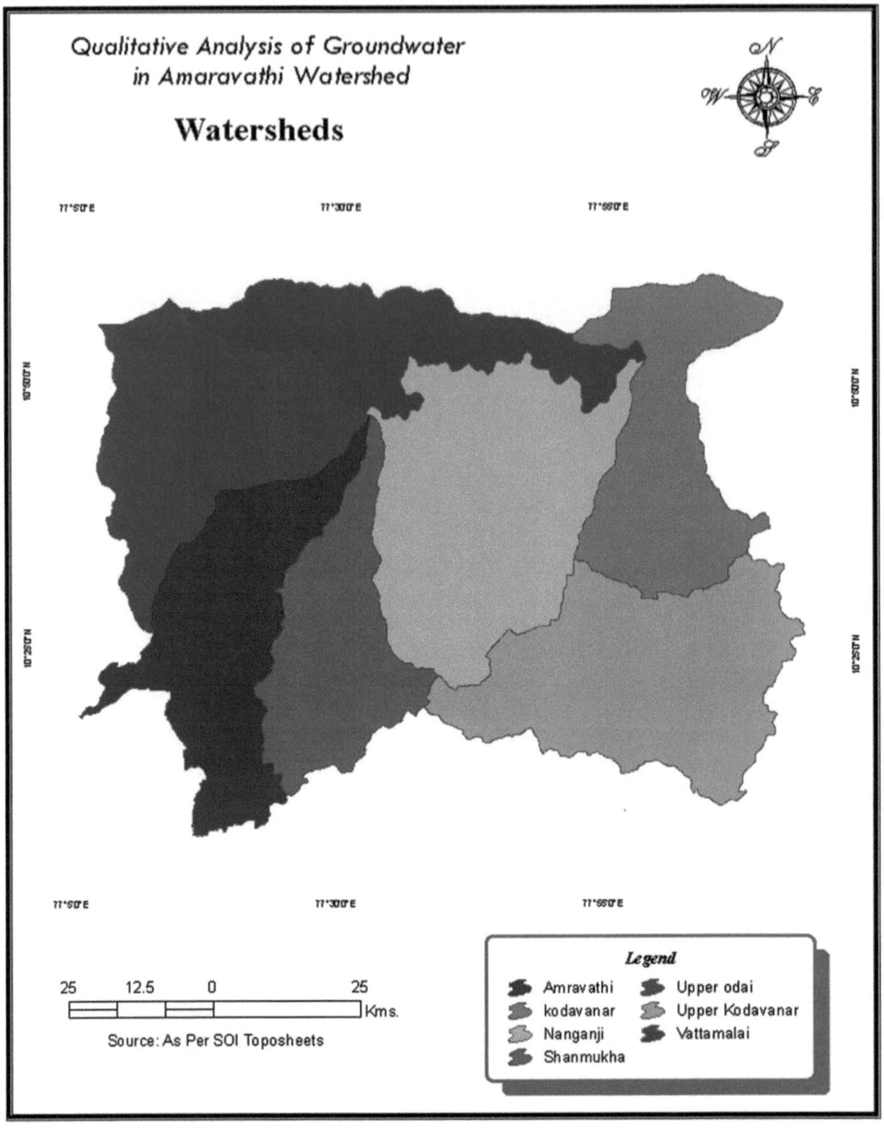

Fig. 7 Watersheds within the Amaravathi River basin

Table 1 List of sampling locations and the water quality parameters

S.No.	Sampling locations	District	Pre monsoon				Post monsoon			
			Ph	TDS	TH	Cl	Ph	TDS	TH	Cl
1	Appanichenakkampatti	Erode	7.75	1,656	665	568	7.3	1,553	550	468
2	Kungampalayam	Erode	7.9	1,213	770	686	7.5	1,152	654	596
3	Alampalayam	Erode	7.85	1,100	550	650	7.5	1,050	458	586
4	Vellamadai	Erode	8.1	3,636	1,456	1,356	7.6	2,500	1,125	1,125
5	Periya Thirumangalam	Erode	7.65	1,980	880	758	7.4	1,500	657	656
6	Elavanur	Erode	7.8	4,598	720	756	7.5	3,500	653	685
7	Edayakniyar	Erode	7.8	3,125	885	785	7.5	2,564	756	696
8	Thurambadi	Erode	7.65	1,256	562	868	7.3	1,136	468	796
9	Raaika Valasu	Erode	7.55	1,563	685	869	7.3	1,100	556	786
10	Athappakavandanpudur	Erode	7.65	1,654	496	546	7.4	1,200	405	486
11	Sitharavu	Erode	6.83	900	450	447	6.5	654	398	403
12	Ottanagampatti	Erode	6.5	1,900	565	658	6.5	1,324	486	558
13	Andippatti	Erode	8.23	5,468	868	768	7.8	3,520	705	686
14	Kadakkottai	Erode	7.65	1,869	586	575	7.4	1,530	485	526
15	Moddkaur	Erode	7	1,756	515	545	6.8	1,564	486	496
16	Theerappadi	Erode	8.2	4,523	868	768	7.8	3,564	750	654
17	Kottampoondi	Erode	8.26	3,245	775	658	7.8	3,051	698	606
18	Semmadipatti	Erode	7.66	5,689	665	621	7.4	4,535	625	605
19	Pithalapatti	Karur	7.8	2,303	840	860	7.5	1,800	786	686
20	Kilakottai	Karur	7.5	3,625	1,100	968	7.3	2,864	984	798
21	Ayyalur	Karur	7.2	885	480	475	7.1	665	398	365
22	Usilamppati	Karur	7.2	1,996	568	562	7	1,564	489	485
23	Ulliyakkottai	Karur	7.2	1,443	489	586	7	1,246	425	468

(continued)

Table 1 (continued)

S.No.	Sampling locations	District	Pre monsoon				Post monsoon			
			Ph	TDS	TH	Cl	Ph	TDS	TH	Cl
24	E.V. Pudur	Karur	7.3	1,834	520	467	7	1,654	452	425
25	Sadiaya Kavandanpudur	Karur	7.7	1,756	563	468	7.4	1,654	468	436
26	Thalappatti	Karur	7.9	4,394	790	896	7.5	3,548	654	768
27	Thottivadi	Karur	7.1	5,072	964	864	6.8	4,503	874	796
28	Salappalayam	Karur	7.4	3,568	775	896	7.2	2,569	686	788
29	Kodayar	Karur	7.4	3,147	1,250	1,275	7.2	2,489	1,056	1,154
30	Karuppampalayam	Karur	7.6	2,332	950	780	7.5	2,561	862	686
31	Veluswamypuram	Karur	7.3	5,968	1,523	1,023	7.2	3,694	1,164	998
32	Kaliyappanur	Karur	7.1	3,808	986	876	7	2,564	751	786
33	Arasu Calany	Karur	7.1	3,452	610	740	7	2,498	554	686
34	Puliyur	Karur	7.4	2,356	864	785	7	2,154	698	697
35	Sellipalayam	Karur	8.6	3,184	1,400	1,275	7.8	2,564	1,245	1,145
43	Periyakottai	Tiruppur	7.09	2,350	610	860	6.9	1,156	554	796
44	Cinnar	Tiruppur	7.3	1,000	450	298	7.1	896	354	205
45	Vellagavai	Tiruppur	7.26	980	369	459	7.1	750	384	428
46	Poolampatty	Tiruppur	7.37	886	396	384	7.1	657	367	338
47	Bolikavandanpalayam	Tiruppur	6.96	699	425	226	6.5	567	410	205
48	Manuppatti	Tiruppur	7.41	869	420	423	7.1	648	405	409
49	Sethampalli	Tiruppur	7.43	890	414	329	7.1	685	389	345
50	Jothiempatti	Tiruppur	6.82	2,059	606	720	6.5	1,152	554	686
51	Ganapathy pudur	Tiruppur	7.25	965	385	268	7	897	325	246
52	Kaungali valasu	Tiruppur	7.02	925	445	336	7	758	368	308
53	Pachalur	Tiruppur	6.65	932	485	236	6.5	785	427	238

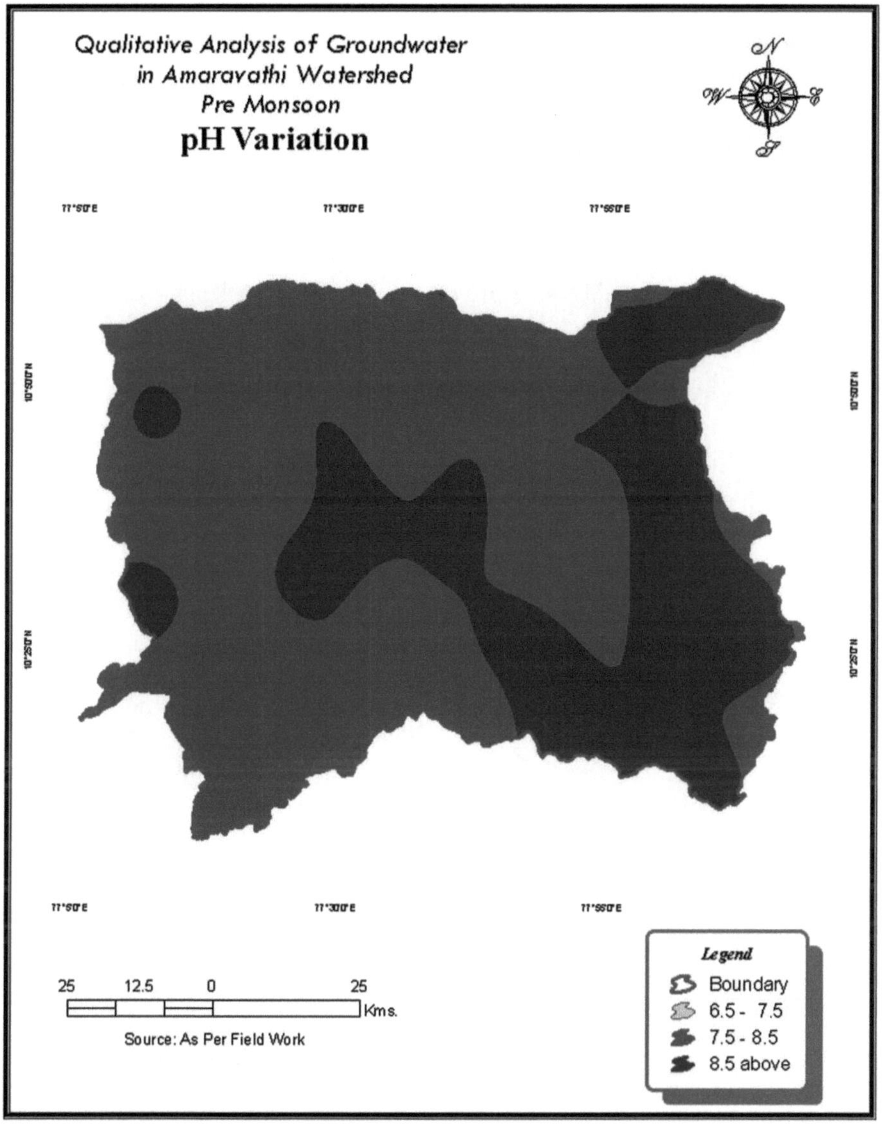

Fig. 8 Spatial distribution of pH during pre-monsoon

Karur and Erode districts. The values show a decreasing trend during the post-monsoon. Higher concentration of Chlorides has been observed in most of the water samples collected. Cl in the natural waters occurs in very low level and enrichments are attributed to the anthropogenic sources. Occurrences of higher levels of Cl regardless of seasons indicate contaminated nature of the groundwater beyond the dilution effect of monsoon rainfall. [Figures 16–19]. A comparative analysis of the

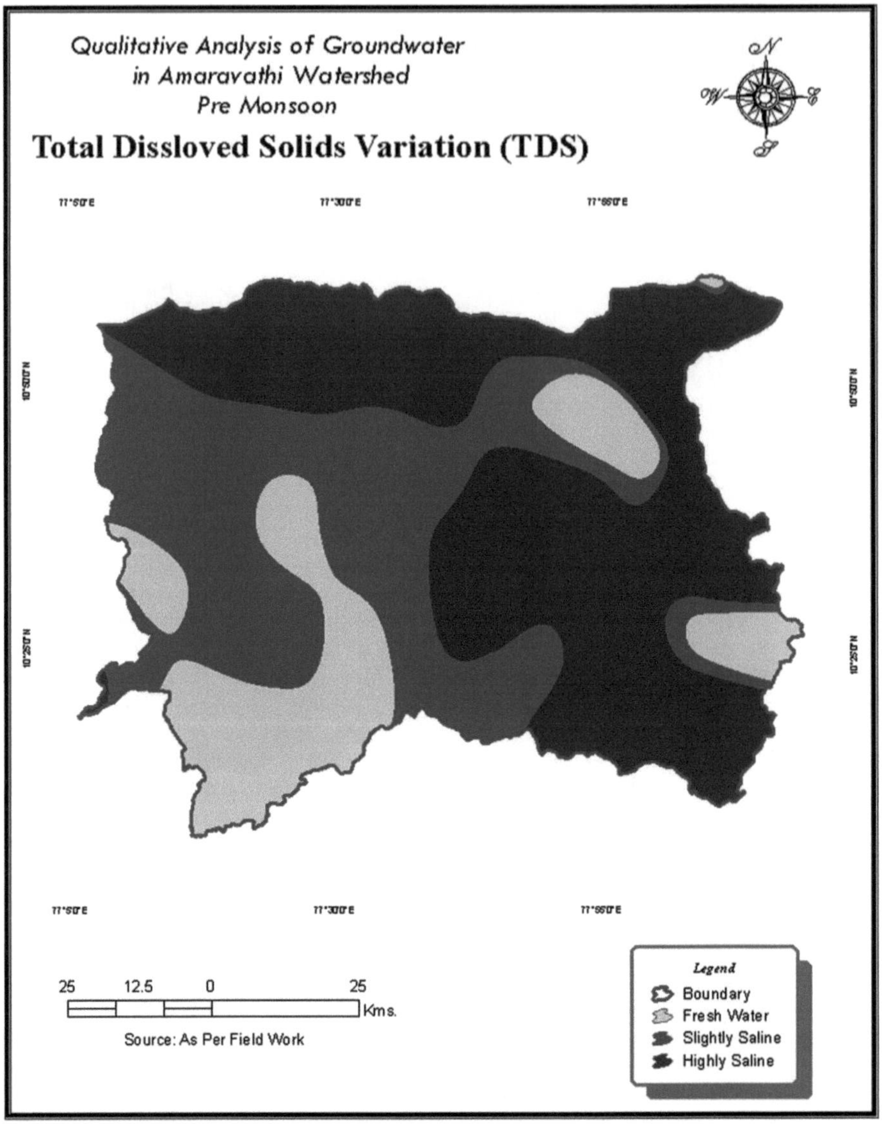

Fig. 9 Spatial distribution of hardness during pre-monsoon

land use/land cover distribution with the maxima and minima of the spatial distributions of water quality parameters shows that the observed highs are always associated with clusters of urban centers, industrial clusters and discharge sites of domestic and industrial wastes. In addition, agricultural return flows have shown

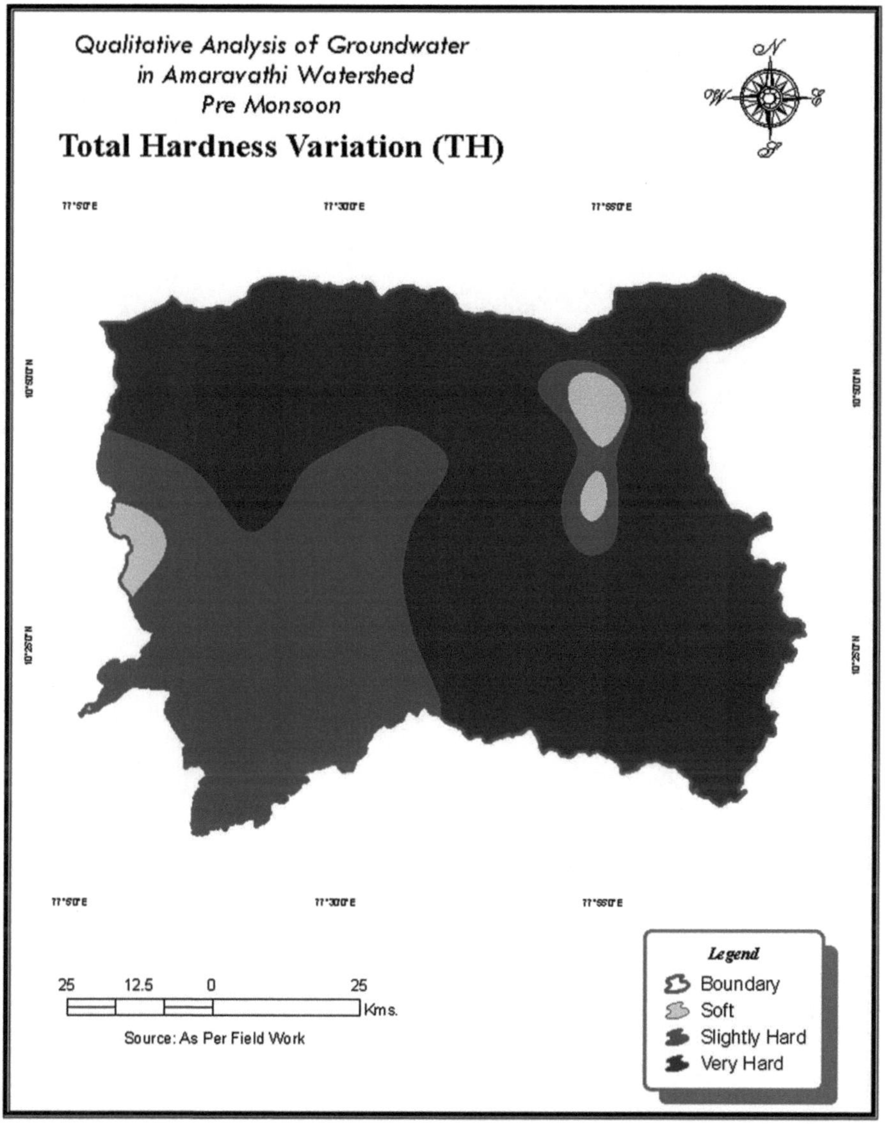

Fig. 10 Spatial distribution of total hardness during pre-monsoon

coincidences with randomly located regions, which may suggest transfer of the contaminants into the sub-surface aquifers through overland flow as well as percolation through the vadose zone.

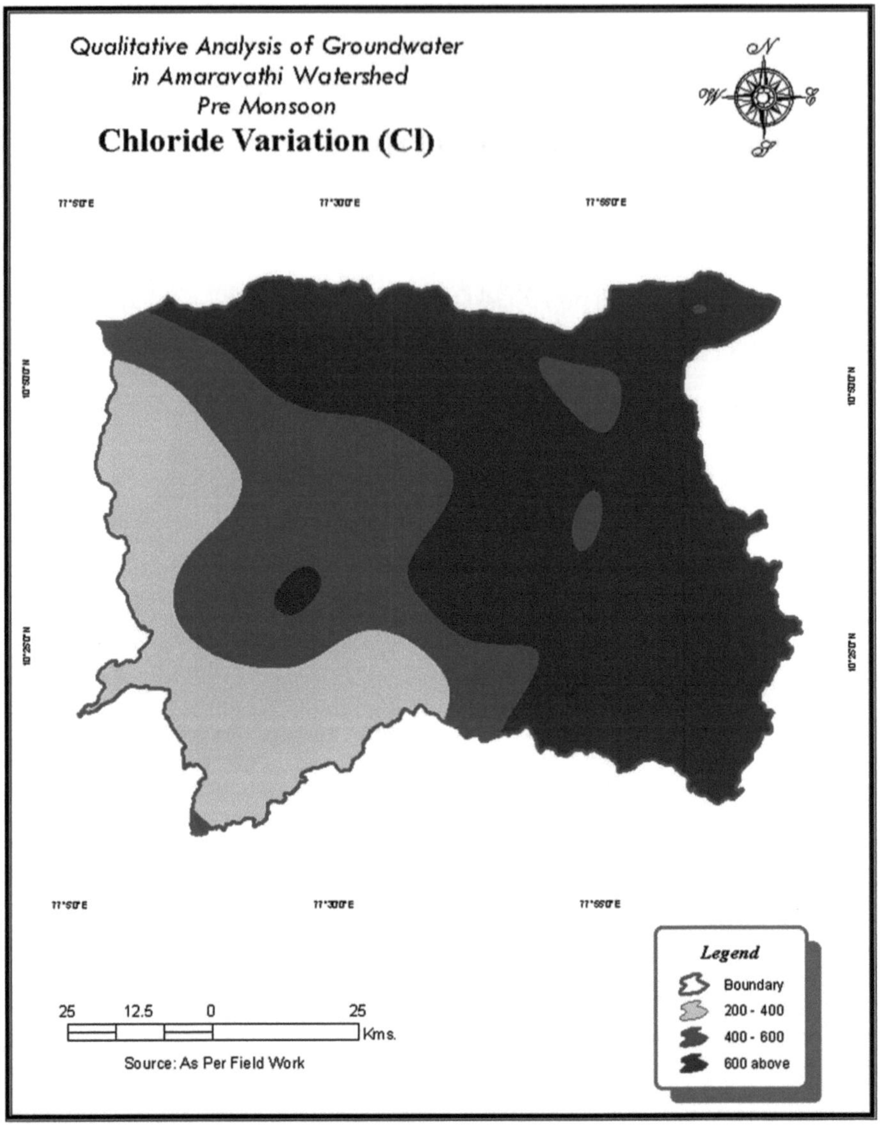

Fig. 11 Spatial distribution of chloride during pre-monsoon

7.2 Environmental Degradation Due to Textile Industry

The primary residual wastes generated from the textile industry are non-hazardous. These include scraps of fabric and yarn, off-specification yarn and fabric and packaging waste. There are also wastes associated with the storage and production of yarns and textiles, such as chemical storage drums, cardboard reels for storing

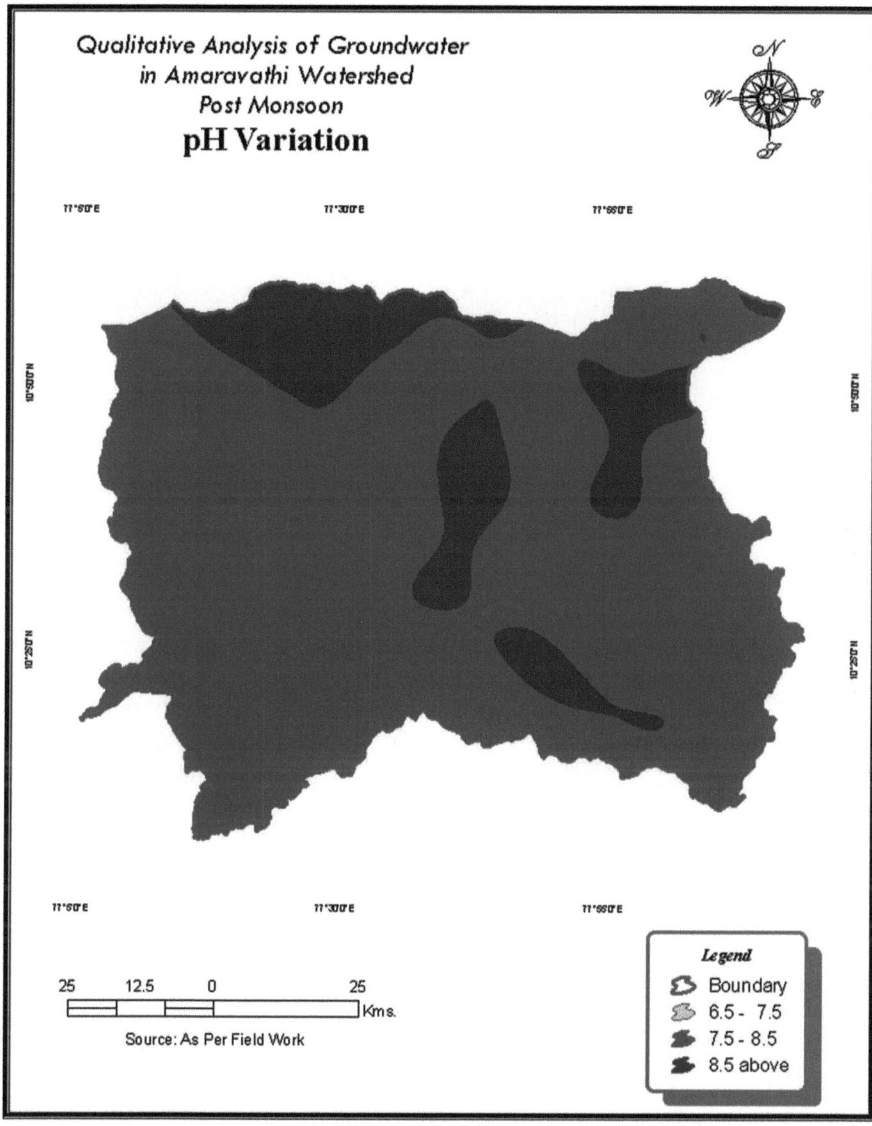

Fig. 12 Spatial distribution of pH during post-monsoon

fabric and cones used to hold yarns for dyeing and knitting. Cutting room waste generates a high volume of fabric scraps, which can often be reduced by increasing fabric utilization efficiency in cutting and sewing. But the major concern is on water due to the limited availability and excessive demand for domestic and industrial purposes and the extreme degradation it causes. The textile industry uses high volumes of water throughout its operations, from the washing of fibers to bleaching,

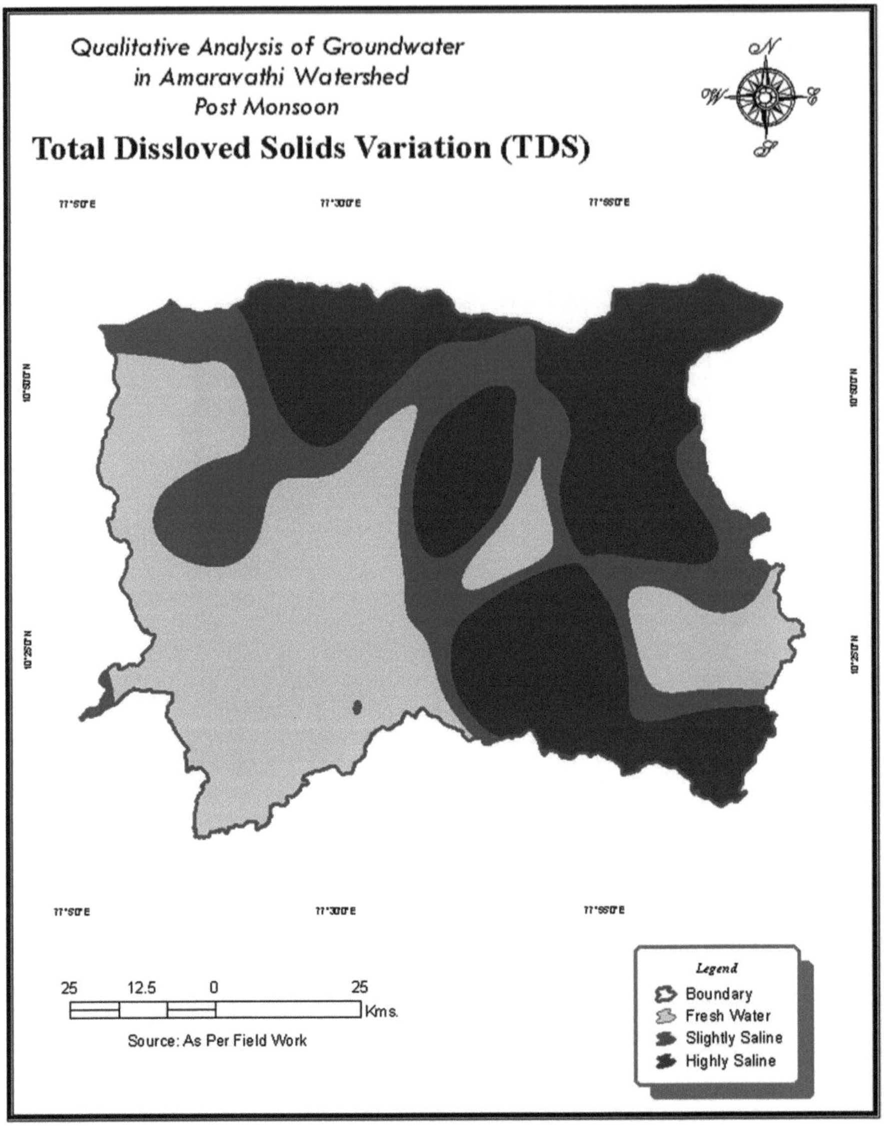

Fig. 13 Spatial distribution of hardness during post-monsoon

dyeing and washing of finished products. On average, approximately 200 l of water are required to produce 1 kg of textile. The large volumes of wastewater generated also contain a wide variety of chemicals. These can cause damage if not properly treated before being discharged into the environment. Of all the steps involved in textiles processing, wet processing creates the highest volume of wastewater. The aquatic toxicity of textile industry wastewater varies considerably among

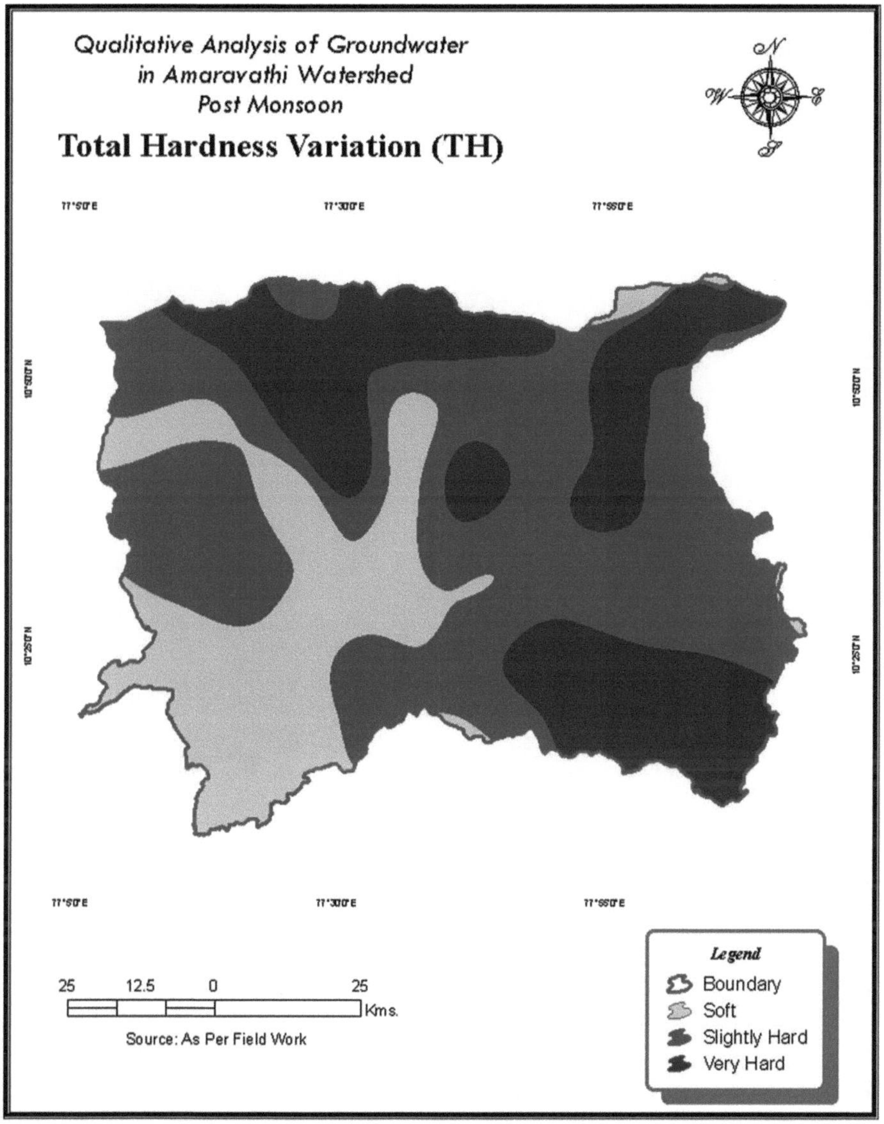

Fig. 14 Spatial distribution of total hardness during post-monsoon

production facilities. The sources of aquatic toxicity can include salt, surfactants, ionic metals and their metal complexes, toxic organic chemicals, biocides and toxic anions. Most textile dyes have low aquatic toxicity. On the other hand, surfactants and related compounds, such as detergents, emulsifiers and dispersants are used in almost each textile process and can be an important contributor to effluent aquatic toxicity, BOD and foaming.

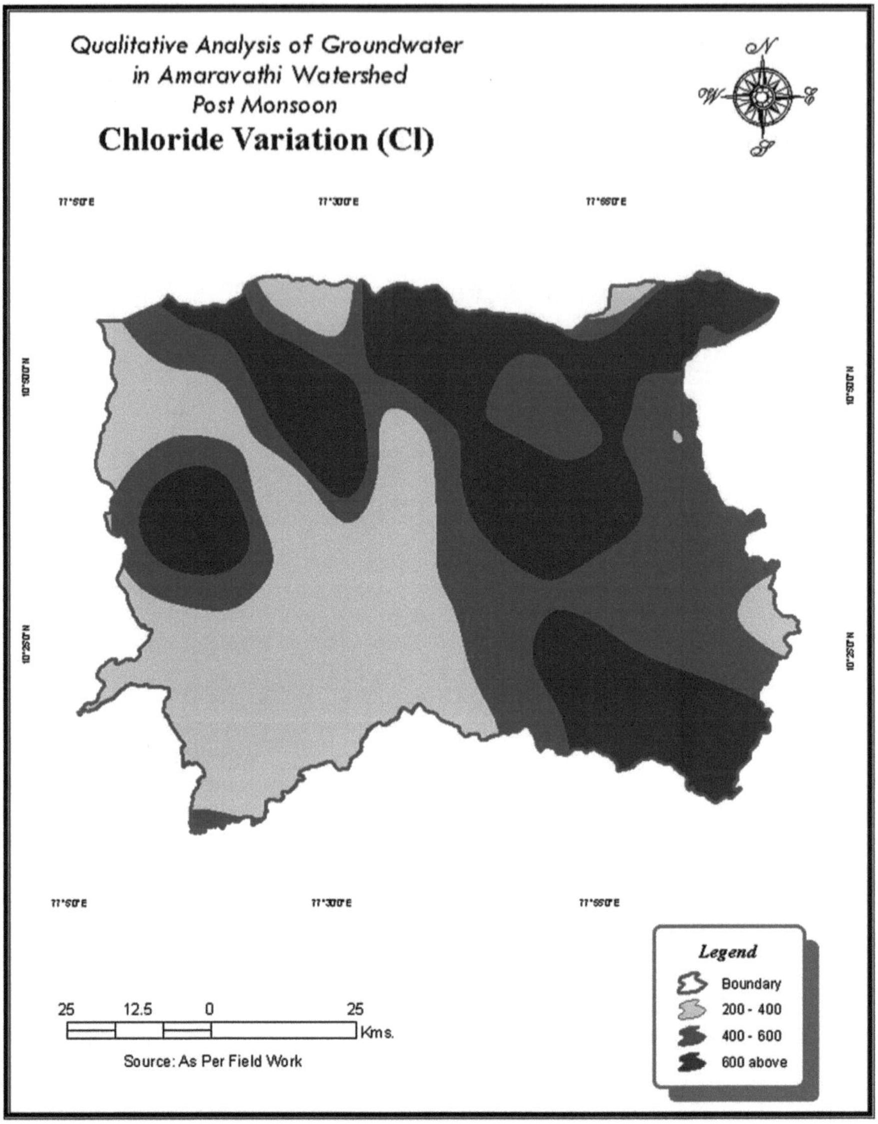

Fig. 15 Spatial distribution of chloride during post-monsoon

7.3 *Effects of Pollution on Various Segments*

Effects on human health: High pH values were reported in some of the sampling sites. High pH will give a metallic taste for human consumption and will cause corrosion of metallic pipelines that are used for supply pipelines. The total hardness is the measure of the capacity of water to precipitate soap. The hardness of more than 50 mg/l in drinking water will cause the formation of kidney stone. The minimum and maximum values recorded were 650 and 5,968 mg/l respectively as a result of higher bicarbonates, drawn into the sub-surface aquifers at levels many orders higher than that can be supplied from the natural weathering sources. The BIS has a suggested level of 500–2,000 mg/l for TDS listed in the Drinking Water Standards. Drinking water supplies with TDS levels greater than 1,200 mg/l are unpalatable to most consumers. An aesthetic objective of 500 mg/l should ensure palatability and prevent excessive scaling. An elevated total dissolved solids (TDS) concentration is not a health hazard. The concentration of the dissolved ions may cause the water to be corrosive, salty or brackish taste, and interfere and decrease efficiency of hot water heaters. Elevated TDS levels in waters are usually associated with elevated levels of nitrate, arsenic, aluminum, copper, lead, etc., which in turn may be toxic. Little is known about the effect of prolonged intake of large amounts of chloride in the diet. As in experimental animals, hypertension associated with sodium chloride intake appears to be related to the sodium rather than the chloride ion. In this regard, ubiquitous presence of elevated Cl levels in almost all the samples collected during both pre and post-monsoon is a cause of worry.

Effects on environment: The effects of hardness on aquatic life depend on which cations are making the water "hard". Concentration of total dissolved solids that are too high or too low may limit the growth and may lead to the death of many aquatic organisms. High concentrations of total dissolved solids may reduce water clarity, which contributes to a decrease in photosynthesis and lead to an increase in water temperature. Many aquatic organisms cannot survive in high temperatures. It is possible for dissolved ions to affect the pH of the body of water, which in turn may influence the overall health of many aquatic species. If TDS levels are high, especially due to dissolved salts, many forms of aquatic life are affected. The salts act to dehydrate the skin of animals. Some freshwater organisms are able to tolerate low dissolved solids levels. If a total dissolved solids increase in the water body, a shift to more salinity-tolerant species can be expected. High salinity may interfere with the growth of aquatic vegetation. Salt may decrease the osmotic pressure, causing water to flow out of the plant to achieve equilibrium. Less water can be absorbed by the plant, causing stunted growth and reduced yields. High salt concentrations may cause leaf tip and marginal leaf burn, bleaching, or defoliation.

Effects on irrigation: The textile units are located in close proximity to the water bodies due to the huge amount of water required for all the processes in the units. It has been revealed by the farmers that during the past years, crops like paddy, cotton, coconut, sugarcane, turmeric, tobacco, vegetables and banana were cultivated in the area. The poor water and soil quality has driven the farmers to cultivate only rain-fed crops and the farmers are maintaining their coconut farms with great difficulty. Even the economically less important crop like "korai" a type of grass extensively used in India for manufacturing bed mats did not grow well in the regions located in the vicinity of effluent discharge. Deposition of carbonates in the supply pipelines of irrigation, often get chocked by a widespread phenomenon of "scaling", a product of higher hardness in the ground water. Inadequate drainage or excessive evaporation from agricultural fields may lead to an accumulation of salts in the soil. Salt in the soil may harm crops. Certain salt constituents alone can prove toxic to some plant varieties. In some cases, rather than destroying a crop, elevated salt levels may simply reduce crop yields and leave the plants prone to disease.

Domestic effects: A vast majority of the households in the affected villages depended on both on-farm and non-farm activities and about 10 % of the households were found employed in non-farm activities. The major non-farm activities of these villages were renting out houses, running own dyeing or bleaching units, knitting units, money lending and water trading. In most of the households in the affected villages, at least one family member has migrated out of the affected area; among those migrated, the permanent migration was found high. The reasons for migration were poor income, desire to earn more, poor water quality, and crop activities that do not fetch any remuneration. Among the reasons, poor income from agricultural operations alone was reported by the farmers. Hard water is objectionable because of the formation of scale in boilers, water heaters, radiators, and pipes with resultant decrease in the rate of flow and heat transfer as well as in increased corrosion. In addition to its effect on soap consumption, excessive hardness can shorten the wearing ability of fabrics and toughen cooked vegetables. The reasons attributed for the sale of croplands were less crop income, migration, and poor water quality and water scarcity. Most of the people in the affected villages had sold their livestock due to various reasons like reduced milk yield, loss in weight, water scarcity and inadequate man power to tend them. Only few people were willing to pay for improving the soil and water qualities.

8 Conclusions

- The study had revealed significant impact of unplanned growth of urban centers and industrial clusters over a once fertile river basin. Though seasonal, the region once was known for cash-crops such as sugarcane, banana, coconut, etc. Now that the agricultural activities were restricted to fodder for cattle. If at all any other crops are sown, it is always associated with excessive application of fertilizers and insecticides which in turn, aggravated the soil and water quality deterioration, leading to conversion of agricultural lands into fallow lands, urban and small-scale industrial clusters. Often the effluents from these agricultural return flows, urban and industrial clusters are discharged into the natural systems without any treatment, leading to a vicious cycle of accelerated pollution.
- Quality monitoring of water courses and point and non-point sources of pollution on a continuous basis may be entrusted to research institutions like universities, colleges etc. for reporting back to Government for remedial measures and designing conservation compliance programs. A working group comprising of farmers, industry, technocrats, policy makers, researchers, enforcement agencies, etc., should be constituted to look into the physical,

economic, social, environmental and ecological issues of industrial pollution, remedial and abatement measures. Proper measures should be undertaken to revive the lost fertility and productivity of the area.

References

Balachandar D, Rutharvel Murthy K, Muruganandam R, Sumathi M, Sundararaj P, Kumaraswamy K (2011) Analysis of landuse/landcover using remote sensing techniques—a case study of Karur district, Tamil Nadu, India. Int J Curr Res 3(12):226–229

Brief industrial profile of Karur district (2012–2013) MSME, Govt of India

District ground water brochure, Karur district (2008) Tamil Nadu, Govt of India, Ministry of Water Resources, CGWB

Geetha A, Palanisamy PN, Sivakumar P, Ganesh Kumar P, Sujatha M (2008) Assessment of underground water contamination and effect of textile effluents on Noyyal River basin in and around Tiruppur Town, Tamil Nadu. E-J Chem 5(4):696–705

Jenifa Latha C, Saravanan S, Palanichamy K (2010) A sem distributed water balance model for Amaravathi River basin using remote sensing and GIS. Int J Geomatics Geosci 1(2):252–263. ISSN NO 0976–4380

Palanisamy PN, Geetha A, Sujatha M, Sivakumar P, Karunakaran K (2007) Assessment of ground water quality in and around Gobichettipalayam Town Erode district, Tamil Nadu. E-J Chem 4 (3):434–439

Pandian M, Kumanan CJ (2013) Geomatics approach to demarcate groundwater potential zones using remote sensing and GIS techniques in part of Trichy and Karur district, Tamil Nadu, India. Arch Appl Sci Res 5(2):234–240

Rajamanickam R, Nagan S (2010) Groundwater quality modelling of Amaravathi River basin of Karur district, Tamil Nadu, Using VISUAL MODFLOW. Int J Environ Sci 1(1):91–108, ISSN 0976–4402

Rajamanickam R, Nagan S (2010) Performance study of common effluent treatment plants of textile dyeing units in Karur, Tamil Nadu (India). J Environ Res Dev 5(3):623–630

Sivakumar KK, Balamurugan C, Ramakrishnan D, Leena Hebsibai L (2011) Studies on physicochemical analysis of ground water in Amaravathi River basin at Karur (Tamil Nadu), India. Water Res Dev 1(1):36–39

Zahir Hussain A, Rajadurai D (2013) Assessment of ground water pollution on the bank of river Amaravathi at Karur district, Tamil Nadu. Pelagia research library, Advances in applied science research (4):6–10. ISSN 0976-8610

Microbial Biodiversity of Selected Major River Basins of India

Ramasamy Balagurunathan and Thanganvel Shanmugasundaram

Abstract Indian subcontinent is one of the richest regions in terms of biodiversity. River and its watersheds are the nurseries and habitats for biodiversity, including the microorganisms. In this paper, the biodiversity and ecological significance of four major Indian river basins namely, the Ganges, the Cauvery, the Krishna and the Godavari are reviewed with special emphasis on microorganisms. Through this review we demonstrate that recording and detecting the microbial biodiversity and ecology of river basins would help in the formulation and implementation of appropriate conservation and management strategies in the river ecosystems. It has also shown that, the Ganges river basin is the major microbial diversity region, consisting of nearly $0.5-2.0 \times 10^6$ cfu/ml followed by the Cauvery river basin ($0.33-2.6 \times 10^5$ cfu/ml), Krishna and Godavari basins $>0.1-1.0 \times 10^4$ cfu/ml.

Keywords Biodiversity · River basin · Microorganisms · Conservation · Ecology

1 Introduction

The purported modern concept of ecology "that people are part of, not separate from the ecosystems in which they live, and are affected by changes in ecosystems, populations of species and genetic changes. In addition, the human health, wealth, security and culture are strongly affected by the changes in biodiversity" has been the way of life in India since Vedic times. Thus, India has a long tradition of biodiversity strategies that are useful to human civilization. Biodiversity plays a fundamental role in maintaining and enhancing the well-being of the world's more than 6.7 billion people, rich and poor, rural and urban alike. The water scenario of India, a subcontinent with population over 1.2 billion, is becoming critical as per

R. Balagurunathan (✉) · T. Shanmugasundaram
Actinobacterial Research Laboratory, Department of Microbiology, Periyar University, Salem 636011, Tamil Nadu, India
e-mail: rbalaguru@yahoo.com

Mu. Ramkumar et al. (eds.), *Environmental Management of River Basin Ecosystems*, Springer Earth System Sciences, DOI 10.1007/978-3-319-13425-3_27

capita availability of water has decreased from 2,309 m^3 in 1990 to 1,820 m^3 in 2001, and will be dropped to 1,140 m^3 by 2,050 (Gupta and Deshpande 2004). Biodiversity through time and space has provided the prospecting view of the genesis and diversification of various life forms. Biodiversity is the most valuable but least indebted resource, and it can be a key factor for maintenance of the world (Wilson 1992). The planning without any developmental activities will decrease the biodiversity of hydrologic regions. It is the right time to understand the aquatic nature and to analyze the diversity of living things, which are present in/associated with water bodies.

Freshwater environments of the tropics and sub-tropics are undergoing rapid deterioration due to developmental pressures, opportunistic exploitation and neglect. One of the challenges being faced by the human race to sustain the current rates of utilization of the natural resources is improving the current knowledge on biodiversity. So that it would, aid in sustainable management of the ecosystem through suitable conservation approaches (Daniels 2003) and provide the ecosystem products and services without any hindrances. The biodiversity and the higher endemism could be credited to the prevailing climate (higher rainfall, evapotranspiration, etc.), location (mid latitude), topographic and geological characteristics of the region (Sreekantha et al. 2007).

2 River Basins

A river basin drains all the land around the major rivers. Basin can be divided into water sheds, or areas of land around a smaller river, stream or lake. A drainage basin or watershed is an extent or an area of land where surface water from rain and melting snow or ice converges to a single point at a lower elevation, and the waters join another water body, such as a river, lake, reservoir, estuary, wetland, sea, or ocean. For example, a tributary stream of a brook, which joins a small river, which is tributary of a larger river, is thus part of a series of successively smaller area but higher elevation drainage basins (watersheds). The landscape is made up of many inter connected watersheds. Within each watershed, all water runs to the lowest point stream, river, lake or ocean. On its way, water travels over the land surface across farm fields, forestland, suburban lawns and city streets, or it seeps into the soil and travels as groundwater. Other terms that are used to describe a drainage basin are catchment, catchment area, catchment basin, drainage area, river basin and water basin (David Lambert 1998).

India is endowed with rich water resources. Approximately 45,000 km long riverine systems crisscross the length and breadth of the country. India has 12 major basins, 46 medium river basins and 14 minor and desert river basins. Figure 1 presents the major rivers and their drainage basins of India. Among these, the Ganga, Cauvery, Godavari and Krishna river basins are major concern. Table 1 presents the data on selected Indian river basins pertaining to their catchment areas and water resources.

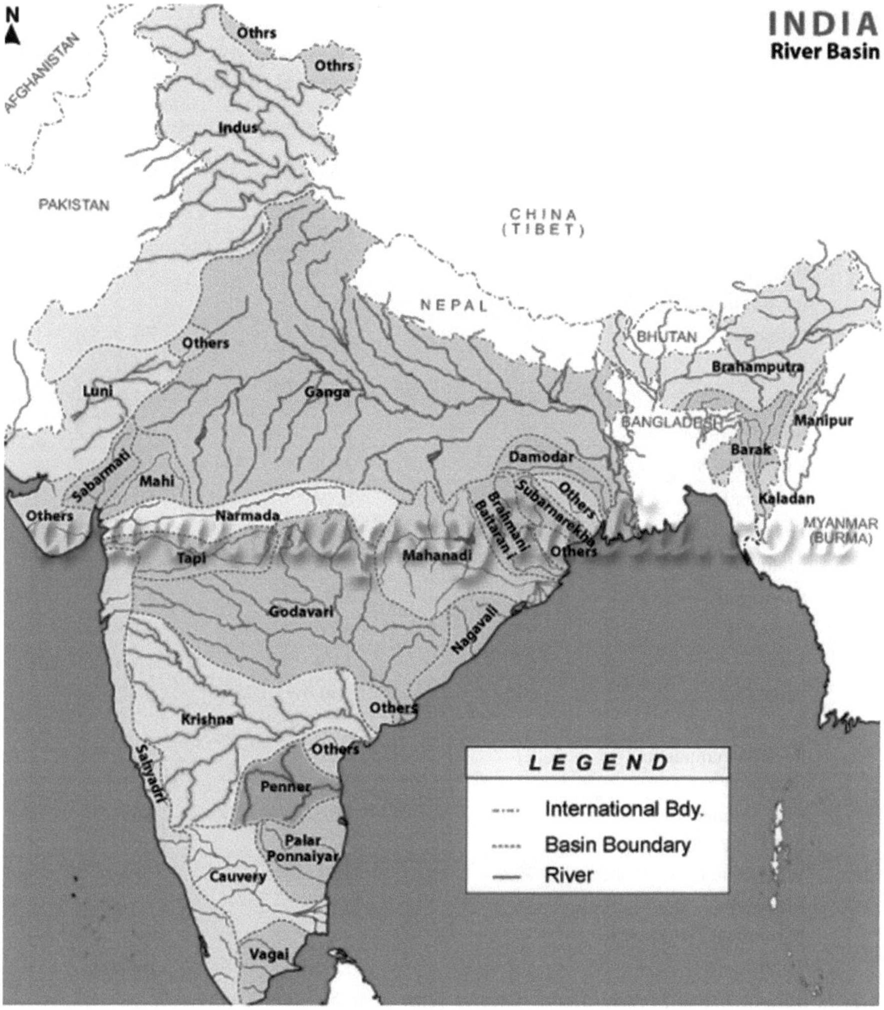

Fig. 1 Major river basins in india *Source* http://www.mapsofindia.com/maps/india/river-basins.html

2.1 Ganges River Basin

The River Ganga (Ganges) originates from the Gangotri Glacier in the Garhwal Himalayas at an elevation of about 4,100 m above the sea level under the name of Bhagirathi. This main stream of the river flows through the Himalayas till another two streams—the Mandakini and the Alaknanda—join it at Dev Prayag, the point of confluence. The combined stream is then known as the Ganga. In India, Ganga River basin is the largest and sprawls over 11 states of Indian Subcontinent. The length of the Ganges is frequently said to be slightly over 2,500 km long, about 2,505 to 2,525 km,

Table 1 Major Indian river basins and its catchment area

S. No	River basin	Catchment area (1,000 sq. km)	Water resources potential (BCM)[a]	Utilizable surface water (BCM)[a]
1	Indus	321	73.3	46.0
2	Ganga-Brahmaputra-Meghna	1,097	1,110.6	274.0
	(a) Ganga	861	525	250.0
	(b) Brahmaputra	194	537.2	24.0
	(c) Barak (Meghna)	42	48.4	
3	Godavari	313	110.5	76.3
4	Krishna	258	78.1	58.0
5	Cauvery	81	21.4	19.0
6	Subarnarekha	29	12.4	6.8
7	Brahmani-Baitarani	52	28.5	18.3
8	Mahanadi	142	66.9	50.0
9	Pennar	55	6.3	6.9
10	Mahi	35	11.0	3.1
11	Sabarmati	21	3.8	1.9
12	Narmada	99	45.6	34.5
13	Tapi	65	14.9	14.5
14	WFRTapi—Kanyakumari	112	200.9	36.2
15	EFR-Mahanadi to Pennar	87	22.5	13.1
16	EFR Pennar to Kanyakumari	100	16.5	16.5
17	WFR Kutch, Saurashtra	322	15.1	15.0
18	Minor rivers flowing into Myanmar, Bangladesh	36	31.0	
	Total		1,869.4	690.0

[a] *BCM* Billions Cubic Metres (*Source* Central Water Commission 2011)

or perhaps 2,550 km (Merriam Webster 1997). In other cases, the length of the Ganges is given for its Hooghly River distributary, which is longer than its main outlet via the Meghna River, resulting in a total length of about 2,620 km, from the source of the Bhagirathi, or 2,135 km, from Haridwar to the Hooghly's mouth. In some cases the length is said to be about 2,240 km, from the source of the Bhagirathi to the Bangladesh border, where its name changes to Padma (Sharad et al. 2007) (Fig. 2)

2.2 Cauvery River Basin

The River Cauvery is an Inter-State river in Southern India (Fig. 3). It is one of the major rivers of the Peninsular India flowing east and running into the Bay of

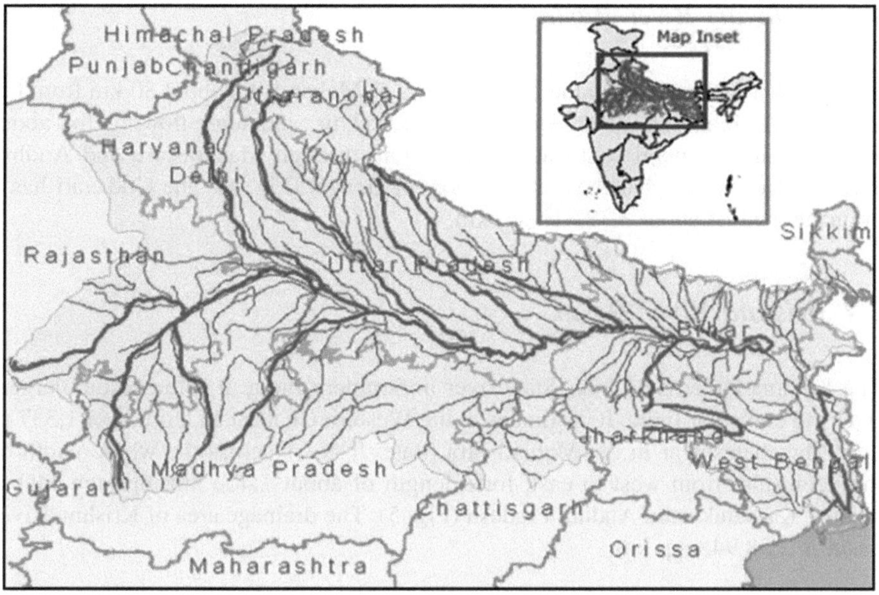

Fig. 2 Ganges River Basin *Source* http://www.indiawaterportal.org/articles/basin-maps-ganga-river-showing-basin-indicators-landcover-classes-and-biodiversity

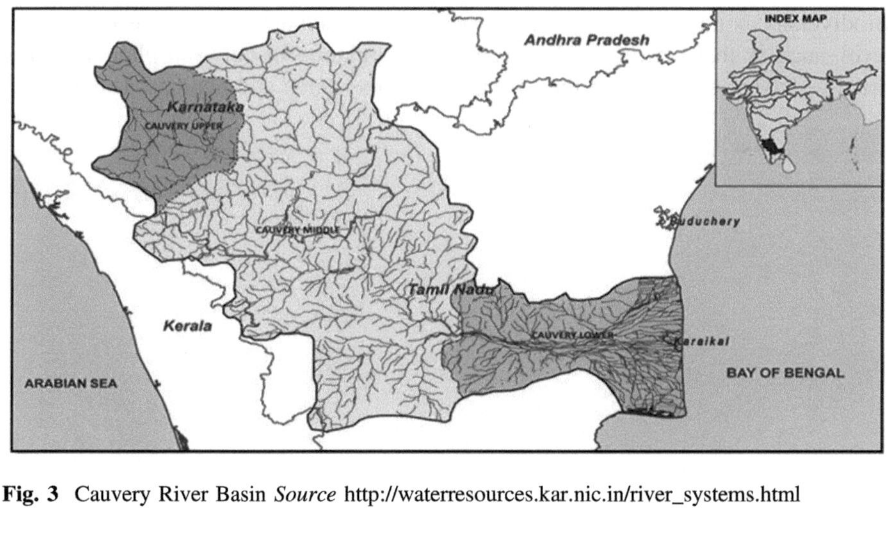

Fig. 3 Cauvery River Basin *Source* http://waterresources.kar.nic.in/river_systems.html

Bengal. The River Cauvery rises at Talakaveri on the Brahmagiri Range in the Western Ghats, presently in the Coorg district of the State of Karnataka, at an elevation of 1.341 m (4,400 ft.) above mean sea level. The drainage area of the Cauvery Basin is 81,155 sq. km.

2.3 Godavari River Basin

The Godavari River rises in the Nasik district of Maharashtra about 80 km from the shore of Arabian sea, at an elevation of 1,067 m, and after flowing for about 1,465 km in a general southeasterly direction, through Maharashtra and Andhra Pradesh, Godavari debauches into the Bay of Bengal (Fig. 4). The Godavari has a drainage area of about 3,12,813 sq. km.

2.4 Krishna River Basin

The Krishna River is an Inter-State river in Southern India. It is the second largest river in Peninsular India. It originates in the Western Ghats at an altitude of 1,337 m near Mahabaleshwar in the Maharashtra State. It flows across the whole width of the peninsula, from west to east, for a length of about 1,400 km, through Maharashtra, Karnataka and Andhra Pradesh (Fig. 5). The drainage area of Krishna River basin is 2,58,948 sq. km.

3 Biodiversity

Biodiversity is the variety of life on the Earth. It includes diversity at the genetic level, such as that between individuals in a population or between plant varieties, the diversity of species, and the diversity of ecosystems and habitats. Biodiversity

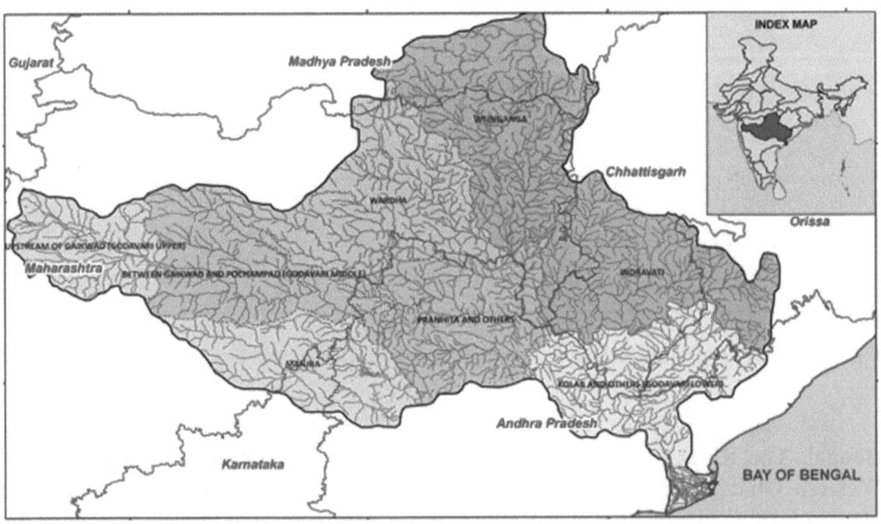

Fig. 4 Godavari River Basin *Source* http://waterresources.kar.nic.in/river_systems.html

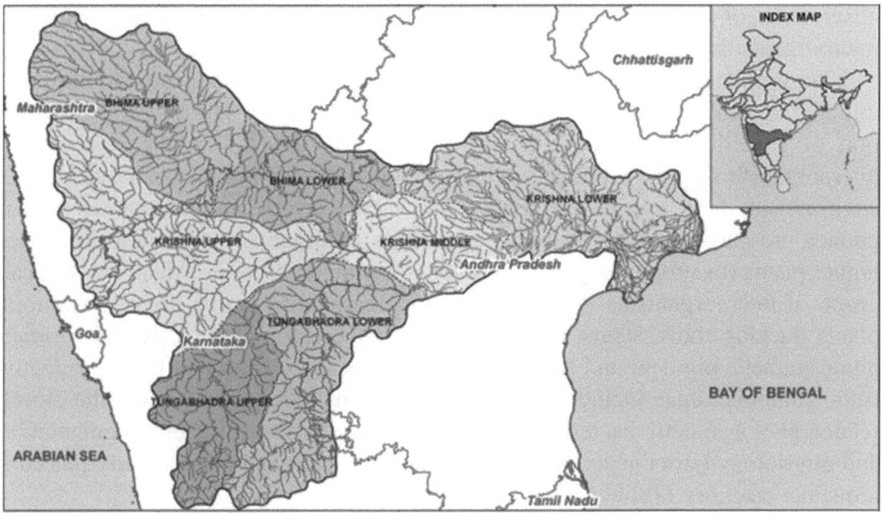

Fig. 5 Krishna River Basin *Source* http://waterresources.kar.nic.in/river_systems.html

encompasses more than just variation in appearance and composition. It includes diversity in abundance (such as the number of genes, individuals, populations or habitats in a particular location), distribution (across locations and through time) and in behavior, including interactions among the components of biodiversity, such as between pollinator species and plants, or between predators and prey. Biodiversity also incorporates human cultural diversity, which can be affected by the same drivers as biodiversity, and which has impacts on the diversity of genes, other species and ecosystems. Biodiversity has evolved over the last 3.8 billion years or so of the planet's approximately 4.6 billion-year history. Although five major extinction events have been recorded over this period, the large number and variety of genes, species and ecosystems in existence today are the ones with which human societies have developed, and on which people depend.

4 Microorganisms in Aquatic Systems

Microorganisms are those living things that are visible as individual organisms only with the aid of magnification. Microorganisms are components of every ecosystem on the Earth. Microorganisms range in complexity from single to multi-cellular organisms. Most microorganisms do not cause disease and many are beneficial. Microorganisms require food, water, air, ways to dispose of waste, and an environment in which they can live. Investigation of microorganisms is accomplished by observing organisms using direct observation with the aid of magnification, observation of colonies of these organisms and their waste, and observation of microorganisms effects on an environment and other organisms.

Life originated between 3.5 and 4 billion years ago in the aquatic environment, initially as self-replicating molecules (Alberts et al. 1962). The subsequent evolution of unicellular microbes, followed by eukaryotes, led to the existence of microorganisms which are highly adapted to aquatic ecosystems. In the freshwater environment, light energy conversion and related synthesis of carbonated compounds are carried out by three major groups of organisms (primary producers) higher plants (macrophytes), algae, and photosynthetic bacteria. Algae are the main group of microorganisms involved in this process and may be defined as simple plants (lacking roots, stems, and leaves) that have chlorophyll-A as their primary photosynthetic pigment and lack a sterile covering of cells around the reproductive cells. Although algae include unicellular organisms (blue-green algae), the closely related photosynthetic bacteria differ in terms of cell size, microbial pigmentation, and physiology (strict anaerobes, not evolving oxygen) and are generally placed in a distinct category (Table 2).

Bacteria occur as one of three major groups of prokaryotes in the freshwater environment, differing from blue-green algae in their heterotrophic mode of nutrition but showing many physiological similarities to the actinobacteria. The majority of bacterial cells have a maximum linear dimension in the picoplankton (0.2–2 mm) range, though some freshwater bacteria fall into the femtoplankton (<0.2 mm) and nanoplankton (2–20 mm) categories. These organisms can be readily observed in water samples by light (dark field, phase-contrast) microscopy or by transmission and scanning electron microscopy. Table 3 presents some of the naturally occurring bacteria, and their phenotypic and habitat diversity in freshwater ecosystems.

Fungi and fungal-like organisms lack chlorophyll and have a major saprophytic (or saprotrophic) role in aquatic environments, where they are important decomposers of both plant and animal detritus. The breakdown of biomass by these organisms is important in the regeneration of soluble materials, and they play a substantial role in the carbon, nitrogen, and phosphorus cycles of lakes, rivers, and other freshwater habitats. As heterotrophic organisms, fungi are a key component of many aquatic food webs, and are in direct competition for organic material with bacteria and protozoa. Each of these groups has adopted a particular strategy to promote their saprophytic existence, which in the case of fungi involves the development of a filamentous branching growth form—the mycelial growth habit. Some researchers estimated that there are approximately 1.7 million fungal species on the earth (Hyde and Bussaban 2007). Of these, only around 3,000 species are known to be associated with aquatic habitats and only 470 species occur in marine waters (Shearer et al. 2007). This small proportion of aquatic fungal taxa is surprising because the aquatic ecosystems are a potentially good habitat for many species. Based on this notion assume that the "real" number of aquatic fungi is much larger than 3,000 and includes a large variety of hitherto under scribed species with unknown ecological function.

Although freshwater viruses (Table 4) have probably been the least researched of all among aquatic microorganisms, their widespread occurrence and general role as parasites gives them an ecological significance equal to the other, more extensively

Table 2 Major divisions of freshwater algae: microscopical appearance, motility, and typical habitat

Algal division (Class)	No. of genera within the group	Typical colour	Typical morphology of freshwater species	Aquatic habitats	Typical examples
Cyanophyta Blue-green algae	124	Blue-green	Microscopic or visible—usuallycolonial	Lakes and streams Planktonic or attached	*Smicrocystisy-nechocystis*
Chlorophyta Green algae	302	Grass-green	Microscopic or visible — unicellular or filamentous colonial	Lakes, rivers, estuaries Planktonic or attached	*Chlamydomonas cladophora*
Euglenoids Euglenophyta	10	Various colours	Microscopic—unicellular	Lakes and ponds Planktonic	*Euglena colacium*
Yellow-green algae: Eustigmatophyta Raphidiophyta Tribophyta	90	Yellow green	Microscopic—unicellular or filamentous	Planktonic, benthic and epiphytic Wide habitat range	*Chlorobotrys vischeria*
Dinoflagellates Dinophyta	37	Red-brown	Microscopic—unicellular	Lakes and estuaries Planktonic	*Ceratium peridinium*
Cryptomonads Cryptophyta	12	Various colours	Microscopic—unicellular	Lakes planktonic	*Rhodomonas cryptomonas*
Chrysophytes[a] Chrysophyta	72	Golden brown	Microscopic—unicellular or colonial	Lakes and streams Planktonic	*Mallomonas dinobryon*
Diatoms Bacillariophyta	118	Golden brown	Microscopic—unicellular or filamentous colonies	Lakes, rivers, estuaries Planktonic or Attached	*Stephanodiscus aulacoseira*
Red algae Rhodophyta	25	Red	Microscopic or	Mainly streams,	*Batrachospermum*
			visible—unicellular or	some lakes	
			colonial	Attached	*bangia*
Brown algae Phaeophyta	4	Brown	Visible—multicellular cushions and crustose thalli	Lakes and streams Attached	*Pleurocladia heribaudiella*

Source Wehr and Sheath (2003), Lee (1997), Wehr and Sheath (2003)
[a] Including haptophyte and synurophyte algae

Table 3 Some naturally-occurring aquatic bacteria, illustrating the variety of phenotypic and habitat diversity in freshwater systems. Individual species are arranged within selected numbered groups, which represent a relatively small fraction of the 35 phenotypic assemblages enumerated in bergey's manual of systematic bacteriology

Bacterium	Some key phenotypic features	Habitat
Group 1: helically shaped, motile bacteria (Spirochaetes); single celled, contain periplasmic flagella enclosed within an outer sheath, gram-negative		
Spirochaeta plicatilis	Helical cells, capable of locomotion both in suspension and on solid surfaces, Anaerobic or microaerophilic	Common in brackish and marine freshwaters rich in H_2S
Group 2: aerobic or microaerophilic; motile helical or vibrioid (comma-shaped) cells, gram-negative		
Spirillum volutans	Rigid helical rods, bipolar tufts of flagella, breakdown of organic material	Microaerophile, eutrophic, stagnant fresh or saline waters
Bdellovibrio bacteriovorus	Small, curved cells Motile via single polar flagella at one or both ends of cells, Parasite of bacteria and algae, Biphasic life cycle	Widely present throughout freshwaters and soil
Group 4: aerobic rod- and coccoid-shaped cells, gram-negative		
Azotobacter chroococcum	Typically occurs as large ovoid cells, but varies in shape (pleomorphic), Motile (with peritrichous flagella) or non-motile, Important N_2 fixer	Range of soil and freshwater systems
Methylomonas methanica	Straight, curved or branched rods, Motile via a single polar flagellum, Methane oxidising bacterium	Aerobic conditions close to anaerobic sediments
Pseudomonas aeruginosa	Motile (polar flagella), rod-shaped, General breakdown of organic material	Very common throughout aerobic freshwater environments
Group 5: facultative anaerobes, rod shaped, gram-negative		
Escherichia coli	General breakdown of organic material	Faecal contaminant of freshwaters
Group 7: strictly anaerobic, morphologically diverse bacteria; sulphate- or sulphur-reducing, gram-negative		
Desulfovibrio desulfuricans	Spiral to vibrioid-shaped cells, Motile by single or tufts of polar flagella, Important sulphur reducing bacterium	Anaerobic sediments and hypolimnia
Group 10: photosynthetic bacteria, not generating oxygen; wide range of shape, gram-negative		
Thiopedia rosea	Spherical to ovoid cells, Non-motile, Obligately anaerobic and phototrophic, Bright purple-red colour	Anaerobic sediments and top of hypolimnia
Rhodospirillum rubrum	Photosynthetic purple non-sulphur bacterium, Spiral cells, motile by	Lakes and mudflats

(continued)

Table 3 (continued)

Bacterium	Some key phenotypic features	Habitat
	polar flagella, Anaerobic to aerobic growth	
Group 13: budding and/or appendaged bacteria; a very diverse group, gram-negative		
Caulobacter vibrioides	Rod shaped or vibrioid cells, with a single flagellum at one end of the cell and a stalk at the other, Non-budding, Motile and attached phases in life cycle, Predatory	Widely present in a range of freshwater systems
Group 14: bacteria contained within a sheath of extracellular material, growing as chains of cells in filaments, gram-negative		
Sphaerotilus natans	Occurs as chains of rod-shaped cells, with prominent sheath and holdfasts for attachment, Important in breakdown of organic material	Attached to submerged plants and stones, in high nutrient flowing waters
Group 15: gliding motility (on solid surfaces); gram-negative, diverse morphology		
Beggiatoa alba	Long filamentous bacterium, occurring as single, cells or in filaments, Cells often contain distinct sulphur inclusions, Oxidises H_2S to S	Occurs in sediments at interface between underlying anoxic high sulphide zone and the overlying oxic zone
Cytophaga hutchinsonii	Rod-shaped, with gliding motility, Important in breakdown of polysaccharides (cellulose and chitin) and proteins	Lake and river sediments, associated with decomposing organic matter
Lysobacter enzymogenes	Thin rods, Able to lyse a variety of microorganisms, including blue-green algae	Widely present in freshwaters
Group 18: endospore-forming gram-positive rods and cocci		
Bacillus pituitans	Rod-shaped, Motile by peritrichous, flagella, decomposition of protein, forming H_2S	Bog lakes of high organic content

Source Holt et al. (1994)

studied microbial groups (Wommak and Colwell 2000). Various ecological studies have been carried out on the role of viruses in freshwater systems (Weinbauer and Hofle 1998), but there is little information on viruses in benthic systems such as biofilm communities. The relative lack of ecological information on freshwater compared with marine viruses also means that information obtained from marine environments may fill the gaps in our understanding of freshwater systems.

Protozoa are important consumers of organic debris and microorganisms in many freshwater bodies, including natural ecosystems of both standing and flowing waters and man-made aquatic systems of economic importance such as wastewater treatment plants. In most of these systems, protozoa compete with other grazing organisms (multi-cellular invertebrates) for food supply, and the relative grazing impact of these two major groups can be considered both in terms of biomass transfer and type of food supply.

Table 4 Freshwater bacteriophages and cyanophages in the families Myoviridae, Siphoviridae and Podoviridae

	Family	Phage species	Host alga or bacterium
Myoviridae	Phages with a central tube and contractile tail, separated from the head by a neck	AS-1	*Anacystis nidulans* *Synechococcus cedrorum*
		N-1 *Pseudomonas* phage D312	*Anabaena* spp.
		Pseudomonas phage UT1	*Pseudomonas aeruginosa* *Pseudomonas aeruginosa*
Siphoviridae	Long, non contractile tails	SM-2	*Synechococcus elongatus*
		S-2L	*Synechococcus* sp. 698 *Microcystis aeruginosa*
		Methanobacterium phage ø F3	*Methanobacterium* sp.
Podoviridae	Short tail	LPP-1	*Lyngbya* *Plectonema* *Phormidium*
		SM-1	*Synechococcus elongatus* *Microcystis aeruginosa* *Microcystis aeruginosa*
		Pseudomonas phage ø PLS27	*Pseudomonas aeruginosa*

Source Martin and Kokjohn (1999) and Tidona and Darai (2002). Each family of phages contains both cyanophages (shaded) and bacteriophages. Large linear viruses (family Inoviridae) that infect bacteria have also been detected in some freshwater environments (Middelboe et al. 2003)

5 Microbial Diversity of River Basins and Environmental Management

River basin management is the process of coordinating conservation, management and development of water, land and related resources across sectors within a given river basin, in order to maximize the economic and social benefits derived from water resources in an equitable manner while preserving and, where necessary, restoring freshwater ecosystems. Therefore, management of river basins must include maintaining ecosystem functioning as a major goal. These kinds of ecosystem approaches are central tenet of the convention on biological diversity. River basins are dynamic over space and time, and any single organization involvement has implications for the system as a whole. Rivers form a hydrological mosaic, with an estimated 263 international river basins covering 45.3 % of the land surface area of the earth, excluding Antarctica.

Earth and its ecosystem components are dominated by two vastly different sets of processes. For at least 75 % of the approximately 4 billion year history of life, microorganisms particularly, bacteria and archaea controlled the elemental cycling, organic matter production and turnover and the planetary climate. In various ways, they still do. Microbes including bacteria, archaea, phytoplankton, protozoans and fungi still catalyze the major transformations of the elements, break down organic matter, and produce and consume oxygen and carbon dioxide in aquatic environments (Smil 2003). Around half the global net primary productivity is by unicellular phytoplankton in the sea and fresh water ecosystems (Falkowski et al. 2000), and most of the universal respiration (terrestrial and marine) is microbial. However, just within the past century (0.00000001 % of the history of life!), many processes on the earth have become dominated by human activities. Vitousek et al. (1997) describe the extent of anthropogenic impact on the earth system processes. To name but a few of the examples given by those authors: over 60 % of all marine and river fisheries are fully exploited, overexploited or exhausted; over 20 % of all bird species on the earth have become vanished; and 50 % of all the accessible surface freshwater is used in human activity. The extent of human perturbation of the chemical composition of the atmosphere is well known: the CO_2 concentration has increased by nearly 40 % since 1,750. The transformation in the nitrogen cycle is even more striking. Human activity now accounts for more than half of all reactive nitrogen entering terrestrial ecosystems, an increase of over 100 % in the global nitrogen cycle. Some researchers suggest the importance of recognizing the shared dominance of the planet by the extreme ends of the evolutionary process: microbes and man. Space limits this to a superficial treatment, with most specific examples emphasizing the bacteria.

5.1 Microbial Diversity in the Selected Indian River Systems

More than 70 % of the earth's surface is covered by various water bodies. Life originated in the water systems about 3.5 billion years ago especially in the oceans and microbes were the only form of life for the first two thirds of the planet's existence. The development and maintenance of all other living things in water system completely depend absolutely on the past and present activities of aquatic microorganisms. Such understanding is vital for knowing of our life in a period of rapid global change. In this section, microbial diversity in few selected major river systems of India namely, the Ganges, Cauvery, Godavari and Krishna are presented.

Ganges River: According to a World Bank sponsored study (State of Environment Report India State of Environment Report, India, 2009), 9–12 % of total disease burden in Uttar Pradesh (U.P) are contributed by the pollution of Ganga River. The coliform bacteria levels are in excess of 2 lakh MPN/100 ml as against the national water quality standards of 5,000 MPN/100 ml. The presence of coliforms in water also hints at the potential presence of pathogenic microbes, especially *E. coli* which might cause water borne diseases (Kulshrestha and Sharma 2006). The Ganga River ecosystem supports 25,000 or more species ranging from microorganisms to mammals in Indian Subcontinent (Munendra Singh and Amit K. Singh 2007). Rai et al. (2010) observed extensive occurrences of *Salmonella, Vibrio, E. coli* and *Clostridium* in the Ganga waters at Varanasi. The antibiotic resistant bacterial strains were isolated from intestine of *Labeo rohita*, which are commonly originated in the Ganges River. The resistant pattern determined that 3 water and 1 clinical isolate belonging to *Pseudomonas, Proteus* and *Klebsiella* species dominated the total microbial diversity (DebMandal et al. 2011). An attempt was made to estimate the water quality of the River Ganga during mass bathing in Haridwar during Maha Kumbha of 2010. The results have shown log 6.79 cfu. ml^{-1} colonies and Most Probable Number (210 and 150 MPN/100 ml) for total and fecal coli form, particularly *Escherichia coli* (Arora et al. 2013).

Cauvery River: The microbiological scrutiny was performed in the Cauvery River basin based on samples collected during monsoon (2007), winter, summer and spring sessions (2008). The samples were analyzed for Total viable counts (TVC), Total coliform counts (TC) and Total Streptococci counts (TS). Total viable counts were found in the range of $6.2–26.0 \times 10^4$ cfu/ml in monsoon, $5.2–20 \times 10^4$ cfu/ml in winter and $3.3–15.5 \times 10^4$ cfu/ml in spring, respectively (Kumarasamy et al. 2009). The mean cell-length of heterotrophic bacterioplankton and their relationships with environmental variables were analyzed for two years based on samples collected from the Cauvery River Channel and four of its upstream tributaries during February 2000–January 2002. In addition, samples were collected from Talakaveri, from where the River Cauvery originates and analyzed for bacteriological aspects. The results show that the MPN counts recorded a maximum of 21 MPN/100 ml. Very less count of *Salmonella* sp. and *Shigella* sp. were recorded in tested water samples (Krishna et al. 2012).

The water quality of Cauvery River at Kudige, Kushalnagar was studied by using bacteriological parameters. The statistical analysis of seasonal bacterial parameters recorded high density of bacterial counts of 296 cfu/ml during summer season from February 2010 to May, 2010. The MPN count of Kudige Cauvery water samples were found to be $\geq$ 1,600 MPN/100 ml, in all the seasons from February 2009 to January 2011. The colonies of *Salmonella* sp. and *Shigella* sp. were found in the range of 1.0–6.0 colonies/100 ml. Maximum count of *Shigella* sp. (6.0 colonies/100 ml) was noted, when compared to *Salmonella* sp. (Krishna and Jayashankar 2012). Rajesh Muthu et al. (2013) studied the actinobacterial diversity of the Cauvery River. These authors isolated five different actinobacterial strains from 25 soil samples of the Cauvery River basin and recorded the presence of actinobacterium, *Isoptericola variabilis* by using 16S rRNA sequencing (Rajesh Muthu et al. 2013). Vijayalakshmi et al. (2013) attempted analysis of bacterial and fungal population present in the Cauvery River water at Pallipalayam region, Tamil Nadu. The results have shown that the maximum and minimum values of population density ranged between $4.3-0.5 \times 10^3$ cfu/ml and $3.2-0.1 \times 10^3$ cfu/ml. The total coliform count of the Cauvery River water samples, taken from Vairapalayam were examined by Sivaraja and Nagarajan (2014). The results revealed that the samples contained coliform count of about 1,800 MPN/100 ml.

Krishna—Godavari Basin: Bacteriological analysis of the river water samples collected from upstream, midstream and downstream regions of the River Krishna revealed higher diversity of *Escherichia coli, Citrobacter freundii, Citrobacter diversus, Enterobacter aerogenes* and *Klebsiella* speiecs (Chitanand et al. 2010). A group of antibiotic resistant bacterial strains were isolated from Krishna–Godavari basin, Bay of Bengal, India. Totally 53 isolates were separated and screened against a sequence of antibiotics. The strains such as *Pseudomonas, Bacillus, E. coli, Enterobacter, Clostridium, Yersinia, Klebsiella, Proteus, Microbacterium*, etc. showed resistant nature against tested antibiotics (Ruban and Gunaseelan 2011). The drinking water quality of the Godavari River at Nanded city, Maharashtra was studied by Rizvi et al. (2013) and the results indicated that the range of < 10–100 cfu/ml bacterial colonies were present in the studied samples. The microbial parameters of the Krishna water samples in and around Vijayawada were studied in the months of January–December, 2012. The biochemical identifications showed the presence of *E.coli, Salmonella typhi, Staphylococcus aureus, Pseudomonas aeruginosa* and *Vibrio cholerae* in Krishna river water ecosystem (Jayalakshmi and Lakshmi 2014).

6 Conclusions

- The microbes came into existence many millions of years ago and form part of every conceivable ecological niches of the Earth. Considering their ecological services, and impacts on the aquatic systems, a better understanding on their diversity and ecological functions, especially in the riverine ecosystems are necessary.
- Currently, extensive researches are being conducted on the microbial community, for the synthesis of various primary and secondary metabolites. Till date, nearly 33,500 bioactive compounds are produced from microbial sources including fungi (15,600), actinobacteria (13,700) and other unicellular—bacteria (4,200).
- With this view the present review is a base to enhance the research in aquatic microbial populations and their bioactive metabolites, in particular, the microbial communities associated with river basins.

References

Alberts B (1962) Molecular biology of the cell. Garland Publishing Inc., New York, USA

Arora NK, Tewari S, Singh S (2013) Analysis of water quality parameters of river ganga during maha kumbha, Haridwar, India. J Environ Biol 34:799–803

Avijit G (2007) Large rivers: geomorphology and management. Wiley. p 347. ISBN 978-0-470-84987-3

Central Water Commission (2011) Major river basins of India-An overview. Ministry of water Resources, Government of India, New Delhi

Chitanand MP, Kadam TA, Gyananath G, Totewad ND, Balhal DK (2010) Multiple antibiotic resistance indexing of coliforms to identify high risk contamination sites in aquatic environment. Indian J Microbiol 50:216–220

Daniels RJR (2003) Biodiversity of the western ghats: An overview. In: Gupta AK, Ajith Kumar, Ramakanthan V (eds) ENVIS bulletin: wildlife and protected areas, conservation of rainforests in India, 4(1):25−40

DebMandal M, Mandal S, Kumar Pal N (2011) Antibiotic resistance prevalence and pattern in environmental bacterial isolates. Open Antimicrob Agents J 3:45–52

Falkowski P, Scholes RJ, Boyle E, Canadell J (2000) The global carbon cycle: a test of our knowledge of earth as a system. Science 290:291–296

Gupta SK, Deshpande RD (2004) Water for India in 2050: first-order assessment of available options. Curr Sci 86:1216–1224

Holt J (1994) Bergey's manual of systematic bacteriology. Williams and Wilkins Company, Baltimore

Hyde K, Bussaban B, Paulus B (2007) Diversity of saprobic microfungi. Biodivers Conserv 16 (1):7–35

Jayalakshmi V, Lakshmi N (2014) Assessment of microbiological parameters of water and waste waters in and around Vijayawada. IOSR J Environ Sci Toxicol Food technol 8(2):53–57

Krishna Jayashankar M (2012) Physicochemical and bacteriological study of Kaveri river at Kudige, Kodagu District. Karnataka. Int J Environ Sci 2(4):2040–2049

Krishna H, Hosmani S, Jayashnakar M (2012) Physico-chemical and bacteriological parameters of Kaveri river at talakaveri region—A comparative study. Natl Mon Refereed J Res Sci Technol 1:6

Kulshrestha H, Sharma S (2006) Impact of mass bathing during Ardhkumbh on water quality status of river Ganga. J Environ Biol 27(2):437–440

Kumarasamy P, Vignesh S, Arthur James R, Muthukumar K, Rajendran A (2009) Enumeration and identification of pathogenic pollution indicators in Cauvery river, South India. Res J Microbiol 1:1–10

Lambert D (1998) The field guide to geology. Checkmark Books. pp 130−13. ISBN 0-8160-3823-6

Lee R (1997) Phycology. Cambridge University Press, Cambridge

Martin E, Kokjohn T (1999) Cyanophages. In: Granoff A, Webster R (eds) Encyclopaedia of virology. Academic Press, San Diego, pp 324–332

Merriam-Webster (1997) Merriam-Webster's geographical dictionary. Merriam-Webster. p 412. ISBN 978-0-87779-546-9

Middelboe M (2003) Distribution of viruses and bacteria in relation to diagenetic activity in an estuarine sediment. Limnol Oceanogr 48:1447–1456

Rai PK, Mishra A, Tripathi BD (2010) Heavy metal and microbial pollution of the river Ganga: a case study of water quality at Varanasi. Aquat Ecosyst Health Manage 13(4):352–361. doi:10.1080/14634988.2010.528739

Rajesh Muthu M, Subbaiya R, Balasubramanian M, Ponmurugan P, Selvam Masilamani (2013) Isolation and identification of actinomycetes *Isoptericola variabilis* from Cauvery river soil sample. Int J Curr Microbiol App Sci 2(6):236–245

Rizvi R, Kamble LH, Kadam AS (2013) Heterotrophic plate count bacteria in drinking water supply of a selected area of Nanded City. Sci Res Report 3(1):66–68

Ruban P, Gunaseelan Chandran (2011) Antibiotic resistance of bacteria from Krishna Godavari Basin, Bay of Bengal, India. Environ Exp Biol 9:133–136

Singh Munendra, Singh Amit K (2007) Bibliography of environmental studies in natural characteristics and anthropogenic influences on the Ganga river. Environ Monit Assess 129:421–432. doi:10.1007/s10661-006-9374-7

Sivaraja M, Nagarajan K (2014) Levels of indicator microorganisms (Total and Fecal Coliforms) in surface waters of rivers Cauvery and Bhavani for circuitously predicting the pollution load and pathogenic risks. Int J Pharm Tech Res 6(2):455–461

Sharad KJ, Pushpendra KA, Vijay PS (2007) Hydrology and water resources of India. Springer. pp 334−342. ISBN 978-1-4020-5179-1

Shearer C, Descals E, Kohlmeyer B, Kohlme yer J, Marvanova L, Padgett D (2007) Fungal biodiversity in aquatic habitats. Biodivers Conserv 16(1):49–67

Smil V (2003) The earth's biosphere: evolution, dynamics and change. MIT Press, Cambridge

Sreekantha, Subash Chandran MD, Mesta DK, Rao GR, Gururaja KV, Ramachandra TV (2007) Fish diversity in relation to landscape and vegetation in central Western Ghats, India. Curr Sci 92:11

State of Environment Report, India (2009) Ministry of environment and forests, Government of India, New Delhi

Tidona C, Darai G (2002) The Springer index of viruses. Springer, Berlin

Vijayalakshmi G, Ramadas V, Nellaiah H (2013) Evaluation of physico-chemical parameters and microbiological populations of Cauvery river water in the Pallipalayam region of Tamilnadu, India. Int J Res Eng Technol 2(3):305–312

Vitousek PM, Mooney HA, Lubchenco J, Melillo JM (1997) Human domination of earth's ecosystems. Sci 277:494–499

Wehr J, Sheath R (2003) Freshwater algae of North America. Academic Press, Amsterdam

Weinbauer M, Hofle M (1998) Significance of viral lysis and flagellate grazing as factors controlling bacterioplankton production in a eutrophic lake. Appl Environ Microbiol 64:431–438

Wilson EO (1992) Diversity of life. W.W. & Company, Norton, p 424

Wommak K, Colwell R (2000) Virioplankton: viruses in aquatic ecosystems. Microbiol Mol Biol Rev 64:69–114

Web References

http://waterresources.kar.nic.in/river_systems.html

http://www.indiawaterportal.org/articles/basin-maps-ganga-river-showing-basin-indicators-landcover-classes-and-biodiversity

http://www.mapsofindia.com/maps/india/river-basins.html

Application of Diatom-Based Indices for Monitoring Environmental Quality of Riverine Ecosystems: A Review

R. Venkatachalapathy and P. Karthikeyan

Abstract Diatoms are a large and diverse group of single-celled algae. Diatom-based indices are increasingly becoming important tools for assessment of environmental conditions in aquatic systems. Diatoms have long been lauded for their use as powerful and reliable environmental indicators. The objective of this paper is to review and explain the application of diatoms in environmental studies of river aquatic system. Review of Diatom Research in India and abroad shows that the diatom study and its applications are at nascent stage when compared to their utilization in Australia, Canada, United States and Brazil. The occurrences of Diatoms in surface waters especially in major rivers of the world are yet to be recorded. However, the review of Diatom Research provides considerable scope and applications in understanding and monitoring of environments. From the analysis of diatoms in surface waters of India, we conclude that the diatom studies can be best used in environmental assessments of water-quality of ecosystem.

Keywords Diatoms · Environment · Monitoring · Management · Aquatic ecosystem

1 Introduction

Diatoms are single celled microscopic algae that posses ornamented cell wall composed of silica (SiO_2). They inhabit all the aquatic environments and are found in great abundances and diversity. The shape, size and pattern of silica frustules form the basis for classification and identification of diatom taxa. Due to their rapid response to environmental changes (Karthick et al. 2008; Venkatachalapathy and

R. Venkatachalapathy (✉) · P. Karthikeyan
Department of Geology, Periyar University, Salem 636 011, India
e-mail: rvenkatachalapathy@gmail.com

Mu. Ramkumar et al. (eds.), *Environmental Management of River Basin Ecosystems*,
Springer Earth System Sciences, DOI 10.1007/978-3-319-13425-3_28

Karthikeyan 2012, 2014), deterioration of water quality especially from impacts such as nutrient enrichment, acidification and metal contamination, diatoms have been used widely for biomonitoring of aquatic ecosystems (Cholnoky 1968; Lowe 1974; Schoeman and Archibald 1984, 1985, 1987, 1988; Somers et al. 2000; Kelly and Whitton 1995a, b). According to Kelly (1998a, b), diatoms are one of the basic components of river bio-monitoring and assessment of ecological status of rivers. These authors have also established many diatom indices for water quality assessment of rivers and lakes. Recent studies have shown that the diatom based indices vary in their capacity to ionic composition and organic pollution in rivers (Gomez and Licursi 2001; Taylor et al. 2007). The strong correlations between diatoms and ionic concentration enable diatoms to be used to reconstruct past changes in lake water salinity driven by hydrologic and climatic change (Fritz et al. 1993, 1999; Blinn 1995). Diatom assemblages in rivers and streams can be analyzed by rigorous statistical techniques to establish their relationship to the environmental factors. The relative abundance of diatom species is used as the most valuable characteristics of diatom assemblages for bio-assessment of river health. River health can be assessed by using diatoms [Prygiel and Coste (1993) in France, Harding and Kelly (1999) in UK and Stevenson and Pan (1999) in the USA].

A diatom has shorter generation times than fish and macro-invertebrates and responds rapidly to environmental changes, thus provides scope as early warning indicators for both pollution increases and habitat restoration success. Cost of sampling and analysis are relatively low when compared to other organisms. Diatom samples can be collected easily for long periods of time for analysis. Thus, the study of diatoms has become an important element of monitoring and assessment programs in countries around the world. Undisturbed core sediments from ecosystems will provide habitat history of surface water bodies (Amoros and van Urk 1989; Cremer. et al. 2004; Gell et al. 2005). Past conditions in streams and rivers can also be assessed utilizing museum collections of diatoms on macrophytes and fish (van Dam and Mertens 1993; Rosati et al. 2003; Yallop et al. 2006). Environmental changes in marine, brackish waters and estuaries can also be inferred by diatom studies, although the techniques and interpretations are more challenging than those used in freshwater lakes and rivers (John 1983; Snoeijs 1999).

Artificial substrata are used for precise assessments in streams with highly variable habitat conditions and natural substrate unsuitable for sampling. The latter may be the case in deep, channelized or silty habitats. Benthic algal communities on artificial substrata are commonly different than those on natural substrata (Tuchman and Stevenson 1980). Suitable laboratory methods and field-sampling methods should be followed to minimize errors in assessments of water quality in surface water bodies.

This paper reviews developments in the diatom studies over the past few decades with special emphasis on the application of diatom indices for monitoring the environmental quality of freshwater ecosystems.

2 Prerequisites for the Study of Diatoms

2.1 Extraction of Diatoms

The attached microalgae, of which diatoms are a dominant component, can be collected by scraping natural substrates in streams. Stones (pebbles) and macrophytes are good sources of diatoms (Fig. 1). Floating filamentous algae or floating wood and plant materials are often full of loosely attached diatoms (Fig. 2). Diatoms grow attached to concrete substrates in urban drains. Epiphytic diatoms grow attached to macrophytes and large algae. Metaphyton includes diatoms loosely attached to algal mats and filamentous algae. Typically diatoms can be seen in the form of brown or black, slimy coatings on stones, submerged plants and mud. The colored coatings may be removed with a brush, 'spoon' or blunt scalpel from hard substrates and by squeezing or washing off from soft substrates like macrophytes and filamentous algae.

Benthic diatoms may be collected by 'scooping' the top 3–5 cm of the sediment of a stream, with a vial, but this method disturbs and mixes the sediments (Fig. 3). An intact 'core' may be collected, by pushing an inverted open vial (with a small hole at the bottom), into the sediment and lifting it up. Using this method one can collect epipelon (growing on mud) and episammon (growing on sand) intact from a specific area. Motile species are common in mud samples.

Tyagi (1985) collected water samples from various water bodies viz., lakes, ponds, wells and drains in and around Delhi and treated them with concentric HCl acid and the supernatant was discarded. It was followed by addition of concentric H_2SO_4 to remove the organic material present in the sample material. The supernatant was cooled and added with solid $NaNO_3$. Bhatt et al. (2005) used a sharp-edged knife to

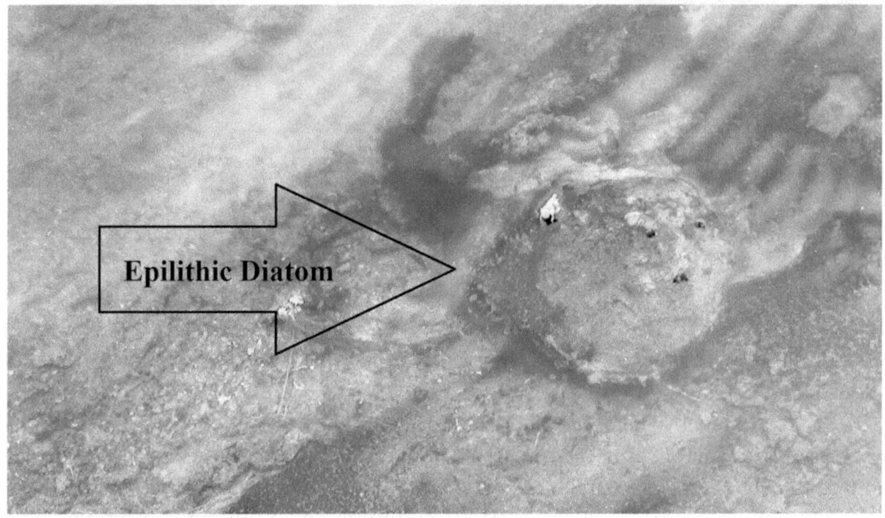

Fig. 1 Epilithic diatom samples collection in Cauvery River

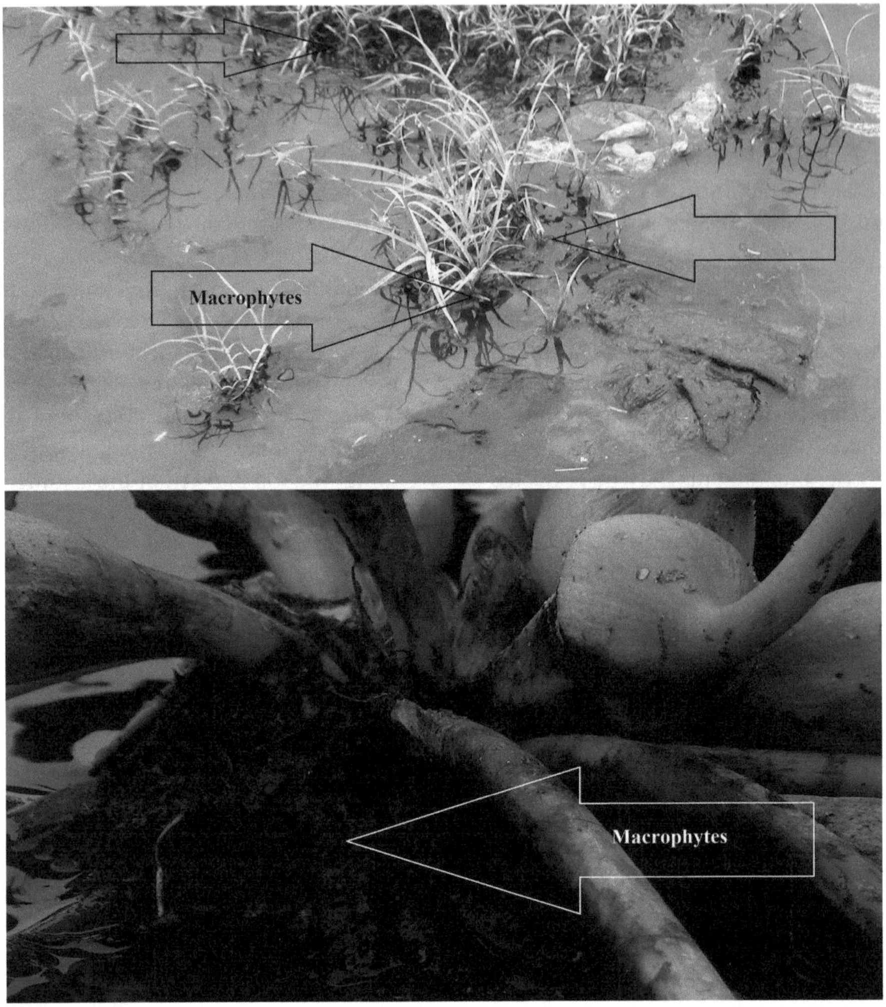

Fig. 2 Macrophytes diatom samples collection in Cauvery River

collect epilithic diatom samples by scraping the rocks and boulder surfaces and collected scraps of 3 mm^2. The epilithic diatom samples obtained thus were cleaned with nitric acid and potassium dichromate. The samples were centrifuged at 10,000 rpm for 10–30 min and the supernatant was discarded. The pellets were washed twice with Isopropanol, Xylene and by distilled water for making permanent slides using Canada balsam.

Venkatachalapathy and Karthikeyan (2014) collected epiphytic samples from the Cauvery River by brushing the undersurfaces and petioles of at least 20 plant leaves and roots. The collected materials were preserved in formaldehyde (4 %). For Polarizing microscopic analysis, a 10 ml epiphytic and Epilithic subsamples were extracted and cleaned using 30 % H_2O_2 and concentrated HNO_3. Singh et al. (2010)

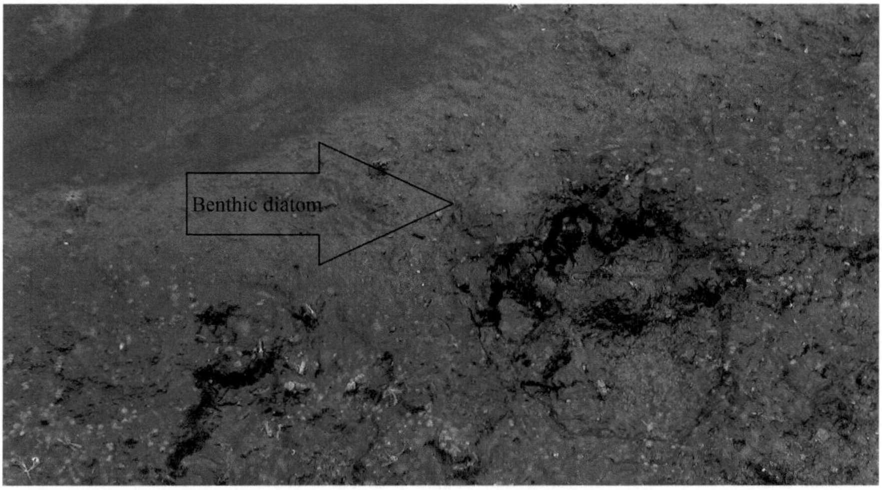

Fig. 3 Benthic diatoms samples collection in Cauvery River

collected diatom samples using planktonic mesh net (pore size 40 μm) and preserved it in Lugol's solution and treated it hot HCl and $KMnO_4$ method. Kociolek and Karthick (2011) collected samples from floating aquatic plants at Kodaikanal Lake by scrubbing the plants with a toothbrush and the resultant suspension was preserved in ethanol. The samples were digested using concentrated nitric acid and centrifuged several times to remove the acid to prepare the diatom slides. A list of procedures involved in collection and segregation of diatoms are listed in the Table 1.

2.2 Preparation Diatom Slides

Fresh samples should be examined, if habit and chloroplasts are to be observed. 'Colonies' are disintegrated by the processing employed for permanent preparation. Chloroplasts are destroyed by the chemical processing required for making permanent slides. If diatom samples are preserved by Lugol's solution, 4 % formalin or Transeau's Algal Preservative (6:3:1 = Water, Ethyl alcohol, Formalin), colonial structure and chloroplasts can be retained.

Concentrated diatom samples are boiled in 50–60 % Nitric Acid to oxidize the organic materials and then the acid is washed off with deionizer water by settling or centrifugation. The final 'cleared' diatom samples-an ash colored material is mounted in a mounting medium e.g. Napthrax with a refractive index close to 1.7 and placed on a hotplate until the solvent in the medium evaporates. The slide is cooled immediately, by removing from the hotplate. The diatom slide thus obtained is permanent. The diatoms cleaned cells in permanent slides are observed under ×40 and photographed using black and white for best results or by digital photography.

Table 1 Methods used by researchers for diatom slides preparation

Author(s)	Material	Methods	Environment	Country
Frederic Rimet (2012)	Epilithic	4 % formaldehyde. In laboratory, the diatom valves were cleaned using 40 % H_2O_2 and HCl. Clean valves were mounted in a resin (Naphrax$^©$)	River	France
Solak (2011)	Epilithic	Diatoms were collected by scraping 20 cm^2 area stones. They were cleaned with acid (HNO_3) and mounted on microscopie for observation with a magnification of ×1,000	River	Turkey
Kociolek and Karthick (2011)	Epiphytic	Using concentrated nitric acid and centrifuged several times to remove the acid	Lake	India
Nenad Jasprica et al. (2005)	Epiphytic	2.5 % neutralized formaldehyde. The material was treated with 10 % HCl add 30 % H_2O_2. 2.5 % neutralized formaldehyde. 10 % HCl and distilled water. Using Naphrax as the mounting medium	Lake	Croatia
Potapova (2003)	Epilithic	Oxidizing organic material in samples with nitric acid in a laboratory microwave oven and mounting cleaned diatoms in Naphrax	River	USA
Jiunn-Tzong Wu (2002)	Epilithic	Lugol's iodine solution immediately after collection and cleaned with acid (acetic acid: sulfuric acid = 9:1) in the laboratory. Washing with deionized water, Naphrax	River	Taiwan
John (2000)	Epiphytic, Epilithic	Lugol's solution, 4 % formalin or Transeau's Algal (6:3:1 = Water, Ethyl alcohol, 50–60 % Nitric Acid, Napthrax	River	Australia

Epilithic growing on stones; *Epiphytic* growing on macrophytes

2.3 Classification and Systematic Descriptions

The Classification and systematic descriptions of diatoms has been almost exclusively based upon frustules characteristics i.e., shape, size, symmetry, structure and density of striae, nature of raphe, copulae and processes on the valves. The biological species concept based upon reproductive isolation is difficult to apply to diatoms (Round et al. 1990). Natural population of large number of a taxon should be studied incorporating as many characters as possible in delineating a species. The classification of Diatoms by Round et al. (1990) recognizes diatoms as a division of Bacillariophyta with three major classes as follows: (a) Coscinodiscophyceae-(Centric diatoms); (b) Fragilariophyceae-(Araphid-pennate diatoms) and

(c) Bacillariophyceae-(Raphid-pennate diatoms). The International Journal of Diatom Research-the official journal of the International Society of Diatom Researchers follows this classification.

2.4 Diatom Studies in India

Diatom research in India has a history of over 100 and 50 years (Ehrenberg 1854). Notable works on diatom taxonomy from this region includes the publications of Skvortzow (1935), Gonzalves and Gandhi (1952, 1953, 1954), Krishnamurthy (1954), Gandhi (1959a, 1966, 1970, 1998), Venkataraman (1957) and Sarode and Kamat (1984). Most of these works focused on the pennate diatoms, with less or almost no attention given to the centric forms.

In a major contribution to diatom studies, Desikachary and Ranjitha Devi (1986), Desikachary (1988, 1989) illustrated and described many recent marine and fossils centric forms from the Indian Ocean in the "*Atlas of the Diatoms*". Similarly Gandhi (1952, 1955, 1956a, b, 1957a, b, c, 1958b, c, 1959b, c, d, 1960a, b, c, 1961, 1962b, c) Gandhi et al. (1983a, b, c, 1986), Trivedi (1982), Jakher et al. (1990), Dadheech et al. (2000), Singh et al. (2006, 2010), Kumar et al. (2008, 2009), Tarar and Bodhke (1998), Bhagat (2002), Mishra and Mishra (2002), Mishra (2006), Patil and Kumawat (2007), Anand (1998) have studied various aquatic systems.

Algal Data Base of Tamil Nadu Environmental Information System (ENVIS) Centre in http://tnenvis.nic.in/tnenvis_old/images/Algal_Database.pdf reported 668 taxa of freshwater algae from Tamil Nadu. Among them 346 green algae, 219 bluegreen algae, 60 diatoms, 32 charophytes and 9 freshwater red algae are listed in Table 7, pp. 85–105. The algae reported in that work were collected from temple tanks, ponds, beach pools, roadside puddles in and around Chennai (Desikachary 1959; Iyengar and Desikachary 1981) and a few places near Madurai (Desikachary 1959), Tanjore (Subramanian 2001) Virudachalam (Subramanian 2000) and Pollachi (Sankaran 1992, 2002, 2005a, b). The only comprehensive work on freshwater diatoms not only from the plains but also from several localities belonging to Kodaikanal and Ootacamund is that of Krishnamurthy (1954).

Diatoms are used as bio-indicators to assess the water quality of Cauvery River in parts of Tamil Nadu, Yercaud lake, Shevaroys hills, Tamil Nadu and surface water bodies in Manipur State NE India (Venkatachalapathy and Karthikeyan 2014; Venkatachalapathy et al. 2014a, b).

2.5 Diatom Studies in Other Parts of the World

In Australia, Schmid (1874–1959) reported occasional records of diatoms in Australia. Wood (1961a) and Wood et al. (1959) reported the occurrence of marine and inland diatoms in Australia. However, John (1983) presented the first systematic treatise with micrographs, descriptions and ecological information of 360 diatom taxa recovered from Swan River Estuary, Western Australia which includes both marine and freshwater species. Diatoms as tools for assessing health of rivers and streams in the south west of Western Australia were carried out by John (1998). Palaeoecological studies using diatoms to infer past environmental conditions were attempted by several workers (John 1993c; Gell 1997, 1998; McBride and Selkirk 1998). Importance of identification of diatoms by comparison of the modern or fossil diatom with descriptions and illustrations in floras for specific geographical areas or particular ecological systems has been emphasized by Battarbee et al. (2001b). Information on floras in an electronic format are increasingly becoming available that can be accessed either by CD-ROM (Kelly and Telford 2007) or through the internet (Battarbee et al. 2001a). Use of electronic monographs is an important contribution in the time of rapidly declining basic training and expertise on diatoms taxonomy (Stoermer 2001). Patrick and Reimer (1966, 1975) are the guides widely used in North America. Krammer and Lange-Bertalot (1986, 1988, 1991a, b) are German language guides to the central European flora.

2.6 Keywords on Diatom Studies

With the wide acceptability of diatom studies in environmental monitoring, there has been a growth of terminologies associated with such studies. Important among them are provided herein for the benefit of the readers: acidity index, acidophilous species, acidophilous algae, diatom index, diatom indicator, diatom indices, pollution indicator, pollution sensitivity index, saprobic index, saprobity, saproby index, tropic diatom index (TDI), trophic indices, trophic state.

3 Diatom Indices for Water Quality Assessment

There are many varieties of Diatom Indices for water quality assessment of surface water bodies (Table 2). They include, but not limited to, Generic Diatom Index or GDI (Coste and Ayphassorho 1991), the Specific Pollution Sensitivity Index (Indice de Polluosensibilite Specifique) or SPI (IPS) (Coste in Cemagref 1982), the Biological Diatom Index or BDI (Lenoir and Coste 1996), the Artois-Picardie Diatom Index or APDI (Prygiel et al. 1996), Sladeceks index or SLA (Sladecek 1986), the Eutrophication/Pollution Index or EPI (DellUomo 1996), Rotts Index or ROT (Rott 1991), Leclercq and Maquets Index or LMI (Leclercq and Maquet 1987), the Commission of Economical Community Index or CEC (Descy and Coste 1991), Schiefele and Schreiners index or SHE (Schiefele and Schreiner 1991), the Trophic Diatom Index or TDI (Kelly and Whitton 1995a) and the Watanabe index or WAT (Watanabe et al. 1986).

Table 2 Biotic diatom indices

Abbreviation	Full name	Reference
IPS	Specific pollution sensitivity metric	Coste (1987)
SLAD	Sládeček's pollution metric	Sladecek (1986)
DESCY	Descy's pollution metric	Descy (1979)
L&M	Leclercq and Maquet's pollution metric	Leclercq and Maquet (1987)
SHE	Steinberg and Schiefele trophic metric	Steinberg and Schiefele (1988)
WAT	Watanabe et al. pollution metric	Lecointe et al. (2003)
TDI	Trophic diatom metric	Kelly and Whitton (1995b)
EPI-D	Pollution metric based on diatoms	DellUomo 1996)
ROTT	Trophic metric	Rott et al. (1999)
IDG	Generic diatom metric	Lecointe et al. (2003)
CEE	Commission for economical community metric	Descy and Coste (1991)
IBD	Biological diatom metric	Prygiel and Coste (1999)
IDAP	Indice Diatomique Artois Picardie	Lecointe et al. (2003)
IDP	Pampean diatom index (IDP)	Gomez and Licursi (2001)

In the twentieth century, many biotic diatom indices were developed in Europe. These include, the trophic diatom index (TDI) by Kelly and Whitton (1995a) in Great Britain, the generic diatom index (GDI) by Rumeau and Coste (1988), the specific pollution-sensitivity index (SPI) by Cemagref (1982) and the biological diatom index (BDI) by Lenoir and Coste (1996) and Prygiel (2002) in France, the eutrophication pollution diatom index (EPI-D) by DellUomo (1996) in Italy, the Rott saprobic index (Rott et al. 1997) and the Rott trophic index (Rott et al. 1998) in Austria, the Schiefele and Kohmann trophic index by Schiefele and Kohmann (1993) in Germany and the CEE by Descy and Coste (1991) in France and Belgium.

The diatom assemblage index of organic pollution (DAIPo) was developed in Japan (Watanabe et al. 1986) and the saprobic index (Pantle and Buck 1955) in the USA. These indices were later tested in neighboring regions. In Spain, Goma et al. (2004, 2005) and Blanco et al. (2008) tested the SPI, BDI, TDI and EPI-D in Cataluna, East Spain and north-West Spain respectively. Torrisi and DellUomo (2006) and Battegazzore et al. (2004) tested the EPI-D in Italian rivers and compared its performance with those of other diatom indices. In Germany, Koster and Hubener (2001) tested the Rott saprobic index, the Rott prophic index, the TDI, the Lange–Bertalot classes, the CEE index, and the Schiefele and Kohmann trophic index in German rivers. Kelly (2002) and Kelly et al. (2009a) tested the TDI for assessment of river quality in English rivers. Diatoms also enabled efficient assessment of the quality of Moroccan rivers (Fawzi et al. 2002).

Several studies reported the use of diatom indices in regions with very different climates from the area they were created. Thus the TDI and GDI were tested in East Africa (Bellinger et al. 2006; Ndiritu et al. 2006) and the saprobic index and the TDI were tested in Malaysia (Maznah and Mansor 2002). The TDI was also tested in Australia (Newall and Walsh 2005), the Himalayas (Juttner et al. 2003) and Iran

(Atazadeh et al. 2007). The GDI, SPI, BDI, and EPI-D were tested in South Africa (Walsh and Wepener 2009), the TDI and Rott saprobic index in Turkey (Gurbuz and Kivrak 2002; Kalyoncu et al. 2009a, b), and the SPI, BDI, and DAIPo in Vietnam (Duong et al. 2006, 2007).

Dela-Cruz et al. (2006) tested the suitability of ecological tolerances/preferences of diatoms (Lange-Bertalot 1979) defined in the northern hemisphere in Australian rivers. In all cases, even if these diatom indices and diatom tolerances were developed and defined in very different regions (Europe, USA and Japan) from those where they were tested, pollution assessment results were good and demonstrated the robustness of diatom biomonitoring.

In some cases diatom indices were applied in situations for which they were not planned: the SPI, GDI, BDI, and EPI-D were tested in springs and did not reflect their hydrochemical characteristics. When a field study is too different from the initially intended scope, the authors prefer to develop their own diatom index for their specific study. After testing European diatom indices, an Australian diatom index based on species determination was developed by Chessman et al. (2007).

Eloranta and Soininen (2002) also tested an already existing trophic index, the TDI and proposed a new phosphorus diatom equation adapted to Finnish rivers. Lavoie et al. (2009) followed IDEC (Quebec diatom index) in Quebec Rivers, Canada and compared it with European and U.S. diatom indices. Lavoie et al. (2009) emphasized the importance of using diatom indices and suitable modifications to the regions of the study to obtain reliable results, especially for extreme conditions (very polluted or reference conditions). In South America, Lobo et al. (2004a, b) developed and integrated a biological water-quality index. Hurlimann and Niederhauser (2002) and Kupe et al. (2008) developed and tested a diatom index in Switzerland (DI-CH).

In Taiwan, Wu (1999) developed and tested the generic diatom index (Wu and Kow 2002) whereas in China, Tang et al. (2006) developed their own multimetric diatom index and compared it with European diatom indices and confirmed that the Diatom indices are more suitable in assessment of river water quality.

In U.S., Potapova and Charles (2003) developed and tested the U.S. diatom metrics for assessment of pollution in rivers. They have also compared the assessment of U.S. river quality with that of diatom metrics developed in Europe. Wang et al. (2005) developed a diatom index of biological integrity based on seven metrics for a particular U.S. ecoregion (interior plateau ecoregion).

In India, Venkatachalapathy and Karthikeyan (2014) used many diatom indices for testing the water quality conditions of the Cauvery River. Based on the results obtained these authors have suggested that the diatom indices can be utilized for water quality assessment of Indian rivers.

Thus, there is a consensus among the workers that the distribution of diatoms can be reliable indicators of ecological conditions of water (Cholnoky 1968; Lowe 1974) (Tables 3 and 4).

Table 3 Class limit values for diatom indices. Eloranta and Soininen (2002)

Index score	Class	Trophy
>17	High quality	Oligotrophy
15–17	Good quality	Oligo-mesotrophy
12–15	Moderate quality	Mesotrophy
9–12	Poor quality	meso-eutrophy
<9	Bad quality	Eutrophy

Table 4 Diversity parameters and indices

Index	Equation	Remarks	References
Abundance	$\frac{\text{No. of Individuals of a Species} \times 100}{\text{No. of Sampling Units}}$		
Shannon Weiner's (H′)	$-\sum_{I=1}^{S} \text{pi In pi}$ Pi: proportion of individuals of ith species	The value ranges between 1.5 and 3.5 and rarely surpasses 4.5	Ludwig and Renolds (1988), Legendre and Legendre (1998)
Simpson's D	$\text{D} = \frac{\sum_{I=1}^{S} \text{ni}(\text{ni}-1)}{N(N-1)}$	The value varies from 0 to 1. A value of 0 indicates the presence of only one species, while 1 means that all species are equally represented	Ludwig and Reynolds (1988)
Dominance	1-Simpson index $\text{D} = \text{sum}\left(\frac{ni}{n}\right)^2$ where ni is number of individuals of taxon i	The occupancy of a species over an area. Ranges from 0 (all taxa are equally present) to 1 (one taxon dominates the Community completely)	
Evenness	$\text{H}' = -\sum_{I=1}^{S} \text{pi In pi}$	The measure of biodiversity which quantifies how equal the community	
Fisher's Alpha	$\text{S} = \text{a} * \ln\left(\frac{1+n}{a}\right)$ where S is number of taxa, n is number of individuals and *a* is the Fisher's alpha	It is a mathematical model used to measure diversity	
Berger-Parker	$d = \frac{N_{max}}{N}$ where N_{max} is the number of individuals in the most abundant species and N is the total number of individuals in the sample	The number of individuals in the dominant taxon relative to n, where n is the total number of species	Berger and Parker (1970)

4 Reliability of Diatom-Based Assessments of Water Quality and Environment

Diatom assemblages have been widely related to specific environmental conditions in different geographical regions. The sensitivity of diatom communities has led them to be used as indicators of environmental conditions, such as water quality and habitat conditions in stream and river systems (Soininen et al. 2004). Round (1991) presented a history of diatom research related to water quality in rivers beginning with the study by Kolkwitz and Marsson (1908). Diatoms are widely used to monitor river pollution because they are sensitive to water chemistry, especially ionic content, pH, dissolved organic matter and nutrients apart from their wide geographic distribution (McCormick and Cairns 1994). According to Whitton and Rott (1996), biological indicators describe water quality and its changes over a long time scale more reliably than a few physicochemical analyses. Biological monitoring has been proven to be useful especially in running waters, where concentrations can fluctuate notably even within a few hours. Benthic diatoms have been found to be valuable for river monitoring purposes in several European countries (Prygiel et al. 1999). In Finland, the applicability of diatoms in water quality assessments has been tested by Eloranta (1995) and Eloranta and Andersson (1998).

Potapova (2007) used U.S. Geological Survey National Water-Quality Assessment program data to create diatom metrics for monitoring eutrophication, and showed that these metrics provide better assessments in U.S. Rivers than similar metrics developed for European inland waters. Van Dam et al. (1994) stated that the trophic diatoms are most commonly used in bio-assessment studies (Fore and Grafe 2002).

The potential of diatoms as indicators of water quality was realized early on in South Africa by Cholnoky (1968), Archibald (1972), Schoeman (1976) , Schoeman (1979) tested Lange-Bertalot's (1979) method in the upper Hennops River and found the method successful, with a good correlation between the species composition of diatom communities and water quality. Taylor (2007) opined that the diatom analyses obtained in South Africa were comparable to those obtained in several European studies. Taylor (2007) stressed the need for diatom-based indices which will provide a valuable tool for the biological monitoring of water quality in South Africa.

In Europe concerted efforts were made to standardize the routine sampling and processing of diatoms for water quality assessments (Kelly et al. 1998; Prygiel et al. 2002). In recent years the methodological tools for bio-assessment have improved significantly. Diatom indices and ecological classifications have been tested along pollution gradients in several countries (Kelly 1998a; Kwandrans et al. 1998; Rott et al. 1998).

Monitoring of nutrient levels in rivers and streams is problematic because of periodic and diffuse input from non-point sources (Cattaneo and Prairie 1995). However, diatom species composition responds directly to nutrients and can be used as indicator of trophic state (e.g. Oligotrophy, Oligo-mesotrophy, Mesotrophy, meso-eutrophy and Eutrophy) than measurements of nutrient concentrations or algal biomass (Stevenson and Pan 1999).

5 Application of Diatoms in Environmental Studies

It appears that in the 21st century, the application and understanding of diatoms are growing, as the roles they play in global nutrient, oxygen and silica cycling, their utility in understanding the status and trends of aquatic ecosystem health, harmful algal blooms, and their potential as a possible source of renewable fuels are better understood and appreciated (Stoermer and Smol 1999).

Assessments of environmental conditions in rivers and streams using diatoms have a long history in which two basic conceptual approaches emerged. First, based on the work of Kolkwitz and Marsson (1908), autecological indices were developed to infer levels of pollution based on the species composition of assemblages and the ecological preferences and tolerances of taxa (Butcher 1947; Fjerdingstad 1950; Zelinka and Marvan 1961; Lowe 1974; Lange-Bertalot 1979). Second, Patrick's early monitoring studies (Patrick 1949; Patrick et al. 1954; Patrick and Strawbridge 1963) relied primarily on diatom diversity as a general indicator of river health (i.e. ecological integrity), because species composition of assemblages varied seasonally and species diversity varied less. The conceptual differences in these two approaches really address two different goals for environmental assessments, one inferring pollution levels and the other determining biodiversity, a more valued ecological attribute. Thus, the concepts and tools for assessing ecosystem health and diagnosing causes of impairment in aquatic habitats, particularly rivers and streams, were established and developed between ~50 and 100 years ago. Today, diatoms are being used to assess ecological conditions in streams and rivers around the world (Asai 1996; Kelly et al. 1998; Wu 1999; Lobo et al. 2004; Wang et al. 2005; Chessman et al. 2007; Taylor et al. 2007; Porter et al. 2008).

Review of published literature on diatom studies shows the emergence of a great diversity of methods and findings, identifying regionally refined tools, and organizing the application of these tools in scientifically sound protocols for solving environmental problems. Using correct diatom assessment tools for correct reason during an assessment is necessary effective environmental management. According to Stevenson et al. (2004) two fundamental questions need to be answered in almost all ecological assessments: "Is there a problem?" and "What is causing the problem?" Understanding the meaning of these questions, and how they will be asked and answered by government agencies or other scientists, is important for determining how diatoms can help answer both of these questions. As one of the most species-rich components of river and stream communities, diatoms are important elements of biodiversity and genetic resources in rivers and streams (Patrick 1961).

In addition, diatoms are the source of many nuisance algal problems, such as taste and odor impairment of drinking water, reducing water clarity, and clogging water filters (Palmer 1962). Diatoms are valuable indicators of environmental conditions in rivers and streams, because they respond directly and sensitively to many physical, chemical, and biological changes in river and stream ecosystems, such as temperature (Squires et al. 1979), nutrient concentrations (Pan et al. 1996; Kelly 1998), and herbivory (McCormick and Stevenson 1989).

John (1998) has recorded 200 diatom species from 46 rivers and streams from Western Australia, which is considered the largest study of river health assessment conducted in Australia using diatoms at species level. He has generated data by sampling over 150 sites in the south-west of Western Australia covering an area of 24,000 km^2. An artificial substrate collector (The JJ Periphytometer) was used for collecting diatom samples. All sites were classified according to the environmental factors and diatom distribution pattern.

Diatoms are used in water-quality assessments in many states in the USA and were included in the National Rivers and Streams Assessment being conducted from 2008 to 2011 by the U.S. Environmental Protection Agency (EPA). Diatoms provide as precise an assessment, as macrophytes, invertebrates, and fish and actually are more sensitive to some stressors than other organisms. Both planktic and benthic diatoms (i.e. phytobenthos or periphyton) are used in assessments of surface water bodies.

In South Africa, Taylor et al. (2007) reported of the study 245 diatom taxa were encountered, comprising 54 genera. A full species list of the taxa encountered (including all synonyms) is given in Taylor (2004). He has concluded that the unique composite picture of ecosystem conditions provided by the diatoms can only be replicated by intensive chemical monitoring studies. In Turkey, Tahir and Olcay (2011) reported a total of 93 taxa from Dam Lake (Ankara) and concluded that though the dam served as the freshwater source, owing to the algal bloom, potential human health risks may occur. In Brazil, Bere (2011) reported a total of 112 diatom species belonging to 44 genera of the families *Achnanthidiaceae*, *Achnanthaceae*, *Bacillariaceae*, *Eunotiaceae*, *Cymbellaceae*, *Gomphonemataceae*, *Fragilariaceae*, *Melosiraceae*, *Naviculaceae*, *Rhoicospheniaceae*, *Rhopalodiaceae* and *Surirellaceae*.

Recently, Venkatachalapathy and Karthikeyan (2014) analyzed and evaluated diatom assemblages for its applicability to assess the water quality of Cauvery River in part of Tamil Nadu, India. Venkatachalapathy et al. (2014b), reported the occurrence of 37 Diatoms species at Manipur River, Nambul River, Waishen River and Loktak Lake in Manipur state, NE India. Venkatachalapathy et al. (2014a) recorded 21 diatom species belongs to 13 genera in Yercaud Lake, Salem district, Tamil Nadu. Venkatachalapathy and Karthikeyan (2014) and Venkatachalapathy et al. (2014) reported the occurrence pollution tolerant diatom species and their dominance in India. Few of the diatom species inhabiting pristine and polluted areas of aquatic systems are shown in the Plates I, II and III.

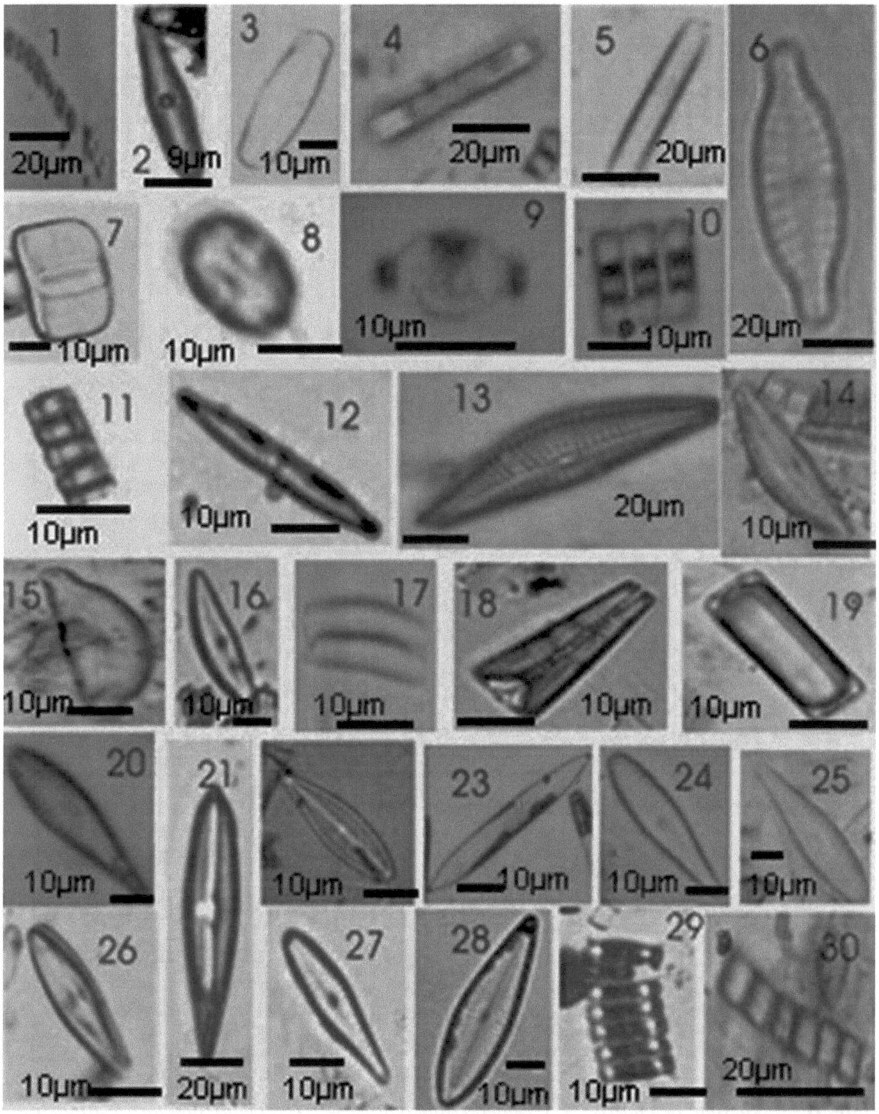

Plate I 1–2 *Achnanthesbrevipes*; 3 *Achnanthidium binodis*; 4–5 *Achnanthidium minutissima*; 6 *Achnanthidium plonensis*; 7 *Amphora holsatica*; 8 *Amphora ovalis*; 9 *Anomoeoneis sphaerophora*; 10–11 *Aulacoseira distans*; 12 *Ctenophora pulchella*; 13 *Cymbella lanceolata*; 14–16 *Cymbella tumida*; 17–18 *Eunotia curvata*; 19 *Eunotia pectinalis*; 20 *Gomphonema affine*; 21–22 *Gomphonema lanceolatum*; 23 *Gomphonema parvulum*; 24–26 *Gomphonema truncatum*; 27 *Gomphonema undulatum*; 28 *Mastogloia braunii*; 29 *Melosira moniliformis*; 30 *Melosira varians* (×40) (Mettur Dam pristine side)

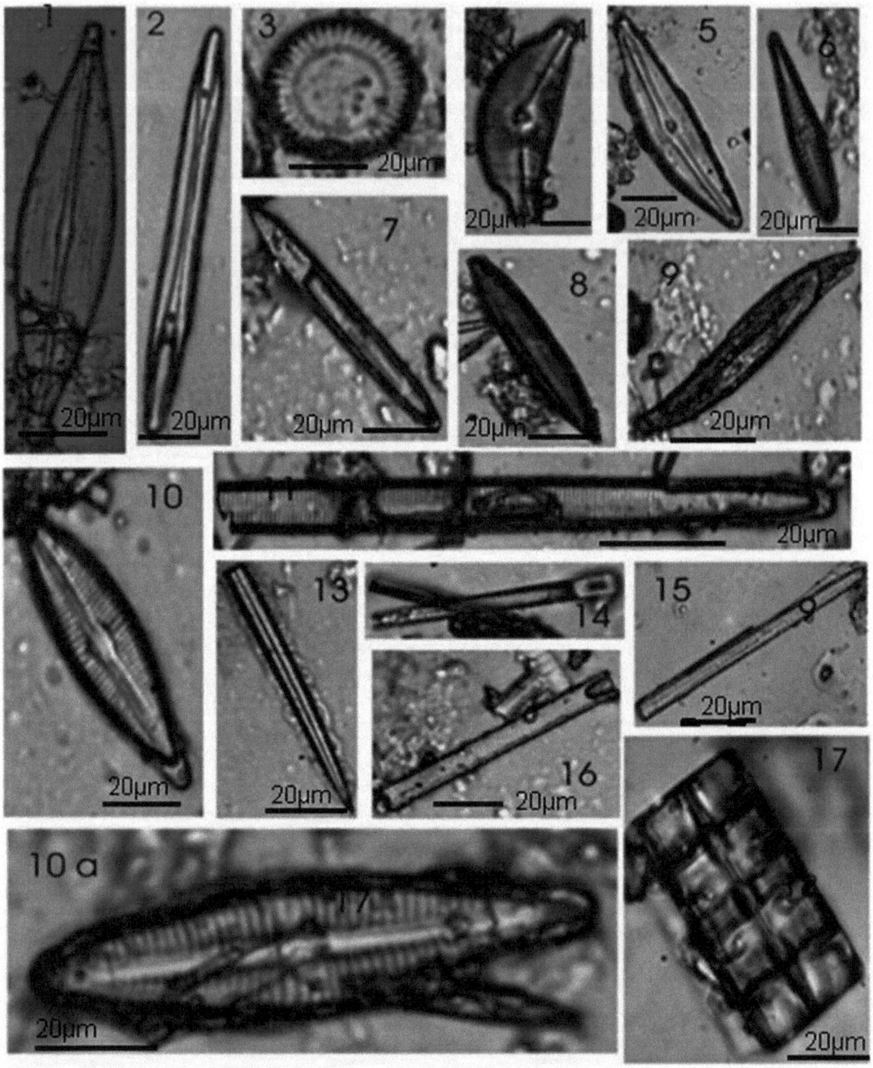

Plate II 1 *Nitzshia acicularis*; 2 *Cymbella lancolata*; 3–5 *Cymbella tumida*; 4 *Aulacoseira granulate*; 6 *Navicula rhychocephala*; 7 *Anomoeoneis sphaerophora*; 8 *Gomphonema undulatum*; 9, 13, 15 *Stauroneis phoenicenteron*; 10, 12, 16 *Cyclotella meneghiniana*; 11 *Tryblionella apiculate*; 12, 16 *Cyclotella meneghiniana*; 14 *Frugulia megaliesmontana*; (×40) (Mettur Dam pristine side)

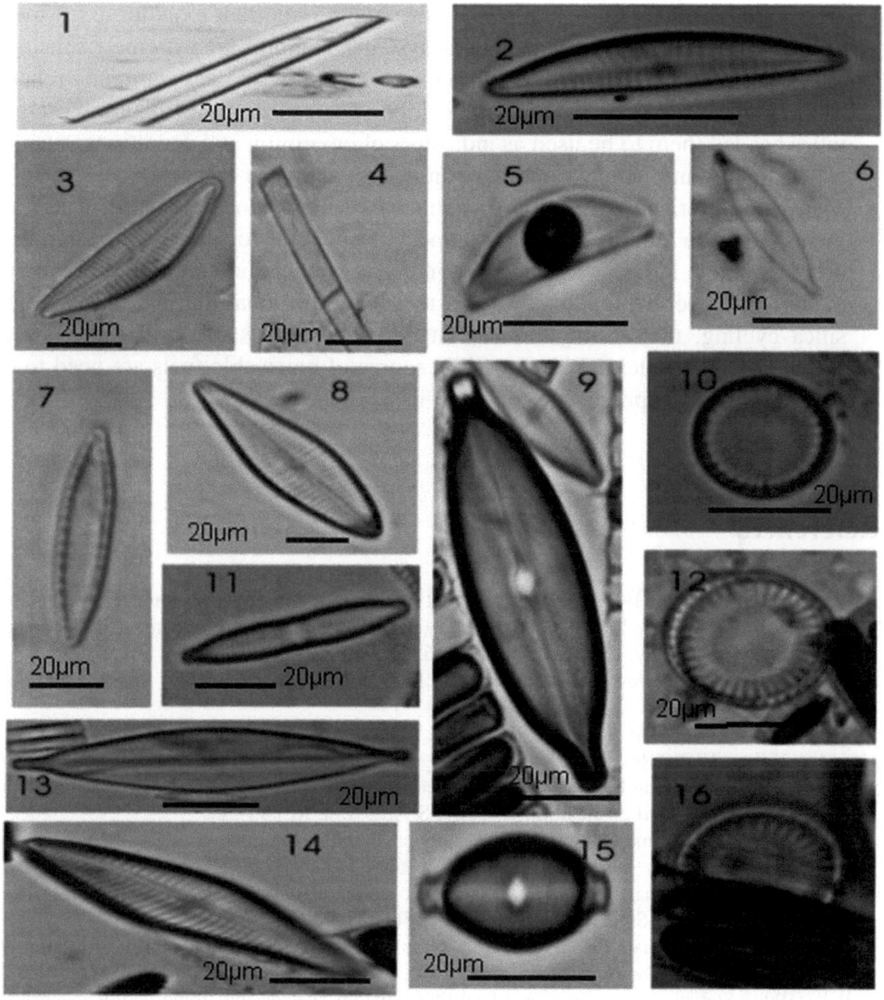

Plate III 1 *Anomoeoneis sphaerophora*; 2 *Bacillaria paxillifer*; 3 *Cyclotella meneghiniana*; 4–5 *Cymbella lanceolata;* 6 *Gomphonema truncatum*; 7 *Fragilaria punctata*; 8 *Navicula elginensis*; 9 *Pleurosigma longum*; 10, 10a *Gomphonema truncatum*; 11 *Stauroneis dubitabilis Hust*; 13–16 *Synedra ulna*; 17 *Tabellaria flocculosa*; (×40) (Bhavani polluted area)

6 Conclusions

- The Classification and systematic descriptions of diatoms has been almost exclusively based upon frustules characteristics i.e. shape, size, symmetry, structure and density of striae, nature of raphe, copulae and processes on the valves. Use of electronic monographs is an important development in the time of rapidly declining basic training and expertise on diatoms taxonomy.

- Diatom-based indices are increasingly becoming important tools for assessment of environmental conditions in aquatic systems. Diatoms are extremely sensitive to pH and salinity. Diatom distribution in surface water bodies is directly related to the ionic concentration and composition. The sensitivity of diatom communities has led them to be used as indicators of environmental conditions, such as water quality and habitat conditions in stream and river systems.
- Many characteristics of diatom assemblages and methods are used in assessments of water quality and environment. Depending on the specific requirements, suitable indices can be employed. The application and understanding of diatoms are everywhere, as the roles they play in global nutrient, oxygen and silica cycling, their utility in understanding the status of aquatic ecosystem health and their potential as a possible source of renewable fuels are need to be understood and appreciated.

References

Amoros C, Van Urk G (1989) Palaeoecological analyses of large rivers: some principles and methods. In: Petts GE, Moller H, Roux AL (eds) Historical change of large alluvial rivers: Western Europe, Wiley, Chichester, pp 143–65

Anand N (1998) Indian fresh water microalgae. Bishen Singh Mahendra Pal Singh, Dehradun, pp 1–94

Archibald REM (1972) Diversity in some South African diatom associations and its relation to water quality. Water Res 6:1229–1238

Archibald REM, Schoeman FR (1984) Amphora coffeaeformis (Agardh) Kützing: a revision of the species under light and electron microscopy. S Afr J Bot 3(2):83–102

Archibald REM, Schoeman FR (1985) Amphora hybrida Grunow (Bacillariophyceae)—its identity and taxonomy. Nova Hedwig 41:159–166

Archibald REM, Schoeman FR (1987) Taxonomic notes on diatoms (Bacillariophyceae) from Great Usutu River in Swaziland. S Afr J Bot 53(1):75–92

Archibald REM, Schoeman FR (1988) A taxonomic review of Amphora hartii in relation to Amphora thermalis. Diatom Res 3(2):181–190

Asai K (1996) Statistical classification of epilithic diatom species into three ecological groups relating to organic water pollution. (1) Method with coexistence index. Diatom 10:13–34

Atazadeh I, Sharifi M, Kelly MG (2007) Evaluation of the trophic diatom index for assessing water quality in River Gharasou, western Iran. Hydrobiologia 589:165–173

Battarbee RW, Jones VJ, Flower RJ (2001a) Diatoms. In: Smol JP, Birks HJB, Last WM (eds) Tracking environmental change using lake sediments. Terrestrial, algal, and siliceous indicators, vol 3. Kluwer Academic Publishers, Dordrecht, pp 155–202

Battarbee RW, Juggins S, Gasse F (2001b) European diatom database (EDDI). An information system for palaeoenvironmental reconstruction. Environmental Change Research Centre, University College London, London, p 210

Battegazzore M, Morisi A, Gallino B, Fenoglio S (2004) Environmental quality evaluation of alpine springs in NW Italy using benthic diatoms. Diatom Res 19:149–165

Bellinger BJ, Cocquyt C, O'Reilly CM (2006) Benthic diatoms as indicators of eutrophication in tropical streams. Hydrobiologia 573:75–87

Bere T, Tundisi JG (2011) Diatom-based water quality assessment in streams influence by urban pollution: effects of natural and two selected artificial substrates, São Carlos-sp, Brazil. Braz J Aquat Sci Technol 15(1):54–63
Berger WH, Parker FL (1970) Diversity of planktonic foraminifera in deep-sea sediments. Science 168:1345–1347
Bhagat P (2002) Limnological investigation on Naukuchia Tal. Ph.D. thesis, Kumaun University Nanital, India
Bhatt JP, Bhaskar A, Pandit MK (2005) Biotic communities of Kishanganga river: a pre-impoundment case study of a Himalayan river. Aqua Ecosyst Health Manage 8:259–265
Blanco S, Ector L, Huck V, Monnier O, Cauchie HM, Hoffmann L, Becares E (2008) Diatom assemblages and water quality assessment in the Duero Basin (NW Spain). Belg J Bot 141:39–50
Blinn DW (1995) Diatom community structure along salinity gradients in Australian saline lakes: biogeographic comparisons with other continents. In: Kociolek JP, Sullivan MJ (eds) A century of diatom research in North America: a tribute to the distinguished careers of Charles W. Reimer and Ruth Patrick. Koeltz Scientific Books, Champaign, pp 163–174
Butcher RW (1947) Studies in the ecology of rivers. IV. The algae of organically enriched water. J Ecol 35:186–191
Cattaneo A, Prairie YT (1995) Temporal variability in the chemical characteristics along the Rivie`re de l'Achigan: how many samples are necessary to describe stream chemistry. Can J Fish Aquat Sci 52:828–835
Cemagref (1982) Etude de Methodes Biologiques Quantitatives Appreciation de la Qualité des Eaux. Rapport Q.E. Lyon-A.F.B. Rhone- Mediterrannée-Corse 218
Chessman BC, Bate N, Gell PA, Newall P (2007) A diatom species index for bioassessment of Australian rivers. Mar Freshw Res 58:542–557
Cholnoky BJ (1968) Die Ökologie der Diatomeen in Binnengewässern. J Cramer, Lehre
Coste M (1987) Study of quantitative biological methods of appreciation of the quality of the water. Report qulities division iyon waters agency i water rhone, 28
Coste M, Ayphassorho H (1991) Etude de la qualité des eaux du bassin Artois Picardie à l'aide des communautés de diatomées benthiques (Application des indices diatomiques), Rapport Cemagref Bordeaux — Agence de l'Eau Artois Picardie, p 227
Cremer H, Gore D, Hultzsch N, Melles M (2004) The diatom flora and limnology of lake in the Amery Oasis, East Antarctica. Polar Biol 27:513–531
Dadheech PK, Srivastava P, Sharma KP (2000) Status of phycodiversity of Rajasthan State (India) in the New Millennium. In: National symposium on phycology in the New Millennium, Centre for Advanced Studies in Botany, University of Madras, Chennai, p 14
Dela-Cruz J, Pritchard T, Gordon G, Ajani P (2006) The use of periphytic diatoms as a means of assessing impacts of point source inorganic nutrient pollution in south-eastern Australia. Freshw Biol 51:951–972
DellUomo A (1996) Assessment of water quality of an Apennine river as a pilot study for diatom-based monitoring of Italian watercourses. In: Whitton BA, Rott E (eds) Studia student. G.m.b. H., Innsbruck, pp 65–72
Denys L (2004) Relation of abundance-weighted averages of diatom indicator values to measured environmental conditions in standing freshwaters. Ecol Ind 4(2004):255–275
Descy JP (1979) A new approach to water quality estimation using Diatoms. Nova Hedwigia, Heft 64
Descy JP, Coste M (1991) A test of methods for assessing water quality based on diatoms. Verh Int Ver Theor Angew Lim 24:2112–2116
Desikachary TV (1959) Cyanophyta. Indian Council of Agricultural Research, New Delhi, p 700
Desikachary TV (1988) Marine diatoms of the Indian Ocean region. In: Desikachary TV (ed) Atlas of diatoms. Fasc.V. Madras Science Foundation, Madras, pp 1–13
Desikachary TV (1989) Marine diatoms of the Indian Ocean region. In: Desikachary TV (ed) Atlas of diatoms. Fasc.VI. Madras Science Foundation, Madras, pp 1–27
Desikachary TV, Prema P (1987) Diatoms from the Bay of Bengal. In: Desikachary TV (ed) Atlas of diatoms. Fasc. III. Madras Science Foundation, Madras, pp 1–10

Desikachary TV, Ranjitha Devi KA (1986) Marine fossil diatoms from India and Indian Ocean region. In: Desikachary TV (ed) Atlas of diatoms. Fasc. I. Madras Science Foundation, Madras, pp 1–77
Desikachary TV, Gowthaman S, Latha Y (1987) Diatom flora of some sediments from the Indian Ocean region. In: Desikachary TV (ed) Atlas of diatoms. Fasc. II. Madras Science Foundation, Madras, pp 78–221
Duong TT, Coste M, Feurtet-Mazel A, Dang DK, Gold C, Park YS, Boudou A (2006) Impact of urban pollution from the Hanoi area on benthic diatom communities collected from the red, Nhue and Tolich rivers (Vietnam). Hydrobiologia 563:201–216
Duong TT, Feurtet-Mazel A, Coste M, Dang DK, Boudou A (2007) Dynamics of diatom colonization process in some rivers influenced by urban pollution (Hanoi, Vietnam). Ecol Ind 7:839–851
Ehrenberg CG (1854) Mikrogeologie. Das Erden und Felsen schaffende Wirken des unsichtbar kleinen selbstständigen Lebens auf der Erde. Leopold Voss, Leipzig, p xxviii
Eloranta P (1995) Type and quality of river waters in central Finland described using diatom indices. In: Marino D, Montresor M (eds) Proceedings of the 13th international diatom symposium, 1994. Biopress, Bristol, pp 271–280
Eloranta P, Kwandrans J (1996) Distribution and ecology of freshwater red algae (Rhodophyta) in some central Finnish rivers. Nord J Bot 16(1):107–117
Eloranta P, Andersson K (1998) Diatom indices in water quality monitoring of some South-Finnish rivers. Verh. Internat. Verein. Limnol. 26:1213–1215
Eloranta P, Soininen J (2002) Ecological status of some Finnish rivers evaluated using benthic diatom communities. J Appl Phycol 14:1–7
Fawzi B, Loudiki M, Oubraim S, Sabour B, Chlaida M (2002) Impact of wastewater effluent on the diatom assemblages structure of a brackish small stream: Oued Hassar (Morocco). Limnologica 32:54–65
Fjerdingstad E (1950) The microflora of the River Molleaa with special reference to the relation of benthic algae to pollution. Folia Limnol Scandanavica 5:1–123
Fore LS, Grafe C (2002) Using diatoms to assess the biological condition of large rivers in Idaho (U.S.A.). Freshw Biol 47:2015–2037. doi:10.1046/J.1365-2427.2002.00948.X
Fritz SC, Juggins S, Battarbee RW (1993) Diatom assemblages and ionic characterisation of lakes of the Northern Great Plains, North America. A tool for reconstructing past salinity and climate fluctuations. Can J Fish Aquat Sci 50:1844–1856
Fritz SC, Cumming BF, Gasse F, Laird KR (1999) Diatoms as indicators of hydrologic and climatic change in saline lakes. In: Stoermer EF, Smol JP (eds) The diatoms: applications for the environmental and earth sciences. Cambridge University Press, Cambridge, pp 41–72
Gandhi HP (1955) A contribution to our knowledge of the freshwater diatoms of Pratapgarh, Rajasthan. J Indian Bot Soc 34:304–338
Gandhi HP (1956a) A contribution to the freshwater Diatomaceae of S. Western India—I. Freshwater diatoms of Dharwar. J Indian Bot Soc 35:194–202
Gandhi HP (1956b) A preliminary account of the soil diatom of Kolhapur. J Indian Bot Soc 35:402–408
Gandhi HP (1957a) A contribution to our knowledge of the diatom genus Pinnularia. J Bombay Nat Hist Soc 54:845–853
Gandhi HP (1957b) Some common freshwater diatoms from Gersoppa-falls (Jog-Falls). J Poona Univ Sci Sect 12:13–21
Gandhi HP (1957c) The freshwater diatoms from Radhanagari—Kolhapur. Ceylon J Sci (Biol Sect) 1:45–47
Gandhi HP (1958b) Freshwater diatoms from Kolhapur and its immediate environs. J Bombay Nat Hist Soc 55:493–511
Gandhi HP (1958c) The freshwater diatoms flora of the Hirebhasgar Dam area, Mysore State. J Indian Bot Soc 37:249–265
Gandhi HP (1959a) Fresh-water diatoms from Sagar in the Mysore State. J Indian Bot Soc 38:305–331

Gandhi HP (1959b) Freshwater diatom flora of the Panhalgarh Hill Fort in the Kolaphur district. Hydrobiologia 14:93–129
Gandhi HP (1959c) Notes on the Diatomaceae from Ahmedabad and its environs-II. On the diatom flora of fountain reservoirs of the Victoria Gardens. Hydrobiologia 14:130–146
Gandhi HP (1959d) The freshwater diatom flora from Mugad, Dharwar District with some ecological notes. Ceylon J Sci (Biol Sect) 2:98–116
Gandhi HP (1960a) On the diatom flora of some ponds around Vasna village near Ahmedabad. J Indian Bot Soc 39:558–567
Gandhi HP (1960b) Some new diatoms from the Jog Falls, Mysore State. J R Microsc Soc 79:81–84
Gandhi HP (1960c) The diatom flora of the Bombay and Salsette Islands. J Bombay Nat Hist Soc 57:78–123
Gandhi HP (1961) Notes on the Diatomaceae of Ahmedabad and its environs. Hydrobiologia 17:218–236
Gandhi HP (1962b) Some freshwater diatoms from Lonawala Hill Station in the Bombay State (Maharashtra). Hydrobiologia 20:128–154
Gandhi HP (1962c) The diatom flora of the Bombay and Salsette Islands. II. Nova Hedwig 3:469–505
Gandhi HP (1966) The fresh-water diatom flora of the Jog-Falls, Mysore State. Nova Hedwig 11:89–197
Gandhi HP (1967) Notes on Diatomaceae from Ahmedabad and its environs. VI. On some diatoms from fountain reservoirs of Seth Sarabhai's Garden. Hydrobiologia 30:248–272
Gandhi HP (1970) A further contribution to the diatom flora of the Jog Falls, Mysore State. Nova Hedwig 31:633–652
Gandhi HP (1998) Fresh-water diatoms of Central Gujarat. Bishen Singh Mahendra Pal Singh, Dehra Dun
Gandhi HP, Vora AB, Mohan DJ (1983a) Fossil diatoms from Baltal, Karewa beds of Kashmir. Curr Trends Geol VI (Climate and Geology of Kashmir) 6:61–68
Gandhi HP, Mohan DJ, Vora AB (1983b) Preliminary observation on Baltal and Ara sediments, Kashmir. In: Proceedings of the Xth Indian colloquium on micropalaeontology and stratigraphy, pp 555–570
Gandhi HP, Vora AB, Mohan DJ (1983c) Review of the fossil Diatomflora, of the Karewa beds of Kashmir. Curr Trends Geol VI (Climate and Geology of Kashmir) 6:57–60
Gandhi HP, Vora AB, Mohan DJ (1986) Ecology of diatoms from the Karewa beds of Baltal area, Kashmir, India. In: Proceedings of the Xth Indian colloquium on micropalaeontology and stratigraphy, bulletin of geological, mining and metallurgical society of India, Part II (Stratigraphy and Microflora) No 54, pp 159–161
Gell PA (1997) The development of a diatom database for inferring lake salinity towards a quantitative approach for reconstructing past climates. Aust J Bot 45:389–423
Gell PA (1998) Quantitative reconstructions of the Holocene palaeosalinity of paired crater lakes based on a diatom transfer function. Palaeoclimates 31(1–3):83–96
Gell P, Bulpin S, Wallbrink P, Hancock G, Bickford S (2005) Tareena Billagong—apalaeolimnological history of an ever-changing wetland, chowilla floodplain, lower murrayDarling Basin, Australia. Mar Freshw Res 56:441–56
Goma J, R Ortiz, Cambra J, Ector L (2004) Water quality evaluation in Catalonian Mediterranean rivers using epilithic diatoms as bioindicators. Vie et Milieu—Life Environ 54:81–90
Goma J, Rimet F, Cambra J, Hoffmann L, Ector L (2005) Diatom communities and water quality assessment in Mountain rivers of the upper Segre basin (La Cerdanya Oriental Pyrenees). Hydrobiologia 551:209–225
Gomez N, Licursi M (2001) The pampean diatom index (IDP) for assessment of rivers and streams in Argentina. Aquat Ecol 35:173–181
Gonzalves EA, Gandhi HP (1952) A systematic account of the diatoms of Bombay and Salsette I. J Indian Bot Soc 31:117–151
Gonzalves EA, Gandhi HP (1953) A systematic account of the diatoms of Bombay and Salsette—II. J Indian Bot Soc 32:239–263

Gonzalves EA, Gandhi HP (1954) A systematic account of the diatoms of Bombay and Salsette—III. J Indian Bot Soc 33:338–350
Gurbuz H, Kivrak E (2002) Use of epilithic diatoms to evaluate water quality in the Karasu River of Turkey. J Environ Biol 23:239–246
Harding JPC, Kelly MG (1999) Recent developments in the use of algae to monitor rivers in the U.K. In: Prygiel J, Whitton BA, Bukowska J (eds) Use of algae for monitoring rivers III. Agence de l'Eau Artois-Picardie, Douai, pp 26–34
Hurlimann J, Niederhauser P (2002) Methode détude et d'appréciation de l'état de santé des cours d'eau: Diatome es - niveau R (région). Office Fédéral de l'Environnment, des Forêts et des Paysages, Berne, 1–111
Jakher GR, Bhargava SC, Sinha RK (1990) Comparative limnology of Sambhar and Didwana Lake (Rajasthan, NW India). Hydrobiologia 197:245–256
Jasprica N, Hafner D, Batistic M, Kapetanovic T (2005) Phytoplankton in three freshwater lakes in the Neretva River delta (Eastern Adriatic, NE Mediterranean). Nova Hedwig 81(1–2):37–54
John J (1983) The diatom flora of the Swan River estuary, Western Australia. Bibl Phycol 64:1–359. J Cramer Veduz 360
John J (1993c) Palaeolimnology of North Lake, Perth, Western Australia, based on diatom assemblages. In: Bowdrey G, Hope GS (eds) Sixth international palaeolimnology symposium, Australia National University, Canberra, p 49
John J (1998) Diatoms: tools for bioassessment of river health. Report to Land and Water Resources Research and Development Corporation, Canberra, 388 pp
John J (2000) Diatom prediction and classification system for urban streams. LWRRDC Canberra
Juttner I, Sharma S, Dahal BM, Ormerod SJ, Chimonides PJ, Cox EJ (2003) Diatoms as indicators of stream quality in the Kathmandu valley and Middle Hills of Nepal and India. Freshw Biol 48:2065–2084
Kalyoncu H, Cicek NL, Akkoz C, Ozcelik R (2009) Epilithic diatoms from the Darioren stream (Isparta/Turkey): biotic indices and multivariate analysis. Fresen Environ Bull 18:1236–1242
Kalyoncu H, Cicek NL, Akkoz C, Yorulmaz B (2009b) Comparative performance of diatom indices in aquatic pollution assessment. Afr J Agric Res 4: 1032–1040
Karthick B, Kociolek JP (2011) Four new centric diatoms (Bacillariophyceae) from the Western Ghats, South India. Phytotaxa 22:25–40
Karthick B, Krithika H, Alakananda B (2008) Short guide to common freshwater diatom genera (Poster). Energy and Wetlands Research Group, CES, IISc, Bangalore
Kelly MG (1998a) Use of community-based indices to monitor eutrophication in rivers. Environ Conserv 25:22–29 (in press)
Kelly MG (1998b) Use of the trophic diatom index to monitor eutrophication in rivers. Wat Res 32:236–242
Kelly MG (2002) Role of benthic diatoms in the implementation of the urban wastewater treatment directive in the river wear, North-East England. J Appl Phycol 14:9–18
Kelly MG, Whitton BA (1995a) The trophic diatom index: a new index for monitoring eutrophication in rivers. J Phys 7:433–444
Kelly MG, Whitton BA (1995b) The trophic diatom index: a new index for monitoring eutrophication in rivers. J Appl Phycol 7:433–444
Kelly M, Telford RJ (2007) Common fresh water diatoms of Britain and Ireland: an interactive identifications key (CD ROM). Environment Agency, UK
Kelly MG, Haigh A, Colette J, Zgrundo A (2009a) Effect of environmental improvements on the diatoms of the River Axe, southern England. Fottea 9:343–349
Kolkwitz R, Marsson M (1908) Okologie der pflanzlichen Saprobien. Berichte der Deuts chen Botanischen Gesellschaft 26A:505–519
Koster D, Hubener T (2001) Application of diatom indices in a planted ditch constructed for tertiary sewage treatment in Schwaan, Germany. Int Rev Hydrobiol 86:241–252
Krammer K, Lange-Bertalot H (1986) Bacillariophyceae. 1. Teil: Naviculaceae. In: Ettl H, Gerloff J, Heynig H, Mollenhauer D (eds) Süsswasser flora von Mitteleuropa, Band 2/1. Gustav Fischer Verlag, Stuttgart, New York, p 876

Krammer K, Lange-Bertalot H (1988) Bacillariophyceae. 2. Teil: Bacillariaceae, Epithemiaceae, Surirellaceae. In: Ettl H, Gerloff J, Heynig H, Mollenhauer D (eds) Süsswasserflora von Mitteleuropa, Band 2/2. VEB Gustav Fischer Verlag, Jena, p 596

Krammer K, Lange-Bertalot H (1991a) Bacillariophyceae. 3. Teil: Centrales, Fragilariaceae, Eunotiaceae. In: Ettl H, Gerloff J, Heynig H, Mollenhauer D (eds) Süsswasserflora von Mitteleuropa, Band 2/3. Gustav Fischer Verlag, Stuttgart, Jena, p 576

Krammer K, Lange-Bertalot H (1991b) Bacillariophyceae. 4. Teil: Achnanthaceae, Kritische Ergänzungen zu Navicula (Lineolatae) und Gomphonema, Gesamtliteraturverzeichnis Teil 1-4. In: Ettl H, Gärtner G, Gerloff J, Heynig H, Mollenhauer D (eds) Süsswasserflora von Mitteleuropa, Band 2/4. Gustav Fischer Verlag, Stuttgart, Jena, p 437

Krishnamurthy V (1954) A contribution to the diatom flora of South India. J Indian Bot Soc 33:354–381

Kumar A, Sharma LL, Aery NC (2008) Physicochemical characteristics and diatoms as indicators of trophic status of Kishore Sagar, Rajasthan. In: Proceedings of Taal 2007: 12th world lake conference, pp 1804–1809

Kumar A, Sharma LL, Aery NC (2009) Physicochemical characteristics and diatom diversity of Jawahar Sagar—a wetland of Rajasthan. Sarovar Saurabh 5(1):8–14

Kupe L, Schanz F, Bachofen R (2008) Biodiversity in the benthic diatom community in the upper river toss reflected in water quality indices. Clean: Soil Air Water 36:84–91

Kwandrans J, Eloranta P, Kawecka B, Wojtan K (1998) Use of benthic diatom communities to evaluate water quality in rivers of southern Poland. J Appl Phycol 10:193–200

Lange-Bertalot H (1979) Pollution tolerance of diatoms as a criterion for water quality estimation. Nova Hedwig 64:285–304

Lavoie I, Hamilton PB, Wang YK, Dillon PJ, Campeau S (2009) A comparison of stream bioassessment in Quebec (Canada) using six European and North American diatom-based indices. Nova Hedwig 35:37–56

Leclerq L, Maquet B (1987) Deux nouveaux indices chimique et diatomique de qualité d'eau courante. Application au Samson et àses affluents (bassin de la Meuse belge). Comparaison avec d'autres indices chimiques, biocénotiques et diatomiques. Institut Royal des Sciences Naturelles de Belgique, documented travail 28

Lecointe C, Coste M Prygiel J (2003) Omnidia 3.2. Diatom Index Software Including Diatom Database with Taxonomic Names, References and Codes of 11645 Diatom Taxa

Legendre P, Legendre L (1998) Numerical ecology, 2nd English edn. Elsevier, Amsterdam

Lenoir A, Coste M (1996) Development of a practical diatom index of overall water quality applicable to the French national water board network. In: Whitton BA, Rott E (eds) Use of algae for monitoring rivers II. Institut für Botanik, Universität Innsbruck, pp 29–43

Lobo EA, Bes D, Tudesque L, Ector L (2004a) Water quality assessment of the Pardinho river, RS, Brazil, using epilithic diatom assemblages and faecal coliforms as biologica indicators. Vie et Milieu—Life Environ 54:115–125

Lobo EA, Callegaro VLM, Hermany G, Gomez N, Ector L (2004b) Review of the use of microalgae in South America for monitoring rivers, with special reference to diatoms. Vie et Milieu—Life Environ 54:105–114

Lowe RL (1974) Environmental requirements and pollution tolerance of freshwater diatoms. Environmental monitoring series. National Environmental Research Center, Cincinnati, Ohio

Ludwig John A, Reynolds JF (1988) Statistical ecology: a primer of methods and computing. Wiley Press, New York, p 337

Maznah WOW, Mansor M (2002) Aquatic pollution assessment based on attached diatom communities in the Pinang River Basin, Malaysia. Hydrobiologia 487:229–241

McBride P, Selkirk J (1998) Palaeo lake diatoms on subarntartic Macquarie Island: possible markers of climate change. In: John J (ed) 15th International diatom symposium abstracts

McCormick PV, Cairns J Jr (1994) Algae as indicators of environmental change. J Appl Phycol 6:509–526

McCormick PV, Stevenson RJ (1989) Effects of snail grazing on benthic algal community structure in different nutrient environments. J North Am Benthol Soc 82:162–172

Mishra KN (2006) Impact of Sugar industries effluent on species diversity and algal productivity at Shahganj (Jaunpur) U.P. Human population natural resources. Jaspal Prakashan Patna (Bihar), pp 103–108
Mishra UC, Mishra KN (2002) Species richness and Shannon diversity indices of diatomic community in Gomati water influenced from Jaunpur city sewage. In: Proceedings of 89th plant science section of Indian science congress Lucknow, p 58
Ndiritu GG, Gichuki NN, Triest L (2006) Distribution of epilithic diatoms in response to environmental conditions in an urban tropical stream, Central Kenya. Biodivers Conserv 15:3267–3293
Newall P, Walsh CJ (2005) Response of epilithic diatom assemblages to urbanization influences. Hydrobiologia 532:53–67
Palmer CM (1962) Algae in water supplies. US Department of Health, Education and Welfare, Washington
Pan Y, Stevenson RJ, Hill BH, Herlihy AT, Collins G B (1996) Using diatoms as indicators of ecological conditions in lotic systems: a regional assessment.J North Am Benthol Soc 15 (4):481–495
Pantle R, Buck H (1955) Die biologischeUberwaswaschung der Gewässer und die Darstellung der Ergebnisse. Gas Wasserfach 96:604
Patil SB, Kumawat DA (2007) Studies on the Centric diatoms from Abhora Dam of Jalgaon District, Maharashtra. Research Link 345(9):154–155
Patrick R (1949) A proposed biological measure of stream conditions based on a survey of the Conestoga basin, Lancaster County, Pennsylvania. Proc Acad Nat Sci Philadelphia 101:277–341
Patrick R (1961) A study of the numbers and kinds of species found in rivers of the Eastern United States. Proc Acad Nat Sci Philadelphia 113:215–258
Patrick R, Reimer CW (1966) The diatoms of the United States, exclusive of Alaska and Hawaii, Volume 1-Fragilariaceae, Eunotiaceae, Achnanthaceae, Naviculaceae. Academy of Natural Sciences of Philadelphia Monograph No 13, p 688
Patrick R, Reimer CW (1975) The diatoms of the United States, exclusive of Alaska and Hawaii, Volume 2, Part 1-Entomoneidaceae, Cymbellaceae, Gomphonemaceae, Epithemaceae. Academy of Natural Sciences of Philadelphia Monograph No 13, p 213
Patrick R, Strawbridge D (1963) Variation in the structure of natural diatom communities. Am Nat 97:51–57
Patrick R, Hohn MH, Wallace JH (1954) A new method for determining the pattern of the diatom flora. Notulae Naturae, No 259
Porter SD, Mueller DK, Spahr NE, Munn MD, Dubrovsky NM (2008) Efficacy of algal metrics for assessing nutrient and organic enrichment in flowing waters. Freshw Biol 53:1036–1054
Potapova M, Charles DF (2003) Distribution of benthic diatoms in U.S. rivers in relation to conductivity and ionic composition. Freshw Biol 48:1311–1328
Potapova M, Hamilton PB (2007) Morphological and ecological variation within the *Achnanthidium minutissimum* (Bacillariophyceae) species complex. J Phycol 43:561–575
Prygiel J (2002) Management of the diatom monitoring networks in France. J Appl Phycol 14:19–26
Prygiel J, Coste M (1993) The assessment of water quality in the Artois-Picardie water basin (France) by the use of diatom indices. Hydrobiologia 269(279):343–349
Prygiel J, Leveque L, Iserentant R (1996) Un nouvel indice diatomique pratique pour l'évaluation de la qualité des eaux en réseau de surveillance. Rev Sci Eau 1:97–113
Prygiel J, Coste M, Bukowska J (1999) Review of the major diatom-based techniques for the quality assessment of rivers-state of the art in Europe. In: Prygiel J, Whitton BA, Bukowska H (eds) Use of algae for monitoring rivers III, pp 224–238
Rimet F (2012) Recent views on river pollution and diatoms. Hydrobiologia 683:1–24. doi:10.1007/s10750-011-0949-0
Rosati TC, Johansen JR, Coburn MM (2003) Cyprinid fishes as samplers of benthic diatom communities in freshwater streams of varying water quality. Can J Fish Aquat Sc 60:117–25

Rott E (1991) Methodological aspects and perspectives in the use of periphyton for monitoring and protecting rivers. In: Whitton BA, Rott E, Friedrich G (eds) Use of algae for monitoring rivers. Institut fur Botanik, Universität Innsbruck, pp 9–16

Rott E, Hofmann G, Pall K, Pfister P, Pipp E (1997) Indikationslisten für Aufwuchsalgen Teil 1: Saprobielle indikation. Bundesministerium für Land- und Forstwirtschaft, Wien, Austria, pp 1–73

Rott E, Pipp E, Pfister P, Van Dam H, Ortler K, Binder N, Pall K (1998) Indikationslisten für aufwuchsalgen in Österrichischen fliessgewässern. Teil 2: Trophieindikation. Arbeitsgruppe Hydrobotanik, Institut für Botanik, Universitat Innsbruck, Austria, pp 1–248

Rott E, Pfister P, Van Dam H, Pipp E, Pall K, Binder N, Ortler K (1999) Indikationslisten fu¨r aufwuchsalgen. Wien. Bundesministerium fu¨r Land - und Forstwirtschaft 248

Round FE, Crawford RM and Mann DG (1990) The diatoms: biology and morphology of the genera. Cambridge University Press. pp747

Round FE (1991) Use of diatoms for monitoring rivers. In: Whitton BA, Rott E, Friedrich G (eds) Use of algae for monitoring rivers I. STUDIA Studentenf˙orderungs-Ges.m.b.H, Innsbruck. pp 25–32

Rumeau A, Coste M (1988) Initiational systematique des diatomées d'eau douce pour l'utilisation pratique d'un indice diatomique générique. Bulletin Franc¸ais de la Pêche et de la Pisciculture 309:1–69

Sankaran V (1992) Blue-green algae from brackish water pools. Seaweed Res Utiln 15(1 and 2):207

Sankaran V (2002) Algae of the Anamalais hill ranges of Western Ghats-Chlorophyceae. Phykos 41(1 and 2):21–26

Sankaran V (2005) Freshwater algal biodiversity of the Anaimalai hill ranges, Tamilnadu-Chlorophyta-Ulotrichales and Ulvales. Indian Hydrobiol 7(Suppl):93–95

Sarode PT, Kamat ND (1984) Freshwater diatoms of Maharashtra. Saikripa Prakashan, Aurangabad, 338 pp

Schiefele S, Kohmann F (1993) Bioindikation der Trophie in Fliessgewassern. Umweltforschungsplan des Bundesministers für Umwelt. Naturschutz und Reaktorsicherheit, Germany, pp 1–211

Schiefele S, Schreiner C (1991) Use of diatoms for monitoring nutrient enrichment acidification and impact salts in Germany and Austria. In: Whitton BA, Rott E, Friedrich G (eds) Use of algae for monitoring rivers. Institüt für Botanik, Universität Innsbruk

Schmidt A (1874–1959) Atlas der diatomaceen-kunde. Band 1, II and III. Tafel 1- 480. Reprint 1984. Koeltz scientific books Koenigstein. Schmidt A, Schmidt M, Fricke F, Heiden H, Muller O, Hustedt F

Schoeman FR (1979) Diatoms as indicator of water quality in the upper Hennops River. J Limnol Soc South Afr 5(2):73–78

Schoeman FR, Archibald REM (1976) The diatom flora of southern Africa. C.S.I.R. special report WAT 50, Pretoria, South Africa

Singh R, Singh R, Singh R, Thakar MK (2006) Diatomological studies from three water bodies of Jaipur. Indian Internet J Forensic Med Toxicol 4:3

Singh M, Lodha P, Singh GP (2010) Seasonal diatom variations with reference to physico-chemical properties of water of Mansagar Lake of Jaipur, Rajasthan. Res J Agric Sci 1(4):451–457

Skvortzow BW (1935) Diatoms from Calcutta. India Phil J Sci 58:179–192

Sladecek V (1986) Diatoms as indicators of organic pollution. Acta Hydrochim Hydrobiol 14:555–566

Snoeijs P (1999) Diatoms and environmental change in brackish waters. In: Stoermer EF, Smol JP (eds) The diatoms: applications for the environmental and earth sciences. Cambridge University Press, Cambridge, pp 299–333

Soininen J, Kononen K (2004) Comparative study of monitoring South Finnish rivers and streams using macroinvertebrate and benthic diatom community structure. Aquat Ecol 38:63–75

Solak CN (2011) The application of diatom indices in the Upper Porsuk River, Kütahya-Turkey. Turk J Fish Aquat Sci 11:31–36

Somers KM, Lavoie I, Paterson AM, Dillon PJ (2000) Assessing scales of variability in benthic diatom community structure. J Appl Phycol 17:509–513
Squires LE, Rushforth SR, Brotherson JD (1979) Algal response to a thermal effluent: study of a power station on the Provo River, Utah, USA. Hydrobiologia 63:17–32
Steinberg C, Schiefele S (1988) Biological indication of trophy and pollution of running waters. Z Wasser Wasser-Forsch 21:227–234
Stevenson RJ, Pan Y (1999) Assessing environmental conditions in rivers and streams with diatoms. In: Stoermer EF, Smol JP (eds) The diatoms: applications for the environmental and earth sciences. Cambridge University Press, Cambridge, pp 11–40
Stevenson RJ, Bailey BC, Harass MC (2004) Designing data collection for ecological assessments. In: Barbour MT, Norton SB, Preston HR, Thornton KW (eds) Ecological assessment of aquatic resources: linking science to decision-making. Society of Environmental Toxicology and Chemistry, Pensacola, pp 55–84
Stoermer EF, Smol JP (eds) (1999) The Diatoms: applications for the environmental and earth sciences. Cambridge University Press, Cambridge, UK
Stoermer EF (2001) Diatom taxonomy for paleolimnologists. J Paleolimnol 25:393–398
Subramanian D (2000) New taxa of Nitella from Tamil Nadu. Indian Hydrobiol 3(2):71–80
Subramanian D (2001) Charophytes from Kannankulam of Thanjavur district of Tamilnadu. Phykos 40(1 and 2):135–141
Tahir A, Olcay O (2011) The diatoms of Asartepe Dam Lake (Ankara), with environmental and some physicochemical properties. Turk J Bot 34(2010):541–548
Tang T, Cai QH, Liu JK (2006) Using epilithic diatom communities to assess ecological condition of Xiangxi River system. Environ Monit Assess 112:347–361
Tarar JL, Bodhke S (1998) Studies on diatoms of Nagpur. Phykos 37(1–2):104–108
Taylor (2004) The application of diatom- based Pollution indices in the vaal catchment Unpublished M. Sc. Thesis, North- West University, Potchefstroom Campus, Potchefstroom
Taylor JC (2007) The application and testing of diatom-based indices in the Vaal and Wilge Rivers, South Africa. Water SA 33(1):51–60
Taylor JC, Janse van Vuuren MS, Pieterse AJH (2007) The application and testing of diatom-based indices in the Vaal and Wilge Rivers, South Africa. Water SA 33(1):51–60
Torrisi M, DellUomo A (2006) Biological monitoring of some Apennine rivers (central Italy) using the diatom based eutrophication/pollution index (EPI-D) compared to other European diatom indices. Diatom Res 21:159–174
Trivedi RK (1982) Some observations on algal flora of Jaipur, Rajasthan. Phykos 21:106–163
Tuchman M, Stevenson RJ (1980) Comparison of clay tile, sterilized rock, and natural substrate diatom communities in a small stream in southeastern Michigan, U.S.A. Hydrobiologia 75:73–79
Tyagi GD (1985) Diatoms of Delhi. J.F.M.S. 2(3):18–23
Van Dam H, Mertens A, Sinkeldam J (1994) A coded checklist and ecological indicator values of freshwater diatoms from The Netherlands. Neth J Aquat Ecol 28:117–133
Venkatachalapathy R, Karthikeyan P (2012) Environmental impact assessment of Cauvery River with Diatoms at Bhavani, Tamil Nadu, India. Int J Geol Earth Environ Sci 2(3):36–42
Venkatachalapathy R, Karthikeyan P (2014) Diatom indices for water quality assessment in Cauvery river, Tamil Nadu. India. Gondwana. Geol Mag 15:109–116
Venkatachalapathy R, Nanthakumar G, Karthikeyan P (2014a) Diatoms and water quality assessment of the Yercaud Lake, Salem District, Tamil Nadu. South India. Gondwana. Geol Mag 15:13–16
Venkatachalapathy R, Chandra Singh M, Karthikeyan P, Somorjit Singh L, Sanjoy Singh L (2014b) Diatoms as indicators of water quality in Imphal, Nambul, Waishen Rivers and the Loktak Lake, Manipur State, North East India. Gondwana. Geol Mag 15:61–66
Venkataraman GS (1957) A contribution to the knowledge of the diatomaceae of Kanya Kumari (Cape Comorin), India. Proc Natl Acad Sci (Sect B) 23:80–88
Walsh G, Wepener V (2009) The influence of land use on water quality and diatom community structures in urban and agriculturally stressed rivers. Water SA 35:579–594

Wang YK, Stevenson RJ, Metzmeier L (2005) Development and evaluation of a diatom-based index of biotic integrity for the interior Plateau ecoregion, USA. J North Am Benthol Soc 24:990–1008

Watanabe T, Asai K, Houki A (1986) Numerical estimation of organic pollution of flowing waters by using the epilithic diatom assemblage-diatom assemblage index (DIApo). Sci Total Environ 55:209–218

Whitton BA, Rott E (eds) (1996) Use of algae for monitoring rivers II. Institut für Botanik, Universität Innsbruck, p 196

Wood EJF (1961a) Studies on Australian and New Zealand diatoms IV. Descriptions of further sedentary species. Trans R Soc NZ 33(4):699–698

Wood EJF, Crosby LH, Cassie V (1959) Studies on Australian and New Zealand diatoms III. Descriptions of further discoid species. Trans R Soc NZ 87(3,4):211–219

Wu JT (1999) A generic index of diatom assemblages as bioindicator of pollution in the Keelung River of Taiwan. Hydrobiologia 397:79–87

Wu J-T, Kow L-T (2002a) Applicability of a generic index for diatom assemblages to monitor pollution in the tropical River Tsanwun, Taiwan. J Appl Phycol 14:63–69

Wu JT, Kow LT (2002b) Applicability of a generic index for diatom assemblages to monitor pollution in the tropical River Tsanwun, Taiwan. J Appl Phycol 14:63–69

Iyengar MOP, Desikachary TV (1981) Volvocales. ICAR, New Delhi, p 532

Yallop M, Hirst H, Kelly M (2006) Validation of ecological status concepts in UK rivers using historic diatom samples. Aquat Bot 90:289–95

Zelinka M, Marvan P (1961) Zur Proazisierung der biologischen Klassifikation des Reinheit fliessenderGewasser. Archiv für. Hydrobiologie 57:389–407

Ecohydrology of Lotic Systems in Uttara Kannada, Central Western Ghats, India

T.V. Ramachandra, M.D. Subash Chandran, N.V. Joshi, B. Karthick and Vishnu D. Mukri

Abstract The Western Ghats is the primary catchment for most of the rivers in peninsular India. Pristine forests in this region are rich in biodiversity but are under environmental stress due to unplanned developmental activities. This has given rise to concerns about land use/land cover changes with the realization that the land processes influence the climate. Rapid and unscientific land-use changes undermine the hydrological conditions, and deteriorate all the components in the hydrological regime. The developmental programs, based on ad-hocdecisions, are posing serious challenges to the conservation of fragile ecosystems. Considerable changes in the structure and composition of land use and land cover in the region have been very obvious during the last four decades. Pressure on land for agriculture, vulnerability of degraded ecosystems, the vagaries of high intensity rainfall and consequent occurrences of accelerated erosion and landslides, lack of integrated and coordinated land use planning become some of the reasons for rapid depletion of natural resource base. These changes have adversely affected the hydrological regime of river basins, resulting in diminished river/stream flows. This necessitates conservation of ecosystems in order to sustain their biodiversity, hydrology and ecology. In this situation, for resolving present problems and to avoid any future crisis, a comprehensive assessment of land use changes, its spatial distribution and its impact on hydrological regime were carried out. Accordingly, appropriate remedial methods have been explored for the sustainable utilization of the land and water resources in the catchment. The current research, focusing on five rivers located in the central Western Ghats, monitors water quality along with that of diatoms, land use in the catchment and threats faced by these ecosystems.

Keywords Western Ghats · Lotic ecosystems · Water quality · Diatoms

T.V. Ramachandra (✉) · M.D. Subash Chandran · N.V. Joshi · B. Karthick · V.D. Mukri
Energy and Wetlands Research Group, Centre for Ecological Sciences, Indian Institute of Science, Bangalore, India
e-mail: cestvr@ces.iisc.ernet.in

Mu. Ramkumar et al. (eds.), *Environmental Management of River Basin Ecosystems*,
Springer Earth System Sciences, DOI 10.1007/978-3-319-13425-3_29

1 Introduction

Freshwater ecosystems are grouped into lotic and lentic systems, that is, systems comprising of flowing or standing water. There are many varieties of plant and animal communities in these ecosystems which have adapted to the physical conditions associated with them. Environmental pollution, mainly pertaining to water, has gained public interest (Niemi et al. 1990) in recent times. Not only the developed countries have been affected by environmental problems, but also the developing nations suffer the impact of pollution (Listori and World-wide Bank 1990) due to unplanned developmental activities. Surface waters are vulnerable to pollution due to their proximity to pollutants on land which get dispersed off as polluted runoff and wastewaters and also due to the sustained inflow of untreated sewage. Quality of the surface waters are altered by the natural processes such as precipitation, erosion, and weathering as well as from the anthropogenic influences such as agricultural activities, urbanization, industrialization and intensive-exploitation of water resources (Jarvie et al. 1998). These impacts reduce both water quality (Sweeting 1996) and biological diversity of aquatic ecosystems (Maddock 1999).

Rivers play a major role in the assimilation or in carrying off the municipal and industrial wastewater and run-off from agricultural land. The surface run-off is a seasonal phenomenon, which is largely influenced by the climate prevailing in the basin. Seasonal variations in precipitation, surface run-off, interflow, groundwater flow and water inflows/outflows have a strong effect on the river discharge and subsequently on the concentration of nutrients/pollutants in the river water (Vega et al. 1998). Rivers are the main inland water resources for domestic, industrial and irrigation purposes and it is imperative to prevent and control river pollution. This necessitates regular monitoring to have reliable information on quality of water for effective management. In view of the spatial and temporal variations in hydrochemistry of rivers, regular monitoring programs are required for reliable estimates of water quality and conservation of riverine biodiversity. An integrated aquatic ecosystem management requires sound understanding of physical, chemical and biological aspects. An attempt is made in the present study to determine the water quality status through diatoms as bio-indicators in the rivers of central Western Ghats.

The Western Ghats of India, one of the global biodiversity hotspots, is a chain of mountains on the Western Coast with about 1,600 km long and about 100 km wide stretch (between 8°N and 21°N). The region has varied forest types from tropical evergreen to deciduous to high altitude sholas. It is also an important watershed for the peninsular India with as many as 37 west flowing rivers, three major east flowing rivers and innumerable tributaries. In this paper, the water quality along with diatoms, land use in the catchment and threats faced by these ecosystems are evaluated based on the study of five rivers in central Western Ghats. Aim of this work is to understand the ecohydrology of west flowing rivers in the central Western Ghats. The work involved exploring the current water quality status of five rivers of the Uttara Kannada District, Karnataka, assessment of the seasonality of diatoms and application of diatoms in bio-monitoring in Western Ghats, understanding the impact of catchment land-use and land-cover on water quality and diatom community in streams, identification of the stretches with major water pollution and provide recommendation for mitigation and conservation of rivers of Uttara Kannada.

2 Lotic Ecosystems of Central Western Ghats: An Overview on the Study Area

Rivers of the central Western Ghats are unique in their geomorphology, due to the presence of 'river capture' in most of the rivers. When the Indian plate moved away from the Gondwanaland, peninsular portion experienced an eastward tilt, which changed the pattern of drainage in many rivers. In many cases, like Sharavathi and Kali rivers in Uttara Kannada (Fig. 1), the western faulting led to 'river capture' and diversion of the easterly drainage to the west (Radhakrishna 1991; Kamath 1985). Five rivers are chosen for the present study. Brief descriptions on them are presented herein.

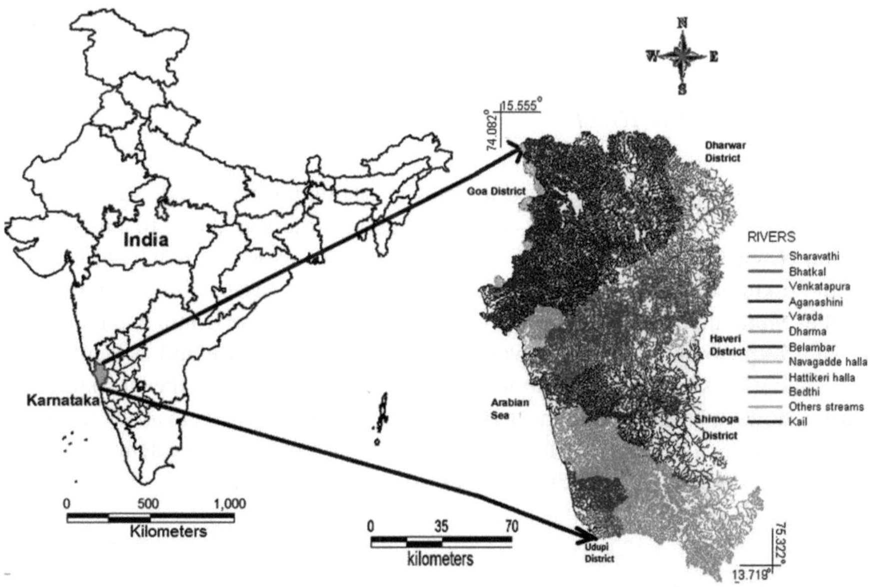

Fig. 1 Study region—Uttara Kannada district with rivers (*Source* Energy and Wetlands Research Group, CES, Indian Institute of Science)

2.1 Kali River

The Kali River (Fig. 2) flows for a length of 184 km. Previously, it originated near the village Diggi in Supataluk, as Karihole. After the construction of the dam near Supa, the entire region is now submerged in the reservoir. Pandri and Ujli are the two main tributaries of this river in the North and the stream Tattihalla also joins near Haliyal. The Kaneri and the Vaki are its two main tributaries that join at Dandeli and Anshi Tiger Reserve respectively. Later near Kadra, Thananala joins the main river. In all, the catchment area of the river is about 5,179 km^2 and the annual river discharge is estimated to be 6,537 million cu. m (Bhat 2002). There are four major dam projects on this river—the Supa reservoir near the headwaters, the Bommanhalli reservoir near the Dandeli Wildlife Sanctuary, the Kodasalli dam near Ganeshgudi and finally, one at Kadra (which is the part of the Kaiga project).

2.2 Bedthi River

The River Bedthi River (Fig. 3) originates near Hubli taluk. The river, has a total length of 152 km with a catchment area of 3,902 km^2. It discharges 4,925 million cu. m of water annually.

2.3 Aghanashini River

The Aghanashini River (Fig. 4) originates at Manjguni near Sirsi. After winding westerly course of about 70 km, it debauches into the sea about 10 km south of Bedthi. The river has two sources—a tributary called Bakurhole, rising at Manjguni, about 25 km west of Sirsi and Donihalla, which is close to Sirsi. These two streams meet at Mutthalli about 16 km south of Sirsi. Under the name Donihalla, it flows for about 25 km south of Sirsi westwards to Sahyadri's west face and at Heggarne in Siddapur, it falls off a height of about 116 m as the Lushington (or the Unchalli) falls. Further down 6 km from Bilgi near Hemanbail, it flows down again as the Burdejog. It finally meets the sea at Uppinpatna. The Aghanashini covers a catchment area of 2,146 km^2. It has an annual discharge of 966 million cu. m.

2.4 Sharavathi River

The 128 km long Sharavathi River (Fig. 5) originates at Ambutirtha in Tirthahallitaluk of Shimoga District. After a northerly course of about 64 km from Sagar, it forms the southeastern border of the Uttara Kannada District for about 13 km and flows a further

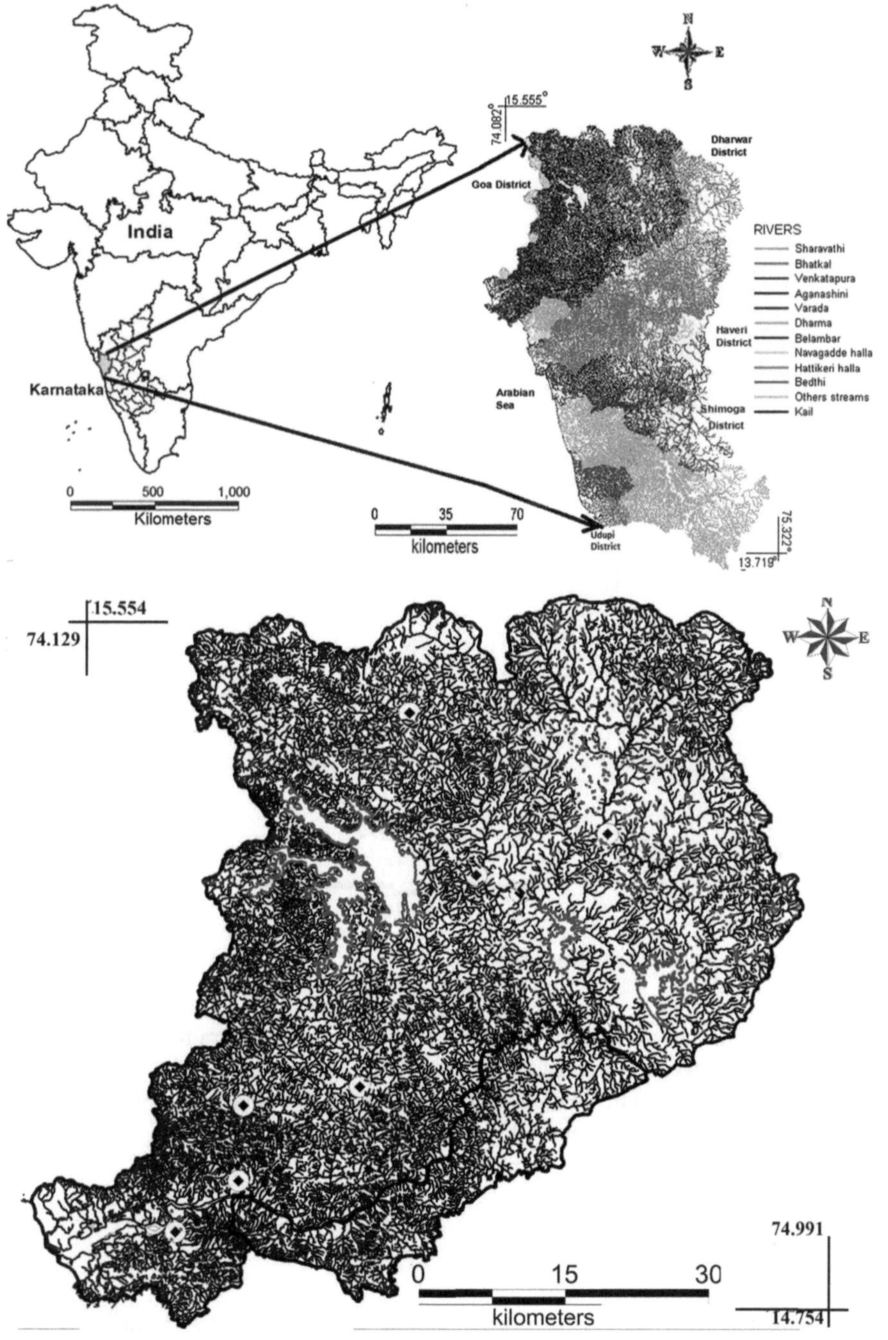

Fig. 2 River Kali with sampling sites

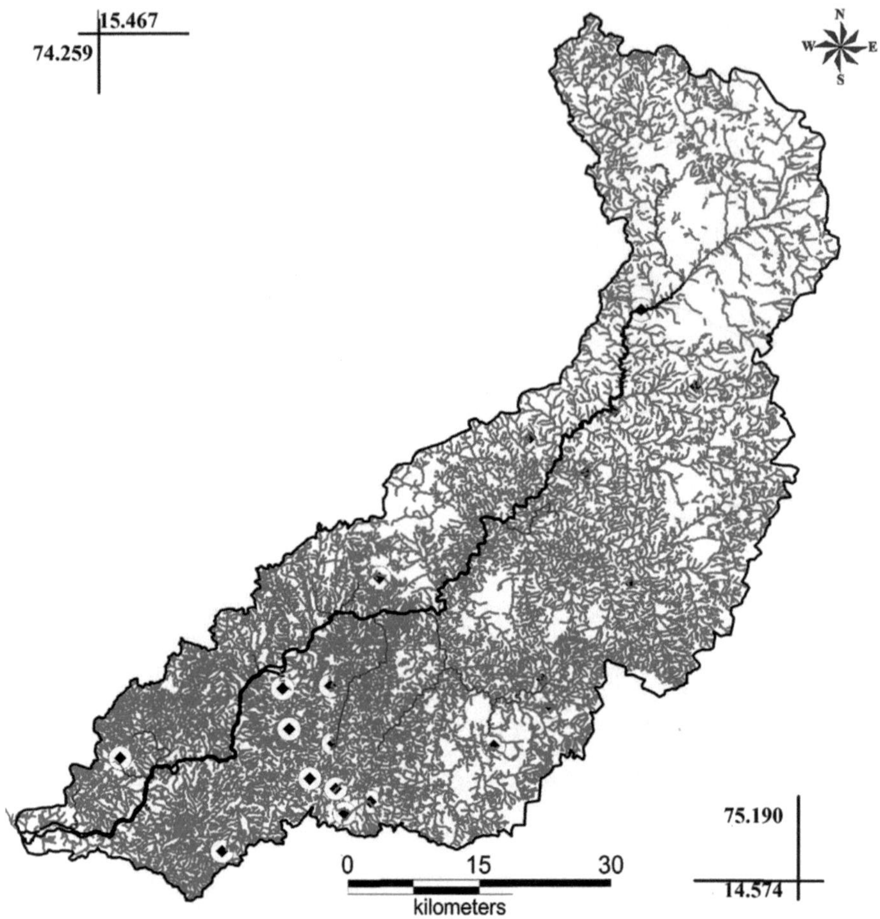

Fig. 3 River Bedthi with sampling sites

32 km to join the sea at Honnavar. Soon after touching the Uttara Kannada border the river falls off the western face of the Ghats in Jog falls at a height of 252 m into a pool 117 m deep. About 30 km west, it reaches Gersoppa. The Sharavathi has a catchment area of 2,209 km^2 and an annual discharge of 4,545 million cu.m.

2.5 Venkatapura River

The Venkatapura River (Fig. 6) originates in Western Ghats and confluence into Arabian Sea after a course of 45 km near Venkatapura with a catchment of 335 km^2. The river basin is divided into sub basins namely Chitihalla, KatagarNala, BastiHalla, Kitrehole and Venkatapura based on major tributaries.

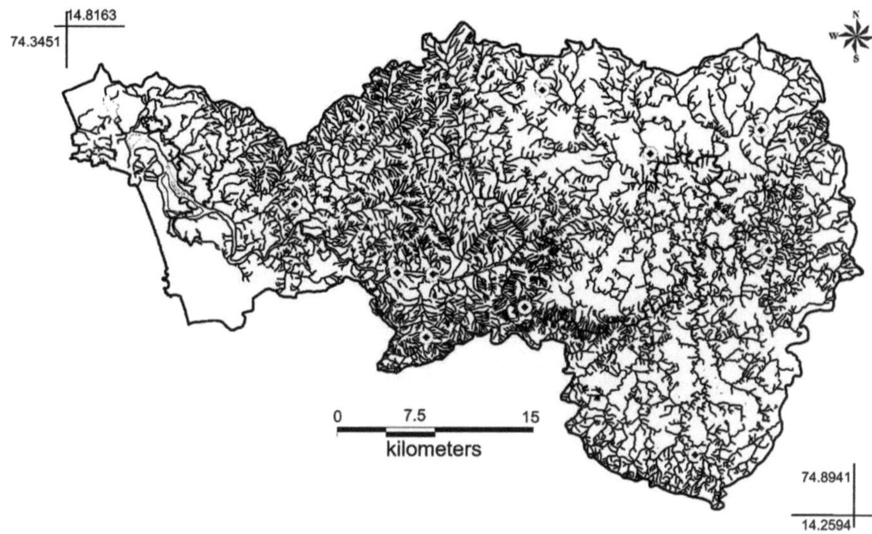

Fig. 4 River Aghanashini with sampling sites

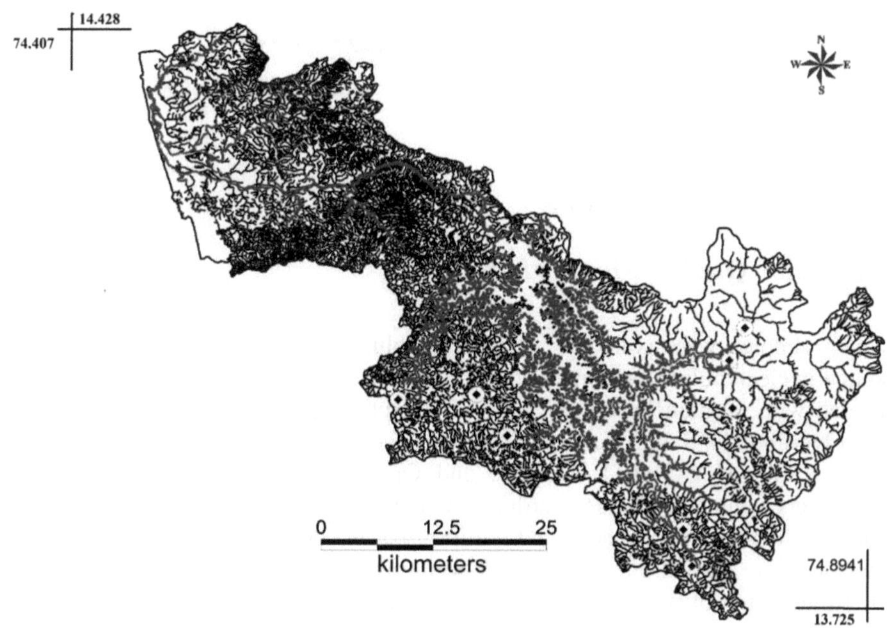

Fig. 5 River Sharavathi with sampling sites

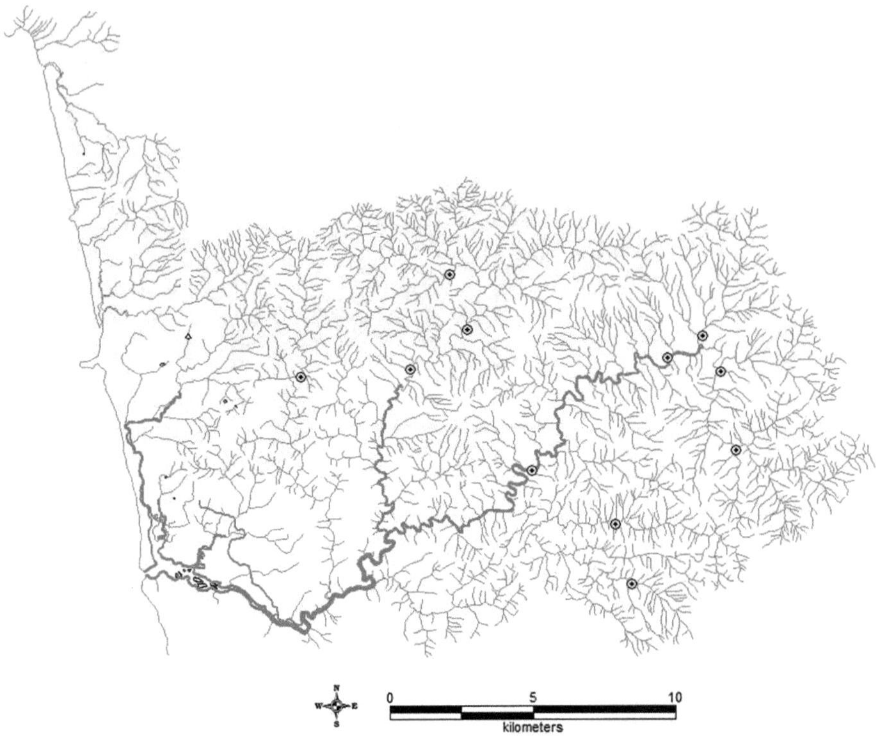

Fig. 6 Venkatapura River with sampling sites

3 Materials and Methods

3.1 Water Quality Monitoring

Water samples were collected at each sampling locations (Table 1) from each source in clean polythene containers of 2.5 L capacity. The sample containers were labeled with a unique code and date of collection. pH, water temperature, Total Dissolved Solids, Salinity and Nitrates were recorded immediately after collection using EXTECH COMBO electrode and Orion Ion Selective Electrode. Other parameters namely, chloride, hardness, magnesium, calcium, sodium, potassium, fluoride, sulphate, phosphates, and coliform bacteria were analyzed in lab. All the analyses were carried out as per the procedures provided in Standard Methods for the examination of water and wastewater (APHA 1998). Details of the methods of water quality determination are presented in the Table 2.

Table 1 Details of the sampling sites (river basin-wise—marked in Figs. 2, 3, 4, 5, and 6)

SITES	CODE	LAT	LON	SITES	CODE	LAT	LON
Aghanashini river basin (ARB)				*Kali river basin (KRB)*			
Sonda	A1	74.4834	14.4868	Beegar	K1	74.5818	14.9163
Nellimadke	A10	74.8431	14.5289	Astolli	K10	74.5383	15.4289
Neralamane	A11	74.8439	14.4554	Kervada	K2	74.6368	15.2454
Balur	A12	74.8098	14.4853	Mavlangi	K3	74.5923	15.2561
Baillalli	A13	74.7920	14.3013	Tatwala	K4	74.7466	15.0879
Hulidevarakodlu	A2	74.6643	14.4040	Sakathi	K5	74.3378	14.9185
Donehole	A3	74.5878	14.4330	Naithihole	K6	74.2593	14.8543
Deevalli	A4	74.5584	14.4332	Kesrolli	K7	74.7412	15.3037
Ullurmatha	A5	74.5823	14.3844	Kaneri	K8	74.4676	15.0247
Yanahole	A6	74.5355	14.5344	Badapoli	K9	74.3560	15.0144
Jalagadde	A7	74.6127	14.5480				
Kurse	A8	74.6900	14.5595				
Sappurthi	A9	74.7562	14.5234				
Bedthi river basin (BRB)				*Sharavathi river basin (SRB)*			
Mathigadda	B1	74.5926	14.6730	Nandiholé	S1	75.1245	14.0418
Vajgadde	B10	74.6154	14.6213	Haridravathi	S2	75.1084	14.0209
Nycti. Site	B11	74.6120	14.6390	Mavinaholé	S3	75.1055	13.9735
Angadibail	B12	74.5332	14.6067	Sharavathi	S4	75.0804	13.8532
Daanandhi	B13	74.8667	14.7358	Hilkunji	S5	75.0896	13.7730
Hemmadi	B14	74.8586	14.7510	Nagodiholé	S6	74.8839	13.9269
Attiveri	B15	75.0357	15.0759	Hurliholé	S7	74.8428	13.9786
Yerebail	B16	74.9395	15.0470	Yenneholé	S8	74.7268	13.9650
Gunjavathi	B17	74.9140	14.9921				
Chitgeri	B18	74.9834	14.8557	*Venkatapura river basin (VRB)*			
Karadrolli	B19	74.8356	14.9918	Badabhag	V1	75.6293	14.0588
Kammani	B2	74.5958	14.7132	Bachochodi	V10	74.6907	14.0901
Dabguli	B20	74.6572	14.8508	Kelanur	V11	74.6959	14.0653
Ramanguli	B21	74.6054	14.1238	Undalakatle	V2	74.5900	14.0910
Kalghatghi	B22	74.9785	15.1586	Midal	V3	74.5543	14.0888
Manchiker	B23	74.7861	14.8910	Arkala	V4	74.6563	14.0415
Apageri	B3	74.5840	14.6389	Galibyle	V5	74.6085	14.1038
Hasehalla	B4	74.5840	14.7551	Nagoli	V6	74.6735	14.0946
Kaleswara	B5	74.6095	14.7587	Ondalasu	V7	74.6028	14.1213
Andhalli	B6	74.8016	14.6701	Hegganamakki	V8	74.6848	14.1018
Makkigadde	B7	74.4299	14.7095	Kurandura	V9	74.6616	14.0227
Kelaginkeri	B8	74.5926	14.6730				
Devanahalli	B9	74.6635	14.6281				
Kurandura	V9	74.6616	14.0227				

Table 2 Methods used for analysing water samples

Parameters	Units	Methods	Section no. APHA 1998
pH	–	Electrode method	4500-H^+ B
Water temperature	°C		2550 B
Salinity	ppm		2520 B
Total dissolved solids	ppm		2540 B
Electrical conductivity	µS		2510 B
Dissolved oxygen	mg/L	Iodometric method	4500-O B
Alkalinity	mg/L	HCl titrimetric method	2320 B
Chlorides	mg/L	Argentometric method	4500-Cl^- B
Total hardness	mg/L	EDTA titrimetric method	2340 C
Calcium hardness	mg/L	EDTA titrimetric method	3500-Ca B
Magnesium hardness	mg/L	Calculation method	3500-Mg B
Sodium	mg/L	Flame emission photometric method	3500-Na B
Potassium	mg/L	Flame emission photometric method	3500-K B
Fluorides	mg/L	SPADNS method	4500-F- D
Nitrates	mg/L	Nitrate electrode method and phenol disulphonic acid method	4500-NO3- D
Sulphates	mg/L	Turbidimetric method	4500-SO_4^{2-} E
Phosphates	mg/L	Stannous chloride method	4500-P D

3.2 Diatom Collection, Preparation and Enumeration

Figure 7 illustrates the habitat of diatoms—diatom colonies on stones, sand, etc. At each site, three to five stones were randomly selected across the stream and diatoms were scraped off the exposed surface of the stones using a tooth brush. Fresh samples were carefully checked to assure that majority of the diatom frustules were alive prior to acid combustion. A hot HCl and $KMnO_4$ method was used to clean frustules of organic materials. The cleaned diatom samples were dried on 18 × 18 mm cover slips and mounted with Pleurax. A total of 400 frustules per sample were enumerated and identified using compound light microscope (Lawrence and Mayo LM-52-series, with 1,000 X magnification) following the methods described by Taylor et al. (2005) and Karthick et al. (2010). Diatoms were identified at species level according to Gandhi (1957a, b, c, 1958a b, 1959a, b, c, 1960a, b, c), Krammer and Lange-Bertalot (1986–1991) and Taylor 2004; Taylor et al. (2007a, b).

Fig. 7 Diatoms on stone in streams

3.3 Land Use Land Cover (LULC) Analysis

The remote sensing data were processed to quantify the land use of respective basins broadly into 6 classes—forest and vegetation; agriculture and cultivated area; open scrub and barren; water bodies; built-up; and others (includes categories like rocky outcrop, etc.). The multi-spectral data of Indian Remote Sensing (IRS) LISS-III with a spatial resolution of 23.5 m were analyzed using IDRISI Andes (Eastman 2006; http://www.clarklabs.org) and GRASS (http://ces.iisc.ernet.in/grass). Land use analysis involved (a) generation of False Colour Composite (FCC) of remote sensing data (bands—green, red and NIR). This helped in locating heterogeneous patches in the landscape (b) selection of training polygons (these correspond to heterogeneous patches in FCC) covering 15 % of the study area and uniformly distributed over the entire study area, (c) loading these training polygons co-ordinates into pre-calibrated GPS, (d) collection of the corresponding attribute data (land use types) for these polygons from the field. GPS helped in locating respective training polygons in the field, (e) supplementing this information with Google Earth (http://www.googleearth.com) (f) 60 % of the training data has been used for classification, while the balance was used for validation or accuracy assessment. Based on these signatures, corresponding to various land features, supervised image classification was carried out using Gaussian Maximum Likelihood Classifier (GMLC) to the final six categories.

3.4 Data Analysis

Compiled data were tested for normality before performing statistical analyses. Statistical analyses comprised Kruskal-Wallis test (H), Principal Component Analysis (PCA) and Non-Metric Multi Dimensional Scaling (NMDS). All the tests were performed using the R-software (R Development Core Team 2006). Box plots were used to visually summarize the data. The line in the box indicates the median value of the data. If the median line within the box is not equidistant from the edges of the box, then the data are skewed. "Gridding" is the operation of spatial interpolation of scattered 2D data points onto a regular grid. Gridding allows the production of a map showing a continuous spatial estimate. The spatial coverage of the map is generated automatically as a square covering the data points. Non-metric multidimensional scaling is based on Bray-Curtis distance matrix was performed for classifying the sites across river basins. In NMDS, data points are placed in 2 or 3 dimensional coordinates system preserving ranked differences.

The non-parametric Kruskal-Wallis test was used to assess whether species richness, species diversity and turnover across water quality regimes were significantly different. Temporal variation in diatom assemblages in each site was analyzed by NMDS using absolute abundance data. NMDS is an ordination method well suited to data that are non-normal or are arbitrary or discontinuous and for ecological data containing numerous zero values (Minchin 1987; McCune and Grace 2002). Results were visualized showing the most similar samples closer together in ordination space (Gotelli and Ellison 2004). A final stress value, typically between 0 and 15, was evaluated as a measure of fitted distances against the ordination distance, providing an estimation of the goodness-of-fit in multivariate space. Changes in species composition or percentage turnover (T) were used to indicate community persistence. T was calculated as $T = (G + L)/(S1 + S2)$ times 100 where G and L are the number of taxa gained and lost between months respectively, and S1 and S2 are the number of taxa present in successive sampling months (Diamond and May 1977; Brewin et al. 2000; Soininen and Eloranta 2004). The relationship between the local population persistence, the local abundance in terms of relative abundances, and the regional occupancy were examined using correlation analysis (Soininen and Heino 2005). For the species distribution model, the species were classified as core species as species that occurred in over 90 % of sites, and satellite species as species that occurred in fewer than 10 % of sites (McGeoch and Gaston 2002; Soininen and Heino 2005). Local occupancy of each diatom species was calculated by their percentage of occurrence at each site across the seasons. Seasonal diatom community was related to the water quality parameters using multiple linear regressions. Finally, water quality variables were used in PCA to elucidate the spatial water quality variation.

4 Results and Discussion

4.1 PH

The pH of river water is the measure of negative logarithm of hydrogen ion concentration that indicates how acidic or basic the water is on a scale of 0–14. Most of the peninsular rivers fall between 6.5 and 8.5 on this scale with 7.0 being neutral. The optimum pH for river water is around 7.4. Water's pH can be altered by industrial and agricultural runoff. Vajgadde at BRB (Bedthi River basin) has recorded low pH of 6.9 and Kalghatgi from the same river has recorded highest pH of 8.27. Low ranges of pH are observed in the forested streams and high alkaline pH are observed in sites contaminated with agricultural and urban runoff (Figs. 8 and 9). A pH of 8.0 should be sufficient to support most river life with the possible

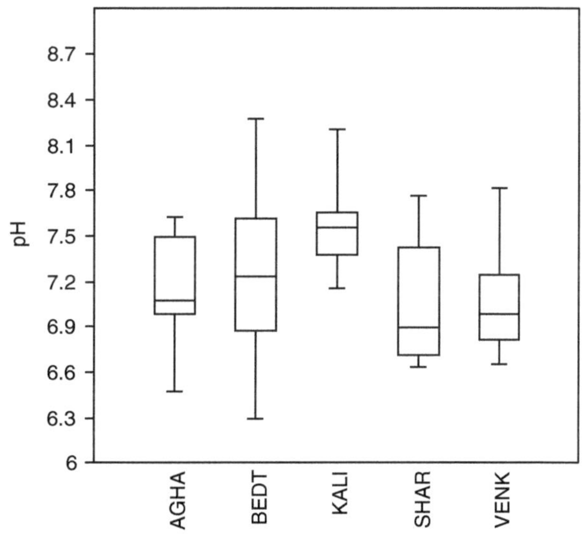

Fig. 8 pH across river basins

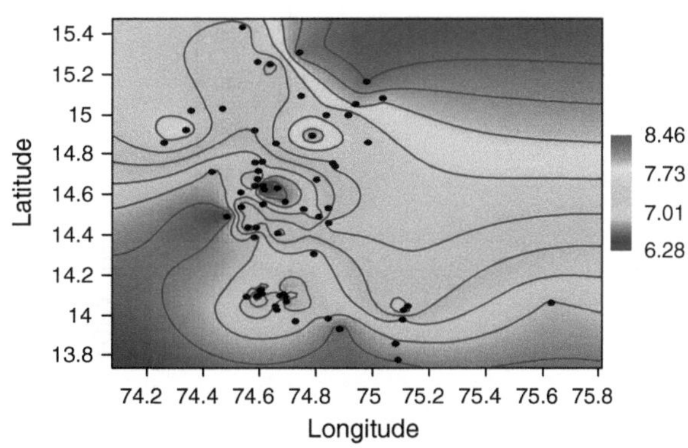

Fig. 9 Spatial representation of pH across sites

exception of snails, clams, and mussels, which usually prefer a slightly higher pH. The average pH in the study was 6.9, a value that is only sufficiently basic for bacteria, carp, suckers, catfish, and insects. BRB (Bedthi River basin) and KRB (Kali River basin) record most of the alkaline nature, whereas ARB (Aghnashini River basin), SRB (Sharavathi River basin) and VRB (Venkatapura River basin) sites record neutral to near acidic nature.

4.2 Electrical Conductivity and Total Dissolved Solids

Pure water does not conduct electricity. Conductivity is a measure of the ability of water to pass an electrical current. Conductivity in water is affected by the presence of inorganic dissolved solids such as chloride, nitrate, sulfate, and phosphate anions (ions that carry a negative charge) or sodium, magnesium, calcium, iron, and aluminum cat ions (ions that carry a positive charge). Organic compounds such as oil, phenol, alcohol, and sugar do not conduct electrical current very well and therefore have a low conductivity when in water. Conductivity is also affected by temperature: the warmer the water, the higher the conductivity. Discharges to streams can change the conductivity depending on their make-up. A failing sewage system would raise the conductivity because of the presence of chloride, phosphate, and nitrate in it; an oil spill would lower the conductivity.

Low level of electrical conductivity was observed at Vajgadde at BRB (22.5 μS/cm) and highest value was recorded from Kalghatghi of the BRB (1038.95 μS/cm), as illustrated in the Fig. 10. Sites of BRB showed high levels of variation when compared to the SRB and VRB. KRB sites recorded comparatively higher conductivity and

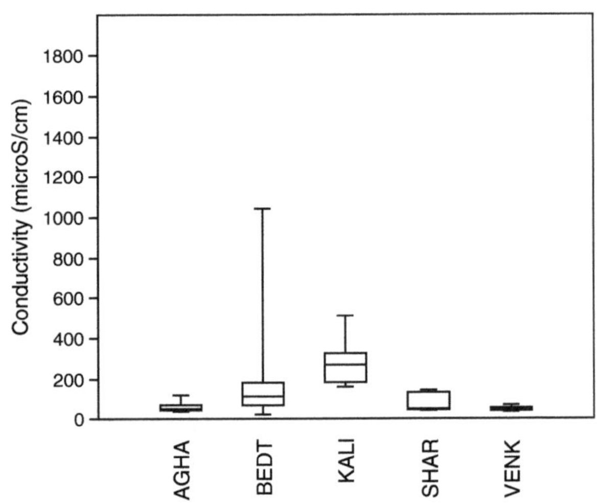

Fig. 10 Electrical conductivity across river basins

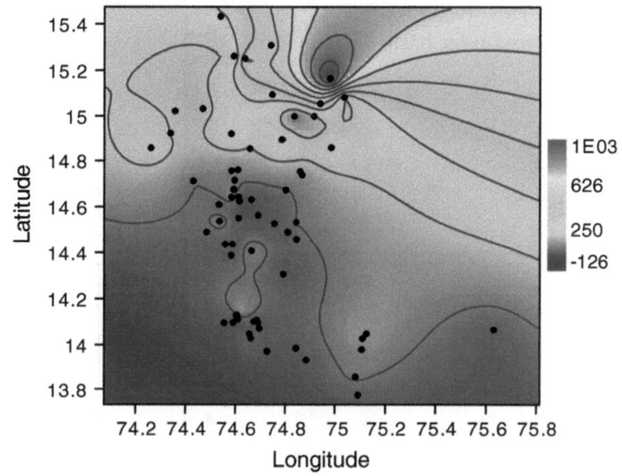

Fig. 11 Spatial representation of electrical conductivity across sites

total dissolved solids, perhaps due to the accelerated erosion associated with the conversion of natural vegetation into monoculture plantations in its catchment area in the north part of the river basin (Fig. 11).

4.3 Dissolved Oxygen

The atmosphere is a major source of dissolved oxygen in river water. Waves and tumbling water mix atmospheric oxygen with river water. Oxygen is also produced by rooted aquatic plants and algae as a product of photosynthesis. An adequate supply of dissolved oxygen (DO) is essential for the survival of aquatic organisms. A deficiency of DO in is a sign of an unhealthy river. There are a variety of factors affecting the levels of dissolved oxygen.

In the present study, lowest dissolved oxygen levels were observed in Kervada (3.67 mg/L) located in the KRB. This site is located adjacent to the effluent discharge point of a Paper mill (Fig. 12). The paper mill effluent is characterized with high levels of organic content, which might consume most of the oxygen for its degradation with the help of bacteria. Bacteria which decompose plant material and animal waste consume dissolved oxygen, and decrease the quantity available to support life. Ironically, it is life in the form of plants and algae that grow uncontrolled due to fertilizer that leads to the masses of decaying plant matter. This site is also infested with marsh crocodiles (*Crocodylus palustris*), which prevails as a major threat to the humans and livestock in the surroundings. Crocodiles are attracted to this particular place due to availability of the solid organic contents present the paper mill effluent. Sites at SRB recorded saturated levels of dissolved oxygen in the streams and all other sites recorded with high variation. Apart from the organic pollution other reason for low levels of dissolved oxygen is lack of mixing in water. Small to medium sized check dams are constructed in the middle

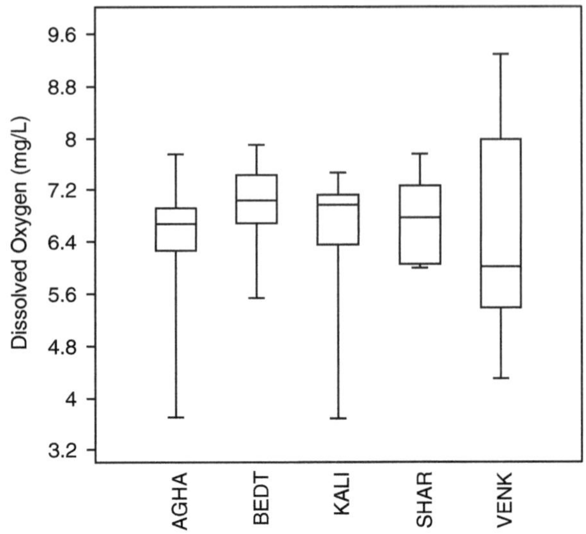

Fig. 12 Dissolved Oxygen levels across river basins

reaches of all the rivers, might have converted the lotic system into lentic system wherein diffusion of atmospheric oxygen into stored water is highly restricted due to the stagnant conditions. Thus, many factors namely organic pollution, active consumption by bacteria, algae and exotic plants and reduced influx due to damming have all contributed towards low levels of dissolved oxygen.

4.4 Nutrients

Unlike temperature and dissolved oxygen, the presence of nitrates usually does not have a direct effect on aquatic insects or fish. However, excess levels of nitrates in water can create conditions that make it difficult for aquatic insects or fish to survive. Nitrate-nitrogen is important because it is biologically available and is the most abundant form of nitrogen in Central Western Ghats streams. Like phosphorus, nitrate can stimulate excessive and undesirable levels of algal growth in water bodies leading to eutrophication. Nitrates come to the streams mainly from the runoff from the agriculture farms and eroded sediments. Runoff from the agriculture farms carries huge amount of fertilizer residues. Among the studied river basins, the KRB and BRB recorded high levels of nitrates from its upstream region (Fig. 13). Both KRB and BRB possess intense agriculture and limited surface water bodies in their upstream regions which leads to high levels of nitrates.

Along with the nitrates, phosphate also plays an important role in the river hydrobiology. Phosphorus is an important nutrient for plant growth. Excess phosphorus in the river is a concern because it can stimulate the growth of algae. Excessive algae growth, death, and decay can severely deplete the oxygen supply in

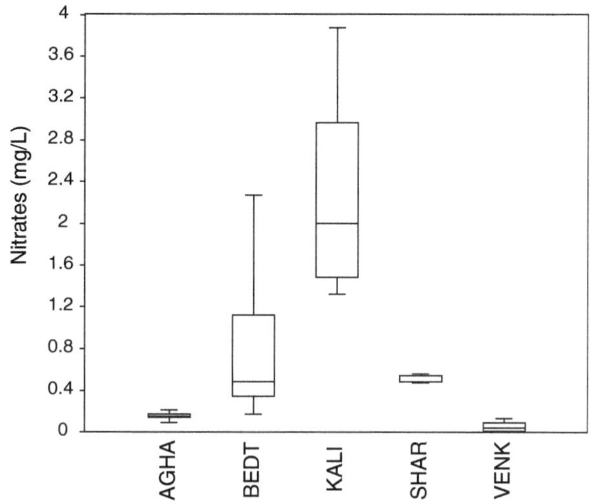

Fig. 13 Nitrate across River Basins

the river, endangering fish and other forms of aquatic life. Urban runoff is the major source for phosphates in the streams. Among the studied basins, BRB receives considerable amount of urban sewage from Hubli city. The impact of high levels of phosphates leads to algal blooms in many reaches of the Bedthi River. Manchikeri site located in between Sirsi and Yellapur has a check dam for pumping water for drinking water supply. Recently another check dam was constructed near the Manchikeri Bridge to store water for Yellapur drinking water supply. Though check dam stores water to support drinking water supply, owing to the intensive agricultural and other activities in the upstream regions, check dam also plays as a reservoir for pollutants and reduces the chances of the accumulated pollutants diffusing away. Stagnated water with heavy nutrient content leads to algal bloom. Preliminary investigations suggest that the algal bloom was created by algal genus *Microcystis,* a blue-green algae (also referred to as Cyanobacteria). It is a common bloom-forming algae found primarily in nutrient enriched river and lake waters. This genus is colonial, which means that single cells can join together in groups which tend to float on the water surface. Colony sizes will vary from a few to hundreds of cells. Any large algal bloom has the potential to result in fish kills by depleting the water of oxygen. The dead algal cells sink down and consume huge amount of oxygen for their decomposition. In such situations, there may not be enough oxygen remaining in the water to support fish in the vicinity. Furthermore, as these large blooms die and sink to the bottom, they commonly release chemicals that can produce a foul odor and musty taste. Some strains of *Microcystis* may produce toxins that have been reported to result in health problems to animals that drink the water, and minor skin irritation and gastrointestinal discomfort in humans that come in contact with toxic blooms. Uncontrolled growth of single species of algae will also lead to death of aquatic invertebrates and fishes due to unavailability of food, which in turn affects the aquatic food chain.

4.5 Lotic Ecosystems: Intra Basin Variations in Quality

Principal component analysis reveals that the BRB contains sites with pristine to heavily polluted waters. Most of the sites in northern part of BRB stand out separately in ordination space due to their very high amount of ions and nutrients. Sites in the SRB, KRB and ARB seem to fall in the same quality of water, whereas the VRB stands out separately with very pristine water quality status (Figs. 14 and 15). NMDS plot of the water quality variables shows that ionic and physical parameters have the same origin, where as nutrients arise from different source (Fig. 15).

4.6 Seasonality of Benthic Diatoms and Water Quality

The water chemistry data along the Bedthi River showed high annual variation across sites. The parameters which showed significant difference among the groups are pH, conductivity, chlorides, hardness, calcium, magnesium, sodium and potassium. All these parameters were found to be high in HPAS, moderate in MPPS and very low in LPFS (Table 3). Irrespective of the pollution status, dissolved

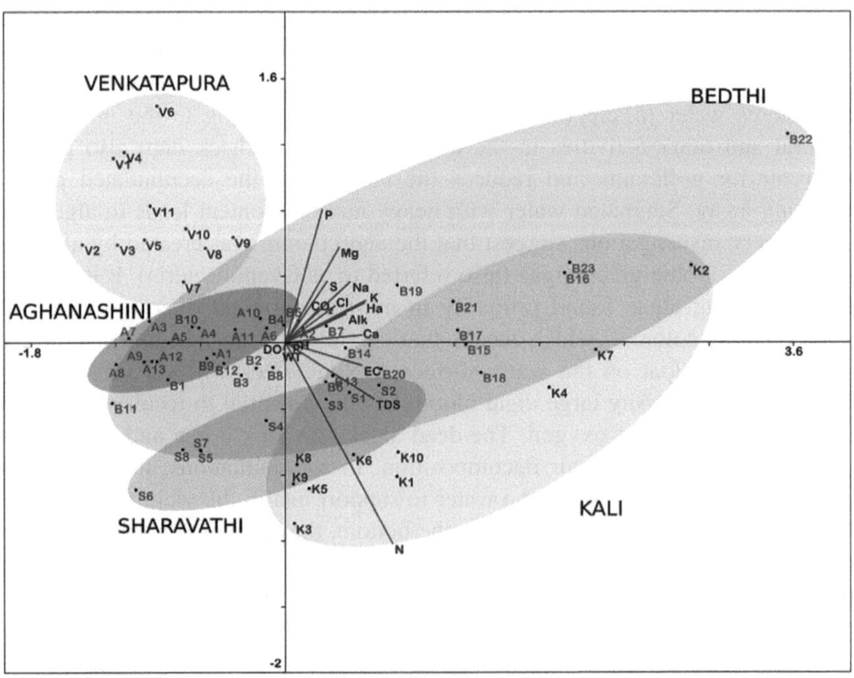

Fig. 14 PCA plot for water quality variables across the River Basins

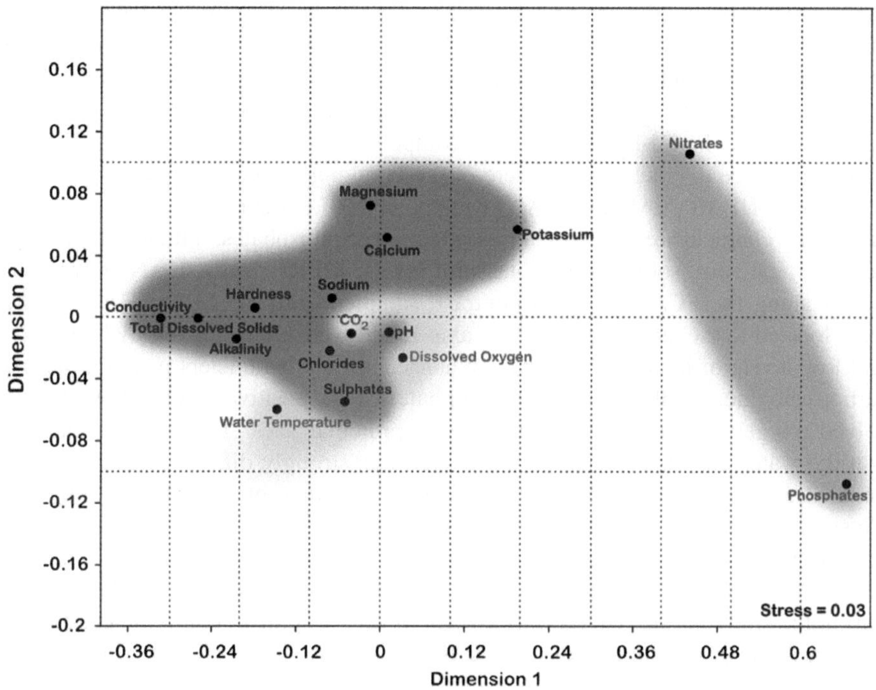

Fig. 15 NMDS plot of water quality variables across River Basins

Table 3 Species richness and diversity across space and time at BRB sites

Months	LPFS		HPAS		MPPS	
	KAM	HAS	KAL	MAN	AND	DAN
Jan	14 (1.10)	10 (1.04)	7 (1.58)	-L-	4 (0.81)	14 (2.27)
Feb	10 (1.58)	6 (1.10)	7 (1.65)	9 (1.63)	4 (1.04)	19 (2.34)
Mar	4 (1.25)	6 (0.85)	8 (1.73)	14 (1.90)	6 (0.95)	-D-
Apr	7 (1.47)	7 (0.67)	9 (1.65)	1 (0)	8 (0.93)	-D-
May	11 (1.40)	4 (0.47)	3 (0.92)	8 (1.34)	11 (1.76)	-D-
Jun	9 (1.16)	3 (0.15)	4 (1.02)	-M-	3 (0.98)	5 (1.14)
Jul	4 (0.77)	-M-	1 (0)	-M-	6 (1.19)	7 (1.46)
Aug	-M-	-M-	-M-	-M-	5 (0.79)	5 (1.35)
Sep	1 (0)	-M-	1 (0)	-M-	2 (0.69)	9 (1.67)
Oct	4 (0.78)	5 (0.77)	5 (1.34)	5 (1.15)	9 (1.29)	9 (1.35)
Nov	4 (0.82)	4 (0.94)	6 (1.54)	-L-	-L-	4 (0.84)
Dec	4 (0.87)	-L-	8 (1.76)	16 (2.02)	2 (0.69)	10 (1.39)

oxygen (DO) levels across water quality regimes were roughly similar with mean DO levels (Mean ± S.D) of 7.51 ± 1.67, 7.08 ± 2.21, 6.43 ± 2.91. However, anoxic DO level of 0.86 mgL^{-1} was observed in one sample from the HPAS (KAL). PCA results indicated that water quality differed markedly among sampling sites and across seasons (Fig. 14) with the first component explaining 84.6 % of the total variation. Three distinct clusters were observed along a pollution gradient. Sample scores from HPAS (KAL and MAN) were positioned to the right along PCA axis 1, and were characterized by higher conductivity, phosphates, nitrates, alkalinity, hardness, calcium, sodium and potassium levels. MPPS (AND, DAN) were positioned along the PCA axis 2. In contrast, samples from the LPFS (HAS, KAM) were located to the left along the PCA axis 1, and were characterized by higher DO and low levels of ions and nutrients. Water chemistry parameters namely, the pH, carbon dioxide, alkalinity, nitrates, sulphates were positively loaded while dissolved oxygen was negatively loaded with principal axes. These results indicate that the water chemistry between these sites was strongly different throughout the year. Stream water chemistry differed between the three groups of sites (Fig. 16). Clusters illustrated in Fig. 17 reveal distinct grouping based on the ion and nutrient concentrations in the respective sampling sites across the river basins.

One hundred and three species of diatoms were recorded from all the six sites during the study period, with a flora typical of oligotrophic to eutrophic conditions. Among the taxa recorded, the *Achnanthidium minutissimum* (KÜtz) Czarn., *Gomphonema gandhii* Karthick and Kociolek, *G. difformum* Karthick and Kociolek, *Nitzschia palea* (KÜtz) W. Sm., *Nitzschia frustulum* (KÜtz) Grun., *Cymbella* sp. and *Navicula* sp. *Achnanthidium minutissimum*, *Gomphonema gandhii* and *G. difformum*, were present throughout the study period in LPFS and MPPS, while *Nitzschia palea* and *Nitzschia frustulum* were dominant in HPAS. In contrast, *Cymbella* sp. was the only diatom present at the MAN site during the month of April. The samples from headwater oligotrophic streams (often with low pH and conductivity) were characterized by the occurrence of *Gomphonema gandhii, Achnanthidium minutissimum and Gomphonema difformum.* Assemblages from eutrophic streams (HPAS) were characterized by dominance of *Nitzschia palea, N. frustulum and* occasionally with *Cyclotella meneghiniana.*

The species richness was highest at three sites; KAM, DAN and MAN, even though each one inherited different water chemistry regimes. All the six sites were characterized with very low species richness during the monsoon season (Table 3). In all the sites during the entire study period, the diversity (H′) ranged between the highest 2.34 in DAN during the month of February to lowest of 0 in KAM and KAL during the monsoon months. Kruskal–Wallis results showed that the species diversity across three water chemistry regimes were significantly different (Kruskal–Wallis, $H = 6.97$; $p = 0.03$).

Species abundances across season suggested trends within community composition in ordination space (Fig. 18). In sites KAM and KAL, the communities of post monsoon season aggregated in ordination space. However, this trend was not recognized in HAS, DAN, MAN and AND sites. In LPFS (KAM and HAS) and MPPS (AND and DAN), the diatom assemblages were identical for pre-monsoon,

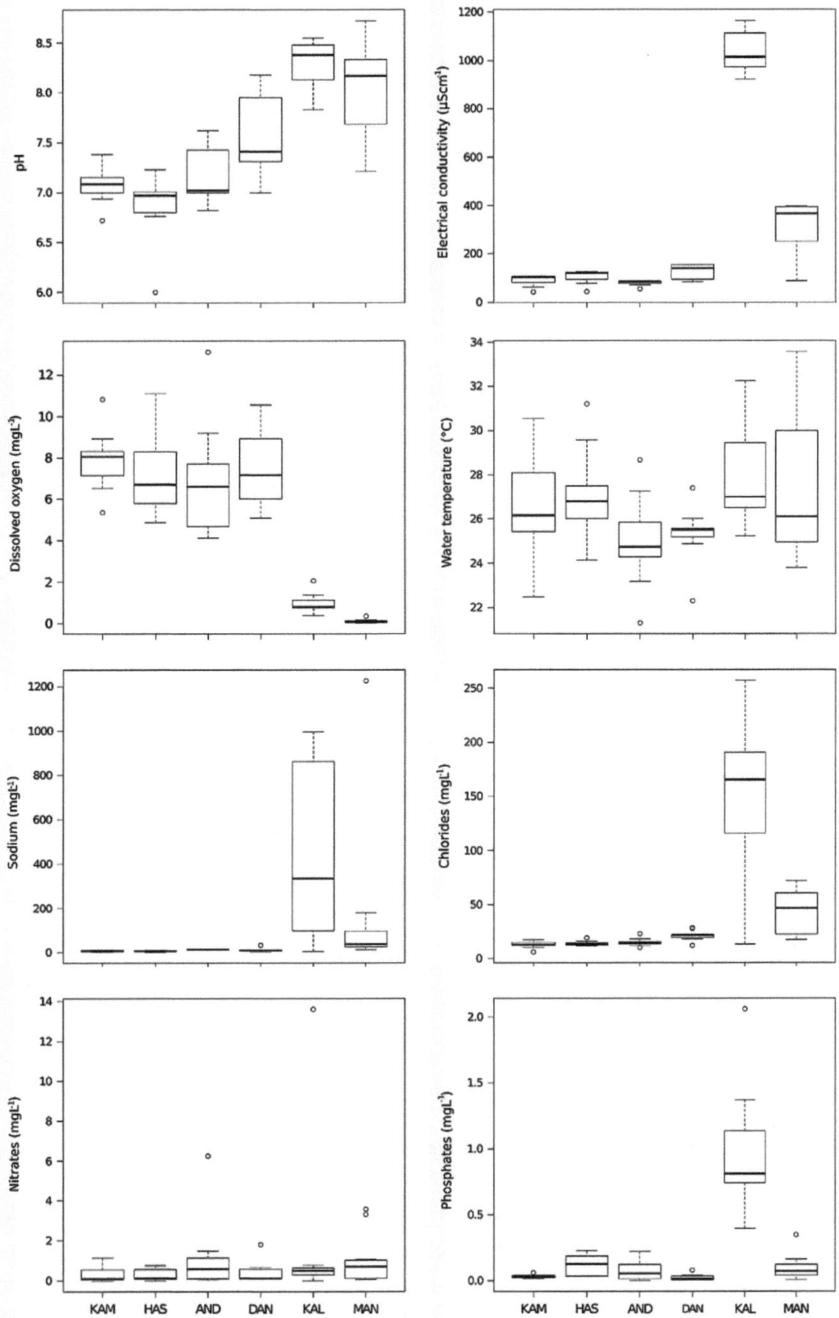

Fig. 16 Water chemistry at sampled sites during the study period at BRB

Fig. 17 Cluster analysis of sampling sites across river basins based on water quality

monsoon and post-monsoon seasons respectively, whereas the assemblages in HPAS (KAL and MAN) were not identical across seasons. Though there are trends on community composition, a strong relation with the seasonally dynamic environmental variables could also be envisaged. The difference in the species richness among sites were not significant (Kruskal–Wallis $H = 6.07$; $p = 0.29$). Species richness from highest to lowest within water quality regimes, followed the order: LPFS > MPPS > HPAS. Overall, species richness was lowest during the monsoon

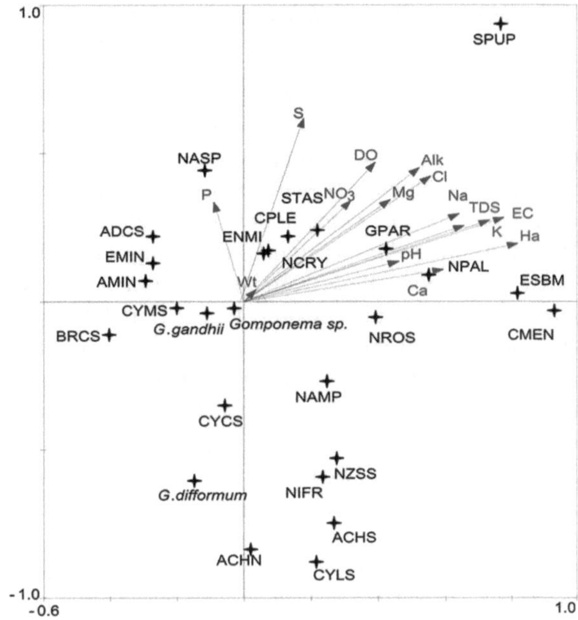

Fig. 18 CCA bi-plot of water chemistry variables and dominant species assemblages

months in all the sites. Changes in species composition or percentage turnover (T) did not follow any trend irrespective of site water chemistry. The highest mean turnover (94.44 % ± 11.11) was observed in MAN, indicating the lowest persistence (Table 3), followed by DAN (79.08 ± 14.47), HAS (70.46 ± 27.64) and AND (64.77 ± 23.71). The mean species turnover was less than 50 % in KAL (47.96 ± 38.1) and KAM (44.03 ± 20.85). Interestingly, KAL showed a wide range of turnover with a minimum of 9 % during the post monsoon and a maximum turnover of 100 % during the monsoon months. In LPFS sites 25 % of the species were persistent across seasons and in MPPS sites 30 % of the species were persistent. However in the HPAS sites, a minimum persistence of 7.14 % was observed for KAL and 80 % persistence in MAN. The differences in turnover were significant across sites (Kruskal–Wallis H = 17.52; p = 0.0036).

Percentage occupancy showed a significant relationship with local maximum species abundance (r = 0.49; P = <0.0001) and local mean species abundance (r = 0.37; P = <0.0001). This positive correlation was slightly stronger for the local maximum abundance. However, it was highly significant for both the local abundance measures. Species that occurred locally with more frequency also tended to be abundant across the sites. The species–occupancy frequency distribution (Fig. 18) followed a "satellite-mode" (Hanski 1982) of species distribution, where a high proportion of species occurred at a small number of sites. Sixty three species occurred at only one site, twenty two species occurred in two sites, eleven species occurred in three sites; five species occurred in four sites, three species occurred in five sites and none of the species occurred in all the six sites.

4.7 Diatom Based Biomonitoring

A total of 140 diatom taxa were identified across sites, 61 of them reaching a relative abundance of over 5 % in at least one site. Appendix 1 provides the checklist of diatoms. The species compositions were dominated by *Gomphonema gandhii* Karthick and Kociolek, *Achnanthidium minutissimum* Kützing, *Achnanthidium* sp., *Gomphonema* sp., *Gomphonema parvulum* Kützing, *Nitzschia palea* (Kützing) W.Smith, *Nitzschia frustulum* (Kützing) Grunow var. *frustulum*, *Navicula* sp., *Navicula cryptocephala* Kützing, *Cyclostephanos* sp., *Cymbella* sp., *Eolimna sub-minuscula* (Manguin) Moser Lange-Bertalot and Metzeltin, *Sellaphora pupula* (Kützing) Mereschkowksy, *Eunotia minor* (Kützing) Grunow in Van Heurck, *Nitzschia amphibian* Grunow f. *amphibia*, *Cyclotella meneghiniana* Kützing, *Gomphonema difformum* Karthick and Kociolek, *Navicula rostellata*Kützing, *Cocconeis placentula* Ehrenberg var. *euglypta* (Ehr.) Grunow, *Brachysira* sp., *Stauroneis* sp., *Encyonema minutum* (Hilse in Rabh.) D.G. Mann, *Cyclotella* sp. and *Nitzschia* sp. The species composition contains cosmopolitan to possible Western Ghats endemic species. In general, species from oligotrophy to highly eutrophic condition were observed. The current study also documents some of the species for the first time in Western Ghats and many new species descriptions are underway. Waters were circumneutral throughout the study area (Table 4), with certain tendency towards alkalinity in the streams drained from agriculture and urban catchment. The highest ionic and nutrient values correspond to the agriculture catchment dominated streams, particularly in the leeward side of the mountains. Oxygenation was generally close to saturation; the lowest values are due to wastewater water inflows in few localities. The most oligotrophic sites were located in mountain watercourses, while downstream sites were generally more polluted, becoming eutrophic in condition. The detailed water chemistry variables are presented in Table 5.

The results of correlation performed between diatom indices and water chemistry variables are presented in the Table 6. It is observed that significant correlations, albeit at varying degrees exist between most of the diatom indices and water chemistry variables. Diatom indices IPS, EPI and SID showed correlation with more number of water chemistry variables when compared to the other indices. TDI and IPS are negatively correlated with pH, EC, TDS, alkalinity, calcium, magnesium,

Table 4 Summary of the canonical correspondence analysis for the stream sites from central Western Ghats

Variables	Axis order			
	1	2	3	4
Eigen value	0.275	0.193	0.162	0.119
Species-environment correlations	0.815	0.755	0.890	0.754
Cumulative percentage variance of species data	10.0	17.0	22.9	27.2
Cumulative percentage variance of species-environment relation	25.8	43.8	59	70.1

Table 5 Waterchemistry variables in 45 sites of CWG streams

Variables	Mean	Std. dev	Median	Min	Max
pH	7.22	0.49	7.14	6.03	8.16
WT (°C)	25.31	2.70	25.07	19.00	33.00
EC (μScm^{-1})	160.55	207.10	107.67	41.55	1164.67
TDS (mg L^{-1})	122.24	204.98	60.30	20.88	1299.67
Alkalinity (mg L^{-1})	54.55	50.32	30.00	6.81	180.00
Chlorides (mg L^{-1})	32.39	40.40	22.72	5.90	220.24
Hardness (mg L^{-1})	51.26	71.05	28.00	10.00	348.00
Calcium (mg L^{-1})	13.88	16.14	8.02	1.60	78.56
Magnesium (mg L^{-1})	16.35	16.73	9.36	1.17	65.95
DO (mg L^{-1})	6.96	1.68	7.23	2.93	10.87
Phosphates (mg L^{-1})	0.36	0.56	0.04	0.00	2.30
Nitrates (mg L^{-1})	0.74	1.10	0.13	0.03	4.30
Sulphates (mg L^{-1})	25.73	20.84	16.87	0.00	74.10
Sodium (mg L^{-1})	25.77	72.18	9.09	4.11	370.00
Potassium (mg L^{-1})	6.33	15.72	1.30	0.19	75.00

sodium and potassium. Percent pollution tolerant diatoms were positively correlated with most of the ionic variables. None of the indices were correlated with water temperature. No correlation of temperature with any of the indices observed that may be due to differing temperature regime in tropical when compared to temperate streams. Similar observation was recorded by Taylor et al. (2007a) from South African rivers. The first four axes of CCA explain 70.1 % variance of species-environment relation and the ordination plot reveals two distinct clusters of species.

Among the species observed in this study, two species were possibly endemic to Western Ghats (*G. gandhii*, *G. difformum* and few other species that are yet to be identified). In few sites, these species were very dominant (>80 % of the total assemblages). The remaining dominant taxa were cosmopolitan and well documented in international literature (Krammer and Lange Bertalot 1986–1991). It is important to note that the indices that were developed and tested in European rivers, lack Western Ghats endemic taxa. Most sites were oligo-mesotrophic and only a few of the streams were eutrophic. The differences in the water quality of these rivers were reflected in the values for the diatom indices, by the relative abundances of indicators of trophic/saprobic stage and by different types of diatom community.

The correlations obtained in the present study are comparable to those demonstrated by Taylor et al. (2007b) in South Africa and by Kwandrans et al. (1998), Prygiel and Coste (1993) and Prygiel et al. (1999) in Europe. Significant correlations emphasize that diatom indices can be used to reflect changes in general water quality (Table 6). Canonical correspondence analysis (Fig. 18) demonstrates that certain widely distributed taxa have similar ecological characteristics in widely separated geographic areas. Species commonly associated with poor water quality in Europe e.g., *Eolimna subminuscula* Lange-Bertalot, *Nitzschia palea* (Kützing)

Table 6 Pearson correlation coefficients between measured water chemistry variables and diatom index scores in 45 sites of CWG streams

INDICES	pH	WT	EC	TDS	Alk	Cl	Ha	Ca	Mg	Na	K
SLA	−0.33*	–	−0.59**	−0.52**	−0.49**	–	−0.62**	−0.43**	–	−0.51**	−0.58**
DESCY	−0.32*	–	−0.54**	−0.46**	−0.41**	–	−0.65**	−0.47**	–	−0.49**	−0.52**
IDSE/5	−0.32*	–	−0.60**	−0.50**	−0.46**	–	−0.63**	−0.45**	–	−0.55**	−0.60**
SHE	–	–	−0.52**	−0.38**	−0.38*	–	−0.56**	−0.41**	–	−0.43**	−0.56**
WAT	–	–	–	–	–	–	−0.36*	–	–	−0.34*	−0.44**
TDI	−0.32*	–	−0.64**	−0.54**	−0.46**	–	−0.69**	−0.53**	−0.30*	−0.52**	−0.58**
%PT	0.36*	–	0.68**	0.62**	0.35*	0.43**	0.66**	0.50**	0.41**	0.65**	0.58**
GENERE	–	–	−0.49**	−0.39**	−0.30*	–	−0.54**	−0.41**	–	−0.41**	−0.41**
CEE	–	–	–	–	–	–	−0.36*	–	–	–	−0.40**
IPS	−0.36*	–	−0.68**	−0.59**	−0.42**	–	−0.66**	−0.46**	−0.31*	−0.56**	−0.58**
IBD	–		−0.56**	−0.43**	−0.34*	–	−0.61**	−0.46**	–	−0.46**	−0.51**
IDAP	–	–	−0.51**	−0.38*	−0.38*	–	−0.56**	−0.40**	–	−0.44**	−0.53**
EPI-D	−0.33*	–	−0.58**	−0.51**	−0.41**	−0.31*	−0.59**	−0.44**	–	−0.53**	−0.55**
DI_CH	–	–	−0.54**	−0.43**	−0.45**	–	−0.58**	−0.39**	–	−0.41**	−0.53**
IDP	–	–	−0.48**	−0.35*	−0.39**	–	−0.58**	−0.42**	–	−0.43**	−0.49**
SID	−0.36*	–	−0.50**	−0.45**	−0.40**	−0.38**	−0.47**	−0.37*	–	−0.43**	−0.46**
TID	–	–	−0.53**	−0.43**	−0.47**	–	−0.59**	−0.41**	–	−0.40**	−0.48**
Evenness	–	–	0.39**	0.40**		–	0.41**	–	–	–	–

WT Water temperature, *EC* Electric conductivity, *TDS* Total dissolved solids, *ALK* Alkalinity, *Cl* Chlorides, *Ha* Total hardness, *Ca* Calcium hardness, *Mg* Magnesium hardness, *Na* Sodium, *K* Potassium. *Diatom Indices SLA* Sládeček's index, *DESCY* Descy's pollution metric, *SHE* Steinberg and schiefele trophic metric, *WAT* Watanabe index, *TDI* Tropical diatom index, *GENRE* Generic diatom index, *CEE* Comission for economical community Index, *IPS* Specific pollution sensitivity metric, *IBD* Biological diatom index, *IDAP* Indice diatomique artois picardie, *EPI-D* Eutrophication/pollution index, *IDP* Pampean diatom index, *%PT* Percentage tolerant

*p<0.1 and **p<0.05

W. Smith, *Sellaphora pupula* (Kützing) Mereschkowksy, *Gomphonema parvulum* (Kützing) ordinate on the right side of the CCA together with elevated levels of ionic and nutrients. Taxa typical of cleaner, less polluted waters ordinate on the left side of the diagram e.g., *Gomphonema difformum* Karthick and Kociolek. However, *Gomphonema gandhii* Karthick and Kociolek seems to have a wider ecological tolerance when compared to its morphologically related species. *Achananthidium minutissimum* group from Western Ghats streams contains morphologically three distinct taxa with wide ecological preferences. Despite the reevaluation of this genus multiple times (Lange-Bertalot and Krammer 1989; Krammer and Lange-Bertalot 1986–1991; Potapova and Hamilton 2007) there are still major gaps in taxonomy and ecology apart from non-inclusion of specimens from tropical rivers. Similar problem holds good for some of the other genus like *Gomphonema*. This analysis had demonstrated that the widely distributed species encountered in the streams of Western Ghats are not only morphologically identical, but also have similar environmental tolerances. *G. gandhii* and *G. difformum* are few among the dominant taxa in this data set but are not included in any of the index calculations. Their omission in the index calculations could result in an under or overestimation of the index scores. Taylor et al. (2007b) cautioned about the associated problems with the usage of European indices in South African rivers. However, the data provided in the present study suggest that European diatom indices can be used in India provided indices address the issues concerned with ecology of endemic species. Hence, the list of taxa included in the indices needs to be adapted according to the study region by providing more importance to the local endemic flora which encourages taxonomic and ecological studies in tropics. The structure of benthic diatom communities and the use of diatom indices yield good results in water quality monitoring in India. However, the occurrence of possible endemic species necessitates a diatom index unique to India.

4.8 LULC Analysis

LULC showed considerable variability among catchments, with forest/vegetation land cover as a dominant class (mean = 64.36 %, range = 0.13–95.45 %), followed by agriculture/cultivation area (mean = 24.27 % range = 2.55–63.63 %), among the 24 catchments. LULC analysis shows that natural vegetation is poor towards the leeward side of the mountains (eastern region), due to the intense anthropogenic activities. This region has more of agriculture, open scrub/barren land, and built-up area. In the entire study region, the class forest/vegetation covers predominantly moist deciduous type, with small isolated patches of semi-evergreen vegetation in the eastern region and the western region (windward side) with rugged hilly terrain and heavier rainfall ($\sim$5,000 mm) having characteristic evergreen to semi-evergreen forests. The detailed LULC for each catchment is given in the Table 7 and the land cover images are given in the Fig. 19. The dendrogram of sites based on LULC obtained by Ward's method is shown in the Fig. 20. Three well differentiated clusters

Table 7 Percentage LULC classes for each catchment in the study area

	Forest/ vegetation	Agriculture/ cultivation	Open scrub/ barren land	Water bodies	Built up	Others
CHI	52.5	34.24	10.9	1.18	0.14	1.04
MEL	83.58	10.88	0.57	2.84	1.52	0.61
KEL	85.83	8.19	1.4	3.03	0.92	0.62
BEE	90.41	2.55	0.16	5.11	1.16	0.61
ANG	81.54	12.67	1.38	0.08	0.18	4.14
MAK	95.45	3.02	0.38	0.01	0.13	1.01
HUR	52.84	28.62	11.89	3.91	0.1	2.64
MAV	48.85	36.26	10.79	2.94	0.43	0.73
YEN	56.15	24.49	10.47	5.61	0.16	3.12
BAI	70.85	23.49	0.99	0.58	0.2	3.89
DEE	64.34	28.64	3.62	0.51	0.2	2.69
YAN	90.83	6.52	0.56	0.02	0.08	1.98
SAP	42.8	50.61	4.5	0.97	0.31	0.81
BAD	88.36	7.29	0.24	0.11	0.06	3.94
NAI	67.43	17.89	3.65	1.03	0.44	9.58
SAK	80.08	14.94	0.26	0.92	0.24	3.58
AND	68.1	24.87	3.36	0.93	0.98	1.77
DAA	86.29	10.84	0.58	1.26	0.41	0.63
HAS	88.6	5.77	0.42	3.96	0.88	0.37
KAM	88.8	5.31	0.6	4.01	0.89	0.39
MAN	23.46	50.42	20.69	0.67	0.52	4.25
KAL	0.53	59.6	30.22	0.02	0.84	8.79
SAN	0.13	63.63	31.48	0.02	0.16	4.59
GUN	36.82	51.72	10.52	0.52	0.35	0.07

can be seen, with forest cover decreasing and agriculture/cultivable land cover increasing from top to bottom. The third cluster from top includes sites SAN, KAL, MAN, SAP and GUN, which are characterized with intensive agricultural activities (>50 %). The group located in the center of the dendrogram is characterized by more forest cover (>50 %) with moderate amount of agricultural land. The topmost group is dominated by forest land cover of more than 80 %.

4.9 Relationship of LULC with Water Chemistry and Diatom Assemblages

A PCA bi-plot of water quality variables and LULC for all sample sites is given in Fig. 21. The two-dimensional bi-plot describes 65 % of the variation in data, where 52 % displayed on the first axis and 13 % is displayed on the second axis. Among

Fig. 19 Land use of study catchments in central Western Ghats.**a** Daanandhi,**b** Deevalli, **c** Badapoli, **d** Mavinahole, **e** Makkegadde, **f** Gunjavathi, **g** Sakathihalla, **h** Angadibail, **i** Hurlihole, **j** Chitgeri, **k** Kalghatghi, *l* Naithihole land use of study catchments in central Western Ghats. **m** Beegar, **n** Andhalli, **o** Yennehole, **p** Kammani, **q** Sangadevarakoppa, **r** Machikere, **s** Melinakeri, **t** Yanahole, **u** Sapurthi, **v** Bailalli, **w** Kelaginakere, **x** Hasehalla

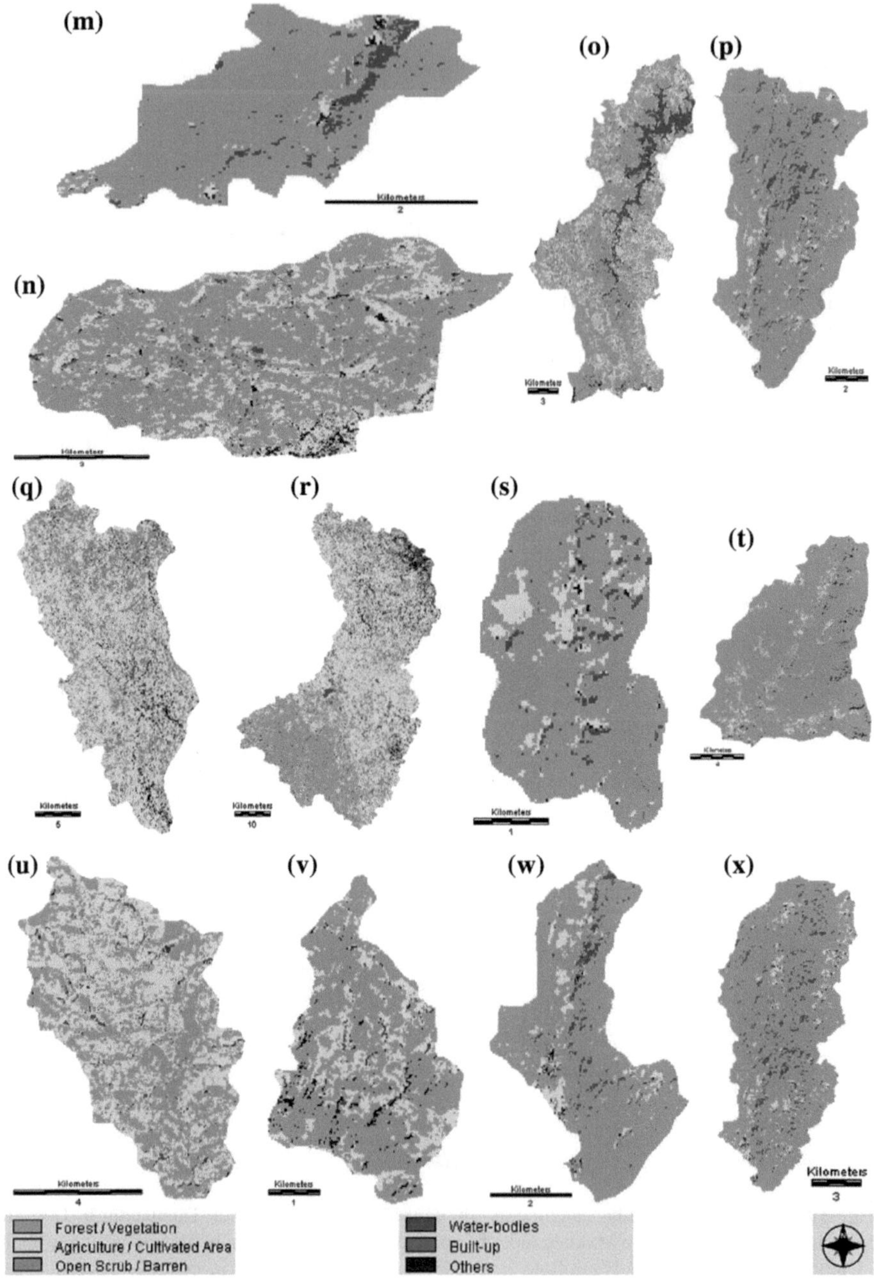

Fig. 19 (continued)

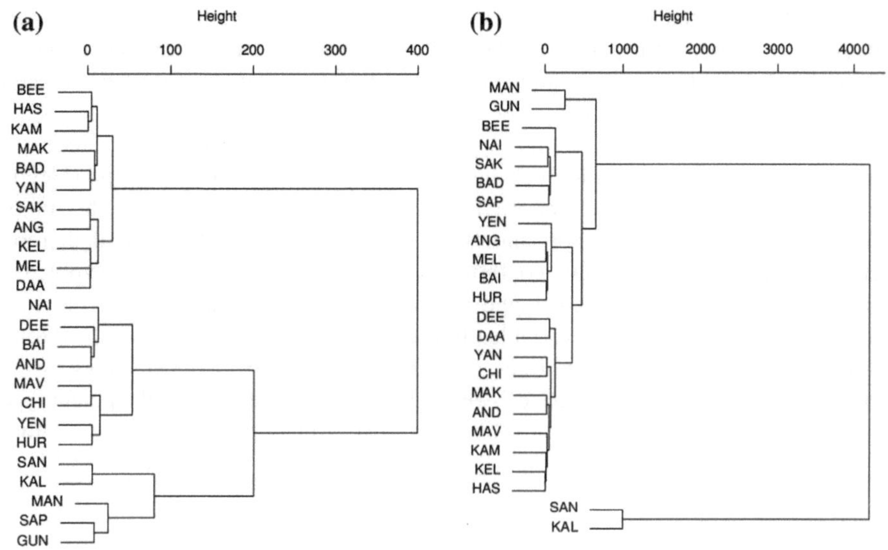

Fig. 20 Dendrogram of the cluster analysis based on **a** LULC and **b** water quality in the 24 sampling sites of the central Western Ghats

Fig. 21 PCA bi-plots of water chemistry and LULC variables in study sites in central Western Ghats Streams

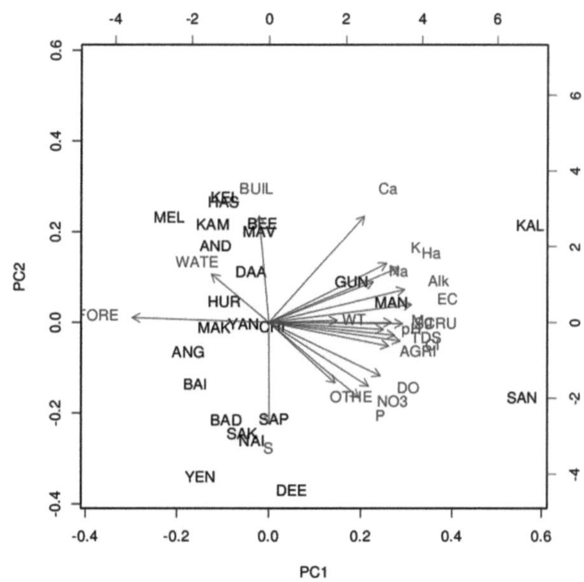

water chemistry variables, ionic variables were positively correlated with first axis and among the LULC variables percentage agriculture and scrub land cover were positively related to the first axis. Sites with more than 50 % of agriculture land cover were separated from other sites on the PC2 axis indicating trends in water quality may be related to land use. Agriculture dominated sites were placed due to the higher conductivity, ionic and nitrates levels relative to the forest dominated sites, which are characterized by low ionic and nutrient in nature.

Correlation between percentage agricultural land cover with water chemistry variables and diatom autecological indices revealed the role of landscape (Hegde et al. 1994). Previous studies reported agricultural expansion as one of the major driver for deforestation (Menon and Bawa 1998) in Western Ghats and thereby determine the environmental condition of streams and diatom assemblages (Fig. 22). The gradient of percentage agriculture land cover were positively correlated with water chemistry variables like electrical conductivity (r = 0.67), total dissolved solids (r = 0.62), nitrates (r = 0.60) and pH (r = 0.52). Gradient of percentage agriculture land cover were positively correlated with percentage pollution tolerant diatoms (r = 0.65, Fig. 23). Relation between the diatom

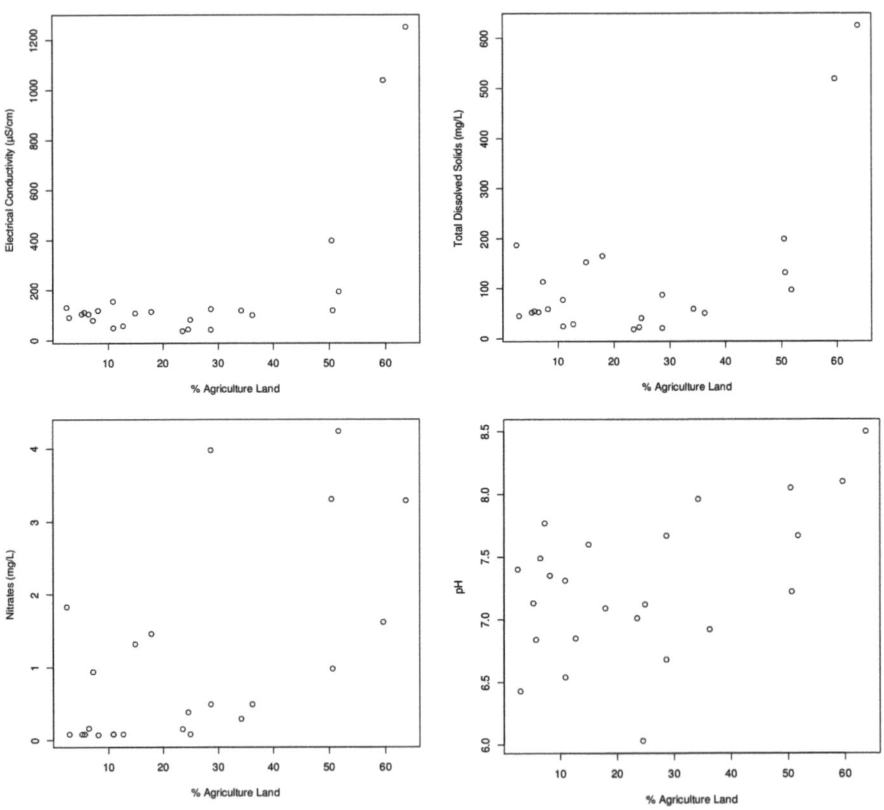

Fig. 22 Changes in water quality variables along a gradient of percentage agricultural land cover in central Western Ghats

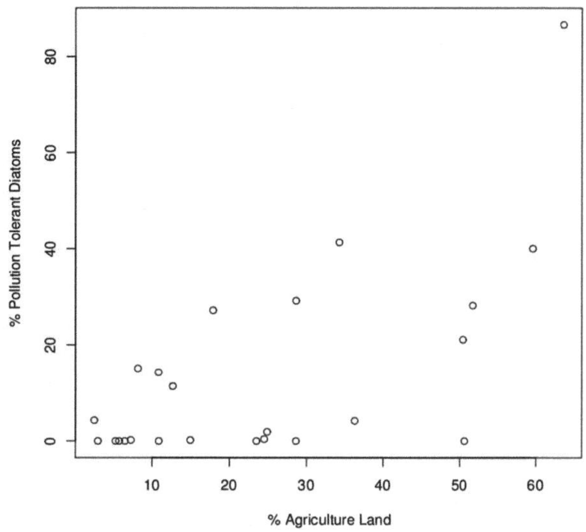

Fig. 23 Pollution tolerant diatoms with gradient of agricultural land cover (%) in central Western Ghats

autecological indices with land cover and water chemistry variables are given in Tables 7 and 8 respectively. Most of the diatom autecological parameters were positively correlated with forest/vegetative cover and negatively correlated with agriculture/cultivable and scrub land cover. All the diatom indices were normalized to a range of 0–20, where <9 indicates bad water quality, 9–12 indicates poor water quality, 12–15 indicates moderate water quality, 15–17 indicates good quality and >17 indicates high quality. The present study shows that within a similar eco-region, the diversity and community composition of diatoms changes with LULC pattern. Among all the 24 catchments, most of the catchments were dominated by forest/vegetation land cover. However, forest cover in the leeward side catchment was very low owing to anthropogenic activities. Hydro power projects commenced in the study area since 1960s seem to have lost.

The streams draining the catchments with agriculture and scrub land cover were characterized with ionic and nutrient rich waters, which highlight that the water chemistry variables are driven by the composition of land cover. Many studies have reported that urban and agricultural land use play a primary role in degrading water quality in adjacent aquatic systems by altering the soil surface conditions, increasing the impervious area and generating pollution (Tong and Chen 2002; White and Greer 2006). The results suggest better water quality tendencies in watersheds having less urbanization with more natural vegetation region. Percent agriculture in the catchment ranged from 2 to 63 % with an average of 24.27 %.

Table 8 Water chemistry variables of the sampling sites in central Western Ghats

Water chemistry variables (units)	Mean ± S.D	Range
pH	7.28 ± 0.58	6.03–8.50
Water temperature (°C)	26.09 ± 2.52	22.10–35.43
Electrical conductivity (μScm^{-1})	199.00 ± 301.44	37.17–250.67
Total dissolved solids (mg L^{-1})	120.26 ± 149.68	18.67–624.67
Alkalinity (mg L^{-1})	70.44 ± 111.46	12.00–421.07
Chlorides (mgL^{-1})	35.25 ± 62.36	4.99–255.92
Hardness (mg L^{-1})	69.49 ± 96.94	12.00–376.00
Calcium (mg L^{-1})	14.76 ± 17.33	1.60–84.97
Magnesium (mg L^{-1})	15.72 ± 18.58	1.17–71.01
Dissolved oxygen (mg L^{-1})	7.58 ± 1.77	4.81–11.52
Phosphates (mg L^{-1})	0.18 ± 0.36	0.01–1.30
Sulphates (mg L^{-1})	19.94 ± 18.07	2.91–67.91
Sodium (mg L^{-1})	52.92 ± 201.09	1.05–996.03
Potassium (mg L^{-1})	11.01 ± 35.01	0.41–168.33
Nitrates (mg L^{-1})	1.06 ± 1.33	0.07–4.24

Thus the sites selected for the present study covered a good range of the land-use gradient and hence the inferences drawn from the statistic may not have been influenced by skewed sampling. An aggregated measure of LULC such as percentage agriculture in catchments may only represent the potential of LULC effects on streams. Percentage agriculture lands in catchments were positively correlated with the ionic and nutrient variables. Studies have shown that the percentage of agriculture at watershed scale is a primary predictor for nitrogen and phosphorus (Ahearn et al. 2005).

Diatom community structure in streams of the central Western Ghats was found to be strongly related to the land use practices as could be observed elsewhere (Stevenson et al. 2009; Walsh and Wepner 2009). The nutritional changes in the streams triggered by the LULC changes stands as a determining factor in structuring the diatom species composition. Effects of nutrient are commonly identified as one of the most important determinants of diatom species composition in lentic and lotic ecosystems (Pan and Stevenson 1996). However, the diatom species composition at CWG streams were controlled more by the ionic variables than the nutrient concentration. More of pollution tolerant species were seen in the streams in agriculture dominated catchments. Blinn (1993, 1995) found that higher salinities (≥35 mScm^{-1}) tend to override other water quality parameters in structuring diatom communities in salt lakes. Agriculture dominated sites represented high pH, TDS and nutrient loads, (Figs. 22 and 23) which is also supported by positive correlation between percentage of agriculture land with percentage of pollution tolerant diatoms (Fig. 23).

5 Conclusions

- The results indicate that (a) the water quality regimes show seasonal variations, (b) diatom species assemblages change accordingly in all the water quality regimes, due to seasonal water quality conditions and (c) the species distribution across the sites followed the satellite-mode due to the specific ecological niches of the diatoms.
- This study concludes that the environmental quality of the Western Ghats streams can be monitored by biomonitoring ventures and compared to other water monitoring programs. This study also suggests that the diatom community in this region is rich with possible endemic taxa; hence considerable amount of importance has to be given for the taxonomy of the lesser-known species before commencing the biomonitoring programs.
- The analyses and results provided insights into the linkages between land use practices and water quality in the streams and the relative sensitivity of water quality variables to alterations in land use. The relationships between the diatom indices and water chemistry variables relation showed the impact of land use on the stream ecosystem.
- It has been evident that the causes and sources of water pollution in the five river basins are due to agricultural land use, anthropogenic activities and industrialization. The major occupation in the study area is agriculture, which is main source of increase in nitrates and ionic components in streams. Domestic and industrial sewage discharges into the rivers are responsible for the observed high concentration of electrical conductivity, total dissolved solids, total hardness and other ionic components. Proper treatment of effluent from the industrial processes to the acceptable levels and discouraging stagnation of water through small dams are the two major recommendations to minimize the damages on the river ecosystem in the central Western Ghats. Table 9 lists the threats and remedial measures.

Table 9 Threats and mitigation measures

River basin	Region	Problem	Remedial measures
Kali	Dandeli	Paper mill effluent	Enforce effluent treatment by the industry (implementation of the control of water pollution, Polluter pays principle)
Kali	Ramnagar	Non-point source pollution in streams and rivers from Agriculture fields	Avoiding intense use of chemical fertilizers and pesticides
Kali	Honkon (Brackish)	Mechanized sand mining	Stopping of sand mining in certain ecologically sensitive region and regulated sand mining in selected localities
Bedthi	Sangdevarkoppa	Non-point source pollution	Avoiding intense use of chemical fertilizers and pesticides
Bedthi	Kalghatghi	Urban domestic sewage, non-point source pollution	Implementation of sewage treatment plant in Hubli town. Sewage should be treated before letting into the river
Bedthi	Kalghatghi	Solid Waste Disposal in River	Setting up Solid waste disposal facility in outskirts of Hubli town
	Manchikeri	Urban domestic sewage, non-point source pollution	Implementation of sewage treatment plant in Hubli town. Sewage should be treated before letting into the river
Sharavathi	Gerusoppa and downstream	Mechanized sand mining	Stopping of sand mining in certain ecologically sensitive region and regulated sand mining in selected localities

Acknowledgment We are grateful to the NRDMS division, the Ministry of Science and Technology (DST), Government of India, The Ministry of Environment and Forests (MoEF), Government of India and Indian Institute of Science for the financial and infrastructure support.

Appendix

Appendix 1 Checklist of epilithic diatoms of Rivers of Uttara Kannada, Karnataka

Taxa	Kali (KRB)	Bedthi (BRB)	Aghan ashini (ARB)	Shara vathi (SRB)	Venkat apura (VRB)
Achnanthes sp. J.B.M. Bory de St. Vincent		+			
Achnanthes minutissima Kützingv. *minutissima* Kützing (*Achnanthidium*)	+	+	+	+	+
Achnanthes sp.		+			
Achnanthidium sp.	+	+	+	+	+
Actinocyclus sp.	+				
Amphora montana Krasske	+				
Amphora pediculus (Kützing) Grunow	+				+
Amphora species		+			
Aulacoseira ambigua (Grunow) Simonsen	+	+		+	
Aulacoseira granulata (Ehr.) Simonsen		+			
Aulacoseira granulata (Ehr.) Simonsenmorphotype *curvata*		+			
Bacillaria paradoxa Gmelin	+				
Brachysira neoexilis Lange-Bertalot	+	+	+	+	+
Brachysira sp.	+	+	+	+	+
Brachysirawygaschii Lange-Bertalot	+		+	+	+
Brassierea sp Hein and Winsborough		+			
Caloneis bacillum (Grunow) Cleve	+	+		+	
Caloneis hyalina Hustedt	+				
Caloneis silicula (Ehr.) Cleve	+	+			+
Caloneis species		+			
Cocconeis placentula Ehrenberg *var.euglypta* (Ehr.) Grunow	+	+	+		+
Craticula sp A. Grunow		+			
Craticulaacco modiformis Lange-Bertalot		+			

(continued)

Appendix 1 (continued)

Taxa	Kali (KRB)	Bedthi (BRB)	Aghan ashini (ARB)	Shara vathi (SRB)	Venkat apura (VRB)
Craticula molestiformis (Hustedt) Lange-Bertalot		+			
Craticula submolesta (Hust.) Lange-Bertalot	+	+			+
Craticula vixnegligenda Lange-Bertalot		+			
Cyclostephanos sp F.E. Round		+			
Cyclostephanos species	+	+			+
Cyclotella sp F.T. Kützingex A de Brébisson		+			
Cyclotella meneghiniana Kützing	+	+			
Cyclotella ocellata Pantocsek		+			
Cyclotella species		+			
Cymbella kolbei Hustedt var. *kolbei*	+	+		+	+
Cymbella species	+	+	+	+	+
Cymbella tumida (Brebisson) van Heurck	+	+			
Cymbopleura (Krammer) Krammer					+
Cymbopleura sp.	+			+	+
Diadesmis contenta (Grunow ex V. Heurck) Mann	+	+			+
Diploneis elliptica (Kützing) Cleve		(+)			
Diploneis oblongella (Naegeli) Cleve-Euler		(+)			
Diploneis ovalis (Hilse) Cleve		+			
Diploneis subovalis Cleve	+	+			+
Encyonema mesianum (Cholnoky) D.G. Mann	+				+
Encyonema minutum (Hilse in Rabh.) D.G. Mann	+	+			+
Encyonema species	+				+
Entomoneis alata Ehrenberg		+			
Eolimna subminuscula (Manguin) Moser Lange-Bertalot and Metzeltin	+	+			

(continued)

Appendix 1 (continued)

Taxa	Kali (KRB)	Bedthi (BRB)	Aghan ashini (ARB)	Shara vathi (SRB)	Venkat apura (VRB)
Eunotia sp C.G. Ehrenberg		+			
Eunotiabi lunaris (Ehr.) Mills var. *bilunaris*					+
Eunotia incisa Gregoryvar. *incisa*	+	+			
Eunotia minor (Kützing) Grunow	+	+	+	+	+
Eunotia rhomboidea Hustedt	+		+	+	+
Eunotia sp.	+	+		+	
Fallaciainsociabilis (Krasske) D.G. Mann		+			
Fallacia pygmaea (Kützing) Stickle and Mann sp. *pygmaea* Lange-Bertalot	+	+			
Fallaciatenera (Hustedt) Mann in Round		+			
Fragilaria biceps (Kützing) Lange-Bertalot	+	+	+	+	+
Fragilaria species		+			
Fragilaria ulna (Nitzsch.) Lange-Bertalotvar. *ulna*	+	+	+	+	+
Fragilari aungeriana Grunow	+				
Frustulia saxonica Rabenhorst				+	
Frustulia species	+			+	+
Geissleriadecussis (Ostrup) Lange-Bertalot and Metzeltin		+			
Gomphonema acuminatum Ehrenberg	+				
Gomphonema difformum Karthick and Kociolek		+	+	+	
Gomphonemadi minutum Karthick and Kociolek	+	+	+		
Gomphonema gandhii Karthick and Kociolek	+	+	+	+	+
Gomphonema parvulum (Kützing) Kützingvar. *parvulum f. parvulum*	+	+	+	+	+
Gomphonema pseudo augur Lange-Bertalot		+			
Gomphonema species	+	+	+	+	+
Gyrosigma acuminatum (Kützing) Rabenhorst	+	+			

(continued)

Appendix 1 (continued)

Taxa	Kali (KRB)	Bedthi (BRB)	Aghan ashini (ARB)	Shara vathi (SRB)	Venkat apura (VRB)
Gyrosigma scalproides (Rabenhorst) Cleve	+				
Gyrosigma species		+			
Hantzschia distincte punctata Hustedt in Schmidt et al.				+	
Hippodontaavittata (Cholnoky) Lange-Bert. Metzeltin and Witkowski	+				+
Luticola species	+	+			
Luticola species (*aff. mutica*)	+				
Navicula species		+			
Navicula antonii Lange-Bertalot		+		+	
Navicula cincta (Ehr.) Ralfs in Pritchard	+				
Navicula cryptocephala Kützing	+	+	+	+	+
Navicula cryptotenella Lange-Bertalot	+				
Navicula elginensis (Gregory) Ralfs in Pritchard					+
Navicula erifuga Lange-Bertalot	+	+			
Navicula gracilis Ehrenberg	+			+	
Navicula hustedtii Krasske					
Navicula hustedtii Krasskevar. *obtusa* Hustedt	+			+	
Navicula leptostriata Jorgensen	+	+	+	+	+
Navicula peregrina (Ehr.) Kützing	+				
Navicula reinhardtii (Grunow) Grunow in Cl. and Möller				+	
Navicula riediana Lange-Bertalot and Rumrich	+			+	+
Navicula rostellata Kützing	+	+		+	+
Navicula sp.	+	+	+	+	+
Navicula symmetrica Patrick	+	+	+	+	+
Navicula viridula (*Kützing*) Ehrenberg	+				
Navigiolum species.					
Neidium affine (Ehrenberg) Pfitzer	+				+

(continued)

Appendix 1 (continued)

Taxa	Kali (KRB)	Bedthi (BRB)	Aghan ashini (ARB)	Shara vathi (SRB)	Venkat apura (VRB)
Nitzschia sp. A.H. Hassall		+			
Nitzschia amphibia Grunowf. *amphibia*	+	+			+
Nitzschia clausii Hantzsch	+	+		+	+
Nitzschia compressa (J.W. Bailey) Boyer		+			
Nitzschia dissipata (Kützing) *Grunowvar.media* (Hantzsch.) Grunow				+	
Nitzschia fonticola Grunow in Cleve et Möller	+	+			
Nitzschia frustulum (Kützing) Grunow var. *frustulum*	+	+			
Nitzschia gracilis Hantzsch				+	
Nitzschia linearis (Agardh) W.M. Smith var. *linearis*					
Nitzschia nana Grunow in Van Heurck	+				
Nitzschia obtusa W.M. Smith var. *kurzii* (Rabenhorst) Grunow	+	+		+	+
Nitzschia palea (Kützing) W. Smith	+	+		+	+
Nitzschia reversa W. Smith	+	+		+	+
Nitzschia sigma (Kützing) W.M. Smith		+		+	+
Nitzschia species	+	+			+
Nitzschia umbonata (Ehrenberg) Lange-Bertalot		+			
Pinnularia acrospheria W.M. Smith var. *acrospheria*		+			+
Pinnularia brebissonii (Kütz.) Rabenhorst var. *brebissonii*	+	+		+	
Pinnularia divergens W.M. Smith. var. *undulata* (M. Perag. and Herib.) Hustedt				+	
Pinnularia gibba Ehrenberg				+	
Pinnularia species		+		+	+
Placoneis sp.	+	+		+	
Planothidium frequentissimum (Lange-Bertalot) Lange-Bertalot	+	+	+	+	+

(continued)

Appendix 1 (continued)

Taxa	Kali (KRB)	Bedthi (BRB)	Aghan ashini (ARB)	Shara vathi (SRB)	Venkat apura (VRB)
Planothidium rostratum (Oestrup) Round and Bukhtiyarova	+	+			+
Planothidium sp. Round and Bukhtiyarova		+	+	+	
Pleurosigma salinarum (Grunow) Cleve and Grunow	+				
Pseudostaurosira brevistriata (Grun. in Van Heurck) Williams and Round		+			
Rhopalodia gibba (Ehr.) O.Mullervar.gibba					+
Rhopalodia operculata (Agardh) Hakansson	+				+
Sellaphora species	+	+			
Sellaphora americana (Ehrenberg) D.G. Mann	+			+	+
Sellaphora laevissima (Kützing) D.G. Mann				+	
Sellaphora nyassensis (O. Muller) D.G. Mann	+	+			
Sellaphora pupula (Kützing) Mereschkowksy	+	+		+	+
Seminavis sp. D.G. Mann		+			
Seminavis species		+			
Skeletonema species					
Stauroneis species	+	+			+
Surirella angusta Kützing	+	+		+	+
Surirella species	+	+		+	+
Synedra sp.		+			
Tryblionella calida (grunow in Cl. and Grun.) D.G. Mann	+	+			
Tryblionella levidensis W.M. Smith		+			
Total	83	95	22	51	55
Total number of taxa reported from all river basins	140				

References

Ahearn DS, Sheibley RS, Dahlgren RA, Anderson M, Jonson J, Tate KW (2005) Land use and land cover influence on water quality in the last free-flowing river draining the western Sierra Nevada, California. J Hydrol 313:234–247

APHA (1998) Standard methods for the examination of water and wastewater, 20th edn. American Public Health Association, Washington

Bhat A (2002) A study of the diversity and ecology of freshwater fishes of four river systems of the Uttara Kannada District, Karnataka, India. Ph.D., dissertation, Indian Institute of Science, Bangalore, India, 178 pp

Blinn DW (1993) Diatom community structure along physicochemical gradients in saline lakes. Ecology 74:1246–1263

Blinn DW (1995) Diatom community structure along salinity gradients in Australia saline lakes: biogeographic comparisons with other continents. In: Kociolek P, Sullivan MA (eds) Century of diatom research in North America. Special symposium for Reimer and Patrick festschrift, Koeltz Scientific Books, pp 163–174

Brewin PA, Buckton ST, Ormerod SJ (2000) The seasonal dynamics and persistence of stream macroinvertebrates in Nepal: do monsoon floods represent disturbance? Freshw Biol 44:581–594

Diamond JM, May RM (1977) Species turnover rates on islands: dependence on census interval. Science 197:266–270

Eastman JR (2006) IDRISI Andes. Clark University, Worcester. http://www.clarklabs.org

Gandhi HP (1957a) The freshwater diatoms from Radhanagari—Kolhapur. Ceylon J Sci Biol Sect 1:45–47

Gandhi HP (1957b) Some common freshwater diatoms from Gersoppa-falls (Jog-Falls). J Poona Univ Sci Sect 12:13–21

Gandhi HP (1957c) A contribution to our knowledge of the diatom genus Pinnularia. J Bombay Nat Hist Soc 54:845–853

Gandhi HP (1958a) Freshwater diatoms from Kolhapur and its immediate environs. J Bombay Nat Hist Soc 55:493–511

Gandhi HP (1958b) The freshwater diatoms flora of the Hirebhasgar Dam area, mysore state. J Indian Bot Soc 37:249–265

Gandhi HP (1959a) The freshwater diatom flora from Mugad, Dharwar District with some ecological notes. Ceylon J Sci Biol Sect 2:98–116

Gandhi HP (1959b) Freshwater diatom flora of the Panhalgarh Hill Fort in the Kolhapur district. Hydrobiologia 14:93–129

Gandhi HP (1959c) Notes on the Diatomaceae from Ahmedabad and its environs-II. On the diatom flora of fountain reservoirs of the Victoria Gardens. Hydrobiologia 14:130–146

Gandhi HP (1960a) On the diatom flora of some ponds around Vasna village near Ahmedabad. J Indian Bot Soc 39:558–567

Gandhi HP (1960b) Some new diatoms from the Jog Falls, Mysore State. J Roy Microscopic Soc 79:81–84

Gandhi HP (1960c) The diatom flora of the Bombay and Salsette Islands. J Bombay Nat Hist Soc 57:7, 123

Gotelli NJ, Ellison AM (2004) A primer of ecological statistics. Sinauer Associates Inc, Sunderland

Jarvie HP, Whitton BA, Neal C (1998) Nitrogen and phosphorus in east-coast British rivers: speciation, sources and biological significance. Sci Tot Environ 210–221:79–109

Hanski L (1982) Dynamics of regional distribution: the core and satellite species hypothesis. Oikos 38:210–221

Hegde VR, Shreedhara V, Hegde VS (1994) Changing landuse/land cover pattern in the Kali River Basin in Western Ghats, south India. Curr Sci 66(2):128–137

Kamath SU (1985) Gazetteer of India, Karnataka State Gazetteer, Uttara Kannada District. Government of Karnataka Publications, Bangalore

Karthick B, Taylor JC, Mahesh MK, Ramachandra TV (2010) Protocols for collection, preservation and enumeration of diatoms from aquatic habitats for water quality monitoring in India. ICFAI Univ J Soil Water Sci 1:1–36
Krammer K, Lange-Bertalot H (1986–1991) Bacillariophyceae. Süsswasserflora von Mitteleuropa, 2(1–4) Fischer, Stuttgart
Kwandrans J, Eloranta P, Kawecka B, Kryzsysztof W (1998) Use of benthic diatom communities to evaluate water quality in rivers of southern Poland. J Appl Phycol 10:193–201
Lange-Bertalot H, Krammer K (1989) Achnanthes. Eine Monographie der Gattung mit Definition der Gattung Cocconeis und Nachtr∼igen zu den Naviculaceae. Bibl Diatom 18:392 pp
Listori JJ, World-wide Bank (1990) Environmental health components for water supply, Sanitation and Urban Projects. World-wide Bank, Washington
Maddock I (1999) The importance of physical habitat assessment for evaluating river health. Freshw Biol 41:373–391
McCune B, Grace JB (2002) Chapter 11—Hierarchical clustering. In: McCune B, Grace JB (eds) Analysis of ecological communities. MJM Publishers, Chandler, pp 86–96
McGeoch MA, Gaston KJ (2002) Occupancy frequency distributions: patterns, artifact, and mechanisms. Biol Rev Cambridge Philos Soc 77:311–331
Menon S, Bawa KS (1998) Tropical deforestation: reconciling disparities in estimates for India. Ambio 27(7):576–577
Minchin PR (1987) An evaluation of the relative robustness of techniques for ecological ordination. Vegetation 69:89–107
Niemi GJ, DeVore P, Detenbeck N, Taylor D, Lima A, Pastor J, Yount JD, Naiman RJ (1990) Over-view of case studies on recovery of aquatic systems from disturbance. Environ Manage 1 (4):571–587
Pan Y, Stevenson RJ (1996) Gradient analysis of diatom assemblages in western Kentucky wetlands. J Phycol 32:222–232
Potapova M, Hamilton PB (2007) Morphological and ecological variation within the Achnanthidium minutissimum (Bacillariophyceae) species complex. J Phycol 43:561–575
Prygiel J, Coste M (1993) Utilisation des indices diatomiques pour la mesure de la qualité des eaux du bassin Artois-Picardie: bilan et perspectives. Ann Limnol 29:255–267
Prygiel J, Coste M, Bukowska J (1999) Review of the major diatom-based techniques for the quality assessment of rivers—State of the art in Europe. In: Prygiel J, Whitton BA, Bukowska J (eds) Use of algae for monitoring rivers III, Agence de l'Eau Artois-Picardie, pp 224–238
R Development Core Team (2006) R: a language and environment for statistical computing, R foundation for statistical computing, Vienna, Austria. ISBN 3-900051-07-0
Radhakrishna BP (1991) An excursion into the past—The Deccan volcanic episode. Current Sci 61:641–647
Soininen J, Eloranta P (2004) Seasonal persistence and stability of diatom communities in rivers: are there habitat specific differences? Eur J Phycol 39:153–160
Soininen J, Heino J (2005) Relationships between local population persistence, local abundance and regional occupancy of species: distribution patterns of diatoms in boreal streams. J Biogeogr 32:1971–1978
Stevenson RJ, Novoveska L, Riseng CM, Wiley MJ (2009) Comparing responses of diatom species composition to natural and anthropogenic factors in streams of glaciated eco-regions. Nova Hedwig Beih 135:1–13
Sweeting RA (1996) River pollution. In: Petts G, Calow P (eds) River restoration. Blackwell, Oxford
Taylor JC (2004) The application of diatom-based pollution Indices in the Vaal catchment. Unpublished M.Sc. thesis, North-West University, Potchefstroom Campus, Potchefstroom
Taylor JC, de la Rey PA, van Rensburg L (2005) Recommendations for the collection, preparation and enumeration of diatoms from riverine habitats for water quality monitoring in South Africa. Afr J Aquat Sci 30:65–75
Taylor JC, Harding WR, Archibald CGM (2007a) An illustrated guide to some common diatom species from South Africa. WRC Report TT. 282

Taylor JC, Prygiel J, Vosloo A, De La Rey PA (2007b) Can diatom-based pollution indices be used for biomonitoring in South Africa? A case study of the crocodile west and marico water management area. Hydrobiologica 592:455–464
Tong STY, Chen W (2002) Modeling the relationship between land use and surface water quality. J Environ Manage 66(4):377–393
Vega M, Pardo R, Barrado E, Deban L (1998) Assessment of seasonal and polluting effects on the quality of river water by exploratory data analysis. Water Res 32:3581–3592
Walsh G, Wepner V (2009) The influence of land use on water quality and diatom community structures in urban and agriculturally stressed rivers. Water SA 35:579–594
White MD, Greer KA (2006) The effects of watershed urbanization on the stream hydrology and riparian vegetation of Los Peñasquitos Creek, California. Landscape Urban Plan 74:125–138

Relationships Among Subaquatic Environment and Leaf/Palinomorph Assemblages of the Quaternary Mogi-Guaçú River Alluvial Plain, SP, Brazil

Fresia Ricardi-Branco, Sueli Yoshinaga Pereira, Melina Mara Souza, Francisco Santiago, Paulo Ricardo Brum Pereira, Fabio C. Branco, Victor Ribeiro and Karen Molina

Abstract Environmental features of the Mogi-Guaçú River fluvial plain located in the northeast portion of the São Paulo State, Brazil as characterized by the dynamics of subaquatic environment, the current pollen rain and climatic analysis of leaf morphologies in selected meander bends. As the area forms a transition between the Cerrado (Wood Savanna) and Mata Atlântica (Atlantic Forest), located in the Ecological Station of Mogi-Guaçú (EEcMG), in the municipality of Mogi-Guaçú, district of Martinho Prado Jr. (between 22° 10′S and 22° 18′S and 47° 08′W and 47° 11′W), the present environmental study involves helped improving the knowledge about the origin and evolution of the Cerrado and Mata Atlântica, and the dynamics of the Quaternary plains that sustain this river ecosystem.

Keywords Pollen rain · Leaf assemblage · Hydrogeology · Hydrochemistry

1 Introduction

The study and evolution of the landscape is a very important tool regarding paleoenvironmental reconstructions. The use and characterization of plant macro debris assemblages assist in reconstruction of past climatic conditions and

---

F. Ricardi-Branco (✉) · S.Y. Pereira · M.M. Souza · F. Santiago · V. Ribeiro · K. Molina
Department of Geology and Natural Resources, Institute of Geosciences,
University of Campinas, 51 J. Pandiá Calógeras Street, Campinas, SP 13083-870, Brazil
e-mail: fresia@ige.unicamp.br

P.R.B. Pereira
Forestry Institute, São Paulo State Environmental Secretariat, 874, 7 de Setembro Street, Mogi Mirim, SP 13801-350, Brazil

F.C. Branco
Environmentality, 177 Michigan Street, São Paulo, SP 04566-000, Brazil

Mu. Ramkumar et al. (eds.), *Environmental Management of River Basin Ecosystems*,
Springer Earth System Sciences, DOI 10.1007/978-3-319-13425-3_30

landscapes. The reconstruction of past environments is essential to understand the modes and mechanisms involved in the preservation of fossil assemblages. These modes, mechanisms and other processes can be better understood by the study of deposition of organic debris and associated characteristics of subaquatic/aquatic environments.

The Cerrado presents a peculiar biome and its plant formations change from open fields to dense forests that reach up to 30 m high (Aguiar LMS 2004). It stands out as a phytophysiognomic unit because of its great expressiveness in the percentage of occupied areas in Brazil—currently about 21 % of the Brazilian territory, the 2nd largest biome in Brazil. The present study employed various tools of environmental analyses to access and better understand the Cerrado Biome evolution during the Quaternary. There are few studies in this specific area for the state of São Paulo (Gouveia et al. 1999; Souza et al. 2013). Most studies in Brazil are more general (Ribeiro and Walter 2008; Salgado-Labouriau 1997; Barberi et al. 2000; Ledru et al. 2001).

In the present study, the approaches followed by Gastaldo et al. (1987), Burnham (1990), Gastaldo (1994, 2001, 2004) and Burk et al. (2005) are employed. These authors attempted a taphonomic research and related those characteristics with the physico-chemical characteristics of the aquatic environment in which the organic material of plant origin accumulated. These studies focused on the dynamics of deposition of macro debris of mangroves that were preserved in environments with slightly oxidant and acidic pH. On the other hand, the hydrological studies often analyse the interaction of groundwater and surface water in semi-arid areas and temperate climates. Lautz and Siegel (2006) highlighted the role of groundwater–surface water interface in hyporheic areas of drainages located in semi-arid environments in the debris preservation at meanders. Rushton (2007) used mathematical models to study the river-aquifer interaction. In the River Plate, Nebraska, Chen (2007) simulated the influence of riparian vegetation over the dynamics of groundwater flow. Rassam et al. (2006) studied the riparian zones of two watercourses, a perennial and an ephemeral, with emphasis on the hydraulic processes that supported the removal of nitrate and nutrients from the water flow in temperate climates. Some studies focussed on the changes of the water table level in riparian areas [for example, Schilling (2007) in Iowa (USA), and Burt et al. (2002a, b) in England and in various parts of Europe].

This research seeks to contribute to the characterization of environmental processes that influence the Cerrado biome in the state of São Paulo and how to identify it by integrating studies of groundwater dynamics reflected in the assemblies of leaves and to understand the climate signal given by the modern pollen rain.

2 Regional Setting

2.1 Geographic Location

The study area is located in the Campininha Farm in Mogi-Guaçú district of Martin Prado Jr. State of Sao Paulo, between the geographical coordinates 22° 10′ and 22° 18′ South latitude, and 47° 08′ 47° 11′ West longitude (Fig. 1). The southern portion of the Campininha Farm, where the Ecological Station of Mogi (EEcMG)

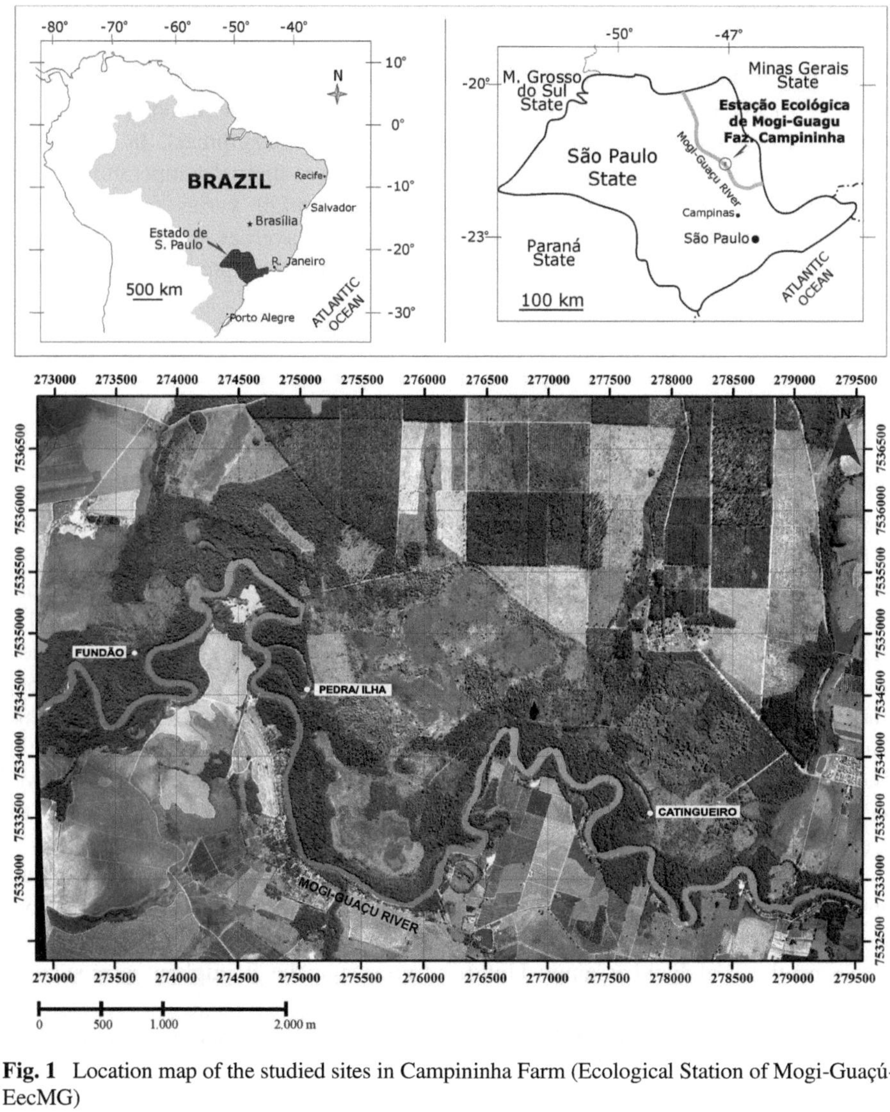

Fig. 1 Location map of the studied sites in Campininha Farm (Ecological Station of Mogi-Guaçú-EecMG)

is, located on the right bank of the Mogi-Guaçú River. The hydrographical basin of the Mogi-Guaçú River has an area of 17,460 km^2 out of which 2,650 km^2 is in the territory of Minas Gerais State and 14,653 km^2 in the territory of São Paulo State. The Mogi-Guaçú River originates in Morro do Curvado, in the municipality of Bom Repouso (MG), at about 1,510 km^2 altitude, and after draining longitudinally for about 530 km, it joins the Pardo River in the NE of São Paulo State (Zancopé et al. 2009).

2.2 Climate

According to Sparovek et al. (2007), and Koppen, the area is classified under Aw Climate (tropical climate, dry in the winter, and rainy in the summer), and Cwa Climate (humid temperate climate, dry winter, and hot summer). In the hydrographical basin of the Mogi-Guaçú River, the average annual temperatures range from 20.5 to 22.5 °C and the average annual rainfall varies from 1,400 to 1,600 mm. In the rainy season (October–March), the average precipitation varies from 1,100 to 1,250 mm, and in the dry season (April to September) from 250 to 300 mm (CBH-Mogi 1999).

2.3 Vegetation

The study area is located in the transition zone between the biomes of Cerrado (Wood Savannah) and Atlantic Forest or Submontane Semideciduous Forest (Mata Atlântica) with characteristics of both, and is considered to be an area of ecotone or ecological tension. The Cerrado biome is peculiar for its vegetation types ranging from open fields to dense forests that reach 30 m height (Aguiar LMS 2004). The Cerrado stands out as a phytophysiognomic unit, because of its great expressiveness in the percentage of occupied areas in Brazil, about 21 % of the Brazilian territory, the 2nd largest biome in Brazil. The Atlantic Forest vegetation occupies most of the Brazilian coastal region. Currently, it occupies an area of 100 km^2 and it is one of the most important rainforests in the world. The Biological Reserve and the EEcMG (Ecological Station of Mogi-Guaçú) are located in an area of priority for the Cerrado biome conservation.

The main plant formation or vegetation types in the study area are the Submontane Semideciduous Forest (Mata Atlântica), the Cerrado Biome (Wood Savannah) represented here by the Alluvial Semideciduous Forest (Riparian Forest), and the Wet Field (Campo Úmido) comprising the floodplains and swamps (Eiten 1963).

According to the studies of Eiten (1963), the Submontane Semideciduous Forest formation occurs mainly in the Ecological Station and the Biological Reserve, where there is a transition between this formation and the Cerrado Biome.

The Alluvial Semideciduous Forest also occurs predominantly in the Ecological Station, but is confined along the Mogi-Guaçú River. According to Montovani (1983), the Cerrado is in remote areas of Mogi-Guaçú River, and small patches occur near the stream with herbaceous plants called Wet Field (Campo Úmido). The wetlands (Riparian Forest) are located in the transition area of forest and the edge of watercourses (Eiten 1963; Pinto et al. 1997). Two areas of Wet Field (Campo Úmido) were delimitated within the EEcMG near the Experimental Station. They are covered by grass.

According to the studies conducted at the EEcMG, 1,090 plant species sampled were among the groups of bryophytes, pteridophytes, and phanerogams, and belong to 134 families. The list of plants at the EEcMG presents families from the most important specimens of plants in the study area.

Passos (1998) conducted a phytosociological survey for the woody families along the Riparian Forest of EEcMG, and highlighted ten families namely, Annonaceae, Euphorbiaceae, Fabaceae, Flacourtiaceae, Lauraceae, Melastomataceae, Meliaceae, Mimosaceae, Myrtaceae and Phytolacaceae as most important in the area.

2.4 Geology

The study area is within the Paraná Basin (Perrotta et al. 2005), which is characterized by the accumulation of thick sedimentary rocks located along the banks of the Mogi-Guaçú River, and some of its main tributaries. The Paraná Basin is a wide sedimentary region in South America covering an area close to 1.5 sq km and includes territorial portions of Southern Brazil, Eastern Paraguay, Northeast Argentina and Northern Uruguay. In Brazil, the basin sprawls in the states of Paraná, Santa Catarina, Rio Grande do Sul, Mato Grosso, Mato Grosso do Sul, Goiás, southern Minas Gerais, and almost all the countryside portion of the State of São Paulo (Petri and Fulfaro 1981). The basin is considered to be the most important hydrogeological province in Brazil with about 45 % of the country's groundwater reserves.

In the study area, the Paraná Basin is represented by rocks from Permian and Jurassic ages. Thus, the Permian strata belong to the groups of Itararé, Guatá, and Passa Dois. The rocks from the Serra Geral Formation are overlapping the previous (Perrotta et al. 2005) and represent the lava flows during the Jurassic, as the result of the rifting that separated Africa from South America. Finally, the Cenozoic alluvial deposits are found predominantly on the banks of the Mogi-Guaçú River, occupying about 68.5 % of the EEcMG.

2.5 Hydrology and Hydrogeology

The EEcMG is located in the Unidade de Gerenciamento de Recursos Hídricos do Mogi-Guaçú-UGRHI 9 (Unit of Water Resource Management of Mogi-Guaçú) with 14,653 km^2 of drainage area. The municipalities of Mogi-Guaçú, Conchal, and Araras belong to the Upper Mogi region, along with the municipalities of Engenheiro Coelho, Espírito Santo do Pinhal, Estiva Gerbi, Leme, Mogi Mirim, Pirassununga, Porto Ferreira, and Santa Cruz da Conceição (CBH-Mogi 1999). The Mogi-Guaçú River limits the EEcMG to the south and in several parts presents the average river flow (Q) variations in portions of the river (CBH-Mogi 1999) viz., Upstream (3D-004 gaging station)—Qmean (between 30 and 125 m^3/s); Qminimum (25 and 75 m^3/s); Qmaximum (55 and 225 m^3/s); Upper Portion (4C-005 gaging station)—Qmean (between 75 and 250 m^3/s); Qminimum (50 and 150 m^3/s); Qmaxim (55 and 225 m^3/s); Lower Portion (4C-007 gaging station)—Qmean (between 30 and 125 m^3/s); Qminimum (25 and 75 m^3/s); Qmaximum (110 and 440 m^3/s); Downstream (5C-025 gaging station)—Qmean (between 160 and 490 m^3/s); Qminimum (110 and 300 m^3/s); Qmaximum (200–700 m^3/s).

The study area is located near the 3D-004 gauging station, and the average flow of the Mogi-Guaçú River lies between 30 and 125 m^3/s. The main aquifers of the study area are the undifferentiated Cenozoic covers, Itararé (Tubarão) and Cristalino Fraturado (CBH-Mogi 1999). According to CBH-Mogi (1999), the undifferentiated Cenozoic covers are characterized by different grained sands, clays and gravels from Itaqueri, Rio Claro, and São Paulo formations. The 100 m thick Tubarão Aquifer (Itararé Group) consists deposits of sandstones, siltstones, diamictites, rhythmites, and mistites from Carboniferous to Permian. The productivity of deep wells has an average flow of 7 m^3/h, in a band ranging from 3 to 30 m^3/h. The Cristalino Fraturado (Precambrian rocks) is composed by granites, gneisses, migmatites, phyllites, schists, quartzites, and metasediments. It s productivity is between 2 and 40 m^3/h, with an average of 5 m^3/h.

3 Materials and Methods

3.1 Dynamics of Subaquatical Environment

To characterize the environment of groundwater, 23 surveys were performed; out of which, 15 in the area called Pedra, and 8 in Catingueiro (Table 1). During the drilling, samples of unconsolidated material were collected for visual description and grain analysis. 50.8 mm-diameter-monitoring wells were drilled. The casing and the screen material are comprised of white PVC, and the screen has 0.25 mm opening. A filter pack material (coarse sand) was filled around the screen, and then the screen was sealed with bentonite. The drilling depth varied according to the

Table 1 Location and depth of monitoring wells

Catingueiro				Pedra lake			
Monitoring well	UTM E	UTMN	Depth (m)	Monitoring well	UTM E	UTMN	Depth (m)
P 1	277,789	7,533,779	5.00	IF 2	274,836	7,535,737	14.32
P 2	277,711	7,533,737	6.63	IF 3	274,830	7,535,739	14.67
P 3	277,690	7,533,722	4.28	IF 6	274,865	7,535,735	15.27
P 6	277,717	7,533,641	4.87	IF 7	274,864	7,535,743	15.37
P 7	277,747	7,533,642	7.12	IF 8	274,901	7,535,794	14.07
P 8	277,777	7,533,657	6.23	IF 9d	274,917	7,535,775	16.28
P 9	277,837	7,533,680	4.01	IF 9s			12.31
P 10	277,915	7,533,631	4.02	IF 10	274,940	7,535,787	13.63
				IF 11d	274,973	7,535,809	13.39
				IF 11s			10.45
				IF 12	274,913	7,535,735	15.76
				IF 13d	274,942	7,535,725	19.61
				IF 13s			15.30
				IF 14	274,960	7,535,692	14.90
				IF 15	274,990	7,535,701	12.16
				IF 16	275,025	7,535,695	10.48
				IF 17d	275,054	7,535,690	12.60
				IF 17s			7.91
				IF 18	275,085	7,535,695	7.00

Legend IF 13d—deep monitoring well, IF 13s—shallow monitoring well

depth of the water table. Thus, the wells drilled in the island were deeper than Catingueiro. Table 1 shows the locations of the wells.

In Catingueiro, the depth of the wells ranged from 4.01 to 7.12 m, while in Pedra the depth of the wells ranged from 7.00 to 15.75 m (the shallower wells), and from 12.60 to 19.61 m (the deeper wells). It was not possible to install multilevel deeper wells in Catingueiro due to the occurrences of crystalline rocks at shallow depths. In Pedra, the well depths ranged from 7.00 to 19.61 m; four were multilevel wells, drilled to study the behavior of deeper portion of the aquifer and groundwater vertical flow in the plain area.

The hydraulic conductivity (K) was estimated through slug-tests, by using Hvorslev (1951) method. The monitoring of water levels was performed in February 2012, and May 2012. The groundwater sampling for physical and chemical analysis was performed in May 2012. The day before the collection, the wells were exhausted to remove the stagnant water. Data were collected by bailer, and the sample underwent filtration through a 0.22-μm filter. 40 mL of water was collected for ion chromatography, and another 400 mL was collected for ICP-MS and alkalinity analyses. All samples were chilled to 4 °C. Immediately on collection, the pH, ORP, EC, and temperature were measured by portable equipment in the field. The samples were analyzed at the Laboratory of Analytical Geochemistry (Institute of Geosciences/UNICAMP) to determine the major constituents by ion chromatography, alkalinity by titration and other elements by ICP—MS (Mass Spectrometry with Inductively Plasma coupled). Collection of blank (deionized water) was used with the same sampling methods as analytical quality parameter.

3.2 Current Pollen Rain

At randomly selected locations in the wetlands of Catingueiro, Pedra, and Fundão, sediment samples from water/sediment interface, and soil samples in the vicinity were collected for documenting the current pollen rain. Sediments in bromeliads located in the trees near each studied site (Table 2) were also collected to evaluate the changes and to obtain the records of the dominant plants. Four samples were processed in each studied pond (Table 2). Cp1-F, Cp2-F, Cp3-F, and Cp4-F are samples of pollen rain from Fundão; Cp1-P, Cp2-P, Cp4-P are samples from Pedra; and Cp1-C, Cp3-C, Cp4-C represent samples from Catingueiro. The samples Cp3-P and Cp2-C could not be analyzed, because pollen grains were not found in the slides. The samples were processed in the laboratory according to standard technique for Quaternary palynology of Faegri and Iversen (1989). After the chemical treatment, pollen slides were prepared from the samples for identification and palinomorphs count. 200–300 spores and pollen grains per slide were counted. In the sequence, diagrams were built with individual percentage of each taxon and habitat to analyse the results of current pollen rain in each location.

3.3 Climatic Analysis of Leaf Morphologies

3.3.1 Litter Sampling

Seven sets of leaves were collected within an area of 1.5 × 1.5 m (Burnham 1994), divided into three specific sites known as Catingueiro, Pedra and Fundão. The sites represent three different evolution stages of oxbow lakes: Catingueiro as the youngest, Pedra the intermediate, and Fundão the most advanced. Out of seven sets, four were located in overbank areas (levees) under a canopy of Cerrado, and three

Table 2 Points of current pollen rain collection at the EEcMG

Ponds	Samples		
Studied site	Sediment	Bromeliads	Sterile
Catingueiro	Cp1 C /Cp3 C	Cp4 C	Cp2 C
Pedra	Cp1 P/Cp2 P	Cp4 P	Cp3 P
Fundão	Cp1 F/Cp2 F	Cp3 F/Cp4 F	

were located in the inner banks of meander bends under a canopy of Riparian Forest. The leaves collected were pressed during collection to prevent deterioration, and they were classified into morphotypes in the laboratory, according to Ellis et al. (2009) and the Denver Museum of Nature and Science (2011).

3.3.2 Foliar Physiognomy

Leaf Margin Analysis (LMA): Evaluation was made with the use of five equations that use the principle of LMA for the calculation of MAT. The first two were proposed by Wilf (1997) (Table 3) based on CLAMP (Climate Leaf Analysis Multivariate Program; Wolfe 1993) data. These equations include locations mainly in the Northern Hemisphere, and both were recommended by Kowalski (2002) for MAT calculation of floras with high temperatures, and low elevation in South America tropical regions. Hinojosa et al. (2011) proposed the three following equations. The first (Table 3) was generated from the CLAMP database

Table 3 Equations based on the Leaf Margin Analysis (LMA) and Leaf Analysis Area (LAA) used for the estimation of mean annual temperature and mean annual precipitation

LMA-Equation number and source	Equation	n[a]	SE[b] (°C)	R^{2c}	*p*-value[d]
Equation 1—Wilf (1997)	MAT = 29.10$(E)^{e}$—0.266	106	3.4	0.760	<0.0005
Equation 2—Wilf (1997)	MAT = 24.40E + 3.250	74	2.1	0.840	<0.0005
Equation 3—Hinojosa et al. (2011)	MAT = 25.00E + 3.420	144	2.2	0.870	<0.0001
Equation 4—Hinojosa et al. (2011)	MAT = 23.42E + 3.600	44	3.5	0.480	<0.0001
Equation 5—Hinojosa et al. (2011)	MAT = 26.03E + 1.310	74	2.8	0.820	<0.0001
LAA-Equation number and source	Equation	n	SE (mm)	R^2	*p*-value
Equation 6—Wilf et al. (1998)	lnMAP = 0.548 $(\text{MlnA})^{f}$ + 0.768	50	0.359 (ln)	0.760	<0.01
Equation 7—Jacobs and Herendeen (2004)	lnMAP = 2.566 + 0.309MlnA	42	39	0.734	<0.001

[a] Number of sites
[b] Standard errors of the model
[c] Coefficients of determination
[d] Calculated statistical value
[e] Proportion of entire-margined species
[f] MlnA, is the average leaf area of all dicot species calculated according to Wilf et al. (1998), where $\text{MlnA} = \sum a_i p_i$, and a_i represents the seven means of the natural log areas of the size classes of Raunkiaer (1934) as modified by Webb (1959), and p_i represents the proportion of species in each of the size class

(http://www.open.ac.uk/earth-research/spicer/CLAMP/Clampset1.html), which also includes locations in North America, Central America and Japan. The second and third equations used data from the tropical and southern floras of South America.

Leaf Area Analysis (LAA): Two equations (Table 3) were used to estimate MAP. The first one was proposed by Wilf et al. (1998), based on 50 modern forests in Africa and Central America, thus, covering a wide variety of different climates and plant types. Jacobs and Herendeen (2004) proposed the second equation based on data from 42 locations of tropical climate in Africa and Bolivia.

Climate Leaf Analysis Multivariate Program (CLAMP): With the help of CLAMP Online (http://clamp.ibcas.ac.cn/) tool two analyses were performed using the set of Physg3brcAZ data. The first one was with Met3brAZ climatic data, and the second was with GRIDMet3brcAZ squared climatic data (http://www.open.ac.uk/earth-research/spicer/CLAMP/Clampset1.html).

4 Results

4.1 Dynamics of Subaquatical Environment

The area of Catingueiro is located 200 m far from Mogi-Guaçú, and lies in a topographically lower region than flood plain, but higher than in the bank of the river or meander. The study area in Pedra is located about 200 m from the main river, in a low land but located higher than the river bank. The aquifers studied are shallow and unconfined, have primary porosity and alluvial origin. The two areas (Catingueiro and Pedra) are close to Mogi-Guaçú River. Figures 2 and 3 present the areas of study and the location of monitoring wells in each area.

There is grain size variation in the texture of sediments with the predominance of sand and little silt. Overall, these sediments ranged from sandy—loamy sand—sandy clay and sand in Catingueiro, and clayey sand—sandy clay—clay in Pedra area. In Catingueiro (Fig. 5), sand fraction predominates (above 60 %), the clay fraction varies and occurs at a higher percentage than the silt fraction (up to 10 %). The clay occurs in larger percentages in the shallowest portion of the soil between 1 and 2 m deep, and ranges from 10 to 30 %. Figure 4 shows the classification of texture of the sediments. In Pedra area, the sediments are sandy with a high percentage of sand fraction (above 40 %) and with greater textural variation. The clay has a better distribution (between 10 and 80 %). As in Catingueiro, the percentage of silt is low and varies between 0 and 20 %. There are occurrences of higher percentages of clay and silt in the superficial portions, especially near the lake. The profile IF 6 (west)—IF 11 (northeast) shows the vertical distribution of fractions. There is a decrease in sand fraction and consequent increase of clay fraction in the eastern portion profile (Fig. 6). In Profile IF 13 (west)—IF 18 (northeast), there is a distribution of the finer fractions along this profile in the northeastern portion (Fig. 7). The other feature points finer fractions (mainly clay) in the superficial

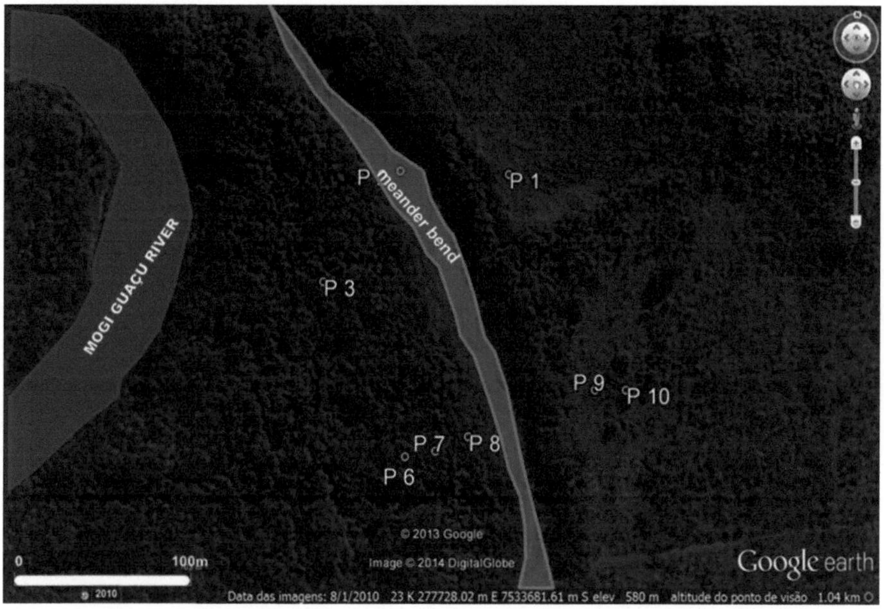

Fig. 2 The wells location in Catingueiro area. Monitoring wells in *blue*. Coordinates: 23 K 27777728.02 mE and 7533681.61 mS; elevation 580 m. *Image* 08/01/2010. Google Earth

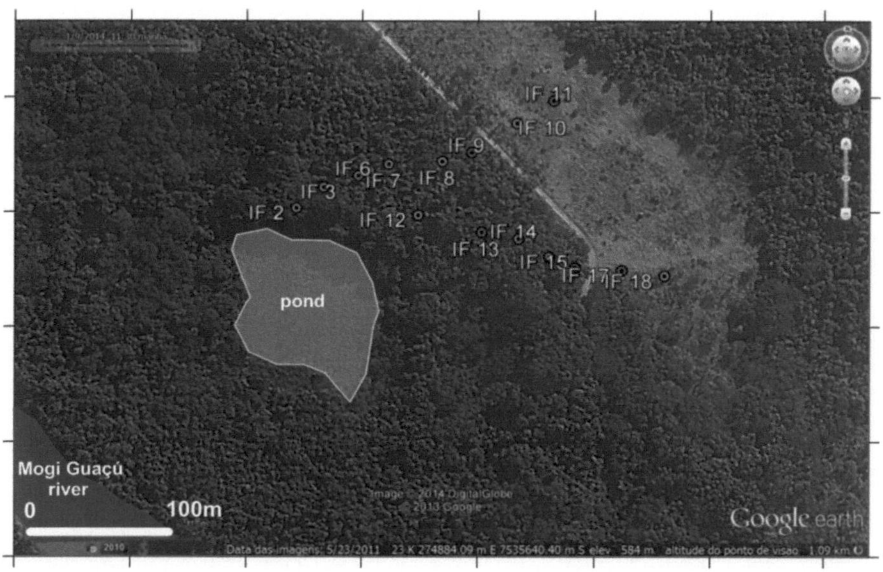

Fig. 3 The wells location in Pedra area. Multilevel wells in *red* and monitoring wells in *blue*. Coordinates: 23 K 274884.09 mE and 7535640.40 mS; elevation 580 m. *Image* 5/23/2011. Google Earth

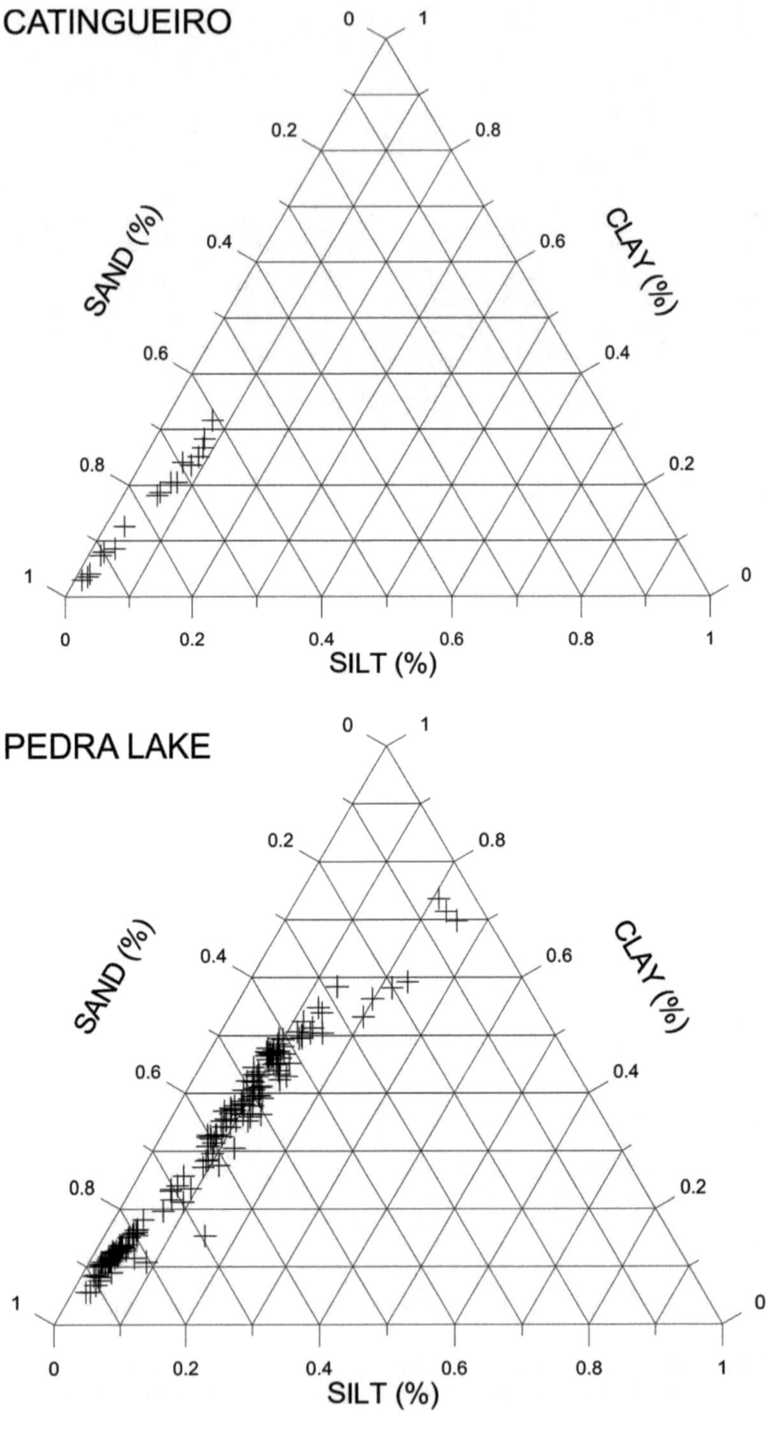

Fig. 4 Texture classification of sediments collected during the drilling of monitoring wells in Catingueiro and Pedra areas

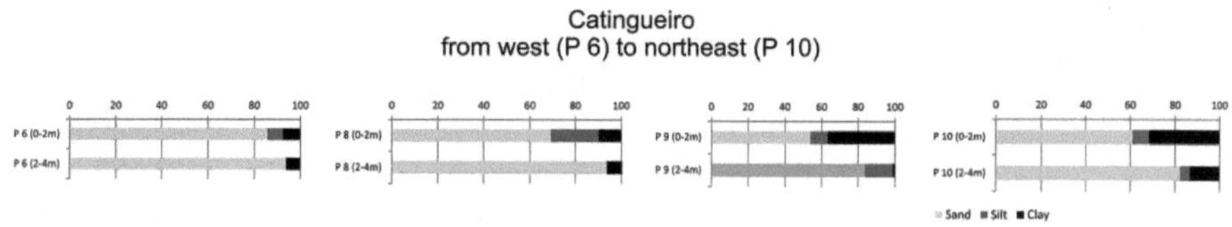

Fig. 5 Cross section (from P 6 to P 10) of vertical distribution in percentage of sand, silt and clay in sediments profiles in Catingueiro area

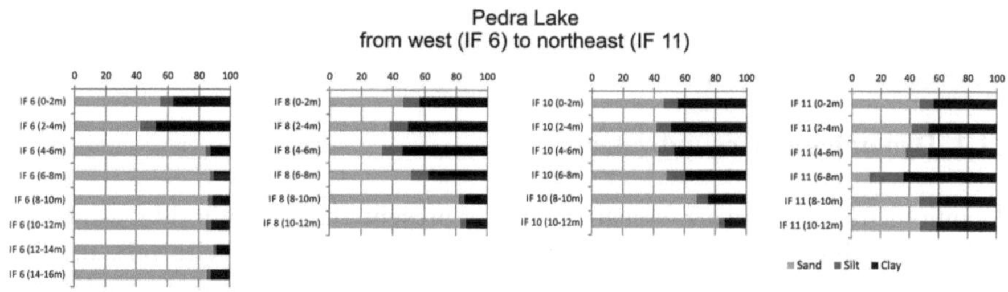

Fig. 6 Cross section (from IF 6 to IF 11) of vertical distribution in percentage of sand, silt and clay in sediments profiles in Pedra area

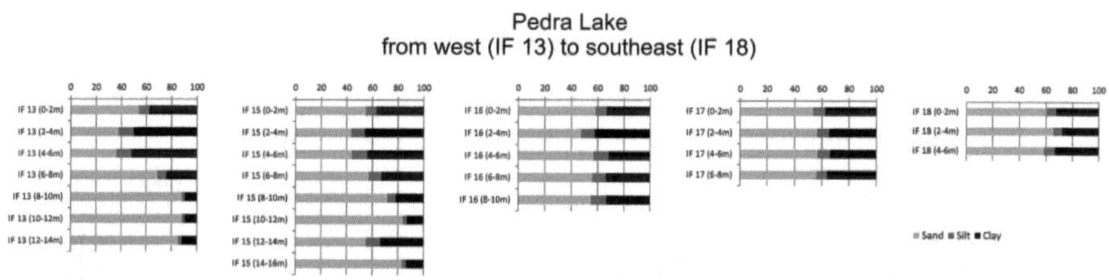

Fig. 7 Cross section (from IF 13 to IF 18) of vertical distribution in percentage of sand, silt and clay in sediments profiles in Pedra area

portions of the profile. An increasing content of sand could be observed towards the water body and river and it gets further enhanced towards the deeper portion.

The aquifer values of Hydraulic Conductivity (K) ranged (Table 4) from 1.03×10^{-1} to 3.83×10^{-3} cm/s (shallow aquifer), a moderate value of 5.9×10^{-3} cm/s (deep aquifer) in Pedra, and 5.8×10^{-3} to 2.1×10^{-1} to 2.1×10^{-1} cm/s in Catingueiro. The values of K indicate an aquifer of high permeability, and the presence of well-sorted sands to fine sand and silt. The presence of gravels in some wells is also significant.

Table 4 Values of hydraulic conductivity of the aquifers in Catingueiro and Pedra areas

Catingueiro		Pedra Lake	
Monitoring well	K (cm/s)	Monitoring well	K (cm/s)
P 1	5.3×10^{-2}	IF 2	2.65×10^{-2}
P 2	9.0×10^{-2}	IF 6	3.83×10^{-3}
P 3	2.1×10^{-1}	IF 8	9.68×10^{-2}
P 6	7.9×10^{-2}	IF 9 s	1.03×10^{-1}
P 7	5.8×10^{-3}	IF 13s	2.58×10^{-2}
P 8	2.0×10^{-2}	IF 17s	4.49×10^{-2}
P 9	7.0×10^{-2}	IF 9d	5.00×10^{-3}
P 10	1.8×10^{-2}	IF 11d	3.78×10^{-2}
		IF 13d	6.82×10^{-3}

4.1.1 Dynamics of Groundwater

Potentiometric maps were constructed to represent three monitoring campaigns during the rainiest period (or when groundwater levels were shallower), the field conditions of the sampling period for physical and chemical analysis, and during the driest period (or when groundwater levels were deeper). The maps how the groundwater flow situation in Catingueiro and Pedra in January 2012, May 2012, and October 2012.

Figures 8, 9 and 10 show the potentiometric maps of Catingueiro with flow directions toward the meander (which has effluent behavior—discharge into the river from groundwater). From these, it is inferred that, the direction of water flow do not change, however, there is a change of hydraulic heads. During the rainy

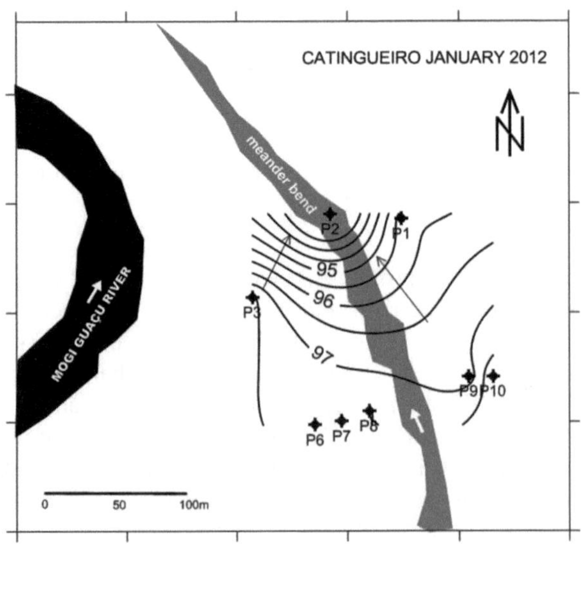

Fig. 8 Potentiometric map to Catingueiro area in January 2011 (rainy season). Legend: *blue lines* equipotential, *black dots* monitoring wells, *arrows* (*white* and *blue*) river and groundwater flow directions

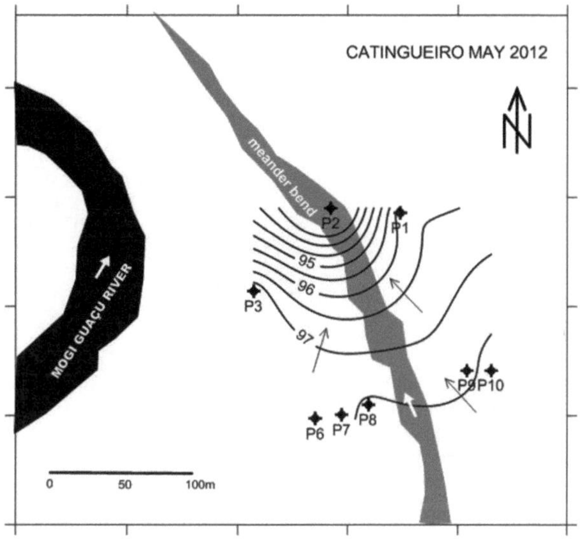

Fig. 9 Potentiometric map to Catingueiro area in the early less rainy season (May 2012). Legend: *blue lines* equipotential, *black dots* monitoring wells, *arrows* (*white* and *blue*) river and groundwater flow directions

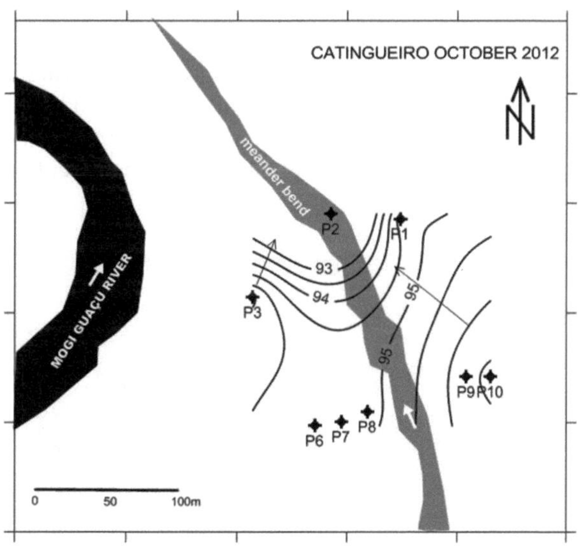

Fig. 10 Potentiometric map to Catingueiro area in the drier season (October 2012). Legend: *blue lines* equipotential, *black dots* monitoring wells, *arrows* (*white* and *blue*) river and groundwater flow directions

season (January 2012), there was an elevation of the hydraulic head indicating higher hydraulic gradient (higher groundwater velocities). In May 2012, early in the less rainy season, water levels started declining and consequently the hydraulic head. In October 2012, the flow direction, and the distribution of groundwater flow changed in the southwest portion of the map. During the dry season, the hydraulic heads had lower values and gradients, resulting in lower flow rates.

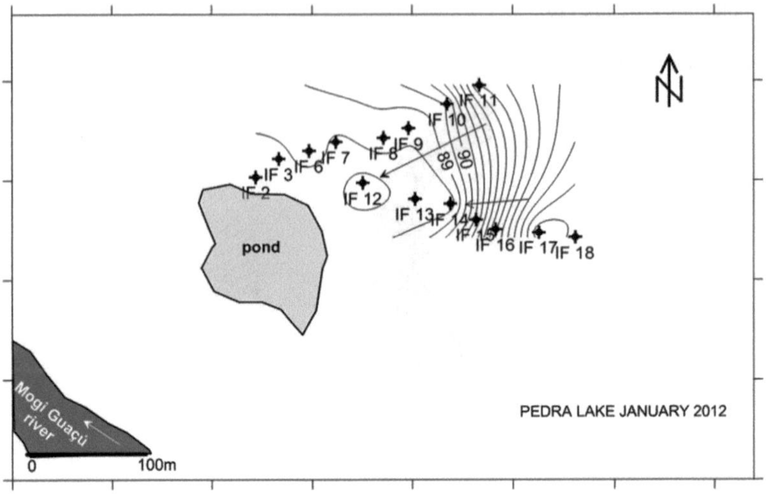

Fig. 11 Potentiometric map to Pedra area in January 2012 (rainy season). Legend: *blue lines* equipotential, *black dots* monitoring wells, *arrows* (*white* and *blue*) river and groundwater flow directions

The potentiometric situations of Pedra during January 2012, May 2012 and October 2012 are presented in the Figs. 11, 12 and 13 respectively. These potenctiometric maps of the shallow aquifer show the flow direction to the lake. It is revealed by these maps that there exists an effluent behavior (groundwater discharge

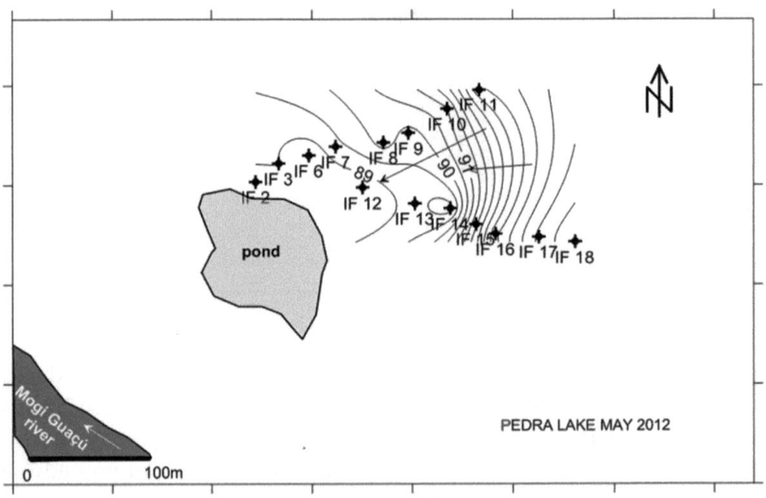

Fig. 12 Potentiometric map to Pedra area in May 2012 (less rainy season) Legend: *blue lines* equipotential, *black dots* monitoring wells, *arrows* (*white* and *blue*) river and groundwater flow directions

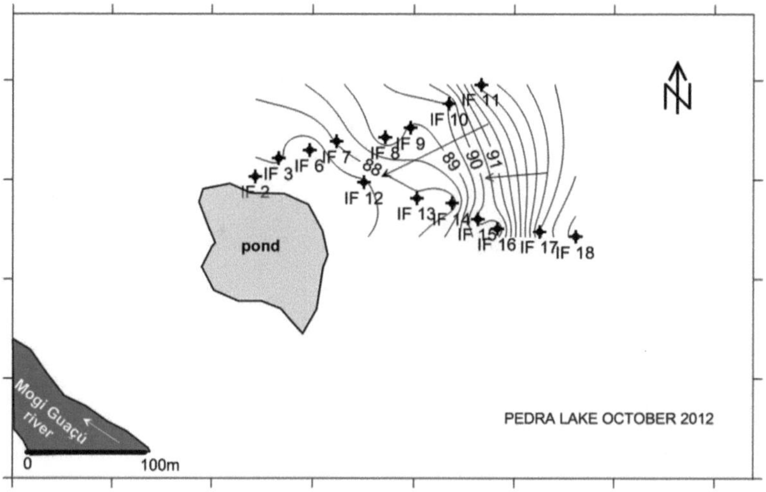

Fig. 13 Potentiometric map to Pedra area in October 2012 (drier season) Legend: *blue lines* equipotential, *black dots* monitoring wells, *arrows* (*white* and *blue*) river and groundwater flow directions

into the lake) regardless of rainy or dry seasons. However, differences occur. During the rainy season (January 2012), the higher and more concentrated equipotential lines show higher hydraulic head and larger hydraulic gradients, indicating a greater dynamic in the aquifer, and increased discharge of water in the water body. In May 2012, the flow has the same behavior as in January 2012, but hydraulic head rises and directions of groundwater move subtly in the northern portion, discharging in the water body (located southwest of the area). In the drier season (October 2012), the equipotential lines are more spaced apart, and have lower values of hydraulic heads and hydraulic gradients, indicating lower flow velocity.

In the deeper portion of the aquifer, the behavior of groundwater flow was studied for the same periods. Figures 14, 15 and 16 show a significant variation during the months monitored. In January 2012, the levels were shallow, and the potentiometric value had higher hydraulic gradients and predominant east-west direction. In May 2012, the hydraulic gradient had southwest flow direction, and the same hydraulic gradient as in January 2012. In October 2012, there was a significant reduction in hydraulic loads, and the flow direction changed (northeast–southwest).

4.1.2 Chemical Composition of Groundwater

Table 5 and Appendix A.1 show the results of the groundwater from Catingueiro and Pedra. It follows from these that the groundwaters from shallower wells were acidic and oxidizing, with pH ranging from 5.20 to 6.94, and Eh (mV) ranging between 242

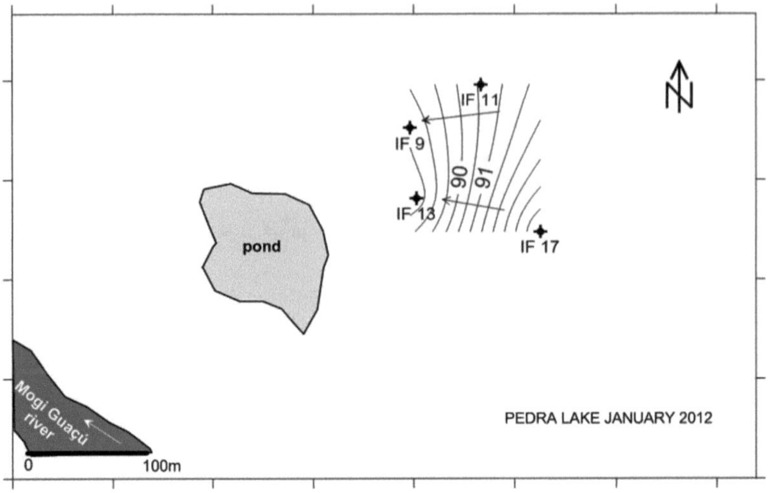

Fig. 14 Potentiometric map of the aquifer in deeper level to Pedra area during the rainy season (January 2012) Legend: *blue lines* equipotential, *black dots* monitoring wells, *arrows* (*white* and *blue*) river and groundwater flow directions

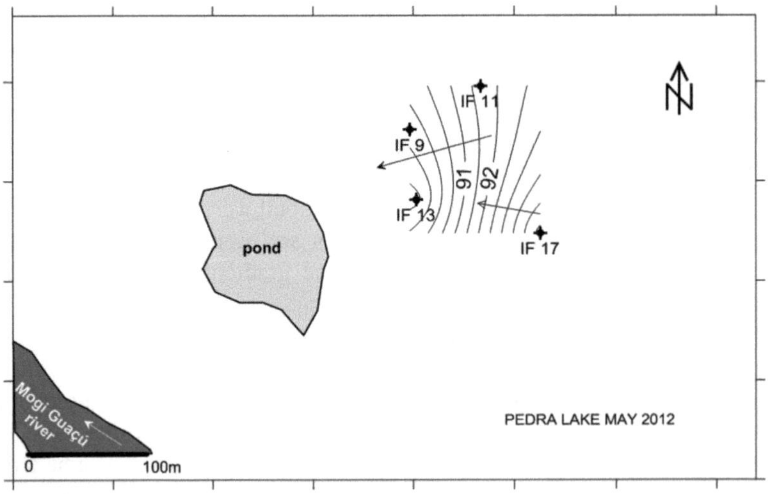

Fig. 15 Potentiometric map of the aquifer in deeper level to Pedra area during the less rainy season (May 2012) Legend: *blue lines* equipotential, *black dots* monitoring wells, *arrows* (*white* and *blue*) river and groundwater flow directions

and 478 mV. The low values of Electrical Conductivity (EC 8.3–28.2 μS/cm), indicated waters with low mineralization. The temperature of the water was between 21.2 and 21.9 °C. Bicarbonate concentration ranged from 4.2 to 49.0 mg L^{-1}. The groundwater samples of IF 11d and IF 17d wells (deeper ones) were different from

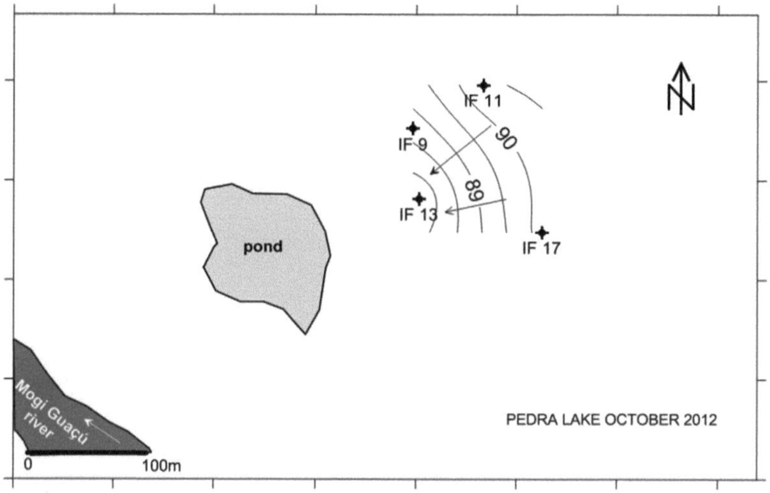

Fig. 16 Potentiometric map of the aquifer in deeper level to Pedra area during the drier season (October 2012) Legend: *blue lines* equipotential, *black dots* monitoring wells, *arrows* (*white* and *blue*) river and groundwater flow directions

the others with greater mineralization (EC 146.2 and 402.0 μS cm^{-1}), neutral pH (6.94 and 7.42), and lower values of Eh (381 and 300 mV). The water temperatures ranged from 20.5 to 22.7 °C, and bicarbonate from 94 to 282 mg L^{-1}.

Table 6 shows the results of chemical analysis of groundwaters from Catingueiro and Pedra areas. They showed low chloride concentrations (between 50 and 910 μg L^{-1}), sulphate (from <10 to 4,640 μg L^{-1}), NO_3^-N (between <3 and 160 μg L^{-1}), calcium (between 55 and 44,280 μg L^{-1}), sodium (between 50 and 61,600 μg L^{-1}), ammonium (between <20 and 60 μg L^{-1}), and potassium (between 40 and 1,180 μg L^{-1}). The fluoride ion occurred only in IF 17d well. The deeper groundwater in Pedra area had higher concentrations of major elements. In the Piper diagram (Fig. 17), Catingueiro groundwater presents calcium bicarbonate and mixed-cation bicarbonate types. Pedra groundwater was classified as calcium bicarbonate, and one sample (IF 17d) was classified as mixed-cation bicarbonate type.

The low mineralization, acidic pH and oxidative environment characterize the Catingueiro groundwater. The shallow aquifer indicates groundwaters with low residence time, and recharge of seasonal waters and sandy aquifers. The acidic pH indicates the contribution of carbon dioxide from the atmosphere into the soil, and the presence of humic acids interacting in the shallow aquifer. In the Pedra area, the low mineralization of water, acidic pH and oxidative environment characterize the groundwater. The shallowest nature of the aquifer indicates a low residence time, and seasonal water recharge. However, the groundwaters are highly mineralized in the deeper wells, and indicate a greater residence time and contact with the aquifer.

Appendix A.1 shows the results of 20 minor elements analysis. In general, the concentrations of these elements are low, and highlight the occurrence of total iron (10.9–385 μg L^{-1}), manganese (6.24–106 μg L^{-1}) and silica (2,298–7,016 μg L^{-1}).

Table 5 Results of physical–chemical parameters of groundwater to Catingueiro and Pedra Lake areas

Monitoring well	Temp. (°C)	pH	Eh (mV)	Electric conductivity (μS/cm)	HCO_3^- (mg/l)
P 1	21.7	5.25	440	11.4	4.3
P 2	21.7	5.80	298	25.1	4.6
P 3	21.8	5.45	353	20.8	7.4
P 6	21.5	5.24	433	21.5	5.2
P 7	21.4	5.71	242	40.3	5.2
P 8	21.7	5.73	316	28.1	5.2
P 9	21.2	5.25	428	8.3	4.2
P 10	21.9	5.20	448	8.7	4.2
IF 2	20.8	5.51	478	19.4	7.2
IF 3	21.4	5.41	390	20.8	–
IF 6	21.2	5.57	416	20.9	–
IF 7	20.7	5.69	424	21.1	–
IF 8	20.9	5.39	453	13.9	–
IF 9s	21.2	5.37	434	10.8	5.2
IF 10	22.3	5.66	425	17.1	–
IF 11s	20.8	5.58	433	12.3	7.4
IF12	21.7	5.40	395	15.9	–
IF 13s	20.5	5.88	378	20.8	13.3
IF 14	22.7	5.17	407	11.5	–
IF 15	21.6	5.16	414	12.9	–
IF 16	22.0	5.18	374	10.3	–
IF 17s	21.4	5.53	370	11.7	7.4
IF 18	22.1	5.24	402	10.3	–
IF 9d	20.9	6.02	397	26.3	49
IF 11d	20.7	6.94	381	146.2	94
IF 13d	21.0	5.91	391	28.2	18
IF 17d	21.2	7.42	300	402.0	282

P1 Monitoring well in Catingueiro; IF 17s—shallow monitoring well in Pedra; IF 17d—deep monitoring well in Pedra

4.2 Current Pollen Rain

Results of the estimates on the current pollen rain are presented in the Fig. 18. The diagram shows the most representative taxa of each point studied, and the taxa index for each kind of plant and habit. Data expressed as percentages are relative to the total amount of pollen grains, including pollen grains of arboreal plants (AP), herbaceous or not arboreal plants (NAP), indeterminate pollen grains, spores and aquatics. Pollen types with less than 1 % presence are represented in the diagram by +sign. Figure 19 shows the habit pollen diagram sampled in points Cp1-C, Cp3-C and Cp4-C (Catingueiro). An increase of arboreal taxa compared to other points is

Table A.1 Results to chemical analyses (ICP-MS) at Catingueiro

	P 1	P 2	P 3	P 6	P 7	P 8	P 9	P 10	IF 2	IF 9s	IF 11s	IF 13s	IF 17s	IF 9d	IF 11d	IF 13d	IF 17d	Limit of Detection (LOD)	Field blank
$\mu g\ L^{-1}$																			
Al	27.1	142	186	349	64.0	63.5	14.4	17.1	19.1	17.1	16.6	37.4	17.8	675	35.7	52.9	18.4	1.3	4
As	0.06	0.15	0.18	0.35	0.11	0.14	<LOD	<LOD	0.06	<LOD	<LOD	<LOD	<LOD	0.15	0.07	0.05	0.15	0.04	<LOD
B	2.08	2.64	2.78	2.55	1.91	2.49	0.67	0.79	1.21	0.45	0.77	0.66	1.84	1.61	1.54	1.13	23.0	0.2	<LOD
Ba	30.0	81.8	100	59.4	99.6	129	15.6	16.8	50.2	17.0	13.1	21.4	12.8	22.4	56.7	32.6	57.3	0.012	0.13
Be	0.05	0.25	0.33	0.18	0.22	0.24	0.04	0.04	0.12	0.04	0.02	0.03	0.02	0.02	0.002	0.02	0.002	0.001	0.001
Cd	0.009	0.028	0.028	0.034	0.036	0.043	0.003	0.004	0.02	0.012	0.005	0.017	0.012	0.012	0.010	0.016	0.026	0.003	0.00
Co	0.39	6.09	2.46	1.37	6.10	4.78	0.17	0.18	0.64	0.31	0.25	0.31	0.15	0.77	0.06	0.43	0.21	0.005	<LOD
Cr	<LOD	1.75	1.87	1.27	1.21	0.71	<LOD	<LOD	0.14	<LOD	<LOD	0.11	<LOD	1.24	0.22	0.17	0.10	0.09	<LOD
Cu	0.35	0.44	2.04	2.70	0.39	0.39	<LOD	0.13	1.70	0.33	0.25	0.39	0.14	2.52	0.43	0.37	2.60	0.1	<LOD
Fe	50.0	168	115	905	58.3	489	12.2	20.6	<LOD	<LOD	<LOD	10.5	<LOD	385	10.9	17.9	57.6	7.7	<LOD
Li	0.29	1.61	1.49	0.91	2.58	2.06	0.30	0.32	0.72	0.57	0.43	0.55	0.38	0.69	0.31	0.52	0.25	0.008	<LOD
Mn	28.1	71.6	46.9	82.3	89.2	106	11.9	21.5	15.6	15.7	11.8	12.8	6.24	25.7	7.43	30.5	84.7	0.14	<LOD
Ni	0.14	1.17	0.71	0.58	0.80	0.72	0.05	0.07	1.36	0.27	0.24	0.81	0.23	5.46	0.35	0.48	0.69	0.04	<LOD
Pb	0.25	0.21	0.27	0.89	0.08	0.13	0.06	0.06	0.72	0.09	0.06	0.27	0.14	1.02	0.07	0.26	0.24	0.003	0.03
Rb	1.09	1.79	2.40	2.73	2.14	1.73	1.21	0.81	4.42	1.42	1.23	2.82	1.42	3.48	2.97	4.13	2.90	0.02	0.09
Si	3,324	4,987	4,193	3,648	5,538	5,103	2,518	2,298	5,618	3,457	3,263	3,625	2,651	4,629	3,565	4,272	7,016	803	<LOD
Sr	6.64	12.3	18.1	12.9	15.4	12.0	2.98	3.98	8.42	4.34	5.22	18.9	3.91	50.5	40.2	21.4	31.9	0.006	0.04
Ti	0.25	1.04	1.18	2.62	0.53	0.43	0.17	<LOD	<LOD	0.06	0.14	0.59	0.13	29.9	1.06	1.36	0.62	0.05	<LOD
U	0.05	0.10	0.11	0.10	0.04	0.04	0.01	0.01	0.06	0.04	0.01	0.04	0.02	0.12	0.12	0.02	3.16	0.0006	<LOD
Zn	8.79	22.6	8.94	7.77	32.3	20.0	3.62	5.79	14.0	21.1	27.0	75.1	18.7	32.7	47.7	60.6	76.2	1.2	<LOD

<LOD—lesser than the limit of detection

Table 6 Results of chemical analyses of groundwater to Catingueiro and Pedra

Sample	F^-	Cl^-	SO_4^{2-}	NO_3^--N	Ca^{2+}	Na^+	NH_4^+	K^+
$\mu g\ L^{-1}$								
P 1	<LOD	530	350	10	570	760	30	260
P 2	<LOD	230	1,310	10	1,028	1,150	20	330
P 3	<LOD	60	300	70	1,526	820	20	530
P 6	<LOD	260	230	10	1,110	690	20	550
P 7	<LOD	210	2,370	50	1,313	1,510	10	320
P 8	<LOD	140	1,670	70	963	1,380	40	220
P 9	<LOD	150	<LOD	10	608	320	<LOD	250
P 10	<LOD	233	50	<LOD	720	340	<LOD	170
Field blank	<LOD	50	<LOD	<LOD	55	50	<LOD	40
IF 2	<LOD	300	70	20	1,555	410	10	910
IF 9s	<LOD	90	<LOD	<LOD	939	310	40	300
IF 11s	<LOD	500	140	90	1,480	320	<LOD	340
IF 13s	<LOD	910	480	110	3,957	560	<LOD	740
IF 17s	<LOD	260	50	110	1,368	500	10	410
IF 9d	<LOD	590	760	20	12,669	570	<LOD	670
IF 11d	<LOD	620	520	150	27,783	980	60	930
IF 13d	<LOD	630	350	120	3,957	900	20	1,180
IF 17d	130	13,900	4,640	160	44,298	61,600	<LOD	1,130
Field blank	<LOD	50	<LOD	<LOD	<LOD	40	<LOD	30
Limit of Detection (LOD)	1	15	10	3	55	20	20	50

<LOD—values lesser than the limit of detection

recognizable from this figure. The herbaceous have higher percentages in points Cp2-P, Cp4-P (Pedra) and Cp1-F (Fundão). The aquatic ones predominated between Cp1-P and Cp1-F. The spores had a significant increase between Cp1-F and Cp4-F. This high frequency may be attributed by the faster development of Pteridophytes than the trees. It may also indicate that the site remained as swamp for a longer time.

4.2.1 Pollen Rain in Catingueiro

The Apocynaceae and Euphorbiaceae families prevail among the herbaceous taxa in Catingueiro (Fig. 20). The *Pinus* sp. Represents an exotic form, and prevails among gymnosperms. The Poaceae and Asteraceae families are the prevailing herbaceous taxa, and the Cyperaceae family represents the aquatics. The percentage of spores was lower against the Pedra and Fundão. The same is represented by the

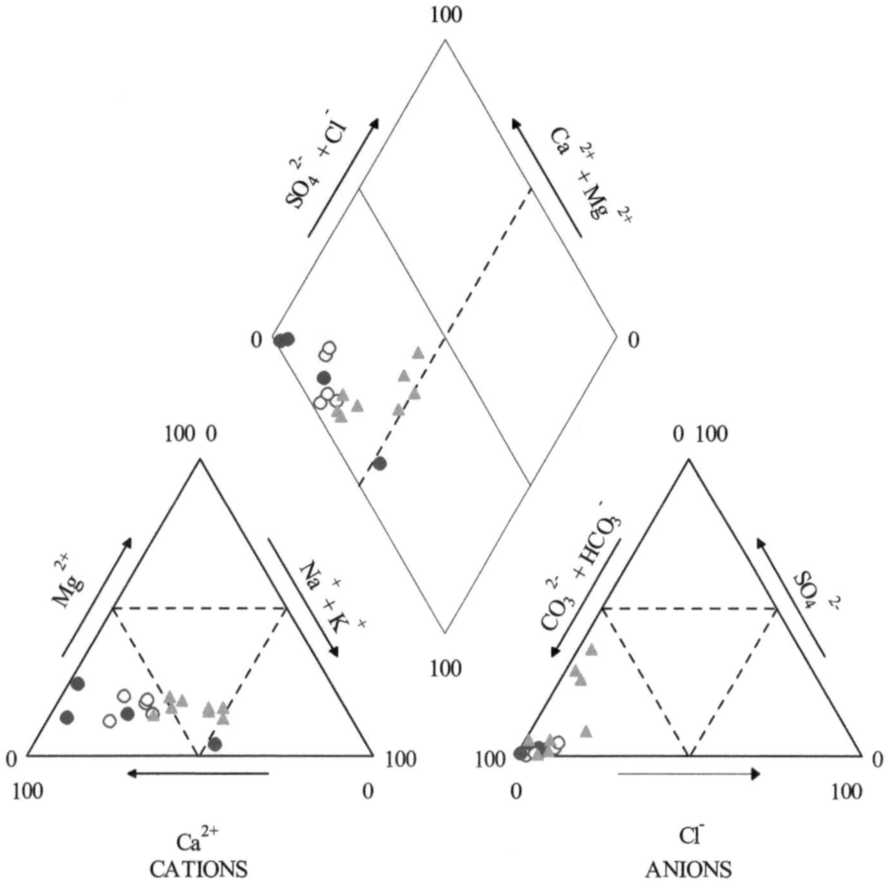

Fig. 17 Classification of groundwater in Catingueiro area (symbol in full *blue circles*) and Pedra area (symbol in full *red circles* deeper aquifer, in *red circles* shallow aquifer) according to Piper diagram

genus *Polipodium* and *Cyathea,* where the frequency of spores ranges from 0.72 to 2.18 %. The pollen signature of the site features an enclosed forest with large trees and we observed an increase of arboreal taxa.

4.2.2 Current Pollen Rain in Pedra

The Apocynaceae, Euphorbiaceae and Araliaceae families are the prevailing arboreal taxa in Pedra (Fig. 21). The *Pinus* sp. represents an exotic form, and prevails among gymnosperms. The Poaceae and Asteraceae families prevail among the herbaceous taxa. An increase in the percentage of aquatics (Cyperaceae family) is recognizable, because it is near the edge of Mogi-Guaçú River, indicating a more wetland condition.

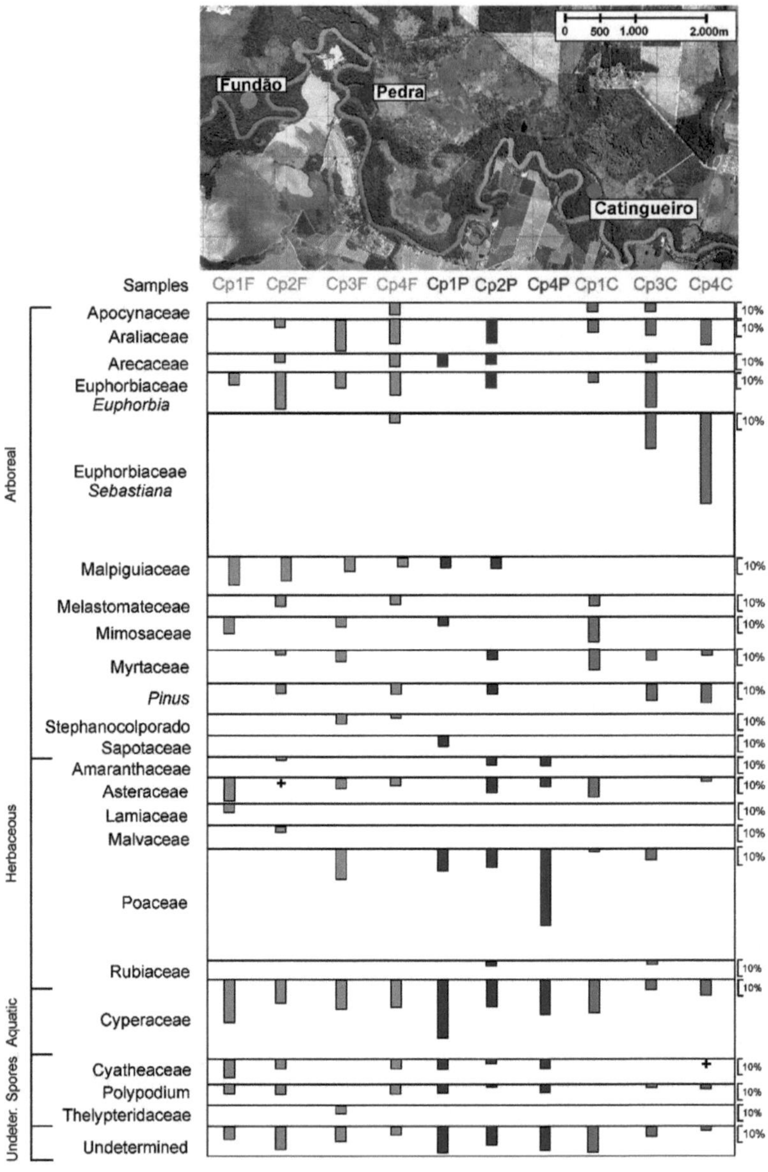

Fig. 18 Diagram of pollen grains percentage of current pollen rain in EEcMG (Ecological Station of Mogi-Guaçú)

Significant increase of spores in Pedra over the Catingueiro, represented by the genus *Polipodium* and *Cyathea* is also recognizable. The frequency ranges from 1.87 to 6.95 %. Similarly, the herbaceous taxa increase and confirm the proximity to a more wet area, given the higher percentage of crawling plants.

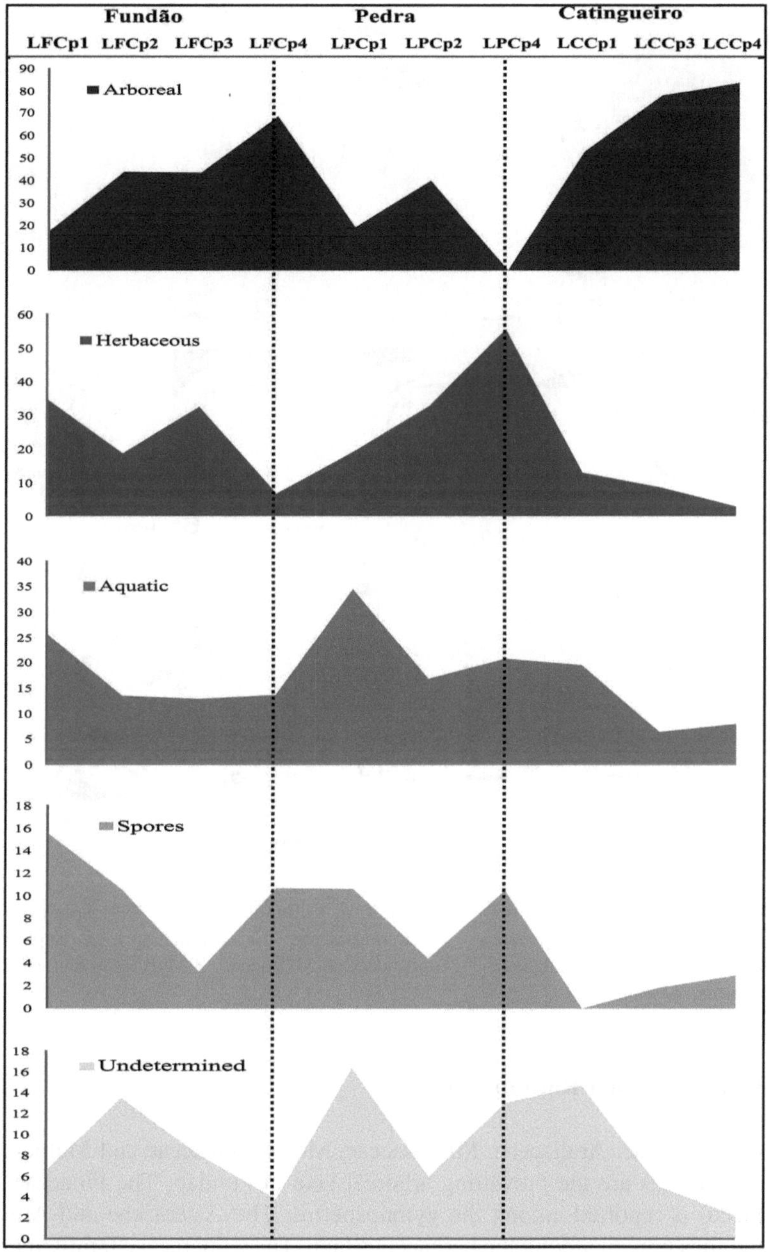

Fig. 19 Diagram of pollen habit of current pollen rain in EEcMG (Ecological Station of Mogi-Guaçú), where the frequency of pollen grains occurrence is represented in percentage

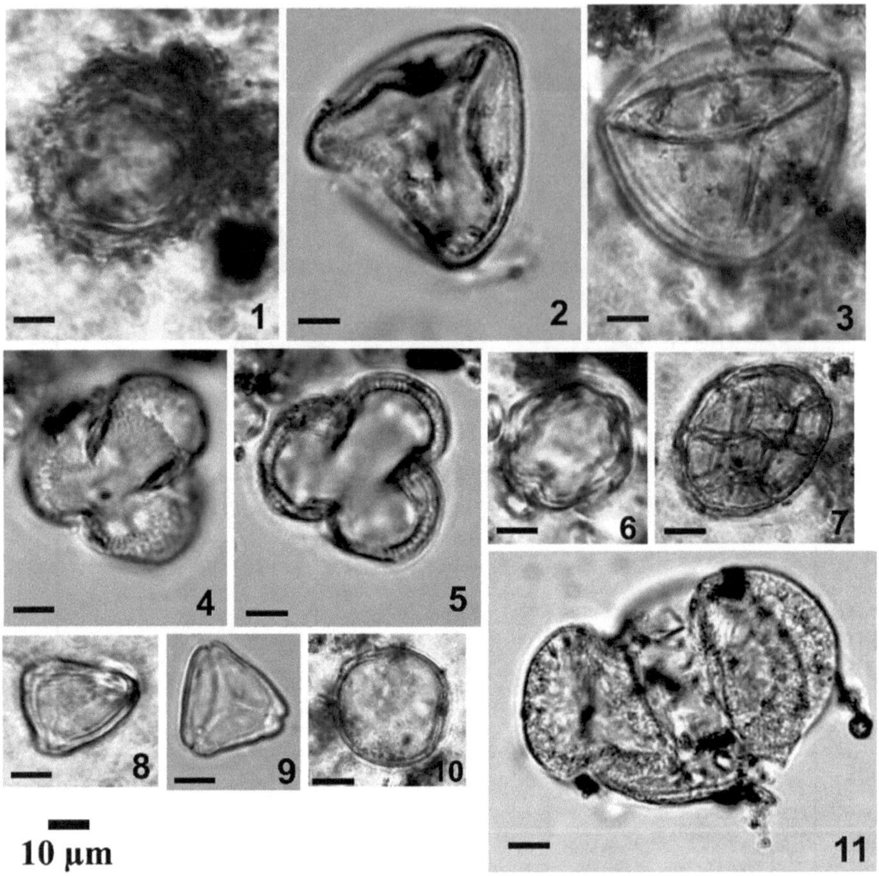

Fig. 20 Angiosperms/Gymnosperms—Pollen rain in Catingueiro pond. Angiosperms: *1* Asteraceae, *2* Cyperaceae, *3* Cyperaceae, *4* Euphorbiaceae (*Sebastiana* sp.), *5* Euphorbiaceae (*Sebastiana* sp.), *6* Melastomataceae, *7* Mimosaceae, *8* Myrtaceae, *9* Myrtaceae and *10* Poaceae. Gymnosperms: *11 Pinus* sp

4.2.3 Current Pollen Rain in Fundão

The Euphorbiaceae, Araliaceae, Mimosaceae, Melastomataceae and Malpiguiaceae families (Fig. 22) are the prevailing arboreal taxa in Fundão. The Pinaceae family (introduced) is reported among the gymnosperms. The Asteraceae and Amaranthaceae families prevail among the herbaceous taxa. The Cyperaceae family represents the aquatic, which appears in a large percentage in analogy to the localities of Pedra and Catingueiro. There is a significant increase of spores (Fig. 23) represented by Cyatheaceae, Polypodiaceae and Lycopodiaceae families. An area with abundant

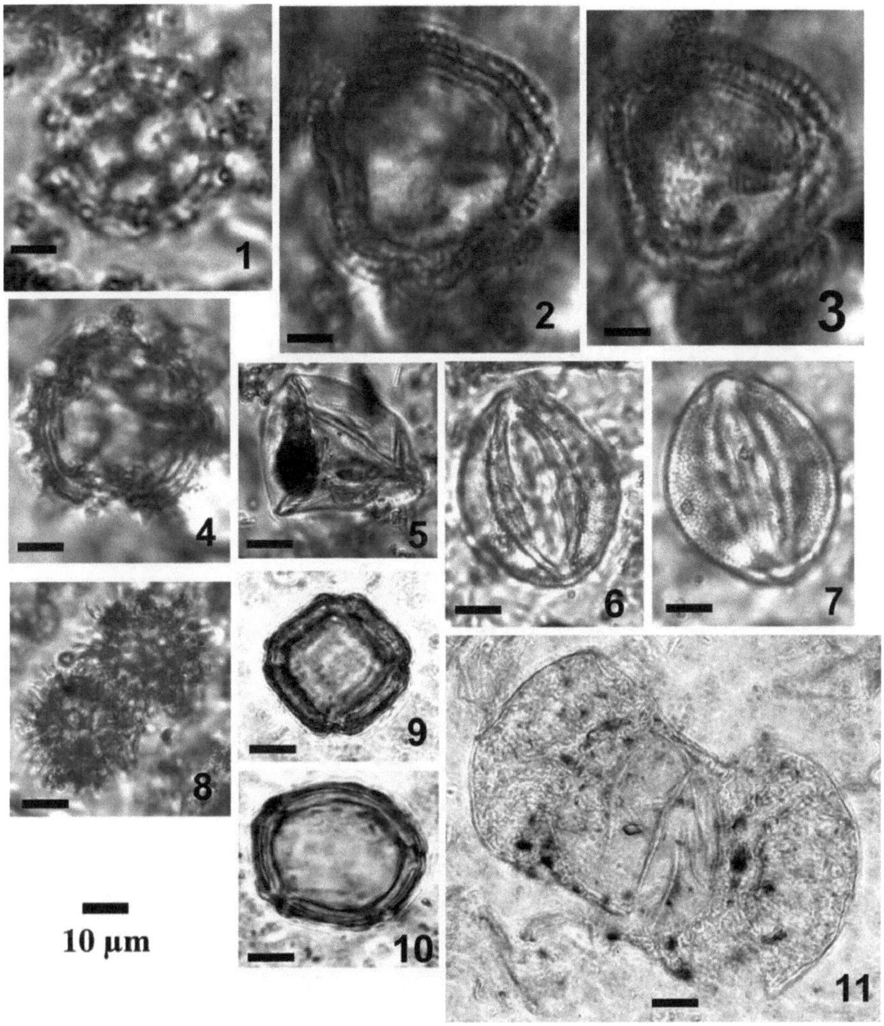

Fig. 21 Angiosperms/Gymnosperms—Pollen Rain in Pedra pond. Angiosperms: *1* Amaranthaceae, *2* Araliaceae, *3* Araliaceae, *4* Asteraceae, *5* Cyperaceae, *6* Euphorbiaceae, *7* Euphorbiaceae, *8* Asteraceae, *9* Malpiguiaceae, *10* Malpiguiaceae. Gymnosperms: *11 Pinus* sp

spores is defined as the result of the data obtained in the site, and related to the proximity of a more wet area. The predominance of herbaceous and arboreal plants determine and represent the biome transition between the Cerrado and Atlantic Forest (Mata Atlântica). The *Mauritia* (Arecaceae) and *Byrsonima* (Malpiguiaceae) are common on the edges of lakes, wetlands and swamps, in the Cerrado region. The record of Cyperaceae in almost all the pollen rain diagram at EEcMG enhances the data of an open vegetation and close to a wetter area.

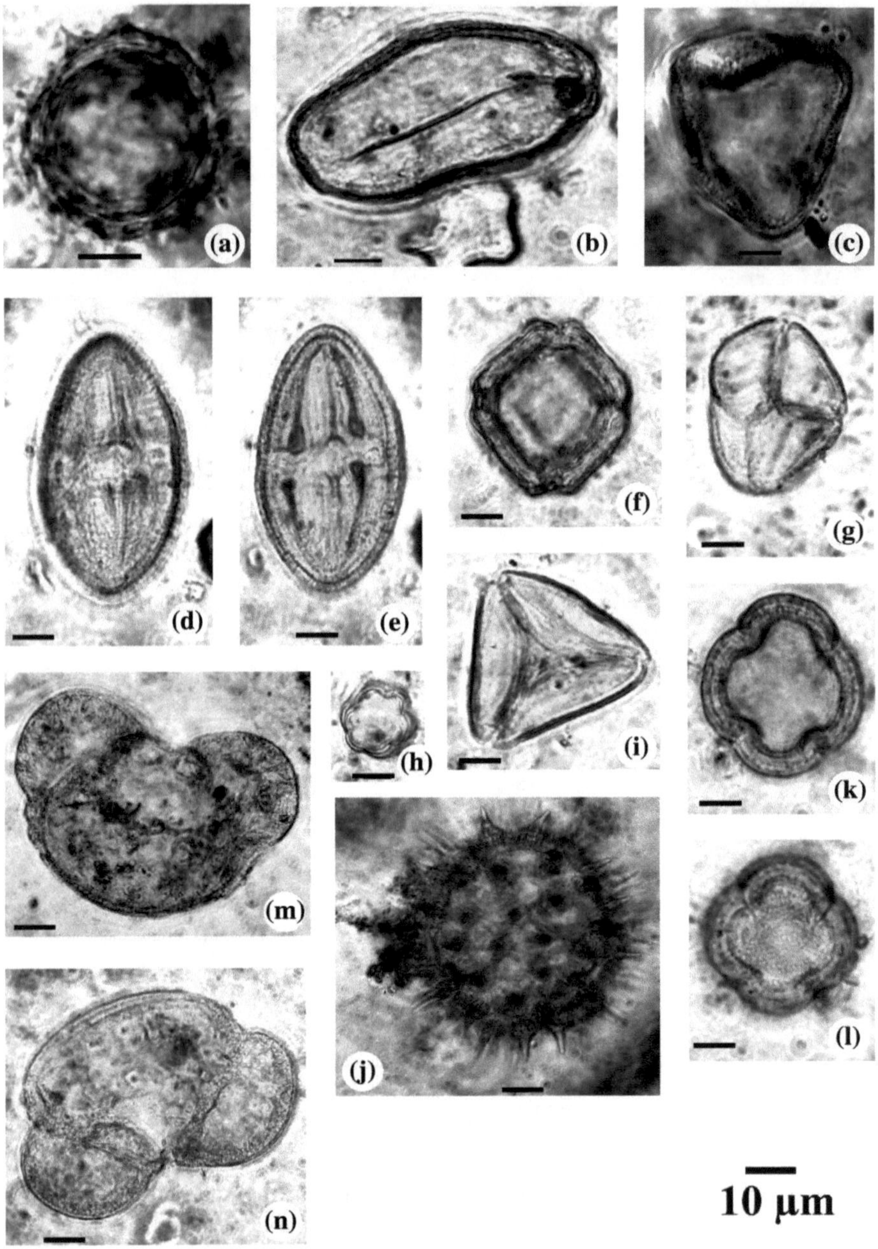

Fig. 22 Angiospermas/Gimnospermas—Pollen rain in Fundão Pond. Angiosperms: **a** Asteraceae, **b** Arecaceae, **c** Cyperaceae, **d** Euphorbiaceae, **e** Euphorbiaceae, **f** Malpiguiaceae, **g** Mimosaceae, **h** Melastomateceae, **i** Myrtaceae, **j** Malvaceae, **k** Stephanocolporate-psilate, **l** Stephanocolporate-psilate. **m** Gymnosperms: and **n** *Pinus* sp

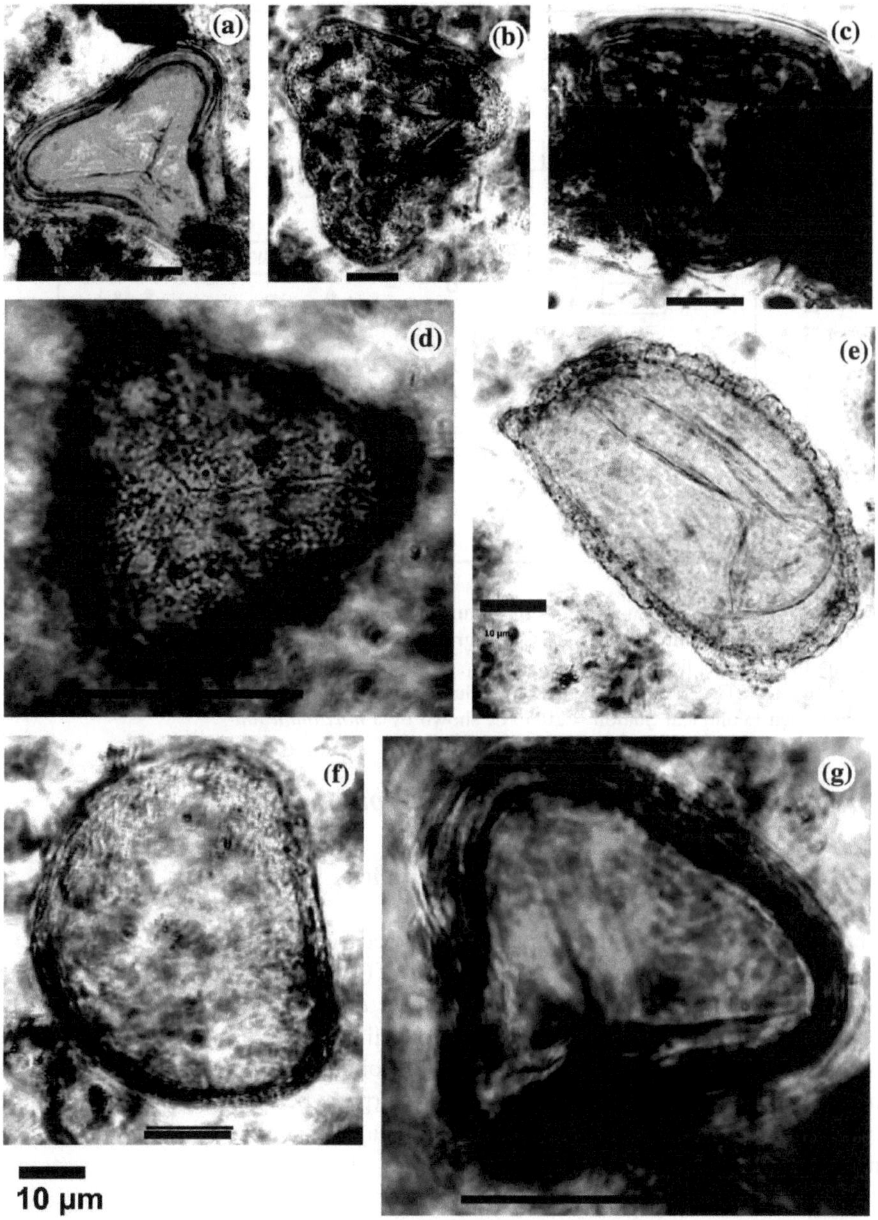

Fig. 23 Spores—Pollen rain in Fundão. **a** Dicksoniaceae, **b** Cyatheaceae, **c** Cyatheaceae, **d** Lycopodiaceae, **e** Thelypteridaceae, **f** Polypodiaceae, and **g** Cyatheaceae

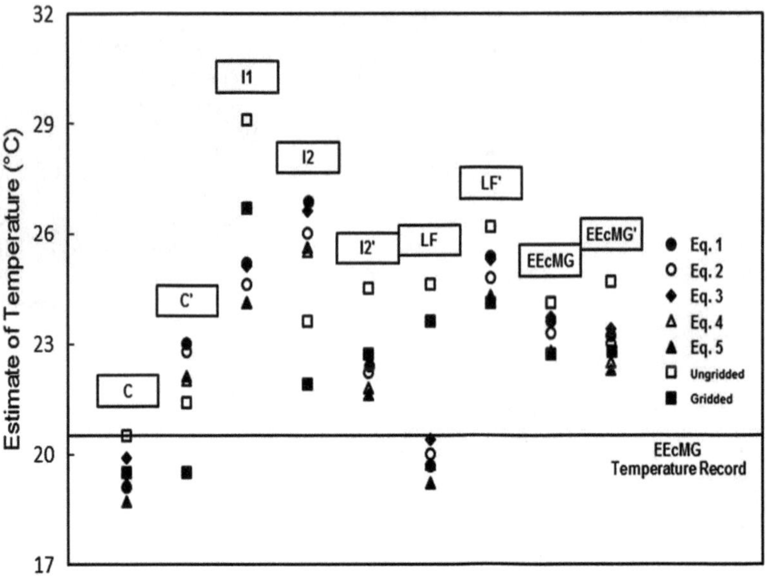

Fig. 24 Estimates of mean annual temperature based on leaf margin analysis (LMA) and two different CLAMP calibration datasets (Ungridded Physg3brcAz and Gridded Physg3brcAZ) starting from litter collected in the levee (C, I_1, I_2, LF, EEcMG) and in the inner banks of meander bends (C', I_2', LF', EEcMG'), of the Mogi-Guaçú Ecological Station. Actual temperature in the Mogi-Guaçú Ecological Station = 20.5 °C, shown by a *horizontal line*

4.3 Climatic Analysis of Leaf Morphologies

4.3.1 Sample Size and Morphotypes Identified

The number of leaves studied in each set varied between 293 and 771 specimens, and between 15 and 31 morphotypes identified (Table 7). The relationship between the number of specimens and morphotypes changes a lot, except the values obtained for Catingueiro (Table 7). The collections were also organized according to geomorphological position, which were collected (levee and inner banks of meander bends) to have a larger number of morphotypes. The result showed that in both geomorphological situations demonstrate a very similar proportion among the number of specimens per morphotypes (Table 7).

4.3.2 Foliar Physiognomy

Estimation of Mean Annual Temperature (MAT): Table 7 and Fig. 25 summarize MAT estimates. The estimates were obtained from sets of leaves collected, and

Table 7 Estimation of mean annual temperature and mean annual precipitation from litter samples taken in the localities of Catingueiro (C-C′), Pedra (I_1-I_2-I_2′), and Fundão (LF-LF′) in Mogi-Guaçú Ecological Station (EEcMG-EEcMG’) using Leaf Margin Analysis (LMA), Leaf Area Analysis (LAA), and two different CLAMP calibration datasets (Ungridded Physg3brcAz and Gridded Physg3brcAZ)

Site	S[a]	m[b]	S:m[c]	E[d]	LMA—MAT (°C)					CLAMP—MAT (°C)		LAA—MAP (mm)		CLAMP—MAP (mm)	
					Eq. 1	Eq. 2	Eq. 3	Eq. 4	Eq. 5	Ungridded	Gridded	Eq. 6	Eq. 7	Ungridded	Gridded
Levee of river and abandoned meander bends															
C	771	18	42.8:1	66.7	19.1	19.5	19.9	19.2	18.7	20.5	19.5	1,179	1,240	2,063	1,897
I_1	485	16	30.3:1	87.5	25.2	24.6	25.1	24.1	24.1	29.1	26.7	1,560	1,455	3,557	3,290
I_2	454	31	14.6:1	93.5	26.9	26.1	26.6	25.5	25.7	23.6	21.9	1,025	1,133	2,587	2,312
LF	293	16	18.3:1	68.7	19.7	20	20.4	19.7	19.2	24.6	23.6	1,343	1,343	2,484	2,572
EEcMG	2,003	61	32.8:1	82	23.6	23.3	23.7	22.8	22.7	24.1	22.7	1,252	1,277	2,580	2,437
Mean absolute errors					2.4	2.2	2.7	1.8	1.6	3.9	2.4	–66	−49	1,319	1,166
Inner banks of meander bends															
C′	649	15	43.2:1	80	23	22.8	23.2	22	22.1	21.4	19.5	1,179	1,240	2,688	2,263
I_2′	528	27	19.5:1	77.8	22.4	22.2	22.7	21.8	21.6	24.5	22.7	1,514	1,426	2,682	2,546
LF′	399	17	23.5:1	88.2	25.4	24.8	25.3	24.3	24.3	26.2	24.1	1,240	1,277	3,350	2,910
EEcMG’	1,576	47	33.5:1	80.8	23.2	23	23.4	22.5	22.3	24.7	22.8	1316	1,316	2,895	2,646
Mean absolute errors					3	2.7	3.1	2.1	2	3.7	1.8	−23	−20	1,569	1,256
Current mean annual temperature based on EEcMG record = 20.5 °C															
Current mean annual precipitation based on EEcMG record = 1,335 mm															

[a] Number of specimens per sample
[b] Leaf morphotypes (assumed to be species)
[c] Relationship between sample size and number of species recovered
[d] Proportion of entire-margined species

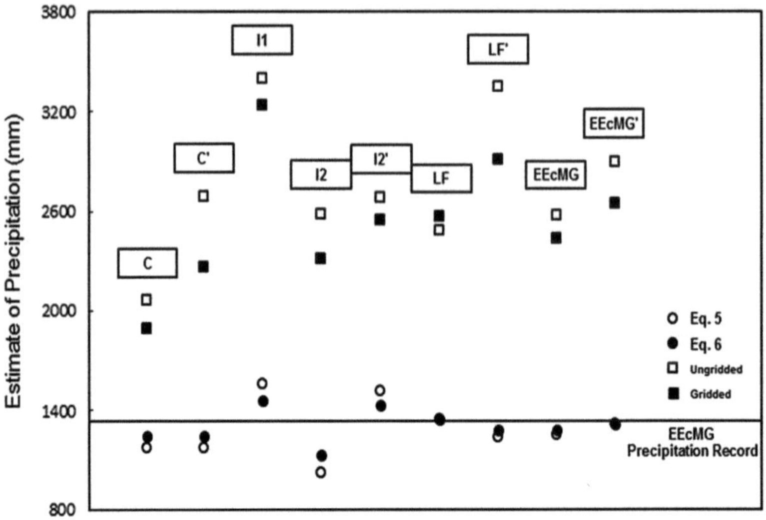

Fig. 25 Estimates of mean annual precipitation based on leaf area analysis (LAA) and two different CLAMP calibration datasets (Ungridded Physg3brcAz and Gridded Physg3brcAZ) starting from litter collected in the levee (C, I_1, I_2, LF, EEcMG) and in the inner banks of meander bends (C′, I_2′, LF', EEcMG′), of the Mogi-Guaçú Ecological Station. Actual precipitation in the Mogi-Guaçú Ecological Station = 1,335 mm, shown by a *horizontal line*

from data matching of each of the selected environments, which allowed maximize the number of specimens and morphotypes. Leave samples collected in the levees (over bank) best reflect the temperature of leaf assemblages collected near the inner banks of meander bends, and better reflect the temperature based on LMA equations (Table 7 and Fig. 24). Errors of 1.6–2.7 °C in the over bank deposits (levee) and associated meander assets (2–3, 1 °C), are observed while considering Eqs. 4 and 5 (Table 7). With respect to the estimated temperature associated with CLAMP data sets, it was observed that the results closer to reality were those obtained in the associated assemblages to inner banks of meander bends (Table 7 and Fig. 24).

Estimation of Mean Annual Precipitation (MAP): Table 7 and Fig. 25 summarize MAP estimates for sets studied. Their combination were according to their geomorphology position (levee and inner banks of meander bends). MAP values for all the assemblages using LAA (Eqs. 6 and 7) were very consistent and much lower than those derived by sets of CLAMP calibration data (Table 7 and Fig. 25). Thus, the set of Gridded data related to levee deposits generated a more reliable result to the current MAP (Table 7 and Fig. 25).

5 Discussion

5.1 Dynamics of Subaquatic Environment

The dynamics of groundwater in alluvial plains is related to the dynamics of the river and the permeable zones where the groundwater flows. Thus, both the shallow and the deep aquifers flow towards the Mogi-Guaçú River, located at west of the areas studied. Locally, the dynamics of groundwater in both areas show flow direction towards the river meanders and pond (effluent character), during the period of present study. These meanders, which dry up during the less rainy season, are also paths of preferential subsurface water flow. The systems of groundwater flow were controlled by the depositional textures of the sediments and structures of sediment layers, slope, and high values of hydraulic conductivity (Woessner 2000). In Catingueiro, the flow system of the shallow groundwater aquifer was influenced by the meander. The flow direction of the aquifer and the meander follow the flow direction of the Mogi River. In Pedra, the groundwater in the shallow aquifer flowed towards a pond to the west of the Mogi River. In the deeper portion of the aquifer, the hydraulic gradient changed significantly during the less rainy period (October 2012), but the flow was toward the river. According to the conceptual models of Vidon and Hill (2004), the study areas, mainly Pedra, characterize a large hillslope. The hydraulic gradient does not change significantly during the year, but the flow direction changes near the banks of the pond.

The study areas have low groundwater residence time and thus the waters had low mineralization, acidic pH, and experienced oxidizing environment. The high values of hydraulic conductivity favor these physico-chemical characteristics, which in turn was influenced by the rainwater recharge, as well as by the the existing water bodies. In the plains, when the paleo-channels were in contact with active meanders, they might have been the preferential flow paths of shallow groundwater aquifers (Sophocleous 2002). In the deeper portions of the aquifer, there was a greater mineralization of water and a pH increase that turned the environment less oxidant, indicating a greater residence time.

The area of Catingueiro showed a porous aquifer of low thickness and shallow water table. The groundwaters had low mineralization, acid pH and Eh oxidant, and mixed bicarbonate type of water. As the porous aquifer in Pedra, was thicker, it was possible to analyze the behavior of two levels of the aquifer. The groundwater level was deeper (greater than 7 m, approximately 13 m), and in the deeper portion (above 7.91 m and average depth of 13 m) the waters were calcium-bicarbonate type, and greatly mineralized.

5.2 Current Pollen Rain

The data integration provided by the analysis of current pollen rain of the three studied areas allows some considerations: In the EEcMG pollen record, 23 types were studied with a total number of 2,083 pollen grains and spores. 158 could not be identified, and remain undetermined. The taxa that most contributed to this record were Poaceae, Cyperaceae, *Euphorbia*, Araliaceae, Malpiguiaceae Arecaceae, *Pinus* and spores (described in the percentage pollen diagram, Fig. 18). Thus, the forest elements (arboreal plants) are dominant compared to the herbaceous.

5.3 Climatic Analysis of Leaf Morphologies

5.3.1 Sample Size and Morphotypes Identified

Samples collected in levees contain 293–771 specimens or between 16 and 31 morphotypes. Samples collected in the inner banks of meander bends fluctuate between 399 and 649 specimens or 27 morphotypes (Table 7). The values of specimens and morphotypes for Campininha Farm are similar to those reported for other rain forests. Burnham (1989) found a variation of 176–994 specimens or 6–24 morphotypes in paratropical forests of southern Mexico. Later, Burnham (1994) reported between 236 and 671 specimens related to 20–30 species, and finally, Ricardi-Branco et al. (2009) reported 585–673 specimens associated with 14–15 species for coastal riparian florests in the State of Sao Paulo, Brazil. All these studies provide only one reference, even using different sampling methods. According to Burnham (1989), a large number of species is determined among the first 350–400 specimens of leaves. Thenceforth, the occurrence of more species decreases with only two new plant species for every 100 leaves. In the same work, the author showed that leaf assemblages corresponding to the levees are the richest in species. Data generated by the present study reinforces the conclusion drawn by Burnham (1989).

5.3.2 Estimation of Mean Annual Temperature (MAT)

For all sample sets, the temperature estimates based on LMA equations show an average error of 1.6 and 3.1 °C (Table 2), which is below the error associated with LMA (generally it is ±5 °C Peppe et al. 2011; Royer 2012). The errors obtained by Eqs. 4 and 5, both for levee assemblages and for inner banks of meander bends, lie between 1.6 and 2.1 °C (Table 7), very close to the error presented by the regional equations calculated with LMA ±2 °C (Royer 2012). Equations 4 and 5 present an average error of 0.8 and 0.7 °C, as evaluated in different types of plants, such as the Cerrado and Atlantic Forest biomes, and the transition between them (Fanton 2013) below the value the obtained in this study. Equation 5 is found to be the most accurate for calculating temperature (Table 7 and Fig. 25). Its accurateness has also been testified by other authors (Hinojosa et al. 2011; Fanton 2013).

It is worth noting that in the study area, the sample sets from levees and associated with Cerrado reflected the best temperature estimates when equations based on LMA were used. Burnham et al. (2001) attributed the fact to the leaves origin, since woody plants that live near water courses and lakes tend to underestimate the MAT between 2.5 and 5 °C. On the contrary, the woods with closed canopy provide very accurate estimates.

The temperature values showed an error range of 1.8 and 3.9 °C during the application of CLAMP data for sheet sets of leaves collected in both geomorphological positions and studied at EEMG. This error is lesser than the associated error, and similar to LMA (±5 °C; Royer 2012), and closer to the error shown in rain forests (±3 °C; Burnham et al. 2005). Though the recent data sets may not be considered relevant to the estimation of climate variables of rain forests or places in the southern hemisphere, more data are required to calibrate the equations in tropical and subtropical regions (Steart et al. 2010; Yang et al. 2011). The Gridded data set is the one that best indicates the temperature for samples collected in levee, than those collected in the inner banks of meander bends. The average error was between 1.8 and 2.4 °C, which is close to the average error of Gridded data set, ±2.1 °C (http://clamp.ibcas.ac.cn/Clampset2.html). While estimating with the CLAMP database, it was observed that samples of levees were those that best indicated the temperature.

5.3.3 Estimation of Mean Annual Precipitation (MAP)

MAP values obtained with the LAA (Eqs. 6 and 7) showed an average error between −66 and −20 mm (Table 7). The values were lower than the errors associated with LAA, which ranged between ±500 mm (Wilf et al. 1998) and ±1,000 mm (Peppe et al. 2011). Samples collected in the inner banks of meander bends better showed rainfall predictions based on LAA. They represented an error around −20 and −23 mm. On the other hand, for levee samples the error was of −49 and −66 mm, and thus, the Eq. 7 was the most accurate. This small precipitation difference between the two environments can be a consequence of the temperature, the type of associated soil or groundwater influences (Royer 2012). The MAP estimates using CLAMP data sets, consistently overestimate the value of precipitation for both sets of leaves collected in the levees than the leaves of the inner banks of meander bends (Table 7). Wilf et al. (1998) previously observed this trend. In this sense, authors such as Burnham et al. (2005) reported that MAP in rainforests may be underestimated with errors greater than 400 mm.

5.4 Comparison with Other Studies

For the first time, integrated analyses of hydrogeological parameters and their influence on the vegetation distribution in riparian areas were studied. The influence was assessed based on the characteristics observed in current pollen rain, and the leaf assemblages of the middle course of the Mogi-Guaçú River. However, in Brazil, Ricardi-Branco et al. (2011) associated the characteristics of subaquatic environments with the preservation of underwater layers of thick leaves in the Itanhaém River estuary. Ricardi-Branco et al. (2011) showed that these assemblies have para-utochthonous origin and reflect the vegetation of its surroundings which were preserved due the oxidizing and acidic environment of the shallow aquifer. Pereira et al. (2013) studied the anthropogenic influences and the preservation of the pollen record in a tropical urban pond. They concluded that the layers with higher content of toxic trace elements (Cu, Ni, Pb) are the ones that showed pollen preservation.

Minckley et al. (2011) studied the ecotone of grasslands from Sonoran and Chihuahuan Desert, Arizona (USA), in the paleoenvironmental records of in-channel wetland sedimentation. The sedimentological studies indicated a transition from strong to weak stream flow, possibly a seasonal flow (~7,000 cal BP). The change in the characteristics of sedimentation and the establishment of wetland vegetation indicates the importance of regional groundwater hydrology from the mid Holocene to Present day.

Thus, it can conveniently be stated that the studies addressing multidisciplinary interpretations of ecological parameters allow a better understanding of history and environmental evolution during the Holocene.

6 Conclusions

- The Cerrado reflects regions with tropical climate (hot and seasonal). The average temperature is 20.5–22.5 °C with few changes and high average annual rainfall, but concentrated in certain months of the year.
- At the levee, the forest reflects the temperature accurately, but not the rainfall. This phenomenon explains the influence of groundwater, type of soil, and temperature on rainfall records. The current pollen rain also shows this trend, since the dominant pollens belong to arboreal plants, and the herbaceous are less represented in pollen sets. The trees have root systems that are better developed during the drier seasons and can reach the groundwater.
- In the inner bank of meander bend, the assemblages of leaves better reflect rainfall than temperature, since the parameter can change with the proximity of water bodies (groundwater discharge and proximity to the river or meander).
- The subsurface waters promote the preservation of organic matter. Owing to the favourable physical environment and water chemistry, plant biomass is preserved.
- These conclusions helped understanding the processes that influence the environmental signatures of fossil records.

Acknowledgments The authors acknowledge the grant 2010/20379-6, from São Paulo Research Foundation (FAPESP), and the collaboration of the Forestry Institute of Sao Paulo State, National Council of Scientific and Technology Development (CNPq), and the Coordination of Personal Improvement for Graduation (CAPES) for the grants given to students and the researchers.

Appendix

Results to chemical analyses (ICP-MS) at Catingueiro. <LOD – lesser than the limit of detection.

References

Aguiar LMS, Camargo AJA (2004) Cerrado: ecologia e caracterização. Embrapa cerrados, Brasilia, Planaltina-DF/

Barberi M, Salgado-Labouriau ML, Suguio K, Martin L, Turq B, Salgado-Labouriau ML, Suguio K (2000) Paleovegetation and Paleoclimate of " Vereda de Águas Emendadas", DF, Central do Brazil. J S Am Earth Sci 13:241–254

Burk D, Uhl D, Walter H (2005) Some aspects of the actuotaphonomy of leaves in stagnant ponds with implications for the formation of fossil leaf deposits—Preliminary results: Neues Neues Jahrbuch fur Geologie und Paleontologie. Monatshefte 12:705–728

Burnham RJ (1989) Relationships between standing vegetation and leaf litter in a paratropical forest: implications for paleobotany. Rev Palaeobot Palynol 58:5–32. doi:10.1016/0034-6667 (89)90054-7

Burnham RJ (1990) Paleobotanical implications of drifted seeds and fruits from modern mangrove litter Twin Cays, Belize. Palaios 5:364–370

Burnham RJ (1994) Patterns in tropical leaf litter and implications for angiosperm paleobotany. Rev Palaeobot Palynol 81:99–113. doi: http://dx.doi.org/10.1016/0034-6667 (94) 90129-5

Burnham RJ, Ellis B, Johnson KR (2005) Modern tropical forest taphonomy: does high biodiversity affect paleoclimatic interpretations? Palaios 20:439–451

Burnham RJ, Pitman NCA, Johnson KR, Wilf P (2001) Habitat-related error estimating temperatures from leaf margins in a humid tropical forest. Am J Bot 88:1096–1102. doi:10. 2307/2657093

Burt TP, Bates PD, Stewart MD, Claxton AJ, Anderson MG, Price DA (2002a) Water table fluctuations within the floodplain of the River Severn, England. J Hydrol 262:1–20

Burt TP, Pinay G, Matheson FE, Haycock NE, Butturini A, Clement JC, Danielescu S, Dowrick DJ, Hefting MM, Hillbricht-Ilkowaska A, Maitre V (2002b) Water table fluctuations in the riparian zone: comparative results from a pan-European experiment. J Hydrol 265:129–148

CBH-Mogi—Comitê da Bacia Hidrográfica do Rio Mogi-Guaçú (1999) Diagnóstico da Bacia Hidrográfica do Rio Mogi-Guaçú—"Relatório Zero", p 252

Chen X (2007) Hydrologic connections of a stream-aquifer-vegetation zone in south-central Platte River valley, Nebraska. J Hydrol 333:554–568

Denver Museum of Nature and Science (2011) Guide to morphotyping fossil floras. http://www. paleobotanyproject.org/morphotyping.aspx. Accessed 20 January 2011

Eiten G (1963) Habitat flora of Fazenda Campininha, São Paulo, Brazil. In: Ferri MG (Coord.). Simpósio sobre o Cerrado. Edgard Blucher e EDUSP, São Paulo, pp 157–202

Ellis B, Daly DC, Hickey LJ, Johnson KR, Mitchell JD, Wilf P, Wing SL (2009) Manual of leaf architecture. Cornell University Press, Ithaca, New York

Faegri K, Iversen J (1989) Textbook of pollen analysis. John Wiley & Sons, LTD, Chichester

Fanton JCM (2013) Reconstruindo as florestas tropicais úmidas do Eoceno-Oligoceno do sudeste do Brasil (Bacias de Fonseca e Gandarela, Minas Gerais) com folhas de Fabaceae, Myrtaceae e outras angiospermas: Origens da Mata Atlântica. Dissertation, Universidade Estadual de Campinas

Gastaldo RA (1994) The genesis and sedimentation of phytoclasts with examples from coastal environments. In: Traverse A (ed) Sedimentation of organic particles. Cambridge University Press, Cambridge, UK, pp 103–127

Gastaldo RA (2001) Plant Taphonomy. In: Briggs DEG, Crowther PR (eds) Palaeobiology II. Blackwell Scientific, Oxford, pp 314–317

Gastaldo RA (2004) The relationship between bedform and log orientation in a paleogene fluvial channel Weibelster Basin, Germany: implication for the use of coarse woody debris for paleocurrent analysis. Palaios 19:587–597

Gastaldo RA, Douglass DP, MacCarroll SM (1987) Origin, characteristics and provenance of plant macrodetritus in a holocene crevasse splay, mobile delta, alabama. Palaios 2:229–240

Gouveia SEM, Pessenda LCR, Boulet R, Aravena R, Scheel-Ybert R (1999) Isótopos do carbono dos carvões e da matéria orgânica do solo em estudos de mudança de vegetação e clima no Quaternário e da taxa de formação de solos no Estado de São Paulo. Anais da Acadêmia Brasileira de Ciências 71:969–980. doi:10.1016/0895-9811(96)00007-7

Hinojosa LF, Pérez F, Gaxiola A, Sandoval I (2011) Historical and phylogenetic constraints on the incidence of entire leaf margins: insights from a new South American model. Global Ecol Biogeogr 20:380–390. doi:10.1111/j.1466-8238.2010.00595.x

Hvorslev MJ (1951) Time lag and soil permeability in groundwater observations. Waterway experiment station. US Army. Bull 36:1–50

Jacobs BF, Herendeen PS (2004) Eocene dry climated and woodland vegetation in tropical Africa reconstructed from fossil leaves from northern Tanzania. Palaeogeogr Palaeoclimatol Palaeoecol 213:115–123. doi:10.1016/j.palaeo.2004.07.007

Kowalski EA (2002) Mean annual temperature estimation based on leaf morphology: a test from tropical South America. Palaeogeogr Palaeoclimatol Palaeoecol 188:141–165. doi:10.1016/j.gloplacha.2007.07.001

Lautz LK, Siegel DI (2006) Modeling surface and groundwater mixing in the hyporheic zone using MODFLOW and MT3D. Adv Water Resour 29:1618–1633

Ledru MP, Campello RC, Landim D, Dominguez JM, Martin L, Mourguiaat P, Sifeddine A, Turcq B (2001) Late-Glacial cooling in Amazonia inferred from pollen at Lagoa do Caçó, Northern Brazil. Quatern Res 55:47–56

Minckley TA, Brunelle A, Blissett S (2011) Holocene sedimentary and environmental history of an in-channel wetland along the ecotone of the sonora and chihuahua desert grasslands. Quatern Int 235:40–47. doi:10.1016/j.quaint.2010.06.031

Montovani S (1983) Composição e similiaridade florística, fenologia e spectro biológico do Cerrado na Reserva Biológica de Mogi-Guaçú, Estado de São Paulo. Campinas. 147 p. Dissertação (Mestrado em Biologia—Ecologia)—Instituto de Biologia, Universidade Estadual de Campinas, Campinas

Passos MJ (1998) Estrutura da vegetação arbórea e regeneração natural em remanescentes de Mata Ciliar do Rio Mogi Guaçú-SP. Piracicaba. 68 p. Dissertação (Mestrado em Ciências Florestais). Escola Superior de Agricultura “Luiz de Queiroz”, Universidade de São Paulo, Piracicaba

Peppe DJ, Royer DL, Cariglino B et al (2011) Sensitivity of leaf size and shape to climate: global patterns and paleoclimatic applications. New Phytol 190:724–739. doi:10.1111/j.1469-8137.2010.03615.x

Pereira SY, de Souza MM, Ricardi-Branco F, Pereira PRB, Cardinale F, Zazera R (2013) Trace elements and palynomorphs in the core sediments of a tropical urban pond. In: Yuanzhi Z, Pallav R (eds) Climate change and regional/local responses, vol. 1. InTech, Croatia, pp 225–233

Perrota MM, Salvador ED, Lopes RC, D’ Agostino LZ, Peruffo N, Gomes SD, Sachs LLB, Meira VT, Garcia MGM, Lacerda Filho JV (2005) Mapa Geológico do Estado de São Paulo, escala 1:750.000. Programa Geologia do Brasil—PGB, CPRM, São Paulo

Petri S, Fúlfaro VJ (1981) Fanerozóico. In: Queiroz TA (ed) Geologia do Brasil. Editora da USP, São Paulo

Pinto MM, Giudice Neto JD, Batista EA, Toledo Filho DV, Mota IS (1997) Vegetação nativa das unidades de conservação e produção de Mogi-Guaçú. In: Coletânea de trabalhos do Congresso

Brasileiro de Unidades de Conservação. Secretária do Meio Ambiente do Estado de São Paulo, São Paulo
Rassam DW, Fellows CS, De Hayr R, Hunter H, Bloesch P (2006) The hydrology of riparian buffer zones: two cases studies in an ephemeral and perennial stream. J Hydrol 325:306–324
Raunkiaer C (1934) The Life forms of plants and statistical plant geography. Clarendon Press, Oxford
Ribeiro JF, Walter BMT (2008) As principais fitofionomias do Bioma Cerrado. In: Sano SM, De Almeida SP, Ribeiro JF Cerrado: Ecologia e Flora. Embrapa Cerrados.- Brasília, DF: Embrapa Informação tecnológica, vol 1, pp 153–212
Ricardi-Branco F, Branco FC, Garcia RF, Faria RS, Pereira SY, Portugal R, Pessenda LC, Pereira PRB (2009) Features of plant accumulations along the Itanhaém River, on the southern coast of the Brazilian state of São Paulo. Palaios 24:416–424. doi:10.2110/palo.2008.p08-079r
Ricardi-Branco F, Pereira SY, Cardinale F, Pereira PBR (2011) Accumulation of bio debris and its relation with the underwater environment in the estuary of Itanhaem river, Sao Paulo State. In: Imran AD, Mithas AD (eds) Earth and environmental sciences/Book2, 1st edn, vol 2. In Tech Publisher, Rijeka, pp 565–590
Royer DL (2012) Climate reconstruction from leaf size and shape: new developments and challenges. In: Ivany LC, Huber BT (eds) Reconstructing Earth's Deep-Time Climate. The State of the Art in 2012, Paleontological society short course, November 3, 2012. (Paleontological Society Papers) 18:195–212
Rushton K (2007) Representation in regional models of saturated river-aquifer interaction for gaining/ losing rivers. J Hydrol 334:262–281
Salgado-Labouriau ML (1997) Late quaternary paleoclimate in the savannas of South América. J Quat Sci 12:371–379
Schilling KE (2007) Water table fluctuations under three riparian land covers, Iowa (USA). Hydrol Process 21:2415–2424
Spavorek G, Van Lier QDJ, Dourado Neto D (2007) Computer assisted koeppen climate classification: a case study for Brazil. Int J Climatol 27:257–266
Sophocleous M (2002) Interactions between groundwater and surface water: the state of the science. Hydrogeol J 10:52–67. doi:10.1007/s10040-001-0170-8
Souza MM, Ricardi-Branco F, Jasper A, Pessenda LCR (2013) Evolução Paleoambiental Holocênica no Nordeste do Estado de São Paulo. Revista Brasileira de Paleontologia. 16:297–308. doi:10.4072/rbp.2013.2.10
Steart DC, Spicer RA, Bamford MK (2010) Is southern Africa different? An investigation of the relationship between leaf physiognomy and climate in southern African mesic vegetation. Rev Palaeobot Palynol 162:607–620. doi:10.1016/j.revpalbo.2010.08.002
Vidon PF, Hill AR (2004) Landscape controls hydrology of stream riparian zones. J Hydrol 292:210–228
Webb LJ (1959) A physiognomic classification of Australian rain forest. J Ecol 47:551–570
Wilf P (1997) When are leaves good thermometers? A new case for leaf margin analysis. Paleobiology 23:373–390
Wilf P, Wing SL, Greenwood DR, Greenwood CL (1998) Using fossil leaves as paleoprecipitation indicators: an Eocene example. Geology 26:203–206
Woessner W (2000) Stream and fluvial plain ground water interactions: rescaling hydrogeologic thought. Groundwater 38:423–429
Wolfe JA (1993) A method of obtaining climatic parameters from leaf assemblages. US Geol Surv Bull 2040:1–71
Yang J, Spicer RA, Spicer TEV, Li CS (2011) 'CLAMP Online': a new web-based palaeoclimate tool and its application to the terrestrial Paleogene and Neogene of North America. Palaeobio Palaeoenv 91:163–183. doi:10.1007/s12549-011-0056-2
Zancopé MHC, Perez-Filho A, Carpi S Jr (2009) Anomalias no Perfil Longitudinal e Migração dos meandros do Rio Mogi Guaçu. Revista Brasileira de Geomorfologia 10:31–42

Diatom Indices and Water Quality Index of the Cauvery River, India: Implications on the Suitability of Bio-Indicators for Environmental Impact Assessment

R. Venkatachalapathy and P. Karthikeyan

Abstract Physico-chemical properties of water are routinely utilized for understanding environmental quality. Diatoms are used as bio-indicators to assess the water quality of surface waters. Independent assessments of environmental quality of the surface waters from the Cauvery River were made and the resultant quality-indicators were assessed in the light of national river water quality standards. Occurrences of sixty (60) species belonging to 21 genera are recorded from the Cauvery River. Water samples were analyzed for various physico-chemical parameters, viz., pH, Electrical Conductivity (μS/cm), Dissolved solid (mg/l), Biochemical Oxygen Demand (BOD) (mg/l), Calcium (Ca) (mg/l), Magnesium (Mg) (mg/l), Sodium (Na) (mg/l), Potassium (K) (mg/l), Chloride (Cl) (mg/l), Bicarbonate (HCO_3) (mg/l) and Sulphate (SO_4) (mg/l). These physico-chemical parameters formed the basis of computing the Water Quality Index. The results of the present study on diatom assemblages in the Cauvery River revealed moderate pollution at Siluvampalayam and Koneripatti and high levels of pollution at Peramachipalayam, Kottampatty, Sanyasipatti and Bhavani. The Water Quality Index revealed pristine nature of the Cauvery River water in upstream regions that became unsuitable for human consumption downstream of Bhavani Town. The values of DO and BOD levels indicated absence of major organic pollution. Comparative validation of the physico-chemistry and bio-indicators suggested sensitive nature of the Diatom indices to environmental variables and thus the diatom indices can be a reliable tool for environmental impact assessment.

Keywords Cauvery river · Diatom indices · Diatom assemblages · Water quality index

---

R. Venkatachalapathy · P. Karthikeyan (✉)
Department of Geology, Periyar University, Salem 636 011, India
e-mail: pkarthikeyangold@gmail.com

Mu. Ramkumar et al. (eds.), *Environmental Management of River Basin Ecosystems*,
Springer Earth System Sciences, DOI 10.1007/978-3-319-13425-3_31

1 Introduction

Rivers form a lifeline associated with people community. Nowadays, rivers are generally amongst the most susceptible water bodies to pollution on account of unprecedented progress. The standard of water is actually followed by its physical, chemical and biological features. Monitoring of river water quality will help in preventing outbreak of diseases which may arise due to contamination of water due to the merger of sewage from domestic and industry in rivers. Surface water environments are vulnerable to contamination because of their easy accessibility for disposal of wastage from domestic and industry. Seasonal variations in precipitation, surface runoff, ground water flow and water interception and abstraction possess a sturdy effect on the river discharge and eventually for the focus of contaminants within river water (Vega et al. 1998; Shrestha and Kazama 2007). Anthropogenic effects (urban, industrial, and agricultural activities, increasing consumption of water resources) and also inorganic operations (precipitation inputs, erosion, weathering connected with crustal materials) degrade surface waters and impair their use for drinking, industrial, agricultural, recreational, or other purposes.

Diatoms are the main producers in rivers. Diatoms are present in all the aquatic environments. Diatoms can be easily collected and preserved. Due to their rapid response to environmental changes, deterioration of water quality especially from impacts such as nutrient enrichment, acidification and metal contamination diatoms have been used widely for biomonitoring of aquatic ecosystems (Kelly and Whitton 1995; Stoermer and Smol 1999). The shape, size and pattern of silica frustules form the basis for classification and identification of diatom taxa. According to Kelly et al. (2008), diatoms are one of the basic components of river bio-monitoring and assessment of ecological status of rivers. These authors also established many diatom indices for water quality assessment of rivers and lakes. Recent studies have shown that the diatom based indices vary in their capacity to ionic composition and organic pollution in rivers (Gomez and Licursi 2001; Taylor et al. 2007). Diatoms are sensitive to environmental changes and can be used as important tool in monitoring environmental conditions of surface waters (Venkatachalapathy and Karthikeyan 2012, 2014).

Water Quality parameters can be broadly classified into three different types: physical, chemical and biological. Physical parameters consist of temperature, turbidity, color and odor; chemical parameters include things like pH, Dissolved Oxygen (DO), Biological Oxygen Demand (BOD) and Chemical Oxygen Demand (COD), Nitrogen and Phosphorus; and Biological parameters consist of occurrences of Fecal Coli-form and other pathogens. The Water Quality Index (WQI) synthesizes complex information about many water quality parameters. The Water Quality Index (WQI) number ranges from 0 to 100. This shows the water quality in terms of a number where a greater number signifies the better water quality. Diatoms can be used for identification of the river sites affected/influenced by urban pollution

(Venkatachalapathy and Karthikeyan 2013a, b, c and d). In the present study an attempt has been made to evaluate the applicability of diatom based indices for water quality assessment in parts of the Cauvery River, south India.

2 Study Area

The Cauvery River is one of the major rivers of South India. It originates in the Western Ghats and flows eastward, passes through the states of Karnataka, Tamil Nadu, Kerala and the Union Territory of Pondicherry before debauching into the Bay of Bengal. The total length of the river is about 800 km. The Cauvery basin extends over an area of 81,155 km^2. The study area is confined to the main channel of the Cauvery River from Mettur Dam to Bhavani town. It lies between 77° 40′E to 77° 42′E longitudes and 11° 25′N to 11° 27′N latitudes (Fig. 1). Within this stretch of the river, there are more than 100 small, medium and large industries including

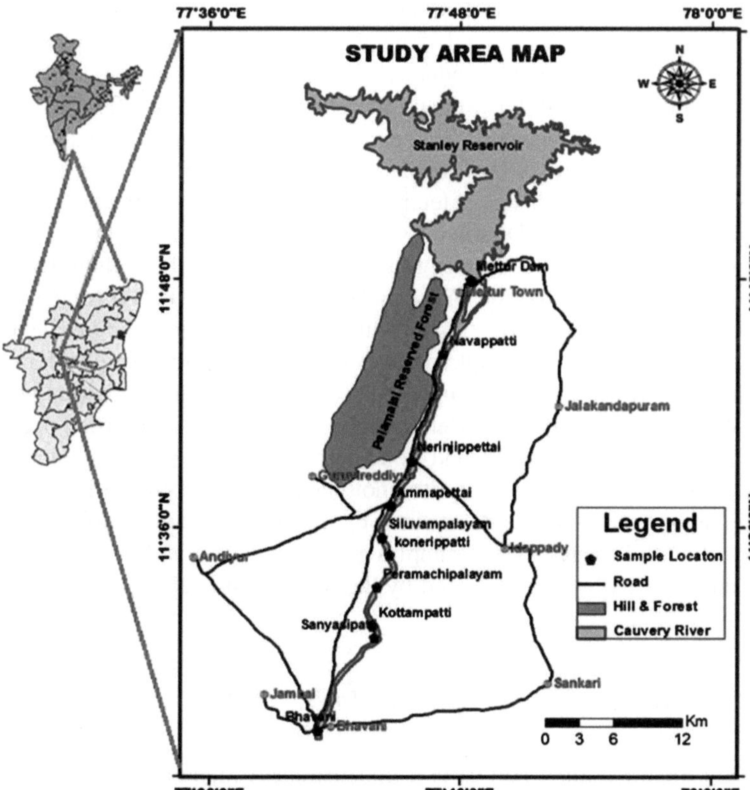

Fig. 1 Map showing the Cauvery River and sample locations

thermal power plant, metal processing units (aluminium, magnesium and iron), dying factories, and foundries. A common feature associated with these industrial and associated urban centers is that they are all located along or in the vicinity of Cauvery River channel for easy access to fresh water that is needed for their processing and easy disposal of effluents without any treatment.

3 Materials and Methods

3.1 Sample Collection and Analytical Procedures

Water samples were collected from 10 locations (Mettur Dam, Navappatti, Nerinjipettai, Ammapettai, Siluvampalayam, Koneripatti, Peramachipalayam, Kottampatty, Sanyasipatti and Bhavani) situated in the Cauvery River channel in a stretch extending from Mettur Dam to Bhavani Town (Fig. 1) during pre and post monsoon seasons in the year 2013. The samples were kept in pre-cleaned (cleaned first with metal-free soap, rinsed repeatedly with distilled water, soaked in 10 % nitric acid for 24 h and rinsed with ultrapure water; washed again at the sampling site with the sampled water) 2 L polyethylene plastic bottles. In the field, immediately on collection, pH, EC (electrical conductivity), hardness, temperature, turbidity, biological oxygen demand (BOD) were measured with the help of a field kit. Then, the samples were stored in a ice-box, maintained at 4 °C until transport to the laboratory for further analyses following standard procedures (APHA 2005). Seventeen parameters namely, pH, Change in temp, Turbidity, Electrical Conductivity (μS/cm), Dissolved Oxygen (mg/l) Dissolved solid (mg/l), Biochemical Oxygen Demand (BOD) (mg/l), Calcium (Ca) (mg/l), Magnesium (Mg) (mg/l), Sodium (Na) (mg/l), Potassium (K) (mg/l), Chloride (Cl) (mg/l), Bicarbonate (HCO_3) (mg/l), Sulphate (SO_4) (mg/l), Total Phosphorus (mg/l), Nitrate Nitrogen (mg/l) and Fecal Coliforms were determined for all the samples collected. While chemical oxygen demand was determined on the same day of sample collection, all other analyses were performed within 48 h of sample collection. Determination of Chloride was carried out by using silver nitrate ($AgNO_3$) titration and potassium chromate (K_2CrO_4) solution as an indicator. SO_4 was determined spectrophotometrically with the barium sulfate turbidity technique. NO_3-N was determined by means of phenol disulfonic acid calorimetry. The acid-treated water samples were analyzed for Ca, Na, K by flame photometry, and Mg was determined by the flame atomic absorption spectrometer.

3.2 Calculation of Water Quality Index (WQI)

The Water Quality Index (WQI) is a dimensionless quantity that brings together many water-quality ingredients by normalizing valuations to subjective rating curves (Miller et al. 1986). Factors being a part of Water Quality Index (WQI)

Table 1 Model for water quality index calculator (excel sheet)

Parameter	Test result	Units	Q-value	Weighting factor	Weighting factor	Subtotal
pH		pH units	NM	0.12	NM	NM
Change in temp		°C	NM	0.11	NM	NM
DO		% saturation	NM	0.18	NM	NM
BOD		mg/L	NM	0.12	NM	NM
Turbidity		NTU	NM	0.09	NM	NM
Total phosphorus		mg/L P	NM	0.11	NM	NM
Nitrate nitrogen		mg/L NO_3-N	NM	0.10	NM	NM
Fecal coliforms[a]		CFU/100 mL	NM	0.17	NM	NM
				Totals	0.00	0.00
				Water quality index		NM
				Water quality rating		NM

[a] Only use one microorganism, not fecal coliforms

design might vary based on the specified water utilizations in addition to local inclinations. A few of these factors include DO, pH, BOD, COD, total coliforms bacteria, temperature and nutrients (nitrogen and phosphorus), and many others. Most of these parameters take place in several ranges and stated in different units. The Water Quality Index (WQI) takes the difficult methodical information on these kinds of variables and synthesizes into a single number. Ranges of numbers are categorized into different quality levels (Table 1). Various researchers used these concepts and models (for example, Bolton et al. 1978; Bhargava 1983; House 1989; Mitchell and Stapp 1996; Pesce and Wunderlin 2000; Cude 2001; Liou et al. 2004; Said et al. 2004; Nasiri et al. 2007). In the present study the Water Quality Index (WQI) parameters such as pH, Change in temp (°C), DO, BOD, Turbidity, Total Phosphorus, Nitrate Nitrogen, Fecal Coliforms involving river water quality from the Cauvery River were analyzed based on the model (Table 1) and categorized according to the quality levels listed in Table 2, and interpreted as per the usage domain as per the scheme listed in the Table 3. Water quality index of water samples collected during pre monsoon and post monsoon are presented in Tables 4 and 5 respectively.

Table 2 Water quality classification based on WQI value

Water quality	Value intervals (%)
Excellent	95–100
Good	75–94
Moderate	50–74
Marginal	25–49
Poor	0–24

Table 3 Interpretation scheme for the values of water quality index, on usage domains

Use score (%)	PWS (Potable water supply)	FAWL (Fish and wildlife)	Industry	Recreation
100	No treatment required	–	Selected uses without treatment	–
90	–	Suitable for all species of fish and wildlife	–	Suitable for all recreation activities
80	Minor purification	–	Minor purification if high quality water is required	–
70	–	Doubtful for game fish. Supports populations of coarse fish	No treatment for most uses	Doubtful for direct contact sports
60	Conventional treatment	–	–	–
50	Advanced treatment	Reasonable coarse fisheries	Advanced treatment required for most uses	Indirect and noncontact activities only
40	Doubtful use	Tolerant species only	–	–
30	–	–	Only industries needing poor quality water	Non-contact uses only
20	Unacceptable	Unacceptable	Unacceptable	Unacceptable
10	–	–	–	–

Source after House and Ellis 1987

3.3 Diatom Sample Collection and Preparation

Macrophytes samples were collected from the same ten locations (from where water samples were collected) in the Cauvery River for water quality assessment using diatom indices. The samples were collected in polythene bottles from all obtainable habitats such as plants (Epiphytic) and stones (Epilithic) following methods of suggested by Taylor et al. (2007) and Karthick et al. (2010). Diatoms are sampled by brushing stones with a tooth brush, following recommendations of Kelly et al. (1998). At least five, pebbles to cobble (5–15 cm) sized stones were collected from the river channel bed. They were also brushed and the diatom suspension was put in a small plastic bottle. Epilithic and epipelic diatoms were sampled at five sampling stations during May 2013. Epiphytic samples were taken by brushing the undersurfaces and petioles of at least 20 plant leaves and roots. All these samples were preserved in formaldehyde (4 %). For Polarizing microscopic analysis, a 10 ml epiphytic and epilithic subsamples were extracted and cleaned using 30 % H_2O_2 and concentrated HNO_3 (Stoermer et al. 1995). Identification of diatoms was carried out

Table 4 Water quality index calculator for pre-monsoon season

Location	pH	Temp (C)	DO	BOD (mg/L)	Turbidity (NTU)	Total P (mg/L)	Nitrate-N (mg/L)	Fecal coliforms (NTU)	Q value	Weight factor	WQ index	Rating
Mettur dam	7.9	20	3.6	1.08	3.62	0.02	1.3	110	61.5	1	56.35	Medium
Navappatti	8.3	24	7.2	0.78	3.08	0.02	0.4	170	63.25	1	57.28	Medium
Nerinjipettai	7.8	22	8.2	1.06	5.36	0.02	0.3	260	63.88	1	57.65	Medium
Ammapettai	8.1	24	5.1	1.08	6.57	0.04	0.6	170	61.25	1	55.73	Medium
Siluvampalayam	8.6	23	7.2	5.21	8.25	0.0055	1.1	110	54.5	1	50.7	Medium
Koneripatti	8.4	23	7.8	4.21	8.94	0.018	0.6	130	57.75	1	52.72	Medium
Peramachipalayam	8.3	21	8.1	7.32	13	0.076	1.1	110	51.25	1	47.15	Bad
Kottampatty	8.6	23	7.2	5.21	8.25	0.05	1.1	110	53.5	1	49.2	Bad
Sanyasipatti	8.7	25	9.2	3.2	7.9	0.4	0.18	280	50.25	1	45.45	Bad
Bhavani	8.7	25	11	3.7	16.8	0.03	0.5	851	53.5	1	48.37	Bad

Table 5 Water quality index calculator for post-monsoon season

Location	pH	Temp (C)	DO	BOD (mg/L)	Turbidity (NTU)	Total P (mg/L)	Nitrate –N (mg/L)	Fecal coliforms (NTU)	Q value	Weight factor	WQ index	Rating
Mettur dam	8.4	24.1	8.5	1.06	2.57	0.03	0.28	0170	63.12	1	57.16	Medium
Navappatti	8.1	25	6.3	1.05	9.31	0.03	0.23	0790	60.38	1	54.36	Medium
Nerinjipettai	7.9	25	11	1.57	5.68	0.06	0.2	3500	59.5	1	52.92	Medium
Ammapettai	8.6	26	8.7	1.57	4.51	0.78	0.22	1700	50.13	1	44.5	Bad
Siluvampalayam	8.2	26	6.6	5.21	9.25	0.59	0.27	2200	46.88	1	41.67	Bad
Koneripatti	8.1	25	14	12.2	29.8	0.02	0.48	1100	49.38	1	44.59	Bad
Peramachipalayam	8.7	26	7.9	7.8	09.1	0.31	0.21	2200	45.00	1	39.63	Bad
Kottampatty	8.7	25	8	7.8	12.4	0.03	0.58	1100	50.50	1	44.97	Bad
Sanyasipatti	8.7	25	11	3.7	16.8	0.03	0.5	0851	53.5	1	48.37	Bad
Bhavani	8.8	26	9.3	7.3	17.9	0.017	0.23	1017	50.88	1	45.38	Bad

using taxonomic guides (Gandhi 1957, 1959a, b, 1961, 1962, 1967, 1998; Karthick et al. 2008). The Diatom species were identified and photographed by using Euromex (Holland) polarizing microscope.

4 Results and Discussion

4.1 Diatom Distributions

A total of 60 diatom species belonging to 21 genera are recorded in the samples collected from the Cauvery River. The Diatom species recorded and identified in the present study are as follows: *Achnanthes brevipes, A. inflata, Achnanthidium binodis, A. minutissima, Amphora holsatica, A. ovalis, Anomoeoneis sphaerophora, Caloneis pulchra, C. silicula, Cocconeis placentula, Ctenophora pulchella, Cyclotella catenata, C. meneghiniana, Cymbella aspera, C. cymbiformis, C. lanceolata, C. tumida, C. tumidula, C. turgida, C. ventricosa, Eunotia curvata, E. fallax, E. pectinalis, Fragilaria intermedia, Gomphonema affine, G. clavatum, G. gracile, G. lanceolatum, G. parvulum, G. olivaceum, G. truncatum, G. undulatum, Mastogloia braunii, Melosira granulata, M. moniliformis, M. varians, Navicula mutica, N. radiosa, N. symmetrica, N. virudila, N. acicularis, N. linearis, N. microcephala, N. palea, N. pseudofonticola, N. recta, N. thermalis, N. sigma, Tabellaria flocculos, Pinnularia acrosphaeria, Pleurosira indica, Pleurosigma salinarum, Stauroneis anceps, Surirella linearis, S. robusta, S. splendida, S. tenera, Synedra rumpens* and *S. ulna*. Selected species are depicted in the Plates I, II and III. Several indices were also computed as per the formula listed in the Table 6.

Dominance is the total of a particular species with respect to different species where an ecological community predominates. It ranges from 0 (all taxa are equally present) to 1 (a taxon dominates the community completely). The Dominance analysis shows that the Nerinjipettai locality in the Cauvery River has 10 species with *Cyclotella meneghiniana* as dominant species (dominance: 0.14), while *Aulocosira granulata* (19.86 %) dominated at Koneripatti (dominance 0.11) Remaining sites showed dominance index value between 0.1 and 0.4 (Table 7). Evenness is a measure of biodiversity which numerically quantifies the equality of the community. *Sellaphora pupula* constitute more than 23.17 % of the total population accounted for low evenness in Koneripatti (Table 7).

Shannon diversity index (H) computed as per Eq. 2 (Table 6) takes into account the number of individuals as well as number of taxa. It varies from 0 for communities with only a single taxon to high values for communities with many taxa, each with few individuals. Low H was recorded in Nerinjipettai (2.03, *C. meneghiniana* representing 92 %) (Table 7). High Fisher's alpha diversity index (Eq. 6, Table 6) was noticed in Sanyasipatti (11.67), Koneripatti (9.55) and Ammapettai (8.28). Nerinjipettai (1.8) shows low index value (Table 7).

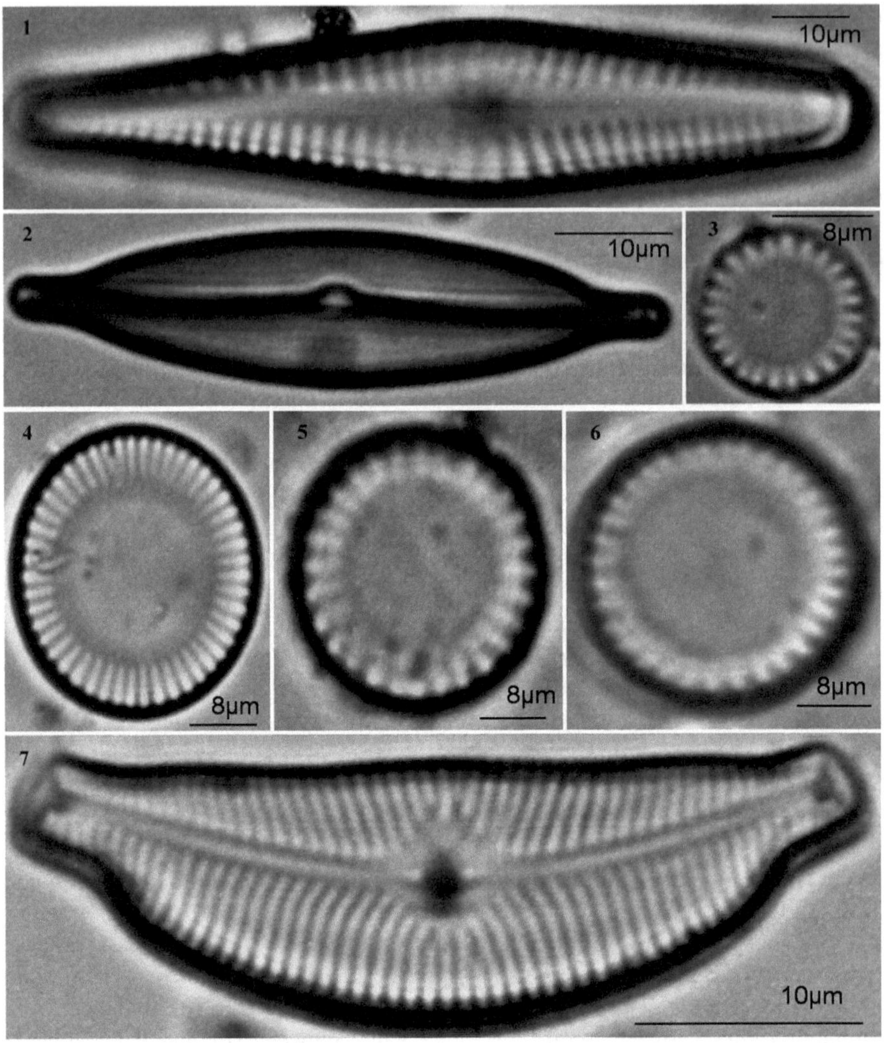

Plate I 1 *Gomphonema undulatum*, 2 *Stauroneis anceps* 3–6 *Cyclotella meneghiniana*, 7 *Cymbella tumida* (40X magnification)

Berger-Parker was calculated (Eq. 7 in Table 6; computed values are presented in the Table 7.) from the number of individuals in the dominant taxon relative to the total number of species. *Cyclotella meneghiniana* is the dominant species (with 92 % abundance) showing a high index value in Koneripatti. In Nerinjipettai *Aulocosira granulata, Cyclotella meneghiniana* and *Nitzschia obtuse* represent 23 % of the population from a macrophytes habitat. *Gomphonema parvulum* represents 32 % and *Nitzschia palea* (67 %) in Bhavani. *Nitzschia palea* (45 %) and *Navicula sp.* (32 %) characterize the Mettur Dam.

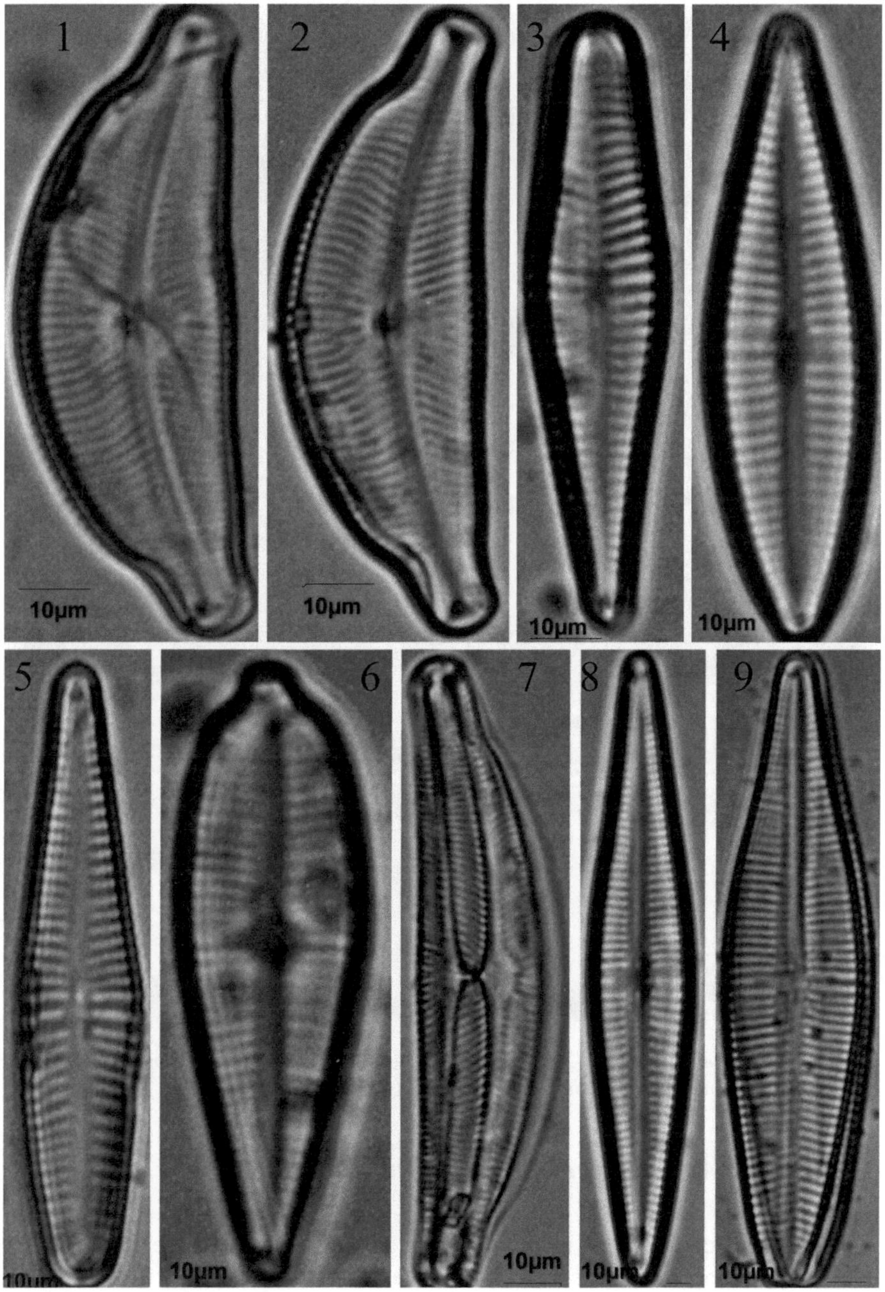

Plate II 1–2 & 7 *Cymbella tumida*, 3–5 *Gomphonema acuminatum*, 6 *G. affine*, 8 *G. auger*, 9 *G. parvulum* (40X magnification)

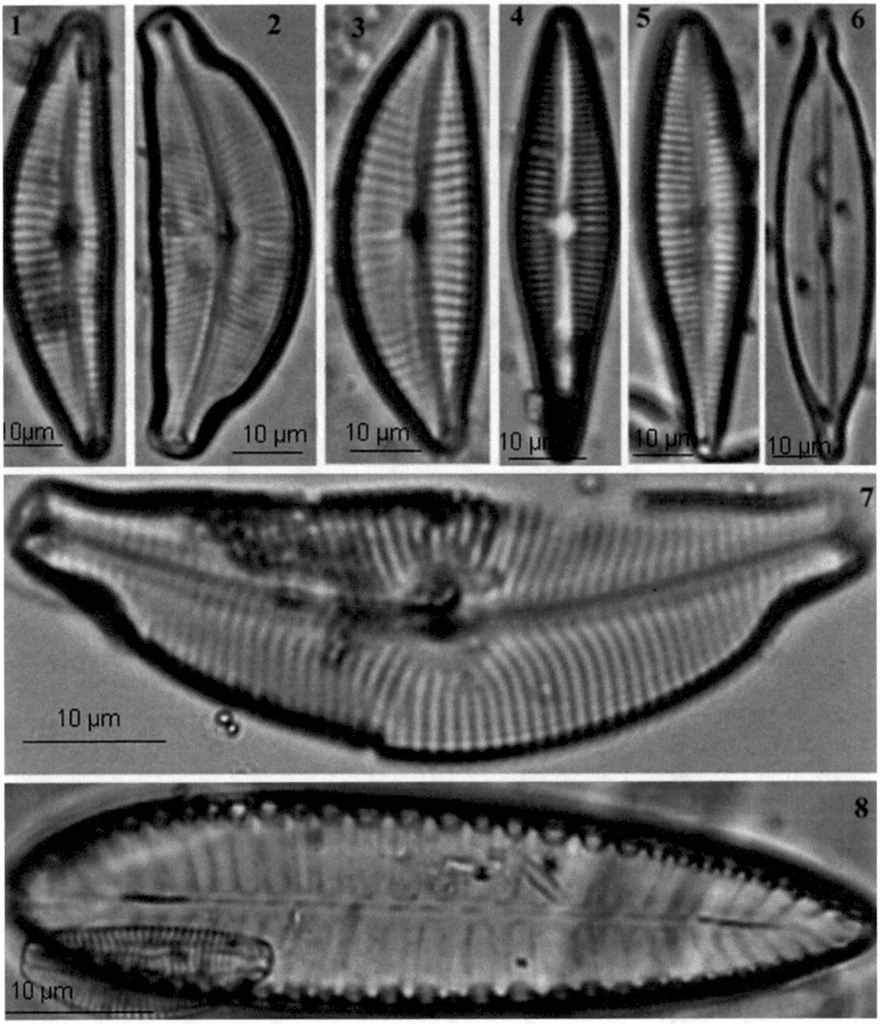

Plate III 1 *Cymbella minuta*, 2–3 & 7 *C. tumida*, 4 *Gomphonema lanceolatum*, 5 *G. olivaceum*, 6 *Stauroneis phoenicenteron*, 8 *Surirella splendida* (40X magnification)

4.2 Diatom Assemblages and Trophic Condition

Among 23 species present in the Cauvery River at Bhavani, the cosmopolitan, extreme pollution resistant species *Achnanthes minutissima* Kutz, *Achnanthidium Plonensis, Aulacoseira distans, Cymbella turgida* (Greg) Cleve, *Cymbella ventricosa* Kutz, *Fragilaria intermedia* Grun var. *robusta*, *Gomponema lanceolatum* Ehr, *Nitzschia sigma* (Kutz) W Smith, *Synedra rumpen, S. ulna* (Nitzsch) Ehr are the most abundant species highlighting the eutrophic status of water with higher

Table 6 Diversity parameters and indices (following Karthick, B. 2009)

Index	Equation	Remarks	References	Eq. No
Abundance	$\frac{\text{No.of Individuals of a Species X 100}}{\text{No.of Sampling Units}}$			1
Shannon Weiner's (H)	$-\sum_{I=1}^{S}$ pi In pi Pi: proportion of individuals of *i*th species	The value ranges between 1.5 and 3.5 and rarely surpasses 4.5	Ludwig and Reynolds (1988); Legendre and Legendre (1998)	2
Simpson's D	$D = \frac{\sum_{I=1}^{S} ni\ (ni-1)}{N(N-1)}$	The value varies from 0 to 1. A value of 0 indicates the presence of only one species, while 1 means that all species are equally represented.	Ludwig and Reynolds (1988))	3
Dominance	1-Simpson index $D = \text{sum}\left(\frac{ni}{n}\right)^2$ where ni is number of individuals of taxon i.	The occupancy of a species over an area. Ranges from 0 (all taxa are equally present) to 1 (one taxon dominates the Community completely)		4
Evenness	$H = -\sum_{I=1}^{S}$ pi In pi	The measure of biodiversity which quantifies how equal the community		5
Fisher's alpha	$S = a^* \text{In}\left(\frac{1+n}{a}\right)$ where S is number of taxa, n is number of individuals and *a* is the Fisher's alpha.	It is a mathematical model used to measure diversity		6
Berger-parker	$d = \frac{Nmax}{N}$ where N*max* is the number of individuals in the most abundant species and N is the total number of individuals in the sample.	The number of individuals in the dominant taxon relative to n, where n is the total number of species	Berger and Parker (1970)	7

electrolyte. *Cyclotella meneghiniana* a cosmopolitan species, resistant to extreme pollution with wide range of distribution including eutrophic, electrolyte rich water, accounts for more than 90 % of 10 species in Bhavani.

4.3 Diatom Indices

Diatom specific indices like Generic Diatom Index or GDI (Coste and Ayphassorho 1991), the Specific Pollution sensitivity Index (Indice de Polluosensibilite Specifique) or SPI (IPS) (Coste in Cemagref 1982), the Biological Diatom Index or BDI

Table 7 Diversity indices for Cauvery river

	No. of species	Dominance	Simpson	Shannon H	Evenness	Margalef	Equitability	Fisher	Berger-parker
Mettur dam	16	0.07	0.93	2.7	0.93	2.61	0.97	3.56	0.1
Navappatti	22	0.05	0.95	3.02	0.93	3.5	0.98	4.99	0.08
Nerinjipettai	8	0.14	0.86	2.03	0.95	1.4	0.97	1.81	0.19
Ammapettai	36	0.03	0.97	3.5	0.92	5.43	0.98	8.28	0.05
Siluvampalayam	21	0.05	0.95	3	0.95	3.39	0.98	4.85	0.09
Koneripatti	12	0.11	0.89	2.34	0.86	3.46	0.94	9.55	0.21
Peramachipalayam	15	0.07	0.93	2.64	0.93	3.41	0.97	6.35	0.08
Kottampatty	19	0.05	0.95	2.92	0.98	4.09	0.99	7.76	0.06
Sanyasipatti	51	0.02	0.98	3.88	0.95	7.34	0.99	11.67	0.04
Bhavani	23	0.05	0.95	3.08	0.94	3.64	0.98	5.22	0.08

(Lenoir and Coste, 1996), the Artois-Picardie Diatom Index or APDI (Prygiel et al.1996), Sladeceks index or SLA (Sladecek 1986), the Eutrophication/Pollution Index or EPI (DellUomo, 1996), Rotts Index or ROT (Rott, 1991), Leclercq and Maquets Index or LMI (Leclercq and Maquet 1987), the Commission of Economical Community Index or CEC (Descy and Coste 1991), Schiefele and Schreiners index or SHE (Schiefele and Schreiner 1991), the Trophic Diatom Index or TDI (Kelly and Whitton 1995) and the Watanabe index or WAT (Watanabe et al. 1986) were also computed and the resultant values of the studied sites are listed in the Table 8. All the diatom indices were calculated using the following Equation (where aj = abundance (proportion) of species j in sample, vj = indicator value and sj = pollution sensitivity of species j (after Zelinka and Marvan 1961) except for the CEC, SHE, TDI and WAT index. A maximum value of 20 for all these indices (except TDI—maximum value of 100), indicates pristine water.

$$index = \frac{\sum_{j}^{n} = 1\, a_j s_j v_j}{\sum_{j}^{n} = 1\, a_j v_j}$$

The performance of the indices depends on the values given to the constants ‘s’ and ‘v’ for each taxon. The values of the index ranges from 1 to an upper limit equal to the highest value of ‘s’. Each diatom species used in the calculation/equation was assigned two values; the first value reflects the tolerance or affinity of the diatom to a certain water quality (good or bad) while the second value indicates how strong (or weak) the relationship is and base on these abundance and weighted average were computed. These would indicate how many of the particular diatoms in the sample occur in relation to the total number counted. Table 9 provides the class limits of diatom indices. Based on these class limits, the scores listed in the Table 8, indicate an increasing level of pollution or eutrophication.

4.4 Physical-Chemical Characteristics of River Water

High pH induces the formation of tri halo methane which is toxic (Trivedy and Goel 1986). pH is an important factor that determines the suitability of water for various purposes, including toxicity to animals and plants. The pH values of water samples of present study ranged from 7.8 to 8.7. The Cauvery River at Bhavani shows highest pH value of 9.8 during post monsoon and high pH values of 8.7 and 8.6 in Sanyasipatti and Siluvampalayam locations respectively, indicating polluted nature. This was supported by the diatom assemblage, dominated by *Gomphonema parvulum* (32 %), and Nitzschia *palea* (67 %) both of which are known to thrive in waters with high pH values. Field investigations revealed that the pollution in these locations is due to the discharges of sewages and industrial effluents. Temperature of water may not be important in surface water because of the wide range of temperature tolerance in aquatic life; but in polluted water, temperature can have profound

Table 8 Diatom indices values for the Cauvery river

Sampling site	IPS	SLAD	DESCY	L&M	SHE	WAT	TDI	EPI-D	ROTT	GDI	CEE	IBD	IDAP
Mettur dam	16.5	14.8	17.4	15.0	16.4	18.1	31.7	12.7	16.1	16.7	19.4	16.2	19.0
Navappatti	14.9	14.4	17.4	14.7	15.5	16.8	45.2	12.4	15.0	15.5	17.3	15.6	16.0
Nerinjipettai	15.9	14.9	16.3	14.5	16.1	19.4	33.6	12.7	14.6	15.9	17.0	16.5	18.1
Ammapettai	15.1	14.3	16.3	14.7	14.5	17.5	38.0	12.6	14.3	15.3	16.4	14.7	18.0
Siluvampalayam	15.2	14.4	15.1	13.6	15.2	18.0	32.8	12.4	13.2	12.2	16.4	15.0	16.0
Koneripatti	12.2	14.3	14.6	13.1	9.8	9.7	31.5	11.5	11.1	12.6	16.0	10.6	12.3
Peramachipalayam	8.1	13.4	16.5	13.8	13.6	13.2	63.7	10.3	13.6	7.2	13.4	11.0	12.8
Kottampatty	7.2	13.9	15.0	14.9	13.6	14.0	26.6	13.3	14.5	7.2	16.8	13.3	16.3
Sanyasipatti	7.1	14.7	15.3	14.5	16.4	18.8	28.4	12.4	14.9	7.4	16.6	16.5	18.2
Bhavani	7.2	15.0	14.7	14.5	15.8	19.0	29.9	12.5	14.8	7.1	16.0	16.7	17.9

Table 9 Class limit values for Diatom indices (Eloranta and Soininen 2002)

Index score	Class	Trophy
>17	High quality	Oligotrophy
15–17	Good quality	Oligo-mesotrophy
12–15	Moderate quality	Mesotrophy
9–12	Poor quality	Meso-eutrophy
<9	Bad quality	Eutrophy

effects on dissolved oxygen (DO) and biological oxygen demand (BOD). The fluctuation in river water temperature usually depends on the season, geographic location, sampling time and temperature of effluents entering the river. The Cauvery River water at Sanyasipatti location shows high temp (25 °C). In this location the temperature tolerant species namely, *Aulocosira granulata* (15 %) ,*Cyclotella meneghiniana* (10 %) and *Nitzschia obtuse* (20 %) represents 45 % of the total population.

Dissolved oxygen content is one of the most important factors in surface waters. Oxygen is the single most important gas for most aquatic organisms; free oxygen (O_2) or dissolved oxygen is needed for respiration. Dissolved oxygen deficiency in water directly affects the ecosystem of the river due to bioaccumulation and biomagnifications. The oxygen content in water samples depends on a number of physical, chemical, biological and microbiological processes. Dissolved oxygen values also changes depending on the industrial and human activity. Dissolved oxygen values were found to be maximum during pre monsoon and minimum during post monsoon, which might be due to natural turbulence. It could also be attenuated by higher algal productivity (produces O_2 by photosynthesis) in rainy period and active utilization in bacterial decomposition of organic matter. The pre monsoon Dissolved oxygen maxima was recorded at Sanyasipatti (9.2) followed by Mettur (8.5). BOD is the parameter used to assess the pollution of surface water and ground water. The values of BOD obtained in the present study are within permissible levels. The Cauvery River at Peramachipalayam shows high value BOD of 7.32, which is supported by the presence BOD tolerant diatom species *Aulacoseira distans* (23 %), *Cymbella ventricosa* Kutz (34 %); whereas at Kottampatti the BOD concentration was 5.21 and found to be dominated by *Gomphonema parvulum* (36 %) and *Nitzschia* (24 %).

Phosphate and Nitrate determinations are important in assessing the potential biological productivity of surface waters. Increasing concentration of phosphorus and nitrogen compounds in river and reservoirs leads to eutrophication. Phosphorus levels greater than 1.0 mg/L may interfere with coagulation in water treatment plants. As a result, organic particles that harbor microorganisms may not be completely removed before distribution. Phosphates and nitrates of all the sites showed post monsoon maxima (0.78 mg/L) and pre monsoon minima (0.005 mg/L). This could be due to the agricultural return flow during rainy season and utilization as nutrients by algae and other aquatic plants. A significant increasing trend was observed at inlet, and could be due to sewage water. The pre monsoon Phosphate maximum in the Cauvery River at Sanyasipatti is associated with *Achnanthidium Plonensis* (34 %) and *Cymbella turgida* (24 %). Enhanced levels of phosphate are associated with the municipal sewage discharge, as observed during the field investigation.

5 Conclusions

- Sixty (60) diatom species belonging to 21 genera in the Cauvery River are recorded in this study. Among these the species such as *Achnanthes minutissima, Achnanthidium plonensis, Aulacoseira distans, Cymbella turgida, C. ventricosa, Fragilaria. intermedia* var. *robusta, Gomponema lanceolatum, Nitzschia sigma, Synedra rumpen, S. ulna* are abundant in the all the locations.
- The presence of diatom species viz. *Cyclotella meneghiniana, Nitzschia sigma, Gomphonema parvulum* and *Synedra rumpen, S. ulna* in good numbers at Mettur dam with IPS 16.5 and GTI 16.7, Navappatti with IPS 14.9 and GTI 15.5, Nerinjipettai with IPS 15.9 and GTI 15.8, Ammapettai with IPS 15.1 and GTI 15.3 indicate good quality class oligo-mesotrophy conditions of the Cauvery River in these areas. The diatom assemblages at Siluvampalayam and Koneripatti with IPS 13.2 and GTI 12.2 and IPS 12.2 and GTI 12.6 respectively indicate moderate quality class and mesotrophy conditions, while Peramachipalayam with IPS 8.1 and GTI 7.2, Kottampatty with IPS 7.2 and GTI 7.1, Sanyasipatti with IPS 7.1 and 7.4 and Bhavani with IPS 7.2 and GTI 7.1 indicate bad quality class and Eutrophy.
- Analysis water quality index from the study area shows that the Water Quality Index (WQI) at Bhavani is 48.38 with Water Quality Rating (WQR) as bad in both pre- and post- monsoon periods indicating that the Cauvery River is under the stress of severe pollution in this region when compared with the values of Mettur.
- Evaluation of the diatom-based indices along with the physic-chemistry based interpretations suggests that, utilization of diatom species and indices for water quality assessment can be a reliable tool for environmental impact assessment, particularly to the river ecosystems.

References

Berger WH, Parker FL (1970) Diversity of planktonic foraminifera in deep-sea sediments, Science 168:1345–1347

Bhargave DS (1983) Use of water quality index for river classification and zoning of Ganga River. Environ Poll Serv B Chem Phys 6:51–76

Bolton PW, Currie JC, Tervet DJ, Welch WT (1978) An index to improve water quality classification. Water Pollut Control 77:271–284

Cemagref (1982) Etude des methods biologiques d'appré- ciation quantitative de la qualitédes eaux. Rapport Q. E. Lyon, Agence de l'eau Rhöne-Me´diterrane´e- Corse-Cemagref. Lyon, France

Coste M, Ayphassorho H (1991) Étude de la qualité deseaux du Bassin Artois-Picardie àl'aide des communautés de diatomées benthiques (application des indices diatomiques). Rapport Cemagref. Bordeaux-Agence de l'Eau Artois- Picardie, Douai

Cude C (2001) Oregon water quality index: a tool for evaluating water quality management effectiveness. J Am Water Resour Assoc 37(1):125–137

DellUomo A (1996) Assessment of water quality of an Apennine river as a pilot study. In: Whitton BA, Rott E (eds), Use of algae for monitoring rivers II. Institut fu¨ r Botanik, Universität Innsbruck, pp 65–73

Descy JP, Coste M (1991) A test of methods for assessing water quality based on diatoms. Verhandlungen der Internationalen Vereinigung für theoretische und angewandte Lim 24:2112–2116

Eloranta P, Soininen J (2002) Ecological status of some Finnish rivers evaluated using benthic diatom communities. J Appl Phycol 14:177

Gandhi HP (1957) A contribution to our knowledge of the diatom genus Pinnularia. Nat Soc J 54:845–853

Gandhi HP (1959) Fresh-water Diatoms from Sagar in the Mysore State. J Ind Bot 38:305–331

Gandhi HP (1998) Freshwater Diatoms of central Gujarat. Bishen Singh Mahendra Pal Singh, Dehra Dun

Gandhi HP (1961) Notes on the Diatomaceae of Ahmedabad and its environs. Hydrobiologia 17:218–236

Gandhi HP (1962) Notes on the Diatomaceae from Ahmedabad and its environs- IV—The diatom communities of some freshwater pools and ditches along Sarkhej Road. Phykos 1:115–127

Gandhi HP (1967) Notes on Diatomaceae from Ahmedabad and its environs. VI. On some diatoms from fountain reservoirs of Seth Sarabhai's Garden. Hydrobiologia 30:248–272

Gomez N, Licursi M (2001) The Pampean Diatom Index (IDP) for assessment of rivers and streams in Argentina. Aquat Ecol 35:173–181

House MA (1989) A Water quality index for river management. J Inst Water Environ Manage. 3:336–344. doi:10.1007/s10653-005-9001-5

House MA, Ellis JB (1987) The development of water quality indices for operational management, Water Sci Technol 19:145–154

Karthick B, Krithika H, Alakananda B (2008) Short guide to common freshwater Diatom Genera (Poster). Energy and Wetlands Research Group, CES, IISc, Bangalore

Karthick B, Alakananda B, Ramachandra TV (2009) Diatom based pollution monitoring in urban wetlands of Coimbatore, Tamil Nadu, Envis Technical Report—560012, India

Karthick B, Taylor JC, Mahesh MK, Ramachandra TV (2010) Protocols for collection, preservation and enumeration of Diatoms from aquatic habitats for water quality monitoring in India. IUP J Soil Wat 3(1):25–60

Kelly MG, Whitton BA (1995) The Trophic Diatom index: a new index for monitoring Eutrophication in rivers. J Phy 7:433–444

Kell MG, Cazaubon A, Coring E, Dell'Uomo A, Ector L, Goldsmith B, Guasch H, Hürlimann J, Jarlman A, Kawecka B, Kwandrans J, Laugaste R, Lindstrom EA, Leitao M, Marvan P, Padisak J, Pipp E, Prygiel J, Rott E, Sabater S, Dam VH, Vizinet J (1998) Recommendations for the routine sampling of diatoms for water quality assessments in Europe. J Phy 10:215–224

Kelly M, Juggins S, Guthrie R et al (2008). Assessment of ecological status in U.K. rivers using diatoms. Freshwater Biol 53:403–422

Leclerq L, Maquet B (1987) Deux nouveaux indices chimique et diatomique de qualite´ d'eau courante. Application au Samson et àses affluents (bassin de la Meuse belge). Comparaison avec d'autres indices chimiques, bioce´notiques et diatomiques. Institut Royal des Sciences Naturelles de Belgique, documented travail 28

Legendre P, Legendre L (1998) Numerical ecology. 2nd English edn. Elsevier, Amsterdam

Lenoir A, Coste M (1996) Development of a practical diatom index of overall water quality applicable to the French National Water Board network. In: Whitton BA, Rott E (eds) Use of algae for monitoring rivers II. Institut für Botanik, Universität Innsbruck, pp 29–43
Liou S, Lo S, Wang S (2004) A generalized water quality index for Taiwan. Environ Monitor Assess 96:35–52
Ludwig JA, Reynolds JF (1988) Statistical ecology: a primer of methods and computing. John Wiley and Sons, xviii, pp 337
Miller WW, Joung HM, Mahannah CN, Garrett JR (1986) Identification of water quality differences Nevada through index application. J Environ Qual 15:265–272
Mitchell MK, Stapp WB (1996) Field manual for water quality monitoring: an environmental educational program for schools. Thomson-Shore Inc, Dextor, p 277
Nasiri F, Maqsiid I, Haunf G, Fuller N (2007) Water quality index: a fuzzy river pollution decision support expert system. J Water Resour Plann Manage 133:95–105
Pesce SF, Wunderlin DA (2000) Use of water quality indices to verify the impact of Cordoba City (Argentina) on Suquira river. Water Res 34:2915–2926
Prygiel J, Leveque L, Iserentant R (1996) Un nouvel indice diatomique pratique pour l'évaluation de la qualité des eaux en réseau de surveillance. Rev Sci Eau 1:97–113
Rott E (1991) Methodological aspects and perspectives in the use of periphyton for monitoring and protecting rivers. In: Whitton, BA, Rott, E, Friedrich G (eds), Use of Algae for Monitoring Rivers. Institut fur Botanik, Univ. Innsbruck, pp 9–16
Said A, Stevens D, Selke G (2004) An innovative index for water quality in streams. Environ Manage 34:406–414
Schiefele S, Schreiner C (1991) Use of diatoms for monitoring nutrient enrichment acidification and impact salts in Germany and Austria. In: Whitton BA, Rott E, Friedrich G (eds) Use of algae for monitoring rivers. Institüt für Botanik, Univ. Innsbruk
Shrestha S, Kazama F (2007) Assessment of surface water quality using multivariate statistical techniques: a case study of the Fuji river Basin, Japan. Environ Model Softw 22(4):464–475
Sladecek V (1986) Diatoms as indicators of organic pollution. Acta Hydrochim Hydrobiol 14:555–566
Stoermer EF, Smol JP (eds) (1999) The diatoms: applications for the environmental and earth sciences. Cambridge University Press, Cambridge, UK
Stoermer EF, Pilskaln CH, Schelske CL (1995) Siliceous microfossil distribution in the surficial sediments of Lake Baikal. J Pal 14:69–82
Taylor JC, Harding WR, Archibald CGM (2007) An illustrated guide to some common diatom species from South Africa. WRC Report TT 282/07. Water Research Commission, Pretoria
Trivedy RK, Goe PK (1986) Chemical and biological methods for water pollution studies. Environmental Publications, Karad, p 7
Vega M, Pardo R, Barrado E, Deban L (1998) Assessment of seasonal and polluting effects on the quality of river water byex ploratory data analysis. Water Res 32(12):3581–3592. doi:10.1016/S0043-1354(98)00138-9
Venkatachalapathy R, Karthikeyan P (2012) Environmental impact assessment of Cauvery river with Diatoms at Bhavani, Tamil Nadu, India. Int J Geol Earth Environ Sci 2(3):36–42
Venkatachalapathy R, Karthikeyan P (2013a) Physical, chemical and environmental studies on Cauvery river in parts of Tamil Nadu (Mettur and Bhavani). Univ J Environ Res 3(3):415–422
Venkatachalapathy R, Karthikeyan P (2013b) Diatoms assemblages distribution in Cauvery Bhavani rivers, Tamil Nadu in relation to chemical and physiographical factors. Res J Chem Sci 3(11):55–59
Venkatachalapathy R, Karthikeyan P (2013c) A taxonomic and morphological study of fresh water Diatom species Synedra ulna (Nitzsch) Ehrenberg in Cauvery river at Bhavani region, Tamil Nadu, India. Int Res J Environ Sci 2(11):18–22
Venkatachalapathy R, Karthikeyan P (2013d) Benthic Diatoms in river influenced by urban pollution, Bhavani region, Cauvery river, South India. Int J Innovative Technol Exploring Eng 2(3):206–210

Venkatachalapathy R, Karthikeyan P (2014) Diatom indices for water quality assessment in Cauvery river, Tamil Nadu, India. Gondwana Geol Mag 15(2014):109–116
Watanabe T, Asai K, Houki A (1986) Numerical estimation of organic pollution of flowing waters by using the epilithic diatom assemblage-Diatom Assemblage Index (DIApo). Sci Total Environ 55:209–218
Zelinka M, Marvan P (1961) Zur Präzisierung der biologischen Klassifikation der Reinheit fliessender Gewässer. Arch Hydrobiol 57:389–407

Species Diversity and Functional Assemblages of Bird Fauna Along the Riverine Habitats of Tiruchirappalli, India

Manjula Menon, M. Prashanthi Devi, V. Nandagopalan and R. Mohanraj

Abstract Riverine ecosystems have complex relationship with human since time immemorial and play an integral role in the socio-economy of a region. However, many riverine habitats, particularly within urban centers in the developing countries are subjected to overexploitation that affects the natural ecological processes and functions of rivers. Such perturbations in riverine habitats are often linked to biodiversity loss. Birds discharge crucial ecosystem services and are closely associated to wetlands and rivers for their survival. This study attempts to document the bird diversity and their community assemblages along the riverine habitats in relation to urban effects, vegetative attributes, seasonal parameters and other anthropogenic pressures. The study finds that bird diversity and species richness were higher in the rural landscape and gradually decreased towards the urban region. A total of 120 bird species consisting of two 'Near Threatened' were recorded along the riverine habitats of the River Cauvery. Few species (*Egretta garzetta* and *Phalacrocorax niger*), were found to be densely populated and adapted to the urban environment while few others declined. Salient features of the results include: seasonality did not affect the variability of the riverine species and the difference in composition of birds during the wet and dry seasons were insignificant. While the factors namely, tree cover, tree height, and number of trees were found to be positively correlated, the anthropogenic factors namely, extent of built up land, noise levels and vehicular traffic contributed negatively towards bird diversity. Among the various riverine stretches/study sites, the Kallanni region recorded highest species richness and diversity. The most important conservation measure would be to declare Kallanai an important bird reserve and the agricultural farmlands on either sides of river Cauvery at Kallanai be declared as 'High Nature Value' (HNV) wetlands/farmlands to protect the overall biodiversity.

M. Menon (✉) · M. Prashanthi Devi · R. Mohanraj
Department of Environmental Management, School of Environmental Sciences, Bharathidasan University, Tiruchirappalli 620 024, India
e-mail: manj.mn@gmail.com

V. Nandagopalan
National College, Tiruchirappalli, India

Mu. Ramkumar et al. (eds.), *Environmental Management of River Basin Ecosystems*, Springer Earth System Sciences, DOI 10.1007/978-3-319-13425-3_32

Keywords Urbanization · Riverine · Diversity · Abundance · Species richness · Corridors

1 Introduction

Rivers are one among the most dynamic ecosystems on the Earth in terms of spatial and temporal complexity (Power et al. 1995; Kingsford 2000). River flow is directly or indirectly linked to recharge patterns of lakes, channels, swamps, marshes and tributaries within the basin. River water acts as a lifeline in the complex ecological processes and services. It also plays a pivotal role in the interactions of a wide range of species within the river ecosystem and also the inter-connected ecosystems including the agricultural ecosystems (Tilman et al. 1996; Daily 2001). Often the river corridor and its adjacent terrestrial areas function as an integrated ecological unit, and are intimately linked through exchanges of energy and materials (Mazeika et al. 2007). The river and river based wetlands may attain bloom during wet conditions fostering the life-forms ranging from bacteria, planktons, fishes, amphibians, birds and mammals which drive the complex food webs (Bunn and Boon 1993; Bunn and Davies 1999). Even smaller water bodies contribute significantly in enhancing and enriching the environments (Gosselin and Johnson 1995; Girling and Helphand 1997) and these smaller wetlands are linked to the riverine systems to avoid isolation and destruction (Titton 1995). More importantly, river corridors act as an abode to water birds as well as upland terrestrial birds (Sullivan et al. 2007). Within the riverine ecosystem, birds play a crucial role in sustenance of the system including its integrity (Carignan and Villard 2002). Bird diversity across the river corridors and their differences in feeding habits are also known to increase the overall vegetation diversity, evenness and species richness (Smith 1995). Some of the birds are greatly recognized for their role as keystone species, which decides the overall health of the ecosystem.

Ever since the human population commenced an exponential growth, the water demand for agricultural, industrial, and domestic purposes has changed the structures and processes of the river basins (Nilsson and Berggren 2000). Such changes have largely impacted the associated ecosystem network like channels, lakes, ponds and waterholes. Rapid urbanization including commercial, residential developments and construction of roads in cities coupled with population growth has impaired river and river based wetlands in developing countries (Azous and Horner 2000; Masero 2003; Tracy et al. 2004). Although such repercussions have been largely overcome in developed countries, many river basins in the developing countries are yet to get restored to their original status. As a consequence, in addition to several ill effects, bird communities along the riparian systems have also been victimized resulting in their population decline and loss.

Kingsford (2000) stated that the construction of dams affects the estuarine and coastal ecology thereby preventing water birds from inhabiting these landscapes. Few studies have documented that invasion of exotic vegetation due to human ramification has dramatically changed the vegetation community structure and composition causing disturbances to the riparian birds (Pavlik and Pavlik 2000; Sogge et al. 2008). Krueper et al. (2003) observed that removing cattle from riparian areas in the southwestern United States had profound benefits for breeding birds. However, along the Colorado River in Grand Canyon National Park, Arizona, bird species appeared to exhibit ecological plasticity in response to anthropogenic increase in prey resources (Yard et al. 2004). Many such studies examining water birds have generally focused on marsh systems in natural environments, with only fewer studies on riverine habitats (Edelson 1990; Hoyer and Canfield 1990, 1994). Few studies have also emphasized the importance of habitat heterogeneity of river corridors at local scales to bird communities (Brotons et al. 2004). Myriad numbers of studies have portrayed the effects of urbanization on native avifauna and concluded that bird survival in urban wetlands and parks is threatened by the risk of collision with man-made objects, changes in the predator assemblage, food supply, and diseases (Chace and Walsh 2006). However, our understanding of the ecological dynamics of bird communities in riverine systems in relation to anthropogenic impacts is still in its infancy (Pearce et al. 2007).

In the Indian scenario, studies highlighting the ecological dynamics of riverine birds and the anthropogenic implications are very limited, however a large number of studies have reported the significance of wetland birds. In Barna wetland of Narmada River basin, a study by Balapure et al. (2012) observed 63 species of water birds including the vulnerable Saras crane and advocated effective measures for the preservation of the wetland. In a review, Balachandran (2012), opined that coastal wetlands in India provide winter refuge for migratory waterfowl from different parts of the world such as north, central and west Asia, Europe and Mediterranean regions. Another study on wetlands of western part of the Kachchh District (Gajera et al. 2012) documented 152 species including 26 migratory species and concluded that diversity of wetland birds was rich and require conservation efforts to sustain it. In an attempt to address the lacuna of information on riverine avifauna in relation to habitat attributes and anthropogenic impacts including their conservation, and management, this study was carried out at Tiruchirappalli along the River Cauvery and the associated wetlands. The aim of this study was to examine the anthropogenic effects of development by studying (a) the seasonal functional assemblages of water birds found along the river channel in urban Tiruchirappalli (b) avifaunal diversity, abundance and the community composition of riverine birds along an urbanization gradient and (c) the factors that closely relate to the riverine bird diversity, and to analyze possible relationships among these habitat attributes.

2 Study Area

The Cauvery basin extends over an area of 81,155 km^2, which is nearly 24.7 % of the total geographical area of the country. Tiruchirappalli District is located at the central part of Tamil Nadu, India, lying between 10° 10′ and 11° 20′ of the Northern

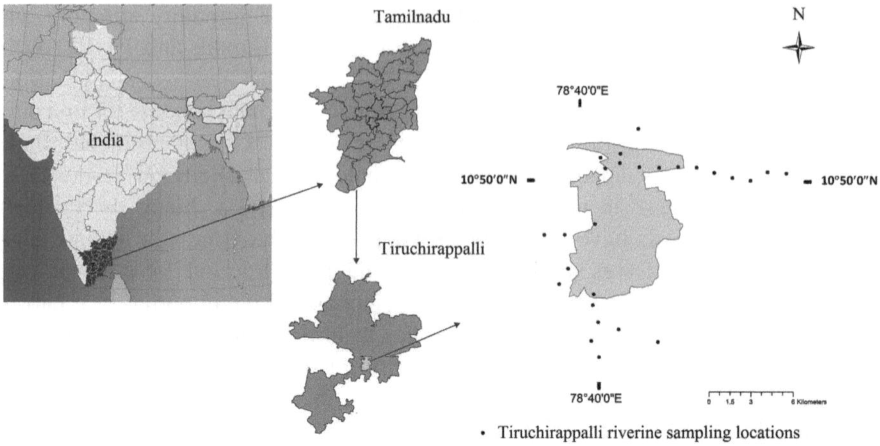

Fig. 1 Location of the study area and the sampling sites

latitudes and 78° 10′ and 79° 0′ of the Eastern longitudes (Fig. 1). The district has an area of 4,404 km^2. It is the 4th largest district and also the fourth largest urban agglomeration in the state. The city of Tiruchirappalli lies on the banks of the river 'Cauvery. The city is located at the delta head of the Cauvery River from where it branches into Cauvery and Coleroon channels and further bifurcates into Cauvery and Vennar' at Kallanai and again a network of 36 major channels which in turn branch off into thousands of distributary channels, canals, etc. The district exhibits hot and dry climate with high temperature and low degree of humidity throughout the year, with a short period of rainy season and winter from September to December. Southwest monsoon generally sets in at the beginning of June and lasts till the end of August. Tiruchirappalli district has a forest coverage area of 27,254 ha, which is 6 % to the total area of the district and the forest areas are classified under Reserve Forest Category. The types of forest predominantly identified in Tiruchirappalli district are Tropical Dry Deciduous forest and Tropical Thorn forest.

Tiruchirappalli is the fourth most populated city in the state of Tamilnadu. As per 2011 census, Tiruchirappalli district as whole had a population of 2,713,858 individuals. There was an increase of 12.22 % in the population compared to 2001. Tiruchirappalli city is one of the progressive industrial areas of the state. There are more than 300 metal fabricating industries, where about 1.5 lakh tones of steel are being processed annually. Tiruchirappalli being one of the oldest inhabited cities in Tamil Nadu, its earliest settlements dates back to the second millennium BC.

3 Methodology

3.1 Design of the Survey

Birds were surveyed along riverine patches of the River Cauvery in Tiruchirappalli at 26 point counts along an urbanization gradient from October 2010 to September 2012 using the point count method. At each point, bird surveys were conducted within 25 m radius, for 10 min from 06:00 AM to 09:00 AM on a monthly basis to maximize count efficiency (Petit et al. 1995). The study points were geo-referenced with a Garmin hand held GPS and all the birds detected visually and acoustically were recorded. Point count data was used to calculate detection probabilities, the abundance of individuals and species richness at each site. Bird community compositions at the different sampling points were expressed by diversity, richness and abundance of each species, obtaining a 'species by site' matrix. Species richness for each sampling point was defined as the total number of species detected during each visit. Abundance for each sampling point was defined as the maximum number of individuals present in the point count. At each point count, spatial and temporal variation in species richness and abundance of birds were also recorded.

When counting birds, special care was taken that individuals were counted only once. Breeders and other visitors were not distinguished as distinction is difficult. Over-flying birds were counted only when they were flying low or showed connection to the ground environment (i.e. searching for food). Bird surveys were not preformed during heavy rains, fog and during strong winds, since these conditions reduce bird activity, deduction and flawless identification (Sutherland 2004). Species identification was carried out using binocular and field guides (Grimmett and Inskipp 2005).

3.2 Survey Site Characterization

Bird abundance and species richness along the gradient were associated with various vegetative attributes (tree cover, shrub cover, herb cover, number of trees, number of shrubs, number of herbs, tree height, shrub height, herb height and DBH) and various scores of urbanization (urban structures and anthropogenic pressure), measured at different spatial scales that surround the riverine patches in the area. Urban structures were evaluated in terms of built up cover, number of buildings, building height, road cover, road width, number of cables and anthropogenic pressure was measured in terms of vehicular movement, (number of vehicles/ minute), pedestrian movement (number of people/minute), road kills (number of birds killed by vehicular movement), and noise levels (recorded using sound level meter and noise recorded/15 min), along the riverine patches. Sound levels were recorded for peak and non peak hours, however only the non peak hour values were used for analysis as this coincided with the bird survey. Tree cover and built up

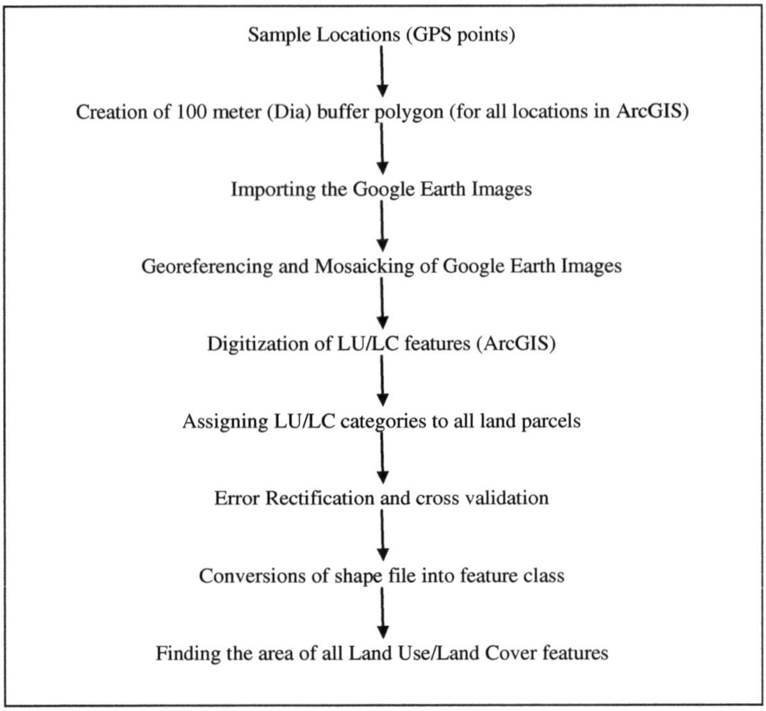

Fig. 2 Habitat attribute estimation using Arc GIS 10

cover were estimated using Arc GIS 10 (Fig. 2). Shrub cover and Herb cover surrounding these riverine patches were estimated using Braun–Blanquet categorical scale (Kent and Coker 1992), within a 50 m radius around each point count station. The cover scale was 0 (<1 %), 1 (1–5 %), 2 (6–25 %), 3 (26–50 %), 4 (51–75 %), and 5 (76–100 %).

4 Results and Discussion

4.1 Species Diversity and Richness of the Riverine Patches

Along the riverine patches of Tiruchirappalli, a total of 120 species of birds were recorded. Of the 120 species recorded, 30.83 % species were exclusively water birds and 69.17 % were land birds. The mean diversity of birds recorded along the riverine patches was: urban 1.13 ± 0.5 (Mean ± SD), suburban 1.59 ± 0.50 and rural 1.82 ± 0.46 (Fig. 3). The mean abundance of individuals recorded along the riverine stretch was: urban 43.41 ± 27.70, suburban 41.08 ± 24.23 and rural 42.69 ± 28.39. Species richness was recorded highest in the the rural landscape (9.73 ± 3.42),

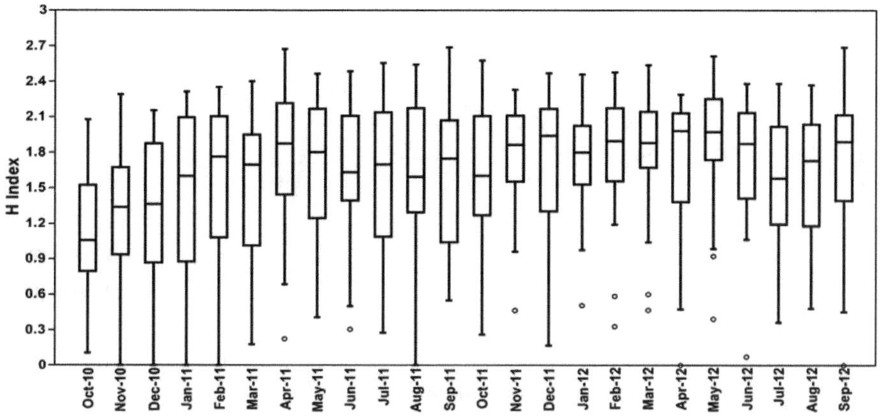

Fig. 3 Mean monthly variation of diversity (H index) recorded in the riverscapes during the study period

Table 1 Diversity indices of the riverine patches

Wetlands (n = 26)	Urban(8)	Suburban(4)	Rural(14)
Taxa_S	5.71 ± 2.79	8.65 ± 3.57	9.73 ± 3.42
Abundance	43.41 ± 27.70	41.08 ± 24.23	42.69 ± 28.39
Dominance	0.46 ± 0.24	0.30 ± 0.17	0.23 ± 0.13
Simpson_1-D	0.53 ± 0.24	0.69 ± 0.17	0.76 ± 0.13
Shannon_H	1.13 ± 0.57	1.59 ± 0.50	1.82 ± 0.46
Evenness_e^H/S	0.65 ± 0.18	0.65 ± 0.15	0.70 ± 0.13
Margalef	1.37 ± 0.80	2.09 ± 0.82	2.42 ± 0.80
Fisher_alpha	2.61 ± 2.76	3.84 ± 1.99	4.97 ± 2.97

followed by the suburban (8.65 ± 3.57), and the urban matrix (5.71 ± 2.79). However species abundance was recorded higher in the urban matrix, (Table 1). The result shows that rural regions are highly species rich when compared to suburban and urban regions. From the rural region towards the urban matrix, species diversity and richness decreases indicating the urban influence.

The mean species diversity recorded in the riverine patches during the wet season was 1.59 ± 0.44 and during the dry season 1.62 ± 0.52. During the wet season a mean taxa of 8.30 was recorded whereas the dry season recorded a mean taxa of 9.05 (Figs. 4, 5 and 6). A total of 72 bird species were common to both seasons, but 8 riverine species were exclusive to the wet season and 4 riverine species were exclusive to the dry season. Among the water birds in the riverine patches of Cauvery, a total of 8,245 individuals were recorded. The most abundant were *Egretta garzetta* (26.74 %), *Phalacrocorax niger* (19.09 %), *Ardeola grayii* (11.17 %), *Mesophoyx intermedia* (8.71 %), *Fulica atra* (5.08 %) and *Bubulcus ibis*

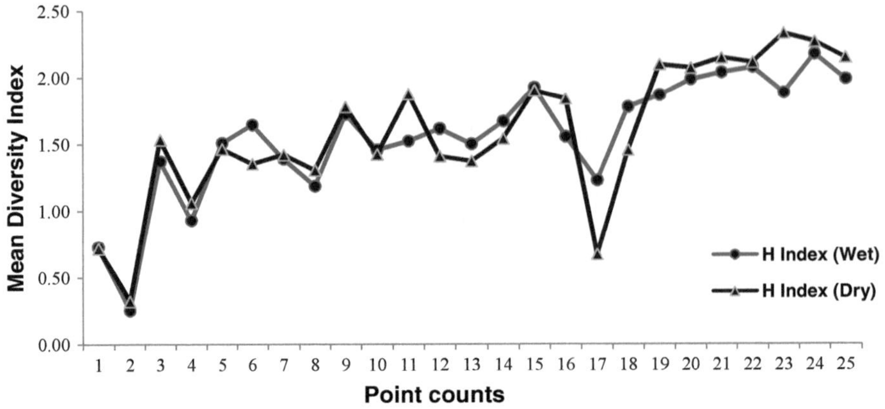

Fig. 4 Mean diversity (H Index) of birds recorded during the wet and dry seasons along the riverine urban-rural stretch

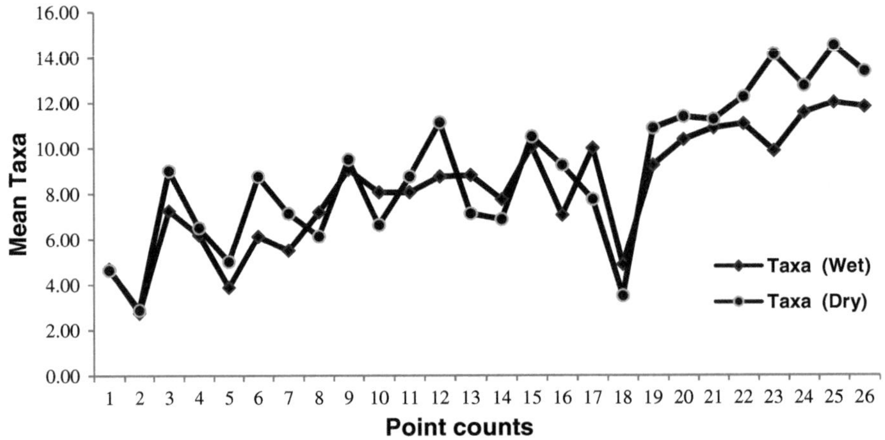

Fig. 5 Mean taxa (species richness) of birds recorded during the wet and dry seasons along the riverine urban-rural stretch

(4.50 %). The six species of water birds accounted for 48.55 % of the total bird abundance (Fig. 7). The mean species diversity during the wet and dry seasons was found to be insignificant (df = 1, f = 0.05, p > 0.05) and is consistent with the studies conducted by Aynalem and Bekele (2008). There were significant differences among landscapes (urban-suburban-rural) within the wet season (df = 2, f = 13.16, p = 0.000) and among landscapes within the dry season (df = 2, f = 7.99, p = 0.002). There were no significant seasonal difference in the species richness recorded (df = 1, f = 0.79, p > 0.05) and in the abundance of species (df = 1, f = 0.82, p > 0.05), as maximum number of species recorded were resident birds.

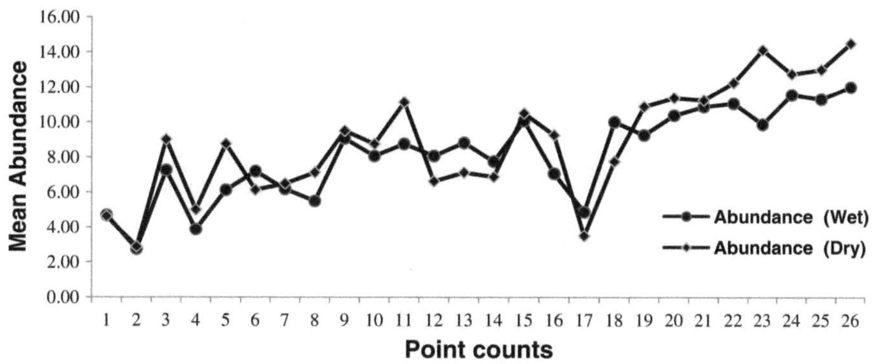

Fig. 6 Mean abundance of birds recorded during the wet and dry seasons along the riverine urban-rural stretch

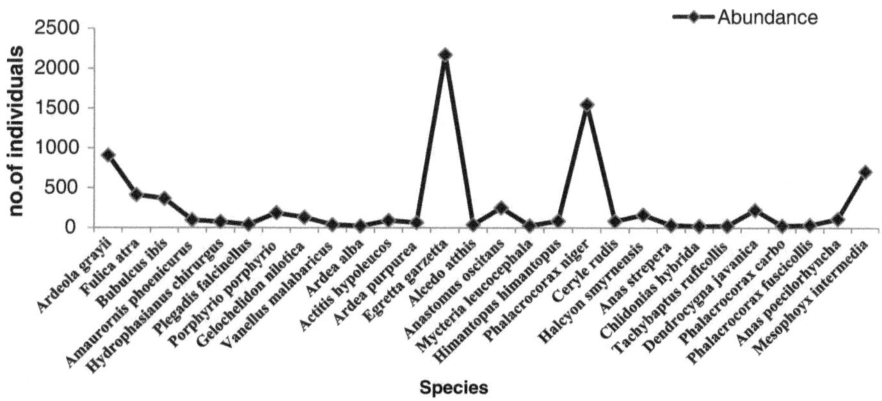

Fig. 7 Most abundant species of the riverine patches

The relative richness and abundance of bird species during the wet and dry seasons may also be related to the availability of food, habitat conditions and breeding season of the species.

Few species namely, the *Egretta garzetta* (Little Egret), with a mean abundance of 97.50 individuals, *Phalacrocorax niger* (Little Cormorant) 69.50, *Ardeola grayii* (Indian Pond Heron) 41.29 and *Mesophoyx intermedia* (Intermediate Egret), 34.46 individuals (Table 2) were found to thrive well in the urban environment due to a broader territorial range. However, few other species namely, *Casmerodius albus* (Great Egret) and *Gallinula chloropus* (Common Moorhen), sensitive to habitat variations showed a decline. The high abundance of fewer species namely, the *Egretta garzetta* (Little Egret), and *Phalacrocorax niger* (Little Cormorant) may set in the processes of biotic homogenization which may in future favor the extinction of native water birds.

Table 2 List of riverine species recorded (U-urban, SU-suburban, R-rural)

Species	Mean abundance	Family	Order	Land use pattern
Actitis hypoleucos	3.33	Scolopacidae	Charadriiformes	R
Alcedo atthis	1.92	Alcedinidae	Coraciiformes	U/SU/R
Amaurornis phoenicurus	7.83	Rallidae	Gruiformes	U/SU/R
Anas poecilorhyncha	6.00	Anatidae	Anseriformes	U/SU/R
Anas querquedula	0.50	Anatidae	Anseriformes	R
Anastomus oscitans	8.63	Ciconiidae	Ciconiiformes	R
Ardea cinerea	0.54	Ardeidae	Pelecaniformes	R
Ardea purpurea	2.83	Ardeidae	Pelecaniformes	U/SU/R
Ardeola grayii	41.29	Ardeidae	Pelecaniformes	U/SU/R
Bubulcus ibis	16.17	Ardeidae	Ciconiiformes	U/SU/R
Casmerodius albus	1.79	Ardeidae	Pelecaniformes	U/SU/R
Ceryle rudis	4.08	Cerylidae	Coraciiformes	U/SU/R
Chlidonias hybrida	2.00	Sternidae	Charadriiformes	R
Ciconia episcopus	0.08	Ciconiidae	Ciconiiformes	R
Dendrocygna javanica	9.54	Anatidae	Anseriformes	U/SU/R
Egretta garzetta	97.50	Ardeidae	Pelecaniformes	U/SU/R
Fulica atra	16.42	Rallidae	Gruiformes	U/SU/R
Gallinula chloropus	0.54	Rallidae	Gruiformes	U/SU/R
Gelochelidon nilotica	5.54	Sternidae	Charadriiformes	R
Halcyon smyrnensis	7.83	Halcyonidae	Coraciiformes	U/SU/R
Himantopus himantopus	3.63	Recurvirostridae	Charadriiformes	R
Hydrophasianus chirurgus	5.21	Jacanidae	Charadriiformes	U/SU/R
Mesophoyx intermedia	34.46	Aredidae	Pelecaniformes	U/SU/R
Metopidius indicus	0.13	Jacanidae	Charadriiformes	R
Mycteria leucocephala	1.13	Ciconiidae	Ciconiiformes	R
Nycticorax nycticorax	0.38	Ardeidae	Pelecaniformes	R
Phalacrocorax carbo	0.88	Phalacrocoracidae	Suliformes	U/SU/R
Phalacrocorax fuscicollis	0.08	Phalacrocoracidae	Suliformes	R
Phalacrocorax niger	69.50	Phalacrocoracidae	Suliformes	U/SU/R
Plegadis falcinellus	1.75	Threskiornithidae	Pelecaniformes	R
Porphyrio porphyrio	8.00	Rallidae	Gruiformes	U/SU/R
Pseudibis papillosa	0.08	Threskiornithidae	Pelecaniformes	R
Tachybaptus ruficollis	0.63	Podicipedidae	Podicipediformes	U/SU/R
Threskiornis melanocephalus	0.04	Threskiornithidae	Pelecaniformes	R
Tringa stagnatilis	0.04	Scolopacidae	Charadriiformes	R
Vanellus indicus	5.25	Charadriidae	Charadriiformes	U/SU/R
Vanellus malarbaricus	1.67	Charadriidae	Charadriiformes	R

Among the various riverine patches in Tiruchirappalli, the Kallanai region recorded high species richness and diversity (H Index = 2.20 ± 0.06), with the minimum and maximum diversity ranging between 1.94 and 2.20. Kallanai Dam in Tiruchirappalli across the River 'Cauvery' is one of the oldest dams in the world, built around 2,000 years ago but still in active service. Today, this riverine stretch serves as an excellent habitat for many species of resident and migratory birds. The large, continuous, and intact patch at Kallanai and the occurrence of dense vegetation in its vicinity provides favorable nesting and roosting sites and perfect habitat for feeding. Among the various riverine patches studied, high species diversity both during the wet (2.17 ± 0.23) and the dry seasons (2.32 ± 0.23) were recorded at Kallanai. Migratory species were recorded from November to February along this riverine stretch. Few important winter visitors and migratory species recorded at this region were *Dendrocygna javanica* (Lesser whistling duck), *Anas querquedula* (Gargeny), *Ardea cinerea* (Grey Heron), *Ardea purpurea* (Purple Heron), *Mycteria leucocephala* (Painted stork), *Ciconia episcopus* (Woolly-necked Stork), *Anastomus oscitans* (Asian open billed stork), *Himantopus himantopus* (Black-winged Stilt), *Plegadis falcinellus* (Glossy Ibis), *Threskiornis melanocephalus* (Black headed Ibis), *Pseudibis papillosa* (Black Ibis), *Chlidonias hybridus* (Whiskered Tern), and *Gelochelidon nilotica* (Gull-billed Tern). High species richness of land birds were also recorded along the riverine stretches of Kallanai. This may be also due to the increased densities of flying insects shown to be associated with the riverine patches. This observation and inference are consistent with the earlier study too (Scheffers et al. 2006). The agricultural farmlands on either side of river 'Cauvery' provides alternate nesting and roosting grounds for birds. The most abundant land birds that were frequently associated with the surrounding of the Cauvery river were *Corvus splendens* (House Crow), *Acridotheres tristis* (Common Myna), *Cypsiurus balasiensis* (Asian Palm Swift), *Dicrurus macrocercus* (Black Drongo), *Psittacula krameri* (Rose-ringed Parakeet), *Merops orientalis* (Green Bee-eater), and *Turdoides affinis* (Yellow-billed Babbler).

Migratory birds also shift their feeding habitats between seasons and look for riverine patches to feed on varied resources (McKinney et al. 2011). The distribution and abundance of migratory birds were also determined by the composition of the vegetation that forms a major element of their habitats. Few migratory species like *Mycteria leucocephala* (Painted Stork), and *Threskiornis melanocephalus* (Black-headed Ibis), enlisted in the IUCN Red List under the 'Near Threatened' category (Figs. 8 and 9) and *Ciconia episcopus* (Woolly-necked Stork) enlisted as 'Vulnerable' were frequent visitors to the riverine patches of Kallanai. The *Anastomus oscitans* (Asian open billed stork), is the third important local migratory winter visitor that frequently visit the riverine patches in the study area. Though they are enlisted in the 'Least Concern' category in the IUCN Red List, there are reports of decline of the species from some parts of the world. The distribution map (Figs. 10 and Plate 1) identifies 18 favored habitats of *Anastomus oscitans* in the study area, mostly agricultural landscapes and riverine patches including Kallanai.

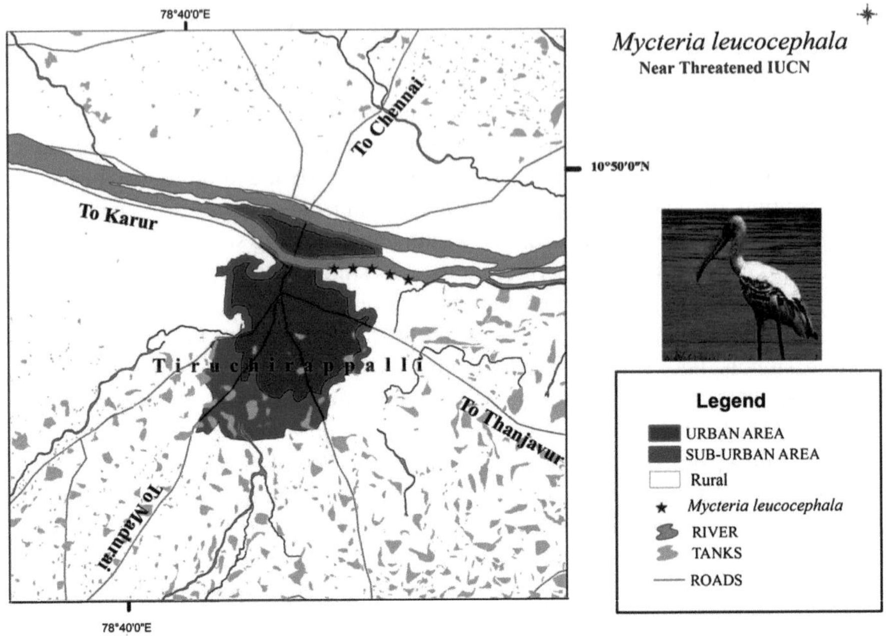

Fig. 8 Habitat selection of *Mycteria leucocephala* in the study area

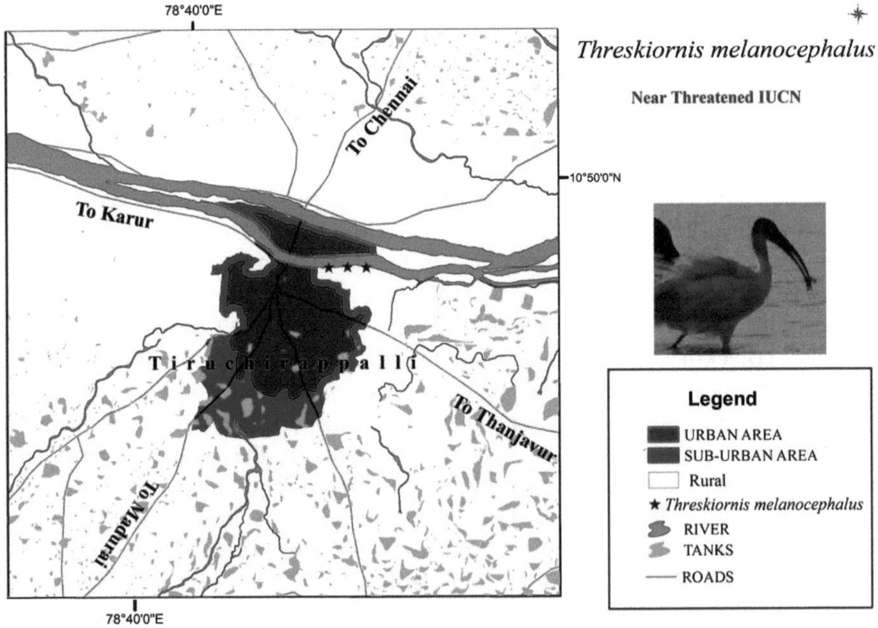

Fig. 9 Habitat selection of *Threskiornis melanocephalus* in the study area

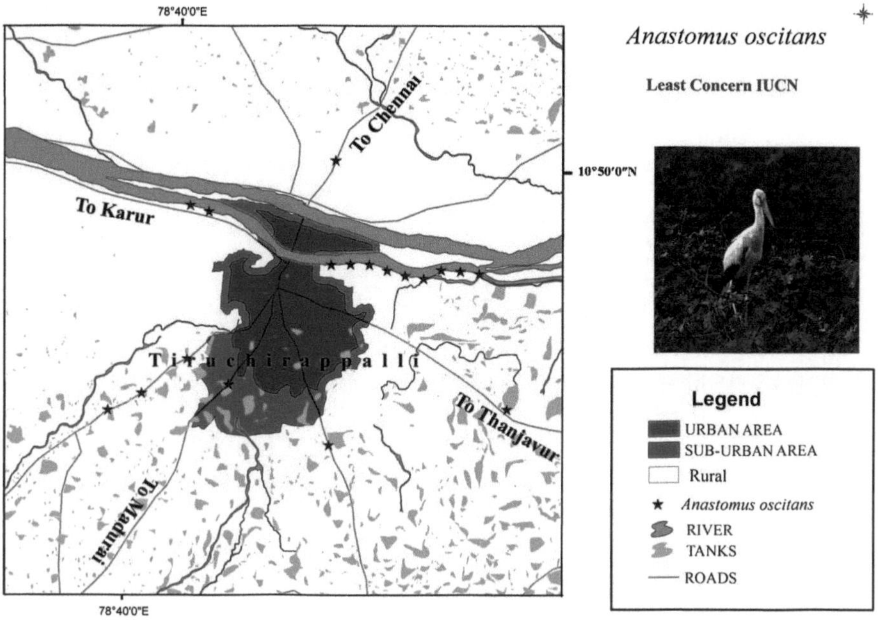

Fig. 10 Habitat selection of *Anastomus oscitans* in the study area

Other riverine patches in Tiruchirappalli are fragmented, polluted and are more exposed to human interventions. Hence, migratory birds less frequent these regions. Unfortunately, the turnover rates of migratory species have dropped dramatically, because of the loss and degradation of riverine patches in these regions.

4.2 Bird Variability and Habitat Attributes

Habitat features affect birds both positively and negatively. The study conducted by Pearce et al. (2007) recommends that adjacent forest cover (and other natural vegetation), which provides both habitat and a buffer for birds from urban activities, be included in the wetland and riverine evaluations and should be a criteria for riverine bird conservation. The adjacent land cover and land use patterns immensely affect the avian movements. Thus, in order to understand the likely factors that determine the species richness and abundance of riverine birds, Principal Component Analysis (PCA) was attempted with the survey site characteristics derived at the point counts surrounding the riverine patches. Vegetative factors like tree cover, shrub cover, herb cover, number of trees, number of shrubs, number of herbs, DBH (Diameter at Breast Height), tree height, shrub height, herb height and anthropogenic factors like built up cover, road cover, building height, number of buildings, road width, cable poles, pedestrian movement, noise levels, road kills and vehicular

Plate 1 Water birds at Kallanai. **a** *Hydrophasianus chirurgus* (Pheasant-tailed Jacana), **b** *Ardea cinerea* (Grey Heron), **c** *Anas querquedula* (Gargeny), **d** *Anastomus oscitans* (Asian Open Billed Stork), **e** *Ardea purpurea* (Purple Heron), **f** *Porphyrio porphyrio* (Purple Swamphen)

traffic were used for analysis. PCA transforms the set of variables into a smaller set of linear combinations that retains the original information as much as possible.

Principal components with an initial Eigen value score greater than 0.5 were considered to show a strong relationship to the principal component. Eigen vector scores that depart from 0 indicate an increasing importance of that variable to the principal component. Once the principal components were identified, the measure

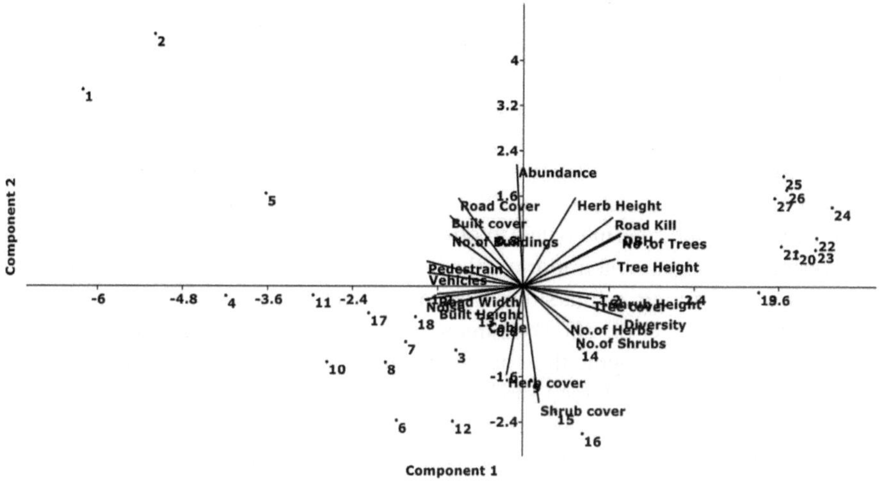

Fig. 11 PCA-Correlating bird diversity to various attributes of vegetation and anthropogenic factors

with the highest Eigen vector score for each component was selected to represent the variability along that component. The PCA analysis between diversity index and habitat attributes including scores of urbanization explained two main factors of the classification analysis, with 60 % of the variance (Factor 1 = 46.36 %, Factor 2 = 14.26 %). Factor 1 represented vegetation components (tree cover, shrub cover, herbaceous plant cover, number of trees, tree height, DBH), at its positive end, and Factor 2 represented urban parameters (built up cover, road width, road cover, building number, building height, pedestrian movement, noise levels and vehicular traffic) at its negative end (Fig. 11). The result shows that vegetative components are favorable attributes for riverine birds and urban development pose threat to riverine diversity.

Pearson's correlation analysis also affirmed that bird species diversity of the riverine patches significantly contributed to the vegetative factors like Tree cover ($r = 0.62, p = 0.001$), DBH ($r = 0.75, p = 0.001$), tree height ($r = 0.79, p = 0.001$), shrub height ($r = 0.57, p = 0.001$), and number of trees ($r = 0.73, p = 0.001$). Diversity index of the riverine patches contributed negatively to built up cover ($r = -0.72, p = 0.001$), building height ($r = -0.65, p = 0.001$), pedestrian movement ($r = -0.74, p = 0.001$), noise levels ($r = -0.83$, $p = 0.001$) and vehicular traffic ($r = -0.81$, $p = 0.001$). As reported previously by several researchers (Munyenyembe et al. 1989; Aurora et al. 2009; Evans et al. 2009; Suarez-Rubio and Thomlinson 2009), the present study also reports that tree cover, DBH, number of trees, tree height and shrub height to be key elements to maintain species richness of riverine birds within urban areas as they provide suitable zones for roosting, nesting, hiding and foraging. Forested and non-forested vegetations are often crucial for riverine bird communities and may provide necessary resources that facilitate the livelihood of the birds in these regions

(Sullivan et al. 2007; Caula et al. 2008, 2010; Murgui 2009). Previous studies also emphasized the importance of vegetation structure and tree composition as important attributes in governing the riverine bird richness (Kreyer and Zerbe 2006). Jobin et al. (2004) stated that the local diversity of the riverine habitats increases with the diversity of wooded habitats adjoining it. Several landscape-scale studies have shown reduced avian diversity in wetlands due to the higher levels of human disturbance (Whited et al. 2000; DeLuca et al. 2004). Increase in built up cover restricts bird movement from one riverine patch to other and prevents alternate foraging grounds for riverine birds. The most important threat to wetland and riverine birds come from the degradation of their habitats that were once ideal zones for feeding and thermo-regulation (Brouwer et al. 2003). Human activities threaten the existence of many riverine species by destroying their natural habitat and thereby affecting their survival and reproductive success (Green and Hirons 1991). Traffic volume and noise level are critical factors in determining the distribution and geographic extent of water birds in the urban and also along the gradient. Above certain threshold levels, traffic noise diminishes the habitat quality for breeding birds than on wintering birds by causing distortion in vocal communication (Brumm and Slabbekoorn 2005). Often breeding birds are very sensitive to their habitats when compared to wintering birds and look for alternate favorable habitats for successful completion of their breeding cycle (Traut and Hostetler 2004; Swaddle and Page 2007). Pedestrians can also limit the availability of resource patches and favorable breeding grounds for birds by reducing their foraging and breeding opportunities. These observations of the present study are consistent with the study conducted by Fernandez-Juricic and Telleria (2000).

Few riverine species namely, *Egretta garzetta* (Little Egret), *Ardeola grayii* (Indian Pond Heron) and *Mesophoyx intermedia* (Intermediate Egret) were found in huge abundance in highly urbanized locations and adapted well to urbanization associated factors. Similar observations were also recorded by many others elsewhere (for example, Butler 1992; Blair 1996; Weller 1999; Savard et al. 2000). Other species, such as the *Porphyrio porphyrio* (Purple Swamphen*), Amaurornis phoenicurus* (White Breasted Water hen) and *Nycticorax nycticorax* (Black-crowned Night Heron), were also found in urbanized areas, but at lower abundance. Few water birds such as *Casmerodius albus* (Great Egret), *Anas poecilorhyncha* (Spot-billed Duck), and *Hydrophasianus chirurgus* (Pheasant-tailed Jacana) were entirely intolerant to disturbance; however occasionally visited urban riverine habitats. The decrease in total number of waders, in contrast to the increase in number of herons and egrets in the study area may probably relate to the decline of suitable breeding and roosting habitats for birds. The result shows that large numbers of bird species are associated with riverine habitats and even small riverine patches may provide valuable habitat and resources. Unfortunately, many of these smaller riverine patches are fragmented and are under constant threat due to urbanization.

The current riverine policies in India do not address issues pertaining to small and isolated riverine patches, re-population of these riverine patches by water birds and the crucial role the surrounding forest cover and land use patterns play in maintaining the biodiversity of these riverine regions. There should be a cultural basis adapted for successfully integrating and conserving the riverine habitats to protect them from disappearing.

5 Conclusions

- Riverine stretches are important reserves for both land birds and riverine species. Urbanization and its associated factors have negative impact on the riverine bird diversity. Species diversity and richness increased from the most to the less urbanized city centers.
- Bird abundance decreased from the urban towards the suburban, however gradually increased towards the rural setting. Fewer riverine species were found to thrive well in the urban environment that had a broader territorial range, where as few others sensitive to habitat variations declined.
- Riverine patches are vital zones for not only the wetland species but also for different species of land birds due to the abundant resources available in these regions. Hence, it becomes important to protect large tracts of forest along riverine stretches to protect and conserve the natural diversity of the river environment.
- Since the Cauvery River experiences water flow only seasonally, it becomes very important to preserve and protect the natural habitat conditions from anthropogenic and mining activities that are highly prevalent in this region.
- The large area covered by Kallanai and surrounding thick patch of vegetation provides alternate feeding and roosting sites for birds contributing to high bird species diversity. Agricultural lands on either side of this river also attract substantial number of birds. As the Kallanai dam and the entire stretch are several centuries old and has not urbanized, it provides refuge for resident water birds and also acts as an excellent corridor for migratory birds. Hence, it is essential that Kallanai be declared an important bird reserve to protect the habitats of the winter visitors to this riverine stretch. As a first step, this stretch shall be demarcated as 'Eco-Sensitive Zone' which ensures protection under the 'Environmental Impact Assessment Act' (2006), of the Ministry of Environment and Forests, Government of India. Conservation measures should ascertain the presence of minimum flow of water in this stretch even during dry seasons. The agricultural farmlands on either sides of river 'Cauvery' at Kallanai shall be declared as 'High Nature Value' (HNV) wetlands/farmlands as in Europe (Doxa et al. 2010), since these farmlands also support high diversity.

Acknowledgments The authors thank the Department of Science and Technology, Government of India, for providing financial assistance under the DST Women Scientists Scheme (WOS-A).

References

Aurora AL, Simpson TR, Small MF, Berder KC (2009) Toward increasing avian diversity: urban wildscapes programs. Urban Ecosyst 12:347–358

Aynalem S, Bekele A (2008) Species composition, relative abundance and distribution of bird fauna of riverine and wetland habitats of Infranz and Yiganda at southern tip of Lake Tana Ethiopia. Trop Ecol 49(2):199–209

Azous AL, Horner RM (2000) Wetlands and urbanization: implications for the future. Lewis Publishers, Boca Raton

Balachandran S (2012) Avian diversity in coastal wetlands of India and their conservation needs. www.upsbdb.org/pdf/Souvenir2012/ch-19.pdf

Balapure S, Dutta S, Vyas V (2012) Avifauna diversity in Barns wetland of Narmada basin in Central India. J Res Biol 2(5):460–468

Blair RB (1996) Land use and avian species diversity along an urban gradient. Ecol Appl 6:506–519

Brotons L, Herrando S, Martin JL (2004) Bird assemblages in forest fragments within Mediterranean mosaics created by wild fires. Landscape Ecol 19:663–675

Brouwer JW, Mullie C, Scholte P (2003) White storks *Ciconia ciconia* wintering in Chad, northern Cameroon and Niger: a comment on P. Berthold, W. Van Den Bossche, W. Fiedler, M. Kaatz, Y. Leshem, E. Nowak & U. Querner. 200. Ibis 145:499–501

Brumm H, Slabbekoorn H (2005) Acoustic communication in noise. Adv Study Behav 35:151–209

Bunn SE, Boon PI (1993) What sources of organic carbon drive food webs in billabongs? A study based on stable isotope analysis. Oecologia 96:85–94

Bunn SE, Davies PM (1999) Aquatic food webs in turbid, arid-zone rivers: preliminary data from Cooper Creek, western Queensland. In: Kingsford RT (ed) A free-flowing river: the ecology of the Paroo river. New South Wales National Parks and Wildlife Service, Sydney, pp 67–76

Butler RW (1992) Great blue heron. The birds of North America: life histories for the 21st century, No. 25. American Ornithologists Union, Washington

Carignan V, Villard MA (2002) Selecting indicator species to monitor ecological integrity: a review. Environ Monit Assess 78(1):45–61

Caula SA, Marty P, Martin JL (2008) Seasonal variation in species composition of an urban bird community in Mediterranean France. Landscape Urban Plan 87:1–9

Caula SA, Sirami C, Marty P, Martin JL (2010) Value of an urban habitat for the native Mediterranean avifauna. Urban Ecosyst 13:73–89

Chace JF, Walsh JJ (2006) Urban effects on native avifauna: a review. Landscape Urban Plan 74:46–69

Daily GC (2001) Ecological forecasts. Nature 411:245

DeLuca WV, Studds CE, Rockwood LL, Marra PP (2004) Influence of land use on the integrity of marsh bird communities of Chesapeake bay, USA. Wetlands 24:837–847

Doxa A, Bas Y, Paracchini ML, Pointereau P, Terres JM, Jiguet F (2010) Low-intensity agriculture increases farmland bird abundances in France. J Appl Ecol 47(6):1348–1356

Edelson NA (1990) Foraging ecology of wading birds using an altered landscape in central Florida. Master's thesis, University of Florida, Gainesville

Evans KL, Newson SE, Gaston KJ (2009) Habitat influences on urban avian assemblages. Ibis 151:19–39

Fernandez-Juricic E, Telleria JL (2000) Effects of human disturbance on blackbird (Turdus merula) spatial and temporal feeding patterns in urban parks of Madrid (Spain). Bird Study 47:13–21

Gajera NB, Mahato AKR, Vijaykumar V (2012) Wetland birds of arid region-a study on their diversity and distribution pattern in Kachchh. Columban J Life Sci 13(1 and 2):47–51

Girling CL, Helphand KI (1997) Retrofitting suburbia: open space in Bellevue, Washington, USA. Landscape Urban Plan 26:301–313

Gosselin H, Johnson B (1995) The urban outback—wetlands for wildlife: a guide to wetland restoration and frog-friendly backyards. Metro Toronto's Adopt-a-Pond Wetland Conservation Programme, Metro Toronto Zoo, Toronto

Green RE, Hirons GJM (1991) The relevance of population studies to the conservation of threatened birds. In: Perrings CM, Lebreton JD, Hirons GJM (eds) Bird population studies. Oxford University Press, New York, pp 594–621

Grimmett R, Inskipp T (2005) Birds of Southern India. D and N publishing, Lowesden Business Park, Hungerford, Berkshire

Hoyer MV, Canfield DE (1990) Limnological factors influencing bird abundance and species richness on Florida lakes. Lake Reservoir Manage 6(2):133–141

Hoyer MV, Canfield DE (1994) Bird abundance and species richness on Florida lakes: influence of trophic status, lake morphology, and aquatic macrophytes. Hydrobiologia 297(280):107–119

Jobin B, Belanger L, Boutin C, Maisonneuve C (2004) Conservation value of agricultural riparian strips in the Boyer River watershed, Québec (Canada). Agric Ecosyst Environ 103:413–423

Kent A, Coker P (1992) Vegetation description and analysis. A practical approach. Wiley, New York

Kingsford (2000) Ecological impacts of dams, water diversions and river management on floodplain wetlands in Australia. Aust Ecol 25:109–127

Kreyer D, Zerbe S (2006) Short-lived tree species and their role as indicators for plant diversity in the restoration of natural forests. Restor Ecol 14:137–147

Krueper D, Bart J, Rich TD (2003) Response of vegetation and breeding birds to the removal of cattle on the San Pedro River, Arizona (USA). Conserv Biol 17(2):607–615

Masero JA (2003) Assessing alternative anthropogenic habitats for conserving waterbirds: salinas as buffer areas against the impact of natural habitat loss for shorebirds. Biodivers Conserv 12:1157–1173

Mazeika SMP, Watzin MC, Keeton WS (2007) A riverscape perspective on habitat associations among riverine bird assemblages in the Lake Champlain Basin, USA. Landscape Ecol 22:1169–1186

McKinney RA, Raposab KB, Cournoyerc RM (2011) Wetlands as habitat in urbanizing landscapes: patterns of bird abundance and occupancy. Landscape Urban Plan 100:144–152

Munyenyembe F, Harris J, Hone J (1989) Determinants of bird populations in an urban area. Aust J Ecol 14:549–557

Murgui E (2009) Influence of urban landscape structure on bird fauna: a case study across seasons in the city of Valencia (Spain). Urban Ecosyst 12:249–263

Nilsson C, Berggren K (2000) Alterations of riparian ecosystems caused by river regulation. Bioscience 50(9):783–792

Pavlik J, Pavlik S (2000) Some relationships between human impact, vegetation, and birds in urban environment. Ekologia-Bratislava 19:392–408

Pearce CM, Green MB, Baldwin MR (2007) Developing habitat models for water birds in urban wetlands: a log-linear approach. Urban Ecosyst 10:239–254

Petit DR, Petit LJ, Saab VA, Martin TE (1995) Fixed-radius point counts in forests: factors influencing effectiveness. In: Ralph CJ, Sauer JR, Droege S (eds.) Monitoring bird populations by point counts. CA: USDA forest service general technical report PSW-GTR-149, Albany pp 49–56

Power ME, Sun A, Parker G, Dietrich WE, Wootton JT (1995) Hydraulic food chain models. Bioscience 45:159–167

Savard JPL, Clergeau P, Mennechez G (2000) Biodiversity concepts and urban ecosystems. Landscape Urban Plan 48:131–142

Scheffers BR, Harris JB, Haskell DG (2006) Avifauna associated with ephemeral ponds on the Cumberland Plateau, Tennessee. J Field Ornithol 77:178–183

Smith JP, Richardson JR, Collopy MW (1995) Foraging habitat selection among wading birds (Ciconiiformes) at Lake Okeechobee, Florida, in relation to hydrology and vegetative cover. In: Aumen NG, Wetzel RG (eds) Ecological studies of lake Okeechobee, Florida. Arch Hydrobiol (Special Issues), Advanc Limnol 45:247–285

Sogge MK, Sferra SJ, Paxton EH (2008) Tamarix as habitat for birds: implications for riparian restoration in the southwestern United States. Restor Ecol 16(1):146–154

Suarez-Rubio M, Thomlinson JR (2009) Landscape and patch-level factors influence bird communities in an urbanized tropical island. Biol Conserv 142:1311–1321

Sullivan SMP, Watzin MC, Keeton WS (2007) A riverscape perspective on habitat associations among riverine bird assemblages in the Lake Champlain Basin, USA. Landscape Ecol 22:1169–1186

Sutherland W (2004) Diet and foraging behaviour. In: Sutherland W, Newton I, Green R (eds) Bird ecology and conservation: a handbook of techniques, Oxford University Press, pp 233–250

Swaddle JP, Page LC (2007) High levels of environmental noise erode pair preferences in zebra finches: Implications for noise pollution. Anim Behav 74:363–368

Tilman D, Wedin D, Knops J (1996) Productivity and sustainability influenced by biodiversity in grassland ecosystems. Nature 379:718–720

Titton DL (1995) Integrating wetlands into planned landscapes. Landscape Urban Plan 32:205–209

Tracy CR, Averill-Murray R, Boarman WI, Delehanty D, Heaton J, McCoy E, Morafka D, Nussear K, Hagerty B, Medica P (2004) Desert tortoise recovery plan assessment. Report to United States Fish and Wildlife Service. University of Nevada, Reno

Traut AH, Hostetler ME (2004) Urban lakes and waterbirds: effects of shoreline development on avian distribution. Landscape Urban Plan 69:69–85

Weller MW (1999) Wetland birds: habitat resources and conservation implications. Cambridge University Press, Cambridge

Whited D, Galatowitsch S, Tester JR, Schik K, Lehtinen R, Husveth J (2000) The importance of local and regional factors in predicting effective conservation, planning strategies for wetland bird communities in agricultural and urban landscapes. Landscape Urban Plan 49:49–65

Yard HK, Riper CV, Brown B (2004) Kearsley MJ (2004) Diets of insectivorous birds along the Colorado River in Grand Canyon. Ariz Condor 106(1):106–115

Ecologically Sound Mosquito Vector Control in River Basins

Tapan Kumar Barik

Abstract Human population growth and resultant environmental modifications across all the landscapes of the Earth, new interactions between human species and the natural processes have emerged, that promote new epidemiological patterns of vector-borne diseases. Among all the vectors that spread epidemics among the humans, mosquitoes stand second to none. Over the decades, several methods have been developed, for protection against mosquito borne diseases albeit with varying rates. No one mosquito-control strategy fits all situations, and management of mosquitoes should be addressed on a case-by-case basis that identifies local and regional differences in mosquito species, geographic setting, and watershed context. Further, socio-economic, environmental and epidemiological factors will determine the choice of vector management strategies. A majority of these methods attempt killing both adult and immature stages of mosquito vector. Currently, mosquito vector control is mainly based on insecticides, but its sustainability has been undermined by the development of resistance and growing concerns about the long-term environmental impact of insecticides. In addition to the environmental damage, anthropogenic activities have created extensive breeding sites for the vectors, especially the mosquitoes. Mosquitoes usually breed in places close to the banks of rivers and creeks where there is protection from currents by obstacles, protruding roots, plants etc. Source reduction is usually the most effective, eco-friendly and economical approach for mosquito control technique which is accomplished by eliminating mosquito breeding sites. This paper provides information of ecologically sound mosquito vector control methods particularly in and near river channels. Usually, it is presumed that the stagnant regions of river channels and adjacent locations are the preferred regions of mosquito production. However, river channel with healthy biodiversity with a diverse community of predators that prey on immature stages of mosquito often produce few to no mosquitoes. Hence, detailed knowledge of the sites in and near the river channels, and promoting and preserving biodiversity in the river basins are necessary for cost-effective, eco-friendly vector

T.K. Barik (✉)
Applied Entomology Laboratory, Post Graduate Department of Zoology,
Berhampur University, Berhampur 760007, India
e-mail: tkbarik@rediffmail.com

Mu. Ramkumar et al. (eds.), *Environmental Management of River Basin Ecosystems*,
Springer Earth System Sciences, DOI 10.1007/978-3-319-13425-3_33

control measure. Further, basic research is recommended to better define the cost-effectiveness of the various tools and management options used currently for the management of mosquito vectors particularly in and near the river channels.

Keywords Mosquito · Vector · Control · River basins

1 Introduction

The human population increasing in a geometric ratio and as a result of environmental modifications to satisfy the necessities of the population, new interactions between the humans and the environment have emerged that lead to new epidemiological patterns of vector borne diseases (Özer 2005). Globally, more than 17 % of infectious diseases are contributed by vector-borne diseases (Townson et al. 2005). Vector-borne disease refers to describe an illness caused by an infectious microbe that is transmitted to human beings by blood-sucking arthropods. The arthropods that most commonly serve as vectors include, blood sucking insects (mosquitoes, fleas, lice, biting flies, bugs) and blood sucking arachnids (mites and ticks). There are several global-scale physical and social environments factors that promote actively many vector borne diseases. Therefore, to address the risk of vector borne diseases efficiently, a comprehensive risk management framework including, the assessment, evaluation of contributing factors and evolving methods of mitigation, control and eradication in a sequential manner are necessary.

Where there is no effective cure for vector borne diseases, vector control remains the only option to protect the population. Vector control is any method that focuses on utilizing preventative methods to control or to eliminate vector populations. To be effective, this method requires education and promotion of methods amongst the population to raise the awareness of vector threats. Removing or reducing areas where vectors can easily breed can help limit their proliferation. Different species of mosquitoes prefer different types of standing water to lay their eggs. The presence of beneficial predators such as fish and dragonfly nymphs in permanent ponds, lakes and streams usually keep these bodies of water relatively free of mosquito larvae. However, portions of marshes, swamps, clogged ditches and temporary pools and puddles are all prolific mosquito breeding sites.

Environmental management is an eco-friendly method for vector control that refers to the planning, organisation, carrying out and monitoring of activities for the modification and/or manipulation of environmental factors, with the aim of preventing or minimising vector breeding and reducing human-vector-parasite contacts (Singer et al. 2005). Source reduction (physical or permanent control) is usually the most effective and economical approach for mosquito control techniques. It is accomplished by eliminating mosquito breeding sites that mainly includes modification such as filling, leveling and drainage of breeding places, and water management so that it exerts its effect not only by reducing the number of adult

Fig. 1 River bed breeding sites of mosquito vector *An. culicifacies s.l.* (*Source* National Institute of Malaria Research, Delhi 110054, India)

mosquitoes, but also by extending generation times by forcing the remaining gravid females to spend more time searching for oviposition sites (Gu et al. 2006).

Usually, streams and rivers are fast flowing water bodies and are not preferred by mosquito larvae. However, there are often fringe area associated with these streams and rivers where the water moves very slowly or is stagnant. These fringe areas can provide favorable breeding habitats (Fig. 1) for mosquito larvae. Further, rivers and streams that are drying up leave stagnant, pooling water behind that can serve as larval habitats. A river basin is the land that water flows across or under on its way to a river. Basins can be divided into watersheds, or areas of land around a smaller river, stream, or lake. Sometimes, pools may form in river beds in the dry season.

2 Classification of Mosquitoes

Mosquitoes have been on Earth for more than 100 million years. They are flies of the Order Diptera having a pair of wings and a pair of halters to aid in flight. There are about 3,500 named species of mosquitoes available worldwide. Each species has its own particular life history, habitat preference, and dispersal ability. These insects complete their life cycle through four morphologically distinct stages such as: egg, larva, pupa, and adult. Eggs can be laid directly onto the water or in areas that will be flooded with water. Larval and pupal stages are spent in water and the adult stage is spent out of water. Each of these stages can be easily recognized by its special morphological appearance. Adults of males feed on nectar or sap from trees, and females feed on vertebrate blood to provide proteins and lipids for egg development. Mosquitoes are known to carry many infectious diseases from several different classes of microorganisms, including viruses and parasites. Some mosquitoes are vectors and responsible for transmission of certain vector borne diseases, like malaria, dengue, chikungunya, japanese encephalitis, yellow fever, lymphatic filariasis etc. (Rozendaal 1997; Barik et al. 2009). Malaria is one of the most common vector-borne diseases transmitted by *Anopheles* mosquitoes widespread in tropical and subtropical regions (WHO 2007) and estimated about 219 million cases of malaria and 6,60,000 malaria deaths globally in 2010 (WHO 2012). Like *Anopheles*, *Culex* mosquitoes are also serve as vectors of important diseases, such as west nile virus, filariasis, japanese encephalitis, st. louis encephalitis etc. *Culex quinquefasciatus*, the vector of bancroftian filariasis is the most widely distributed mosquito, having around 31 million microfilaraemics, 23 million cases of symptomatic filariasis, and about 473 million individuals potentially at risk of infection in India (Agrawal et al. 2006). Similarly, *Aedes* mosquitoes are vector for dengue, chikungunya and, yellow fever. It is estimated that about 50 million dengue infections occur annually and approximately 2.5 billion people live in dengue endemic countries (WHO 2009, 2011). Females of different mosquito species have selective oviposition sites based on the environmental factors and hydrologic conditions. Depending on their egg-laying and hatching behavior, mosquitoes can be classified into two groups, namely (a) floodwater-ephemeral water habitat mosquitoes and (b) permanent and semi-permanent aquatic habitat mosquitoes (Knight et al. 2003).

2.1 Floodwater-Ephemeral Water Habitat Mosquitoes

Floodwater mosquitoes are primarily daytime feeding mosquitoes in the genera *Aedes*, *Ochlerotatus*, and *Psorophora*. These mosquitoes are prevalent in sites receiving agricultural runoff and also along wetland edges. For example, if the runoff from a constructed wetland is let out to a disposal or recharge area in a floodplain, then suitable microhabitats for these mosquitoes may be created, especially if the soils become impermeable. In some cases, the salt content of soils in the receiving basins could increase to levels that would attract certain mosquitoes. Eggs are deposited by these mosquitoes on moist substrate. These eggs do not hatch until the regions are subsequently inundated.

2.2 Permanent and Semi-permanent Aquatic Habitat Mosquitoes

Permanent and semi-permanent aquatic habitat mosquitoes lay eggs singly (*Anopheles* spp.) or in large rafts (*Culex* spp.) on the water surface. These eggs hatch within a few days without an external hatching stimulus. Each egg is equipped with special structures (floats), which enable the egg to remain on the water surface. Depending upon the species and the environmental conditions, essentially all the eggs may hatch shortly after egg deposition, hatching may be staggered over a period of several weeks, or some eggs may become stranded on moist substrate above the water and not hatch until after a flooding episode.

3 Breeding Sites of Mosquitoes

Mosquito species differ in their breeding habitat preferences and usually breed in non or slow flowing, shallow water bodies, unaffected by waves. The riparian (adjacent to a river) zones are generally forested, narrow, and have only small areas of inundation. Usually, these areas do not produce significant mosquito population densities. However, due to various diversities of these wetlands, their distribution, and the presence of tree falls, water holding debris, and other elements, can produce pestilential mosquito populations. *Anopheles gambiae* s.s. and *An. arabiensis* prefer to breed in temporary sunlit puddles, including those produced by rain and irrigation, as well as in shallow shoreline puddles adjacent to rivers. Further, floodplain systems can be major producers of several species of mosquitoes in natural and modified settings particularly in the spring following snowmelt in the case of snow-fed rivers and following monsoon in the case of monsoon-fed rivers. For example, the Sudbury and Concord River floodplains near Boston city are very productive sites for *Aedes* mosquitoes.

In addition to the riparian zone and floodplain systems, man-made dams are also sometimes provide important breeding sources for malaria vectors such as *An. gambiae* complexes, *An. funestus* in Africa, *An. sacharovi* in Turkey (Mather and That 1984).

4 Common Strategies for Mosquito Vector Control

Mainly, chemotherapy, vaccination, and mosquito vector control have played major roles in reduction of mosquito borne disease transmission in many parts of the world (Reiter 2001). The most important facet of controlling mosquito borne diseases is to control mosquito vectors and many control strategies are devised in order to control and eliminate mosquitoes in specific habitats under different environmental conditions (Reynolds and Hellenthal 2003). The options available for vector control efforts include chemical, biological, phytochemicals, and environmental management. Immense literature is available for vector control, and a WHO manual on vector control prepared by Rozendaal (1997) is highly informative, in which various methods are given on use of insecticides, insecticide-treated materials, insect growth regulators, biological control agents, environmental management, and personal protection methods against mosquito vectors. Depending on the situation, control of larvae (larvicide) or control of adults (adulticide) may be used to manage mosquito populations. In brief, control of adult mosquitoes is mainly accomplished by application of chemical pesticides particularly in the form of indoor residual spray (IRS), space spray and insecticide treated material for the control of malaria vectors. However, the emergence of widespread insecticide resistance and the potential environmental issues associated with some synthetic insecticides has necessitated additional approaches to control the proliferation of mosquito population.

It is generally agreed that larvicide is more effective than adulticide because the aquatic immature stages are restricted within a water body where they are easily accessible, relatively immobile, and unable to escape. Flying adults on the other hand tend to be widely dispersed, often inaccessible, and highly mobile, and usually require adulticides to be applied over a larger area than larvicides. Adults often tend to exist in closer proximity to humans than their larval counterparts (Kumar and Hwang 2006). As larvicides, chemicals like petroleum oils (Gratz and Pal 1988), Paris Green (Rozendaal 1997), temephos and fenthion (Organophosphate) (Sharma et al. 1996) are used in mosquito vector control. In addition to these insecticides, insect growth regulators are also used for the control of mosquito vector which are relatively safe to non-target organisms (Mulla and Darwazeh 1979; Schaefer et al. 1984; Mulla et al. 1985, 1986). Biological control methods include use of natural enemies like larvivorous fish, copepods, nematode, predators, toxic products of bacterial agents and products derived from plant origins are used for mosquito vector control (Walker 2002; Lloyd 2003). Some microbial pathogens of mosquitoes like protozoa, microsproidia, fungal pathogens and insect killing viruses have been recognized to be promising larvicides (Scholte et al. 2003; Ren et al. 2008) although their commercial formulations are not still available for mosquito control (Scholte et al. 2004). Toxins of *Bacillus thuringiensis* (Bt) have been widely demonstrated to be effective larvicides against mosquitoes (Poopathi and Abidha 2010). However, recently, resistance to Bt toxins are also reported in certain species of mosquito vectors. Although the biological and chemical control methods show some notable success in mosquito control but in many cases, it has not been sustainable in the long term due to their insecticide resistance, re-invasion, environmental damage etc. (Wilke et al. 2009). Further, methods like genetic engineering, sterile insect technique, cytoplasmic incompatibility, incompatibility due to chromosomal factors, chromosome translocations, conditional lethal, meiotic drive, compound chromosome, etc., are proposed for alternative vector control strategies (Pal and LaChance 1974) but were not extensively tested and need further research.

5 Mosquito Vector Control Strategies in River Basins

Mosquitoes do not need big swamps, ponds and big water bodies to lay their eggs and complete their life cycle. Usually, mosquitoes breed in places close to the banks of rivers and creeks where there is protection from currents by obstacles, protruding roots, plants and so on. Further, humans have always altered the environment to suit their need and these modifications resulted in the development of extensive breeding sites. It is very difficult to find out the exact location of the breeding sites and also to cover the large such areas for effective mosquito vector control during their larval stage. Therefore, for effective management of mosquito vector population, planned recurrent activities, like proper drainage, water level management, stream flushing, construction of dams, removal of man-made obstructions, vegetation removal is necessary which could produce temporary unfavourable conditions for breeding of vectors in their habitats (WHO 1982) especially in the river basins.

5.1 Drainage

Leakages, obstructions and small pools or puddles of residual water in drainage ditches often provide suitable breeding sites for mosquitoes. Proper drainage reduces mosquito breeding sites and can be accomplished by constructing open waterways and dykes with tidal gates, subsoil drainage and pumping. Planning and construction of some drainage systems are complicated and require the expertise of engineers. However, some small-scale drainage works intended to control mosquitoes can be carried out by community members using simple equipments. The impact of accumulated sediment within the stream channels may require restoration of the waterway prior to management of the floodplain mosquito habitat. The removal of up to 2 ft. of channel sediment is generally a reasonable standard for waterway management. It has been reported that, application of multiple interventions along with draining flooded areas of Luanshya River and its tributaries and swamps reduced the abundance of larval stages of *An. gambiae* and *An. funestus* drastically (Utzinger et al. 2001).

5.2 Water-Level Management

Water level management is an effective environmental control approach for mosquitoes (Cardarelli 1976; Collins and Resh 1989) and the incorporation of design features that reduce mosquito production. Channelization to increase the water flow, to steepen banks and provide access to predators of mosquitoes will reduce the likelihood that isolated pools and marshy areas, which are favorable for mosquito proliferation (Service 1993). Further, management practices that create depressions and collect standing water should be avoided. Utzinger et al. (2001) reported that application of multiple interventions like vegetation clearance along the Luanshya River and its tributaries, modification of river boundaries, removal of man-made obstructions, and draining flooded areas and swamps reduced the abundance of larval stages of *An. gambiae* and *An. funestus* drastically. After the first year of implementation of these interventions it was observed that, the water level of the main river had decreased substantially and the velocity was high enough to interrupt larval development.

5.3 Stream Flushing

Flushing (increasing water flow in streams) is employed in small streams where there is a continuous and abundant supply of water flowing slowly enough to permit mosquitoes to breed in quiet places along the margins. A periodic discharge of a large volume of water washes away the immature stages of the mosquitoes from the edges, or strands them on the banks. Flushing also stirs up sediment at the bottom of the stream which can bury aquatic mosquito stages and can help slow the growth of new marginal vegetation. Although flushing is a long-lasting method for the elimination of immature stages of mosquito from the river banks but needs high initial investment and require little maintenance. It has been successfully used in South-East Asia to control malaria vectors particularly, *An. maculatus* and *An. minimus*. Flushing streams to control *A. culicifacies* Giles was examined in five river systems in rural Sri Lanka in the late 1930s (Konradsen et al. 2004), and again, in one stream in the mid 1990s (Konradsen et al. 1998) and the results demonstrated that the costs of periodical river flushing to eliminate mosquito breeding habitats compared favorably with impregnated bednets (Konradsen et al. 1999) used to avoid disease transmission.

5.4 Construction of Dams

Mainly in Africa, Asia and the Pacific and the Americas, dams have extensively proliferated to cover the food and energy demands of these regions (WHO 2000). Further, micro-dams are developed at much greater rate which have a higher potential for mosquito production and may become an important source of health risk (Tubaki et al. 1994; Molyneux 1997). Lautze et al. (2007) reported that the construction of the Koka dam across the Awash River in Ethiopia was associated with increased levels of malaria transmission in communities adjacent to the dam site. The Manantali Dam in Mali was found to have increased malaria risk (King 1996) and proximity to a micro-dam in northern Ethiopia was found to be an important risk factor for malaria (Ghebreyesus et al. 1999). Similarly, in south-western Ethiopia, a higher prevalence of malaria was found in villages closer to the Gilgel-Gibe Dam (Yewhalaw et al. 2009). However, these results are inconsistent with the findings of a study on the construction of the Diama Dam in the Senegal River basin, which found that the dam had not affected malaria incidence in the region (Sow et al. 2002). Furthermore, pre and post-intervention data demonstrated that the construction and operation of two large dams in the Senegal River, increased *Anopheline* densities but without exerting any significant influence on malaria transmission (Ribas et al. 2012). In India, construction of dam in San Dulakudar village, Sundargarh District, Odisha resulted in impounding of water in a small reservoir, thereby preventing flow of water both above and below the dam and making it unfavorable for breeding of *A. fluviatilis* (Sharma et al. 2008).

5.5 Maintenance of Irrigation Systems and Irrigated fields

Power production, water supply and irrigation utilized the natural interconnectedness of water within a river basin more efficiently. Irrigation may create suitable breeding places for some mosquitoes like *Anopheles* and *Culex*. Breeding may occur in irrigation channels and ditches between vegetation growing along the margins. Holes in the beds of channels provide breeding places when there is no water flow. Further, leaks may provide breeding sites in puddles outside the channels. The prevention of breeding in irrigated fields is difficult because of very large surfaces of standing water are available to mosquitoes. The bottoms of channels and ditches should be smooth with a gentle slope so that no stagnant water remains when they dry out. Proper maintenance should prevent leakages from sluices and the dykes and linings of fields and channels. Breeding in ditches and irrigation channels filled with water can be reduced by making the banks steeper and removing water plants. This helps to speed up the flow of water and expose the larvae to larvivorous fish and other predators.

5.6 Removal of Man-Made Obstructions

Prohibiting the entry of people and livestock in marshy areas and cultivated fields, together with filling the crossing points of cattle and humans along the riverbed with rocks and gravel, prevented the destruction of the plants and the creation of breeding sites in hoof and footprints. It has been reported that, in Luanshya River and its tributaries, removal of man-made obstructions along with other interventions reduced larval density of mosquito vectors (Utzinger et al. 2001).

5.7 Vegetation Control

Vegetation provides food for mosquitoes in the form of plant detritus and also fosters the production of other mosquito food such as bacteria, algae, and protozoa (Clements 1992; Jiannino and Walton 2004). Thick vegetation stands also may reduce water flow and thus reduce the physical inhibitors of mosquito disturbances such as high currents, eddies, and waves that can negatively impact developing mosquito larvae (Jiannino and Walton 2004). Vegetation management, as it relates to mosquito control, is undertaken to create open water areas that are unfavorable for development of breeding sites for mosquitoes (Walton and Workman 1998) and to increase predation pressure on mosquito larvae (Orr and Resh 1989). Reduction of vegetation coverage can significantly reduce mosquito populations (Workman and Walton 2000; Thullen et al. 2002; Lawler et al. 2007), but the plant species, the method of vegetation thinning, and the spatial configuration of the remaining vegetation can have significant impacts on the magnitude of the resulting mosquito population reduction (Jiannino and Walton 2004). Vegetation clearance along the Luanshya River and its tributaries with other interventions reduced the larval densities of *An. gambiae* and *An. funestus* (Utzinger et al. 2001). Reduction in vegetation density had a positive effect on mosquito-fish density and mosquito control. It was noted that plant re-growth quickly replaced the removed vegetation and negated the effects of thatching, combing, and edging. The need for annual vegetation maintenance equates to a relatively high cost for these vegetation management methods. In addition to the vegetation management, plantation of different varieties of trees like Eucalyptus that grow rapidly is also another method of reduction of mosquito breeding site near the river bank as these trees use a lot of water and dry the land by allowing water to evaporate through their leaves. However, it may not be a good choice for regions that reel under scarce water resources.

6 Conclusions

- Mosquito species have evolved to exploit a wide variety of habitats. Because they are a natural part of aquatic ecosystems, permanent and total elimination of mosquitoes is not a realistic or achievable goal. However, current scientific understanding supports the position that we can take environmentally compatible measures to help minimize mosquito production from natural, created, or restored ecosystems.
- Responsible mosquito management strategies should start with an effective surveillance program with mosquito abundance and mosquito-transmitted disease. Ideally, both types of surveillance programs should be integrated to formulate risk assessments that help make informed decisions for mosquito control before a public health emergency occurs. No one mosquito-control strategy fits all situations, and mosquito management should be addressed on a case-by-case basis that identifies local and regional differences in mosquito species, geographic setting, and watershed context.
- Socio-economic, cultural and environmental and epidemiological factors will determine the choice of vector management strategies. Therefore, effective mosquito vector control requires interdisciplinary efforts and an integrated vector management approach.
- People often assume that all site nearer to river produce disease carrying mosquito but healthy basins with a diverse community of predators that prey on immature stages of mosquito often produce few to no mosquitoes. Hence, detailed knowledge of sites in the river basin is highly essential for the management of mosquito vectors.

References

Agrawal Lt, Col VK, Sashindran Wg, Cdr VK (2006) Lymphatic filariasis in India: problems, challenges and new initiatives. Med J Armed Forces India 62:359–362
Barik TK, Sahu B, Swain V (2009) A review on *Anopheles culicifacies*: from bionomics to control with special reference to Indian subcontinent. Acta Trop 109:87–97
Cardarelli N (1976) Controlled release pesticide formulations. CRC Press Inc., Cleveland, p 210
Clements AN (1992) The biology of mosquitoes, vol 1. Chapman and Hall, New York
Collins JN, Resh VH (1989) Guidelines for the ecological control of mosquitoes in non-tidal wetlands of the San Francisco Bay area. In: California Mosquito Vector Control Association, Inc., and University of California Mosquito. Research Program, Sacramento
Ghebreyesus TA, Haile M, Witten KH, Getachew A, Yohannes AM, Yohannes M, Teklehaimanot HD, Lindasy SW, Byass P (1999) Incidence of malaria among children living near dams in northern Ethiopia: community based incidence survey. Br Med J 319:663–666
Gratz NG, Pal R (1988) Malaria vector control: larviciding. In: Wernsdorfer WH, McGregor IA (eds) Malaria: principle and practices of malariology. Churchill Livingstone, Edinburgh, pp 1213–1226
Gu W, Regens JL, Beier JC, Novak RJ (2006) Source reduction of mosquito larval habitats has unexpected consequences on malaria transmission. Proc Natl Acad Sci 103:17560–17563
Jiannino JA, Walton WE (2004) Evaluation of vegetation management strategies for controlling mosquitoes in a southern California constructed wetland. J Am Mosq Control Assoc 20:18–26
King C (1996) The incorporation of health concerns into Africa river basin planning. MIT Ph.D. thesis. Cambridge, USA: Massachusetts Institute of Technology
Knight RL, Walton WE, O'Meara GF, Reisen WK, Wass R (2003) Strategies for effective mosquito control in constructed treatment wetlands. Ecol Eng 21:211–232
Konradsen F, Matsuno Y, Amerasinghe FP, Amerasinghe PH, Hoek WV (1998) *Anopheles culicifacies* breeding in Sri Lanka and options for control through water management. Acta Trop 71:131–138
Konradsen F, Steele P, Perera D, van derHoek W, Amerasinghe PH, Amerasinghe FP (1999) Cost of malaria control in Sri Lanka. Bull World Health Organ 77:301–309
Konradsen F, van der Hoek W, Amerasinghe FP, Mutero C, Boelee E (2004) Engineering and malaria control: learning from the past 100 years. Acta Trop 89:99–108
Kumar R, Hwang JS (2006) Larvicidal efficiency of aquatic predators: a perspective for mosquito biocontrol. Zool Stud 45:447–466
Lautze J, McCartney M, Kirshen P, Olana D, Jayasinghe G, Spielman A (2007) Effect of a large dam on malaria risk: the Koka reservoir in Ethiopia. Trop Med Int Health 12:982–989
Lawler SP, Reimer L, Thiemann T, Fritz J, Parise K, Feliz D, Elnaiem DE (2007) Effects of vegetation control on mosquitoes in seasonal freshwater wetlands. J Am Mosq Control Assoc 23:66–70
Lloyd L (2003) Best practice for dengue prevention and control in the Americas. Environmental Health Project, USAID (Agency for International Development), Washington, DC
Mather TH, That T (1984) Environmental management for vector control in rice fields. FAO irrigation and drainage paper, No. 41, p 152
Molyneux D (1997) Patterns of change in vector-borne diseases. Ann Trop Med Parasitol 91:827–839
Mulla MS, Darwazeh HA (1979) New insect growth regulators against flood and stagnant water mosquitoes effect on non-target organisms. Mosq News 39:746–755
Mulla MS, Darwazeh HA, Ede L, Kennedy B (1985) Laboratory and field evaluation of the IGR fenoxycarb against mosquitoes. J Am Mosq Control Assoc 1:442–448
Mulla MS, Darwazeh HA, Kennedy B, Dawson MW (1986) Evaluation of new insect growth regulators against mosquitoes with notes on non-target organisms. J Am Mosq Control Assoc 2:314–320
Orr BK, Resh VH (1989) Experimental test of the influence of aquatic macrophyte cover on the survival of *Anopheles* larvae. J Am Mosq Control Assoc 5:579–585

Özer N (2005) Emerging vector-borne diseases in a changing environment. Turk J Biol 29:125–135
Pal R, LaChance LE (1974) The operational feasibility of genetic methods for control insects of medical and veterinary importance. Ann Rev Entomol 19:269–291
Poopathi S, Abidha S (2010) Mosquitocidal bacterial toxins (*Bacillus sphaericus* and *Bacillus thuringiensis* serovar israelensis): mode of action, cytopathological effects and mechanism of resistance. J Phys Pathophysiol 1(3):22–38
Reiter P (2001) Climate change and mosquito borne diseases. Environ Health Perspect 109 (1):141–161
Ren X, Hoiczyk E, Rasgon JL (2008) Viral paratransgenesis in the malaria vector *Anopheles gambiae*. PLoS Pathog 4:1000135
Reynolds M, Hellenthal Mc. (2003) Environmental influences on mosquito adult and larvae abundance. University of Notre Dame Environmental Research Center
Ribas JS, Henao GP, Guimaraes AE (2012) Impact of dams and irrigation schemes in Anopheline (Diptera: Culicidae) bionomics and malaria epidemiology. Rev Inst Med Trop Sao Paulo 54 (4):179–191
Rozendaal JA (1997) Vector control: methods for use by individuals and communities. World Health Organization, Geneva
Schaefer GH, Miura T, FFJr D, Stewart RJ, Wilder WH, Juird L (1984) Biological activity of J-2931 against mosquitoes (Diptera: Culicidae) and selected non-target organisms and assessments of potential environment impact. J Econ Entomol 77: 425–429
Scholte E-J, Takken W, Knols BGJ (2003) Pathogenecity of six East African entomopathogenic fungi to adult *Anopheles gambiae* s.s (Diptera: Culicidae) mosquitoes. Proc Exp Appl Entomol NEV Amsterdam 14:25–29
Scholte E, Knols BGJ, Samson RA, Takken W (2004) Entomopathogenic fungi for mosquito control: a review. J Insect Sci 4:1–19
Service M.W (1993) Mosquito ecology: field sampling methods, 2nd edn. Elsevier, NewYork, p 988
Sharma RS, Sharma GK, Dhillon GPS (1996) Planning of malaria control operations. In: Epidemiology and control of malaria in India. National malaria eradication programme. Directorate General of Health Services, New Delhi, p 251
Sharma SK, Tyagi PK, Upadhyay AK, Haque MA, Adak T, Dash AP (2008) Building small dams can decrease malaria: a comparative study from Sundargarh District, Orissa. India. Acta Trop 107:174–178
Singer B, Teklehaimanot A, Spielman A, Schapira A, Tozan Y (2005) Coming to grips with malaria in the new millennium. Earthscan, London
Sow S, De vlas SJ, Engels D, Dryseels B (2002) Water-related disease pattern before and after the construction of the Diama dam in northern Senegal. Ann Trop Med Parasitol 96:575–586
Thullen JS, Sartoris JJ, Walton WE (2002) Effects of vegetation management in constructed wetland treatment cells on water quality and mosquito production. Ecol Eng 18:441–457
Townson H, Nathan MB, Zaim M, Guillet P, Manga L, Bos R, Kindhauser M (2005) Exploiting the potential of vector control for disease prevention. Bull World Health Organ 83:942–947
Tubaki R, Hashimoto S, Domingos M, Berenstein S (1994) Abundance and frequency of culicids, emphasizing Anophelines (Diptera, Culicidae), at Taquaruçu dam in the Paranapanema basin, southern Brazil. Rev Bras Entomol 43(3/4):173–184
Utzinger J, Tozan Y, Singer BH (2001) Efficacy and cost-effectiveness of environmental management for malaria control. Trop Med Int Health 6(9):677–687
Walker K (2002) A review control methods for African malaria vectors, Ph.D. thesis, Environmental health project, US Agency for international. development. Washington DC. 20525,p 54
Walton WE, Workman PD (1998) Effect of marsh design on the abundance of mosquitoes in experimental constructed wetlands in southern California. J Am Mosq Control Assoc 14:95–107
WHO (1982) Manual on environmental management for mosquito control with special emphasis on malaria vectors. WHO Offset Publication 66, Geneva

WHO (2000) Water sanitation and health team, world commission on dams. Human health and dams: the World Health Organization's submission to the World Commission on Dams (WCD). Geneva

WHO (2007) Anopheline species complexes in South and South-East Asia. Reg off South East Asia, New Delhi 57:22–32

WHO (2009) Chapter 1: Epidemiology, burden of disease and transmission. In: Dengue: guidelines for diagnosis, treatment, prevention and control. WHO, Geneva

WHO (2011) Chapter 26: Progress and prospects for the use of genetically modified mosquitoes to inhibit disease transmission. In: Dengue bulletin. vol 35. p 234–235

WHO (2012) World Malaria Report 2012 Summary. http://www.who.int/malaria/publications/world_malaria_report_2012/wmr2012_summary_en.pdf. Accessed 21 Dec 2013

Wilke AB, Nimmo DD, St John O, Kojin BB, Capurro ML, Marrelli MT (2009) Mini-review: genetic enhancements to the sterile insect technique to control mosquito populations. AsPac J Mol Biol Biotechnol 17:65–74

Workman PD, Walton WE (2000) Emergence patterns of *Culex* mosquitoes at an experimental constructed treatment wetland in southern California. J Am Mosq Control Assoc 16:124–130

Yewhalaw D, Legesse W, Van Bortel W, Gebreselassie S, Kloos H, Duchateau L, Speybroeck N (2009) Malaria and water resource development: the case of Gilgel-Gibe hydroelectric dam in Ethiopia. Malar J 8:21. doi:10.1186/1475-2875-8-21